Hydraulics and Pneumatics in Environmental Engineering

Hydraulics and Pneumatics in Environmental Engineering

S. David Graber
P.E. Consulting Engineer, Stoughton, MA, USA

WILEY

Published by John Wiley & Sons, Inc., Hoboken, New Jersey.
Published simultaneously in Canada.

Library of Congress Cataloging-in-Publication Data applied for:

Hardback ISBN: 9781394266142

Cover Design: Wiley
Cover Image: © Northern Lights, taken by Keven L. Graber from Underhill, Vermont, May 10, 2024.

Set in 9.5/12.5pt STIXTwoText by Straive, Pondicherry, India

SKY10089594_103024

Contents

About the Author

S. David Graber is a consulting engineer with over 50 years of experience. He holds the degrees of Bachelor of Science from the University of Miami in Florida, and Master of Science and Professional Degree of Civil Engineer from the Massachusetts Institute of Technology. He has professional engineering registrations in New York and elsewhere. Recognitions include being a fellow of the American Society of Civil Engineers (ASCE), life member of the American Society of Mechanical Engineers (ASME), and a member of the Engineering Honor Society Tau Beta Pi. His awards include the ASCE Samuel A. Greeley Award for 1969, ASCE J.C. Stevens Award for 1972, ASCE Samuel A. Greeley Award for 2005, ASCE Standards Development Council Merit Award for 2014, and the ASCE Standards Development Council Award for 2016. He is the author of over 45 journal publications and presentations. David is blessed by his lovely wife of 58 years, four sons, and five grandchildren. He is an avid reader of nonfiction and is an outdoorsman, who particularly enjoys spending summers on his family's island camp on a remote freshwater lake in Maine.

Preface

This book is intended to fill a large gap in environmental engineering education which has become apparent to the author in his 60 years of close association with practicing environmental engineers. It is written both with the student and the practicing engineer in mind. In the current educational scheme of things, the book should find its greatest value at the graduate level. As regards mathematical background, an understanding of ordinary differential equations, partial differentiation (including chain rules), and elementary properties of vectors (including the scalar or dot product) is assumed. Some of the mathematical background could be introduced or expounded upon as part of courses utilizing this text. More sophisticated vector analysis is avoided to maximize accessibility and, in the case of the critical review of G-*value*, to allow parallels to be drawn with historical developments in the environmental engineering literature. More general vector relations are noted and referenced by footnote where appropriate. A sound undergraduate fluid mechanics course has also been assumed. A student having had strong undergraduate training in fluid mechanics should find some of the material in Chapters 2–5 to be review, but should benefit from the particular perspective.

It is hoped that some environmental engineering students will be inspired to continue with further study of hydraulics, particularly in the area of the natural systems touched upon in the final chapter of this book. It is further hoped that the book will improve communication between the hydraulic and environmental engineer, not only by making the latter more aware of the methods of the former but also by making the former more aware of technical needs in the environmental engineering field. The practicing engineer should find the book valuable both as a general self-study source and as a reference resource. For the latter purpose in particular, the author has endeavored to provide practical summary or overview sections where appropriate. Example problems and solutions are given throughout the book which are, without exception, taken from real-life situations encountered by the author.

This book includes original work not previously reported. Since publication in the present form will largely preclude publication in professional journals, it seems appropriate that material which otherwise might have appeared in current engineering literature be listed, as follows:

- G-value
- Detention basins
- Generalized numerical solution and graphs for horizontal collection conduits of constant top width
- Generalized solutions and graphs for horizontal, frictionless collection conduits of varying top width

The author was privileged to associate with eminent engineers of great influence early in his professional career. He enjoyed a special relation with the late Thomas R. Camp with whom he spent many hours in stimulating discussion. Camp's youthfulness of spirit, bold willingness to risk error, and the tenacity with which he held to his convictions made a deep and lasting impression. The more cautious guidance of the late Richard L. Woodward and the sharing of his immense knowledge and resource material was inspirational and of great educational benefit to the author. Others too numerous to mention are to be thanked for their contribution to this endeavor. They include many professors and professional colleagues and, in particular, the author's clients who have provided the practical experience necessary for this endeavor. Special thanks are due Janet H. Reardon for her excellent typing of much of this manuscript, and the reviewers selected by the publishers, who made helpful suggestions. My last, but by no means least, acknowledgment is to my wonderful wife Arlene; to our children Steven, Brian, Keven, and Allen; and our grandchildren Josephine, Keenan, Thomas Joseph (TJ), Sophia, and Maizey whose generation this book is ultimately intended to serve.

S. David Graber
Stoughton, Massachusetts

Acknowledgments

I began this book about 50 years ago, added a great deal along the way, and once retired could devote the time necessary to see it to completion. Initial inspiration for this work began through my affiliation with the late Dr. Thomas Ringgold Camp of the then engineering firm of Camp, Dresser and Mckee, with whom I was privileged to work, learn from, receive inspiration, and together publish numerous articles. I was fortunate earlier to benefit from the inspiration provided by numerous exceptional teachers and professors at various levels. My very ability to live long enough to see this work to completion is due to the expertise and care of Dr. Margarita T. Camacho, M.D. of Newark Beth Israel Medical Center, who performed my heart transplant surgery with little time to spare in May 2015. My editors at Wiley, Nandhini Karuppiah and Kalli Schultea, were wonderful to work with and very supportive. For many years, I served as a member and corresponding editor of the American Society of Civil Engineers Environmental & Water Resources Institute, to which I owe thanks both for the colleagues with whom I was privileged to work and for exposing me to needs in our field which prompted numerous of my publications. My lovely wife Arlene is thanked for meticulously editing the book and suggesting many changes that greatly improved the book. And thank you to Danijel Katic of KaticCAD in Norwood, Massachusetts, for providing expert assistance in CAD drafting many of the figures that appear in the book.

1

Introduction

1.1 Role of Hydraulics and Pneumatics in Environmental Engineering Design

The environmental engineer has historically been the leading professional in water supply and wastewater collection and disposal. In that capacity, he or she was earlier known as a *sanitary engineer*, a reflection of the public health significance of those endeavors. As this spectrum of activities broadened to deal with the environmental problems facing an increasingly complex society and as the broader environmental objectives of the sanitary engineer's endeavors became apparent, the term *environmental engineer* became more popular within the profession. The aforementioned objectives include protection of drinking water supplies, protection and propagation of aquatic life, aesthetic and recreational aspects, and agricultural (livestock water supply and irrigation) and industrial water supply. In 1972, the Sanitary Engineering Division of the American Society of Civil Engineers formally changed its name to the Environmental Engineering Division. The environmental engineer's areas of activity were at that time considered to include the following: air pollution, environmental quality, water supply, municipal wastes, industrial wastes, agricultural wastes, urban (and rural) land runoff, thermal pollution, nuclear energy, and solid wastes. It was noted that all civil engineers (and other engineers as well) deal with the environment. However, the sole focus of the environmental engineer is on *environmental protection* as opposed to *production* or *environmental modification*. Whereas other branches of engineering deal with activities of potential impact on the environment, that is not their primary interest. Environmental engineering has a singular concern and responsibility for environmental protection.

Hydraulics is the engineering art and science dealing with the macroscopic, physical behavior of liquids. Relating this definition to the above areas of activity, *environmental engineering hydraulics* deals with the hydraulic aspects of: the environmental quality of water in the hydrologic cycle: municipal and industrial water supply; municipal, industrial, and agricultural wastewaters and stormwater runoff plus thermal and nuclear pollution and solids wastes. Environmental quality is impacted by the above wastewater and stormwater runoff plus thermal and nuclear pollution and solid waste.

Pneumatics is the art and science of dealing with the movement of air, generally in human-made systems. (Its inclusion makes this book's title a slight misnomer.) In environmental engineering applications, pneumatic systems assumed initial importance in the aeration of activated sludge. Now air and other gases (mentioned in Chapter 9) play an important role in a wider range of the abovementioned areas of activity. The emphasis in this book is on pneumatic (air) systems, but many of the principles are more generally applicable.

In providing the services mentioned above, the environmental engineer integrates various natural sciences and engineering disciplines. This integration is the result of his or her own interdisciplinary training and experience, and the support of other professionals. The water-oriented environmental engineering curriculum usually provides a smattering of chemistry, microbiology, aquatic ecology, and process theory. The latter entails a mixture of fundamental science and empiricism relating to influent and effluent characteristics of unit processes. Practicing engineers apply this knowledge with the consulting support of specialists within their own field (most particularly process specialists and aquatic chemists). The supporting role of the aquatic ecologist is relatively recent and increasing. The support of the chemist and aquatic ecologist is most significant during the preliminary design phase when treatment processes are being selected and environmental impact assessed.

During the design phase of a project, the support of other design professionals is commonly provided. These include architects, structural engineers, electrical and instrumentation engineers, and mechanical engineers. The latter generally design, as a minimum, the HVAC (heating, ventilating, and air conditioning) and plumbing systems. In some large consulting

Hydraulics and Pneumatics in Environmental Engineering, First Edition. S. David Graber.

firms, mechanical engineers may design process piping, pump systems, and the like, although the environmental engineer is called upon to do this in many offices. Through his training (both formal and on-the-job) and the professional assistance of equipment manufacturers, the environmental engineer is usually capable of handling many aspects of the "mechanical" design.

Although of major significance, the hydraulic engineering aspects present a particular challenge and somewhat of an anomaly in this scheme of things. The environmental engineer is not often trained beyond the undergraduate level in fluid mechanics; many graduate environmental engineering curricula do not include a hydraulic engineering course. Equipment manufacturers are only a limited source of professional support since hydraulic structures (including process tanks) are generally made of concrete with mechanical equipment constituting a small fraction of the cost. Environmental engineering textbooks and design manuals provide some practical hydraulics, but they are often oversimplified and erroneous. Hydraulic engineers are often appalled at the "hydraulics" presented in the environmental engineering literature. Ironically, most hydraulic and environmental engineers have civil engineering backgrounds, and yet the methods of these two branches of civil engineering have not been well integrated. Engineering firms, particularly larger ones, may have one or more engineers with graduate-level training in hydraulics whose advice and review are solicited. This serves both to improve the design and the hydraulic knowledge of the environmental engineers who are exposed to this input. However, hydraulic engineering expertise is often not available; environmental engineering systems do not commonly incorporate the major hydraulic structures that justify a hydraulic engineering staff such as might be found in a hydroelectric design firm.

Furthermore, even when available, the hydraulic engineer may have had little experience with the unique environmental systems.

The availability of hydraulic engineers with experience in environmental engineering systems would reasonably resolve the problems referred to above. However, there is an additional dimension to consider. All water-oriented environmental engineering processes, both in constructed facilities and in the environment, are largely hydraulic processes, and many are predominantly so. However, there is a large gap between hydraulic engineering and environmental engineering technology. The environmental engineer's understanding of the processes with which he is concerned would greatly benefit from an understanding of hydraulic fundamentals. This benefit would extend to research efforts as well as practical applications.

The author has become of the opinion that hydraulics should be as much a part of the environmental engineer's fundamental background as chemistry and microbiology. It is hoped that the chapters which follow speak for themselves in this regard and that the reader will eventually share this opinion.

1.2 Scope, Organization, and Approach

1.2.1 Scope and Organization

Following this introductory chapter are four chapters (Chapters 2 through 5), which deal with basic hydraulic concepts and provide the theoretical foundation for the subsequent chapters. They do, however, incorporate some practical information, which stands by itself. Chapter 2 rigorously employs system and control volume concepts in the presentation and application of mass, momentum, and energy relations. This has been a sound trend in engineering education, and the author believes it to be unifying and of practical importance in the broader range of applications that includes, e.g., open-channel flows. The importance of the Navier–Stokes equations and dissipation function (Chapter 3) lies partially in their relation to "G-value," a seriously misapplied notion that has gained undue prominence. This book attempts to correct the notion of G-value and provide (in later chapters) suitable alternatives. The Navier–Stokes equations are also used (Chapter 4) as a point of departure for the unified discussion of various types of flow. Dimensional analysis (Chapter 5) is dealt with as an important concept that deserves greater attention, both for the perspective it provides and as an analytic tool.

From Chapter 6 on, the book deals with specific applications of the basic concepts presented earlier. Chapters 6 through 14 deal essentially with man-made environmental engineering systems. Although Chapters 6 through 13 fall within the conventional provinces of hydraulics and pneumatics, they are considered here from the standpoint of the environmental engineer's applications. This seems very useful and is done here for the first time known to the author. Chapter 14 deals with process applications from what the author believes is a sound and useful hydraulic perspective. In the design and research aspects of treatment processes, the significance of hydraulics needs greater appreciation. There is a significant disparity between existing knowledge in the field of hydraulics and its applications to treatment processes. The author believes strongly that experimental and theoretical hydraulics, in fact, provide the best framework for properly understanding certain treatment processes and help to elucidate all treatment processes. Hydraulics plays a major, if not dominant, role in essentially all unit processes of water and wastewater treatment. The breadth and utility of the author's interpretation of *environmental engineering hydraulics and pneumatics* should be made clearer by the inclusion of Chapter 13.

Chapter 15 deals with what has loosely been termed "natural systems," and provides an introduction to the hydraulics of such systems. That large subject area is only dealt with in an introductory way for several reasons: (1) natural systems have benefited from more complete treatment by hydraulic engineers; (2) the literature on natural systems (as cited in Chapter 15) is technically sound and reasonably consolidated; (3) the study of natural systems is generally more complex and specialized; and (4) the need for involving specialists in this area is generally recognized by environmental engineers.

This book is intended to prepare the environmental engineer to handle many but not all of the hydraulic aspects that will be encountered. The author has endeavored to point out those areas where the complexity of a problem warrants the hydraulic specialist's involvement. A serious effort has also been made to point out those areas in which hydraulic design errors are often made and to point out where application of methods presented in this book can avoid those pitfalls. Areas that could benefit from research efforts are pointed out; the encouragement of such research is among the purposes of this book.

There are some areas of hydraulics that the environmental engineer may deal with that are not covered in this book. Dam design is a specialty drawing from the disciplines of structural, soil, and hydraulic engineering and is dealt with in books devoted largely to that subject (Davis and Sorensen 1969, USBR 1965). It is not addressed here. Municipal storm drainage systems, largely because of their similarity to sanitary sewer systems, have traditionally been within the province of the environmental engineer. They have become even more proper with the recognition of the pollution associated with urban stormwater runoff. The hydraulic principles of municipal storm sewer design are covered here. Land drainage [as distinguished from municipal storm drainage (Linsley and Franzini 1964, pp. 52–54)] is not covered in any purposeful detail. The principles presented in this text should, however, permit ready transition via self-study or additional course work.

1.2.2 Approach

The author believes strongly in the practical value of a fundamental understanding as opposed to "cookbook" ("add water and stir") approaches. Armed with a fundamental understanding, errors can be better avoided, and problems associated with new applications can be more successfully formulated and solved. There has been a tendency, particularly in large design firms, toward standardization of design. Unfortunately, the reasons behind many such standards have been left with the engineers of earlier days. They are sometimes applied more as a matter of faith without necessary regard to their current applicability. The dissemination of computer programs to those not conversant with the techniques embodied in the programs has been a related concern. The author's observations in this regard are supported by Beck (1979) and Orenstein (1984). Partly to deal with these concerns, the emphasis here is placed on the "why" as well as the "how." Sewer design is a case in point. Too often sewer designers are unfamiliar with the meaning and nature of uniform flow, subcritical vs. supercritical, etc., and run the risk of erring. Most books treating sewer design are of little help in this regard. Unfortunately, hydraulic errors may go undiscovered because the design capacity of sewers (and treatment plants) is years into the future. And conduits receiving storm flows are rarely observed during storm conditions. This book attempts to provide concepts necessary to rectify such design inadequacies. The emphasis is on understanding and design of components for integration into systems described in the popular texts on water supply and wastewater systems.

A rigorous yet practical text incorporating principles for applications to new systems has been the goal. The author has strived to provide a clear exposition of all results. Choices of analytical tools have been made, which are likely to be of practical use. Theoretically oriented hydraulicians have studied phenomena such as turbulence and pipe resistance on a fine scale. The approach here will be a more large-scale and empirical one. A better merger between theory and practice is an intended result.

Examples given throughout this book are from projects successfully designed by the author. References include recent publications, but also older ones of lasting value.

1.3 Fluid Flow Phenomena

Hydraulic engineering is very much an art as well as a science. Many of the systems with which the hydraulic engineer deals are not amenable to complete theoretical analysis. The understanding of a hydraulic system or proportioning of a hydraulic structure must often be arrived at by a "feeling" for the problem, unaided or only partially aided by numerical calculation. A physical feel for the problem is always important. A qualitative understanding of the flow phenomena, aided by general experimental results and observations, will provide the engineer with the ability to proportion many hydraulic structures.

Furthermore, good qualitative understanding coupled with design experience can often avoid the uncertainties of complicated designs, undue analysis, or the need for physical modeling.

Indeed, many present-day environmental engineering devices would not exist if engineers had waited until a satisfactory theory was available to explain the operation of the device. As stated by Stenning and Shearer, "Few systems are pure enough or simple enough to conform very closely to basic theoretical concepts. Most fluid-flow systems are sufficiently complex to require a great deal of judgment in applying the theoretical concepts to them. Experience in the form of empirical data taken by others and in the form of personal observations ... and engineering practice is a vital factor in most successful engineering analyses" (Stenning and Shearer 1960).

A significant characteristic of the hydraulic components of man-made environmental engineering systems is that taken individually they are often somewhat "minor" hydraulic structures. These hydraulic components, although vital to the performance of the overall system, often cannot, due to time and budget constraints, be subjected to the degree of analysis or physical modeling that might be performed for a major hydraulic structure. Furthermore, physical models involving two phases (e.g., liquid–solid as in flocculation, or gas–water mixtures as in aeration) that the environmental engineer often encounters will not necessarily yield results which can be extrapolated to the prototype (see Chapter 14). Thus, one must often do the best that can be done with simplified analytic methods and judicious design. This theme underlies much of this book.

A functional, economical design can be best achieved by intelligent combination of theoretical and empirical knowledge. Regardless of the particular combination of theory and empiricism employed, it is essential to identify and understand the phenomena of concern. In addition to those phenomena imparting the desirable and more obvious properties of fluid flow, some fluid flow phenomena may sometimes impart undesirable characteristics. It behooves the designer or analyst to be thoroughly aware of the range of fluid flow phenomena. Such phenomena and a partial indication of the location of their coverage in this book are listed below:

Fluid Flow Phenomena	Reference
Viscosity	Section 1.5
Zero Slip at Solid Boundary	Section 4.2
Momentum Exchange	Sections 2.4, 2.5, 2.7, 4.4, 4.7, 6.4, 12.1, 12.2, 12.5, 12.6
Boundary Layers, Velocity Profiles	Sections 2.5, 4.2, 4.3, 4.7, 6.2, 6.4, 7.7, 7.8, 9.4, 10.3, 12.1, 13.2, 14.3, 15.2, 15.3, 15.6
Jet Flow, Entrainment	Sections 6.2, 6.3, 7.3, 7.5, 7.7, 7.8, 8.4, 8.6, 8.7, 9.4, 9.5, 11.4, 13.3, 14.7, 15.2, 15.5
Laminar and Turbulent Flow	Sections 3.3, 4.2, 4.4, 4.5, 4.6, 4.7, 6.2, 6.5, 7.1, 7.4, 7.5, 7.8, 8.2, 8.6, 9.3, 9.5, 12.7, 13.2, 13.4, 14.3, 14.4, 14.5, 14.6, 14.7, 14.8, 15.6
Dispersion	Sections 14.2, 15.2, 15.4
Cavitation	Sections 6.5, 8.2, 8.6, 11.4, 12.5
Separation and Reattachment	Sections 4.7, 6.3, 6.4, 6.5, 7.7, 8.1, 8.6, 11.5, 14.3, 14.5
Frictional Drag	Sections 4.5, 7.5, 11.3, 13.4, 13.5,14.8
Form Drag	Sections 4.5, 6.5, 7.5, 10.3, 11.3, 11.5, 14.8
Coanda Effect	Sections 7.8, 15.2
Asymmetric Flow in Symmetric Expansions	Section 15.2
Vortices	Sections 4.7, 6.2, 6.4, 6.5, 7.2, 7.6, 7.7, 8.6, 8.7, 8.8, 11.5, 13.3, 14.9, 15.2, 15.3, 15.4
Secondary Currents	Section 4.7
Surface Waves; Subcritical, Critical, and Supercritical Flow	Chapter 7
Hydraulic Jump, Cross Waves	Sections 6.2, 6.5, 7.2, 7.5, 7.6, 7.7, 7.8, 15.2, 15.4
Compressibility; Subsonic, Sonic, Supersonic, and Choked Flow	Chapter 9; Sections 11.2, 15.4
Sediment Transport	Sections 4.7, 13.2, 13.3, 13.4
Flocculation, Floc Breakup	Sections 3.4, 4.4, 14.2, 14.6
Two-Phase Flow	Sections 6.5, 13.3
Density Currents	Section 14.3
Langmuir Circulation	Section 15.5

The fluid flow films prepared under the sponsorship of the National Science Foundation (National Committee for Fluid Mechanics Films 1972) are excellent educational aids dealing graphically with many of these phenomena.

1.4 Definitions and Classifications of Flows

Certain terms are used throughout much of this book and are defined below:

Fluid – a material that deforms continuously and permanently under the application of a shearing stress, which can be in a liquid or gaseous state. Fluids can be Newtonian or non-Newtonian as discussed in Section 3.2.

Conduit – A natural or constructed boundary that confines flow to move in a particular general direction and which has a total dimension in the general direction of flow much larger than its dimensions transverse to the direction of flow.

Prismatic Conduit – A conduit (generally human-made) with constant cross-sectional geometry.

Pipe – A prismatic conduit of circular cross section.

Pressure Conduit – A conduit flowing full so that the minimum pressure in the cross sections is other than atmospheric.

Open Channel – A liquid-conveying conduit having a free water surface.

Steady Flow – Fluid flow in which no fluid properties (other than those associated with high-frequency turbulent fluctuations) vary with time ($\partial/\partial t = 0$), although they may vary spatially.

Constant Flow – Steady flow in a conduit in which the total mass rate of flow does not vary in the direction of flow. For an incompressible flow (constant density – see below), it is only necessary that the volumetric rate of flow not vary in the direction of flow ($\partial Q/\partial s = 0$).

Uniform Flow – Constant flow in which fluid properties do not vary with distance in the direction of flow ($\partial/\partial s = 0$); such flow occurs in sloped open channels under certain conditions (Section 7.3).

Spatially Varied Flow – Flow in a conduit in which the mass rate of flow varies with distance along the conduit.

Unsteady or Time-Varying Flow – Flow in which fluid properties vary with time (in addition to high-frequency turbulent fluctuations).

Incompressible Flow – Flow in which density changes are negligible.

Compressible Flow – Flow having significant density changes.

The significance of the above definitions will become more apparent as the terms are used in the text.

Numerous additional flow classifications are defined in precise terms in the text. These classifications and the section of the text in which they are defined include the following:

One-dimensional flow (Section 2.6)

Parallel flow (Section 2.6)

Potential flow (Section 4.3)

Laminar flow (Section 4.4)

Creeping flow (Section 4.5)

Turbulent flow (Section 4.7)

Critical flow (Sections 7.2, 7.4)

Subcritical flow (Section 7.4)

Supercritical flow (Sections 7.4, 7.5)

Adiabatic flow (Sections 2.6, 9.3)

Nonadiabatic flow (Section 9.3)

Isothermal flow (Sections 2.6, 9.3)

Isentropic flow (Section 9.4)

Choked flow (Section 9.4)

1.5 Viscosity

Viscosity is a very important feature in environmental fluid mechanics. It also has a particularly illuminating history, as recounted here. We will first define the term "law" as used in science. A scientific law (or "first principle") is the description of an observed phenomenon; it does not explain why the phenomenon exists or what causes it. Examples of laws are the Law of Conservation of Mass, Newton's Second Law of Motion, and the First Law of Thermodynamics, all discussed in Chapter 2. Newton's "Law of Viscosity" is not actually a law because it can be derived from the kinetic theories of gases and liquids (Frenkel 1955, Kirshner undated); its mathematical formulation is provided in Chapter 3. Newton provided the first complete proof of the revolution of the Earth and planets around the Sun and correct explanation of the causes of the tides, as discussed in Section 14.5. Newton's discoveries of some of the most fundamental laws of nature are believed to have occurred primarily during a two-year period of seclusion during a Bubonic Plague (Black Death) outbreak just after he had finished college.

Newton's Laws of Motion are embodied in what has been called the single most important equation in the universe: $F = ma$. Force equals mass times acceleration. Newton's Laws of Motion are commonly taught as three separate laws because that is the way Newton presented them in his *Principia* (Newton 1999) in order to directly address misconceptions of the time. The Second Law of Motion is expressed by the famous equation $F = ma$. But the first and third laws are actually corollaries of $F = ma$ (Graber 2004c). The First Law of Motion says that a body at rest tends to remain at rest, and a body in motion tends to remain in motion. It was and is a profound statement, but it simply says that it takes force to accelerate or decelerate a mass, which is quantified by the Second Law. The Third Law of Motion says that for every action there is an equal and opposite reaction, which also derives from the Second Law. $F = ma$ is actually a somewhat simplified statement of the Second Law of Motion, and the law has many ramifications (e.g., conservation of linear and angular momentum). A more generalized form of the equation that applies to much of the work in environmental and water resources engineering is presented in Section 2.4, and an additional discussion of Newton's Laws of Motion is provided in Chapter 10.

Scientists constantly scrutinize laws to test their accuracy and generality. Newton's Laws of Motion are fundamental laws of nature that have stood up to countless experiments and observations in nature and engineered projects. And, although Albert Einstein's Special Theory of Relativity is sometimes said to have modified Newton's Second Law of Motion when objects move at speeds that approach the speed of light because the mass of the objects changes, the fact is that in the form that Newton originally presented his Second Law, i.e., as force equals the time rate of change of momentum, it is completely compatible with Einstein.

In the fourth century B.C.E., Plato and Aristotle believed that the Earth was immobile and that the Sun, planets, and stars revolved around the Earth. However, also in the fourth century B.C.E., the Greek philosopher Philolaus of Croton, in his *Treatise of the Sky*, expressed the opposite, and as it is now known, correct opinion held by the Pythagoreans that the Earth revolves around the Sun and "its circular motion about its own center produces the day and night." (Aczel 2003, p. 12) Then a superb mathematician and the greatest astronomer of the ancient world, Claudius Ptolemy in Alexandria, Egypt, in the second century C.E., came along and proposed a sophisticated mathematical theory of an Earth-centered solar system. This was consistent with the church's interpretation of the Bible (Joshua 10:12–13; Isaiah 38:7–8; Ecclesiastes 1:5), so the church used Ptolemy's theory to justify the Aristotelian view and attacked what came next. What came next were the scientific endeavors of the sixteenth and seventeenth centuries: Copernicus in Poland, Kepler in Germany, and Galileo in Italy, who advanced the modern heliocentric (Sun-centered) view of the solar system, culminating in the work of Newton in England and Foucault in France. During this period, the premature science of Ptolemy became pseudoscience, as the Catholic Church took strong measures to protect its long-held point of view despite new and overwhelming evidence. Copernicus, despite being in Catholic Poland, escaped the Inquisition because his book, *On the Revolution of Heavenly Spheres*, was published the year of his death (and was according to legend delivered to him on his deathbed). However, Galileo was not so fortunate. In "1600, the Inquisition [had] brought the Italian monk and teacher Giordano Bruno ... in chains to ... the center of Rome, tied him to an iron stake, and burned him alive. One of his crimes was his belief that the Earth rotated." (Aczel 2003, p. 9) That same Inquisition threatened Galileo with torture, forced the great scientist, now regarded as the "Father of Physics," to kneel before his prosecutors, forced him to recant his belief about the solar system to avoid a painful death, and sentenced him to house arrest for the rest of his life. (Aczel 2003, p. 10, Hecht 2003) Léon Foucault, incidentally, gave irrefutable proof that the Earth rotated on its axis with his great pendulum demonstration in 1851 at the Paris Observatory. A five-story-tall replica of Foucault's pendulum can be seen today at the Boston Museum of Science.

It should be noted that Ptolemy's incorrect Earth-centered theory was accepted for more than 1300 years. So another lesson in this is that the presumed validity of an idea should not be based on its longevity. Nor, for that matter, should validity be assumed based exclusively on the credentials of the proponent.

The author has addressed erroneous concepts that have been applied to the design of numerous types of environmental engineering facilities. The following table (Table 1.1) (from Graber 2010c) lists numerous concepts of "established practice" that the author has overturned and for which he has provided corrective methods. This is offered as encouragement to others to not simply accept published design methods but rather to critically review their validity and applicability and modify them for specific applications and more generally when appropriate. The first column of the table references papers by the author, which are among those listed in the References section at the end of this book. The second column gives the concept overturned for which corrective methods are provided in the cited papers. The third column lists the original proponents of the erroneous concepts plus some comments. The reader is referred to the cited papers for details. Subsequent chapters will address a number of these concepts. An interesting paper in this context is "Creativity vs. Skepticism within Science" (Ramachandran 2006).

Viscosity is further discussed in Section 3.2, including explanations of kinematic, dynamic, and absolute viscosity.

Table 1.1 Overturned Concepts of "Established Practice"

Reference	Concept Overturned and Corrective Methods	Proponents/Comments
Graber 1974a, 2005	Weir loadings for primary clarifiers	National Research Council/Acknowledged in 1998 Manual of Practice of American Society of Civil Engineers/Water Environment Federation (WEF/ASCE 1998)
Graber 1994, 1997, 1998, 2004d	G-value or RMS velocity gradient for flocculation, floc breakup, mixing, induced circulation in channels, short-circuiting in tanks, and cleansing of filter media during backwashing	Many, including cited references. One major textbook called this "an important unifying concept in modern water or wastewater operations or processes."
Graber 1972, 1974a, 1975a, 2005	Role of short-circuiting in primary sedimentation	Numerous cited references
Graber 2004a	Assumption of uniform flow for spatially varying flow in subsurface drains	American Iron and Steel Institute, U.S. Soil Conservation Service, U.S. Department of Agriculture, U.S. Bureau of Reclamation, American Society of Civil Engineers, U.S. Environmental Protection Agency/Received ASCE's Samuel A. Greeley Award. Author's replacement method now part of ASCE Standard Guidelines (ASCE 2005)
Graber 2004b, 2004c	Conceptual and practical errors relative to spatially varied flows	Various cited references
Graber 2007a, 2007b	Erroneous methods for design of aggregate, geopipe, and geocomposite subsurface drains	Federal Highway Administration, National Cooperative Highway Research Board, Forrester (2001), Koerner (2005)
Graber 2009a	TR-55 curves can substantially underpredict required detention basin storage, or conversely underpredict the peak outflow from the detention	U.S. Soil Conservation Service (now Natural Resources Conservation Service)
Graber 2009b	Standards underpredict the depth	Factory Mutual Insurance Company; American Society of Civil Engineers; Fairfax County, VA
Graber 2010a	Stormwater pumping station wet-well storage is underpredicted and can result in upstream flooding under design storm conditions	American Society of Civil Engineers, Water Environment Federation, Federal Highway Administration
Graber 2010b, 2010c	Erroneous concepts for design of wastewater pressure-distribution systems, rapid sand filter underdrains, and dividing manifolds more generally	Various cited references
Graber 2013a	Assumption of uniform flow for spatially varying flow in street gutters	Federal Highway Administration, ASCE, WEF/ASCE

1.6 Dimensional Units and Notation

1.6.1 Dimensional Units

The four physical units employed in engineering mechanics are those of force, mass, length, and time. Thermal units and electromagnetic units must be introduced to embrace the full range of physical phenomena. Thermal units will be discussed in Chapter 9; electromagnetic phenomena will not be considered. All four are involved in Newton's Second Law of Motion, which thus provides a useful basis for discussing systems of units in many texts, including this one. Newton's Second Law in its simplest form is given by:

$$F = \frac{1}{g_o} ma$$

in which F is the force acting on a system of particles of mass m, a is the system acceleration (length per time squared), and g_o is a proportionality constant that depends on the units of force, mass, and acceleration.

Sets of units in use and the corresponding proportionality constant are given as follows:

Mass	Length	Time	Force	g_o
slug	ft	sec	lbf	1 slug ft/(lbf sec^2)
lbm	ft	sec	poundal	1 lbm ft/(poundal sec^2)
lbm	ft	sec	lbf	32.174 lbm ft/(lbf sec^2)
kilogram	meter	sec	newton	1 kilogram meter/(newton sec^2)
gram	centimeter	sec	dyne	1 gram centimeter/(dyne sec^2)

in which lbm denotes pound-mass and lbf denotes pound-force. The third set is traditionally used in American industry and is known as the foot-pound-second (FPS) system. The fourth is the basis for the SI units (systeme international d'unites) of the standardized metric system (also known as mks units). Mass, length, and time in the fifth system (known as cgs units) are merely multiples of those in the fourth system.

Dimensionless equations are widely used in fluid mechanics so that units of choice may be used and properly checked. In presenting such equations, a common convenience is the assumption that $g_o = 1$ so that $F = ma$ and the term g_o does not have to be employed. However, the slug and poundal, which make $g_o = 1$ in the first two systems above have received little practical acceptance. Nonetheless, the first system has been widely used in fluid mechanics texts, with conversions to lbm made according to 1 slug = 32.174 lbm.

The tradition of assuming $g_o = 1$ will be continued in this text, and equations will be derived and presented without g_o. The English units used in examples will be those of the third system above, with lbm and lbf used to clearly indicate pound-mass and pound-force, respectively, and the above conversion from slugs to lbm employed.

Reid (2000) and Harish (2023) have noted some major errors due to inconsistent use of dimensional units. One is the unintended crashing of NASA's $125 million Mars Climate Orbiter onto the surface of Mars. Others reported by Reid and the running out of fuel by an Air Canada Boeing 747 jet in mid-flight because of a conversion error in calculating the plane's fuel storage, administering an incorrect dosage of a sedative to a hospital patient, and an American International Airways cargo plane landing 13,000 pounds overweight due to a conversion error.

1.6.2 Notation

It is impossible in a text with the breadth and scope of this one to employ notation that is consistent between subject areas and, at the same time, consistent with the notation common to a given subject area. The author has attempted to strike a reasonable balance in this regard.

2

Mass, Momentum, and Energy

2.1 Introduction

Three basic laws form the basis for the large majority of the analytical work in environmental engineering hydraulics and pneumatics. Before starting and elaborating upon these laws, this introductory statement will be emphasized. The statement is somewhat akin to the statement that *photosynthesizing plants form the basis for all higher life on earth*, in requiring understanding and reflection to appreciate. The importance of the three laws and the value of understanding them thoroughly cannot be overstated. Even the experienced analyst may find occasion to chastise himself or herself for tardiness in returning to these basics when attempting to solve a perplexing problem.

The three basic laws are the Law of Conservation of Mass, Newton's Second Law of Motion, and the First Law of Thermodynamics. Each is presented below, along with pertinent discussion and exemplification. For the reader wanting a review of the basics, excellent discussions of those laws as well as excellent introductory examples of applications are given in Shames' *Mechanics of Fluids* (Shames 1962, Chapter 5). The introductory applications presented in the present chapter, in addition to exemplifying applications of the techniques presented, are important to the practical considerations of subsequent chapters. Portions of this chapter are taken from Graber (2023a).

2.2 Closed Systems and Control Volumes

2.2.1 Closed Systems

The term *closed system* is used in thermodynamics, whereas in fluid mechanics texts the term *system* alone is often used to have the same meaning. However, because of the varied usage of the word "system" in engineering work, the term closed system is less ambiguous and will be employed here.

A *closed system* is defined as a prescribed mass of material, in our case, a prescribed fluid mass. Mass cannot be added to or taken from a closed system. The closed system moves with the flowing fluid and changes shape. For example, a closed system might be defined as the fluid within a particular closed surface at a given instant flowing in a pipe at time t. After a time interval, the closed system would be distorted from the original shape (and possibly broken into disconnected shapes) at time $t + \Delta t$.

2.2.2 Control Volumes

The concept of control volume (CV) is of vital importance in fluid mechanics. Many problems owe their solution to the analyst's ability to judiciously select a CV. The *CV* is defined as a prescribed volume fixed in space. The *CV* is identical to the *open system* of thermodynamics. See Shames (1962) and additional discussion below regarding non-fixed, i.e., moving, CVs. The boundary of the CV is called the *control surface*. Liquid flows into and out of the CV, and the amount and other characteristics of fluid in the CV may change with time (as, for example, if a pipe is alternately drained and filled). The shape and position of the CV once defined remain fixed for all time. A closed system coinciding with the CV one instant would later be found displaced downstream.

Hydraulics and Pneumatics in Environmental Engineering, First Edition. S. David Graber.

The characteristics of the CV with which we will be concerned include differential elements of the CV denoted by dV and differential elements of the control surface denoted by dA. The orientation of the differential area dA assumes importance and is described by the direction of a unit vector normal to dA and pointing outward from the CV interior; the direction and magnitude of the differential area are fully defined by the product of dA and the unit vector denoted by $d\bar{A}$. The total volume of the CV is equal to the integral of $d\mathcal{V}$ over the CV, denoted by the triple integral $\iiint_{c.v.} d\mathcal{V}$. The total area of the control surface is equal to the integral of $d\mathcal{V}$ over the CV, denoted by the surface integral $\oiint_{c.s.} dA$. The reader should not feel at all intimated by these integral notations here or in the equations given below. They merely represent a convenient and precise means of expressing the simple operations to be performed, as explained below, and *serve to emphasize the importance of thinking explicitly about the CV.*

The CV approach is generally more useful than the closed system approach in fluid mechanics *applications*, and the examples presented in this chapter will emphasize the CV approach. The closed system approach is useful in deriving certain fundamental relations in fluid mechanics, for which purpose the closed system concept will be used in the next section and in Chapter 3.

2.3 Conservation of Mass

The Law of Conservation of Mass states that the mass of a substance is conserved in the absence of chemical and nuclear reactions. The word "substance" is meant here to mean a single chemical species or compound or a combination thereof. Chemical and biochemical reactions will be considered in Chapter 14, wherein a more general statement of mass conservation will be given. Diffusion and dispersion concepts will be brought into the mass conservation relation in Chapters 13 and 14, respectively.

2.3.1 Closed Systems

For a *closed system*, conservation of mass is expressed simply by $Dm/Dt = 0$, where m is the mass of fluid in the closed system. The notation D/Dt denotes the *substantial* or *total* derivative with respect to time, which indicates that the derivative is being carried out as a system of fluid particles is being followed. Since $m = \rho\mathcal{V}$ in which ρ is the fluid density (mass per unit volume, e.g., in lbm/ft^3, slug/ft^3, or kgm/l) and $\mathcal{V}$ is the system volume, we have:

$$\frac{D}{Dt}(\rho\mathcal{V}) = \rho\frac{D\mathcal{V}}{Dt} + \mathcal{V}\frac{D\rho}{Dt} = 0 \tag{2.1}$$

In order to derive the differential equation expressing conservation of mass at a point in a flow, we will apply the above relation to the system described in rectangular coordinates in Figure 2.1. The initial position of the system and the position after time Δt are shown by solid and dashed lines, respectively (a similar expansion occurs in the z-direction). The terms v_x, v_y, and v_z denote velocities (e.g., in ft/sec or meter/sec) in the x, y and z directions, respectively. The change in volume with respect to time can be written as:

$$\frac{D\mathcal{V}}{Dt} = \frac{\left[\Delta x\left(1 + \frac{\partial v_x}{\partial x}\Delta t\right)\right]\left[\Delta y\left(1 + \frac{\partial v_y}{\partial y}\Delta t\right)\right]\left[\Delta z\left(1 + \frac{\partial v_z}{\partial z}\Delta t\right)\right] - \Delta x\Delta y\Delta z}{\Delta t} \tag{2.2}$$

Neglecting terms of higher order, the above equation reduces to:

$$\frac{D\mathcal{V}}{Dt} = \left(\frac{\partial v_x}{\partial x} + \frac{\partial v_y}{\partial y} + \frac{\partial v_z}{\partial z}\right)\Delta x\Delta y\Delta z \tag{2.3}$$

The density at a point is functionally expressed by:

$$\rho = f(x, y, z, t) \tag{2.4}$$

From which the differential is:

$$D\rho = \frac{\partial \rho}{\partial x}dx + \frac{\partial \rho}{\partial y}dy + \frac{\partial \rho}{\partial z}dz + \frac{\partial \rho}{\partial t}dt \tag{2.5}$$

The change in density with respect to time becomes:

$$\frac{D\rho}{Dt} = \frac{\partial\rho}{\partial x}\frac{dx}{dt} + \frac{\partial\rho}{\partial y}\frac{dy}{dt} + \frac{\partial\rho}{\partial z}\frac{dz}{dt} + \frac{\partial\rho}{\partial t} \tag{2.6}$$

Since $dx/dt = v_x$, etc., the above equation can be rewritten as:

$$\frac{D\rho}{Dt} = \frac{\partial\rho}{\partial t} + v_x\frac{\partial\rho}{\partial x} + v_y\frac{\partial\rho}{\partial y} + v_z\frac{\partial\rho}{\partial z} \tag{2.7}$$

Substituting Equations 2.3 and 2.7 into Equation 2.1 and canceling terms gives:

$$\rho\left(\frac{\partial v_x}{\partial x} + \frac{\partial v_y}{\partial y} + \frac{\partial v_z}{\partial z}\right) + \frac{\partial\rho}{\partial t} + v_x\frac{\partial\rho}{\partial x} + v_y\frac{\partial\rho}{\partial y} + v_z\frac{\partial\rho}{\partial z} = 0 \tag{2.8}$$

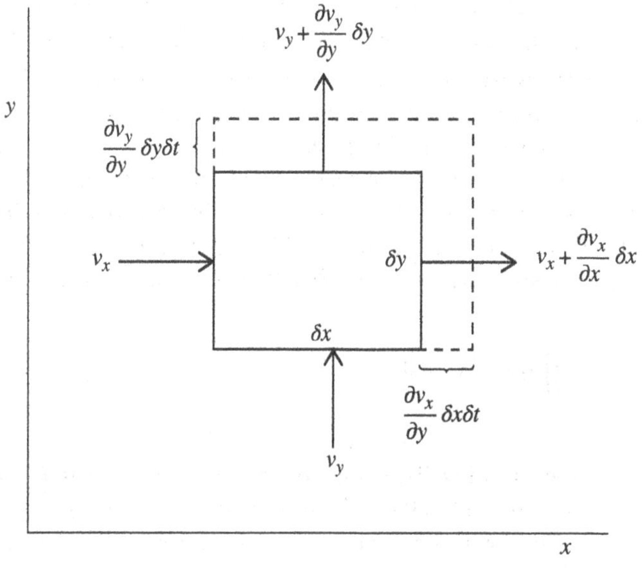

Figure 2.1 Closed System for Conservation of Mass Derivation

The general vector relationship is $\nabla \cdot (\rho\bar{v}) + \partial\rho/\partial t = 0$ in which $\nabla\cdot$ is the divergence operator. This general relation, from which Equations 2.8 and 2.10 (below) may be obtained, is derived by applying Gauss' Theorem (the divergence theorem) to the CV expression of conservation of mass (Equation 2.12) (Kuethe and Schetzer 1959).

Equation 2.8 and simplified versions of it presented below are known as *continuity* equations. For steady flow, the $\partial\rho/\partial t$ term vanishes. For *incompressible flow of a single substance*, whether steady or not, the above equation reduces to:

$$\frac{\partial v_x}{\partial x} + \frac{\partial v_y}{\partial y} + \frac{\partial v_z}{\partial z} = 0 \tag{2.9}$$

The continuity equation can be similarly derived for cylindrical coordinates (see Figure 2.2), giving:

$$\frac{1}{r}\frac{\partial(r\rho v_r)}{\partial r} + \frac{1}{r}\frac{\partial(\rho v_\phi)}{\partial\phi} + \frac{1}{r}\frac{\partial(\rho v_z)}{\partial z} + \frac{\partial\rho}{\partial t} = 0 \tag{2.10}$$

in which v_r, v_ϕ, and v_z are velocities in the r, ϕ, and z directions, respectively. For incompressible flow of a single substance, the above equation reduces to:

$$\frac{\partial v_r}{\partial r} + \frac{v_r}{r} + \frac{1}{r}\frac{\partial v_\phi}{\partial\phi} + \frac{\partial v_z}{\partial z} = 0 \tag{2.11}$$

The continuity equation can also be expressed in spherical coordinates (Rouse 1959).

2.3.2 Control Volumes

The Law of Conservation of Mass can be expressed for a CV by the following equation:

$$\oiint_{c.s.}(\rho\bar{v}\cdot d\bar{A}) + \frac{\partial}{\partial t}\iiint_{c.v.}(\rho d\mathcal{V}) = 0 \tag{2.12}$$

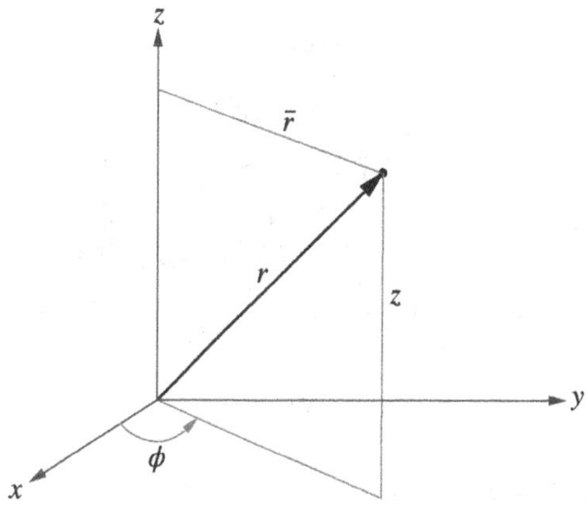

Figure 2.2 Derivation of Continuity Equation for Cylindrical Coordinates (Shames 1962/McGraw-Hill Education)

In the above equation $\bar{v} \cdot d\bar{A}$ is the dot product of the velocity vector and the differential area vector. The dot, scalar, or inner product of two vectors, denoted by $\bar{a} \cdot \bar{b}$, is defined as the product of the length of the two vectors and cosine of the angle θ between the positive directions of the two vectors (Hildebrand 1962). Thus $\bar{a} \cdot \bar{b} = |a||b| \cos\theta$ in which $|a|$ and $|b|$ are the lengths (or absolute values) of their respective vectors.

The first term in the above equation represents the net rate at which mass is passing out of the control surface, while the second term represents the net rate at which mass is accumulating within the CV. The equation says in essence that the rate of mass flow into the CV equals the rate of mass flow out of the CV plus the rate at which mass accumulates inside the CV. The above equation and the simplified forms presented below are also known as *continuity* equations.

For steady flow, we have $\partial/\partial t = 0$, and the above equation simplifies to:

$$\oiint_{c.s.} (\rho\bar{v} \cdot d\bar{A}) = 0 \tag{2.13}$$

For the following discussions, reference will be made to Figure 2.3, which depicts an end-suction centrifugal pump taking suction from a wet well and discharging it to a tank or channel. A small line tees off the suction line for purposes of chemical dosing. The dosed chemical is mixed with the aid of the pump, resulting in a thorough mixture by the time the sampling tee in the figure is reached on the discharge side.

Considering CV (control volume) 1 in Figure 2.3, Equation 2.13 gives:

$$\iint_1 \rho\bar{v} \cdot d\bar{A} + \iint_2 \rho\bar{v} \cdot d\bar{A} + \iint_3 \rho\bar{v} \cdot d\bar{A} = 0 \tag{2.14}$$

or

$$-\iint_1 \rho v dA - \iint_2 \rho v dA + \iint_3 \rho v dA = 0 \tag{2.15}$$

in which the velocities v are parallel to the pipe axes. The above equation may alternately be written as:

$$-w_1 - w_2 + w_3 = 0 \tag{2.16}$$

in which $w = \iint \rho v dA$ is the mass rate of flow (e.g., in lbm/sec, slug/sec, or kgm/sec).

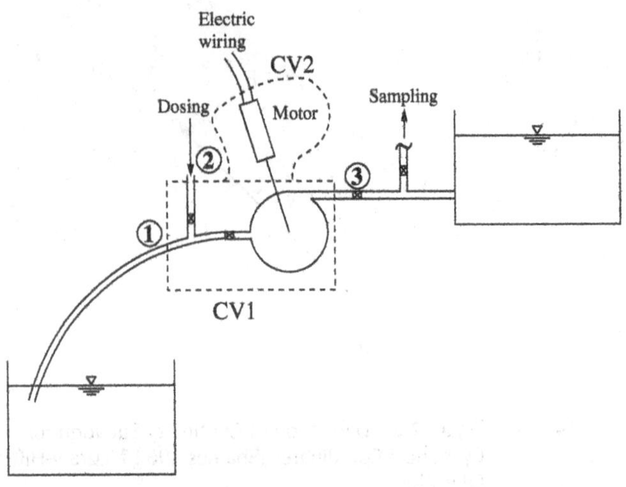

Figure 2.3 Wet Well, Pump, Tank/Channel System

For a homogeneous fluid, the densities are constant across the pipe cross sections, so the densities may be taken outside of the integral signs. Thus:

$$\rho_1 \iint_1 v dA + \rho_2 \iint_2 v dA = \rho_3 \iint_3 v dA \tag{2.17}$$

Since $\iint v dA = Q$, the volumetric rate of flow (e.g., in ft^3/sec or liters/sec), the above equation can alternately be written as:

$$-\rho_1 Q_1 - \rho_2 Q_2 + \rho_3 Q_3 = 0 \tag{2.18}$$

The *average* or *bulk mean velocity* is defined as:

$$V = \frac{\iint v dA}{\iint dA} = \frac{1}{A}\iint v dA = \frac{Q}{A} \tag{2.19}$$

Capital V will henceforth be used to denote average velocities as defined above, whereas small v will refer to point velocities. (Other localized uses of V and v may occur and will be clearly stated.) In terms of the average velocity, Equation 2.18 becomes:

$$\rho_1 V_1 A_1 + \rho_2 V_2 A_2 = \rho_3 V_3 A_3 \tag{2.20}$$

Note that the above relations would all be equally valid if CV1 encompassed a centrifugal air blower and connected piping (i.e., for compressible flow).

Clearly, if the flow streams in Figure 2.3 are of the *same incompressible fluid*, then the densities are equal and Equation 2.20 becomes:

$$V_1 A_1 + V_2 A_2 = V_3 A_3 \tag{2.21}$$

or

$$Q_1 + Q_2 = Q_3 \tag{2.22}$$

In a large proportion of environmental engineering applications, we are concerned with liquid mixtures in which water is the solvent. If the solutes or suspensions are in sufficiently small concentrations (defined below), then the mixture density approximates that of water. In such a case, Equation 2.20 again reduces to Equation 2.21 or 2.22. Such multicomponent flows will be examined in additional detail next.

2.3.3 Multicomponent Flows

The flows of concern are often multicomponent flows, in which we are interested in certain characteristics of individual components. We will introduce some of the pertinent concepts for liquid flows in this section.

For a multicomponent liquid flow, the *component density* is defined as the mass of the component divided by the volume of the mixture. The *component concentration* is defined as the mass of the component divided by the mass of the mixture [e.g., in lbm/lbm, kgm/kgm, or ppm (parts per million by mass)]. For component "i," we thus have:

$$c_i = \rho_i / \rho \tag{2.23}$$

in which ρ is the density of the mixture.

We will most often be concerned with liquid mixtures in which the predominant component is water. Two-component mixtures will be considered, although the concepts may be easily extended to mixtures of larger numbers of components.

We shall designate the two components by "A" and "B." For concreteness, think of component "A" as being fluoride and component "B" as being water, with the outgoing flow being represented without a subscript. The reader should be able to verify that:

$$\rho_A + \rho_B = \rho \tag{2.24}$$

$$c_A + c_B = 1 \tag{2.25}$$

For steady flow, Equation 2.13 applies to the individual components as well as the mixture, in which the control surface encompasses the incoming flows and the outgoing flow. Thus:

$$\oiint_{c.s.} (\rho \bar{v} \cdot d\bar{A}) = 0 \tag{2.26a}$$

$$\oiint_{c.s.} (\rho_A \bar{v} \cdot d\bar{A}) = 0 \tag{2.26b}$$

$$\oiint_{c.s.} (\rho_B \bar{v} \cdot d\bar{A}) = 0 \tag{2.26c}$$

For the CV, we can thus write:

$$\rho_{A1} Q_1 + \rho_{A2} Q_2 = \rho_{A3} Q_3 \tag{2.27a}$$

$$\rho_{B1} Q_1 + \rho_{B2} Q_2 = \rho_{B3} Q_3 \tag{2.27b}$$

$$\rho_1 Q_1 + \rho_2 Q_2 = \rho_3 Q_3 \tag{2.27c}$$

or

$$w_{A1} + w_{A2} = w_{A3} \tag{2.28a}$$

$$w_{B1} + w_{B2} = w_{B3} \tag{2.28b}$$

$$w_1 + w_2 = w_3 \tag{2.28c}$$

From Equation 2.23, Equation 2.27a can be written as follows:

$$c_{A1} \rho_1 Q_1 + c_{A2} \rho_2 Q_2 = c_{A3} \rho_3 Q_3 \tag{2.29}$$

Note from Equations 2.23 and 2.25 that $\rho = \rho_B/(1 - c_A)$. Thus, when the concentrations c_A are small ($c_A \ll 1$), we have $\rho_1 \cong \rho_2 \cong \rho_3 \cong \rho_B$ and the above equation can be approximated by:

$$c_{A1} Q_1 + c_{A2} Q_2 \cong c_{A3} Q_3 \tag{2.30}$$

Solving for c_{A3}, we obtain the following familiar equation:

$$c_{A3} = \frac{c_{A1} Q_1 + c_{A2} Q_2}{Q_1 + Q_2} \tag{2.31}$$

From Equation 2.27c, for small concentrations c_A, we also have:

$$Q_1 + Q_2 \cong Q_3 \tag{2.32}$$

The reader who has prepared treatment plant solids balances should find Equations 2.29–2.32 familiar. In that context, the term $w = c\rho Q$ is referred to as a *loading*, with units of lbm/day or kgm/sec. In the environmental engineering field, units of ppm are commonly used to represent component concentrations, such as the number of pounds of dry solids contained in one million pounds of water. Since the density of water is 1 kgm/liter and 1 kgm = 10^6 g, "concentration" in ppm (mg/10^6 mg) is commonly expressed as mg/l. Note that mg/l is properly regarded as an expression of component density and that the common conversion from ppm to mg/l neglects the density of the solute or suspension and is thus only accurate for dilute concentrations.

In Chapter 9, we will consider compressible air flows, which are actually mixtures of dry air and water vapor. In Chapters 13 and 14, more generalized versions of the point continuity equations (Equations 2.8–2.11) will be presented and applied as they pertain to individual components of a liquid mixture and include chemical (and biochemical) reactions, diffusion, and dispersion.

2.4 Linear Momentum

2.4.1 Closed Systems

Newton's Second Law of Motion states that the forces acting on a system equal the rate of change of momentum of the system. This is expressed for an infinitesimal system by:

$$\sum d\overline{F} = \frac{D[(dm)\overline{v}]}{dt} \tag{2.33}$$

in which $\sum d\overline{F}$ is the sum of forces (e.g., in lbf or newtons) acting on the system of mass dm and velocity $\overline{v}$. The product $(dm)\overline{v}$ is the *linear momentum* of the system.

2.4.2 Control Volumes

For a CV, the relation becomes:

$$\left(\sum \overline{F}\right)_{c.v.} = \oiint_{c.s.} \overline{v}(\rho\overline{v}\cdot d\overline{A}) + \frac{\partial}{\partial t}\iiint_{c.v.}(\rho\overline{v})d\text{V\kern-0.6em-} \tag{2.34}$$

This equation states that the sum of forces acting on the fluid momentarily occupying the CV equals the net rate of outflow of momentum across the control surface plus the rate of increase of momentum within the CV. It is important to understand that the above is a *vector equation*. In the x-direction of a rectangular coordinate system, for example, it is written as:

$$\sum F_x = \oiint_{c.s.} v_x(\rho\overline{v}\cdot d\overline{A}) + \frac{\partial}{\partial t}\iiint_{c.v.}(\rho v_x)d\text{V\kern-0.6em-} \tag{2.35}$$

For steady-state conditions, the second term on the right-hand side of the above two equations is zero, and the total force equals the net outflow of momentum from the CV.

The forces in the above equation may be regarded as being of two types: surface forces and body forces. Surface forces are those that result from pressure and shear stresses acting on the *surface* of the system or CV. (Surface tensions are an additional type of surface force that occurs when interfaces between phases are involved; they are generally not of concern for our applications of systems and CVs.) Body forces arise from force fields and are developed on the system or CV without surface contact. They include gravitational, magnetic, electrodynamic, centrifugal, and Coriolis forces. Here, we shall limit our concern to gravitational forces. {Centrifugal and Coriolis body forces need to be considered explicitly only for noninertial CVs; only inertial CVs [i.e., CVs fixed in, or translating at constant velocity relative to, an inertial reference (Shames 1962)] are considered here.}

2.4.2.1 Tee with Small Inflow Branch

Consider the case of a tee with a "branch" through which flow enters, such as the tee shown in Figure 2.4. The diameter and flow of the branch are assumed to be small compared to those of the "run" of the tee so that the branch flow does not impinge on the wall opposite the branch (these assumptions are discussed further in Chapter 6). It is desired to determine the pressure change in the run across the tee.

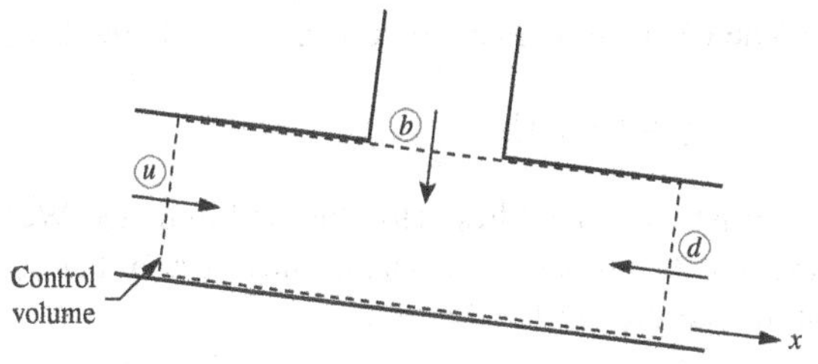

Figure 2.4 Tee with Small Inflow Branch

The CV is defined as shown in Figure 2.4. The flow is assumed steady, so $\partial/\partial t = 0$. The densities are assumed to be approximately equal, and the continuity equation is applied to the control surface to obtain:

$$Q_u + Q_b = Q_d \tag{2.36}$$

in which the subscripts u, d, and b denote the flow in the upstream run, downstream run, and branch, respectively. From Equation 2.35, the steady-flow momentum equation in the x-direction becomes:

$$\sum F_x = \oiint_{c.s.} v_x(\rho\bar{v}\cdot d\bar{A}) = \iint_u v_x(\rho\bar{v}\cdot d\bar{A}) + \iint_d v_x(\rho\bar{v}\cdot d\bar{A}) + \iint_b v_x(\rho\bar{v}\cdot d\bar{A}) \tag{2.37}$$

The surface forces include the pressure forces p_uA_u and $-p_dA_d$, and the frictional force F_f exerted by the pipe walls on the fluid. The pressure variations perpendicular to the direction of flow are essentially hydrostatic. Thus, from fluid statics [see, e.g., Shames (1962, §3–6)] the pressures p_1 and p_2 may be regarded as centroidal pressures or pressures acting at the centroids of their respective body force is the x-component of the weight of the fluid in the CV, which may be neglected in comparison to the net pressure force. Thus:

$$\sum F_x = p_uA_r - p_dA_r \tag{2.38}$$

in which $A_r = A_u = A_d$ is the cross-sectional area of the run.

Variations of velocity perpendicular to the direction of flow are usually small in our applications, so in evaluating momentum flux terms, we will find it sufficient to simply introduce the momentum coefficient attributed to Boussinesq, defined by:

$$\beta = \frac{\iint v^2 dA}{V^2 A} \tag{2.39}$$

The first momentum flux term in Equation 2.37 is given by $\iint_u v_x(\rho\bar{v}\cdot d\bar{A}) = -\iint_u v_x\rho v_x dA = -\rho\iint_u v_x^2 dA = -\beta_u\rho V_u^2 A_r = -\beta_u\rho V_u Q_u$. The term $\iint_d v_x(\rho\bar{v}\cdot d\bar{A})$ similarly becomes $\beta_d\rho V_d^2 A_r = \beta_d\rho V_d Q_d$. The branch flow enters the tee at right angles to the main flow so that its velocity component in the x-direction and, hence, its associated momentum flux term is zero.

Substituting Equation 2.38 and the momentum flux terms into Equation 2.37 gives:

$$(p_u - p_d)A_r = -\beta_u\rho V_u Q_u + \beta_d\rho V_d Q_d \tag{2.40}$$

Employing the continuity equation (Equation 2.36 and $V = Q/A$), we may rewrite the above equation as follows:

$$\frac{p_u - p_d}{\gamma} = 2\left[\beta_d - \beta_u\left(1 - \frac{Q_b}{Q_d}\right)^2\right]\frac{V_d^2}{2g} \tag{2.41}$$

The above equation (with $\beta_u = \beta_d = 1$) will be shown in Section 6.4 to give reasonably good agreement with experiment in certain cases.

2.4.2.2 Tee with Small Outflow Branch

Consider now a tee similar to that of Figure 2.5, except the flow is now dividing through a small (in the sense noted earlier) outflow "branch," as shown in Figure 2.5. The pressure change across the run of the tee is to be determined.

The CV is initially defined by CV1 in Figure 2.5. The flow is assumed to be steady, and the continuity equation gives:

$$Q_u = Q_b + Q_d \tag{2.42}$$

Equation 2.37 applies as for the combining tee. We shall neglect friction and body-term forces as before. The surface forces include the pressure forces p_uA_u and $-p_dA_d$ as before. However, there is now a force $-T_x$ required to turn and accelerate the flow into the outflow branch. Thus:

$$\sum F_x = p_uA_r - p_dA_r - T_x \tag{2.43}$$

For CV1, the x-direction momentum flux terms are the same as for the combining tee (note that the x-direction momentum flux through the branch is again zero). The momentum equation then becomes:

$$(p_u - p_d)A_r - T_x = -\beta_u \rho V_u Q_u + \beta_d \rho V_d Q_d \tag{2.44}$$

One can go no further along these lines using CV1 because the term T_x remains unknown.

This is a case where a judicious choice of CV at the flow division allows circumvention of the force T_x associated with the change in direction of flow (Camp and Graber 1968, 1969). Consider CV2 as shown in Figure 2.5, which is bounded by sections "u" and "d" and the conduit walls except in the vicinity of the branch, at which region the boundary is defined by the imaginary surface separating the branch flow Q_b from the inflow at section "u" (Q_u) and the continuing flow in the run (Q_d). Pressure variations and shear stresses along the control surface in the vicinity of the branch are assumed to be negligible. For this CV, which avoids the forces that turn the flow into the branch, there is an efflux of momentum $\beta_b \rho Q_b V_u$ in the x-direction. The momentum equation then becomes:

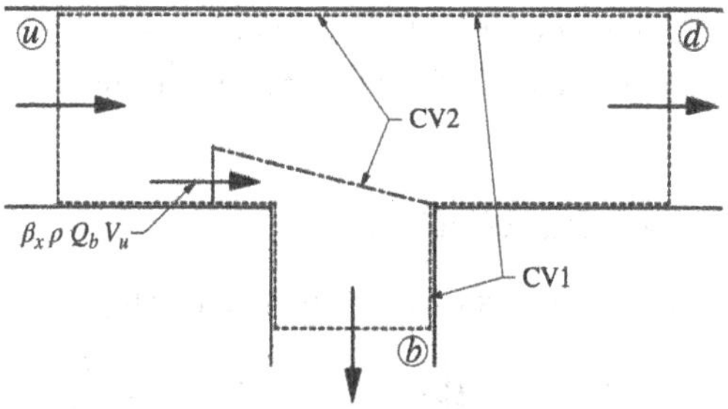

Figure 2.5 Tee with Small Outflow Branch

$$(p_u - p_d)A_r = -\beta_u \rho V_u Q_u + \beta_d \rho V_d Q_d + \beta_b \rho Q_b V_u \tag{2.45}$$

Employing the continuity equation (Equation 2.42 and $V = Q/A$) we may rewrite the above equation for $\beta_u = \beta_d = \beta_b = 1$ as follows:

$$\frac{p_u - p_d}{\gamma} = 2\frac{V_u^2}{2g}\left[\frac{Q_b}{Q_u}\left(\frac{Q_b}{Q_u} - 1\right)\right] \tag{2.46}$$

The above relation indicates that the pressure in the run rises across the tee. As will be shown in Chapter 6, the above relationship approximately predicts the pressure rise for small values of Q_b/Q_u and A_b/A_r, and gives the qualitative shape of the pressure rise relation over a larger range.

Note that by considering CV3, which is the CV between CV1 and CV2, shown in Figure 2.5, we can show the magnitude of the turning force to be:

$$-T_x = -\rho Q_b V_u \tag{2.47}$$

Eliminating T_x from Equations 2.47 and 2.44 results again in Equation 2.45 from which Equation 2.46 is obtained. The reader should be able to demonstrate that Equation 2.46 would be the same if the takeoff were angled, indicating the theoretical independence of the pressure change on the angle of takeoff.

This example demonstrates the value of choosing a CV that includes those forces and velocities that will contribute to the solution of a problem, to the exclusion of extraneous ones.

2.5 Moment of Momentum

2.5.1 Closed Systems

Newton's Second Law provides the basis for the moment of momentum relation, expressed for an infinitesimal system as follows:

$$\sum d\overline{M} = \frac{D}{Dt}[\overline{r} \times d(m\overline{v})] \tag{2.48}$$

in which $\sum d\overline{M}$ is the sum of moments (e.g., in ft-lbf or meter-newton) acting on the system of mass dm, and $\overline{r} \times d(m\overline{v})$ is the cross product of the position vector to the infinitesimal system and the linear momentum vector $d(m\overline{v})$. [The *cross*, *vector*, or *outer product* of two vectors, denoted by $\overline{a} \times \overline{b}$, is defined as the *vector* having: (1) a length, which is the product of the lengths of the vectors $\overline{a}$ and $\overline{b}$ and the sine of the angle θ between the positive directions of the two vectors; and (2) a direction perpendicular to the plane of $\overline{a}$ and $\overline{b}$ and oriented so that $\overline{a}$ is rotated into $\overline{b}$ about $\overline{a} \times \overline{b}$ by the right-hand thumb rule

(Hildebrand 1962). The absolute value of this product is given by $|\bar{a} \times \bar{b}| = ab \sin\theta$ in which a and b are the lengths (or absolute values) of their respective vectors.]

2.5.2 Control Volumes

For a CV, the corresponding relationship is:

$$\sum \overline{M} = \oiint_{c.s.} (\bar{r} \times \bar{v})(\rho\bar{v} \cdot d\bar{A}) + \frac{\partial}{\partial t} \oiiint_{c.v.} (\bar{r} \times \bar{v})(\rho d\Psi) \tag{2.49}$$

This equation states that the sum of moments acting on the control system equals the net rate of outflow of moment of momentum across the control surface plus the rate of increase of moment of momentum within the CV.

As is the linear momentum relation for a CV, the above equation is a *vector* equation. Our concern will be with its application to pumps and rotary blowers, for which Equation 2.49 will be applied in the z-direction of a cylindrical coordinate system:

$$\sum M_z = \oiint_{c.s.} (rv_\phi)(\rho\bar{v} \cdot d\bar{A}) + \frac{\partial}{\partial t} \oiiint_{c.v.} (rv_\phi)(\rho d\Psi) \tag{2.50}$$

For steady-state conditions, the second term on the right-hand side of the above equation is zero, and the total moment equals the net rate of outflow of moment of momentum.

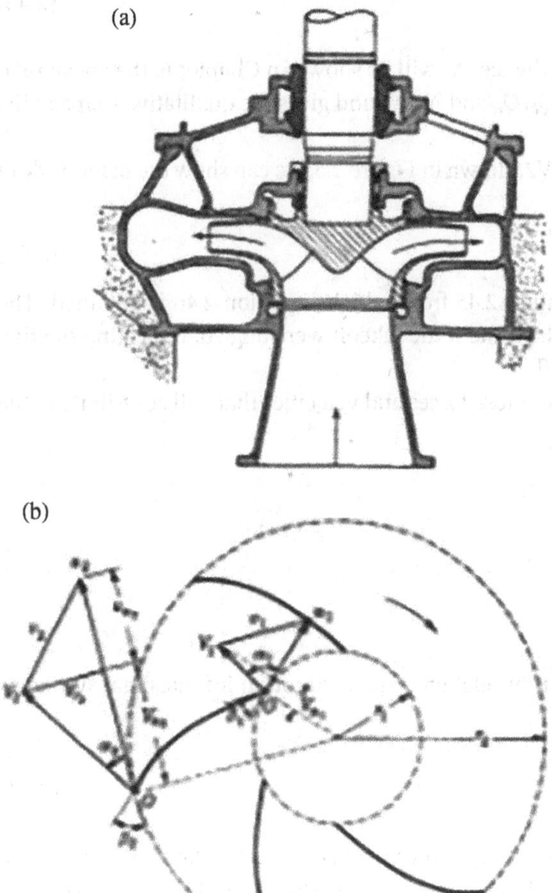

Figure 2.6 (a) Sectional Elevation of Centrifugal Pump; (b) Velocity Relationships for Flow through Centrifugal Pump Impeller (Streeter 1962/McGraw-Hill Education)

An important relation of *turbomachine theory* is derived using the moment of momentum equation given above. That theory is basic to understanding the performance of centrifugal pumps, centrifugal blowers, and turbines. The pertinent equation will be derived below and picked up on in Sections 8.2 and 9.5 in connection with centrifugal pumps and centrifugal blowers, respectively. The following development is patterned after that of Streeter (1962).

For concreteness, consider the end-suction centrifugal pump of Figure 2.3, having the internal configuration shown in Figure 2.6(a). The CV within the dashed lines on Figure 2.6(b) is selected for analysis. This stationary CV just surrounds the impeller and cuts the impeller shaft. The flow entering and leaving the CV is idealized as shown in Figure 2.6(b).

On the entrance and exit vector diagrams shown in Figure 2.6(b), u is the peripheral velocity of the runner and V is the absolute velocity of the fluid. The impeller is rotating at a constant N revolutions per unit time, which is related to the peripheral impeller speeds by:

$$2\pi N = \frac{u_2}{r_2} = \frac{u_1}{r_1} \tag{2.51}$$

The angle α is the angle that the absolute velocity V makes with the peripheral velocity u. Recalling from elementary kinematics that the velocity of B relative to A is given by the vector difference of the velocity of B minus the velocity of A (mathematically: $\bar{v}_{B/A} = \bar{v}_B \rightarrow \bar{v}_A$), one sees that the fluid velocity relative to the impeller is given by v in Figure 2.6(b). Then assume that the impeller blades guide the flow perfectly so that the angle β of the entering and exiting fluid velocity relative to the impeller (measured from $-u$ is the blade angle as shown in Figure 2.6(b). The components of the absolute velocity in the tangential and radial directions are designated by V_u and V_r, respectively.

The flow crossing the control surfaces is not truly steady, as each rotating blade has a velocity pattern or profile. However, the high-frequency variations in torque arising from this unsteadiness are small and not of concern here. On average, the flow may be assumed to be steady so that Equation 2.50 reduces to:

$$\sum M_z = \oiint_{c.s.} (rv_\phi)(\rho\overline{v}\cdot d\overline{A}) = \iint_1 (rv_\phi)(\rho\overline{v}\cdot d\overline{A}) + \iint_2 (rv_\phi)(\rho\overline{v}\cdot d\overline{A}) \tag{2.52}$$

If we neglect friction, the stresses on the control surface are all normal stresses and do not contribute to the moment. The torque from body forces is also zero. Only the torque on the shaft, which is cut by the control surface, contributes to the moment. The above equation thus gives:

$$T_{th} = \rho Q(r_2V_2\cos\alpha_2 - r_1V_1\cos\alpha_1) = \rho Q(r_2V_{u2} - r_1V_{u1}) \tag{2.53}$$

in which T_{th} is the average theoretical torque acting on the fluid within the CV. This equation is in a form of Euler's pump and turbine equation.

2.6 Conservation of Energy

2.6.1 Closed Systems

The First Law of Thermodynamics states that, in the absence of nuclear reactions, energy can neither be created nor destroyed. This law, also known as the Law of Conservation of Energy, may be expressed for a system as follows:

$$\frac{DE}{Dt} = \dot{Q} - \dot{W} \tag{2.54}$$

in which DE/Dt is the rate at which energy (e.g., in BTU or joules) increases in the system, $\dot{Q}$ is the rate of heat flow (e.g., in BTU/hr or watts) *into* the system, and $\dot{W}$ is the rate at which work is done *by* the system (same units as $\dot{Q}$). If the heat flows *out of* the system, it is given a negative sign, as is work done *on* the system. Heat and work represent the means by which energy can be transferred across the system boundaries.

The energy E may be categorized as (Shames 1962): (1) internal energy; (2) kinetic energy associated with the motion of the system; and (3) potential energy associated with the position of the system in external force fields. The specific energy or energy per unit mass e (e.g., in BTU/lbm or joule/gram) may be expressed in terms of these energy categories as follows:

$$e = E/m = u + v^2/2 + (\text{p.e.}) \tag{2.55}$$

in which u is the specific internal energy, $v^2/2$ is the kinetic energy with v being the velocity of the system, and (p. e.) denotes the potential energy per unit mass due to force fields (see Section 2.4). Gravity is the only external force field with which we shall be concerned; thus, the potential energy per unit mass due to gravity is denoted by gz, in which g is the acceleration of gravity and z is the elevation above a defined datum. The specific energy may then be written in terms of its three components as:

$$e = E/m = u + v^2/2 + gz \tag{2.56}$$

The work done by or on a system may be categorized as: (1) work due to pressure stresses at the boundaries of the system; (2) work done by shear stresses at the boundaries of the system; and (3) electric work. These categories are elaborated upon below. We shall omit discussion of capillarity and magnetic forces. The use of *pressure* rather than the more general *normal stresses* is valid for gases and liquids under conditions of thermodynamic equilibrium, except when the gas or liquid approaches the thermodynamic critical point (Shames 1962, §2–4). This assumption suffices for the applications considered in this text.

2.6.2 Control Volumes

The First Law of Thermodynamics may be written for a CV as follows:

$$\dot{Q} - \dot{W}_{shear} - \dot{W}_{elec} = \oiint_{c.s.} \left(e + \frac{p}{\rho}\right)\rho\overline{v}\cdot d\overline{A} + \frac{\partial}{\partial t}\iiint_{c.v.} (\rho e)d\forall \tag{2.57}$$

in which $\dot{Q}$ is the rate of heat flow *into* the CV, and $\dot{W}_{shear}$ and $\dot{W}_{elec}$ are, respectively, the rate at which shear and electric work are done *by* the fluid in the CV on its environment. This formulation is based on Shapiro (1963) and is similar to (and readily deduced from) Shapiro (1953) and Shames (1962).

The work done by pressure stresses at the boundary of the CV is fully incorporated in the term $\oiint_{c.s.}(p/\rho)\rho\bar{v}\cdot d\bar{A}$. From Equation 2.56, the sum $e + p/\rho$ is:

$$e + p/\rho = u + p/\rho + v^2/2 + gz \tag{2.58}$$

When dealing with gases, the first two terms on the right-hand side of the above equation are commonly combined as follows:

$$\tilde{H} = u + p/\rho \tag{2.59}$$

and are together known as *enthalpy* (the notation $\tilde{H}$ is used in this book to avoid conflict with other similar notations; the enthalpy $\tilde{H}$ will be found useful in Chapter 9). In terms of enthalpy, the sum $e + p/\rho$ in Equation 2.58 becomes:

$$e + p/\rho = \tilde{H} + v^2/2 + gz \tag{2.60}$$

The shear work is of two types (Shapiro 1953): (1) shaft work, which is the work done *by* the part of the shaft inside the system or CV on the part of the shaft outside the system or CV. This work is due to the torque in the rotating shaft resulting from the shear stress in the plane cut by the boundary of the system or the control surface; and (2) the work done at the boundaries of the system or CV on the adjacent fluid, which is in motion. The latter is nonzero for a system of particles in a fluid flow (as will be considered in Chapter 3), but becomes zero for a CV in which the region outside the boundary is stationary (as when the control surface coincides with the stationary wall of a duct or casing).

Equations 2.57 and 2.58 give:

$$\dot{Q} - \dot{W}_{shear} - \dot{W}_{elec} = \oiint_{c.s.}(u + p/\rho + v^2/2 + gz)\rho\bar{v}\cdot d\bar{A} \tag{2.61}$$

This equation will be applied to the CVs of Figure 2.3, denoted CV 1 (used previously in this chapter) and CV 2. For CV 1, the control surface cuts the shaft connecting the motor and the pump, and the motor is otherwise outside the CV. Shaft work is done *on* CV 1. For this case, we have $\dot{W}_{elec} = 0$ and $\dot{W}_{shear} = \dot{W}_{sh}$ which is the negative of the shaft power (such as in brake horsepower or kw) supplied to the pump. If the CV is enlarged as CV 2 to include CV 1 and the motor and intersect the electric wiring, then electric work is done on the CV. For this case, $\dot{W}_{shear} = 0$ and $\dot{W}_{elec} = -P_{elec}$ in which $-P_{elec}$ is the electric power input (e.g., in kw) to the motor. Electric work could also be done by an electric heater within the CV with wires running through the control surface to the heater.

Focusing on CV 1, the energy equation becomes:

$$\dot{Q} - \dot{W}_{sh} = \iint_1 (u + p/\rho + v^2/2 + gz)\rho\bar{v}\cdot d\bar{A} + \iint_2 (u + p/\rho + v^2/2 + gz)\rho\bar{v}\cdot d\bar{A} \tag{2.62}$$

Note that no assumption of incompressibility has been made. The above equation is valid for compressible flow, such as could occur if the CV encompassed a centrifugal air blower and connected piping.

2.6.3 One-Dimensional Flow

One-dimensional flow assumptions may now be introduced in the context of the example (CV 1 in Figure 2.3) to which Equation 2.62 applies. (For incompressible flow, the less restrictive and more understandable assumption of parallel flow is usually preferable, as discussed below.) *One-dimensional flow is flow in which all fluid properties (u, p, ρ, v) are assumed to vary only in the mean direction of flow.* Such an assumption is of value if property variations perpendicular to the direction of flow are much smaller than those in the direction of flow, and only the latter are of interest. As a consequence of this assumption, we have $v = V$, which is the average velocity at each cross section. The first energy flux term of Equation 2.62 then becomes:

$\iint_1 (u+p/\rho+v^2/2+gz)\rho\bar{v}\cdot d\bar{A} = -\iint_1 (u+p/\rho+v^2/2+gz)\rho v dA = -\left[(u_1+p_1/\rho_1+V_1^2/2)\rho Q_1+\rho V_1 g\iint z dA\right]$. Noting that $\iint z dA = z_c A$, in which z_c is the z-coordinate of the centroid of the area A, one can write this first energy flux term as $-\left[u_1+p_1/\rho_1+V_1^2/2+g(z_c)_1\right]\rho_1 Q_1]$. Applying similar reasoning to the second energy flux term, Equation 2.62 becomes:

$$\dot{Q}-\dot{W}_{sh} = -\left[u_1+p_1/\rho_1+V_1^2/2+g(z_c)_1\right]\rho_1 Q_1+\left[u_2+p_2/\rho_2+V_2^2/2+g(z_c)_2\right]\rho_2 Q_2 \tag{2.63}$$

Since by continuity $\rho_1 Q_1 = \rho_2 Q_2$, the above equation may be rearranged as follows:

$$\frac{\dot{Q}}{\rho_1 Q_1}-\frac{\dot{W}_{sh}}{\rho_1 Q_1} = -\left[u_1+p_1/\rho_1+V_1^2/2+g(z_c)_1\right]+\left[u_2+p_2/\rho_2+V_2^2/2+g(z_c)_2\right] \tag{2.64a}$$

or

$$\frac{\dot{Q}}{\rho_1 Q_1}-\frac{\dot{W}_{sh}}{\rho_1 Q_1} = -\left[\tilde{H}_1+V_1^2/2+g(z_c)_1\right]+\left[\tilde{H}_2+V_2^2/2+g(z_c)_2\right] \tag{2.64b}$$

in which $\tilde{H}$ is enthalpy as defined by Equation 2.59. The above equation is the *one-dimensional energy equation for a single stream*, applicable to steady flow in a CV in which $\dot{W}_{elec} = 0$ and all the shear work is in the form of shaft work (see above). It is applicable to compressible as well as incompressible flow and will be employed in the former context in Chapter 9. An analogous relation for multiple flow streams may be derived. However, when dealing with pipe networks with friction, it is preferable to apply the *single-stream* equation between nodes. For other multiple flow-stream problems, it is generally best to go back to the CV formulation (e.g., Equation 2.62) and proceed from there.

2.6.4 Energy Loss Due To Friction

The nature of the heat flow and internal energy terms merit a close look. Considering again CV 1 of Figure 2.3, the heat flow and internal energy terms of Equation 2.64a (which applies to that control volume) may be grouped on the right-hand side of the equation to give $\left[-\dot{Q}/\rho Q_1+(u_2-u_1)\right]$. This represents the heat transferred *out of* the CV per unit mass of fluid plus the increase in specific internal energy of the flow. From physics or thermodynamics, $\Delta u = c\Delta T$ in which c is the specific heat of the fluid. [More precisely, c is the specific heat in the case of a liquid and specific heat at constant volume (see Section 9.2) in the case of a perfect gas.] The terms of interest may thus be written as $\left[-\dot{Q}/\rho Q_1+c(T_2-T_1)\right]$.

If T_1 is equal to the ambient temperature outside the CV, then any and all heat transfer and temperature rise must be due to frictional effects within the piping and pump casing. (If T_1 differs from the ambient temperature, then additional transfers of heat and/or temperature rises are superimposed; the general reasoning, however, remains the same.) If the piping and pump were perfectly insulated, then $\dot{Q} = 0$ (adiabatic flow), and all the frictional heating would manifest itself as a temperature rise in the fluid. If the piping and pump freely transferred heat to the environment, there could be little or no temperature rise (isothermal flow), with all the frictional heating lost via $\dot{Q}$. The Second Law of Thermodynamics may be employed to show that such frictional heating is irreversible (meaning, roughly, irrecoverable) and thus represents a *loss* of energy. [The irreversibility of frictional heating (fluid or solid) or pressure drop due to friction may be demonstrated by methods given in basic thermodynamics texts, e.g., Mooney (1959).]

Since, for a liquid (incompressible fluid), the other fluid properties (p, ρ, and V) are not affected by temperature, and since the temperature rise is small and usually not of concern, it is convenient to express $\left[-\dot{Q}/\rho Q_1+(u_2-u_1)\right]$ for this case by the energy loss per unit mass gh_ℓ. Equation 2.64a can thus be written:

$$-\frac{\dot{W}_{sh}}{\rho Q_1} = -\left[p_1/\rho+V_1^2/2+g(z_c)_1\right]+\left[p_2/\rho+V_2^2/2+g(z_c)_2\right]+gh_\ell \tag{2.65}$$

Dividing the above equation through by g, the following is obtained:

$$-\frac{\dot{W}_{sh}}{\rho g Q_1} = -\left[p_1/(\rho g)+V_1^2/(2g)+(z_c)_1\right]+\left[p_2/(\rho g)+V_2^2/(2g)+(z_c)_2\right]+h_\ell \tag{2.66}$$

in which h_ℓ is *energy loss per unit weight.* Note that h_ℓ has units of length (e.g., feet or meters) and is termed *head loss.*

The concepts presented above may be generalized in the following specialization of Equation 2.61:

$$-\dot{Q}_f - \dot{W}_{sh} = \oiint_{c.s.} (p/\rho + v^2/2 + gz)\rho\bar{v}\cdot d\bar{A} \tag{2.67}$$

The above equation is the steady-flow energy equation for a CV, applicable to incompressible flow in which $\dot{W}_{elec} = 0$, and all the shear work is in the form of shaft work (see above); the term $\dot{Q}_f$ represents the rate of energy loss due to friction.

In subsequent chapters, relationships for determining the energy loss or head loss will be presented. Note Hynes' interesting observation that the potential energy of the water flowing in natural streams is converted to heat generated by friction, "a fact that is well demonstrated when meltwater at 0°C carves channels in ice over which it flows" (Hynes 1972, p. 9).

2.6.5 Parallel, Incompressible Flow

A relation similar to the one-dimensional energy equation can be derived for incompressible flow by making assumptions that are less restrictive and for which the significance of the assumptions is clearer than those for one-dimensional flow. This is for the case of *parallel flow,* defined here as *flow in which the fluid velocity vectors at all points in a cross section perpendicular to the mean direction of flow are assumed to be parallel to the mean direction of flow.* For flow in a conduit, the velocity components perpendicular to the direction of flow are neglected. It should be apparent that the conduit (and flow cross section in the case of open channels) must be prismatic or have only gradual variations at sections where these assumptions are to be valid. Furthermore, changes in flow direction (including those resulting from changes in slope) must be gradual so that centrifugal forces do not cause significant transverse velocity components. Further discussion of the parallel-flow assumption and the assumption of one-dimensional flow in the contexts of laminar vs. turbulent flow and secondary currents will be presented in Section 4.7. *As a consequence of the parallel-flow assumption, a hydrostatic pressure distribution occurs over the cross section (see Figure 2.8).*

2.6.5.1 Pressure Conduits

Equation 2.67 may be applied to the pressure-conduit CV bounded by cross sections "1" and "2" and the pipe wall to obtain:

$$-\dot{Q}_f = \iint_1 (p/\rho + v^2/2 + gz)\rho\bar{v}\cdot d\bar{A} + \iint_2 (p/\rho + v^2/2 + gz)\rho\bar{v}\cdot d\bar{A} \tag{2.68}$$

in which $\dot{Q}_f$ is the rate of energy loss due to friction.

Let us investigate the term $\iint(p/\rho + gz)dA$ over a cross section with hydrostatic pressure distribution. Let p_c be the pressure acting at the centroid of the cross section, and let z' be distance measured opposite to z from the centroid. (Actually, any unambiguous point in the cross section could be selected, but the centroid is selected here because of convention for pressure conduits stemming from one-dimensional flow assumptions. For open-channel flow discussed below, a different convention is employed.)

The hydrostatic pressure distribution is then given by $p = p_c + \rho gz'$. It follows that $p/\rho + gz = p_c/\rho + gz' + gz$. Note, however, that $z' + z = z_c$, which is constant for the given datum and cross section. Thus $(p/\rho + gz) = (p_c/\rho + gz_c)$, which is constant for a given cross section. Deleting u from Equation 2.62 and incorporating it in the head loss term, as discussed above, the first flux term in Equation 2.62 becomes $\iint_1(p/\rho + v^2/2 + gz)\rho\bar{v}\cdot d\bar{A} = (p_{c1}/\rho + gz_{c1})\iint_1 \rho\bar{v}\cdot d\bar{A} + (\rho/2)\iint_1 v^2\bar{v}\cdot d\bar{A} = -\left[(p_c + \rho gz_c)Q_1 + (\rho/2)\iint v^3 dA\right]$. An energy coefficient, attributable to Coriolis and analogous to the Boussinesq momentum coefficient (Equation 2.39), is defined by:

$$\alpha = \frac{1}{V^3 A}\iint v^3 dA \tag{2.69}$$

Incorporating the energy coefficient, the first energy flux term thus becomes $\iint_1(p/\rho + v^2/2 + gz)\rho\bar{v}\cdot d\bar{A} = -(p_{c1} + \alpha\rho V_1^2/2 + \rho gz_{c1})Q_1$. Applying similar reasoning to the second energy flux term, Equation 2.68 becomes:

$$-\dot{Q}_f = -\left[p_{c1}/\rho + \alpha_1 V_1^2/2 + gz_{c1}\right]\rho Q_1 + \left[p_{c2}/\rho + \alpha_2 V_2^2/2 + gz_{c2}\right]\rho Q_2 \tag{2.70}$$

Since by continuity $\rho Q_1 = \rho Q_2 = \rho Q$ and since $\dot{Q}_f/\rho Q = gh_\ell$, as discussed above, Equation 2.70 may be rearranged as follows:

$$[p_{c1}/(\rho g) + \alpha_1 V_1^2/(2g) + z_{c1}] = [p_{c2}/(\rho g) + \alpha_2 V_2^2/(2g) + z_{c2}] + h_\ell \quad (2.71a)$$

The above equation is the *parallel-flow energy equation for steady flow in pressure conduits with a single flow stream.* By convention, the elevations are measured above datum to the centroids of the cross sections, and the pressures are those occurring at those centroids. With this understanding, the subscript "c" may be deleted, and the above equation may be expressed in its more common form:

$$\frac{p_{c1}}{\rho g} + \alpha_1 \frac{V_1^2}{2g} + z_1 = \frac{p_{c2}}{\rho g} + \alpha_2 \frac{V_2^2}{2g} + z_2 + h_\ell \quad (2.71b)$$

2.6.5.2 Open-Channel Flow

A development similar to that for pressure conduits can be carried out for open-channel flow. Equation 2.67 may be applied to the open channel shown in Figure 2.7 to obtain an equation identical to Equation 2.68. Subject to the parallel-flow assumptions, that equation reduces to:

$$\frac{p_1}{\rho g} + \alpha_1 \frac{V_1^2}{2g} + z_1 = \frac{p_2}{\rho g} + \alpha_2 \frac{V_2^2}{2g} + z_2 + h_\ell \quad (2.72)$$

in which p and z are respectively measured at and to the same unambiguous point in the cross section, and α is as defined by Equation 2.69. Noting that the pressure at the lowest point of the cross section is equal to $\rho g d \cos\theta$ in which d is the depth of flow measured perpendicular to the channel bottom and θ is the slope angle (Figure 2.7), then $p/(\rho g) = d\cos\theta$ at that lowest point. Since the depth of flow is of primary interest, it is convenient to take $p/(\rho g)$ in the above equation equal to $d\cos\theta$ and use the lowest point in the cross section as the point to which z is measured. The above equation then becomes:

$$d_1\cos\theta_1 + \alpha_1 \frac{V_1^2}{2g} + z_1 = d_2\cos\theta_2 + \alpha_2 \frac{V_2^2}{2g} + z_2 + h_\ell \quad (2.73)$$

The above equation is the *parallel-flow energy equation for steady flow in an open channel with a single flow stream.* The elevations are measured above datum to the low point on the channel bottom, and depths are measured perpendicular to the channel bottom.

Depth of flow is most commonly measured vertically below the water surface rather than perpendicular to the channel bottom. Denoting the vertical depth of flow by y (see Figure 2.7), then $d \cong y\cos\theta$. [The approximation (rather than equality) results from any small differences in d between the section (perpendicular to the flow) of interest and the section below the surface from which y is measured. The notation y is used here because of common convention for open-channel flow and should not be confused with y also used in this book as one of the Cartesian coordinates.] In terms of y, the term $d\cos\theta$ in Equation 2.73 becomes $y\cos^2\theta$. Our usual interest is in *channels of small slope* for which $\cos\theta \cong 1$ and $\cos^2\theta \cong 1$ so that $d\cos\theta \cong d \cong y$, in which case the above equation simplifies to:

$$y_1 + \alpha_1 \frac{V_1^2}{2g} + z_1 = y_2 + \alpha_2 \frac{V_2^2}{2g} + z_2 + h_\ell \quad (2.74)$$

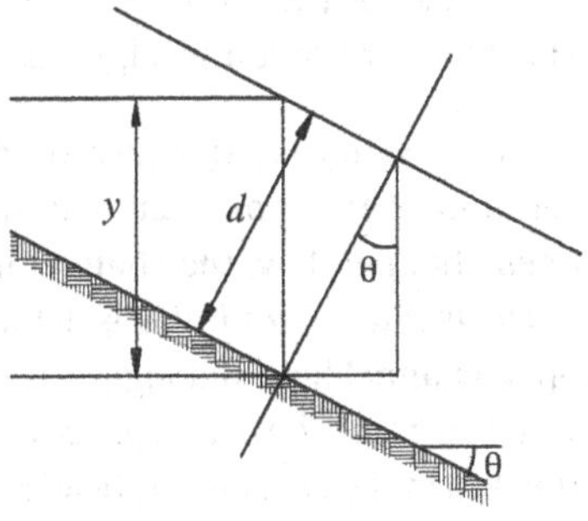

Figure 2.7 Open-Channel Flow Notation

Note that the pressure correction term $\cos^2\theta$ is within about 1% of unity for slopes ($\tan\theta$) up to 0.1 (1 in 10) corresponding to a slope angle of 6°. Thus, the above equation is considered applicable to channels with slopes less than 0.1 (Chow 1959), and the phrase "small slope" will be used hereafter to refer to such channels. Note that the approximation $d\cos\theta \cong d$ (but not $\cong y$) may be used to the same degree of accuracy and d used in place of y in the above equation for channel slopes up to about 0.15, corresponding to a slope angle of 8.5°.

2.6.5.3 Modified Bernoulli Equation

Equation 2.71 may be thought of as being applied between two *points* in a piping system, one point in cross section "1" and one point in cross section "2," without explicit reference to a CV. This concept will be extended by investigating conditions on the suction side of the pump in Figure 2.3 (with the dosing line valve closed). The energy equation will then be applied to a CV bounded by the interior walls of the suction pipe, by a cross section perpendicular to the flow at the pump suction flange, and in the wet well by the portion of the control surface shown in Figure 2.3. The control surface in the wet well is extended out from the end of the pipe to a distance at which the velocities are very small and the kinetic energy of the flow is negligible. The pressure distribution outside of and on the surface of the wet well CV is hydrostatic so that $p = p_{atm} + \rho g(z_a - z)$ in which p_{atm} is the atmospheric pressure and z_a is the elevation above datum of the wet well water surface. The energy flux term at the wet well control surface is thus given by $\iint_1 (p/\rho + v^2/2 + gz)\rho\bar{v}\cdot d\bar{A} \cong \iint_1 (p/\rho + gz)\rho\bar{v}\cdot d\bar{A} = \iint_1 [p_{atm}/\rho + g(z_a - z) + gz]\rho\bar{v}\cdot d\bar{A} = (p_{atm}/\rho + gz_a)\rho\iint \bar{v}\cdot d\bar{A} = -(p_{atm}/\rho + gz_a)\rho Q_1$. (Similar reasoning may be applied to the discharge). The energy equation for the CV of interest becomes:

$$\frac{\dot{Q}_f}{\rho Q_1} = -(p_{atm}/\rho + gz_a) + (p_s/\rho + \alpha V_s^2/2 + gz_s) \tag{2.75}$$

in which the subscript "s" denotes the section at the pump suction flange. The above equation may be thought of as having been applied between a *point* on the water surface of the wet well and a *point* (such as the centroid) in the cross section at the pump suction flange. If gauge pressures rather than absolute pressures are used, then p_{atm} may be dropped.

More generally, we may write:

$$\frac{p_1}{\rho g} + \frac{\alpha_1 V_1^2}{2g} + z_1 = \frac{p_2}{\rho g} + \frac{\alpha_2 V_2^2}{2g} + z_2 + h_\ell \tag{2.76}$$

This equation is referred to as the *modified Bernoulli equation*, based on the similar form derived by Bernoulli for the α's = 1 and $h_\ell = 0$. It is applicable to a single flow stream, in the absence of shear work or electric work, between points of parallel flow or negligible kinetic energy.

2.6.5.4 Heads and Grade Lines

The term "head" has various meanings in connection with fluid flow. The various "heads" are most commonly expressed in a form having units of length (feet or meters). The terms in common usage may be defined with reference to Equation 2.76 as follows:

Pressure head, $p/\rho g$
Elevation head, z
Velocity head, $h_v = V^2/(2g)$
Piezometric head, so called because it is the head measured by a piezometer (Rouse 1950, p. 38), $h_p = p/(\rho g) + z$
Total head or Bernoulli head, $H = p/(\rho g) + V^2/(2g) + z$
Head loss or friction head, h_ℓ, as defined above

The elevation head is measured from a specified datum to a *point* in the flow cross section. For pressure conduits, the centroid of the cross-sectional area is the *point* used here (with the pressure being that at the centroid), whereas for open-channel flow, the channel invert is used as the *point.*

The *hydraulic grade line* is defined as the locus of points representing the piezometric head, as shown in Figure 2.8(a) for open-channel flow and Figure 2.8(b) for closed-conduit flow. For open channels, the hydraulic grade line coincides with the water surface. The *energy grade line* [see Figure 2.8(a)] is defined as the locus of points representing the total energy head or the sum of the piezometric head and velocity head. The energy grade line is lowered in the direction of flow by friction losses. The negative slope of the energy grade line at any point is called the *friction slope* denoted by S_f. Mathematically, $S_f = -dH/dx$ in which x is the direction of flow. The friction slope is the energy loss per unit mass per unit length of conduit.

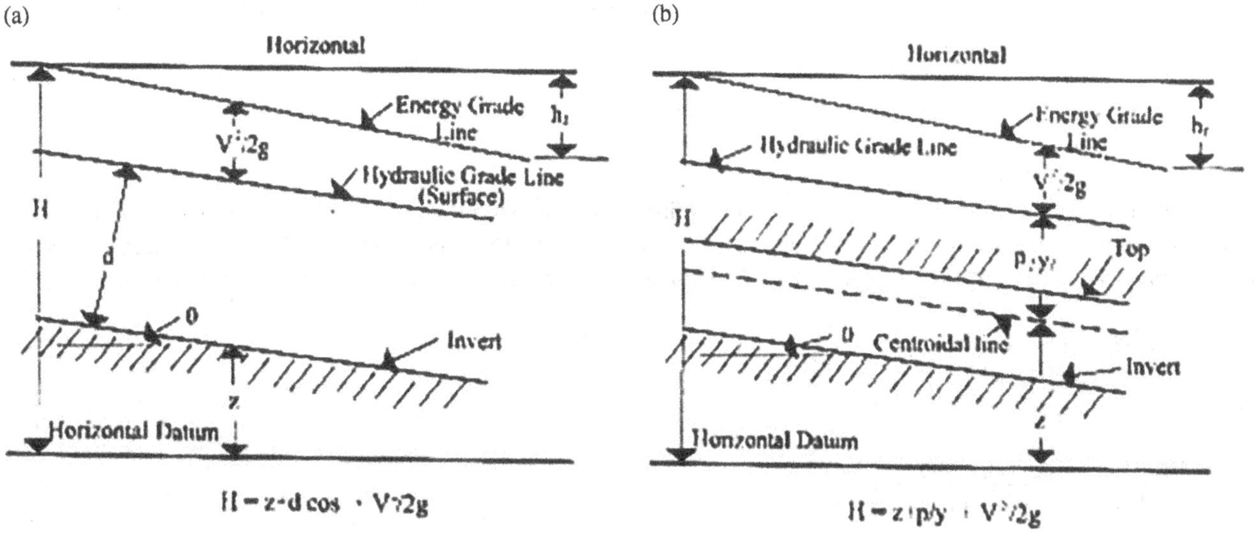

Figure 2.8 Definition of Terms for Total Energy (a) Open-Channel Flow and (b) Closed-Conduit Flow (Graber 2017a / American Society of Civil Engineers)

2.6.5.5 End-Suction Centrifugal Pump

The steady-flow energy equation (Equation 2.67) may be applied to the pump, neglecting energy losses, to obtain:

$$-\dot{W}_{sh} = \rho g Q(H_2 - H_1) = \rho g Q \Delta H \tag{2.77}$$

in which ΔH is the increase in total head (as defined above) across the impeller. The theoretical work done on the impeller shaft equals the actual work done by the pump drive (motor, engine, or whatever) divided by an efficiency factor of less than one, which reflects various energy losses (discussed in Chapter 8). Noting that $\dot{W}_{sh}$ is the negative of the theoretical work done *on* the impeller shaft, one can relate $\dot{W}_{sh}$ to the average theoretical torque acting on the fluid within the CV by:

$$-\dot{W}_{sh} = 2\pi N T_{th} \tag{2.78}$$

Eliminating $\dot{W}_{sh}$ between the above two equations, we obtain:

$$2\pi N T_{th} = \rho g Q \Delta H \tag{2.79}$$

Eliminating T_{th} between Equations 2.53 and 2.79, substituting N from Equation 2.51, and solving for ΔH results in:

$$\Delta H = \frac{u_2 V_{u2} - u_1 V_{u1}}{g} \tag{2.80}$$

Pumps are generally designed so that the angular momentum of the fluid entering the impeller is zero. Thus:

$$\Delta H = \frac{u_2 V_{u2}}{g} \tag{2.81}$$

A theoretical head–discharge curve may be obtained by means of the exit vector diagram in Figure 2.6b, mass continuity, and Equation 2.81. From the vector diagram:

$$V_{u2} = u_2 - V_{r2} \cot \beta_2 \tag{2.82}$$

If impeller vane thickness is neglected and the depth of the impeller at r_2 is denoted by b_2, mass continuity gives the total flow through the impeller as:

$$Q = 2\pi r_2 b_2 V_{r2} \tag{2.83}$$

Substituting Equations 2.82 and 2.83 into Equation 2.81 and expressing u_2 in terms of N from Equation 2.51 gives:

$$\Delta H = \frac{(2\pi)^2 N^2 r_2^2}{g} - \frac{NQ \cot \beta_2}{b_2 g} \tag{2.84}$$

2.6.5.6 Tee with Small Inflow Branch

The steady-flow energy equation alone gives results of little practical value if applied to the CV of Figure 2.4. However, having determined the pressure loss across the run by momentum methods (see Equation 2.41), the change in energy per unit mass across the run can be determined. This change in energy (neglecting potential energy changes) is given by:

$$H_u - H_d = \left(\frac{p_u}{\rho g} + \frac{V_u^2}{2g}\right) - \left(\frac{p_d}{\rho g} + \frac{V_d^2}{2g}\right) \tag{2.85}$$

or:

$$\frac{H_u - H_d}{V_d^2/(2g)} = \frac{(p_u - p_d)/(\rho g)}{V_d^2/(2g)} + \left(\frac{V_u}{V_d}\right)^2 - 1 \tag{2.86}$$

Substituting Equation 2.41 into the above, employing continuity, and rearranging gives:

$$\frac{H_u - H_d}{V_d^2/(2g)} = 2\left[\beta_d - \beta_u\left(1 - \frac{Q_b}{Q_d}\right)^2\right] + \left(1 - \frac{Q_b}{Q_d}\right)^2 - 1 \tag{2.87}$$

For $\beta_u = \beta_d = 1$, the above equation simplifies to:

$$\frac{H_u - H_d}{V_d^2/(2g)} = 1 - \left(1 - \frac{Q_b}{Q_d}\right)^2 \tag{2.88}$$

Since $Q_b < Q_d$, energy per unit mass is lost across the run of the tee.

The above demonstrates a very important phenomenon, too often overlooked in practice: the *exchange of momentum* required to accelerate the branch flow (or any lateral inflow) to the velocity of the flow in the run causes a loss in pressure and energy per unit mass in the flow through the run. The pressure and energy differences between the branch and the run require experimental determination (see Chapter 6).

2.6.5.7 Tee with Small Outflow Branch

As for the inflow tee considered above, the steady-flow energy equation gives results of little value by itself if applied to the CVs of Figure 2.5. As before, however, having determined the pressure loss across the run by the momentum principle, one can determine the change in energy per unit mass across the run. This change in energy is given by:

$$\frac{H_u - H_d}{V_u^2/(2g)} = \frac{(p_u - p_d)/(\rho g)}{V_u^2/(2g)} + 1 - \left(\frac{V_d}{V_u}\right)^2 \tag{2.89}$$

Substituting Equation 2.46 into the above, employing continuity, and rearranging gives:

$$\frac{H_u - H_d}{V_u^2/(2g)} = \left(\frac{Q_b}{Q_u}\right)^2 \tag{2.90}$$

The above relation may also be written as:

$$\frac{H_u - H_d}{V_u^2/(2g)} = \frac{(V_u - V_b)^2}{2g} \tag{2.91}$$

which will be shown in Chapter 6 to be the head loss associated with an abrupt expansion from velocity V_u to reduced velocity V_b.

Thus, in the case of the outflow tee, a pressure rise and head loss across the run of the tee occur. As discussed in Section 6.4, the pressure and energy difference between the branch and the run are amenable to approximate theoretical analysis by a combination of methods presented in the present chapter and Chapter 4.

2.7 Differential Equations for Pipe and Channel Flows

In the next chapter, we will apply the principles given above based on the *system* concept to derive differential equations for three-dimensional laminar flow (defined in Section 4.7). Subsequent chapters will apply the above principles using *CVs* "in-the-large" to derive other useful relations. In the present section, we will apply *CV* principles to derive differential equations for *one-dimensional* flow in pipes and channels. These applications are instructive and provide important results, which will be used in later chapters.

2.7.1 Pipes with Spatially-Increasing Flow

Consider a pipe (or any prismatic pressure conduit) with spatially increasing flow due to a lateral inflow. Such pipes are used for applications including well screens, submerged effluent collectors, and the manifolds in hypochlorite generation systems (these applications are discussed in Chapter 12). The inflow is assumed to enter through many small ports so that it may be regarded as continuously distributed along the pipe. A CV is defined by sections "1" and "2" perpendicular to the direction of flow in the pipe and the inside boundaries of the pipe.

2.7.1.1 Continuity

The flow is assumed to be steady and incompressible. We will use q_l to denote the lateral inflow per unit length. Continuity is expressed by:

$$-VA + (V + \Delta V)A - q_i\Delta x = 0 \tag{2.92}$$

Canceling like terms, passing to differentials, and solving for q_l

$$q_l = A\frac{dV}{dx} = \frac{dQ}{dx} \tag{2.93}$$

2.7.1.2 Momentum

From Equation 2.35, the steady-flow momentum equation in the x-direction becomes:

$$\sum Fx = \iint_1 v_x(\rho\bar{v}\cdot d\bar{A}) + \iint_2 v_x(\rho\bar{v}\cdot d\bar{A}) + \iint_i v_x(\rho\bar{v}\cdot d\bar{A}) \tag{2.94}$$

The surface forces include the pressure forces and the frictional force F_f exerted by the pipe walls on the fluid. The pressure forces are given by p_1A and $-p_2A$. The friction force acts opposite to the direction of flow and will be discussed in some detail in Chapters 4 and 6. For now, the reader is asked to tentatively accept the following relationships:

$$F_f = \bar{\tau}P\Delta x \tag{2.95a}$$

$$\bar{\tau} = C_f\frac{\rho V^2}{2} \tag{2.95b}$$

in which $\bar{\tau}$ is the average shear stress exerted by the pipe walls on the fluid over the pipe perimeter P, C_f is a friction coefficient, and other terms are as defined previously. The body force is the x-component of the weight of the fluid in the CV, equal to $\rho g S_o A\Delta x$ in which $S_o = -dz/dx$ is the local slope of the pipe. Thus:

$$\sum F_x = p_1A - p_2A - C_f\frac{\rho V^2}{2}P\Delta x + \rho g S_o A\Delta x \tag{2.96}$$

Denoting p_1 by p and p_2 by $p + \Delta p$, we have:

$$\sum F_x = pA - (p + \Delta p)A - C_f\frac{\rho V^2}{2}P\Delta x + \rho g S_o A\Delta x = -A\Delta p - C_f\frac{\rho V^2}{2}P\Delta x + \rho g S_o A\Delta x \tag{2.97}$$

The first momentum flux term in Equation 2.94 is given by $\iint_1 v_x(\rho\bar{v}\cdot d\bar{A}) = -\beta\rho V^2A = -\beta\rho VQ$. The term $\iint_2 v_x(\rho\bar{v}\cdot d\bar{A})$ similarly becomes $+\beta\rho(V + \Delta V)^2A$ or $+\beta\rho(V + \Delta V)(Q + \Delta Q)$. For the applications of interest, the lateral inflow enters the conduit at right angles to the main flow. Its velocity component is in the x-direction and hence the associated momentum flux term is zero. Thus, the momentum flux terms become:

$$\iint_1 v_x(\rho\bar{v}\cdot d\bar{A}) + \iint_2 v_x(\rho\bar{v}\cdot d\bar{A}) + \iint_i v_x(\rho\bar{v}\cdot d\bar{A}) = -\beta\rho VQ + \beta\rho(V+\Delta V)(Q+\Delta Q) + 0 \tag{2.98}$$

Substituting Equations 2.97 and 2.98 into Equation 2.96, passing to differentials, dropping second-order terms, and rearranging results in:

$$\frac{dp}{dx} = \rho g S_o - C_f\frac{\rho V^2}{2R} - \frac{\beta\rho}{A}\left(Q\frac{dV}{dx} + V\frac{dQ}{dx}\right) \tag{2.99}$$

in which $R = A/P$ is the hydraulic radius. The continuity relation (Equation 2.93 and $Q = VA$) allows the above equation to be rewritten as follows:

$$\frac{dp}{dx} = \rho g S_o - C_f\frac{\rho V^2}{2R} - \frac{2\beta\rho Q}{A^2}\frac{dQ}{dx} \tag{2.100}$$

We may also write the above equation in terms of the piezometric head h_p as defined above. Noting that $dh_p/dx = [1/(\rho g)]\,dp/dx + dz/dx = [1/(\rho g)]dp/dx - S_o$, Equation 2.100 may be written in terms of h_p as follows:

$$\frac{dh_p}{dx} = -C_f\frac{V^2}{2gR} - \frac{2\beta Q}{gA^2}\frac{dQ}{dx} \tag{2.101}$$

Since $C_fV^2/(2gR)$ equals the friction slope S_f, as defined above. The above equation can be written in the following alternate form:

$$\frac{dh_p}{dx} = -S_f - \frac{2\beta Q}{gA^2}\frac{dQ}{dx} \tag{2.102}$$

2.7.2 Pipes with Spatially-Decreasing Flow

We will now consider a pipe (the analysis applies to any prismatic pressure conduit) with spatially decreasing flow due to a continuous lateral outflow. Such *distribution conduits* are used for a multitude of practical purposes, as discussed in Chapter 11.

2.7.2.1 Continuity

The flow is assumed to be steady and incompressible. The lateral outflow per unit length is denoted by q_o. Considering a CV with upstream area and velocity denoted by A and V entering velocity, respectively, and length of CV and change in velocity denoted by Δx and ΔV, respectively, we have:

$$-VA + (V+\Delta V)A + q_o\Delta x = 0 \tag{2.103}$$

Canceling like terms, passing to differentials, and solving for q_o yields:

$$q_o = -A\frac{dV}{dx} = -\frac{dQ}{dx} \tag{2.104}$$

2.7.2.2 Momentum

From Equation 2.35, the steady-flow momentum equation in the x– direction becomes:

$$\sum F_x = \iint_1 v_x(\rho\bar{v}\cdot d\bar{A}) + \iint_2 v_x(\rho\bar{v}\cdot d\bar{A}) + \iint_o v_x(\rho\bar{v}\cdot d\bar{A}) \tag{2.105}$$

The surface and body forces are the same as for the pipe with spatially-increasing flow analyzed above, and Equation 2.96 applies.

The first momentum flux in Equation 2.105 is given by $\iint_1 v_x(\rho\bar{v}\cdot d\bar{A}) = -\beta\rho V^2 A = -\beta\rho VQ$. The term $\iint_2 v_x(\rho\bar{v}\cdot d\bar{A})$ similarly becomes $+\beta\rho(V+\Delta V)^2 A$ or$+\beta\rho(V+\Delta V)(Q+\Delta Q)$. The momentum flux associated with the outflow may be expressed by $\iint_o v_x(\rho\bar{v}\cdot d\bar{A}) = \beta_o\rho(q_o\Delta x)V$. The momentum flux terms then become:

$$\iint_1 v_x(\rho\bar{v}\cdot d\bar{A}) + \iint_2 v_x(\rho\bar{v}\cdot d\bar{A}) + \iint_o v_x(\rho\bar{v}\cdot d\bar{A}) = -\beta\rho VQ + \beta\rho(V+\Delta V)(Q+\Delta Q) + \beta_o\rho(q_o\Delta x)V \tag{2.106}$$

Substituting Equations 2.96 and 2.106 into Equation 2.105, passing to differentials, dropping second-order terms, taking $\beta_o = \beta$, and rearranging results in:

$$\frac{dp}{dx} = \rho g S_o - C_f\frac{\rho V^2}{2R} - \frac{\beta\rho}{A}\left(Q\frac{dV}{dx} + V\frac{dQ}{dx} + q_o V\right) \tag{2.107}$$

The continuity relation (Equation 2.104) enables the above equation to be rewritten as follows:

$$\frac{dp}{dx} = \rho g S_o - C_f\frac{\rho V^2}{2R} - \frac{\beta\rho Q}{A^2}\frac{dQ}{dx} \tag{2.108}$$

The similarities and differences between the above equation and Equation 2.101 should be noted. As for Equation 2.101, Equation 2.108 can be written in terms of the piezometric head:

$$\frac{dh_p}{dx} = -\frac{C_f}{R}\frac{V^2}{2g} - \frac{\beta Q}{gA^2}\frac{dQ}{dx} \tag{2.109}$$

or in terms of the piezometric head and friction slope:

$$\frac{dh_p}{dx} = -S_f - \frac{\beta Q}{gA^2}\frac{dQ}{dx} \tag{2.110}$$

The above equation provides the basis for the practical results of Chapter 11.

2.7.2.3 Energy

Application of the parallel-flow energy equation is instructive in this case. Equation 2.71a applied to the CV encompassing an open-channel flow in which entering (upstream) conditions are denoted by p, V, z_c which are centroidal pressure, velocity, and elevation, respectively, and corresponding exiting (downstream) conditions are denoted by $p+\Delta p$, $V+\Delta V$, $z_c + \Delta z_c$, respectively, and h_f is frictional head loss:

$$[p/(\rho g) + \alpha V^2/(2g) + z_c] = [(p+\Delta p)/(\rho g) + \alpha(V+\Delta V)^2/(2g) + (z_c + \Delta z_c)] + h_f \tag{2.111}$$

Passing to differentials, canceling like terms, dropping second-order terms, and rearranging results in:

$$\frac{dp}{dx} = -\rho g\frac{dz_c}{dx} - \rho g\frac{dh_f}{dx} - \alpha\rho V\frac{dV}{dx} \tag{2.112}$$

Noting $S_o = -dz_c/dx$, $S_f = dh_f/dx$, and $V = Q/A$, the above equation can be rewritten in terms of the piezometric head as:

$$\frac{dh_p}{dx} = -S_f - \frac{\alpha Q}{gA^2}\frac{dQ}{dx} \tag{2.113}$$

The similarity between the above equation and Equation 2.110 should be apparent. Note also from Equation 2.46 that the pressure drop (hence head loss) across a tee run (and by analogy across an outflow port in a distribution conduit) approaches zero as the ratio of the branch outflow Q_b to the upstream conduit flow Q_u approaches zero. Since for the present case $Q_b/Q_u = q_o dx/Q = dQ/Q$ in which dQ is infinitesimal, we see that the head loss across the ports associated with the outflow is theoretically negligible. This indicates that the energy loss is almost entirely due to pipe friction.

2.7.3 Open-Channel Flow

An instructive and important example pertains to the general open-channel flow of a liquid encompassed by a CV. The main flow enters and exits the CV at the upstream and downstream ends of the channel, respectively. Spatially-varied inflow and outflow occur along the channel. For concreteness, the channel may be thought of as representing: (1) a collection-distribution conduit in which inflow comes over a weir from an upstream tank and is distributed over a weir to a downstream tank; or (2) sheet flow of runoff with inflow of rainfall and outflow to an underlying pervious surface (Figure 10.1). The CV is initially defined by sections "1" and "2" parallel to the y-axis (perpendicular to the main flow) separated by short channel length Δx, and the lateral boundaries of the main flow within this length immediately inside the channel boundaries (whether confining walls or the ground).

2.7.3.1 Continuity

The flow is assumed to be steady so that $\partial/\partial t = 0$. The continuity equation (Equation 2.12) is applied to the control surface to obtain:

$$\iint_1 \rho\bar{v}\cdot d\bar{A} + \iint_2 \rho\bar{v}\cdot d\bar{A} + \iint_i \rho\bar{v}\cdot d\bar{A} + \iint_o \rho\bar{v}\cdot d\bar{A} = 0 \tag{2.114}$$

in which the integrals are taken over sections "1" and "2," the top surface "i" over which the inflow enters, and the bottom surface "o" over which the outflow leaves the CV. Using the notation in Figure 10.1, the term $\iint_1 \rho\bar{v}\cdot d\bar{A}$ becomes $-\rho V_1A_1 = -\rho VA$ and the term $\iint_2 \rho\bar{v}\cdot d\bar{A}$ becomes $+\rho V_2A_2 = \rho(V+\Delta V)(A+\Delta A)$. The inflow and outflow per unit length of channel are denoted by q_i and q_o, respectively, in terms of which we can write $\iint_i \rho\bar{v}\cdot d\bar{A} = -\rho q_i\Delta x$ and $\iint_o \rho\bar{v}\cdot d\bar{A} = +\rho q_o\Delta x$. Equation 2.114 then becomes:

$$-\rho VA + \rho(V+\Delta V)(A+\Delta A) - \rho q_i\Delta x + \rho q_o\Delta x = 0 \tag{2.115}$$

Passing to differentials, canceling second-order terms, and rearranging yields:

$$V\frac{dA}{dx} + A\frac{dV}{dx} - q_i + q_o = 0 \tag{2.116}$$

The above is the differential continuity equation for the general open-channel flow depicted in Figure 10.1.

2.7.3.2 Momentum

From Equation 2.35, the steady-flow momentum equation in the x-direction becomes:

$$\sum F_x = \oiint_{c.s.} v_x(\rho\bar{v}\cdot d\bar{A}) = \iint_1 v_x(\rho\bar{v}\cdot d\bar{A}) + \iint_2 v_x(\rho\bar{v}\cdot d\bar{A}) + \iint_i v_x(\rho\bar{v}\cdot d\bar{A}) + \iint_o v_x(\rho\bar{v}\cdot d\bar{A}) \tag{2.117}$$

The surface forces include the pressure forces F_{p1}, F_{p2}, and F_{px} (the x–component of pressure forces acting on the sides of the CV); the frictional force exerted by the channel walls on the fluid; and a force denoted by T_x which is necessary to turn and accelerate the outflow (see below). The frictional force is equal to $\bar{\tau}P\Delta x$, in which $\bar{\tau}$ is the average shear stress and P is the wetted perimeter. This is in turn equal to $\rho gAS_f\Delta x$ in which S_f is the friction slope. The body force is the x–component of the weight of the fluid in the CV, equal to $\rho gAS_o\Delta x$ (the slope S_o is assumed constant between x and $x+\Delta x$). Thus:

$$\sum F_x = F_{p1} - F_{p2} + F_{px} + \rho gS_oA\Delta x - \rho gS_fA\Delta x - T_x \tag{2.118}$$

2.7.3.3 Pressure Forces

The pressure forces merit detailed discussion not readily available elsewhere. Parallel flow, as defined above, is assumed, which implies a hydrostatic pressure distribution at any cross section perpendicular to the x axis. The z axis is parallel to the gravity vector and not necessarily perpendicular to the x axis. The pressure at a point in any such cross section is given by $p = \rho gz''$ in which z'' is the distance measured opposite to z from the free surface. The pressure force is then given by

$F_p = \iint p dA = \iint \rho g z'' dA = \rho g \iint z'' dA = \rho g \tilde{z} A$ in which $\tilde{z}$ is the z''-coordinate to the centroid of area A. The pressure forces F_{p1} and F_{p2} are given by:

$$F_{p1} = \rho g \tilde{z} A \tag{2.119a}$$

$$F_{p2} = \rho g(\tilde{z} + \Delta\tilde{z})(A + dA) \tag{2.119b}$$

For a prismatic channel, the term $z + \Delta z$ may be obtained by the method of composite sections for centroids (Pletta 1951) as follows:

$$\tilde{z} + \Delta\tilde{z} = \left(\frac{\sum \tilde{z}A}{\sum A}\right)_2 = \frac{(\tilde{z} + \Delta y)A + \frac{\Delta y}{2}\Delta A}{A + \Delta A} \tag{2.120}$$

in which y is depth of flow. Substituting Equation 2.120 into Equation 2.119b

$$F_{p2} = \gamma(\tilde{z} + \Delta y)A \tag{2.121}$$

Neglecting second-order terms, the difference between F_{p1} and F_{p2} is obtained from Equations 2.121 and 2.119a as follows:

$$F_{p1} - F_{p2} = -\gamma A \Delta y \tag{2.122}$$

Since, for a prismatic channel, $F_{px} = 0$, the net effect of the three pressure terms is given by:

$$F_{p1} - F_{p2} + F_{px} = -\gamma A \Delta y \tag{2.123}$$

Interestingly, Equation 2.123 also applies to nonprismatic channels (for which $F_{px} \neq 0$) provided that they satisfy the assumptions of parallel flow. The reason for this can be simply demonstrated by deriving Equation 2.123 for the case of a tapered (nonprismatic) rectangular channel. The pressure forces in this case are given by:

$$F_{p1} - F_{p2} + F_{px} = \frac{1}{2}\gamma b_1 y_1^2 - \frac{1}{2}\gamma b_2 y_2^2 + 2\left(\frac{1}{2}\right)\gamma \frac{\Delta x}{\cos\theta}\left(\frac{y_1 + y_2}{2}\right)^2 \sin\theta \tag{2.124}$$

Substituting $b_1 = b$, $y_1 = y$, $b_2 = b + \Delta b$ and $y_2 = y + \Delta y$ into Equation 2.124 and dropping second-order terms gives:

$$F_{p1} - F_{p2} + F_{px} = -\frac{1}{2}\gamma b_2 y \Delta y - \frac{1}{2}\gamma \Delta b y_1^2 + \gamma\left(\frac{2y + \Delta y}{2}\right)^2 \frac{(\Delta x)\sin\theta}{\cos\theta} \tag{2.125}$$

Since $(\Delta x)\sin\theta = \Delta b/2$ and assuming θ is sufficiently small that $\cos\theta \cong 1$ $(\theta < 10^\circ)$ Equation 2.125 becomes:

$$F_{p1} - F_{p2} + F_{px} = -\gamma b y \Delta y - \frac{1}{2}\gamma y^2 \Delta b + \gamma y^2 \frac{\Delta b}{2} + \gamma y \Delta y \frac{\Delta b}{2} \tag{2.126}$$

Noting that the second and third terms on the right-hand side cancel each other and that the last term is second order, Equation 2.126 reduces to:

$$F_{p1} - F_{p2} + F_{px} = -\gamma b y \Delta y \tag{2.127}$$

which, for a rectangular channel, is identical to Equation 2.123. The energy principle can be employed (see below) to demonstrate the validity of Equation 2.127 for more general nonprismatic, one-dimensional flows.

2.7.3.4 Momentum Flux Terms

The first momentum flux term in Equation 2.117 is given by $\iint_1 v_x(\rho\bar{v} \cdot d\bar{A}) = -\iint_1 v_x \rho v_x dA = -\rho\iint_1 v_x^2 dA = -\beta\rho V^2 A$ or $-\beta\rho VQ$. The term $\iint_2 v_x(\rho\bar{v} \cdot d\bar{A})$ similarly becomes $\beta\rho(V + \Delta V)^2(A + \Delta A)$ or $\beta\rho(V + \Delta V)(Q + \Delta Q)$. Thus, the first two momentum flux terms become $\iint_1 v_x(\rho\bar{v} \cdot d\bar{A}) + \iint_2 v_x(\rho\bar{v} \cdot d\bar{A}) = \beta\rho VQ + \beta\rho(V + \Delta V)(Q + \Delta Q)$. Passing to differentials, canceling second-order terms, and rearranging yields:

$$\iint_1 v_x(\rho\bar{v} \cdot d\bar{A}) + \iint_2 v_x(\rho\bar{v} \cdot d\bar{A}) = \beta\rho V dQ + \beta\rho Q dV \tag{2.128}$$

The momentum flux associated with the lateral inflow is $\iint_i v_x(\rho\bar{v}\cdot d\bar{A}) = -\int_x^{x+\Delta x} U_i(\cos\phi_i)\rho q_i \Delta x = -U_i(\cos\phi_i)\rho q_i\Delta x$, in which U_i is the inflow velocity and ϕ_i is the angle of the inflow, as shown in Figure 10.1. Passing to the differentials, we have:

$$\iint_i v_x(\rho\bar{v}\cdot d\bar{A}) = -\rho q_i U_i(\cos\phi_i)dx \tag{2.129}$$

Particular care must be exercised in evaluating the momentum flux associated with the lateral outflow. We expect this outflow to leave the CV at some angle ϕ_o (see Figure 10.1), but there is no basis for directly determining that angle or the force T_x referred to above. The problem can be circumvented in an approximate way by a judicious choice of CV similar to that employed above in deriving the relations for dividing tees and pipes with spatially decreasing flow. The CV is shown in Figure 10.1, and results in $\iint_o v_x(\rho\bar{v}\cdot d\bar{A}) = \rho q_o V\Delta x$.

2.7.3.5 Energy

If we neglect the energy loss associated with the spatial outflow (as suggested by the analysis of spatially-decreasing pipe flow above), then for the CV and notation of Figure 10.1 for spatially-decreasing flow, development analogous to that leading to Equation 2.73 gives:

$$d\cos\theta + \frac{\alpha V^2}{2g} + z = d\cos\theta + \Delta(d\cos\theta) + \frac{\alpha(V+\Delta V)^2}{2g} + (z+\Delta z) + h_f \tag{2.130}$$

Dividing by Δx, passing to differentials, canceling like terms, dropping second-order terms and rearranging results in:

$$\cos\theta\frac{dd}{dx} - d\sin\theta\frac{d\theta}{dx} + \frac{\alpha V}{g}\frac{dV}{dx} + \frac{dz}{dx} + S_f = 0 \tag{2.131}$$

Substituting $S_o = -dz/dx$, and employing the continuity relation (Equation 2.116 for $q_i = 0$), the above equation may be rewritten:

$$\cos\theta\frac{dd}{dx} - d\sin\theta\frac{d\theta}{dx} + \frac{\alpha V}{gA}(q_i - q_o) - \frac{\alpha V^2}{gA}\frac{dA}{dx} - S_o + S_f = 0 \tag{2.132}$$

Note that the flow area A is a function of x and d so that:

$$dA = \frac{\partial A}{dx}dx + \frac{\partial A}{dd}dd \tag{2.133}$$

Therefore:

$$\frac{dA}{dx} = \frac{\partial A}{dx} + \frac{\partial A}{dd}\frac{dd}{dx} = \frac{\partial A}{dx} + T\frac{dd}{dx} \tag{2.134}$$

in which $T = \partial A/\partial d$ is the top width of the flow. Substituting Equation 2.134 into Equation 2.132 and solving for dd/dx gives:

$$\frac{dd}{dx} = \frac{S_o - S_f + \frac{\alpha Q}{gA^2}(q_o - q_i) + \frac{\alpha Q^2}{gA^3}\frac{\partial A}{\partial x} + d\sin\theta\frac{d\theta}{dx}}{\cos\theta - \frac{\alpha Q^2 T}{gA^3}} \tag{2.135}$$

Our usual interest is in channels of small slope (as discussed above) for which $\cos\theta \cong 1$, $\sin\theta \cong 0$, and $d \cong y$, in which case the above equation takes on the following simpler form:

$$\frac{dd}{dx} = \frac{S_o - S_f + \frac{\alpha Q}{gA^2}(q_o - q_i) + \frac{\alpha Q^2}{gA^3}\frac{\partial A}{\partial x}}{1 - \frac{\alpha Q^2 T}{gA^3}} \tag{2.136}$$

3

Navier–Stokes, Dissipation Function, and G-Value

3.1 Introduction

In the previous chapter, mass, momentum, and energy principles were presented and applied to *control volumes* to derive useful "macroscopic" relationships for fluid flow. The same principles applied to infinitesimal *systems* together with certain "phenomenological" relationships yield "microscopic" differential equations of fluid flow, including the Navier–Stokes equations and the energy dissipation equation. These fundamental equations and their derivations, when viewed from a proper perspective, provide valuable practical insight and tools for understanding and analyzing many types of fluid flow. The foundation provided by the present chapter will be built upon in Chapter 4 in the context of analytical solutions and further perspectives and in Chapter 5 in the context of dimensional analysis. Chapters 4 and 5 together with the present chapter will provide useful techniques for the applied chapters that follow.

The Navier–Stokes equations and dissipation function are also prerequisites to a proper derivation and understanding of the concept of "G-value," which has been widely used (and often misused) by environmental engineers. The particular derivations of the Navier–Stokes equations and dissipation function presented herein (from Graber 1994) have been selected because of their relation to the historical development (shown herein to be incorrect) of G-value.

3.2 Navier–Stokes Equations

The Navier–Stokes equations are differential momentum equations applicable to a *point* in a fluid flow. The derivation of these equations requires the concept of stress at a point, Newton's Second Law of Motion (Section 2.4), the concept of strain at a point, and stress–strain relationships. These concepts and the overall derivation are presented below in summary fashion for the most part, with ample citation of references with which the interested reader can fill in the details.

3.2.1 Stress at a Point

The concept of stress at a point in a fluid is developed by analyzing the stresses acting on elemental volumes of fluid. The development, which is identical for fluids (Streeter 1948, §3, Shames 1962, Chapter 2) and solids (Wang 1953, Chapter 1), demonstrates that the state of stress at a point in *any* continuous medium is fully characterized by nine scalar components of stress on three orthogonal surfaces intersecting at the point. In the rectangular (Cartesian) coordinate system (Figure 3.1), these components are denoted as $\sigma_x, \sigma_y, \sigma_z, \tau_{xy}, \tau_{xz}, \tau_{yx}, \tau_{yz}, \tau_{zx}$ and τ_{zy}, in which σ and τ are normal and shear stresses, respectively. The subscripts for normal stresses denote the coordinate direction to which the stress is parallel. {The sign conventions are discussed elsewhere [see, e.g., Shames (1962)] and will not be addressed here other than to say that σ and τ are positive as

Hydraulics and Pneumatics in Environmental Engineering, First Edition. S. David Graber.

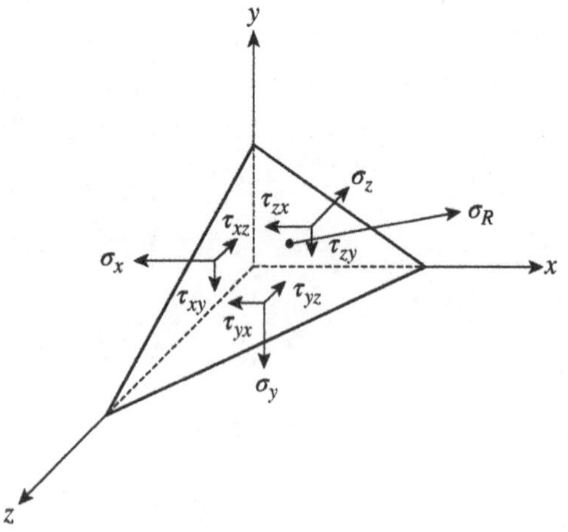

Figure 3.1 General Stress Tetrahedron

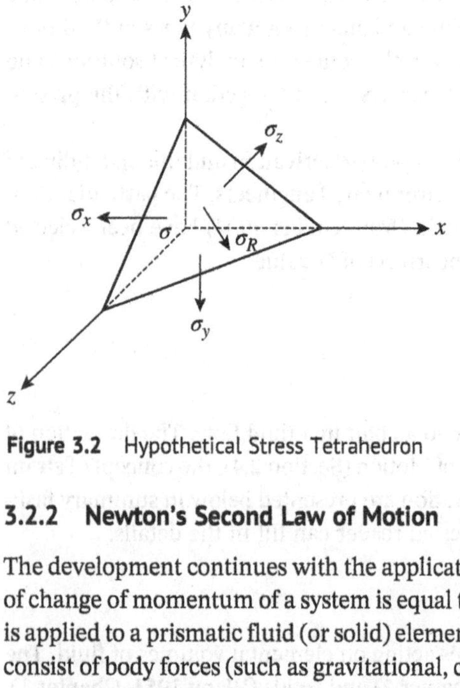

Figure 3.2 Hypothetical Stress Tetrahedron

shown in Figure 3.1}. The subscripts for shear stresses employ double indices, in which the first subscript indicates the direction of the normal to the plane on which the stress acts and the second subscript denotes the coordinate direction to which the stress is parallel (see the parenthetic note above).

One proof (Streeter 1948; the details are included here because of their relation to a proof given in Section 3.4) that the nine stress components fully characterize the state of stress at a point consists of simply showing that the resultant stress σ_R on the arbitrarily inclined face of Figure 3.1 is completely defined by the nine stress components. Letting l, m, and n be the direction cosines (relative to the x, y, and z axes, respectively) of the normal to the inclined face of area A, and letting the components of σ_R be σ_{Rx}, σ_{Ry}, and σ_{Rz}, respectively, forces in the x, y, and z directions are summed to obtain:

$$\sigma_{Rx} = \sigma_x l + \tau_{yx} m + \tau_{zx} n \tag{3.1a}$$

$$\sigma_{Ry} = \tau_{xy} l + \sigma_y m + \tau_{zy} n \tag{3.1b}$$

$$\sigma_{Rz} = \tau_{xz} l + \tau_{yz} m + \sigma_z n \tag{3.1c}$$

Thus, the stress resultants' components, and hence the stress resultant, are completely given in terms of the nine stress components.

The development then demonstrates (see, e.g., Shames 1962, §2.4) the equality of shear stresses having reversed indices, i.e., $\tau_{xy} = \tau_{yx}$, $\tau_{xz} = \tau_{zx}$, $\tau_{yz} = \tau_{zy}$. This reduces to six the number of scalar components characterizing the state of stress at a point.

The next step in the development of the momentum equations consists of demonstrating the existence of three "principal planes." These are planes intersecting at a point, on each of which there is a normal stress but no shear stress. The normal stresses are called principal stresses, and the coordinate axes parallel to these stresses are referred to as principal axes. The principal axes are proven to be orthogonal (Streeter 1948, §109).

3.2.2 Newton's Second Law of Motion

The development continues with the application of Newton's Second Law of Motion (Section 2.4), which states that the rate of change of momentum of a system is equal to the sum of external forces acting on the system (see Equation 2.33). This law is applied to a prismatic fluid (or solid) element having sides Δx, Δy, Δz in a Cartesian coordinate system. The external forces consist of body forces (such as gravitational, centrifugal, and electromagnetic – see Section 2.4) and surface forces, the latter consisting of the appropriate three of the six stress components acting at the midpoint on each face of the prismatic element. The resulting equations are as follows (Streeter 1948, Schlichting 1960):

$$X + \left(\frac{\partial \sigma_x}{\partial x} + \frac{\partial \tau_{xy}}{\partial y} + \frac{\partial \tau_{xz}}{\partial z} \right) = \rho \frac{Dv_x}{Dt} \tag{3.2a}$$

$$Y + \left(\frac{\partial \tau_{xy}}{\partial x} + \frac{\partial \sigma_y}{\partial y} + \frac{\partial \tau_{yz}}{\partial z} \right) = \rho \frac{Dv_y}{Dt} \tag{3.2b}$$

$$Z + \left(\frac{\partial \tau_{xz}}{\partial x} + \frac{\partial \tau_{yz}}{\partial y} + \frac{\partial \sigma_z}{\partial z} \right) = \rho \frac{Dv_z}{Dt} \tag{3.2c}$$

in which X, Y, and Z are the x, y, and z components of body forces; v_x, v_y, and v_z are the x, y, and z velocity components; ρ is the mass density of the medium; and D/Dt denotes the substantial or total derivative (as defined in Section 2.3). In the case of a solid, the acceleration terms drop out, and Equations 3.2 become the equilibrium equations of elasticity theory (Wang 1953, Chapter 11).

3.2.3 Strain at a Point

Since Equations 3.2 contain nine unknown variables in the case of a fluid (six for a solid), the three equations are not sufficient to permit a solution. In order to solve the equations, additional relationships must be provided. In classical mechanics of fluids or elastic solids, the concept of strain (more precisely, rate of strain in the case of fluids) at a point provides some of these additional relationships. In developing this concept, it is demonstrated that the strain or rate of strain at a point is completely characterized by nine scalar components (Wang 1953, Schlichting 1960) as follows:

$$\varepsilon_x = \partial v_x/\partial x \tag{3.3a}$$

$$\varepsilon_y = \partial v_y/\partial y \tag{3.3b}$$

$$\varepsilon_z = \partial v_z/\partial z \tag{3.3c}$$

$$\gamma_{xy} = \frac{\partial v_x}{\partial y} + \frac{\partial v_y}{\partial x} \tag{3.3d}$$

$$\gamma_{xz} = \frac{\partial v_x}{\partial z} + \frac{\partial v_z}{\partial x} \tag{3.3e}$$

$$\gamma_{yz} = \frac{\partial v_y}{\partial z} + \frac{\partial v_z}{\partial y} \tag{3.3f}$$

$$\omega_x = \frac{\partial v_z}{\partial y} - \frac{\partial v_y}{\partial z} \tag{3.3g}$$

$$\omega_y = \frac{\partial v_x}{\partial z} - \frac{\partial v_z}{\partial x} \tag{3.3h}$$

$$\omega_z = \frac{\partial v_y}{\partial x} - \frac{\partial v_x}{\partial y} \tag{3.3i}$$

in which ε_x, ε_y, and ε_z represent linear strain rates; γ_{xy}, γ_{xz}, and γ_{yz} represent angular strain rates; and ω_x, ω_y, and ω_z represent angular rotations. Since the concern is with fluid (or solid) *deformation*, the rotation terms (Equations 3.3g–i) may be deleted. The remaining six components (Equations 3.3a–f) fully characterize the state of strain at a point.

The existence of principal planes of strain, i.e., planes intersecting at a point whose normal (principal) axes do not rotate, may be proven. Along these axes, there will be only linear strain. As with the principal stress axes, the axes of principal strain may be proven to be orthogonal. The development for an elastic solid is identical, the only difference being in the substitution of displacement derivatives for the velocity derivatives in Equations 3.3.

3.2.4 Stress–Strain Relationships

Equations 3.2 and 3.3 comprise a total of 9 equations in 15 unknowns. Six relationships between the stress and strain terms will result in an equal number of equations and unknowns and give a theoretically solvable formulation of the respective problems of fluid or solid mechanics. The formulation of these relations revolves around some of the most important names in the history of engineering mechanics (Streeter 1948, p. 223; Rouse and Ince 1957, pp. 194–197; Schlichting 1960, p. 53).

Isaac Newton, in the seventeenth century, discovered the linear proportionality that exists between shear stress and rate of strain for many fluids in one-dimensional shear flow. (He did so as an intermediate step in his eventually successful attempt to explain the motion of the planets in the solar system.) In terms of the notation defined previously, this proportionality may be written as follows:

$$\tau_{xy} = \mu \frac{dv_x}{dy} \tag{3.4}$$

and similarly for τ_{xz} and τ_{yz}. The constant μ is the dynamic viscosity (also called absolute viscosity) of the particular fluid and was discussed in historical context in Section 1.5. Dynamic viscosity has units of lbm/(sec ft), dyne sec/sq cm (=1 poise), etc. Fluids that behave in accordance with the above relationship are known as Newtonian fluids and include virtually all fluids of concern in environmental engineering. Although viscosity will change with temperature, it does not change with the strain rate. The many fluids whose viscosity changes with the strain rate (i.e., do not undergo strain rates proportional to the applied shear stress) are called non-Newtonian fluids and are the subject of the science of rheology; a common example is paint, which when being applied requires a nonzero shear to flow; this allows the paint to be applied without running down non-horizontal surfaces.

In addition to dynamic or absolute viscosity as defined above, the term kinematic viscosity is used, denoted by ν and defined as $\nu = \mu/\rho$ and with units of ft^2/sec or cm^2/sec. Section 9.2 gives a formula for dynamic viscosity of dry air as a function of absolute temperature. A table of kinematic viscosity vs. temperature for various gases and liquids is given by Shames (1962, Figure B-2).

Newton's contemporary, Robert Hooke, discovered the analogous one-dimensional relationship for an elastic solid. Poisson, in the early nineteenth century, extended Hooke's stress–strain relationship to three dimensions. In the course of deriving general momentum equations, Navier in 1827 and Poisson in 1831, reasoning on the molecular level, demonstrated the plausibility of analogous three-dimensional relationships for a fluid.

Saint-Venant in 1843 derived the same stress–strain relationships from considerations based on the macroscopic properties of fluids. This derivation was considered an improvement over those of Navier and Poisson because Saint-Venant avoided assumptions about certain indeterminate characteristics of molecular motion (Rouse and Ince 1957). In addition, Saint-Venant explicitly presented general stress–strain relationships, which are given below.

Finally, G. G. Stokes (1845) presented an ingenious derivation of the stress–strain relationships based on the property of isotropy and taking advantage of the existence of principal axes. That derivation is considered the classical derivation of what are now known as Stokes' stress–strain relationships.

Stokes' derivation has particular significance to the development of G-value and will, accordingly, be summarized here. Stokes hypothesized that a relationship analogous to that of Equation 3.4 exists in the three-dimensional case. The most general linear relationship would be of the following form:

$$\sigma_x = \mu_1 \varepsilon_x + \mu_2 \varepsilon_y + \mu_3 \varepsilon_z + \mu_4 \gamma_{xy} + \mu_5 \gamma_{xz} + \mu_6 \gamma_{yz} \tag{3.5}$$

with five additional such equations for the other stress terms. A total of $6 \times 6 = 36$ constants results from Stokes' hypothesis.

Taking advantage of the existence of principal axes of stress and strain and the isotropic nature of most Newtonian fluids, Stokes showed that the six linear stress–strain relationships could alternately be written as three equations for three normal stresses and three linear strain rates. Thereby, the number of constants was reduced to nine. By rotating coordinates and taking further advantage of isotropy, the number of independent fluid properties was reduced to two. Then, transforming from principal axes back to x, y, z axes (Streeter 1948, §112), Stokes' stress–strain relationships were obtained. In the case of a fluid, the two constants are the first and second coefficients of viscosity. For an elastic solid, they are known as the Lamé constants (Wang 1953, Chapter 3), which can be related to two physical constants: Young's modulus and either the modulus of elasticity or Poisson's ratio. For incompressible fluid flow, the terms involving the second coefficient of viscosity drop out; the resulting Stokes' stress–strain relationships are as follows (Schlichting 1960):

$$\sigma_x = -p + 2\mu \frac{\partial v_x}{\partial x} \tag{3.6a}$$

$$\sigma_y = -p + 2\mu \frac{\partial v_y}{\partial y} \tag{3.6b}$$

$$\sigma_z = -p + 2\mu \frac{\partial v_z}{\partial z} \tag{3.6c}$$

$$\tau_{xy} = \mu \left(\frac{\partial v_x}{\partial y} + \frac{\partial v_y}{\partial x} \right) \tag{3.6d}$$

$$\tau_{yz} = \mu \left(\frac{\partial v_y}{\partial z} + \frac{\partial v_z}{\partial y} \right) \tag{3.6e}$$

$$\tau_{xz} = \mu \left(\frac{\partial v_z}{\partial x} + \frac{\partial v_x}{\partial z} \right) \tag{3.6f}$$

It is interesting to note that some modern textbook derivations of Equations 3.6 are based, at least partially, on analogy with the related equations of elasticity theory (Prandtl 1952, Schlichting 1960, Shames 1962). Insofar as all of the concepts presented up to this point are concerned, Newtonian fluids and elastic solids are analogous, with the difference that in the latter case, the stresses are proportional to the strain, and in the former case, they are proportional to the rate of strain. By reducing Equations 3.6 to the case of one-dimensional shear flow, it becomes apparent that μ is the absolute viscosity defined by Equation 3.4.

Having eliminated the linear and angular strain rate terms by combining Equations 3.3 and 3.5, there remain (in Equations 3.2 and 3.6) a total of nine equations in nine unknowns. These relationships are theoretically sufficient to solve the general problem of Newtonian fluid flow or elastic solid deformation. They may be reduced to a still smaller number of equations by substituting Equations 3.6 for the stress terms in Equations 3.2. The resulting three equations in three unknowns are known as the Navier–Stokes equations (for incompressible fluid flow), given as follows:

$$\rho\frac{Dv_x}{Dt} = X - \frac{\partial p}{\partial x} + \mu\left(\frac{\partial^2 v_x}{\partial x^2} + \frac{\partial^2 v_x}{\partial y^2} + \frac{\partial^2 v_x}{\partial z^2}\right) \tag{3.7a}$$

$$\rho\frac{Dv_y}{Dt} = Y - \frac{\partial p}{\partial y} + \mu\left(\frac{\partial^2 v_y}{\partial x^2} + \frac{\partial^2 v_y}{\partial y^2} + \frac{\partial^2 v_y}{\partial z^2}\right) \tag{3.7b}$$

$$\rho\frac{Dv_z}{Dt} = Z - \frac{\partial p}{\partial z} + \mu\left(\frac{\partial^2 v_z}{\partial x^2} + \frac{\partial^2 v_z}{\partial y^2} + \frac{\partial^2 v_z}{\partial z^2}\right) \tag{3.7c}$$

(The general vector relationship is: $\rho D\overline{v}/Dt = \overline{F} - \nabla p + \mu\nabla^2\overline{v}$, in which $\overline{v}$ is the velocity vector, $\overline{F}$ is the force field vector, ∇ is the gradient or del operator, and ∇^2 is the Laplacian operator.) The analogous equations for an elastic solid (Wang 1953, p. 34) are called the Lamé equations. For a gravitational force field with the gravity vector in the z direction, we have $X = Y = Z$ and $Z = -gz$.

3.3 Dissipation Function

The derivation of the dissipation function requires general differential energy equations applicable to fluid flow. The First Law of Thermodynamics and the concepts of energy and work as they apply to fluid systems are necessary to derive the energy equations.

The First Law of Thermodynamics (Section 2.6) is applied to the same prismatic fluid element to which Newton's Second Law was applied above. The First Law of Thermodynamics stipulates that the rate of increase of energy in the element (DE/Dt) equals the rate at which heat is added to the element by its environment ($\dot{Q}$) minus the rate at which work is done by the element on its environment ($\dot{W}$). This is expressed mathematically by Equation 2.54.

The energy of a system moving through a force field (in this case, gravitational) is relative to the observer (Hatsopoulus and Keenan 1965). For an observer at location x_o, y_o, z_o who is stationary relative to an elemental system of fluid particles about the point x, y, z moving with velocity components v_x, v_y, v_z, the energy is given by:

$$E = \rho\Delta x\Delta y\Delta z\left[u + \frac{v_x^2 + v_y^2 + v_z^2}{2} + g(z - z_o)\right] \tag{3.8}$$

in which the terms on the right-hand side represent the internal energy, kinetic energy, and potential energy of the system.

The derivation of the differential energy equations may be substantially simplified by expressing the energy in terms of an observer moving with the fluid element (Rohsenow and Choi 1961). Relative to that observer, Equation 3.8 gives the following relationship for the rate of increase of energy:

$$\frac{dE}{dt} = \rho\Delta x\Delta y\Delta z\frac{Du}{Dt} \tag{3.9}$$

in which D/Dt represents the *substantial* or *total* derivative as before.

The rate at which work is done by the element is equal to minus the rate at which work is done by normal and shear stresses acting on the element. Mathematically, the rate of work done by each of the stress components on the element is given by the product of the force resulting from that component and the collinear velocity. Thus, to a stationary observer, the normal forces on the x-faces of the element do work at a rate given by:

$$-\dot{W}_{\sigma_x} = \left[-\sigma_x v_x + \left(\sigma_x + \frac{\partial\sigma_x}{\partial x}\Delta x\right)\left(v_x + \frac{\partial v_x}{\partial x}\Delta x\right)\right]\Delta y\Delta z \tag{3.10}$$

For the moving observer $v_x = 0$, and dropping second-order products of differentials Equation 3.10 reduces to:

$$-\dot{W}_{\sigma_x} = -\sigma_x \frac{\partial v_x}{\partial x}\Delta x \Delta y \Delta z \tag{3.11}$$

Similarly, the shear stresses in the y-direction on the x-faces do work on the system relative to the stationary observer as follows:

$$-\dot{W}_{\tau_{xy}} = \left[-\tau_{xy}\left(v_y - \frac{\partial v_y}{\partial x}\frac{\Delta x}{2}\right) + \left(\tau_{xy} + \frac{\partial \tau_{xy}}{\partial x}\Delta x\right)\left(v_y + \frac{\partial v_y}{\partial x}\frac{\Delta x}{2}\right)\right]\Delta y \Delta z \tag{3.12}$$

For the moving observer ($v_y = 0$) Equation 3.12 reduces to:

$$-\dot{W}_{\tau_{xy}} = -\tau_{xy}\frac{\partial v_y}{\partial x}\Delta x \Delta y \Delta z \tag{3.13}$$

The rate at which work is done by the system on its environment due to all shear and normal forces is given (relative to the moving observer) by:

$$\frac{\dot{W}}{\Delta x \Delta y \Delta z} = -\left[\sigma_x\frac{\partial v_x}{\partial x} + \tau_{yx}\frac{\partial v_x}{\partial y} + \tau_{zx}\frac{\partial v_x}{\partial z} + \sigma_y\frac{\partial v_y}{\partial y} + \tau_{zy}\frac{\partial v_y}{\partial z} + \tau_{xy}\frac{\partial v_y}{\partial x} + \sigma_z\frac{\partial v_z}{\partial z} + \tau_{yz}\frac{\partial v_z}{\partial y} + \tau_{xz}\frac{\partial v_z}{\partial x}\right] \tag{3.14}$$

Combining Equations (2.54, 3.9, and 3.14), and rearranging gives:

$$\left(\frac{\dot{Q}}{\Delta x \Delta y \Delta z} - \rho\frac{Du}{Dt}\right) = -\sigma_x\frac{\partial v_x}{\partial x} - \sigma_y\frac{\partial v_y}{\partial y} - \sigma_z\frac{\partial v_z}{\partial z} - \tau_{xy}\left(\frac{\partial v_x}{\partial y} + \frac{\partial v_y}{\partial x}\right) - \tau_{xz}\left(\frac{\partial v_x}{\partial z} + \frac{\partial v_z}{\partial x}\right) - \tau_{yz}\left(\frac{\partial v_y}{\partial z} + \frac{\partial v_z}{\partial y}\right) \tag{3.15}$$

Equation 3.15 states mathematically that the energy expended by fluid stresses in *distorting* elements of fluid is converted to internal energy and heat, which are then exchanged with the environment. The Second Law of Thermodynamics may be employed to show that this conversion is irreversible (the reader is referred to the Section 2.6 discussion of "Energy Loss Due to Friction.") and thus represents a *dissipation* of energy. For later use, it is important to note that both shear *and normal* stresses are involved in the distortion associated with energy dissipation, contrary to some statements found in the literature (Camp and Stein 1943, Fair *et al.* 1968).

The Stokes stress–strain relationships Equations (3.6) may now be substituted into Equation 3.15, resulting in the differential energy equations, given (for incompressible flow) as follows:

$$\rho\frac{Du}{Dt} - \frac{\dot{Q}}{\Delta x \Delta y \Delta z} = \mu\Phi \tag{3.16a}$$

$$\Phi = \left\{2\left[\left(\frac{\partial v_x}{\partial x}\right)^2 + \left(\frac{\partial v_y}{\partial y}\right)^2 + \left(\frac{\partial v_z}{\partial z}\right)^2\right] + \left(\frac{\partial v_y}{\partial x} + \frac{\partial v_x}{\partial y}\right)^2 + \left(\frac{\partial v_z}{\partial y} + \frac{\partial v_y}{\partial z}\right)^2 + \left(\frac{\partial v_x}{\partial z} + \frac{\partial v_z}{\partial x}\right)^2\right\} \tag{3.16b}$$

The term Φ in Equation 3.16b was first given by Stokes (Stokes 1845; reported by Lamb 1945, p. 580), and is referred to as the *Stokes dissipation function* in modern literature. Equation 3.16b is the form given by Lamb (1945), Schlichting (1960), Rohsenow and Choi (1961), and others. Equation 3.16 states that velocity gradients (angular strain rates) and spatial acceleration (linear strain rates) will be converted through the action of viscosity into internal energy and heat, the conversion representing an irreversible dissipation of energy. The term Φ represents the dissipation of energy per unit volume brought about by fluid friction (viscosity). Equation 3.16 is valid at a *point* in any incompressible fluid flow (laminar or turbulent, the latter as defined in Section 4.7) provided that the fluid properties are continuous (excluding only such systems as rarefied gases, which are of no practical interest for our purposes). It is at the microscopic level, through the action of viscosity, that energy is ultimately dissipated in fluid flows.

By selecting the moving frame of reference, kinetic energy, body force, and stress derivative terms were eliminated in the above development. Had the stationary frame of reference been employed, those terms would remain in Equation 3.8. The differential momentum equations (Equations 3.7) could have then been used to eliminate those terms (Schlichting 1960). When conservative (non-dissipative) energy changes are involved, the momentum equation will fully characterize the system.

Equation 3.16 applies at a *point* in a fluid system, as was noted. For *laminar flow* (see Section 4.7), Equation 3.16 can be integrated over the whole volume of the system. For a closed container in which the fluid velocity is zero at every point of the boundary, the integral is as follows (Lamb 1945, p. 580):

$$2F = \iiint \Phi dxdydz = \mu \iiint \left(\omega_x^2 + \omega_y^2 + \omega_z^2\right) dxdydz \tag{3.17}$$

The terms ω_x, ω_y, ω_z are the rotational terms given in Equations 3.3g–i. Equation 3.17 was derived by Rayleigh in 1873, and the term $2F$ is called the "Rayleigh dissipation function" (Lamb 1945, p. 568), which has since been applied to both Stokes' form (Equation 3.16b) and that of Rayleigh (Equation 3.17). Reference will be made to the above relationship again in the Section 14.6 discussion of flocculation.

3.4 Critical Review of G-Value Theory

Since the 1940s, the concept of G-value [also called the root-mean-square (rms) velocity gradient] has been sporadically applied to various water and wastewater treatment processes. It has been used to quantify flocculation, floc breakup, mixing, induced circulation in channels, short-circuiting in tanks, and cleansing of filter media during backwashing. That usage has led the authors of one major reference book (Fair *et al.* 1968) to call the G-value "an important unifying concept in modern water and wastewater treatment operations or processes."

The lack of experimental support for most of the applications of G-value leads one to expect that a powerful basic principle is involved that puts the concept beyond reproach. Unfortunately, that is not the case. It will be demonstrated herein that the basic derivation on which the concept was founded is incorrect. Although a useful parameter in certain circumstances, its applicability to environmental engineering processes is quite limited. The theoretical basis of G-value is reviewed below.

The original derivation of G-value was presented in 1943 by Camp and Stein (1943), who suggested its application to the process of flocculation. They related the G-value to the dissipation function of fluid mechanics but presented their "dissipation function" in the following form:

$$\Phi' = \mu\left[\left(\frac{\partial v_x}{\partial y} + \frac{\partial v_y}{\partial x}\right)^2 + \left(\frac{\partial v_x}{\partial z} + \frac{\partial v_z}{\partial x}\right)^2 + \left(\frac{\partial v_y}{\partial z} + \frac{\partial v_z}{\partial y}\right)^2\right] \tag{3.18}$$

The above form of the "dissipation function" or simplifications thereof are repeated in other environmental engineering references.

Camp and Stein (1943) noted that the bracketed terms in the above equation are the square of the absolute velocity gradient at a point (recall Equations 3.3d–f). This led them to suggest an equation for the rms velocity gradient or G-value G_m, defined as follows:

$$G_m = \sqrt{\Phi'_m/\mu} \tag{3.19}$$

Since the total energy dissipated per unit volume by a system is equal to the work input per unit volume to the system (in the absence of heat addition), Φ'_m was taken as the time rate of power per unit volume added to the system. Noting the significance of velocity gradients to flocculation (as discussed in Section 14.6, under certain conditions the absolute velocity gradient at a point is proportional to the rate of flocculation), Camp, Stein, and later Camp (1955) proposed the use of Equation 3.19 in the design of flocculation systems. The common form in which Equation 3.19 appears in the literature is as follows:

$$G_m = \sqrt{P/(\mu V)} \tag{3.20}$$

in which P/V is the power input per unit volume (Camp 1955, AWWA 1969, WPCF/ASCE 1977). Since its initial application to flocculation, the G-value has been applied to other environmental processes, as cited later.

Equation 3.18 differs from the classical dissipation function (Equation 3.16b) in that the acceleration terms ($\partial v_x/\partial x$, $\partial v_y/\partial y$, $\partial v_z/\partial z$) are missing from the former. Camp and Stein acknowledged the difference between their function and the dissipation function of classical hydrodynamics. They argued that contrary to the derivation summarized in the previous section, the dissipation of energy accompanying the distortion of a fluid element is due only to shear distortion, with normal stress not contributing to this dissipation. If that were true, then Equation 3.18 would indeed be the correct form of the dissipation function.

To demonstrate that all the energy lost as heat is accounted for by Equation 3.18, Camp and Stein (1943) attempted to prove the existence of planes on which there is no extension or contraction of the fluid element bounded by the planes. On such planes, only shear stresses and no normal stresses would exist; these planes are, in this respect, the opposite of the principal planes previously discussed on which only normal stresses exist.

Camp and Stein's derivation was incorrect. That can be demonstrated most directly by proving that planes having only shear stress and no normal stress do not exist in general. We do this by hypothesizing the existence of such planes and demonstrating that a contradiction results. The proof is similar in certain respects to that employed in proving the existence of principle planes (Streeter 1948), as was required for the above derivation of the Navier–Stokes equations.

Considering the tetrahedron shown in Figure 3.2, assume that the arbitrarily inclined face has only three shear stress components so that the resultant stress σ_R lies in the plane of the inclined face. Since the orthogonal x, y, z axes may be arbitrarily oriented, they are chosen to be parallel to principal axes so that σ_x, σ_y, and σ_z are principal stresses and there are no shear stresses acting on these faces. The areas on which σ_x, σ_y, and σ_z act are Al, Am, and An, respectively, in which A is the area of the inclined face and l, m, and n are the direction cosines defined previously. Then, summing forces in the direction of the normal to the inclined face on which the normal stress is zero, we have:

$$\sigma_x l \bullet l + \sigma_y m \bullet m + \sigma_z n \bullet n = 0 \tag{3.21}$$

The principal stresses σ_x, σ_y, and σ_z are roots of the following equation (Streeter 1948, p. 215):

$$\sigma^3 - I_1\sigma^2 + I_2\sigma - I_3 = 0 \tag{3.22}$$

where I_1, I_2, and I_3 are invariant quantities (independent of axes orientation) given by:

$$I_1 = \sigma_x + \sigma_y + \sigma_z \tag{3.23a}$$

$$I_2 = \sigma_x\sigma_y + \sigma_y\sigma_z + \sigma_z\sigma_x \tag{3.23b}$$

$$I_3 = \sigma_x\sigma_y\sigma_z \tag{3.23c}$$

Since $\sigma_x = \sigma_y = \sigma_z$ are admissible roots of Equations 3.23, the existence of a plane of zero normal stress requires, in this special case, that

$$l^2 + m^2 + n^2 = 0 \tag{3.24}$$

However, since $l^2 + m^2 + n^2 = 1$ by definition of direction cosines, Equation 3.24 is contradictory, and values of l, m, and n satisfying Equation 3.24 do not exist in this case. Therefore, planes of zero normal stress do not exist in general. Accordingly, Equation 3.18 is not valid in the general three-dimensional case.

From the above, we may also conclude that Equations 3.19 and 3.20 are not generally true. Only a portion of the power input per unit volume P/V in Equation 3.20 causes velocity gradients. This realization and experimental observations discussed in Section 14.6 led the author to conduct a critical assessment of the various processes to which G-value theory has been applied. The results of that assessment are presented in the following sections along with suggested alternatives to the use of G-value where appropriate:

Section 13.3 – Induced Circulation in Channels
Section 14.2 – Basic Reactor Concepts (short-circuiting)
Section 14.4 – Mixing
Section 14.6 – Flocculation and Floc Breakup
Section 14.8 – Filtration and Fluidized Beds

4

Analytical Solutions of the Navier–Stokes and Related Equations

4.1 Introduction

A general analytical solution of the Navier–Stokes equations has not been obtained. However, specialized solutions for various simplifications of those equations are attainable and useful both for the theoretical insight they provide and for certain practical applications.

4.2 Potential Flow, Zero Slip, Boundary Layers, and Prandtl's Synthesis

4.2.1 Potential Flow

During the eighteenth and nineteenth centuries, eminent mathematicians devoted much time to obtaining solutions to the equations of *potential flow theory*. Potential flow refers to the *irrotational* (defined below) flow of a fluid. In the present section, an attempt is made to provide a clear overall perspective of the concepts and applicability of potential flow theory. Some of the mathematical details will be summarized rather than presented fully; in such cases, reference will be made to details available elsewhere. In any event, the reader interested in gaining further understanding of potential flow theory should consult one or more of the available references (Kuethe and Schetzer 1959, Chapter 4; Shames 1962, Chapters 8 and 9).

The *stream function* is a scalar (as opposed to vector) function, which is of great utility in potential flow theory. The derivation of the stream function given here is essentially that of Kuethe and Schetzer (1959). Consider a steady, incompressible, two-dimensional flow. In Figure 4.1, ab and cd represent streamlines in the flow. A *streamline* is a path whose direction at every point coincides with that of the mean velocity. In a turbulent flow, the mean velocity refers to the time-averaged velocity, as discussed in Section 2.3. There can be no net (time-averaged) transfer of fluid across ab or cd. Because incompressible, steady flow has been assumed, the same volumetric flow of fluid must cross ef that crosses gh or any other path connecting the streamlines. If the streamline ab is arbitrarily chosen as a base, every other streamline in the flow field can be identified by assigning to it a number equal to the volumetric flow rate of fluid passing between it and the base streamline. This identifier is the *stream function* denoted by ψ. Note that the stream function has a value at any point in the flow field, which represents the volumetric flow rate passing between that point and the base streamline (for which $\psi = 0$). The streamlines are isolines of this parameter.

The component of velocity u_s in any direction is determinable from ψ as follows: Let Δn be an incremental length in the flow field perpendicular to the direction s. Let ψ_1 and ψ_2 be the values of ψ at the endpoints of Δn as shown in Figure 4.2. The volumetric flow rate of fluid crossing the line ij must be equal to that crossing the lines kl and ik. By definition of ψ we have:

$$\psi_2 = \psi_1 + u_s \Delta n \tag{4.1}$$

or:

$$u_s = \frac{\psi_2 - \psi_1}{\Delta n} \tag{4.2}$$

Hydraulics and Pneumatics in Environmental Engineering, First Edition. S. David Graber.

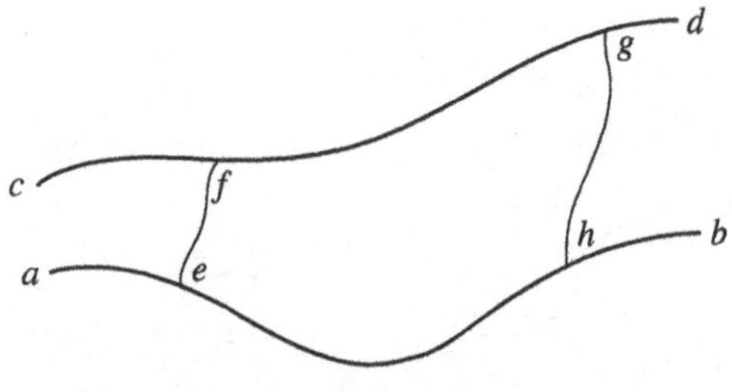

Figure 4.1 Derivation of Stream Function (Kuethe and Schetzer 1959/John Wiley & Sons)

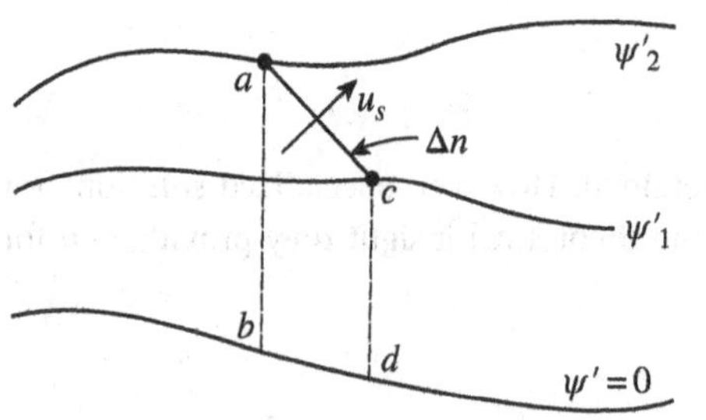

Figure 4.2 Velocity Components from Stream Function (Kuethe and Schetzer 1959/John Wiley & Sons)

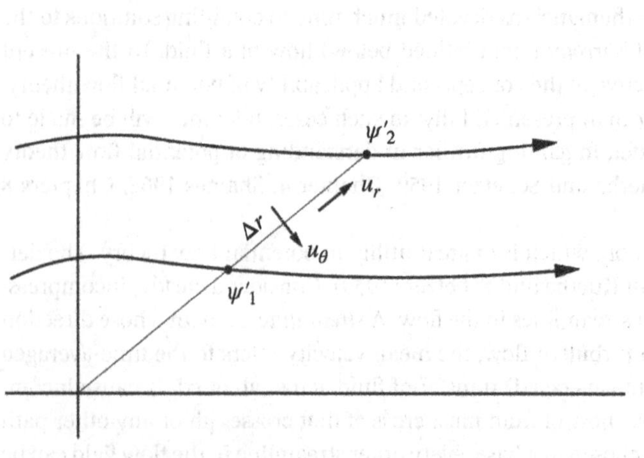

Figure 4.3 Velocity Components from Stream Function in Polar Coordinates

Taking the limit, we obtain:

$$u_s = \frac{d\psi}{dn} \tag{4.3}$$

The velocity component in any direction is found by differentiating the stream function at right angles to that direction. (The sign convention varies in the literature, and caution must be exercised in using results.) Referring to Figure 4.3, in Cartesian coordinates, the velocity components are derived by first noting:

$$\psi_2 - \psi_1 \cong u\Delta y \cong -v\Delta x \tag{4.4}$$

and then taking limits to obtain:

$$u = \frac{\partial \psi}{\partial y} \tag{4.5a}$$

$$v = -\frac{\partial \psi}{\partial x} \tag{4.5b}$$

In polar coordinates, with increasing values of θ and r measured counterclockwise and outward from the origin, the analogous relations are:

$$\psi_2 - \psi_1 \cong -u_\theta \Delta r = u_r r \Delta\theta \tag{4.6}$$

and:

$$u_r = \frac{1}{r}\frac{\partial \psi}{\partial \theta} \tag{4.7a}$$

$$u_\theta = -\frac{\partial \psi}{\partial r} \tag{4.7b}$$

An irrotational flow is one in which the fluid elements are subjected to pure strain with zero rotation; this is depicted in Figure 4.4. For such a flow, the angular velocities ω_x, ω_y, ω_z given by Equations 3.3g–i are all zero. Taking $\omega_z = 0$ in Equation 3.3i (recognizing the differing notation) and combining the result into differentials of Equations 4.5 results in:

$$\frac{\partial^2 \psi}{\partial x^2} + \frac{\partial^2 \psi}{\partial y^2} = 0 \tag{4.8}$$

In terms of the Laplace operator ∇^2 and rectangular, cylindrical, or axially symmetric, and polar or spherical coordinates, the stream function may be expressed for three-dimensional incompressible flows in the following respective forms (Shapiro 1954, Chapter 9; Kuethe and Schetzer 1959, Chapter 7):

$$\nabla^2 \psi = 0 \tag{4.9a}$$

$$\frac{\partial^2 \psi}{\partial x^2} + \frac{\partial^2 \psi}{\partial y^2} + \frac{\partial^2 \psi}{\partial z^2} = 0 \tag{4.9b}$$

$$\frac{1}{r}\frac{\partial}{\partial r}\left(r\frac{\partial \psi}{\partial r}\right) + \frac{1}{r^2}\frac{\partial^2 \psi}{\partial \omega^2} + \frac{\partial^2 \psi}{\partial z^2} = 0 \tag{4.9c}$$

$$\frac{1}{r}\frac{\partial^2}{\partial r^2}(r\psi) + \frac{1}{r^2 \sin\theta}\frac{\partial}{\partial \theta}\left(\sin\theta \frac{\partial \psi}{\partial \theta}\right) + \frac{1}{r^2 \sin\theta}\frac{\partial^2 \psi}{\partial \omega^2} = 0 \tag{4.9d}$$

A stream function may also be defined for compressible flow (Shapiro 1954, Chapter 9; Kuethe and Schetzer 1959, Chapter 7; Shames 1962, Chapter 14).

For irrotational flow, the existence of another useful scalar function can be demonstrated (Kuethe and Schetzer 1959, Chapter 2; Shames 1962, Chapter 6), which is known as the velocity potential ϕ. In rectangular coordinates, the velocity components are related to the velocity potential as follows:

$$u = \frac{\partial \phi}{\partial x} \tag{4.10a}$$

$$v = \frac{\partial \phi}{\partial y} \tag{4.10b}$$

$$w = \frac{\partial \phi}{\partial z} \tag{4.10c}$$

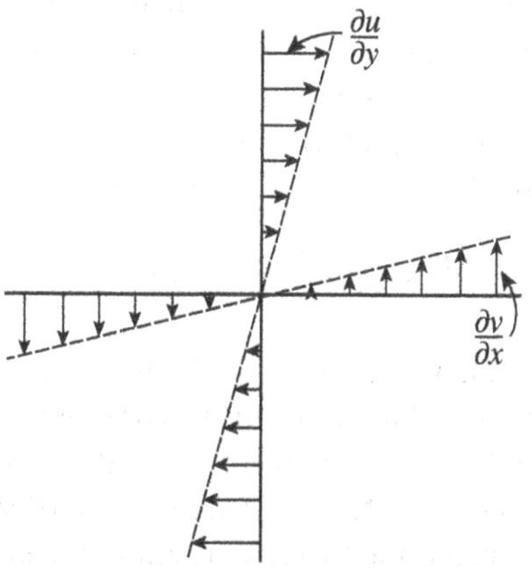

Figure 4.4 Pure Strain (Kuethe and Schetzer 1959/John Wiley & Sons)

Substituting the above equations into the continuity equation (Equation 2.9 or 2.11) results in:

$$\frac{\partial^2 \phi}{\partial x^2} + \frac{\partial^2 \phi}{\partial y^2} + \frac{\partial^2 \phi}{\partial z^2} = 0 \tag{4.11}$$

In terms of the Laplace operator ∇^2 and rectangular, cylindrical or axially symmetric, and polar or spherical coordinates, the velocity potential may be expressed in the following respective forms analogous to Equations 4.9:

$$\nabla^2 \phi = 0 \tag{4.12a}$$

$$\frac{\partial^2 \phi}{\partial x^2} + \frac{\partial^2 \phi}{\partial y^2} + \frac{\partial^2 \phi}{\partial z^2} = 0 \tag{4.12b}$$

$$\frac{1}{r}\frac{\partial}{\partial r}\left(r\frac{\partial \phi}{\partial r}\right) + \frac{1}{r^2}\frac{\partial^2 \phi}{\partial \theta^2} + \frac{\partial^2 \phi}{\partial z^2} = 0 \tag{4.12c}$$

$$\frac{1}{r}\frac{\partial^2}{\partial r^2}(r\phi) + \frac{1}{r^2 \sin\theta}\frac{\partial}{\partial \theta}\left(\sin\theta\frac{\partial \phi}{\partial \theta}\right) + \frac{1}{r^2 \sin\theta}\frac{\partial^2 \phi}{\partial \omega^2} = 0 \tag{4.12d}$$

The general relation between the velocity potential and the velocity vector $\overline{V}$ at a point is given by $\overline{V} = \text{grad}\phi$ in which $\text{grad}\,\phi = \bar{i}\frac{\partial \phi}{\partial x} + \bar{j}\frac{\partial \phi}{\partial y} + \bar{k}\frac{\partial \phi}{\partial z}$ with $\bar{i}, \bar{j}$, and $\bar{k}$ being unit vectors in the x, y, and z directions.

The relation between the equipotential (ϕ = constant) lines and streamlines may now be established. Since ϕ and ψ are both point functions, their differentials may be written in Cartesian coordinates as follows:

$$d\phi = \frac{\partial \phi}{\partial x}dx + \frac{\partial \phi}{\partial y}dy \tag{4.13a}$$

$$d\psi = \frac{\partial \psi}{\partial x}dx + \frac{\partial \psi}{\partial y}dy \tag{4.13b}$$

Setting $d\phi$ and $d\psi$ equal to zero and using Equations 4.10a and 4.10b and 4.5a and 4.5b, the slopes of lines of constant ϕ and ψ are obtained as follows:

$$\left(\frac{dy}{dx}\right)_\phi = -\frac{u}{v} \tag{4.14a}$$

$$\left(\frac{dy}{dx}\right)_\psi = \frac{v}{u} \tag{4.14b}$$

From the above equations, the relation between the slopes of equipotential lines and streamlines is found as follows:

$$\left(\frac{dy}{dx}\right)_\phi = -1/\left(\frac{dy}{dx}\right)_\psi \tag{4.15}$$

From the above equation, it is concluded that equipotential lines are normal (perpendicular) to streamlines. The two sets of lines are said to form an orthogonal network. The orthogonality can also be demonstrated for other two-dimensional coordinate systems.

For frictionless flow with no head loss and momentum coefficient α equal to unity, Equation 2.76 gives the following equation for flow along a streamline:

$$\frac{p}{\rho} + gz + \frac{V^2}{2} = \text{const} \tag{4.16}$$

The above relation is known as Bernoulli's equation. It can alternatively be derived by applying the energy equation (Section 2.6) to a control volume encompassing a "streamtube" (Shames 1962, §5–16). Note the similarity of Equation 4.16 to the terms of Equation 2.58. It can be seen that for frictionless, steady, incompressible flow, the First Law of Thermodynamics and Newton's Second Law of Motion reduce to the same equation. For flow with friction or compressibility effects, however, the two laws result in independent equations that provide two distinct analytical tools.

A specialized version of a theorem variously attributed to Thomson (Lord Kelvin) and Helmholtz states that a fluid subjected to conservative body forces (which include gravity) that is initially irrotational remains irrotational. The proof of this theorem makes use of Euler's equations and other concepts and is given in various fluid mechanics texts. The proofs preferred by the author are those given by Shapiro (1954, §9.4) and Shames (1962, App. A-2). One of the most practical categories of flow patterns satisfying this theorem is that which in some regions is uniform and parallel. Such a flow is approximated ahead of an object (such as a sphere) moving through the fluid by the uniform flow rising from a fluidized filter bed and by the idealized flow entering a rectangular settling basin. (All of these practical situations and others will be discussed below.) In such cases, the entire flow field is irrotational, subject to an important qualification discussed below.

One of the simplest examples of the velocity potential and stream function is in the case of uniform parallel flow in the +x direction, for which:

$$\phi = V_\infty x \tag{4.17a}$$

$$\psi = V_\infty y \tag{4.17b}$$

Equations 4.17a and 4.17b are seen to satisfy the pertinent Laplace equations (Equations 4.8 and 4.11) for ψ and ϕ, and Equations 4.5 and 4.10 give $u = V_\infty$ and $v = 0$. Other examples and means of determining ϕ and ψ are given below.

4.2.2 Zero Slip at Solid Boundaries

The practical value of potential flow theory, as outlined above, hinges largely on the reasonableness of the assumption of frictionless flow. Differences of opinion in this regard caused a significant rift between engineers and mathematicians of the nineteenth century and probably retarded the application of theoretical principles of fluid mechanics. One of the conceptual differences centered around D'Alembert's "paradox." This paradox refers to potential flow theory's predicted absence of drag on a cylinder in a uniform flow, which is contrary to observation. This paradox is ultimately related to the boundary conditions employed in solving potential flow problems.

For concreteness, consider the boundary conditions associated with the flow about a cylinder, as depicted in Figure 4.5. (Such flow may result from any combination of motion by the fluid and cylinder, which results in the relative velocity V between the fluid and the cylinder.) The boundary conditions at infinity are given by:

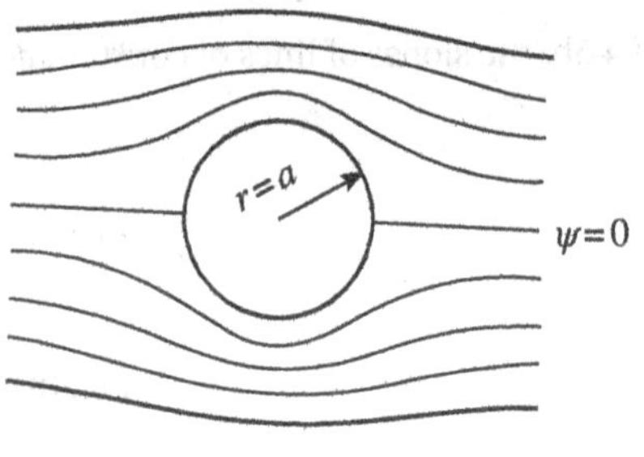

Figure 4.5 Circular Cylinder in a Uniform Stream (Kuethe and Schetzer 1959/John Wiley & Sons)

$$u = \frac{\partial \psi}{\partial y} = \frac{\partial \phi}{\partial x} = V_\infty \quad @x = \pm\infty \tag{4.18a}$$

$$v = -\frac{\partial \psi}{\partial x} = \frac{\partial \phi}{\partial y} = 0 \quad @x = \pm\infty \tag{4.18b}$$

At the cylindrical boundary, the velocity normal to the boundary must be zero. This may be expressed in polar coordinates by:

$$u_r = \frac{\partial \psi}{r\partial \theta} = \frac{\partial \phi}{\partial r} = 0 \quad @r = R \tag{4.19}$$

The stream function satisfying Laplace's equation and the above boundary conditions is given as follows:

$$\psi = V_\infty y\left(1 - \frac{R^2}{r^2}\right) \tag{4.20}$$

Noting that $r = \sqrt{x^2 + y^2}$ and $y = r\sin\theta$, we find that the boundary conditions are all satisfied.

The velocity at the cylindrical boundary can only be in the θ-direction and is obtained by first using Equations 4.7b and 4.20 to obtain:

$$u_\theta = -\frac{\partial\psi}{\partial r} = -V_\infty \sin\theta\left(1 + \frac{R^2}{r^2}\right) \tag{4.21}$$

and then setting $r = R$ to obtain:

$$u_\theta = -2V_\infty \sin\theta \quad @\, r = R \tag{4.22}$$

Potential flow theory thus predicts that the velocity will increase from zero at the leading edge of the cylinder ($\theta = \pi$) to $2V_\infty$ on the sides of the cylinder at $\theta = \pi/2$ and$\theta = 3\pi/2$ and then decrease to zero at $\theta = 0$. The magnitude of the velocity is symmetrical about a vertical axis through the center of the cylinder.

From Bernoulli's equation (Equation 4.16), we find the pressure along the cylindrical surface to be $p = p_\infty + V_\infty^2/(2g) - u_\theta^2/(2g)$. Since p_∞ (the pressure at $x = \pm\infty$) and V_∞ are constants and the magnitude of u_θ is symmetrical, the pressure distribution is also symmetrical about the vertical axis. The pressure force on the upstream half of the cylinder is thus balanced by the pressure force on the downstream side of the cylinder. The result is the paradoxical prediction of zero drag, as noted earlier.

For virtually all potential flow problems, the prescription of zero velocity normal to the solid boundary, together with the other prescribed boundary conditions, will result in a tangential velocity at the boundaries (such as u_θ on the cylindrical surface above). David Bernoulli, in the mid-1700s, first suggested that a fluid could not slip freely over a solid boundary. This clearly raised serious questions about the validity of potential flow theory and began a debate that continued for more than a century. The debate was not only fascinating, but it should serve to boost the spirits of students who are struggling with new concepts to learn of the illustrious names and opposing viewpoints of the participants in this lengthy debate.

Coulomb contended that the slip at the boundary is zero. Navier held that there was slip at the boundary and, strangely, deduced this from the same molecular hypothesis that led him to his equations of motion (see Section 3.2). Poisson adopted a viewpoint similar to Navier's. Stokes studied the various hypotheses and decided to join the no-slip group. Hagen and Poiseuille noted that the fluid at the walls of a tube seemed to have very small velocity but did not comment further. Darcy held to an idea developed earlier by Girard that a layer of fluid adheres to the boundary and the rest of the fluid slips over this layer. Helmholtz adopted Navier's belief.

Various investigators continued experimental work initiated by Stokes. Gradually, the hypothesis of no-slip gathered a following. Maxwell, at the request of the Royal Society, carried out some calculations on the molecular level and concluded that slipping takes place but that the length of slip is a moderate multiple of the mean free path, L, of a gas molecule–probably about $2L$. Thus, at atmospheric pressure, the slip would be negligible (Goldstein 1938, pp. 676–680). Modern work with rarefied gases shows good agreement with Maxwell's prediction and subsequent theories. This fact and the excellent agreement between experiment and theory employing the no-slip condition (Rohsenow and Choi 1961, Chapter 11) resulted in its acceptance for liquids and gas flows at ordinary pressures.

The enhanced applicability of potential flow theory is discussed next.

4.2.3 Boundary Layers and Prandtl's Synthesis

In 1904, Prandtl introduced the important concept and equations of the *boundary layer*. To introduce (or review) this concept, we will consider a uniform flow along a flat plate as depicted in Figure 4.6 (the same general arguments apply to the flow about a streamlined body). As the flow first encounters the plate, there is an extremely abrupt gradient from the velocity V_o to the zero velocity at the plate surface. As one proceeds downstream along the plate, it is found that the thickness of this transitional region increases. This region of high-velocity gradient is called the *boundary layer*.

Prandtl showed that, at high Reynolds Number (defined in Section 4.7), the Navier–Stokes equations could be simplified to derive what are now known as Prandtl's boundary layer equations. The simplest form of these equations, as derived by

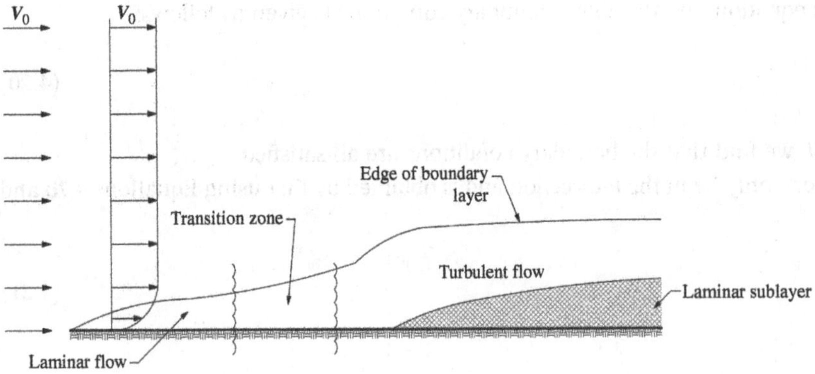

Figure 4.6 Flow Over a Flat Plate Depicting Boundary Layer (Shames 1962/McGraw-Hill Education)

order of magnitude analysis, applies to the steady, two-dimensional flow over a flat plate parallel to the flow. For this case, the applicable Navier–Stokes equations (Equations 3.7a and 3.7b) reduce to:

$$\rho\left(u\frac{\partial u}{\partial x} + v\frac{\partial u}{\partial y}\right) = X - \frac{\partial p}{\partial x} + \mu\left(\frac{\partial^2 u}{\partial x^2} + \frac{\partial^2 u}{\partial y^2}\right) \tag{4.23a}$$

$$\rho\left(u\frac{\partial v}{\partial x} + v\frac{\partial v}{\partial y}\right) = Y - \frac{\partial p}{\partial y} + \mu\left(\frac{\partial^2 v}{\partial x^2} + \frac{\partial^2 v}{\partial y^2}\right) \tag{4.23b}$$

Prandtl's order of magnitude analysis showed that the pressure in the y direction is nearly constant across the boundary layer and may be assumed to be equal to that at the outer edge of the boundary layer. The pressure within the boundary layer is thus imposed by the flow outside of the boundary layer. If that pressure is separately determinable (as it is – see below), then the number of boundary layer unknowns reduces from three (u, v, and p) to two (u and v). Equation 4.23a and the continuity equation (Equation 2.9) then provide the necessary number of equations, and Equation 4.23b may be deleted. Prandtl also showed that the term $\partial^2 u/\partial x^2$ in Equation 4.23a could be neglected. The pertinent equations thus become:

$$u\frac{\partial u}{\partial x} + v\frac{\partial u}{\partial y} = -\frac{1}{\rho}\frac{dp}{dx} + \upsilon\frac{\partial^2 u}{\partial y^2} \tag{4.24a}$$

$$\frac{\partial u}{\partial x} + \frac{\partial v}{\partial y} = 0 \tag{4.24b}$$

Blasius, in 1908, was able to reduce the above two partial differential equations to a single ordinary differential equation (interestingly by using the stream function and a "similarity" variable) and obtain a solution for the laminar boundary layer on a flat plate ("laminar" is defined in Section 4.4). Nikuradse and others provided experimental verification. Boundary layer theory has since been extended to bodies of many shapes and has become an important branch of fluid mechanics. It is of particular quantitative value in the field of aerodynamics.

Of key importance is Prandtl's observation that the flow outside of the boundary layer is one of small velocity gradients in which the influence of viscosity is unimportant. If the flow outside the boundary layer was initially irrotational in addition to being frictionless, then that portion of the flow can be regarded as potential flow. An idea of the thickness of the boundary layer can be obtained by considering the relationship for the *displacement thickness* of a laminar boundary layer on a flat plate based on Howarth's refinement of Blasius' solution. The displacement thickness δ^* is the distance by which the external potential flow field is displaced outwards as a result of the decrease in velocity in the boundary layer (Schlichting 1960, Chapter VII). The relationship is:

$$\delta^* = 1.72\sqrt{\frac{\upsilon x}{U_\infty}} \tag{4.25}$$

In which x is the distance from the leading edge of the plate. For example, for $U_\infty = 1$ ft/sec and $\upsilon = 10^{-5}$ ft^2/sec, we have $\delta^* = 0.00544$ ft at $x = 1$ ft, $\delta^* = 0.0172$ ft at $x = 10$ ft, etc. The boundary of the potential flow may be regarded as the outer limit of the boundary layer (based on a suitable definition of that outer limit). In theory, the potential flow pattern about the body itself can first be obtained. Then the boundary layer dimensions can be determined for the pressure distribution imposed by this first-cut potential flow. The potential flow can then be adjusted for the new boundary conditions, and so forth. In many cases, the boundary layer will be thin enough to obviate the need for adjustment of the potential flow. This procedure is the essence of Prandtl's synthesis of potential flow and boundary layer theory, which established the practical utility of the former.

The notion of "frictionless" flow can now be rigorously qualified. Noting that Equation 4.11 is a result of irrotationality, we may separately take the partial derivatives of that equation with respect to x, y, and z. For the partial derivative with respect to x we obtain:

$$\frac{\partial}{\partial x}\left(\frac{\partial^2 \phi}{\partial x^2} + \frac{\partial^2 \phi}{\partial y^2} + \frac{\partial^2 \phi}{\partial z^2}\right) = 0 = \frac{\partial^2}{\partial x^2}\frac{\partial \phi}{\partial x} + \frac{\partial^2}{\partial y^2}\frac{\partial \phi}{\partial x} + \frac{\partial^2}{\partial z^2}\frac{\partial \phi}{\partial x} \tag{4.26}$$

From the above and Equation 4.9a, we find:

$$\frac{\partial^2 u}{\partial x^2} + \frac{\partial^2 u}{\partial y^2} + \frac{\partial^2 u}{\partial z^2} = 0 \tag{4.27}$$

Note that the above equation eliminates the viscous term in the x-direction Navier–Stokes equation (Equation 3.7a). Similar steps taken for the partial derivatives with respect to y and z eliminate the viscous terms in Equations 3.7b and 3.7c. The general vector relation is:

$$\mathtt{grad}(\nabla^2 \phi) = \nabla^2(\mathtt{grad}\phi) = \nabla^2 \overline{V} \tag{4.28}$$

in which $\nabla^2 \phi = 0$ and hence $\nabla^2 \overline{V} = 0$. The Navier–Stokes equations thus reduce to the Euler equations when the flow is irrotational. In this sense, potential flow may be thought of as a special solution of the Navier–Stokes equations in regions of irrotationality (which excludes the boundary layer).

4.3 Potential Flow Solutions

The preceding section presented the basic concepts of potential flow theory and indicated the conditions to which they reasonably apply. A flow satisfying those conditions has a stream function and velocity potential that satisfy Laplace's equation (Equations 4.9 and 4.12). The problem becomes one of solving Laplace's equation for the boundary conditions appropriate to the geometrical configuration of interest. Solution techniques range from relatively simple to complex. One of the simplest and easiest to visualize, and which has probably been of the greatest utility in environmental engineering applications, is based on the superposition of flows. The superposition method owes its validity to the fact that Laplace's equation is *linear*. Thus, the sum of two or more solutions to Laplace's equation is also a solution. [Such "superposition" is demonstrated to apply to incompressible potential flows (Shapiro 1953, p. 289; Kuethe and Schetzer 1959, §4.2) and supersonic potential flows for slender bodies of revolution (Shapiro 1954, §17.3) but not supersonic potential flows more generally (Kuethe and Schetzer 1959, §7.5).] Pertinent examples of the superposition method are discussed in some detail below, followed by brief mention of other solution techniques.

The uniform parallel potential flow pattern was given previously. Source and sink potential flows are now discussed. Source flow originates at a *point* from which fluid flows with uniform angular symmetry along radial paths. Letting q represent the volumetric flow issuing from the source (called the *source strength*), continuity gives the following relationship:

$$\rho(2\pi r)u_r = \rho q \tag{4.29}$$

or:

$$u_r = \frac{q}{2\pi r} \tag{4.30}$$

The tangential velocity component u_θ is zero, and the stream function is found by direct integration of Equation 4.30:

$$u_r = \frac{1}{r}\frac{d\psi}{d\theta} = \frac{q}{2\pi r} \tag{4.31}$$

$$\psi = \frac{q}{2\pi}\theta + C_1 \tag{4.32}$$

The velocity potential is similarly found by direct integration of Equation 4.30:

$$u_r = \frac{d\phi}{dr} = \frac{q}{2\pi r} \tag{4.33}$$

$$\phi = \frac{q}{2\pi}\ln r + C_2 \tag{4.34}$$

The terms C_1 and C_2 are constants of integration. By choosing the line from which θ is measured ($\theta = 0$) as the zero streamline ($\psi = 0$), C_1 becomes zero. Let $r = R_1$ be the circle on which $\phi = 0$ so that $C_2 = -\frac{q}{2\pi}\ln R_1$. Then the stream function and velocity potential are:

$$\psi = \frac{q}{2\pi}\theta \tag{4.35}$$

$$\phi = \frac{q}{2\pi}\ln\frac{r}{R_1} \tag{4.36}$$

The above ψ and ϕ satisfy the appropriate Laplace's equations, and thus represent potential flows. A negative q corresponds to an inward flow known as a *sink*.

A useful flow pattern is obtained by combining a source and a sink, as shown in Figure 4.7. The source and sink are at coordinates $(-x_o, 0)$ and $(x_o, 0)$, and the radial angles θ_1 and θ_2. The x-axis is chosen as the zero streamline for both the source and the sink, so Equation 4.35 applies to both. The stream function of the combined flow is given by:

$$\psi = -\frac{q}{2\pi}(\theta_1 - \theta_2) \tag{4.37}$$

The angles θ_1 and θ_2 may be expressed in terms of the $x-y$ coordinates, and the difference obtained and rearranged with the aid of the following trig identity:

$$\tan^{-1}\alpha - \tan^{-1}\beta = \tan^{-1}\left(\frac{\alpha-\beta}{1+\alpha\beta}\right) \tag{4.38}$$

to give:

$$\theta_1 - \theta_2 = \tan^{-1}\frac{y}{x-x_o} - \tan^{-1}\frac{y}{x+x_o} = \tan^{-1}\left\{\frac{y/(x-x_o) - y/(x+x_o)}{1+[y^2/(x-x_o)(x+x_o)]}\right\} = \tan^{-1}\frac{2x_o y}{x^2+y^2-x_o^2} \tag{4.39}$$

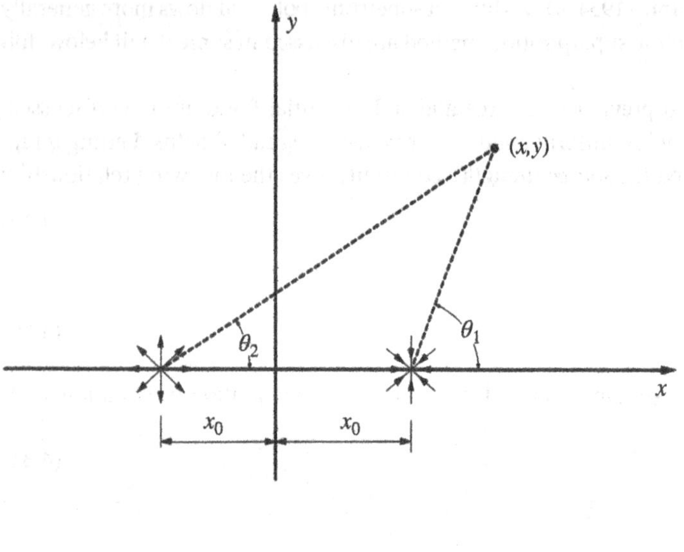

Figure 4.7 Source–Sink Pair (Kuethe and Schetzer 1959/John Wiley & Sons)

The stream function is given by:

$$\psi = \frac{q}{2\pi}\tan^{-1}\frac{2x_o y}{x^2 + y^2 - x_o^2} \tag{4.40}$$

The streamlines (ψ=constant) of the source–sink pair are a series of circles centered on the y-axis as shown in Figure 4.8(a).

Of particular interest is the flow pattern that results when a source and sink of equal strength are allowed to approach each other while holding the product of their strength and distance ($2x_o q \doteq \mu$) constant. Mathematically:

$$\psi = \lim_{x_o\to 0}\left[-\frac{\mu/(2x_o)}{2\pi}\tan^{-1}\frac{2x_o y}{x^2+y^2-x_o^2}\right] = \lim_{x_o\to 0}\left[-\frac{\mu}{4\pi x_o}\frac{2x_o y}{x^2+y^2-x_o^2}\right]$$
$$= -\frac{\mu}{2\pi}\frac{y}{x^2+y^2} \tag{4.41}$$

The resulting streamlines (ψ=constant) are a series of circles passing through the origin with centers on the y-axis, as depicted in Figure 4.8(b). This particular flow pattern is known as a *doublet.*

A number of useful flow patterns can be synthesized by superposition of the potential flows given above. The stream function for flow about a cylinder given earlier is obtained by superimposing the stream function of the doublet (Equation 4.41) on that of uniform parallel flow in the $+x$ direction (Equation 4.17b) to obtain:

$$\psi = -\frac{\mu}{2\pi}\frac{y}{x^2 + y^2} + V_\infty y \tag{4.42}$$

Setting $\psi = 0$ at $r = R$ in the above equation gives $\mu = 2\pi R^2 V_\infty$ which, substituted into the above equation, results in Equation 4.20. The maximum velocity adjacent to the cylinder occurs along the diametral line perpendicular to the flow and is given by Equation 4.22 with $\theta = \pm\pi/2$. The velocity profile along that diametric line has important implications for filter design (Section 14.8).

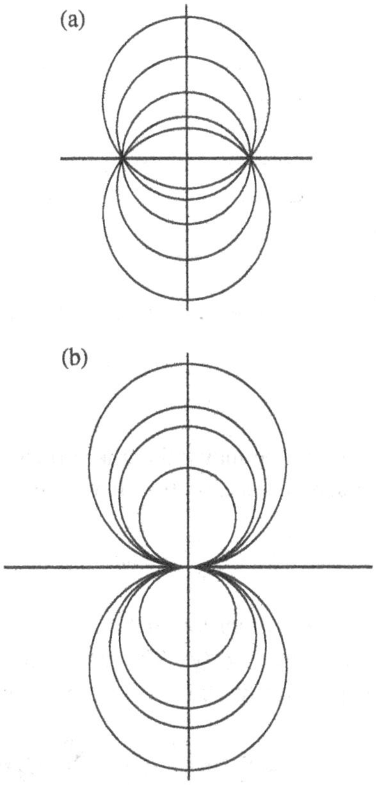

Figure 4.8 (a) Streamlines of a Source–Sink Pair; (b) Streamlines of a Doublet

Another useful potential flow pattern results from a series of sinks equally spaced along the y-axis for which Lamb (1945, p. 71) and Streeter (1948, §60) used complex variables to derive the stream function, which is given by:

$$\psi = -C\tan^{-1}\frac{\tan(\pi y/a)}{\tanh(\pi x/a)} \tag{4.43}$$

in which $2\pi C$ is the strength of each sink and a is the spacing between sinks. [The sign for a sink here is opposite to that of Lamb (1945) and Streeter (1948) due to the different sign conventions used here.] The flow net is depicted in Figure 4.9. The above equation is built upon for practical uses in Section 14.3 (sedimentation) and Section 14.8 (filtration and fluidized beds).

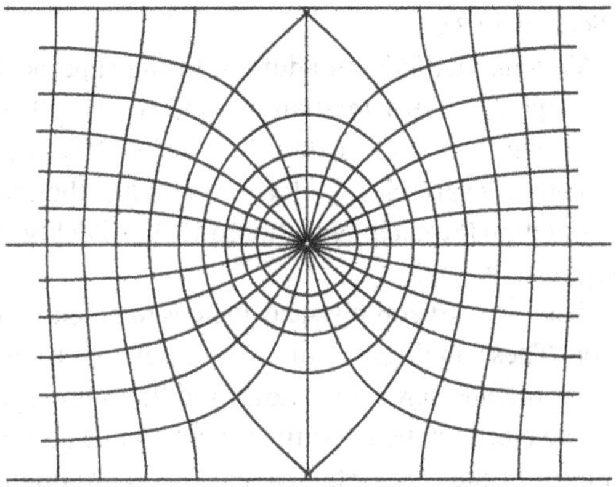

Figure 4.9 Flow Net for One of a Series of Equal and Equidistant Sources (Streeter 1948/McGraw-Hill Education)

The author (Graber 1974a) used the above solution to study, in a qualitative fashion, the effects of end weirs on sedimentation efficiency (Section 14.3). Tesařík (1967) used a similar solution in his studies of flow in sludge-blanket clarifiers (Section 14.3).

Although the method discussed above will be of significant qualitative value for the purposes of Chapter 14, it clearly cannot account for the boundary layer, which may achieve significant proportions in long-settling tanks. Yih (1959) developed an interesting solution that may be considered to account for the effect of the boundary layer in a qualitative way. For two-dimensional flow, the vorticity can be expressed in terms of the stream function such that:

$$\frac{\partial^2\psi}{\partial x^2} + \frac{\partial^2\psi}{\partial y^2} = -2\omega_z \tag{4.44}$$

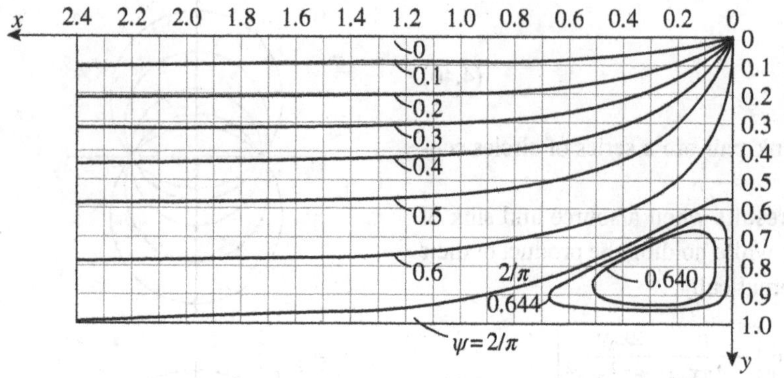

Figure 4.10 Streamline Pattern for Solution (Adapted from Yih 1959)

It may be shown that the vorticity is constant along a streamline in frictionless, two-dimensional flow. For this somewhat incongruous case, the above equation can be written as follows:

$$\frac{\partial^2 \psi}{\partial x^2} + \frac{\partial^2 \psi}{\partial y^2} = f(\psi) \tag{4.45}$$

[The incongruous nature of this case stems from the fact that the vorticity owes itself to the friction, which is neglected in assuming $\omega_z = f(\psi)$. The result is nevertheless of value.] Yih assumed an upstream velocity distribution in which u increases continuously, starting from 0 at the tank floor. His resulting solution of Equation 4.45 is given by:

$$\psi = \frac{2}{\pi}\sin\left(\frac{1}{2}\pi y\right) + \sum_{n=1}^{\infty} C_n \sin(n\pi y)\exp\left[\left(n^2 - \frac{1}{4}\right)^{1/2}\pi x\right] \tag{4.46a}$$

$$C_n = \frac{4}{n\pi^2}\left(1 + \frac{\cos n\pi}{4n^2 - 1}\right) \tag{4.46b}$$

The streamlines are shown in Figure 4.10. A corner eddy arises as shown, which more closely corresponds to physical reality than does Figure 14.1a. This flow pattern enables an important qualitative assessment of sedimentation theory (Section 14.3).

Various methods, in addition to the superposition method discussed above, exist for solving potential flow problems. A graphical approximation method (Rouse 1959, §37) may be useful in some cases and was employed by Ingersoll *et al.* (1956) in their studies of settling tanks. Other methods include some relatively sophisticated mathematical methods, such as complex variables, conformal mapping, the Schwarz–Christoffel transformation, the method of images, and Monte Carlo simulation (Streeter 1948, Rouse 1959, Hildebrand 1962, Webber 2021); and experimental methods relying on physical analogs (see below).

There are several physical analogs to potential flow for which scalar potentials may be defined satisfying Laplace's equation. These analogs are in the areas of heat transfer, gravitational fields, electrostatics, and other types of fluid flows. The latter include "creeping" flows and flows through porous media, which are discussed in Sections 4.5 and 4.6, respectively.

The electrostatic potential provides an experimental analog (Rouse 1959, §38). The few direct applications known to the author of the electrostatic analog to environmental engineering applications could have been accomplished as readily by simple mathematical analysis. However, the potential does exist (pun intended), and the reader should be aware of this technique.

4.4 Laminar Flow Solutions

The flows discussed in this section are **laminar flows**, defined as flows in which water molecules move in lamina or thin layers in which momentum exchange between adjacent layers is by molecular motion, as distinct from **turbulent flow** described most fully in Section 4.7.

Six years after Stokes published his derivation of the general momentum equation, he published its first exact solution. Solutions are said to be exact when terms in the differential equation are eliminated naturally rather than by assumption.

Stokes' solution, published in the year 1851, was for the case of a suddenly accelerated plane wall. Later in the same year, he published his solution for flow near an oscillating flat plate (Schlichting 1960, Chapter V).

Numerous exact solutions of the Navier–Stokes equations for laminar flow are now available. The more classical of these are presented by Streeter (1948, Chapter 11); Schlichting (1960, Chapter 5), and others. The fact that these solutions agree well with experiment proves the validity of the stress–strain assumptions in these simple cases. However, laminar flow occurs in relatively few situations encountered in environmental engineering applications, primarily because the length scales commonly encountered favor turbulent flow. Among the few applications of interest where laminar flow does occur are small tubing, inclined-plate settlers, and model flocculators. Solutions pertinent to those applications are given below. A subclass of laminar flows known as "creeping flows" is discussed in the next section.

Although such solutions are outside of the scope of this book, it should be noted that exact solutions to the differential energy equation are also available. In order to obtain exact solutions to that equation, it is necessary that the corresponding hydrodynamic problem first be solved. Due to the increased complexity, not many of the exact solutions of the differential momentum equations have corresponding thermal flow solutions. Most of the existing ones are given by Schlichting (1960, Chapter XIV, §f).

4.4.1 Steady, Fully-Developed, Prismatic Flows

The solutions of the Navier–Stokes equations (Equations 3.7), which are of immediate interest here, are for steady, fully developed, laminar flow in prismatic channels. (Here the notation u, v, and w are used for velocities in the x, y, and z direction.) For steady flow, the $\partial/\partial t$ terms can be deleted, and for fully developed prismatic flow, it is convenient to align one of the coordinate axes with the direction of flow so that partial derivatives of velocity terms with respect to that direction vanish. Let the flow be in the x–direction, so that $\partial/\partial x = 0$. We make the further assumption that $v = w = 0$. The latter assumption is not as obvious as it may seem, as discussed under secondary currents in Section 4.7.

Subject to the above assumptions, Equation 3.7a reduces to:

$$0 = X - \frac{dp}{dx} + \mu\left(\frac{\partial^2 u}{\partial y^2} + \frac{\partial^2 u}{\partial z^2}\right) \tag{4.47}$$

The only body force of interest here is that due to gravity. With the gravity vector in the $x - z$ plane, $X = -\rho g\, dz/dx$. The above equation can then be rewritten as follows:

$$\frac{\partial^2 u}{\partial y^2} + \frac{\partial^2 u}{\partial z^2} = \frac{1}{\mu}\left(\frac{dp}{dx} + \rho g\frac{dz}{dx}\right) \tag{4.48}$$

Using the piezometric head $h = p/(\rho g) + z$, the above equation may be expressed in the following alternate form:

$$\frac{\partial^2 u}{\partial y^2} + \frac{\partial^2 u}{\partial z^2} = \frac{\rho g}{\mu}\frac{dh}{dx} \tag{4.49}$$

The analogous relation in cylindrical coordinates with flow in the x–direction is:

$$\frac{d^2 u}{dr^2} + \frac{1}{r}\frac{du}{dr} = \frac{\rho g}{\mu}\frac{dh}{dx} \tag{4.50}$$

Several examples of useful flows that satisfy the assumptions inherent in the above equation are presented below:

4.4.1.1 Flow through Circular Tubes

In 1856, Wiedemann published his solution to the Navier–Stokes equations for fully developed flow through a circular tube. Hagen and Poiseuille had earlier analyzed this type of flow experimentally, and such flow often bears their name. It occurs in environmental engineering practice in long tubing of low Reynolds Number such as in sampling lines. Such flow also provided a historical point of departure for a derivation of the equations for flow through filters (Section 4.6). The solution has also been used in the analysis of tube settlers (Yao 1971).

Starting with Equation 4.50, which is a linear nonhomogeneous ordinary differential equation of second order, and letting $D = du/dr$, the equation becomes:

$$\frac{dD}{dr} + \frac{1}{r}D = \frac{\rho g}{\mu}\frac{dh}{dx} \tag{4.51}$$

The differential equation has thus been reduced to a linear one of first order, which may be solved by the method of integrating factor (Hildebrand 1962, §1.4), to obtain:

$$D = \frac{\rho g}{2\mu}\frac{dh}{dx}r + \frac{C_1}{r} \tag{4.52}$$

in which C_1 is an integration constant. Thus:

$$du = \frac{\rho g}{2\mu}\frac{dh}{dx}rdr + \frac{C_1 dr}{r} \tag{4.53}$$

which is integrated to give:

$$u = \frac{\rho g}{4\mu}\frac{dh}{dx}r^2 + C_1 \ln r + C_2 \tag{4.54}$$

in which C_2 is a second integration constant. Since u must be finite at $r = 0$, C_1 must be zero. The boundary condition $u = 0$ at $r = R$ allows one to solve for C_2, and the solution becomes:

$$u = \frac{\rho g}{4\mu}\frac{dh}{dx}(r^2 - R^2) \tag{4.55}$$

The rate of flow is obtained from:

$$Q = \int udA = \int_0^R u2\pi rdr = \pi\int_0^{R^2} ud(r^2) = \pi\frac{\rho g}{4\mu}\frac{dh}{dx}\int_0^{R^2}(r^2 - R^2)d(r^2) = \frac{\pi R^4}{8\mu}\rho g\left(-\frac{dh}{dx}\right) \tag{4.56}$$

or

$$-\frac{dh}{dx} = \frac{8\mu Q}{\rho g\pi R^4} \tag{4.57}$$

The mean velocity is then given by:

$$V = \frac{Q}{A} = \frac{Q}{\pi R^2} = \frac{R^2}{8\mu}\rho g\left(-\frac{dh}{dx}\right) \tag{4.58}$$

The relation of the above solution to the full range of laminar, transitional, and turbulent flows in pipes is discussed in Chapter 6.

4.4.1.2 Flow between Moving and Stationary Parallel Plates

Around 1885, Couette obtained the solution for fully developed flow between moving parallel plates, often referred to as Couette flow. Such flow has been created in experimental flocculators (Swift and Friedlander 1964). The special case of stationary parallel plates has possible application in tube settlers (Yao 1971).

The flow is considered to be two-dimensional so that $\partial/\partial z = 0$ and Equation 4.49 reduces to:

$$\frac{d^2u}{dy^2} = \frac{\rho g}{\mu}\frac{dh}{dx} \tag{4.59}$$

This equation may be integrated once to obtain:

$$\frac{du}{dy} = \frac{\rho g}{\mu}\frac{dh}{dx}y + C_1 \tag{4.60}$$

and then again, giving:

$$u = \frac{\rho g}{\mu}\frac{dh}{dx}\frac{y^2}{2} + C_1 y + C_2 \tag{4.61}$$

Allowing the coordinates to move with the lower plate (if necessary) so that $u = 0$ at $y = 0$, gives $C_2 = 0$. The velocity of the upper plate relative to the lower plate is U. This gives a second boundary condition ($u = U@y = H$) with which we can solve for C_1 and obtain the solution as follows:

$$u = \frac{y}{H}U + \frac{\rho g H^2}{2\mu}\left(-\frac{dh}{dx}\right)\left[\frac{y}{H}\left(1 - \frac{y}{H}\right)\right] \tag{4.62}$$

For $U = 0$ we have flow through a slot, referred to as simple shear flow or simple Couette flow. The relations for flow Q and average velocity V can be obtained as before.

4.4.1.3 Flow in Wide Open Channels

Flow in wide open channels also has possible applications in tube settlers (Yao 1971). Equation 4.61 also applies in this case. The boundary condition $u = 0$ at $y = 0$ gives $C_2 = 0$. At the free surface, there is no confining wall, and the shear stress is assumed to be negligible (the shear stress between the free surface and the air will be quite small in laminar flow). This provides as a secondary boundary condition $\tau = \mu du/dy = 0$ at $y = H$. Referring to Equation 4.61, this gives $C_1 = 0$. The solution becomes simply:

$$u = \frac{\rho g}{\mu}\frac{dh}{dx}\frac{y^2}{2} \tag{4.63}$$

4.4.1.4 Flow through Rectangular Channels

Laminar flow in rectangular channels occurs in tube settlers used to improve sedimentation efficiency (Section 14.3). Yao (1971) presented a solution for the flow in such a channel based on the assumptions discussed above. The solution is instructive, and the author's version is presented below.

Equation 4.61 applies and is a special case of Poisson's equation (Hildebrand 1962, §9.1). That equation can be converted to Laplace's equation in the new variable u_1 as follows:

$$u_1 = u - \frac{\rho g}{4\mu}\frac{\partial h}{\partial x}\left(Ay^2 + Bz^2 + Cy + Dz + E\right) \tag{4.64}$$

We then find that:

$$\frac{\partial^2 u_1}{\partial y^2} + \frac{\partial^2 u_1}{\partial z^2} = 0 \tag{4.65}$$

for $A = 2$ and $B = 0$, or $A = 1$ and $B = 1$, or $A = 0$ and $B = 2$, etc., with C, D, and E arbitrary.

We will assume a product solution of the form:

$$u_1 = Y(y)Z(z) \tag{4.66}$$

in which $Y(y)$ and $Z(z)$ denote functions of y and z, respectively. Substituting this product into Equation 4.65, we obtain:

$$Z\frac{d^2Y}{dy^2} + Y\frac{d^2Z}{dz^2} = 0 \tag{4.67}$$

The above equation may be separated into:

$$-\frac{1}{Y}\frac{d^2Y}{dy^2} = \frac{1}{Z}\frac{d^2Z}{dz^2} = k^2 \tag{4.68}$$

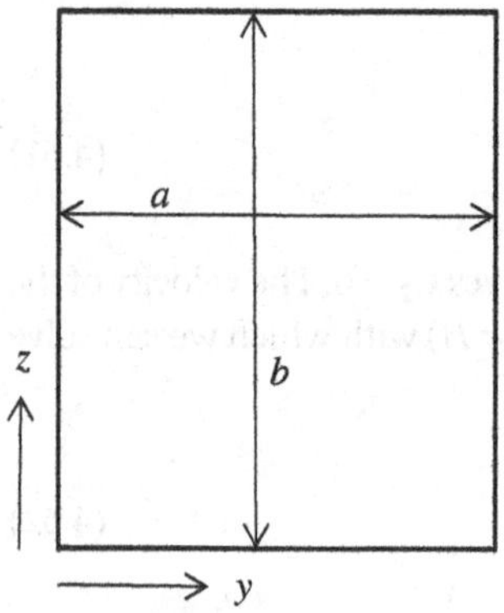

Figure 4.11 Rectangular Cross Section

The separation is based on the first member being independent of z and the second equal member being independent of y. From the above equation, the following two ordinary differential equations with constant coefficients are obtained:

$$\frac{d^2Y}{dy^2} + k^2Y = 0 \tag{4.69a}$$

$$\frac{d^2Z}{dz^2} - k^2Z^2 = 0 \tag{4.69b}$$

The solutions to the above equations are readily obtained (Hildebrand 1962, §1.5) as follows:

$$Y = C_1 \cos ky + C_2 \sin ky \tag{4.70a}$$

$$Z = C_3 \cosh kz + C_4 \sinh kz \tag{4.70b}$$

Referring to the cross section of Figure 4.11, the boundary conditions for $u(y, z)$ are:

$$u(0, z) = 0 \tag{4.71a}$$

$$u(a, z) = 0 \tag{4.71b}$$

$$u(y, 0) = 0 \tag{4.71c}$$

$$u(y, b) = 0 \tag{4.71d}$$

Equations 4.71a and 4.71c give $C_1C_3 = 0$ and $C_2C_3 \neq 0$, respectively, which requires $C_1 = 0$. By virtue of Equations 4.71a–d, the requirement that $C_1 = 0$, and the linearity of Equation 4.70, the solution for u may be expressed in terms of the following series:

$$u = \frac{\rho g}{4\mu}\frac{dh}{dx}(Ay^2 + Bz^2 + Cy + Dz + E) + \sum_{k=1}^{\infty} \sin ky(A_m \cosh kz + B_m \sinh kz) \tag{4.72}$$

The constants in the polynomial portion of the above equation may be determined from the boundary conditions expressed by Equations 4.71a and 4.71b. The first of these gives $B = D = E = 0$. For $B = 0$, one must have $A = 2$ (see above). The second requires:

$$u = \frac{\rho g}{4\mu}\frac{dh}{dx}(2a^2 + ca) + \sum_{k=1}^{\infty} \sin ka(A_m \cosh kz + B_m \sinh kz) = 0 \tag{4.73}$$

from which we obtain $C = -2a$ and $ka = m\pi$ (in which m is an integer). Thus we have:

$$u = \frac{\rho g}{2\mu}\frac{dh}{dx}(y^2 - ay) + \sum_{m=1}^{\infty} \sin\frac{m\pi y}{a}\left(A_m \cosh\frac{m\pi z}{a} + B_m \sinh\frac{m\pi z}{a}\right) \tag{4.74}$$

The remaining boundary conditions enable A_m and B_m to be determined. Substituting Equations 4.71c and 4.71d into Equation 4.74 and employing the theory of Fourier sine series (Hildebrand 1962, §5.10) gives:

$$A_m = -\frac{2\rho g a^2}{\mu m^3\pi^3}\frac{dh}{dx}(\cos m\pi - 1) \tag{4.75a}$$

$$B_m = -\frac{A_m(\cosh m\pi b/a - 1)}{\sinh m\pi b/a} \tag{4.75b}$$

Equations 4.74, 4.75a, and 4.75b represent the complete solution for u.

For purposes of the Section 14.3 discussion of tube settlers, the average velocity and the ratio of the velocity u along the vertical central plane ($y = a/2$) to the average velocity are desired. The average velocity V is obtained from $V = \left(\int u dA\right)/A = \left(\int_0^a \int_0^b u dy dz\right)/(ab)$, which results in:

$$V = -\frac{\rho g a^2}{12\mu}\frac{dh}{dx} - \sum_{m=1}^{\infty}\frac{a}{b}\frac{\cos m\pi - 1}{(m\pi)^2}\left[A_m \sinh\frac{m\pi b}{a} + B_m\left(\cosh\frac{m\pi b}{a} - 1\right)\right] \tag{4.76}$$

The desired ratio can then be obtained from Equations 4.74 to 4.76, giving:

$$\frac{(u)_{y=a/2}}{V} = \frac{-\frac{1}{8} - \sum_{m=1}^{\infty}\frac{2}{m^3\pi^3}(\cos m\pi - 1)\sin\frac{m\pi}{2}\left(\cosh m\pi\frac{z}{a} - \frac{\cosh m\pi\frac{b}{a} - 1}{\sinh m\pi\frac{b}{a}}\sinh m\pi\frac{z}{a}\right)}{-\frac{1}{12} + \frac{a}{b}\sum_{m=1}^{\infty}\frac{2}{m^5\pi^5}(\cos m\pi - 1)^2\left[\sin m\pi\frac{b}{a} - \frac{\left(\cosh m\pi\frac{b}{a} - 1\right)^2}{\sinh m\pi\frac{b}{a}}\right]} \tag{4.77}$$

For a square conduit, $a/b = 1$, the above equation becomes the same as that given by Yao (1971).

4.4.2 Additional Useful Solutions

4.4.2.1 Flow between Concentric Rotating Cylinders

Experimental devices consisting of two concentric cylinders rotating at constant speeds have been used to study flocculation and floc breakup (Bradley and Krone 1971). Oseen and G. Hamel published solutions for such cylinders in 1911 and 1916, respectively, as given by Schlichting (1960).

In this case, it is assumed that $v_r = v_z = 0$. Gradients in the ϕ-direction vanish, and if the axis of the cylinders is vertical $F_\phi = 0$. Subject to the above, the Navier–Stokes equations in cylindrical coordinates reduce to:

$$\frac{d^2 v_\phi}{dr^2} + \frac{1}{r}\frac{dv_\phi}{dr} - \frac{v_\phi}{r^2} = 0 \tag{4.78}$$

With the notational change $u = v_\phi$, this may be rewritten as:

$$r^2\frac{d^2u}{dr^2} + r\frac{du}{dr} - u = 0 \tag{4.79}$$

The above form is a linear, equidimensional differential equation of second order (Hildebrand 1962, §1.6).
The transformation $r = e^z$, $z = \ln r$ (in which z is a dummy variable) changes Equation 4.79 to:

$$\frac{d^2u}{dz^2} - u = 0 \tag{4.80}$$

which is a linear differential equation with constant coefficients (Hildebrand 1962, §1.5). The solution to the above equation is given by:

$$u = C_1 e^z + C_2 e^{-z} = C_1 r + C_2/r \tag{4.81}$$

The boundary conditions are $u = r_1\omega_1$ at $r = R_1$ and $u = r_2\omega_2$ at $r = R_2$. Substituting the boundary conditions into the above equations gives two equations, which may be solved for C_1 and C_2 to give:

$$C_1 = \omega_1 - \frac{r_2^2(\omega_1 - \omega_2)}{r_2^2 - r_1^2} \tag{4.82a}$$

$$C_2 = \frac{r_1^2 r_2^2(\omega_1 - \omega_2)}{r_2^2 - r_1^2} \tag{4.82b}$$

Substituting these equations for C_1 and C_2 into Equation 4.81 results in:

$$u = \frac{1}{r_2^2 - r_1^2}\left[r(\omega_2 r_2^2 - \omega_1 r_1^2) - \frac{r_1^2 r_2^2}{2}(\omega_2 - \omega_1)\right] \tag{4.83}$$

For the special case of $\omega_1 = 0$ the above equation reduces to:

$$u = \frac{\omega_2 r_2^2}{r_2^2 - r_1^2}(r - r_1^2/r) \tag{4.84}$$

Further use of the above equation will be made in Section 14.6.

4.4.2.2 Others

Other exact solutions of the Navier–Stokes equations have been obtained, many of which are given by Schlichting (1960, Chapter V). Approximate solutions for limiting cases are also well covered by Schlichting, including solutions to the boundary layer equations (Section 4.2) and the equations for creeping flow (Section 4.5). It is recommended that the reader peruse Schlichting's book, as use for one of those solutions may be found in the future.

4.5 Creeping Flow

Creeping flow refers to very slow laminar flow in which the inertial and body force terms of the Navier–Stokes equations can be neglected. Inertial and body force terms become negligible when an appropriately defined Reynolds Number (discussed in Section 4.7) becomes sufficiently small. For now, we will simply make the assumption that the above-mentioned terms may be neglected, reducing the Navier–Stokes equations (Equations 3.7) to:

$$\frac{\partial p}{\partial x} = \mu\left(\frac{\partial^2 u}{\partial x^2} + \frac{\partial^2 u}{\partial y^2} + \frac{\partial^2 u}{\partial z^2}\right) \tag{4.85a}$$

$$\frac{\partial p}{\partial y} = \mu\left(\frac{\partial^2 v}{\partial x^2} + \frac{\partial^2 v}{\partial y^2} + \frac{\partial^2 v}{\partial z^2}\right) \tag{4.85b}$$

$$\frac{\partial p}{\partial z} = \mu\left(\frac{\partial^2 w}{\partial x^2} + \frac{\partial^2 w}{\partial y^2} + \frac{\partial^2 w}{\partial z^2}\right) \tag{4.85c}$$

By adding appropriate differentials of the above equations, we obtain:

$$\frac{\partial^2 p}{\partial x^2} + \frac{\partial^2 p}{\partial y^2} + \frac{\partial^2 p}{\partial z^2} = \mu\left[\frac{\partial}{\partial x}\left(\frac{\partial^2 u}{\partial x^2} + \frac{\partial^2 u}{\partial y^2} + \frac{\partial^2 u}{\partial z^2}\right) + \frac{\partial}{\partial y}\left(\frac{\partial^2 v}{\partial x^2} + \frac{\partial^2 v}{\partial y^2} + \frac{\partial^2 v}{\partial z^2}\right) + \frac{\partial}{\partial z}\left(\frac{\partial^2 w}{\partial x^2} + \frac{\partial^2 w}{\partial y^2} + \frac{\partial^2 w}{\partial z^2}\right)\right] \tag{4.86}$$

which may be rewritten as follows:

$$\frac{\partial^2 p}{\partial x^2} + \frac{\partial^2 p}{\partial y^2} + \frac{\partial^2 p}{\partial z^2} = \mu\left[\left(\frac{\partial^2}{\partial x^2} + \frac{\partial^2}{\partial y^2} + \frac{\partial^2}{\partial z^2}\right)\left(\frac{\partial u}{\partial x} + \frac{\partial v}{\partial y} + \frac{\partial w}{\partial z}\right)\right] \tag{4.87}$$

By continuity (Equation 2.9), the term on the right-hand side vanishes, and the following is obtained:

$$\frac{\partial^2 p}{\partial x^2} + \frac{\partial^2 p}{\partial y^2} + \frac{\partial^2 p}{\partial z^2} = 0 \tag{4.88}$$

(More generally, in vector notation, $\nabla^2 p = 0$.) Therefore, the pressure satisfies Laplace's equation in a way analogous to the velocity potential of Section 4.2.

For two-dimensional flow, since $\partial/\partial y(\partial p/\partial x) = \partial/\partial x(\partial p/\partial y)$, Equations 4.85a and 4.85b give:

$$\frac{\partial}{\partial y}\left(\frac{\partial^2 u}{\partial x^2} + \frac{\partial^2 u}{\partial y^2}\right) = \frac{\partial}{\partial x}\left(\frac{\partial^2 v}{\partial x^2} + \frac{\partial^2 v}{\partial y^2}\right) \tag{4.89}$$

Substituting Equations 4.5 into the above gives the creeping flow equation in terms of the stream function:

$$\frac{\partial^4 \psi}{\partial x^4} + 2\frac{\partial^4 \psi}{\partial x^2 \partial y^2} + \frac{\partial^4 \psi}{\partial z^4} = 0 \tag{4.90}$$

(The more general vector form is $\nabla^4 \psi = 0$ which is known as a bipotential or biharmonic equation.) Stokes obtained a solution to the above equation for the case of parallel flow about a sphere. The mathematics is quite elaborate and may be found in Stokes' original publication and in Lamb (1945, pp. 594–617). With the coordinate origin at the center of the sphere of radius R, the solution is given as follows:

$$\psi = \frac{3}{4} V_\infty R \left(1 - \frac{R^2}{3r^2}\right) \sin^2\theta \tag{4.91a}$$

$$u = V_\infty \left[\frac{3}{4}\frac{Rx^2}{r^3}\left(\frac{R^2}{r^2} - 1\right) + 1 - \frac{1}{4}\frac{R}{r}\left(3 + \frac{R^2}{r^2}\right)\right] \tag{4.91b}$$

$$v = V_\infty \frac{3}{4}\frac{Rxy}{r^3}\left(\frac{R^2}{r^2} - 1\right) \tag{4.91c}$$

$$w = V_\infty \frac{3}{4}\frac{Rxz}{r^3}\left(\frac{R^2}{r^2} - 1\right) \tag{4.91d}$$

$$p - p_\infty = -\frac{3}{2}\frac{\mu V_\infty Rx}{r^3} \tag{4.91e}$$

in which p_∞ is the pressure at infinity.

Flow about a sphere satisfying the above equations is known as Stokes flow. It is of interest in connection with the settling of solid particles (Section 13.2) and the rising of small gas bubbles (Section 13.3). Most pertinent are the shear stresses in the vicinity of the sphere and the drag force on the sphere. The shear stress tangent to the sphere surface in a plane of axial symmetry is obtained by using the appropriate derivatives of Equations 4.91b and 4.91c to obtain (Streeter 1948, §120):

$$\tau_{r\theta} = -\frac{3\mu V_\infty}{2R} \sin\theta \tag{4.92}$$

The maximum shear stress occurs along a circumferential line perpendicular to the direction of flow ($\theta = \pm\pi/2$) and has a magnitude $\tau_{max} = 3\mu V_\infty/(2R)$ (this will be used in Section 13.3). The drag force on the sphere is obtained by integrating the x −components of the pressure force and shear force over the surface of the sphere to obtain (Streeter 1948, §120):

$$D = 6\pi R \mu V_\infty \tag{4.93a}$$

The pressure and shear forces contribute 1/3 and 2/3 of the total drag force, respectively.

Summing the drag given by Equation 4.93a and the weight and buoyant force acting on a small particle gives the settling velocity in the form referred to as "Stokes Law" (although, with reference to the discussion in Section 1.3, it is not actually a "law") acting on a small particle as (Streeter 1948, §120):

$$V_\infty = \frac{2}{9}\frac{R^2}{\mu}(\gamma_s - \gamma) \tag{4.93b}$$

which experiments show to hold for Reynolds Numbers below 1, i.e., $R = 2R\rho V_\infty/\mu < 1$. Examples in Chapter 13 for settling sand-sized particles in our applications are found to exceed this Reynolds Number.

Creeping flows also occur in oil-lubricated bearings and in the so-called Hele-Shaw flow. The latter refers to the very slow flow of a viscous fluid between parallel flat plates separated by a short distance. If a body is inserted between the flat plates, the streamlines can be shown to be the same as those for potential flow about a body of similar shape. This is the basis of a physical analogy that may be used to solve potential flow problems. However, the electrostatic analogy, discussed in Section 4.3, is generally more convenient.

4.6 Flow through Porous Media

Flow through porous media defied mathematical representation until 1856, when Henry Darcy published his experimental results in the following form:

$$v = -k\frac{dh}{ds} = ki \tag{4.94}$$

in which v is the *discharge velocity* or *face velocity* defined as volumetric rate of flow per unit of *total* media area, k is the coefficient of permeability (with dimensions of velocity), h is piezometric head, s is distance in the direction of flow, and $i = -dh/ds$ represents the hydraulic gradient. The above equation, commonly known as Darcy's Law, indicates a linear relation between the discharge velocity and hydraulic gradient.

Before discussing the significance and limitations of Darcy's Law, an alternative means of obtaining it will be presented, which should add to the physical feeling of the problem. As reported by Camp (1964), Kozeny and Fair and Hatch (1933), derived an equation for the hydraulic gradient through a porous medium by starting with Poiseuille's equation for laminar flow through a circular tube:

$$i = -\frac{dh}{ds} = \frac{32\nu}{g}\frac{1}{D_t^2}V \tag{4.95}$$

The term D_t denotes the tube diameter. Kozeny noted that D_t is equal to 4 times the hydraulic radius and reasoned by analogy with the circular tube that the average hydraulic radius in a unit volume of filter equals the *pore* volume divided by the surface area of the sand grains. The porosity or void ratio ϕ is related to the pore volume and grain volume by ϕ = pore volume/(pore volume + grain volume), so that:

$$\text{pore volume} = \frac{\phi}{1-\phi}\text{grain volume} \tag{4.96}$$

The volume and area per grain are expressed by βd^3 and αd^2, respectively, with d denoting the grain diameter. The ratio α/β is termed the shape factor s, which is six for spheres and increases to nine or more for angular grains (Camp 1964). The equivalent D_t and hydraulic radius thus become:

$$D_t \propto \text{hydraulic radius} = \frac{\phi}{1-\phi}\frac{\beta d^3}{\alpha d^2} = \frac{\phi}{1-\phi}\frac{d}{s} \tag{4.97}$$

Interpreting v in Equation 4.94 as the velocity through the pores, so that $V = \nu/\phi$ in which v is the discharge velocity, one can substitute Equation 4.97 into Equation 4.95 to obtain:

$$i = \frac{js^2\upsilon}{g}\frac{(1-\phi)^2}{\phi^3}\frac{v}{d^2} \tag{4.98}$$

in which j is a proportionality constant. The relation between Equations 4.94 and 4.98 is apparent, and the latter elucidates the nature of the permeability term in the former.

More will be said about the terms in the above equations in later sections (principally Sections 14.8 and 15.6, in connection with filtration and groundwater, respectively). Here we will concentrate on further implications of Darcy's Law. We will first generalize Equation 4.94 for a homogeneous, isotropic porous medium to three-dimensional flow:

$$u = -k\frac{dh}{dx} \tag{4.99a}$$

$$v = -k\frac{dh}{dy} \tag{4.99b}$$

$$w = -k\frac{dh}{dz} \tag{4.99c}$$

(In general vector notation, $V = -k\,\text{grad}\,h$.) Substituting the above relations into the continuity equation (Equation 2.9) results in:

$$\frac{d^2h}{dx^2} + \frac{d^2h}{dy^2} + \frac{d^2h}{dz^2} = 0 \tag{4.100}$$

(In general vector notation $\nabla^2 h = 0$.) Thus, the piezometric head h represents a scalar potential satisfying Laplace's equation; note the analogy with Equation 4.88 and equations in Section 4.2. The analogy with Equation 4.88 for creeping flow is of particular significance since it suggests that potential flow in a porous medium is limited to slow, viscous flows in which inertial terms may be neglected.

Principles initiated in the present section will be extended to non-Darcian flows and used in Section 15.6 in connection with subsurface drains and water wells.

4.7 Turbulence, Separation, and Secondary Currents

In this section, the phenomena of *turbulence*, *separation*, and *secondary currents* will be discussed. These phenomena are very common and important in environmental engineering applications. They substantially limit the range of applications to which the quantitative concepts presented in prior sections of this chapter may be applied.

4.7.1 Turbulence

There are fundamental differences between *laminar* and *turbulent* flow. The terms "laminar" and "turbulent" are, unfortunately, used somewhat indiscriminately at times. A clear, accurate understanding of their differences is quite important. The term "laminar flow" is defined in Section 4.4.

The term "turbulent" properly denotes fluid motion "in which an irregular fluctuation (mixing or eddying motion) is superimposed on the main stream" (Schlichting 1960, p. 457). The velocity and pressure at a fixed point in a turbulent flow do not remain constant with time but exhibit irregular fluctuations of high frequency, which can be detected by responsive sensors. The measured fluctuations are the result of the erratic motion of macroscopic "lumps" of fluid. The fluctuations of these fluid "lumps" are the principal agents of momentum, heat, and mass transfer in turbulent flow. In laminar flow, no such macroscopic fluctuations exist; it is the motion of individual molecules that affects the transfer of properties. The gross behavior of laminar and turbulent flows is quite distinct.

Progressively larger values of an appropriately defined **Reynolds Number** determine the laminar, transitional, and turbulent regimes of flow. Reynolds Number is one of the most important dimensionless parameters in fluid mechanics, denoted by R and defined as $R = \rho VL/\mu$, in which ρ is fluid density, V is velocity, L is characteristic length, and μ is dynamic viscosity. The Reynolds Number is a measure of the ratio of inertial forces (ρV) to viscous forces (μ/L). At low Reynolds Numbers, viscous effects dominate and the flow is laminar, whereas at high Reynolds Numbers inertia predominates and the flow may be turbulent. Reynolds Number is used, sometimes along with other dimensionless numbers, to provide a criterion for determining dynamic similitude, as discussed in Chapter 5. It is also used in conjunction with the Moody Diagram, discussed in Section 5.3, and for other purposes discussed in subsequent chapters.

With the exception of flow in certain solution lines, instrument air pipes, low-velocity sludge pipes, small contact tanks, porous media, and the like, most fluid flows in environmental engineering applications have relatively large characteristic dimensions and turbulent conditions exist. This is most certainly true of flow through mixers, flocculators, process piping, the vast majority of process tanks, and open channels (natural and man-made).

The energy dissipated in turbulent flow is not extracted directly from the mean flow (as in laminar flow). Rather, energy is extracted from the mean flow by the largest eddies and cascades through increasingly small eddies, until it ultimately dissipates as heat by the action of viscosity. At a point in a turbulent flow, the laminar momentum and energy equations, derived in Chapter 3, are applicable provided that the velocity and pressure terms are considered to be the instantaneous values that exhibit high-frequency variations with time. In order to apply these equations to the mean properties of the flow, the terms in the equations have been separated into terms representing the mean motion and a fluctuating or eddy motion. This has resulted in the Reynolds momentum equations (Rouse 1950, Schlichting 1960, Shames 1962) and the equations of energy (Rouse 1959) for turbulent flow.

4.7.2 Separation

When fluid flow cannot follow a solid boundary smoothly, such as in the case of an airfoil at a high angle of attack, the flow separates from the boundary. Vortices will form in the separated region. This causes a substantial increase in the drag on the object from which the flow separates.

Section 8.2 discusses the separation phenomenon associated with surge in centrifugal pumps; Section 9.5 discusses related phenomena in blowers and compressors; Section 11.5 discusses separation in the context of the Kármán Vortex Street; and Section 14.3 discusses boundary layer separation in settling tanks.

4.7.3 Secondary Currents

Secondary currents are classified as being of the first and second kinds. The terms "secondary flow of the first kind" and "secondary flow of the second kind" were coined by Prandtl (see below). Secondary currents of the first kind are the result of centrifugal forces and occur at bends, whereas those of the second kind occur in noncircular cross sections with turbulent flow in straight (or bent) conduits (Prandtl 1952, Chapter 3, §8; Schlichting 1960, Chapter XX, §e and §h). Secondary currents of the second kind are also associated with Langmuir circulation, which occurs in lakes and the ocean, as discussed in Section 14.3. The Helmholtz vortex theorems are useful in providing a predictive, qualitative understanding of secondary currents. They are expressed below in two different forms to provide a fuller understanding, including a form given by Shapiro (1963).

Helmholtz's first theorem: The strength of a vortex filament is constant along its length. The strength of a vortex tube does not vary with time.

Helmholtz's second theorem: A vortex filament cannot end in a fluid; it must extend to the boundaries of the fluid or form a closed path. Fluid elements lying on a vortex line at some instant continue to lie on that vortex line. More simply, vortex lines move with the fluid. Also, vortex lines and tubes must appear as a closed loop, extend to infinity, or start/end at solid boundaries.

Helmholtz's third theorem: In the absence of rotational external forces, a fluid that is initially irrotational remains irrotational. Fluid elements initially free of vorticity remain free of vorticity.

Helmholtz's theorems strictly apply to inviscid flows but have more general utility. In observations of vortices in real fluids, the strength of the vortices always decays gradually due to the dissipative effect of viscous forces.

Consider the case of a wide, open channel flowing from left to right with constant slope and fully developed flow. If there is no change in channel alignment (i.e., no curvature in the channel), the flow will be one-dimensional in the sense that the velocity varies only in the vertical direction. The flow will have rotation due to the velocity profile, which may be represented according to the right-hand thumb rule as a vorticity vector into the page.

Consider now the same channel with curvature. The outward centrifugal force causes an increase in pressure on the outward wall and a decrease in pressure on the inward wall. This manifests as a decrease in elevation on the inward wall and an increase in elevation on the outward wall, the latter being referred to as superelevation. This is accompanied by a decrease in velocity in the direction of flow near the outward wall and an increase in the velocity near the inward wall. Because vorticity is "locked" in the fluid (Helmholtz's second theorem), the vorticity vector then rotates, as sketched in Figure 15.3. The rotated vorticity vector has a component in the direction of flow, which, by the right-hand thumb rule, corresponds to a secondary current as shown in Figure 15.3. This is observed in actual channels. Such secondary currents have an important influence on sediment transport in channels (Section 13.2) and on sedimentation patterns at river bends (Section 15.2). Similar reasoning will show the secondary current pattern in a curved pipe, as shown in Figure 4.12. (The reader should verify this.)

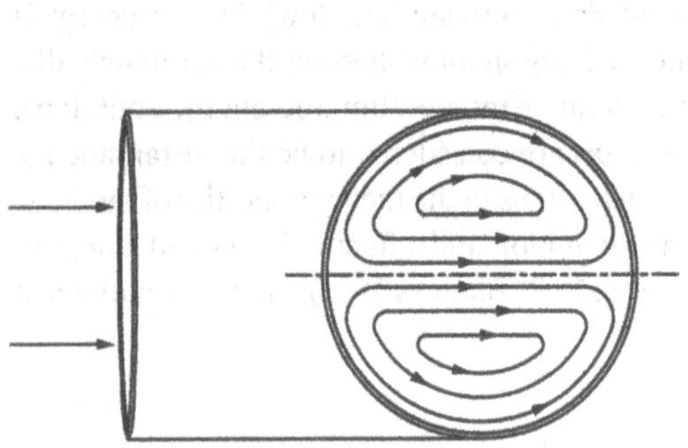

Figure 4.12 Secondary Currents of the Second Kind at Pipe Bend (Prandtl 1952/Hafner Publishing Company)

Example 4.1

Two sedimentation basin influent channels (Deer Island Wastewater Treatment Plant, Boston, Massachusetts) are planned to temporarily convey grit to primary sedimentation influent channels for removal there while grit chambers are taken out of service for replacement. Spiral currents associated with a curved channel (75° bend) conveying the grit-laden wastewater to the influent channels would result in unequal distribution of the grit to those channels despite an equal flow division.

Solution

The author proposed that a splitter vane be installed along the channel centerline before the curved section to maintain equality of grit loading in the channels and reduce secondary currents that would cause an unequal grit distribution. The solution was successfully implemented. A similar problem solved with vanes at the Pittsburgh Sewage Treatment Plant was discussed by Yoshimi and Stelson (1963).

The same type of reasoning as given above with respect to the vorticity vector will give the secondary flow pattern. Gerard (1978) showed with particular clarity (also see Einstein and Li 1958, Naot and Rodi 1982) that secondary currents of the second kind can only occur in turbulent flows. The reader should be able to predict the observed secondary current patterns in the triangular conduit and open rectangular channel depicted in Figure 4.13. In the latter case, the shear between the liquid and the gas at the free surface causes a velocity gradient such that the maximum velocity occurs below the free surface (see Figure 4.13). (Gerard cites references claiming that secondary currents are responsible for that position of the maximum velocity, but it would occur in laminar flow due to the air–water interface in the absence of secondary currents.)

An important ramification of secondary currents of the second kind can be understood by noting that fluid is impelled from areas near the channel boundary where the longitudinal velocities would otherwise be higher toward the center of the channel and, conversely, fluid is impelled from the center of the channel to areas near the channel boundary where the longitudinal velocities would otherwise be lower. This causes the boundary shear stresses to be evened out, a factor that provides the basis for the use of the hydraulic radius for determining the boundary resistance of pipes and channels (Prandtl 1952, p. 149) (see Chapters 6 and 7).

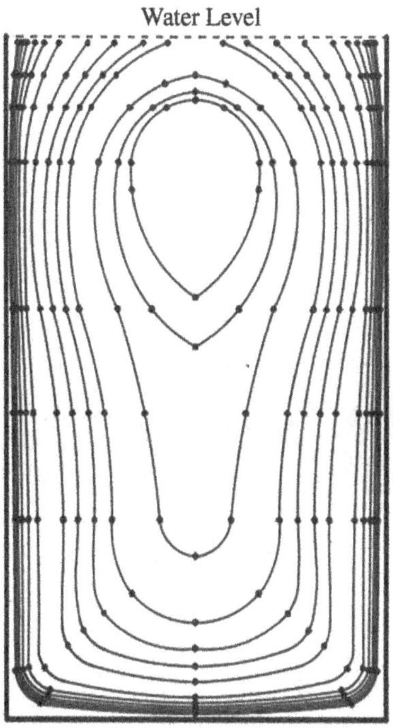

Figure 4.13 Curves of Constant Velocity for a Rectangular Open Channel (Schlichting 1960/McGraw-Hill Education)

5

Dimensional Analysis and Similitude

5.1 Introduction

Dimensional analysis and the related concept of similitude are powerful tools in environmental engineering hydraulics, both in concept and as an analytical technique. Other engineering professions (including chemical and hydraulic engineers) have used dimensional analysis more extensively than the environmental engineer has, although its potential applicability in environmental engineering hydraulics and related aspects of process design is probably as great.

The value of dimensional analysis lies in two areas. The first is in the design, organization, and maximum utilization of experimental results. This may include pilot plant studies as well as more traditional hydraulic experiments. The second area lies in the design and maximum utilization of numerical analyses of hydraulic problems, which may be thought of as numerical analogs of experimental studies.

Dimensional analysis is of greatest value in studying processes in which the variables involved are known, but the relationship between the variables is not known. By dimensional analysis, the process variables may be expressed as a set of dimensionless groups, with the groups numbering less than the number of variables. The advantages of this reduction to a smaller number of dimensionless variables are twofold: (1) available experimental data may be used to establish a relationship between variables extending beyond the range of the actual data, or, conversely, considerably less experimentation is required to establish the relationship between variables over a desired range; (2) the results of experimental (and sometimes analytical) studies can be presented in much more compact and generalized form.

This chapter will present fundamental concepts of dimensional analysis using the Buckingham π-Theorem as a point of departure. Subsequent sections will then present practical applications in the two areas referred to above. Other examples of their use are presented in later chapters. The practical applications, very useful in their own right, should also serve to exemplify techniques that may lead to further applications.

5.2 Fundamentals

The review of fundamentals presented here will be limited to presentation of the basic concepts, followed by a demonstration of their use by example. Derivations of the Buckingham π-Theorem utilize concepts of linear algebra, as developed by Buckingham (1914) and presented more recently by, e.g., Bluman and Kumei (1996) and Palmer (2008).

The basis for reducing variables to a minimum number of dimensionless parameters is the Buckingham π-Theorem, which states (Shames 1962, p. 191) that "the number of independent dimensionless groups that may be employed to describe a phenomenon known to involve n variables is equal to the number $n - r$, where r is usually the number of basic dimensions [e.g., length, mass, and time] needed to express the variables dimensionally." [More precisely, r is the rank of the dimensional matrix (Shames 1962, Section 7.3)]. In formulating dimensionless groups, each must contain one variable that does not appear in the other dimensionless groups. These groups can usually be selected by trial and error with reference to the common dimensionless groups of hydraulics (Reynolds Number, Froude Number, etc.). Formal procedures for their selection are also available (Shames 1962, Section 7.5).

Similitude refers to the relationships between the variables of a process at different scales having certain similarities. For complete similitude, the similarity must be both geometric and dynamic. These concepts are basic to the scale-up of bench or pilot plant studies, or scaling between prototypes of different sizes.

Hydraulics and Pneumatics in Environmental Engineering, First Edition. S. David Graber.

Geometric similarity refers to the need to have complete similarity between all dimensions in order to properly relate process variables at one scale to those at another. For example, geometric similarity requires (among other things) that a bench-scale mixer or flocculator have the shape and proportional dimensions of the envisioned prototype.

Dynamic similarity refers to the conditions, in addition to geometric similarity, that must be met to properly relate process variables at different scales. The physical meaning of dimensionless groups is made clear in discussions of dynamic similarity (e.g., Shames 1962, Section 7.3), and the relationship between these groups and the parameters in the applicable equations (such as the Navier–Stokes equations discussed previously) is brought out. The reader is referred to the references for the details; the important final result, tying the ideas of dimensional analysis and similitude together, is that to achieve dynamic similarity in geometrically similar processes at least all but one of the dimensionless groups yielded by dimensional analysis must be equal.

5.3 Reduction of Experimental Data

One of the classical examples of the use of dimensional analysis to reduce experimental data is the work of Moody (1944), who cast the data for friction in a conduit in terms of the dimensionless parameters Darcy–Weisbach friction factor f, Reynolds Number $\mathbf{R} = \rho VD/\mu$, in which ρ = fluid density, V = bulk mean velocity, D = pipe diameter, μ = absolute viscosity (representing the ratio of inertial forces to viscous forces); and ε/D = roughness ratio with ε = roughness height.

The Darcy–Weisbach friction factor f is related to the friction slope (head loss per unit length of conduit h_f) according to (Streeter 1962, §5.7):

$$h_f = f\frac{L}{D}\frac{V^2}{2g} \tag{5.1}$$

In the case of a noncircular cross section, the term D in the Reynolds Number may be replaced by $4R$ with R being the hydraulic radius, which equals the cross-sectional area A divided by wetted perimeter P [note that for a full conduit, $R = A/P = (\pi/4)D^2/(\pi D) = D/4$ so $D = 4R$]. The Moody Diagram is shown in Figure 5.1. The variable ranges are $\varepsilon/D = \varepsilon/(4R)$ from 0.00001 to 0.5, $\mathbf{R}$ from 5000 to 100,000,000, and f from 0.008 to 0.1.

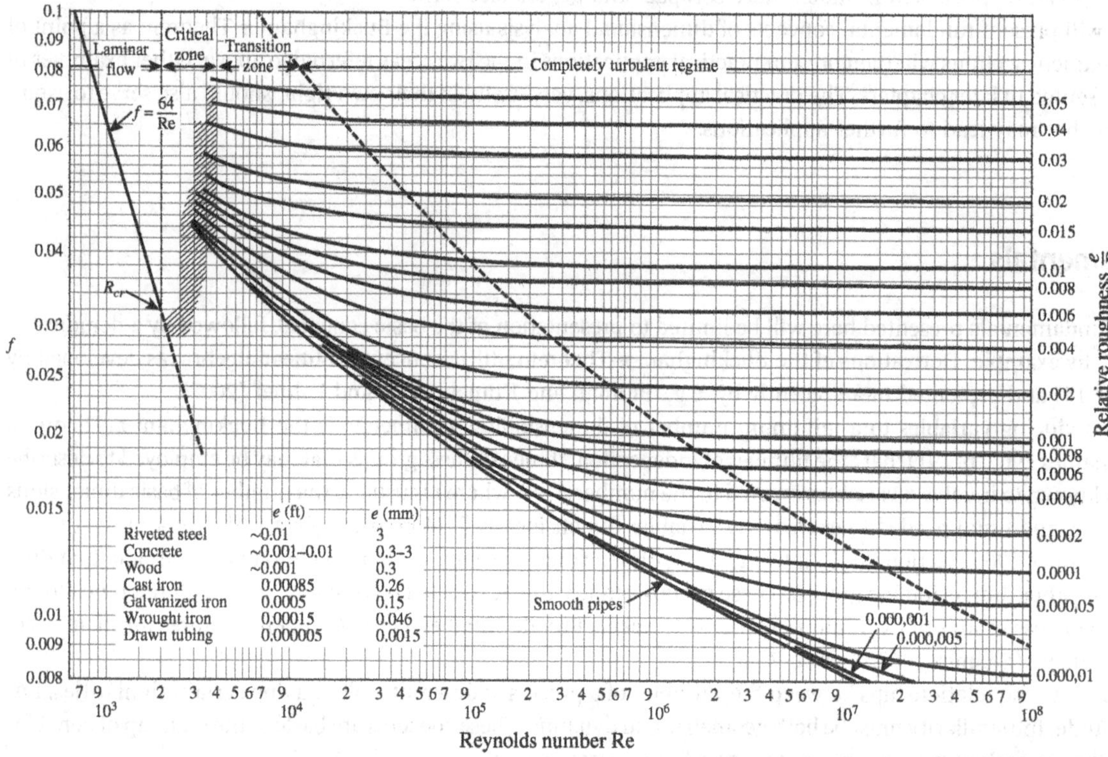

Figure 5.1 Moody Diagram (Moody 1944 / ASME)

Another classical example is that of McNown (1954), who presented experimental data for the change in piezometric head for combining and dividing flow along the run of a tee and between the run and branch of the tee. Further discussion and related plots are presented in Section 6.4.

5.3.1 Curve Fitting

We occasionally wish to fit a mathematical expression to a set of data points. The reasons for doing so include the incorporation of data into a computer program, among others. The fitting procedure entails: (1) selecting one or more general mathematical relations (often aided by a plot of the data); (2) solving for coefficients and/or exponents in each relationship in a way that provides the "best" possible fit; (3) testing how good the particular fit is; and (4) deciding which is the best fit and whether that fit is adequate.

The process of fitting a curve to a data set is known as regression analysis, and entire books have been written on the subject. Here, we shall present simple curve-fitting techniques that may be readily used. A few useful types of regression relationships will be given that are amenable to simple automatic computation (such as on a programmable calculator or using computer programs such as Microsoft Excel).

For bivariate data (data involving two variables – one dependent and one independent), the specific relationships are:

LINEAR IN $f(x)$ and $f(u)$

$$g(y) = b' + mf(x)$$

Nth-ORDER POLYNOMIAL

$$y = a_o + a_1x + a_2x^2 + a_3x^3 + \cdots + a_nx^n$$

The terms b, m, and a_i are the only coefficients to be fitted to the data. The functions $f(x)$ and $g(x)$ can be nonlinear but involve none of the regression coefficients. Examples of expressions linear in $f(x)$ and $g(y)$ are:

Straight Line

$$y = b + mx \quad [g(y) = y, f(x) = x, b' = b]$$

Exponential Curve

$$y = be^{mx} \quad [g(y) = \ln y, f(x) = x, b' = \ln b]$$

Power Curve

$$y = bx^m \quad [g(y) = \ln y, f(x) = \ln x, b' = \ln b]$$

Logarithmic Curve

$$y = b + m\ln x \quad [g(y) = y, f(x) = \ln x, b' = b]$$

Comparable trivariate relationships (for data involving three variables – one dependent and one or two independent) are:

$$z = a_o x^{a_1} y^{a_2}$$

$$y = a_o + a_1\sqrt{x} + a_2x^2$$

in which the regression coefficients and/or exponents are a_o, a_1, and a_2.

An example of a useful trivariate regression relationship is one developed by Wood (1966) to explicitly represent the dependency of the friction factor f on the Reynolds Number $\mathbf{R}$ and relative roughness ε/D in accordance with the Moody Diagram shown in Figure 5.1 (Moody 1944). Wood's relationship (with his D changed to $4R$) is as follows:

$$f = a + b\mathbf{R}^{-c} \tag{5.2a}$$

$$a = 0.094[\varepsilon/(4R)]^{0.225} + 0.53[\varepsilon/(4R)] \tag{5.2b}$$

$$b = 88[\varepsilon/(4R)]^{0.44} \tag{5.2c}$$

$$c = 1.62[\varepsilon/(4R)]^{0.134} \tag{5.2d}$$

These relationships can be readily programmed. Note from the above equations and Figure 5.1 that the coefficient a represents the value of the friction factor at high Reynolds Number.

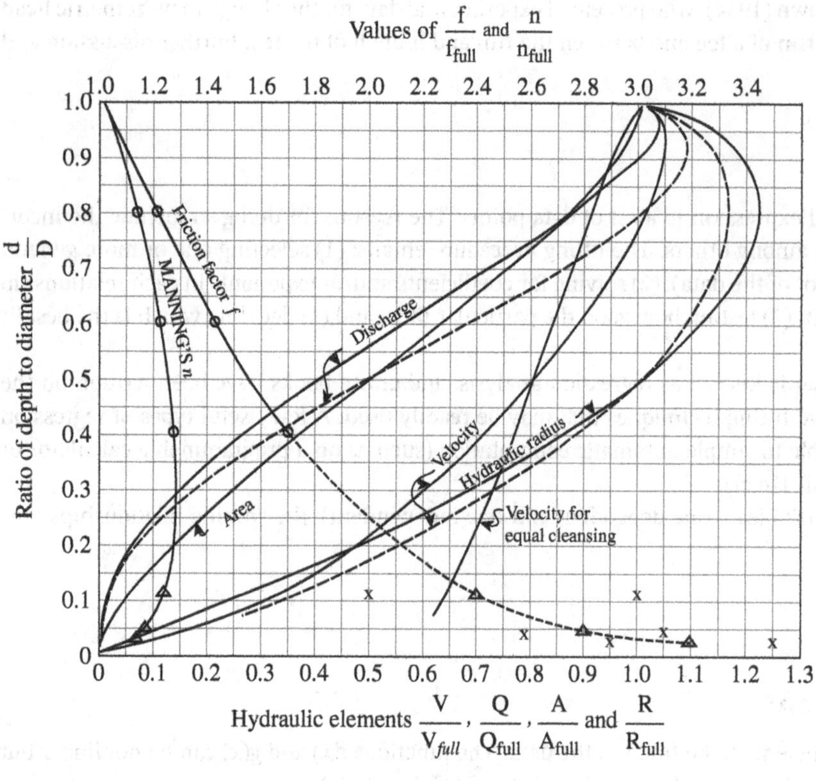

Figure 5.2 Hydraulic Elements Chart (from Camp 1952)

We will here introduce the hydraulic elements chart [e.g., from Camp 1946a, 1952 or Chow's (1959) Figure 6-5 from Camp] shown in Figure 5.2. The variation of Manning's n (see Section 6.2) with depth is shown in that figure, and the discharge and velocity curves are plotted both for constant n (solid curves) and variable n (dashed curves).

In some cases (particularly when no derivatives are required – see below), two or more functions applied to different portions of the curve may provide a more acceptable fit than a single function. Such a relationship will be called a multi-functional regression. The different functions may have the same functional form with different coefficients and exponents or be of different functional forms. For example, the following equation derived by the author provides a good fit to the friction factor curve of the hydraulic elements chart:

$$\frac{f}{f_f} = \delta_1\left(1.2134 - 0.5292\ln\frac{y}{d_o}\right) + \delta_2\left[2.416\exp\left(-0.8801\frac{y}{d_o}\right)\right]$$

$$\delta_1 = \begin{cases} 1, & y/d_o < 0.4 \\ 0, & y/d_o \geq 0.4 \end{cases}$$

$$\delta_2 = \begin{cases} 0, & y/d_o < 0.4 \\ 1, & y/d_o \geq 0.4 \end{cases} \tag{5.3}$$

in which f_f is the Darcy–Weisbach friction factor in a pipe of diameter d_o flowing full and f is the friction factor in the pipe flowing partially full at depth y. A comparison of values computed by means of the above regression relationship against actual data values indicates a maximum difference of less than one-half percent for $y/do \geq 0.4$ and less than 2% for $y/do < 0.4$.

When derivatives of the function are required, a single differentiable function is usually preferred over a multi-functional regression. That is because the derivatives often change abruptly across the boundary values of the function and can require considerable "smoothing" to obtain meaningful results. Thus the relation $K_v = b\exp(mt/T_c)$ (see Section 6.7 for meanings of the terms) provides a fit to the valve curve of Figure 6.15, which is not as good as a multi-functional regression might be. But, when $dK_v/d(t/T_c)$ is important as in waterhammer analysis (Section 8.6), a single differentiable relationship has compensatory advantages along the lines just discussed.

5.4 Generalized Numerical Solution of Differential Equations

Many of the physical situations encountered in environmental engineering can be described by differential equations for which solutions may reasonably be attempted. In some cases, a solution can be obtained in terms of a mathematical expression that can be used directly to calculate a desired variable (or variables) given numerical values of the other problem variables. Several such solutions are presented in Chapter 4, and others will be presented in subsequent chapters. In other cases, a numerical solution to the differential equation can be obtained. Such solutions typically involve the use of a computer or programmable calculator to handle the computational effort of the numerical method and provide a numerical value(s) of the desired variable(s) for the particular given values of the other problem variables (i.e., a case-by-case numerical solution).

In certain cases, numerical solutions can be obtained that involve sufficiently few variables that the results can be presented in a generalized form (i.e., a once-and-for-all numerical solution). Dimensional analysis of the pertinent differential equation enables a determination of the feasibility of obtaining a generalized numerical solution. The basic concepts will be demonstrated by two particularly useful examples, the first pertaining to the design of ground-level detention basins and the second to rooftop detention basins.

The utility of such generalized solutions can be quite substantial in the design office and in the field. Additional generalized solutions, which the author and others have developed and found valuable, are given in later chapters. Others are undoubtedly possible, and the engineer who is prepared to assess the feasibility and develop such generalized solutions will have a substantial capability in his or her bag of tricks.

5.4.1 Ground-Level Detention Basins

The following material is based primarily on Graber (2009a). Detention basins are often provided to receive stormwater runoff from newly developed areas, commonly for the purpose of limiting storm flows to their pre-development levels. Land development causes an increase in runoff due to the increase in imperviousness and decrease in time to peak flow (concentration time). In terms of the Rational Method (Graber 2017a), the result is an increase in the runoff coefficient and average rainfall intensity. [Since this is the only part of the book in which the flow-attenuating characteristics of detention basins are discussed explicitly, the discussion is expanded to include some of the practical aspects of the hydraulic design of detention ponds.]

The water-quality significance of detention basins stems from the concomitant reduction in sediments and certain other stormwater contaminants (see, e.g., Whipple and Hunter 1981), the reduction in downstream erosion potential, and their use in outlying areas to reduce stormwater overflows in downstream combined sewer areas (as in Chicago, for example).

Before getting into more technical details, important safety aspects of detention basins and other considerations are discussed. Detention basins can be a necessary part of developments but should be reviewed carefully as to technical adequacy and maintainability, including responsibility for the latter. Safety aspects of infiltration basins and detention basins must also be considered. Locking manhole covers for manholes and the surface extensions of underground valve pits can be important, particularly in less-frequented areas. The author specified a locking manhole cover for a location in Winthrop, Massachusetts, where kids were using the underground structure for a clubhouse causing risk to themselves and equipment. Ergonomics for manholes is also important (Graber 2017b, §7.3, Buchan 2006). Surface detention basins can pose a drowning hazard, particularly if slopes around the basin become icy ("Drowning site won't be filled," Patriot Ledger, March 12, 1996, page 1; "Boy drowns in icy pond," Stoughton Journal, March 14, 1996, page 1). Limiting side slopes, constructing a flat shelf of ample width around the basin, and fencing all merit consideration.

The conservation of mass relationship for a detention basin is given (from Equation 2.12 for constant density) by:

$$\frac{dS}{dt} = I - Q \tag{5.4}$$

in which I is the volumetric inflow rate, Q is the volumetric outflow rate, and dS/dT is the time rate of increase of effective storage volume. We will limit our attention here to passive outflow controls for which the effective storage is that storage above the minimum level at which outflow commences. The inflow is assumed to have the form of the trapezoidal hydrograph depicted in Figure 5.3, with peak inflow I_p, linear rising limb to the time of concentration t_c, duration t_r, and the linear receding limb of duration Rt_c (Graber 2009a).

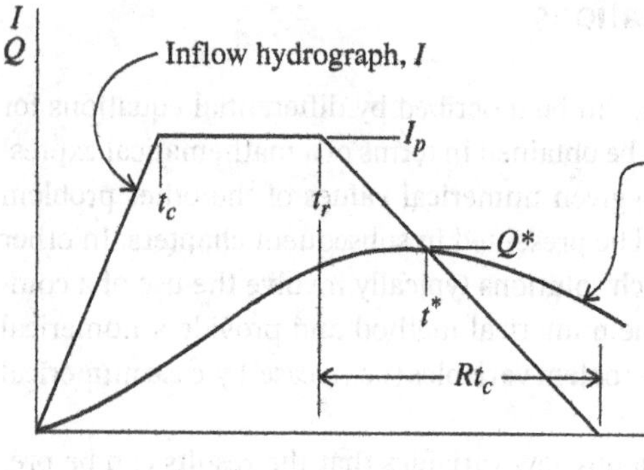

Figure 5.3 Modified Rational Method Hydrograph

We will assume that the effective storage and outflow can be expressed as follows:

$$S = NH^{\ell} \tag{5.5}$$

$$Q = MH^{m} \tag{5.6}$$

in which H is elevation measured upward from the elevation at which effective storage and outflow begin. The storage is thus related to the outflow by:

$$S = N(Q/M)^{\ell/m} \tag{5.7}$$

Therefore:

$$dS = \frac{N}{M^{\ell/m}} d\left(Q^{\ell/m}\right) \tag{5.8}$$

which can be substituted into Equation 5.4 to give:

$$\frac{N}{M^{\ell/m}} \frac{d(Q^{\ell/m})}{dt} = I(t) - \left(Q^{\ell/m}\right)^{\frac{1}{\ell/m}} \tag{5.9}$$

in which $I(t)$ denotes that the inflow is a known function of time.

The above equation is a *nonlinear* differential equation for which a general solution is not known. A numerical solution can be obtained [such as by the Runge–Kutta method (Hildebrand 1962, §3.5)], and the possibility of obtaining a *generalized* numerical solution merits investigation. Equation 5.9 can be nondimensionalized with respect to the peak inflow I_p and time of peak inflow t_c, and rewritten as follows:

$$\frac{d(Q/I_p)^{\ell/m}}{dt/t_c} = B\left\{ I(t)/I_p - \left[(Q/I_p)^{\ell/m}\right]^{\frac{1}{\ell/m}} \right\} \tag{5.10a}$$

$$B = \frac{I_p t_c M^{\ell/m}}{N I_p^{\ell/m}} \tag{5.10b}$$

We retain $(Q/I_p)^{\ell/m}$ as the dependent variable rather than express the differential equation in terms of $d(Q/I_p)/d(t/t_c)$. The latter form requires that we divide by $(Q/I_p)^{\ell/m-1}$, which is inadmissible when $Q/I_p = (Q/I_p)^{\ell/m-1} = 0$, an important initial condition.

We want to solve for the maximum outflow, which occurs when $dQ/dt = d(Q^{\ell/m})/dt = 0$ and (from Equation 5.4) $I = Q$ (see Figure 5.3). Denoting the magnitude of the maximum outflow and the time at which it occurs by Q^* and t^*, respectively, we have:

$$I(t^*) = Q^* \tag{5.11}$$

For a given *mathematically continuous* inflow hydrograph relationship with $Q = 0$ at $t = 0$ as the initial condition, Equations 5.10 imply the following functional relationship:

$$\frac{Q^*}{I_p} = \phi\left(\frac{t^*}{t_c}, B, \frac{\ell}{m}\right) \tag{5.12}$$

From Equation 5.11:

$$\frac{Q^*}{I_p} = \frac{I(t^*)}{I_p} = \psi\left(\frac{t^*}{t_c}\right) \tag{5.13}$$

The term t^*/t_c can theoretically be eliminated from Equations 5.12 and 5.13 to give the following functional relationship:

$$\frac{Q^*}{I_p} = \phi\left(B, \frac{\ell}{m}\right) \tag{5.14}$$

Denoting the maximum required storage volume by $\forall$, we have from Equation 5.7:

$$\forall = N(Q^*/M)^{\ell/m} \tag{5.15}$$

From the above equation and Equation 5.10b, we obtain:

$$B = \left(\frac{Q^*}{I_p}\right)^{\ell/m} \frac{I_p t_c}{\forall} \tag{5.16}$$

The functional relationship of Equation 5.12 can thus be expressed, for a given shape of inflow hydrograph, in the following convenient form:

$$\frac{\forall}{I_p t_c} = \psi\left(\frac{Q^*}{I_p}, \frac{\ell}{m}\right) \tag{5.17}$$

A relationship between three variables, as implied by the above functional equation (or Equation 5.14), can easily be presented graphically, giving a convenient generalized solution.

The inflow hydrographs employed in practice are often *piecewise continuous*. A common example is the trapezoidal hydrograph shown in Figure 5.3 (of which the triangular hydrograph is a special case), which has one continuous relationship applicable for $t \le t_c$, another applying for $t_c < t \le t_r$, and another for $t_r < t < (t_r + Rt_c)$. Specifically:

$$\frac{I}{I_p} = \begin{cases} t/t_c, & t \le t_c & (5.18a) \\ 1, & t_c < t \le t_r & (5.18b) \\ 1 - \dfrac{1}{R}\left(\dfrac{t}{t_c} - \dfrac{t_r}{t_c}\right), & t_r < t \le (t_r + Rt_c) & (5.18c) \end{cases}$$

We will demonstrate how the functional relationship of Equations 5.12 and 5.17 can also be extended to a piecewise continuous inflow hydrograph, focusing specifically on the trapezoidal hydrograph.

For a given R, Equations 5.10 implies the following relationship for a trapezoidal hydrograph for $t \le t_c$:

$$\frac{Q(t)}{I_p} = \phi\left(\frac{t}{t_c}, \frac{\ell}{m}, B\right), \quad t \le t_c \tag{5.19}$$

For $t = t_c$:

$$\frac{Q(t_c)}{I_p} = \phi\left(\frac{\ell}{m}, B\right), \quad t = t_c \tag{5.20}$$

For $t_c < t \le t_r$, since $Q = Q(t_c)$ at $t = t_c$:

$$\frac{Q(t)}{I_p} = \phi\left[\frac{t}{t_c}, \frac{Q(t_c)}{I_p}, \frac{\ell}{m}, B\right], \quad t_c < t \le t_r \tag{5.21}$$

The term $Q(t_c)$ can be eliminated between the above two functional relationships, giving:

$$\frac{Q(t)}{I_p} = \phi\left(\frac{t}{t_c}, \frac{\ell}{m}, B\right), \quad t_c < t \le t_r \tag{5.22}$$

For $t = t_r$:

$$\frac{Q(t_r)}{I_p} = \phi\left(\frac{t_r}{t_c}, \frac{\ell}{m}, B\right), \quad t = t_r \tag{5.23}$$

For $t_r < t \le (t_r + Rt_c)$, since $Q = Q(t_r)$ at $t = t_r$:

$$\frac{Q(t)}{I_p} = \phi\left[\frac{t}{t_c}, \frac{Q(t_r)}{I_p}, \frac{\ell}{m}, B\right], \quad t_r < t \le (t_r + Rt_c) \tag{5.24}$$

The term $Q(t_r)$ can be eliminated between the above two functional relationships, giving:

$$\frac{Q(t)}{I_p} = \phi\left(\frac{t}{t_c}, \frac{t_r}{t_c}, \frac{\ell}{m}, B\right), \quad t_r < t \le (t_r + Rt_c) \tag{5.25}$$

Noting that t^* must fall within the range $t_r < t^* \le (t_r + Rt_c)$, the above equation gives:

$$\frac{Q^*}{I_p} = \phi\left(\frac{t^*}{t_c}, \frac{t_r}{t_c}, \frac{\ell}{m}, B\right) \tag{5.26}$$

in which t^*/t_c is unknown. However, from Equations 5.18a–c, the relationship analogous to Equation 5.13 can be given as:

$$\frac{Q^*}{I_p} = 1 - \frac{1}{R}\left(\frac{t^*}{t_c} - \frac{t_r}{t_c}\right) \tag{5.27}$$

The term t^*/t_c can be eliminated between Equations 5.26 and 5.27 to give the following functional relationship:

$$\frac{Q^*}{I_p} = \phi\left(B, \frac{\ell}{m}, \frac{t_r}{t_c}, R\right) \tag{5.28}$$

Noting that t_r/t_c and R simply characterize the inflow hydrograph, it can be seen that Equation 5.28 is essentially the same functionally as Equation 5.14.

Equations 5.15 and 5.16 still apply and allow Equation 5.28 to be expressed in the form analogous to Equation 5.17 as follows:

$$\frac{\mathcal{V}}{I_p t_c} = \psi\left(\frac{Q^*}{I_p}, \frac{\ell}{m}, \frac{t_r}{t_c}, R\right) \tag{5.29}$$

The inflow volume $\mathcal{V}_i$ resulting from the trapezoidal hydrograph is given by

$$\mathcal{V}_i = I_p t_c\left(\frac{R+1}{2} + \frac{t_r}{t_c} - 1\right) \tag{5.30}$$

From Equations 5.29 and 5.30, one may then obtain

$$\frac{\mathcal{V}}{\mathcal{V}_i} = \frac{\psi\left(\frac{Q^*}{I_p}, \frac{\ell}{m}, \frac{t_r}{t_c}, R\right)}{\left(\frac{R+1}{2} + \frac{t_r}{t_c} - 1\right)} \tag{5.31}$$

An advantage of Equation 5.30 is that the variables $\mathcal{V}/\mathcal{V}_i$ and Q^*/I_p both lie within the range 0–1, facilitating a convenient plot of the results. No new variables have been introduced in going from Equations 5.28 to 5.31, and the functional relationship $\mathcal{V}/\mathcal{V}_i$ can be expressed as follows:

$$\frac{\mathcal{V}}{\mathcal{V}_i} = \psi'\left(\frac{Q^*}{I_p}, \frac{\ell}{m}, \frac{t_r}{t_c}, R\right) \tag{5.32}$$

Generalized solutions in the form of Equation 5.32 for the trapezoidal inflow hydrograph give curves of the type shown in Figure 5.4 for a particular value of ℓ/m. Curves for additional values of $\ell/m = 1.0$, 2.0, and 3.0 are given in Graber (2009a). The value of $R = 1.67$ is employed for the slight conservatism it provides. Others have used the trapezoidal hydrograph with

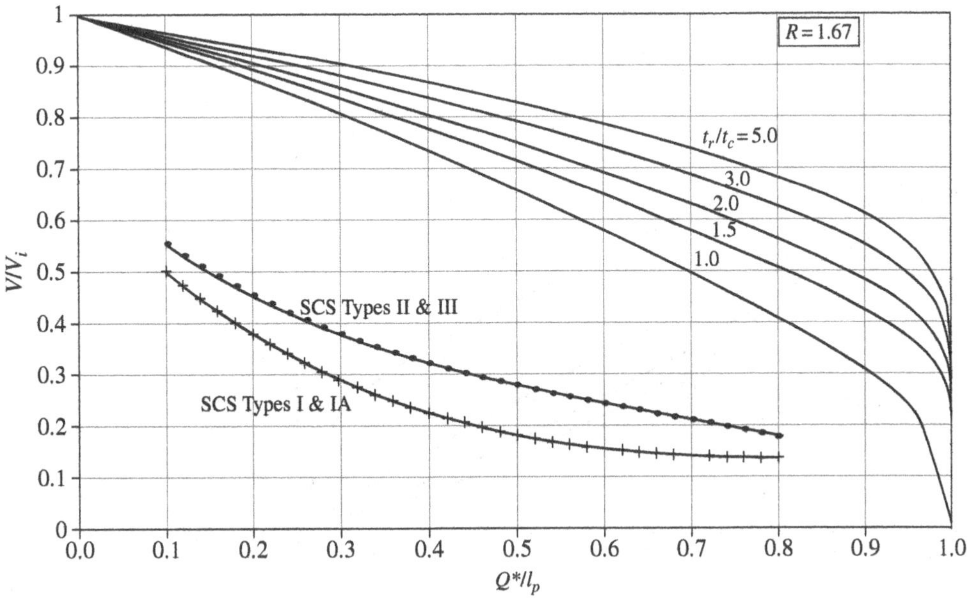

Figure 5.4 V/V_i vs. Q^*/I_p for $\ell/m = 0.2$

various values of R ranging from 1 (triangular hydrograph) to 1.75 [Kaltenbach 1963; USBR 1965, pp. 42–44; Burton 1980; Tourbier and Westmacott 1981; Aron and Kibler 1990; Basha 1994; SCS (McCuen 1982, SCS 1986)]. The use of $R > 1$ in Graber (2010a) for the trapezoidal hydrograph was questioned in a discussion of that paper, prompting the following response (Graber 2011a): "The writer's use of $R > 1$ is a compromise between (1) the long-known fact that the receding limb of the outflow hydrograph has a 'time base longer than that for [the rising limb]' (Williams 1950), the rising limb in our case being linear with time base equal to the time of concentration; and (2) the recognition that the slightly conservative assumption of a linear receding limb with $R = 1.67$ will give an outflow volume greater than the inflow volume. A more complex curve could be used for the receding limb to eliminate the latter conflict but does not seem warranted and would of necessity be arbitrary and complicate the analysis. The discusser notes that, for the writer's example, the discusser's required storage volume is 'close to [and less than] the writer's solution ... in spite of the runoff hydrograph differences [referring to the discusser's $R = 1$ vs. the writer's $R = 1.67$].' The discusser's $R = 1$ solution for storage volume is within 2% of the solution given in the writer's paper ...."

The solutions mentioned in the previous paragraph were obtained using a fourth-order Runge–Kutta method (Hildebrand 1962, Sections 3.5 and 7.10) over most of the range. For Q^*/I_p very close to 1 the numerical solution becomes intractable; in that situation, Equations 5.16 and 5.30 can be combined to give

$$\frac{\mathcal{V}}{\mathcal{V}_i} = \frac{1/B}{\left(\frac{R+1}{2} + \frac{t_r}{t_c} - 1\right)}, \quad Q^*/I_p \cong 1 \tag{5.33}$$

This value of $\mathcal{V}/\mathcal{V}_i$ is then calculated using Equation 5.33 to complete the curve for $Q^*/I_p \cong 1$.

Figure 5.5 gives similar curves for a fixed $t_r/t_c = 1$ and the same four values of ℓ/m shown or mentioned above. This plot demonstrates that a larger ℓ/m gives a smaller V/V_i for a given t_r/t_c and Q^*/I_p, as has been noted by others for other inflow hydrographs (e.g., McEnroe 1992).

Plots can also be presented in the form of Equation 5.28. For rectangular inflow hydrographs ($t_r/t_c \rightarrow \infty$ or $t_c = 0$), applicable especially to roof drainage (see below), the curves for all ℓ/m values can be represented on a single such chart for the forms analogous to Equations 5.28 and 5.32, as given in Graber (2009b). Also for rectangular inflow hydrographs, analytical solutions can be obtained for certain cases (Graber 2009b and references cited therein), including those discussed for roof drainage below.

Graber (2009a) discusses numerous other published methods in relation to the above. They are all limited to the case of storm duration equal to time of concentration or are otherwise unable to consider different storm durations. Only one of those methods is discussed here. Added to Figure 5.4 [and to the other similar figures in Graber (2009a)] are curves taken from Figure 6-1 of TR-55 (SCS 1986), calculated using the formulas in TR-55's Appendix F (i.e., using the formula therein for

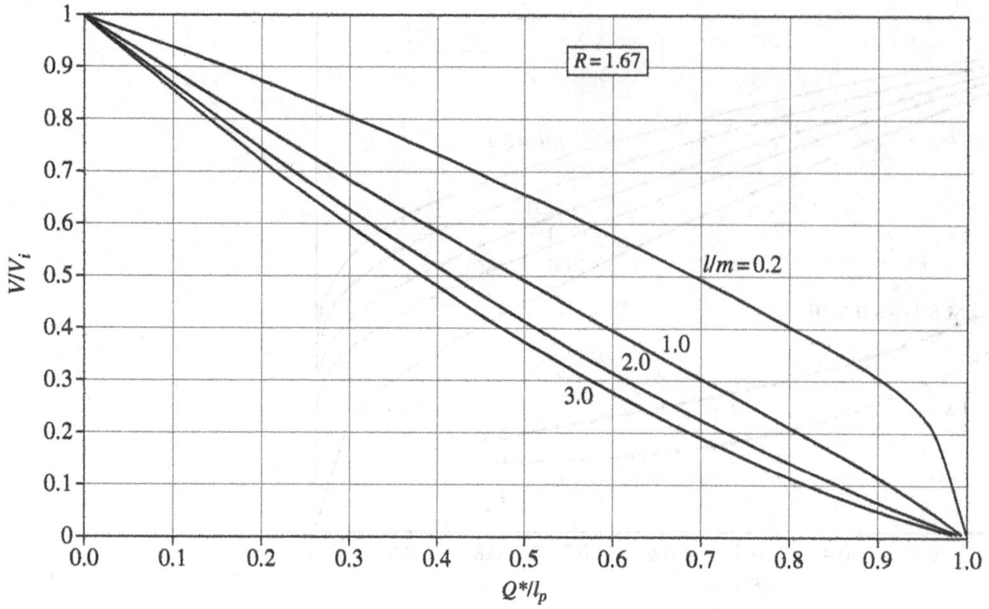

Figure 5.5 V/V_i vs. Q^*/I_p for $t_r/t_c = 1$

Figure 6-1 and coefficients in Table F-2). TR-55's Chapter 6 characterizes the curves as follows: "The method [used in developing the SCS curves] is based on average storage and routing effects for many structures [using the method found in TR-55 Chapter 5]. The relationships ... were determined on the basis of single stage outflow devices. Some were controlled by pipe flow, others by weir flow. Verification runs were made using multiple-stage outflow devices and the variance was similar to that used for both single- and multiple-stage outflow devices ... It is ... adequate ... for final design of small detention basins ... Figure 6-1 is biased to prevent undersizing of outflow devices, but it may significantly overestimate the required storage capacity." For a given ratio of peak outflow to peak inflow, the required ratio of storage volume to runoff volume is actually, over most of the range of variables, substantially higher than predicted by TR-55. Conversely, over most of the range, for a given ratio of storage volume to runoff volume, the ratio of peak outflow to peak inflow is actually substantially higher than predicted by TR-55.

Example 5.1 An example (Westborough, Massachusetts) of the use of the above is shown in Table 5.1, which should be self-explanatory. The time of concentration t_c is 15 minutes, and successively larger storm durations t_c are considered. The rainfall intensity i is taken from intensity–duration–frequency data for each storm duration for the location and return period of interest.

Table 5.1 Detention Basin Design Example

t_r		i m/h	I_p m³/s			$\ell/m = 2$		$\ell/m = 2/3$		$\ell/m = 2$		$\ell/m = 2/3$	
(min)	t_r/t_c	(in./h)	(cu ft/s)	Ꝟ$_i/(I_pt_c)$	Q^*/I_p	Ꝟ/Ꝟ$_i$	Ꝟ$/(I_pt_c)$	Ꝟ/Ꝟ$_i$	Ꝟ$/(I_pt_c)$	Ꝟ(m³)	Ꝟ(ft³)	Ꝟ(m³)	Ꝟ(ft³)
15 (t_c)	1	0.102(4.0)	0.572(20.2)	1.335	0.225	0.716	0.956	0.790	1.055	492.1	17,378	542.9	19,174
30	2	0.071(2.8)	0.400(14.1)	2.335	0.322	0.68	1.588	0.765	1.786	571.6	20,206	643.1	22,732
45	3	0.056(2.2)	0.315(11.1)	3.335	0.409	0.635	2.118	0.735	2.451	600.4	21,175	694.9	24,510
60	4	0.048(1.9)	0.272(9.60)	4.335	0.474	0.601	2.605	0.713	3.091	637.8	22,498	756.6	26,691
75	5	0.043(1.7)	0.243((8.59)	5.335	0.530	0.568	**3.030**	0.691	**3.686**	**662.7**	**23,413**	**806.2**	**28,484**
90	6	0.038(1.5)	0.214(7.58)	6.335	0.600	0.521	3.301	0.656	4.156	635.7	22,501	800.4	28,332
150	10	0.025(1.0)	0.143(5.05)	10.335	0.901	0.263	2.718	0.431	4.454	349.8	12,354	573.3	20,245

Note: Bold font corresponds to critical duration.

The basin peak inflow decreases in proportion to the decrease in rainfall intensity as the storm duration increases. The desired peak outflow Q^* is fixed, so the ratio Q^*/I_p increases in proportion to the decrease in I_p. The ratio of required storage volume to inflow volume $\forall/\forall_i$ is obtained from generalized curves prepared as described above, in this case for $\ell/m = 2$ and $\ell/m = 2/3$. The required storage volume, given by the product $(\forall/\forall_i)\forall_i$ has a maximum value at a storm duration of approximately 75 minutes. The required storage volume can be seen to be greater for the lower value of ℓ/m.

Another example, of an application by the author in Sudbury, Massachusetts, is discussed in Graber (2011b); the methods discussed therein can now be done more expeditiously using software such as HydroCAD®. Other computer models for modeling the hydraulics of stormwater impoundment outlet works are given in Graber (2017b, §9.5).

5.4.2 Rooftop Detention Basins

The following material is based on Graber (2009b). Calculation of the depth of rainwater on roofs is necessary for the determination of rain loads as a component of the total roof load for structural design and analysis. An additional consideration is that of storing rainwater on roofs as a means of attenuating stormwater runoff (Wells and Schmid 1961, Rice 1971, Tourbier and Westmacott 1981). This has more recently been extended to the use of green (vegetated) roofs, which include stormwater reduction among their potential benefits (Carter and Rasmussen 2006, Jarrett *et al.* 2006, Villarreal 2007), and which are being encouraged in the U.S. through LEED (Leadership in Energy and Environmental Design) (U.S. Green Building Council 2005) and other initiatives. At the very least, green roofs reduce the stormwater runoff coefficient compared to impervious roofs. Water storage on ordinary roofs provides more certain capacity for stormwater attenuation, however, because the roof storage usually drains much more quickly than on green roofs, and storage becomes available sooner for the next rainstorm. Some regulatory agencies require that detention be adequate for two back-to-back 24-hour storms or that the maximum time of storage on the roof not exceed 24 hours (e.g., NRCS 2007), which can be accomplished by ordinary roof detention but not generally by a green roof. The focus here is on depth of rainwater and water storage on non-vegetated roofs.

Current U.S. design standards (ASCE 2006, FMIC 2006) are based on the provision of primary and secondary roof drainage. Secondary drainage is required to provide roof protection in the event of clogging of the primary drain (ASCE 2006, sections 8.3 and 8.5; FMIC 2006, sections 2.1.1.1.10, 2.1.1.3.2.4, 2.1.1.3.5, and 2.1.1.3.5.3). Roof collapse has been attributed to clogging of primary-drainage systems by hail (Changnon 1978) as well as clogging by debris. A more recent example attributed to heavy wind and rain was the collapse of a 5200 square meter roof at a General Motors assembly plant in Kansas City, Kansas (Building Design and Construction 1979).

The ASCE and FMIC standards require that the depth of water be that of the secondary-drainage system based on a prescribed rainfall condition. The FMIC standard (2.1.1.3.3) requires use of the local average rainfall intensity for the locale's 100-year, 1-hour storm, except that the design peak flow rate is assumed to be doubled for a controlled drainage system (2.1.1.3.5.4). The ASCE standard refers (section C8.2) to a U.S. plumbing code that uses the 1-hour and 15-minute duration events with 100-yr return periods for the primary and secondary-drainage systems, respectively, and a Canadian building code that uses the 15-minute event with a 10-year return period. The ASCE standard (section C8.5) notes that, in some areas of the U.S., ordinances limit the rate of stormwater flow from roofs and that many roofs designed with controlled-flow drains have a design rain load of 30 lb/ft^2 (1.44 kN/m^2) and are equipped with a secondary-drainage system intended to prevent total water depths greater than 5.75 in. (145 mm) on the roof. The examples in the ASCE standard (Section C8.5) use the 100-yr, 1-hour rainfall for the secondary drainage. An assumption implicit in the ASCE (2006) and FMIC (2006) standards is that an equilibrium condition has been reached such that outflow equals inflow. The present work evaluates these standards and provides recommended improvements. It also provides compatible design methodology for designing for stormwater attenuation.

5.4.2.1 Basic Relationships and Concepts

With zero initial inflow and outflow, the inflow and outflow hydrographs will have the form shown in Figure 5.3, on which the inflow hydrograph is represented as a trapezoid for purposes to follow. Characteristics of the inflow (runoff) hydrograph in Figure 5.3 are the peak inflow I_p, linear rising limb to the time of concentration t_c, duration t_r, and linear receding limb of duration Rt_c. The outflow hydrograph has peak flow Q^* at time t^*. Because storage is a maximum at the maximum outflow, then $dS/dt = 0$, which from Equation 5.4 requires that $Q^* = I$, i.e., the maximum outflow must occur at an intersection of the inflow and outflow hydrograph at $t \geq t_r$.

For roof drainage, and particularly for analysis of roof drainage for the secondary-drainage system, the coefficient C in the Rational Method will be unity, and the time of concentration t_c can be taken as zero. Taking $t_c = 0$ is a reasonable assumption and slightly conservative for these applications. When $t_c = 0$ then Rt_c also equals 0, corresponding to a rectangular inflow hydrograph specialized from Figure 5.3. In that case, the maximum outflow will occur at $t^* = t_r$, knowledge of which allows checking of numerical solutions discussed below. If it is desired to consider cases with $t_c > 0$, that can be done by using the method of trapezoidal hydrographs discussed above.

The storage and outflow terms in Equation 5.4 are expressible as functions of water surface elevation so that with a prescribed inflow, one has a single equation in a single unknown (water surface elevation), which can be solved in concept; numerical methods are commonly required to obtain the solution, and those solutions can sometimes be generalized as discussed below. In the simplest cases, the effective storage and outflow in Equation 5.4 can be expressed as in Equations 5.5 and 5.6.

Cases to which Equation 5.5 applies include a flat roof for which $\ell = 1$ and $N =$ constant roof area; a troughed roof of constant longitudinal slope S_l and width W for which $N = (1/2)W/S_l$ and $\ell = 2$; the inverted-pyramid portion of a roof pitched to a centrally located drain for which $N = (1/3)A_o/H_o^2$ and $\ell = 3$ where $H_o =$ depth corresponding to base area A_o; and the inverted right-wedge portion of a roof pitched with transverse slope S_t and length L to a centrally located drain for which $N = L/S_t$ and $\ell = 2$. In some other cases, the roof contours may enable a regression on an equation in the form of Equation 5.5.

Cases to which Equation 5.6 applies include: $m = 1$ for the unsubmerged proportional weir (Sutro 1915) [see Figure 5.6(a)] with M as given by Howe (1950); linear controlled-flow roof drains for which $m = 1$ and M can be adjustable and obtained from manufacturer's data (Zurn Industries Inc. 2007) (the inverted "parabolic" notches of such drains are essentially Sutro weirs); $m = 1.5$ for an unsubmerged rectangular weir or channel-type scupper with M as discussed below; $m = 2$ for the Lauritson/Greve parabolic weir (Greve 1921, Howe 1950) [see Figure 5.6(b)]; and $m = 2.5$ for an unsubmerged triangular or V-notch weir such as can be placed on a scupper with M as given by Howe (1950). The hydraulic design to attain these m values for the weir configurations is reasonably straightforward.

For the Sutro weir Q is actually proportional to $(H - a/3)$ with H being the depth of water relative to the weir crest and a being the height of the rectangular slot, which is surmounted by the single-sided or symmetrical curved portion(s) (Howe 1950). The direct proportionality is best approximated by making a small. Sandvik (1985) discusses an application of a proportional weir for a detention basin and claims certain advantages of the proportional weir for such applications. Other variations of the proportional weir are given by Sreenivasulu and Raghavendran (1970), Venkataraman and Subramanya (1973), and Chandrasekaran and Rao (1976). Some of the variations of the Sutro type of proportional weir that have been proposed may be of theoretical interest but have no apparent practical advantage over the basic Sutro weir. They may actually be detrimental by employing small acute angles near the weir bottom, which would be more likely to foul.

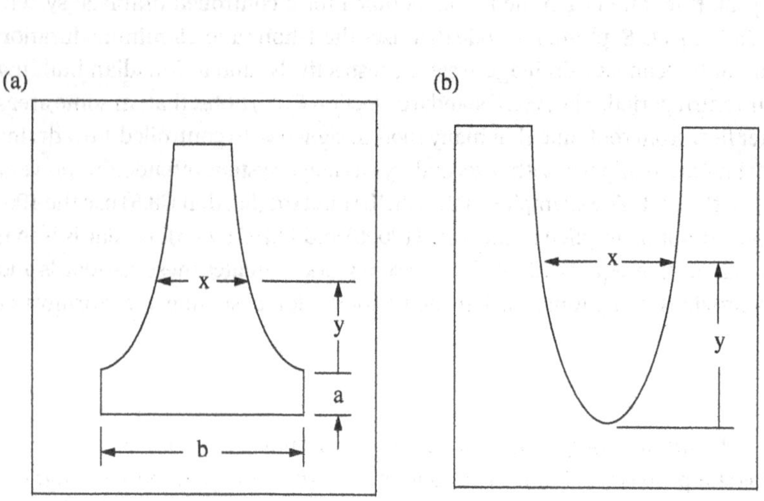

$$\frac{x}{b} = 1 - \frac{2}{\pi}\tan^{-1}\sqrt{\frac{y}{a}} \qquad x^2 = 4py$$

Figure 5.6 (a) Sutro Weir; (b) Parabolic Weir

The author designed a Sutro weir for outlet control on an effluent pond at the Bridgewater Massachusetts Correctional Facility Wastewater Treatment Plant (in conjunction with the dosing siphon discussed in Section 11.4). Constructed of 1/4-inch aluminum plate 18 inches wide by 16-1/2 inches high, the weir dimensions with reference to Figure 5.6(a) were $b = 6$ inches, $a = 1.59$ inches, 10.50 inches overall height, and 1.52 inches wide at the weir top.

In some cases, the hydraulic characteristics may enable a regression on an equation in the form of Equation 5.6. The source of the data should be scrutinized, however; e.g., roof drain data from Tables C8-1 and C8-2 of ASCE (2006) and Table 8 of FMIC (2006) are given without source (nor has the author been able to duplicate the results using weir or orifice formulae) and clearly cannot apply to all manufacturers' roof drains (as exemplified below). The formula and tabulations for channel-type scuppers in Tables C8-1 and 2 of ASCE (2006) and Table 6 of FMIC (2006), respectively, are for an aerated nappe and neglect end contractions; they should be refined as necessary for broad-crested weirs based on, e.g., Brater and King (1976, pp. 5–23 to 5–40); Chow (1959, pp. 52–53, 80, 360–36); Howe (1950); Ippen (1950, pp. 525–528); and Rao and Shukla (1971). The formula and tabulations for closed-type scuppers in those two sources should be refined as appropriate to account for roof width based on, e.g., Bauer *et al.* (1969). More complex relationships should not necessarily be avoided. A numerical routing analysis of the type discussed below is advisable if simplified methods are not applicable or their applicability is uncertain.

5.4.2.2 Analytical Solutions

The storage is related to the outflow by Equation 5.7, which leads to Equations 5.8, 5.9, and 5.10. Using Equation 5.10 or an alternate form expressed in terms of S with the use of Equations 5.7 and 5.8, analytical solutions can be obtained for certain cases of interest here. In the following, constant $I(t) = I_p$, $\mathcal{V}$ = maximum storage volume, $\mathcal{V}_i$ = inflow volume, here equal to $I_p t_r$, and $B' = I_p t_r M^{\ell/m} / \left(N I_p^{\ell/m}\right)$. Using Equation 5.7, a solution for S can be expressed in terms of Q or vice versa. For $\ell/m = 1$ the solution is:

$$\frac{Q^*}{I_p} = 1 - \exp(-B') \tag{5.34a}$$

$$B' = \frac{M t_r}{N} \tag{5.34b}$$

$$\frac{\mathcal{V}}{\mathcal{V}_i} = -\frac{Q^*/I_p}{\ln(1 - Q^*/I_p)} \tag{5.34c}$$

Equation 5.34a is mathematically equivalent to the solution given by Wells and Schmid (1961) specifically for design of roof drains with $t_c = 0$.

For cases with $\ell/m \neq 1$ it is useful, especially for secondary drainage, to include situations with initial storage $\mathcal{V}_o$. For $\ell/m = 1/2$, the author's solution is given by:

$$-2\sqrt{B'}\left[\left(\frac{\mathcal{V}}{\mathcal{V}_i}\right)^{1/2} - \left(\frac{\mathcal{V}_o}{\mathcal{V}_i}\right)^{1/2}\right] - 2\ln\frac{\left[1 - R'(\mathcal{V}/\mathcal{V}_i)^{1/2}\right]}{\left[1 - R'(\mathcal{V}_o/\mathcal{V}_i)^{1/2}\right]} = B' \tag{5.35a}$$

$$B' = \frac{I_p^{1/2} t_r M^{1/2}}{N} \tag{5.35b}$$

$$\frac{Q^*}{I_p} = \sqrt{B'}\left(\frac{\mathcal{V}}{\mathcal{V}_i}\right)^{1/2} \tag{5.35c}$$

The root $\mathcal{V}/\mathcal{V}_i$ must be found numerically (e.g., by bisection). For $\mathcal{V}_o = 0$, the solution can be expressed as:

$$\frac{Q^*}{I_p} = \left[\frac{\exp(2B') - 1}{\exp(2B') + 1}\right]^2 \tag{5.35d}$$

$$\frac{\mathcal{V}}{\mathcal{V}_i} = -\frac{2\sqrt{Q^*/I_p}}{\ln\left[\left(1 + \sqrt{Q^*/I_p}\right)/\left(1 - \sqrt{Q^*/I_p}\right)\right]} \tag{5.35e}$$

For $\ell/m = 2$, the solution is given by:

$$\frac{\forall}{\forall_i}=\frac{U\exp(2B')-1}{B'[U\exp(2B')+1]} \tag{5.36a}$$

$$B'=\frac{t_r M^2}{NI_p} \tag{5.36b}$$

$$U=\frac{1+B'(\forall_o/\forall_i)}{1-B'(\forall_o/\forall_i)} \tag{5.36c}$$

$$\frac{Q^*}{I_p}=\sqrt{B'}\left(\frac{\forall}{\forall_i}\right)^2 \tag{5.36d}$$

For $\forall_o = 0$, the solution can be expressed as:

$$-2\frac{Q^*}{I_p}-2\ln\left(1-\frac{Q^*}{I_p}\right)=B' \tag{5.36e}$$

$$\frac{\forall}{\forall_i}=-\frac{(Q^*/I_p)^2}{-2Q^*/I_p-2\ln(1-Q^*/I_p)} \tag{5.36f}$$

Equations 5.36e and f can be expressed in the form of a solution given by Basha (1994).

5.4.2.3 Generalized Numerical Solution

Full details of the generalized numerical solution are presented in Graber (2009a) for the more general case of a trapezoidal hydrograph depicted in Figure 5.3. Specializing that to the case of the rectangular hydrograph with $t_c = Rt_c = 0$ or $t_r/t_c \to \infty$ (rectangular inflow hydrograph of height I_p and width t_r), the curves can be conveniently plotted as Q^*/I_p vs. ℓ/m and the modified abscissa suggested by the following functional relationship from Equation 5.10b:

$$\frac{Q^*}{I_p}=\Phi\left(\frac{I_p t_r M^{\ell/m}}{NI_p{}^{\ell/m}},\frac{\ell}{m}\right),\quad t_r/t_c\to\infty \tag{5.37a}$$

An applicable form specialized by Graber (2009a) becomes:

$$\frac{\forall}{\forall_i}=\psi'\left(\frac{Q^*}{I_p},\frac{\ell}{m}\right),\quad t_r/t_c\to\infty \tag{5.37b}$$

The form of Equation 5.37a is obtained numerically (see Graber 2009a). The form of Equation 5.37b can be generated by using the numerical solutions for Equation 5.37a and using Graber (2009a) to obtain:

$$\frac{\forall}{\forall_i}=\frac{NI_p^{\ell/m}}{I_p t_r M^{\ell/m}}\left(\frac{Q^*}{I_p}\right)^{l/m} \tag{5.38}$$

Figures 5.7 and 5.8 give these numerical solutions. The analytical solutions given above are in the form of Equations 5.37. Those analytical solutions, for m=½, 1, and 2, are plotted in Figures 5.7 and 5.8 and can be seen to match the numerical solutions.

Outlet control with $m = 1$ can be advantageous because smaller values of m and, hence, larger ℓ/m give lower storage requirements ($\forall$) for a given flow attenuation (Q^*/I_p); this is easily seen in Figure 5.7, and has also been noted for other inflow hydrographs by others [e.g., McEnroe (1992)]. For roof drains, $m = 1$ can be achieved with the unsubmerged proportional (Sutro) weir and linear controlled-flow roof drains as discussed above.

5.4.2.4 Difference Formulation

The most accurate way of performing an event-based analysis of a detention basin or network of detention basins uses accurate formulations of the inflow hydrograph(s), storage–elevation relationship(s), and outflow–elevation relationship(s). Equation 5.4 can be written in different forms for successive time steps as follows:

$$F(y_2)=\frac{S_2(y_2)-S_1}{t_2-t_1}-\frac{I_1+I_2}{2}+\frac{Q_1+Q_2(y_2)}{2}=0 \tag{5.39}$$

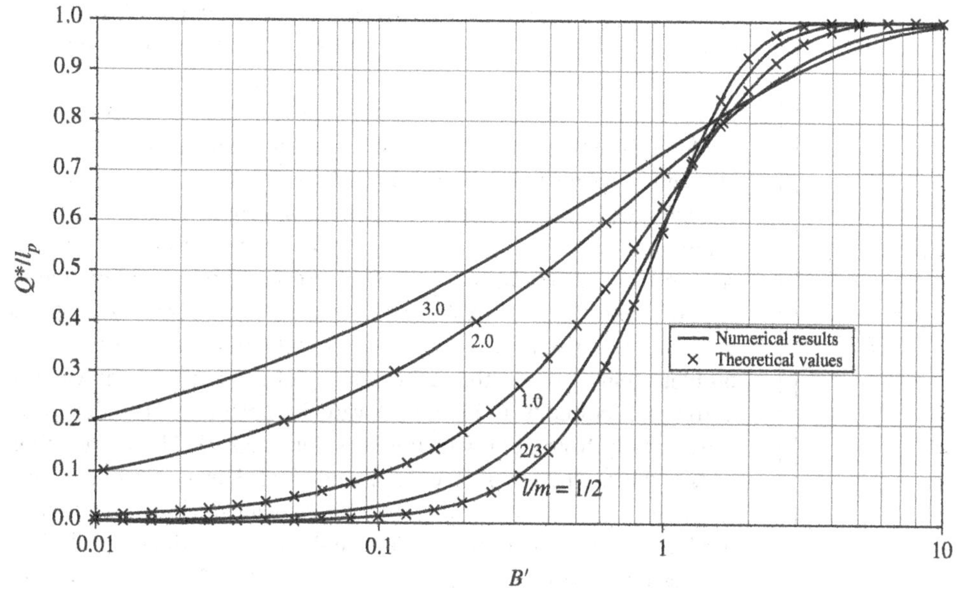

Figure 5.7 Q^*/I_p vs. B' for rectangular inflow hydrograph

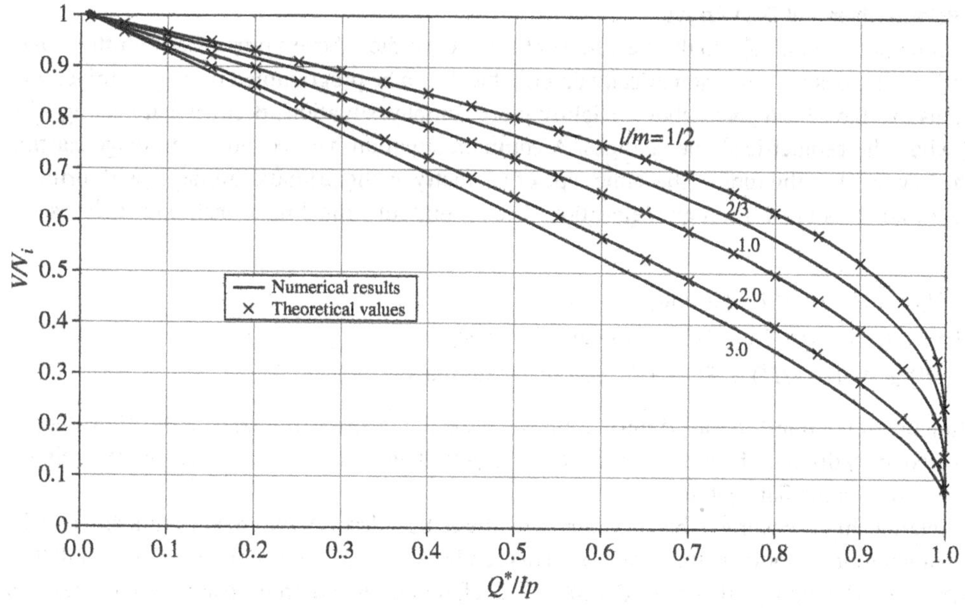

Figure 5.8 V/V_i vs. Q^*/I_p for Rectangular Inflow Hydrograph

in which y_2 = water surface elevation relative to a prescribed datum; I_1 and I_2 = inflow at times t_1 and t_2, respectively; S_1 and Q_1 respectively = storage and outflow calculated at the previous time step t_1; $S_2(y_2)$ = storage at y_2 from the known storage-elevation relationship; $Q_2(y_2)$ = outflow at y_2 from the known outflow–elevation relationship; and $F(y_2)$ = the function whose root is sought. For combined detention-percolation basins, the term $(p_1 + p_2)/2$ is added after the Q terms, in which p_1 and p_2 are the rates of percolation as functions of water surface elevation. Tailwater conditions can be taken into account in the $Q_2(y_2)$ relationship, as discussed in Graber and Elkerton (1999).

The approximate root of Equation 5.38 is sought numerically by finding the value of y_2 within error ε, i.e., $|y_{2L} - y_{2R}| < \varepsilon$ as shown in Figure 5.9. Figure 5.9's point "a" is a location where a zero slope error would occur using the Newton–Raphson method of numerical integration (Hildebrand 1962, Hornbeck 1975, Gullberg 1997), for which outflow then nearly equals

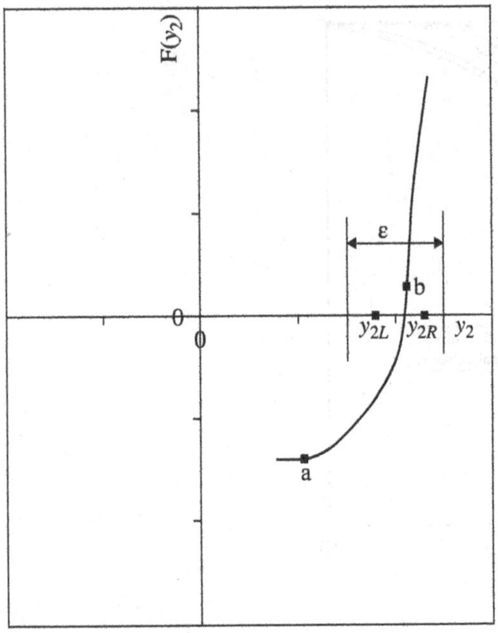

Figure 5.9 $F(y_2)$ vs. y_2

inflow, and in Equation 5.39 $S_1 = S_2 \cong 0$, $Q_1 = I_1$, and $Q_2 = I_2$ so that $dF(y_2)/dy_2 \cong 0$. The Newton–Raphson method is sufficiently robust for multiple basins of the type discussed in Graber and Elkerton (1999) because a zero slope situation at any one of numerous such basins is not fatal. However, fatal zero slope errors can occur with single basins or series of basins with unidirectional flow. Such can occur, for example, with depressed culvert outlets with which appreciable storage only begins above the culvert crown. Therefore, the author uses the bisection method (Hornbeck 1975, Gullberg 1997) or a combination of bisection and Newton–Raphson for case-by-case analyses of the types of basins discussed here. The latter combination addresses another problem that arises in that a steep slope of the curve in the vicinity of the root (point "b" in Figure 5.9) can give a value of $F(y_2)$ departing substantially from zero with the bisection method. The author thus adds a check whereby if $|F(y_2)| > k\varepsilon$, in which k is a prescribed value less than or equal to one, the "root" thus determined by bisection is used as the seed value for the Newton–Raphson method, which then iterates until both $|y_{2L} - y_{2R}| < \varepsilon$ and $|F(y_2)| < k\varepsilon$. The overall methodology yields prompt solutions with current computer speeds. Examples of the use of the methodology are given below.

5.4.2.5 Storage and Outflow Relationships and Data Arrays

The storage–elevation relationship relates the storage to the elevation of the water surface above a prescribed datum (Graber 1999). When the storage–elevation relationship is mathematically describable, it can be programmed into appropriate software. Examples of such relationships have been given above. Slightly more complex situations occur when dealing with secondary roof drainage and when the geometric shape changes. A common situation has secondary drainage starting at an elevation higher than the low point in the roof in the same type of geometry as the primary drainage (with primary drainage clogged), surmounted by a flat roof at a still higher elevation. The general functional relationship in such cases is given by:

$$S = \begin{cases} \Phi(H_p)\ ,\ 0 \le H_p < H' \ \text{(primary drainage)} \\ \Phi(H' + H_s) - \Phi(H')\ ,\ 0 \le H_s < H'' \ \text{(secondary drainage)} \\ \Phi(H' + H'') - \Phi(H') + A_f(H_s - H'')\ ,\ H_s \ge H'' \ \text{(secondary drainage)} \end{cases} \tag{5.40}$$

in which Φ is the geometric functional relationship for primary drainage, H_p is the primary-drainage depth, H' is the primary-drainage depth at which secondary drainage begins, H_s is the secondary-drainage depth, H'' is the secondary-drainage depth to the flat roof elevation, and A_f is the flat roof area.

When the storage–elevation relationship is not amenable to accurate mathematical description, that relationship may be conveniently calculated from known roof contours by interpolation. This can be necessary, for example, when roof deflection due to the water load (referred to as "ponding" in, e.g., ASCE 2006, and references cited therein) results in complex roof surfaces. Alternatively, $\ell = 1$ can be used in Equation 5.5, which would generally be conservative; note from Figure 5.8 that a larger ℓ and, hence, larger ℓ/m gives a smaller $\forall/\forall_i$ for given t_r/t_c and Q^*/I_p.

There is a unique outflow–elevation relationship for a particular outlet geometry. When the outlets are computationally determinable and mathematically simple (some relationships have been given above, and another example is given below), the outflow–elevation relationship can be programmed into appropriate design software. When the outlets are of moderate hydraulic complexity and computationally determinable, an efficient procedure is to initially generate outflow–elevation points. That avoids the computation of a great deal of wasted information in the form of many outflow–elevation points that are not saved for subsequent use. Computationally, the outflow–elevation relationship can then take the form of a two-column 2D array with flow in one column and corresponding basin elevation values in the other. Having set up the array, values of outflow and headwater can be calculated during routing by simple interpolation as part of the solution algorithm (Graber 1999). Interpolation can also be used when the outflow–elevation relationship is based on measured data points.

FMIC (2006) mentions (3.1.1) "backup of storm drainage systems" as a cause of roof failure. If it is necessary to consider such backup as part of the storage analysis, the data array can be set up as described in Graber (1999).

5.4.2.6 Examples

The first of the examples that follow are revisions of those given by ASCE (2006) and FMIC (2006). Sources of each example and computed results are given in Table 5.2. Those references can be consulted for more complete details of roof dimensions, etc. Because only 60-minute rainfalls were provided for the published examples, published Topsfield, Massachusetts, rainfall data (Graber 1992) for different durations (with interpolation for durations between 5-minute intervals) were used for the examples, except as noted. Except as otherwise noted, the examples were analyzed using the difference formulation described above, with error limit $\varepsilon = 0.00000328$, $m = 0.000001$ ft, and time interval $\Delta t = t_2 - t_1 = 0.01$ min. It is important to test successively smaller values of ε and Δt to establish values necessary for appropriate accuracy; as a check on the numerical solutions, it was observed that the time at which the maximum outflow occurred was very nearly equal to the storm duration (at larger values than the ε stated above, that was not always the case) except when equilibrium conditions are achieved (i.e., $Q^* \cong I_p$), in which case maximum outflow can be reached at a time less than the storm duration and persist at that level up to $t = t_r$.

The Example 1 secondary-drainage case in Table 5.2 uses a linear fit to the 102 mm (4 in.) roof drain data for the outflow–elevation relationship; the tabulated drain characteristics [Table C8-1 of ASCE (2006); Table 8 of FMIC 2006] are fit closely over the range of interest by Equation 5.6 with $m = 1$ and $M = 0.211$ L/sec-mm (85 gpm/in.). Table 5.3 gives the calculated results for Example 1, demonstrating the use of successive rectangular hydrographs to ascertain the duration at which the maximum outflow and depth occur. The rainfall intensity i is taken from IDF (intensity-duration-frequency) data for each storm duration for the location and return period of interest. The peak inflow, equal to i times tributary roof area, decreases in proportion to the decrease in rainfall intensity as storm duration increases.

Example 2 has secondary drainage provided by a 305 mm (12 in.) wide channel scupper, for which the drain characteristics are given by the formula in Table 6 of FMIC (2006). The tabulated results for both Examples 1 and 2 show that equilibrium conditions ($Q = I_p$) are achieved at the 60-minute duration and that the depths at the critical durations, including storage attenuation, are greater than the 60-minute equilibrium depths. The secondary-drainage depths are measured above the elevation of the secondary roof drains, which are 51 mm (2 in.) above the roof low points for Examples 1 and 2.

The Example 5a case in Table 5.2 uses a linear fit to the 152 mm (6 in.) roof drain data for the outflow–elevation relationship [see the first line of Equation 5.41 below]. Although the numerical method discussed above was used in the analysis, the results of this example can be compared to the generalized chart in Figure 5.8, which could have been used in this case. From the Table 5.2 results, $Q^*/I_p = 0.323/0.573 = 0.564$ and $\forall/\forall_i = 898.3/[0.573(38)(60)] = 0.688$. With, $\ell = 1$ and $m = 1$, giving $\ell/m = 1$, these ratios can be seen to comport with Figure 5.8. This example can also be compared to the analytical solution given by Equation 5.34, for which $M = 0.236$ L/sec-mm (95 gpm/in.) and $N = 656\ \text{m}^2 = 7056$ sq ft giving $B' = Mt_r/N = 95(38)(12)/[7056(7.48)] = 0.821$, $Q^*/I_p = 1 - \exp(0.821) = 0.560$, and $\forall/\forall_i = -0.560/\ln(1-0.560) = 0.682$; these results are within less than 1% of the corresponding numerical-solution results.

For the Example 5b primary-drainage case, the following more general regression relationship is used for that same 152 mm (6 in.) roof drain:

$$Q = \begin{cases} MH & , \ H \le H_l \\ be^{cH} & , \ H > H_l \end{cases} \tag{5.41}$$

in which $M = 0.236$ L/sec-mm (95 gpm/in.), $b = 2.98$ L/sec (47.2 gpm), and $c = 0.01969\ \text{mm}^{-1}$ (1/2 in.$^{-1}$). The tabulated results for Example 5b show that equilibrium conditions are achieved at the 60-minute duration and that the depths at the critical durations, including storage attenuation, are greater than the 60-minute equilibrium depths. It was necessary to extrapolate the outflow–elevation relationship for the roof drain at the critical duration; in an actual application, larger roof drains should be used that are within the range of the roof drain manufacturer's data. As an additional check of the example, the 9-minute, 100-year rainfall was derived for Dallas (the city to which the example applies) using regional rainfall curves presented by Gilman (1964) and the method given by Froehlich (1993); for the resulting 14.5 in./hr (368 mm/hr) rainfall, the depth was calculated to be 111 mm (4.36 in.), which would result in an unintended overflow of the scuppers set 89 mm (3½ in.) above the roof low point. For the Example 5b, secondary drainage results summarized in Table 5.2, the calculated depths are measured above the (89 mm) 3½-in. depth of the overflow scuppers.

Table 5.2 Summary of Roof Table Examples

	Source of example	Identification	Roof storage geometry	Outflow device	60 min duration			Critical duration				
					I_p (m³/s)	Q (m³/s)	h^* (mm)	t^* (min)	I_p (m³/s)	Q (m³/s)	h^* (mm)	V (m³)
SI units	ASCE (2006)	Example 1: secondary drainage	Inverted obelisk	Roof drain	0.00426	0.00426	20.2	9	0.01172	0.00987	46.7	2.45
	ASCE (2006)	Example 2: secondary drainage	Frustum of sloped triangular trough	Channel scupper	0.01962	0.01962	109.4	12	0.04859	0.03870	172.0	17.22
	FMIC (2006)	Example 5a: primary drainage	Flat rectangular roof	Roof drain	0.01204	0.00884	37.4	38	0.01623	0.00916	38.8	25.44
	FMIC (2006)	Example 5b: primary drainage	Inverted right wedge	Roof drain	0.01204	0.01204	51.0	9	0.03308	0.02767	81.3	8.13
	FMIC (2006)	Example 5b: secondary drainage	Frustum of inverted right wedge	Channel scupper	0.01204	0.01175	48.9	22	0.02234	0.01611	60.5	17.70
	FMIC (2006)	Example 6: primary drainage	Inverted rectangular pyramid	Root dram	0.02274	0.02274	74.2	3	0.07618	0.07420	117.3	3.72
	FMIC (2006)	Example 6: secondary drainage	Inverted obelisk	Roof drain	0.02274	0.02274	74.2	8	0.06485	0.05919	109.2	13.64
	Fairfax County (2007)	Example: secondary drainage	Flat rectangular roof	Clogged detention ring	0.01707	0.01687	21.3	22	0.03165	0.02421	27.1	25.19
	See text	Modified Example 2: primary drainage	Flat rectangular roof	Roof drain	0.01962	0.01076	45.6	67	0.01832	0.01078	45.7	48.82
	See text	Modified Example 2: primary drainage	Rat rectangular roof	Flow-control roof drain	0.01962	0.00436	58.5	132	0.01176	0.00499	67.0	72.81
					I_p (cfs)	Q (cfs)	h^* (in.)	t^* (min)	I_p (cfs)	Q^* (cfs)	h^* (in.)	V (cu ft)
Customary units	ASCE (2006)	Example 1: secondary drainage	Inverted obelisk	Roof drain	0.1506	0.1506	0.795	9	0.414	0.348	1.839	86.4
	ASCE (2006)	Example 2: secondary drainage	Frustum of sloped triangular trough	Channel scupper	0.693	0.693	4.31	12	1.716	1.367	6.77	608.0
	FMIC (2006)	Example 5a: primary drainage	Flat rectangular roof	Roof drain	0.425	0.312	1.474	38	0.573	0.323	1.528	898.3
	FMIC (2006)	Example 5b: primary drainage	Inverted right wedge	Roof drain	0.425	0.425	2.01	9	1.168	0.977	3.20	286.9
	FMIC (2006)	Example 5b: secondary drainage	Frustum of inverted right wedge	Channel scupper	0.425	0.415	1.927	22	0.789	0.569	2.38	625.2
	FMIC (2006)	Example 6: primary drainage	Inverted rectangular pyramid	Roof drain	0.803	0.803	2.92	3	2.69	2.62	4.62	131.3
	FMIC (2006)	Example 6: secondary drainage	Inverted obelisk	Roof drain	0.803	0.803	2.92	8	2.29	2.09	4.30	481.8
	Fairfax County (2007)	Example: secondary drainage	Hat rectangular root	Clogged detention ring	0.603	0.596	0.839	22	1.118	0.855	1.067	889.5
	See text	Modified Example 2: primary drainage	Rat rectangular roof	Roof drain	0.693	0.380	1.796	67	0.647	0.381	1.799	1724.0
	See text	Modified Example 2: primary drainage	Hal rectangular roof	Flow-control roof drain	0.693	0.1539	2.30	132	0.415	0.1762	2.64	2570.9

Note: Depths for secondary drainage are relative to overflow elevations—see text.

The Example 6 cases in Table 5.2 require interpretation of the roof slope: the roof sections are rectangular in plan rather than square, so the single slope of 1/4 in./ft (21 mm/m) given in FMIC (2006) cannot apply in both directions. The stated slope is thus taken as a minimum slope, giving a slope in the other direction of 1/3 in./ft (28 mm/m). The tabulated results show that equilibrium conditions are achieved at the 60-minute duration and that the depths at the critical durations, including storage attenuation, are greater than the 60-minute equilibrium depths. The secondary-drainage depths are measured above the elevation of the secondary roof drains which are 76 mm (3 in.) above the roof low points. This example used the same roof drain formula as was used for Example 5b, primary drainage (see above). It was necessary to extrapolate the outflow–elevation relationship for the roof drain at the critical duration; in an actual application, larger roof drains should be used that are within the range of the roof drain manufacturer's data.

Fairfax County (2007) and City of Manassas (2007) requirements differ from those of ASCE (2006) and FMIC (2006) in requiring that rooftop storage be designed to detain the 10-year, 2-hour storm and that emergency overflow provisions be adequate to discharge the 100-year, 30-minute storm. Those two storms correspond to a rainfall depth of 76 mm (3 inches) (Fairfax County 2007). However, in an example (Fairfax County 2007), overflow depth on a clogged detention ring is inexplicably based on a 5-minute, 100-year storm. In all cases, storage attenuation is neglected. Table 5.2 includes the Fairfax example with attenuation considered, giving the critical-duration values in comparison to 60-minute-duration values. The depths are measured above the 76 mm (3 in.) depth to the overflow elevation of the clogged detention ring. The Fairfax example gives a 51 mm (2 in.) depth above the detention ring for the Fairfax 250 mm/hr (9.84 in./hr) 5-minute, 100-year rainfall; a run was made considering storage attenuation for that rainfall, which gave $Q/I_p = 0.219$ and depth over the detention ring of 18.92 mm (0.745 in.), demonstrating the value of considering storage attenuation. Although the numerical method was used in the analysis, the results of this example can also be compared to the generalized chart in Figure 5.8, which could have been used in this case. From the Table 5.2 results, $Q/I_p = 0.855/1.118 = 0.765$ and $\forall/\forall_i = 889.5/[1.118(22)(60)] = 0.603$. With, $\ell = 1$ and $m = 1.5$, giving $\ell/m = 2/3$, these ratios can be seen to comport with Figure 5.8.

A variation of Example 2 in Table 5.2 is instructive. Consider first that the 2137 m^2 (23,000 sq ft) roof is flat, with each half flowing to the same 152 mm (6 in.) primary drain as in the original example (ASCE 2006). The calculated values for the half-roof primary drainage of modified Example 2 are given in Table 5.2. Now assume that it is desired that the 100-year rate of runoff from the roof not exceed the runoff from the pre-development land with Soil Conservation Service (SCS – now Natural Resources Conservation Service) classifications of forested in fair condition with Hydrologic Soil Group A, for which the Runoff Curve Number $CN = 36$ (McCuen 1982). Using the conversion method of Graber (1992) with a time of concentration $t_c = 5$ minutes and the Topsfield rainfall data (Graber 1992), we get a Rational Method runoff coefficient $C = 0.082$. With the 5-minute, 100-year rainfall intensity $i = 8.2$ in./hr (208 mm/hr) (Table 5.3), the flow objective for each roof half (11,500 sq ft = 1068 m^2) becomes $CiA = 0.082(8.2)(11{,}500)/[12(60)(60)] = 0.1790$ cfs = 80.34 gpm = 0.00507 m^3/sec. For the flat roof, a flow-control roof drain with a linear flow–head relationship is to be used for which the manufacturer gives the flow rate as 10 gpm per notch per inch of head (0.0248 L/sec-mm). By trial and error using the analytical solution, 3 notches are found to give the desired flow attenuation, with the results shown in Table 5.2. By comparison with the variation first discussed in this paragraph, it can be seen that a more restrictive drain can give a lower maximum depth of water. This is because a greater volume of inflow is stored over a longer period of time to peak outflow associated with a longer critical

Table 5.3 Example 1: Secondary-Drainage Data

Tr (min)	100-year i (in./hr)	100-year i (mm/hr)	I_p (m^3/s)	I_p (cfs)	Q^* (m^3/s)	Q^* (cfs)	Q^* (gpm)	Depth (mm)	Depth (in.)
60	2.603	66.12	0.004266	0.1506	0.004266	0.1506	67.6	20.20	0.795
20	5.049	128.2	0.008275	0.2922	0.008158	0.2881	129.3	38.64	1.521
15	5.779	146.8	0.009471	0.3344	0.009027	0.3188	143.1	42.75	1.683
10	6.889	175.0	0.011290	0.3987	0.009810	0.3464	155.5	46.46	1.829
5	8.2	208	0.013439	0.4745	0.009155	0.3233	145.1	43.36	1.707
9	**7.151**	**181.6**	**0.011720**	**0.4138**	**0.009865**	**0.3484**	**156.4**	**46.72**	**1.839**
8	7.413	188.3	0.012150	0.4290	0.009848	0.3478	156.1	46.64	1.836

Note: Emboldened rows give maximum outflow and depth.

duration and a smaller rate of increase of outflow. It is commonly necessary to consider various return periods to assure adequate runoff attenuation.

Tables 5.2 and 5.3 demonstrate the importance of considering storm duration for determination of maximum depth, storage, and outflow. Several authors (Burton 1980, Tourbier and Westmacott 1981, Aron and Kibler 1990) have noted the importance of storm duration more generally in determining the maximum required storage for detention design. Tables 5.2 and 5.3 are also consistent with FMIC's (2006, Section 3.1.1) mention of "rainfall intensities greater than predicted" as a cause of excessive buildup of rainwater leading to collapse. By adding the critical-duration depths in Table 5.2 to the starting depths of secondary roof drains as given above, total depths are obtained: the largest values are for Example 2 secondary drainage at 223 mm (8.77 in.), and Example 6 secondary drainage at 178 mm (7.02 in.). These values are greater than the 145 mm (5.75 in.) total water depth attributed by ASCE (2006), as mentioned above, to some U.S. ordinances. By doubling the 60-minute outflows in Table 5.2, flows are obtained that are less than Q^* in Table 5.2, exemplifying the inadequacy of the doubling method for controlled drainage in the FMIC (2006) standard as mentioned above.

5.4.2.7 Summary and Conclusions

Basic relationships are derived for calculating the flow attenuation and depth of stormwater on roofs. Hyetographs and corresponding hydrographs are specialized for the case of roof drainage and incorporated in various methods that are presented for analysis and design of roof drains. Roof storage and outflow are characterized by various roof geometries and outflow devices. The methods presented for carrying out the integrated analysis and design of roof drainage systems include analytical solutions for certain cases, a difference formulation, generalized numerical solutions, and the use of data arrays. The methods are applied to examples found in the literature, including those in standards of Factory Mutual Insurance Company (FMIC), ASCE, and others. The following assumptions implicit in those standards are tested: (1) that the maximum outflow and, hence, depth of water on the secondary-drainage system with the 100-year rainstorm occurs for a storm of 1 hour or other specified duration; and (2) that an equilibrium condition has been reached such that outflow equals inflow. The limitations of those standards are demonstrated. For accurate results, a critical duration at which the greatest water depth and outflow occur must be found on a case-by-case basis. The critical durations can differ significantly from the fixed durations specified in standards. The corresponding maximum water depths and peak outflows can also differ accordingly. Equilibrium conditions may not be achieved at the critical duration, reflecting the benefit of flow attenuation. Any conservatism in neglecting attenuation (i.e., assuming equilibrium) is offset by failure to consider the critical storm duration. For determination of roof loads, these results will have their greatest importance in warmer climates in which the rain loads govern over snow loads. An example of design specifically for flow attenuation is also presented.

6

Incompressible Flow in Pressure Conduits

6.1 Introduction

This chapter deals with the flow of fluids in pressure conduits under conditions in which the fluid may be regarded as incompressible (constant density). This includes liquids under conditions of steady (temporally constant) flow and, under certain circumstances, time-varying liquid flows, and the flow of gases. Chapter 9 addresses compressible fluid flow, and the conditions under which gas flows may be approximated as incompressible.

The present chapter is limited to spatially-constant flows, which are considered to include branching flows; branching flows may be regarded as spatially constant in the sense that the flows are constant in the pipe components immediately upstream and downstream of the junction, a distinction which should become more apparent in Chapters 11 and 12. Distribution conduits and collection conduits are considered in Chapters 11 and 12, respectively. The initial sections of the present chapter consider only steady flows, that is, flows which are (or may be assumed to be) temporally constant. A later section addresses time-varying liquid flows under conditions in which compressibility (elasticity) effects can be neglected. Time-varying liquid flows in which elasticity effects are important are considered in Section 8.6 Waterhammer.

The present chapter is further limited in scope to fluids having low enough solids or sediment concentration that the fluid flow properties remain essentially those of the base fluid (e.g., the density, viscosity, and other properties of municipal wastewater are essentially those of clean water). Chapter 13 will address the flow of liquids having high solids content, including sludges.

6.2 Steady Flow in Pressure Conduits

Of prime concern in steady-flow pressure, conduits are the associated loss of energy due to friction. Chapter 5 introduced that subject in the context of dimensional analysis, the Darcy–Weisbach friction factor, and the Moody Diagram (Figure 5.1). That subject will be pursued further, with the discussion of additional friction factor relations. However, to emphasize the application of such relations to *developed flows* and provide perspective as to their accuracy in practical piping systems, we shall discuss them under the heading of "DEVELOPED FLOW" after a subsection entitled "FLOW DEVELOPED". Following that, certain problems associated with high-velocity pipe flows, namely erosive velocities and air binding, will be addressed.

6.2.1 Flow Development

Stenning and Shearer (1960, pp. 59–61) provided relationships for the distance to fully-developed flow in a circular pipe. Their analysis is based on water flowing into a conduit from a large chamber such that the velocity distribution at the entrance cross section is nearly radially constant. The strong shearing action between the pipe wall and the fluid at the entrance causes the formation of a boundary layer leading to a fully-developed velocity profile with zero velocity at the wall. For laminar flow in the pipe, the distance required to attain fully-developed flow (parabolic profile) is given by $\ell = 0.058\,D\mathbf{R}$ in which D is pipe diameter and Reynolds Number $\mathbf{R} = \rho VD/\mu$ in which ρ is fluid density, V is cross-sectional bulk velocity, and μ is dynamic viscosity. For turbulent flow in the pipe, we can derive from Stenning and Shearer the equation

Hydraulics and Pneumatics in Environmental Engineering, First Edition. S. David Graber.

$\ell = 1.428D\mathbf{R}^{1/4}$ as the distance to attain fully-developed flow. As a comparative example using these two equations, at a Reynolds Number of 2000 fully-developed flow is attained in about 115 diameters with laminar flow and about 10 diameters with turbulent flow.

Richards and Graber (1965) provided a related discussion of flow development for one-dimensional flow in a rectangular channel (i.e., large depth-to-width ratio) such as used in fluid jet amplifiers; that discussion is based on the work of Schlichting (1960, p. 168) and Sparrow (1955). Fluid jet amplifiers are also discussed in Section 15.2 in the context of fluidic devices used in environmental engineering and for other purposes. The related topic of approach conditions for differential pressure flow meters is discussed in Section 6.3.

6.2.2 Developed Flow

The Darcy–Weisbach equation was discussed in Section 5.3. Additional relationships are discussed below.

Hazen–Williams Equation. The following equation, applicable to full-pipe flow, is applied principally for water transmission (Fair and Geyer 1954, §12-2):

$$Q = 405Cd^{2.63}S^{0.54} \tag{6.1}$$

in which Q is flow rate in gpd, C is the Hazen–Williams coefficient, d is pipe internal diameter in inches, and S is friction slope or slope of the energy grade line equal to head loss h (ft) divided by pipe length L (ft). The above equation can be rearranged and expressed in terms of the head loss:

$$h = \frac{10.50LQ^{1.835}/d^{4.87}}{C^{1.835}} \tag{6.2}$$

Although the Hazen–Williams Equation is used herein under certain appropriate conditions, its range of applicability is limited and the Darcy–Weisbach equation (see Section 5.3) is much preferred due to its sound theoretical basis and extensive database (Liou 1998).

Example 6.1

Calculate the maximum and minimum head losses based on minimum and maximum C values of 100 and 130, respectively, for 12-in. I.D. water pipe with $L = 285$ ft and $Q = 1300$ gpm (Admiralty Hill, Chelsea, Massachusetts).

Solution

Using the above equation with $C = 100$, $Q = 1300\text{ gpm} \times 60 \times 24 = 1{,}872{,}000$ gpd, and other terms as given above:

$$h = \frac{10.50(285)(1,872,000^{1.835})}{(12^{4.87})(100^{1.835})} = 1.839\text{ ft}$$

For $C = 130$ and other terms unchanged,

$$h = \frac{10.50(285)(1,872,000^{1.835})}{(12^{4.87})(130^{1.835})} = 1.136\text{ ft}$$

Manning's Equation. The following equation was proposed by Manning in 1889 for open-channel flows (Yen 1992a, 1992b); it has since been applied successfully to pressure conduits:

$$Q = \frac{K_n A\, R^{2/3} S_f^{1/2}}{n} \tag{6.3}$$

in which $K_n = 1.486$ (customary units) or 1.0 (metric units); Q = flow rate in cubic feet per second (cubic meters per second); A = cross-sectional area of the water stream in square feet (square meters); R = hydraulic radius $R = A/WP$, in feet (meters); S_f = friction slope in feet per foot (meters per meters) (which equals the slope of the drain only for steady uniform flow); WP = wetted perimeter, which is the length of the periphery of the cross-sectional shape of the liquid in contact with the pipe wall surface in feet (meters); and n = Manning's roughness coefficient for the drain product, such as given in Table 6.1 (from Graber 2017a, Table 6-1).

Table 6.1 Recommended Design Values of Manning Roughness Coefficients for Closed Conduits and Open Channels

Conduit Material	Manning's n
Asbestos-cement pipe	0.011–0.015
Brick	0.013–0.017
Cast iron pipe	
Cement-lined and seal-coated	0.011–0.015
Concrete (monolithic)	
Smooth forms	0.012–0.014
Rough forms	0.015–0.017
Concrete pipe	0.011–0.015
Corrugated-metal pipe	
Plain annular	0.022–0.027
Plain helical	0.011–0.023
Paved invert	0.018–0.022
Spun asphalt lined	0.011–0.015
Spiral rib metal pipe (smooth)	0.012–0.015
Ductile iron pipe (cement-lined)	0.011–0.014
Plastic pipe (corrugated)	
3–8 in. (75–200 mm) diameter	0.014–0.016
10–12 in. (250–300 mm) diameter	0.016–0.018
Larger than 12 in. (300 mm) diameter	0.019–0.021
Plastic pipe (smooth interior)	0.010–0.013
Vitrified clay	
Pipes	0.011–0.015
Liner plates	0.013–0.017

Another pertinent relationship is that between the Darcy–Weisbach friction factor and Manning's n:

$$f = \frac{8g}{K_n^2}\frac{n^2}{R^{1/3}} \tag{6.4a}$$

$$n = \left(\frac{fR^{1/3}}{8g/K_n^2}\right)^{1/2} \tag{6.4b}$$

in which, following Yen (1992a, 1992b), $K_n = 1$ for SI units (length terms in meters, time in seconds) and $K_n = 1.486$ for customary units (length terms in feet, time in seconds). For SI units $8g/K_n^2 = 78.5$, and for customary units $8g/K_n^2 = 117$.

Because relative roughness changes as flow depth changes, the Darcy–Weisbach friction factor (which thus also changes with flow depth) is not as convenient for open-channel flow as it is for full-conduit flow. Robert Manning's contribution was to derive the equation named after him with a coefficient (Manning's n) which changes little with flow depth. Since at large Reynold's number f is a function only of relative roughness within the range of the Moody Diagram, over that same range Equation 6.4b infers that Manning's n varies with relative roughness and hydraulic radius. For a given roughness height ε, as R increases the relative roughness $\varepsilon/(4R)$ decreases, causing (referring to the Moody Diagram) f to decrease. Thus, as the term $R^{1/3}$ increases in Equation 6.4b, that increase is offset by the concurrent decrease in f. The effect is to keep n relatively independent of hydraulic radius. This can be shown by an example. For a roughness height ε of 0.01 foot, the relative roughness can be calculated as R varies using Equation 5.2 (for large Reynold's numbers) to determine the corresponding values of f.

Equation 6.4b then gives the corresponding values of Manning's n. The calculations show that n varies by less than ±3% about an average over a tenfold increase in R from 0.2 to 2 ft. (Manning was a genius!) However, at the upper end of the Moody Diagram, for a roughness height ε of 0.1 foot, the calculations show that n varies by about ±8% over R ranging from 0.2 to 2.

An issue that arises concerns the effect of conduit size on Manning's n. Numerous publications have suggested that there is a significant effect requiring consideration for large conduits, as implied by titles such as "FACTORS INFLUENCING FLOW IN LARGE CONDUITS" (Committee on Hydraulic Structures 1965). Eriksen (1974) addressed this concern, from which the following example is taken.

Example 6.2

Determine Manning's n for a 9-ft diameter sewer assuming that the surface irregularities for the 9-ft diameter and a 3-ft diameter reference sewer are identical. Assume $n = 0.013$ and velocity $V = 4$ ft/sec.

Solution

We first determine parameters for the 3-ft sewer. From Equation 6.4a, the Darcy–Weisbach friction factor with hydraulic radius $R = D/4 = 3/4 = 0.75$ is:

$$f = 8(32.2)(0.013)^2/\left[(1.486)^2(0.75)^{(1/3)}\right] = 0.0217$$

The corresponding Reynold's Number is $\mathbf{R} = VD/\nu = 4(3)/10^{-5} = 1.2(10)^6$. The relative roughness is then determined from the Moody Diagram (Figure 5.1 or Equations 5.2) as $\varepsilon/D \cong 0.0013$, from which $\varepsilon = 0.0013(3) = 0.0039$ ft. For the 9-ft sewer then we have $\varepsilon/D \cong 0.0039/9 = 0.000433$ and $\mathbf{R} = 9(4)/10^{-5} = 3.6(10)^6$, from which $f = 0.0162$. From Equation 6.4a and using subscripts to denote the two pipe sizes:

$$\left(\frac{n_9}{n_3}\right)^2 = \frac{f_9 D_9^{1/3}}{f_3 D_3^{1/3}} = \frac{0.0162(9)^{1/3}}{0.0217(3)^{1/3}} = 1.077$$

and then $n_9/n_3 = \sqrt{1.077} = 1.038$ and $n_9 = 0.013(1.038) = 0.0135$. Thus the value of n for the for the 9-ft diameter pipe is only $[(0.0135–0.013)/0.013]\bullet 100 = 3.85\%$ higher than for the 3-ft diameter pipe, an insignificant difference.

6.2.3 Erosive Velocities

Knowledge of erosive velocities is derived from experience. For clear water in "hard-surfaced" conduits, the limiting velocity is very high, with velocities greater than 40 fps (12 m/sec) having been found harmless in, for example, concrete channels. However, when sand or other gritty materials are conveyed, erosion of inverts may result from much lower velocities. In sanitary sewers where flow is continuous, the limiting velocity is commonly taken to be about 10 ft/sec (ASCE/WPCF 1969, pp. 129–130). Recommended maximum velocities for storm sewers, such as determined in Cincinnati, Ohio, are 12 ft/sec with no special measures required; for sewers with velocities greater than 12 ft/sec but less than or equal to 20 ft/sec, the subject sewer must be reinforced concrete pipe or ductile iron pipe. These recommendations should be applicable to free-surface flow conditions as well as to full flow in closed conduits.

6.2.4 Air Binding and Gas Pockets

"Air binding" refers to the presence of stabilized air pockets in liquid-conveying pipelines which can seriously reduce the capacity of the pipe, even to the extent of entirely preventing flow. Wisner *et al.* (1975) did an extensive literature review and conducted experiments on air binding in water lines. The following taken from their Abstract succinctly states most of their findings: "Air is introduced into water transmission lines through vortices at intakes, pumps, vents, and other causes. This results in loss of capacity, surges and blow backs, difficulties in filtering operations, corrosion of lines, and reduction of pump efficiency to name a few. The best solution to air problems is to prevent the introduction of air at the intake. Other means of air removal are air relief valves and by the inertia of the flowing water. No conventional similitude exist to study the time of clearance of isolated air pockets and results of Froudian scale model tests are very conservative." In essence, they found predictive methods for the removal of air pockets to be inadequate and that prevention of air introduction and the use of air relief valves were most reliable.

The use of the hydraulic jump to remove air pockets from water-supply lines and thus prevent air locking is discussed in Section 7.5. Also in Section 7.5 is the discussion of hydraulic jumps in sewers (including free-surface situations), where it is noted that it may be necessary to take air entrainment into account insofar as it affects both the density and volumetric rate of flow. Also, per Kalinske and Robertson (1943) cited in that Section, air entrainment is a byproduct of the hydraulic jumps and cross waves which can accompany supercritical flows; the entrainment of air may result in special requirements for venting of sewers to prevent increased resistance to flow due to formation of air pockets, bulking, and other phenomena. Section 7.7 discusses the formation and removal of air pockets in Type II culvert flow. For certain stilling basins (e.g., Bureau of Reclamation Basin VI discussed in Section 7.8), provision of air vents is important when the approach flow is supercritical to prevent harmful pressure fluctuations. The role of air-release valves in connection with waterhammer control is discussed in Section 8.6.

Gas pockets can form in sewage force mains and cause the exposure of pipe crowns to hydrogen sulfide and attendant corrosion. Walski *et al.* (1994) note that such gas pockets can exist because: (1) air may be pulled into the pipe when the hydraulic grade line falls below the pipe crown, and (2) gas pockets cannot be transported downstream due to the downward slope of the pipe and the buoyancy of the gas pocket. Those conditions are preferably avoided. When unavoidable, methods presented by Walski *et al.* help identify when such conditions may occur and present design (air-release and check valves) and operational steps to prevent force main failure due to the attendant sulfide corrosion. Section 8.6 discusses the use of such valves and their application by the author.

Kohler and Ball (1969, pp. 22–75 to 77) provide an important discussion of air vents which: (1) "act primarily as 'breather' lines to vent or admit air during filling or draining a pipe, and (2) ... deliver a continuous supply of air to a discharging gate or valve."

6.3 Simple Changes in Cross Section

In this section, we will concentrate on cross-sectional changes in which the geometry is relatively simple and reasonably amenable to theoretical hydraulic analysis. Sudden and gradual contractions and expansions, orifices, and venturi tubes are among the configurations considered. The theoretical analyses should provide tools for dealing with these practical appurtenances plus provide insight helpful to a qualitative understanding of more complex configurations considered subsequently.

6.3.1 Sudden Contractions and Expansions: Orifices

Among the types of sudden changes in cross section are sudden contractions, sudden expansions, and pipe orifices. In order to obtain the greatest useful information with the briefest derivation, we will combine these configurations by considering a concentric pipe orifice in a closed conduit having different concentric, prismatic sections upstream and downstream (see Figure 6.1). The results will then be specialized to various practical configurations.

Disregarding initially the depiction at section o, the flow of Figure 6.1 accelerates from Section 1 to Section c and decelerates from Section c to Section 2. The flow is unable to negotiate the abrupt changes in geometry and separates at planes a and b. Section 1 is taken a short distance upstream of plane a, while section c is taken at the plane of minimum flow cross section. Circulating eddies occur between planes a and b and b and d. Due to the assumed concentricity of the orifice, each eddy zone is annular, so that the flow is axisymmetric.

The plane of minimum jet (main flow) cross section is called the *vena contracta*. At this section, the streamlines are parallel and the pressure is nearly uniform over the full cross section within the conduit (including the eddy region). The energy

Figure 6.1 Depiction of Sudden Contraction, Sudden Expansion, and Concentric Pipe Orifice

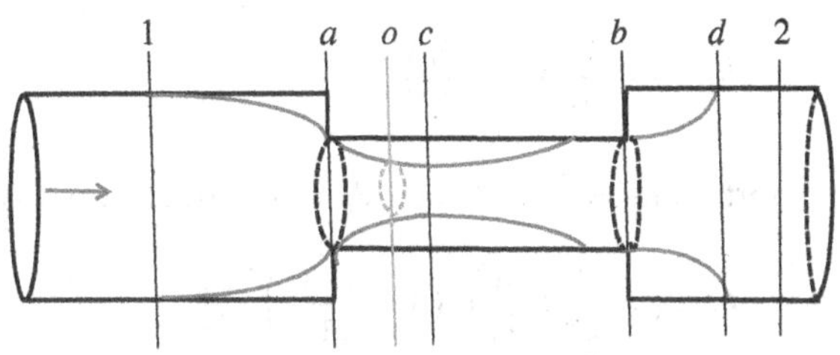

loss in the accelerating flow from station 1 to station c is small, and it is conventional and convenient to momentarily neglect that energy loss. Then the one-dimensional energy equation (from Section 2.6) may be applied to the control volume bounded by stations 1 and c and the conduit boundaries:

$$\frac{p_1}{\rho} + \frac{V_1^2}{2g} = \frac{p_c}{\rho} + \frac{V_c^2}{2g} \tag{6.5}$$

in which V_c is the jet velocity at section c. The jet velocity V_c can be related to the *nominal* orifice velocity (defined by $V_o = Q/A_o$ where A_o is the orifice area) by a contraction coefficient defined by $C_c = A_c/A_o$, where A_c is the cross-sectional area of the vena contracta. Thus, by continuity and the definitions of C_c and V_o:

$$Q = V_1A_1 = V_oA_o = V_cA_c = V_cC_cA_o \tag{6.6}$$

With the above equation, Equation 6.5 can be expressed as:

$$\frac{p_1}{\rho} + \left(\frac{C_cA_o}{A_1}\right)^2 \frac{V_c^2}{2g} = \frac{p_c}{\rho} + \frac{V_c^2}{2g} \tag{6.7}$$

The differential pressure can therefore be expressed in terms of the vena contracta velocity head as follows:

$$\frac{p_1 - p_c}{\rho} = \left[1 - \left(\frac{C_cA_o}{A_1}\right)^2\right] \frac{V_c^2}{2g} \tag{6.8}$$

or in terms of the upstream velocity head by substituting $V_c = V_1A_1/(C_cA_o)$ from Equation 6.6 into Equation 6.8 to obtain:

$$\frac{p_1 - p_c}{\rho} = \left[\left(\frac{A_1}{C_cA_o}\right)^2 - 1\right] \frac{V_1^2}{2g} \tag{6.9}$$

The use of a flow coefficient to correct for the relatively small effect of friction between sections 1 and c is discussed later in this section, in conjunction with flow metering.

Energy losses are significant in the decelerating (expanding) flow between sections c and 2 (again, understood to have no expansion between b and 2). Here the one-dimensional momentum equation (from Section 2.4) may be applied to the control volume bounded by sections c and 2 and the conduit boundaries, giving:

$$p_cA_2 - p_2A_2 = \frac{Q\rho}{g}(V_2 - V_c) \tag{6.10}$$

(It is assumed that the net momentum flux in the eddy zone is negligible.) Using Equation 6.6, the above equation can be rewritten as:

$$\frac{p_c - p_2}{\gamma} = \frac{\rho V_c^2}{2g}\left[2\frac{A_c}{A_2}\left(\frac{A_c}{A_2} - 1\right)\right] \tag{6.11}$$

Application of the energy equation between sections c and 2 gives:

$$\frac{p_c}{\rho} + \frac{V_c^2}{2g} = \frac{p_2}{\rho} + \frac{V_2^2}{2g} + h_{l,c-2} \tag{6.12}$$

Elimination of the pressure terms between the above two equations results in the following relationship for the energy loss per unit weight of fluid:

$$h_{l,c-2} = \frac{V_c^2}{2g}\left(1 - \frac{A_c}{A_2}\right)^2 \tag{6.13}$$

Substituting $A_c = C_cA_o$ and $V_c = A_1V_1/C_cA_o$ (from Equation 6.6) into the above results in:

$$h_{l,c-2} = \frac{1}{C_c^2}\left(\frac{A_1}{A_o}\right)^2\left(1 - \frac{C_cA_o}{A_2}\right)^2 \frac{V_1^2}{2g} \tag{6.14}$$

Equations 6.11 and 6.14 are the principle relationships stemming from the above development. The former gives the pressure differential that can be used for differential pressure-flow metering by appropriate specializations and modifications of the geometry of Figure 6.1. The latter gives the energy loss resulting from the use of the general geometry of Figure 6.1 for the various practical purposes discussed below.

6.3.2 Expansion Transitions

Expansion transitions are used in going from smaller to larger conduit sections. The sudden expansion or increaser is sometimes used when transition length is limited, cost of the transition is to be minimized, and/or head loss is relatively unimportant. Comparing geometries of Figure 6.1 for $A_1 = A_o$ and $C_c = 1$, Equation 6.14 specializes to:

$$h_l = \left(1 - \frac{A_1}{A_2}\right)^2 \frac{V_1^2}{2g} \tag{6.15a}$$

Using continuity, this may also be expressed in the following forms:

$$h_l = \left(\frac{V_1 - V_2}{2g}\right)^2 \tag{6.15b}$$

$$h_l = K\frac{V_1^2}{2g}, K = \left[1 - \left(\frac{D_1}{D_2}\right)^2\right]^2 \tag{6.15c}$$

in which V_1 and V_2 are the upstream and downstream velocities, respectively. Equation 6.15c is obtained by applying continuity (conservation of mass) to the upstream and downstream sections respectively; D_1 and D_2 are the diameters of the smaller and larger pipes, respectively; and V_1 is the velocity in the smaller pipe. Since $D_1 < D_2$, the value of K will be less than one.

6.3.3 Contraction Transitions

Contraction transitions are used to go from larger to smaller conduit sections. The sudden contraction (see Figure 6.1) is sometimes used when transition length is limited, cost of the transition is to be minimized, and/or head loss is relatively unimportant. Comparing geometries of In this case, $A_o = A_2$ and Equation 6.14 specializes to:

$$h_l = \left(\frac{A_1}{A_2}\right)^2 \left(\frac{1}{C_c^2} - 1\right) \frac{V_1^2}{2g} \tag{6.16a}$$

or, in terms of the downstream velocity V_2:

$$h_l = K\frac{V_2^2}{2g}, \quad K = \left(\frac{1}{C_c} - 1\right)^2 \tag{6.16b}$$

The contraction coefficient C_c such as tabulated by Streeter (1962, pp. 222–224) is a function of the downstream-to-upstream pipe area. The author has fit that tabulation with the following equation:

$$C_c = 0.601556 + 0.362452\left(\frac{D_2}{D_1}\right)^4 \tag{6.17}$$

In this case, D_2 and D_1 are the diameters of the smaller and larger pipes, respectively and V_2 is the velocity in the smaller pipe. The value of K will again always be less than one.

Using the above formulas, the increaser and decreaser K values should be consistently assigned to the smaller diameter section irrespective of flow direction. Figure 6.2 plots the expansion and contraction coefficients as functions of the ratio of smaller to larger diameter (D_s/D_l).

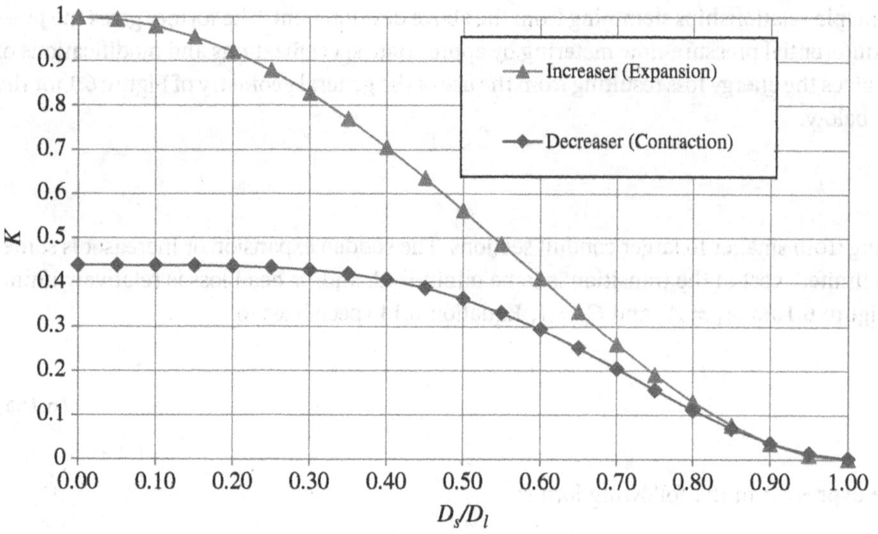

Figure 6.2 Expansion and Contraction Coefficients

6.3.4 Orifice Ports

Ports (including sluice gates) are commonly used to introduce flow from a conduit to a tank. If the downstream cross-sectional area is large compared to A_o, then Equation 6.13 reduces to:

$$h_l = \frac{V_c^2}{2g} \tag{6.18}$$

Thus the full velocity head of the *contracted* jet is dissipated. In terms of the nominal velocity ($V_o = Q/A_o$), the above relationship becomes:

$$h_l = K\frac{V_o^2}{2g} = K\frac{(Q/A_o)^2}{2g}, \quad K = (1/C_c)^2 \tag{6.19}$$

If the upstream cross-sectional area is also large compared to A_o, then $C_c \cong 0.6$ (see Equation 6.17), and $K \cong 2.78$. The effect of C_c on the head loss is thus quite significant. The author knows of more than one case in which the design head loss has been taken as $V_o^2/(2g)$ when it should have been taken as 2.78 times that value.

When upstream and downstream velocities are negligible compared to V_o, Equation 6.19 remains valid even though the flow through the orifice is not aligned with upstream and downstream flows.

The head loss through orifice ports can be reduced by increasing A_o or by smoothing the port entrance to bring C_c closer to 1. The methods for doing the latter are similar to the methods discussed above for contraction transitions. However, it is often desired to maintain a certain minimal head loss for purposes of distribution uniformity (see Chapter 11).

It is emphasized that the orifice or gate must be fully submerged for the above relations to apply. Otherwise, the vertical pressure distribution must be considered as discussed in Chapter 7.

6.3.5 Differential Pressure-Flow Meters

Fluid flow meters have two distinct parts: a primary element and a secondary element. The primary element is in direct contact with the fluid, with which it interacts in some way. The secondary element translates the interaction between fluid and primary element into rates of flow and indicates or records the result and/or provides other required "signals" (such as for pacing chemical feed equipment). Our concern here will be restricted to primary elements of the differential pressure type used to measure the flow of liquids in closed conduits. Section 6.5 will address certain positive-displacement flow meters, Chapter 7 will address open-channel flow measuring devices, Section 9.4 will consider flow metering of compressible fluids (gases), and Section 11.5 will discuss vortex flow meters.

With differential pressure meters the stream of fluid creates a pressure differential as it flows through the primary element (ASME 1971). The magnitude of the pressure difference depends on the velocity and density of the fluid and characteristics of the primary element. Common examples of such primary elements are the venturi tube, flow nozzle, and orifice plate. There are, of course, other categories of flow meters (see, e.g., ASME 1971), but here we will limit ourselves to those of the differential pressure type because: (1) such devices are by far the most widely used in environmental engineering applications, and (2) they require an understanding of the *hydraulic* principles for their *application* (as opposed to, for example, rotameters, nutating-disk meters, and magnetic meters). A good discussion of the history and other details of differential measuring devices and instruments is provided by Babcock (1966, Part 1); also WPCF (1978). Part 2 of Babcock discusses magnetic, ultrasonic, and propeller-type flow meters, as does WPCF (1978); the latter also discusses rotameters and the Dall tube variation on venturi tubes. More on the Dall flow tube is given by Jorinsen (1956). The author specified two 16-in. Dall tubes for the Tallman Island WWTP, New York City, for metering of process air with operating flow range of 1000–10,000 scfm.

The basic relationship between differential pressure and flow is obtained by setting $V_c = Q/A_c = Q/(C_cA_o)$ in Equation 6.8 and solving for Q. For accurate metering purposes, a flow adjustment factor (which we will call C_Q) is applied to the right-hand side of the equation to account for the small friction losses that were neglected in deriving Equation 6.8. In practice, this factor is incorporated into a *discharge coefficient* $C = C_QC_c$ so that:

$$Q = \frac{CA_o}{\sqrt{1-(C_cA_o/A_1)^2}}\sqrt{2g(p_1-p_c)/\rho} \tag{6.20}$$

The flow is proportional to the square root of the difference in pressure between sections 1 and c. The application of this relationship to venturi tubes, flow nozzles, and orifice plates is discussed below.

6.3.5.1 Venturi Tubes and Flow Nozzles

With venturi tubes and flow nozzles, the inlet up to the throat (section o) is shaped to: (1) effectively eliminate the contraction due to flow separation, and (2) condition the incoming flow to reduce the effects of upstream variations. Thus C_c is unity and Equation 6.20 becomes:

$$Q = \frac{CA_o}{\sqrt{1-(A_o/A_1)^2}}\sqrt{2g(p_1-p_c)/\rho} \tag{6.21}$$

in which p_c is now the pressure at the throat of the venturi tube or flow nozzle. Letting d and D represent the diameter of the venturi or nozzle throat and the pipe, respectively, and defining $\beta = d/D$, we can write the above equation as:

$$Q = \frac{CA_o}{\sqrt{1-\beta^4}}\sqrt{2g(p_1-p_c)/\rho} \tag{6.22}$$

An alternative form of the above equation encountered in flow metering practice is:

$$Q = K_fA_o\sqrt{2g(p_1-p_c)/\rho} \tag{6.23a}$$

Recalling that C is the *discharge coefficient* as defined above, the term $K_f = C/\sqrt{1-\beta^4}$ is called the *flow coefficient*. (For which the notation used here is to distinguish that coefficient from K used for form losses as discussed next.)

The head loss for venturis with flow in the normal direction is proportional to the upstream velocity head. Knowing the maximum rated flow Q_R and the head loss h_{LR} at that flow, the equivalent K value can be calculated in terms of those values and the upstream pipe diameter D (this is elaborated upon in Section 9.4):

$$K = \frac{h_{LR}}{V^2/(2g)} = \frac{h_{LR}}{Q_R^2/(2gA^2)} = 2gh_{LR}\left(\frac{\pi}{4}D^2\right)/Q_R^2 \tag{6.23b}$$

For reverse flow through a venturi meter, a conservative approach is to assume sudden expansion from the throat section to the full-pipe section. That can be done by using a modified version of the equation given above for increasers (Equation 6.15c). That modification is derived by noting that the velocity in the increaser equation now becomes the throat velocity which can be expressed in terms of the pipe velocity using continuity. The resulting equation is:

$$H_L = K\frac{V_1^2}{2g}, K = \left(\frac{D_1}{D_T}\right)^4\left[1-\left(\frac{D_T}{D_1}\right)^2\right]^2 \tag{6.23c}$$

in which D_T is the throat diameter and subscript "*1*" refers to the pipe.

An excellent reference discussing the history and additional details about venturi meters is by Briggs and Zimmerman (1999).

6.3.5.2 Pipe Orifices

Equation 6.20 gives the flow equation for orifice meters. According to theory, the pressure taps should be located at a section just upstream of the orifice and at the vena contracta on the downstream side. However, the taps may be located at other convenient locations, with the discharge coefficient C adjusted accordingly (see below). The dominant portion of the head loss across the orifice is obtained by setting $A_2 = A_o$ in Equation 6.15a and applying the contraction coefficient C_c, yielding:

$$h_l \cong \frac{1}{C_c^2}\left(\frac{A_1}{A_o} - 1\right)^2 \frac{V_1^2}{2g} \tag{6.24}$$

The approximate ratio of head loss to pressure differential is obtained by dividing h_l in the above equation by $(p_1 - p_2)/\rho$ as given by Equation 6.9, to obtain:

$$\frac{h_l}{(p_1 - p_c)/\rho} \cong \frac{(A_1/A_o - 1)^2}{(A_1/A_o)^2 - C_c^2} \tag{6.25}$$

Orifices create a relatively large energy loss and are therefore not used frequently for permanent flow metering. However, they are very convenient for temporary test purposes, as they are inexpensive, easily made in local machine shops, and can often be readily inserted between couplings of existing piping. The author has found orifices to be most convenient in this connection in air lines, and they are discussed in that context in some detail in Chapter 9. Alternative pressure tap locations and other practical details are included in the Chapter 9 discussion, and the relations presented there can be easily specialized for incompressible flow situations.

The author specified numerous venturi tubes or insert nozzles for multiple purposes at the Tallman Island facility. A total of eighteen units were applied to metering systems for raw sewage, activated sludge, excess activated sludge, mixed liquor, mixed flow, primary sludge, and return sludge. Capacities ranged from 1.5 to 450 mgd.

Although venturi meters for compressible gas flows are discussed in Section 9.4, ones applied by the author for Tallman Island are for low-pressure situations in which compressibility effects are negligible; they are therefore discussed here. Eight 4-in. venturi insert nozzles (held in place between pipe flanges) were applied to meter process air to air lift pumps, each with an operating flow range of 80–500 scfm with maximum differentials not to exceed nor be significantly less than 60 in. of water.

6.3.5.3 Approach and Following Conditions

Meter accuracy requires suitable approach (upstream) conditions and, in some cases, following (downstream) conditions. These are expressed as distances in terms of diameters of straight pipe required (i.e., length of pipe divided by pipe internal diameter) vs. (in the case of orifices) the ratio of orifice diameter to pipe internal diameter. For orifices, the distances are dependent on the type of preceding fittings (e.g., a single elbow or two elbows or bends and whether the elbows or bends are in the same plane). For some meters, for example, venturi tubes, only upstream conditions pertain; conditions downstream from the venturi have practically no effect on its performance. Straightening vanes can be installed upstream of meters to reduce helical flow and thus decrease the approach conditions and, if applicable, decrease the following conditions. These conditions are given by meter manufacturers, such as BIF®, Akron, Ohio. Howe (1950, p. 204) recommends 25 diameters of straight pipe upstream of orifice meters, which would be conservative per BIF's charts for all except with upstream elbows or bends not in the same plane when the diameter ratio exceeds 0.68. Lefkowitz (1958, pp. 16–20) recommends the same upstream 25 diameters and 5 downstream diameters, the latter being conservative for all conditions discussed above.

6.4 Branching Flows

Branches at which flows combine or divide in piping systems are commonly encountered in environmental engineering practice. Tees, laterals, and wyes are common and are available as standard fittings. Other configurations are also found useful, such as may be fabricated of sheet steel. Similar pressure-flow configurations may be found at junctions in surcharged wastewater and stormwater collection systems. It is often necessary to determine the head losses across such branches. Test data are available for some configurations and are commonly expressed as a K-coefficient times a velocity head in one of the lines. The K-coefficient is a function of geometry, flow ratios, etc. Test data are not available for all configurations of interest, and analytical methods may then be helpful. In any event, the analytical methods are helpful in understanding the phenomena involved and may be of value in extrapolating to untested configurations.

6.4.1 Tees

Isolated tees in pipes with straight upstream and downstream runs were studied in a classical paper by McNown (1954). His comprehensive experimental study, together with data from tests run in Munich, Germany which he included, are given in Figure 6.3 for combining flow and Figure 6.4 for dividing flow. The subscripts "u" and "d" respectively denote the upstream and downstream sides of the conduit (the run of the tee), and the subscript "ℓ" denotes the lateral (the branch of the tee). Blaisdell and Manson (1963) provide additional related information.

Figure 6.3(a) gives the change in piezometric head in the conduit (the run of the tee) for combining flow in multiples of the downstream velocity head. The change in piezometric head Δh equals $(p_u - p_d)/(\rho g)$, and is a function of Q_ℓ/Q_d and D_ℓ/D, in which D_ℓ and D are the diameters of the lateral and conduit, respectively. The pressure loss is primarily a result of momentum exchange as discussed in Section 2.4. Equation 2.41 with $\beta_1 = \beta_2 = 1$ is equivalent to the "$F = 0$" curve in Figure 6.4(a) and is seen to agree well with the data curves for Q_ℓ/Q_d less than 0.1. At higher values of Q_ℓ/Q_d, velocity profile effects (causing β_1 and β_2 to deviate from 1) and possibly frictional effects due to the lateral flow impinging on the opposite wall cause departures from the simple momentum theory.

Figure 6.3(b) gives the loss of head between the lateral and the conduit for combining flow, both as multiples of the lateral velocity head and the downstream conduit velocity head. The loss of head is defined by:

$$h_f' = \left(\frac{p_\ell}{\rho g} + \frac{V_\ell^2}{2g}\right) - \left(\frac{p_d}{\rho g} + \frac{V_d^2}{2g}\right) \tag{6.26}$$

and is seen to be a function of Q_ℓ/Q_d and D_ℓ/D. Note that energy per unit mass is lost at smaller values of Q_ℓ/Q_d and gained at larger values of Q_ℓ/Q_d. Note also that for the particular case of $Q_\ell/Q_d = 1$ and $D_\ell/D = 1$, the lateral (branch) velocity V_1 is equal to the conduit (run) velocity V so Figure 6.3(b) can be used to obtain the K for that case [which is not given on Figure 6.3(a)].

Figure 6.4(a) gives the change in piezometric head in the conduit for dividing flow in multiples of the upstream velocity head. This change in piezometric head Δh equals $(p_d - p_u)/(\rho g)$, corresponds to a rise in pressure, and is a function of Q_ℓ/Q_u

(a)

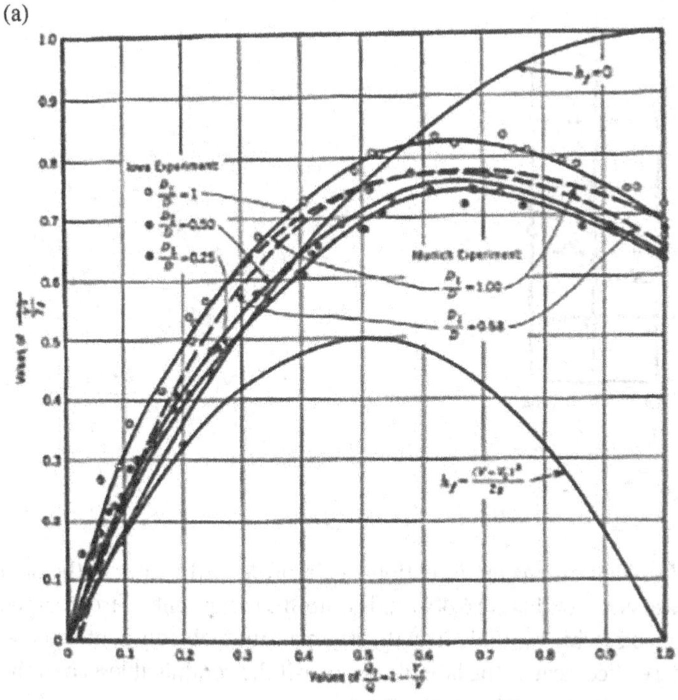

Figure 6.3 (a) Change in Piezometric Head in the Conduit for Dividing Flow; (b) Head Loss between the Conduit and the Lateral for Dividing Flow (McNown 1954/American Society of Civil Engineers)

(b)

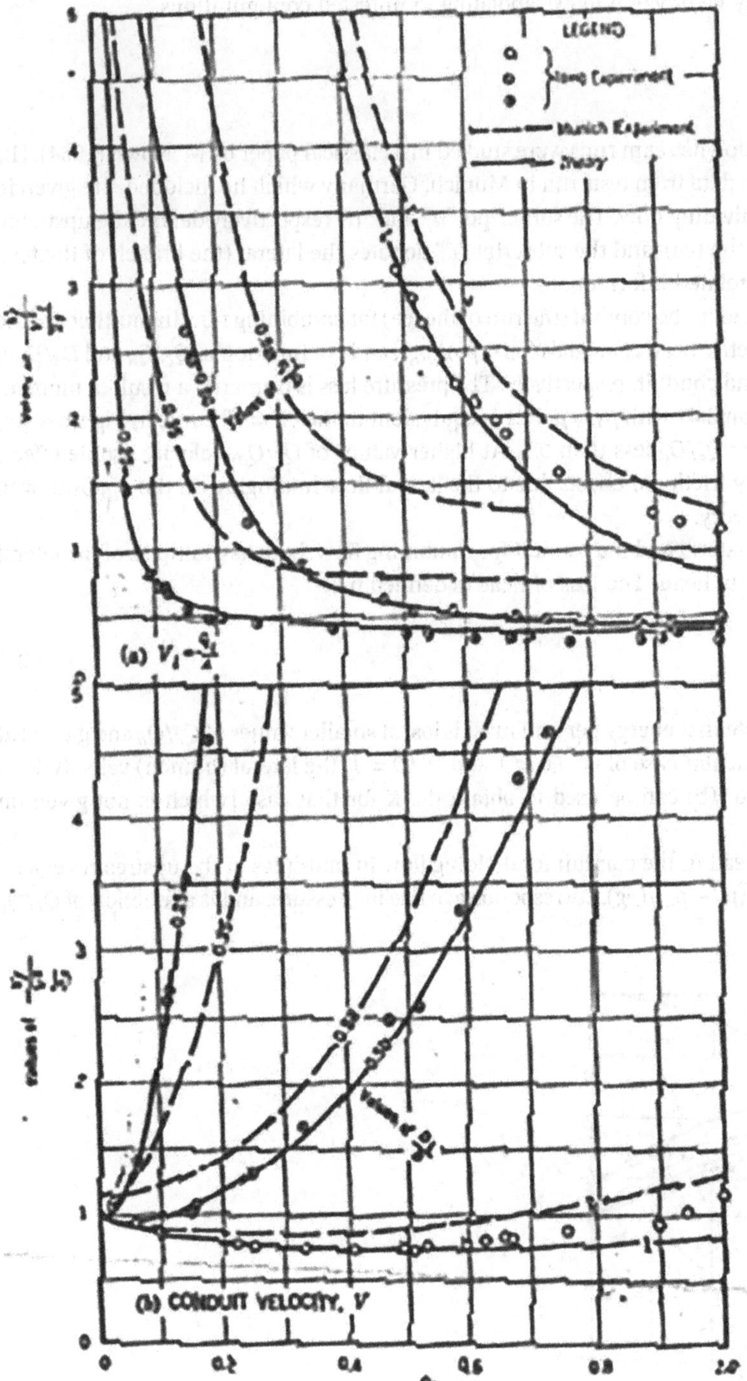

Figure 6.3 (Continued)

and D_e/D. The pressure rise is akin to that occurring in a sudden expansion. Equation 2.46 represents the limit to the data curves for small Q_e/Q_u and D_e/D; it corresponds to the lower curve on Figure 6.4(a) and conforms to the general trend of the data curves over the full range of values. Velocity profile effects (which would affect the β values) undoubtedly contribute to the departures between simple theory and experiment; these effects cause the lateral flow to exit the conduit at less than the average velocity in the conduit (see C.V. 2 on Figure 2.5 and the associated derivation).

The change in piezometric head between the conduit and lateral for dividing flow is amenable to approximate theoretical analysis (McNown 1954). Referring to Figure 6.4, the principle phenomenon is the contraction and subsequent expansion of the flow entering the lateral due to separation at the sharp corner. The flow at the entrance to the lateral is thus similar to that represented by Figure 6.4 with $A_1 = A_2 = A_o$. For this case, Equation 6.14 gives:

$$h_\ell = \frac{1}{C_c^2}(1 - C_c)^2 \frac{V_\ell^2}{2g} = \left(\frac{1}{C_c} - 1\right)^2 \frac{V_\ell^2}{2g} \tag{6.27}$$

McNown and Hsu (1951) applied potential flow theory using a conformal mapping technique (see Chapter 4) to derive a relation for the coefficient of contraction C_c for a two-dimensional idealization of flow into the branch of a pipe (Figure 6.4). They used this relation for C_c in the above equation to obtain the theoretical curves shown on Figure 6.4. The figure presents the head loss in question in multiples of the lateral and upstream conduit velocity heads as a function of Q_ℓ/Q_u and D_ℓ/D. The agreement between experimental data and theory appears to be good except for $Q_\ell/Q_u > 0.5$ with $D_\ell/D = 1$. However, the equation published by McNown and Hsu (1951), which can be readily solved numerically by, for example, graphical bisection, does not give their theoretical curves depicted on Figure 6.4. Nevertheless, their notion that a two-dimensional idealization of the dividing flow problem could apply to a circular branch does seem valid for the limiting case of small $D_\ell/D$for which the theoretical solution for the flow issuing from a large vessel by a two-dimensional aperture gives $C_c = \pi/(\pi + 2)$ (Lamb 1945, Streeter 1948); that limiting value, substituted into Equation 6.27, gives $h'_f/[V_i^2/(2g)] = (2/\pi)^2 \cong 0.405$, which can be seen from Figure 6.3(a) to be accurately reached for $D_\ell/D = 0.25$ when $Q_\ell/Q_u > 0.5$.

McNown (1954) reported that the branch measurements for his data were made at a point five branch diameters along the branch. Hudson *et al.* (1979) noted the difference in measured data for short and long branches, and defined "long" branches as "substantially greater than three pipe diameters in length." Hudson *et al.* also presented data for short branches (without quantifying the branch lengths) which indicated significantly higher branch entrance losses than for long branches

(a)

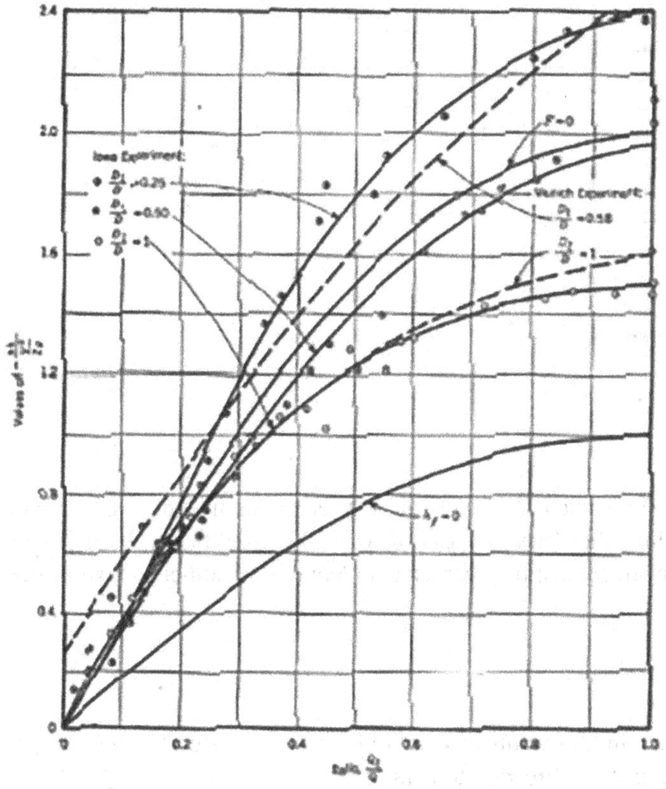

Figure 6.4 (a) Change in Piezometric Head in the Conduit for Combining Flow; (b) Head Loss Between the Conduit and the Branch for Combining Flow (Adapted from McNown 1954)

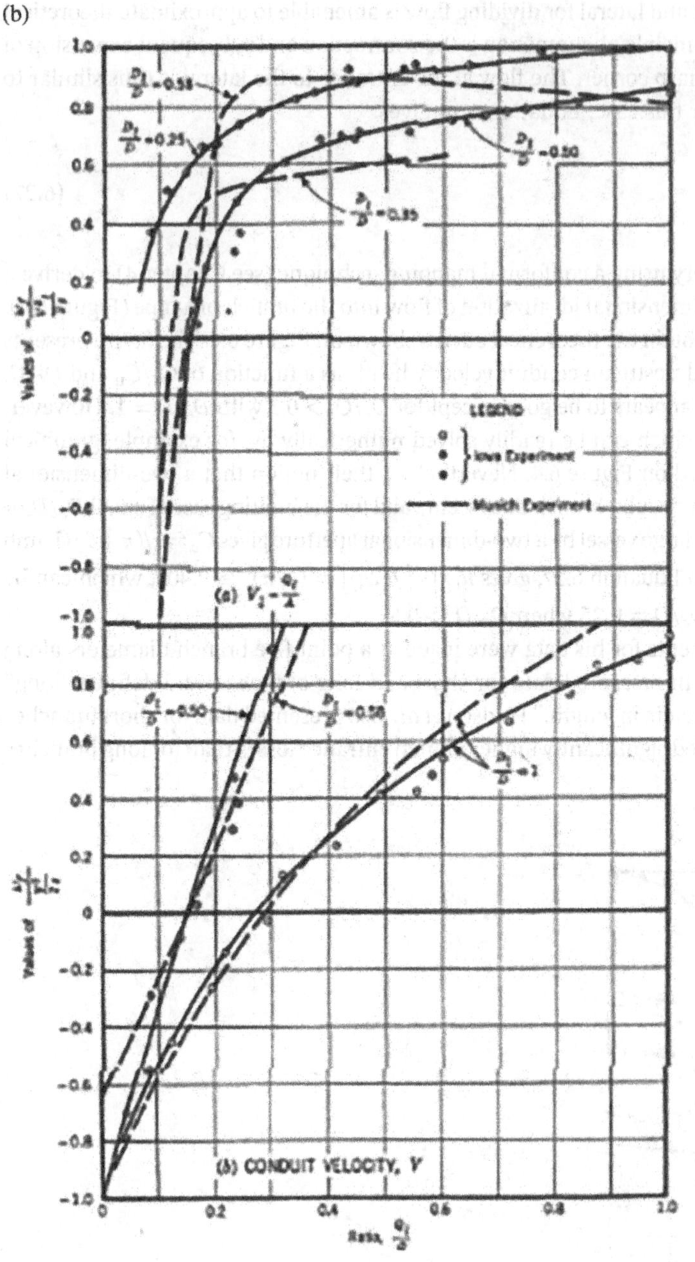

Figure 6.4 (Continued)

(contracted velocity head is not recovered). French (1980) reported observing stability problems in which the branch entrance loss will fluctuate in short branches. Except for branches (risers) with orifices as discussed below, branches for pressure distribution systems should be "long," taken to mean at least five diameters in length (but sufficiently short that entrance losses dominate over wall friction).

6.4.2 Laterals

Blaisdell and Manson (1963, 1967a, 1967b) conducted an investigation of combining flows in laterals. Referring to Figure 6.5, they initially applied the momentum equation in the x-direction to obtain:

$$p_u A - p_d A = \rho Q_d V_d - \rho Q_u V_u - \rho Q_b V_b \cos\theta \tag{6.28}$$

in which $A = A_u = A_d$. They subsequently acknowledged (Blaisdell and Manson 1967a, 1967b) that their analysis neglected the pressure component of the incoming branch flow and the external force necessary to balance the system. However, continuing to ignore those components and employing the continuity relations ($Q_u = Q_d - Q_b$, $V_d = Q_d/A$, $V_u = Q_u/A$, $V_b = Q_b/A_b$), Blaisdell and Manson noted that the above equation may be rewritten as follows:

$$\frac{(p_u - p_d)/\gamma}{V_d^2/(2g)} = 2 - 2\left(1 - \frac{Q_b}{Q_d}\right)^2 - 2\left(\frac{Q_b}{Q_d}\right)^2 \frac{A}{A_b} \cos\theta \tag{6.29}$$

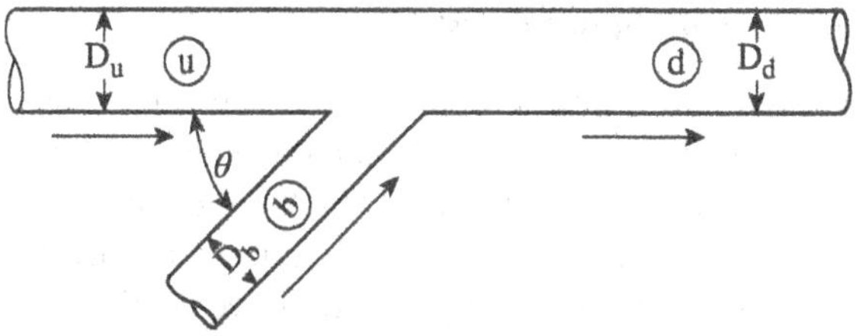

Figure 6.5 Combining Flow in Lateral (Blaisdell and Manson 1967a/American Society of Civil Engineers/).

The total head loss across the main pipe is $H_u - H_d = (p_u/\gamma - p_d/\gamma) + V_u^2/(2g) - V_d^2/(2g)$, which may then be used with the above equation and the continuity relations to give:

$$\varsigma_u = \frac{H_u - H_d}{V_d^2/(2g)} = 2\frac{Q_b}{Q_d} - \left(\frac{Q_b}{Q_d}\right)^2 \left(1 + 2\frac{A}{A_b} \cos\theta\right) \tag{6.30}$$

Blaisdell and Manson (1963) provided experimental results over the full range of Q_b/Q_d values, A_b/A ranging from 1 to 16, and angles θ of 15–165°. Those results did not agree well with the above equation, which they acknowledged was because they did not measure the neglected terms mentioned above. However, the author notes that the neglected terms are subtracted from the right-hand side of Equation 6.30. Therefore that equation can be regarded as giving conservatively high results for the change in head in the main conduit.

Blaisdell and Manson (1967a) presented a theoretical equation for the head loss from the branch to the main conduit. However, they found that a least-squares fit of their experimental data was preferable and gave the equation of that fit as follows:

$$\varsigma_b = \frac{H_b - H_d}{V_d^2/(2g)} = -0.76 + 2.16\frac{Q_b}{Q_d} - \left[1.61 + 0.56\frac{A}{A_b} \cos\theta - 0.84\left(\frac{A}{A_b}\right)^2\right] \left(\frac{Q_b}{Q_d}\right)^2 \tag{6.31}$$

6.4.3 Wyes and Trifurcations

A wye splits flow from one incoming conduit into two outflowing conduits. Ruus (1970) conducted experimental studies of conical and spherical wyes, which Williamson and Rhone (1973) presented in more convenient form. The conical-wye results presented by Williamson and Rhone were for wyes with a 10° conical taper, 60° bifurcation angle, and unequal branches with one branch sum:main diameter ratio of 0.7. The head-loss coefficient for the branch was expressed as a multiple of the velocity head in the main versus the ratio of the velocity in one branch to that in the main and ranged from about 0.5 at zero ratio decreasing to about 0.1 at a velocity ratio of 1.4 then increasing to about 0.5 at a velocity ratio of two. The spherical-wye results were for wyes with a 90° bifurcation angle, and one branch sum:main diameter ratio of 0.7. The head-loss coefficient for the branch was expressed as a multiple of the velocity head in the main versus the ratio of the velocity in one branch to that in the main and ranged from about 0.5 at zero ratio decreasing to about 0.4 at a velocity ratio of 0.4 then increasing to about 1.0 at a velocity ratio of two. Appel and Yu (1966) studied flow dividing into opposing laterals (circular and rectangular) in which a major feature was strong cyclic vortices.

Williamson and Rhone note that there are "wide variations in the loss coefficients from different experiments" and that "[h]ead loss coefficients for prototypes are likely to be less than those observed in model tests" on which the above-mentioned data are based. Of practical significance is that the head losses are relatively small compared to other losses in the systems in which such wyes are usually found.

A trifurcation divides the flow into three branches. Gladwell and Tinney (1965) in ASCE's *Journal of the Power Division* presented hydraulic studies of a large penstock trifurcation. They experimentally determined the coefficient of head loss $K = H_f/H_{VL}$ where H_f is the energy loss per pound of fluid between the main conduit and branch, and H_{VL} is the velocity head in the main tube for various flow configurations (single tube flowing, flow equally divided, etc.). Their study was augmented by numerous discussions in the same journal for January 1966, with authors' closure in March 1967. Rao, Rao, and Ramaswamy in ASCE's *Journal of the Hydraulics Division* (1967) presented related theoretical and experimental studies for trifurcations, which was augmented by discussions in the same journal for January 1968 and March 1968 with authors' closure in November 1968. A related paper was presented by Rao, Rao, and Shivaswamy in ASCE's *Journal of the Hydraulics Division* (1968), with discussion and closure in the same journal for September 1969 and March 1970.

6.4.4 Side Outlet Ports

The term "side outlet ports" is used here to refer to ports along the side(s) of a conduit such that the flow exiting the port is at an angle to the main flow. A perforated outfall diffuser is an example. Such outlet ports usually occur in multiples and as such are discussed in Chapter 11 (Distribution Conduits) in the context of spatially-decreasing flow. However, individual side outlet ports do resemble hydraulically the dividing flow configurations discussed above. Certain hydraulic characteristics of individual outlet ports are conveniently discussed here within the context of the present section.

Just as the approach velocity influences the hydraulic performance of an in-line orifice (such as discussed above in connection with Figure 6.1), so does the velocity adjacent to a side outlet port influence its hydraulic performance.

6.5 Valves and Other Appurtenances

With the above as background, we shall now discuss valves and pipe appurtenances in which the flow patterns are usually more complex than in the configurations discussed above. The discussion of valves is intended to provide an overview, and supplement rather than repeat information readily available elsewhere. Our concern will be chiefly with the flow control, head loss, and cavitation characteristics of valves. The forces at bends and other appurtenances which require anchorage design will be analyzed. Certain other pipe appurtenances including certain positive-displacement flow meters will also be discussed.

6.5.1 Valves

There are a number of excellent references on the types and uses of valves in environmental engineeringh applications (e.g., Thielsch 1967, pp. 7–61 to 112; Symons 1968; Camp and Lawler 1969b, pp. 37–29 to 37; Kohler and Ball 1969, pp. 22–38 to 66). The "Water and Wastes Engineering Manual of Practice on Valves, Hydrants, and Fittings" (Symons 1968) is particularly worthy of note; Table 6.2 lists the valve classifications and types of valves as given in that manual. The standards of AWWA, ANSI, and ASTM are important adjuncts in the standardization and specification of valves; most of the pertinent

Table 6.2 Types of Valves

Air release and air/vacuum	Hydrant
Altitude	Mud
Automatic control	Multiport
Backflow preventer	Needle
Balanced	Pinch*
Ball	Pivot
Butterfly	Plug
Check	Pressure regulating
Cone*	Pressure reducing*
Corporation cock	Pressure Relief
Curb stop	Safety
Diaphragm	Shear
Flap	Sleeve
Float	Sluice
Foot	Surge control
Gate	Tide
Globe	Vacuum release/pressure air release*
	Wafer

* Added.
Source: Symons (1968).

standards are given in the references just cited. Valve manufacturers are the original source of much of that information, and their catalogs and data sheets are valuable sources of additional information. Here the emphasis is on flow-pressure drop characteristics of valves. Various types of valves and practical uses of valve characteristics will be discussed here and in subsequent sections of the text. Valves used on air lines are discussed in Section 9.4.

Automatic valves are used particularly in the waterworks industry. Detailed analyses of such valves can be conducted for each type or variation of automatic valves, as the author has performed for pressure-reducing valves. Such analyses can be detailed and lengthy. The discussion here will be limited to presenting some useful references: Symons (1968, pp. M 29–M 30); Thielsch (1967, pp. 7-97 to 7-98); Schiff (1991); Pérez *et al.* (1993); Bengtson (1998); Prescott and Ulanicki (2003, 2008); Jeppson (1976, pp. 85–86, 129–134) discuss incorporation of pressure-reducing valves in pipe network analyses.

The flow-pressure drop characteristics of most types of pipeline valves are similar to those of pipe orifices as discussed in Section 6.3. In the most common case of valves in pipelines having the same diameter upstream and downstream, the pressure drop and head loss across the valve are identical. The valve openings, analogous to the orifice diameter, are often sufficiently small that head losses across the valve are a multiple of the pipe velocity head. In some types of valves, the geometry of the valve opening is sufficiently simple that its area can be determined and used in conjunction with the pipe orifice equation in an attempt to predict the pressure drop vs. flow rate and valve position. That is particularly common in studies of hydraulic transients. Although not advocated by the author, even for hydraulic transient analyses if avoidable (see below), such geometric relations are useful in some of the developments presented later. The open area and related relations for certain valve geometries are derived below. Wood and Jones (1973) give relations for additional valve geometries.

The rotary ball or cone valve with circular port, as depicted on Figure 6.6(a), has an opening whose projected area perpendicular to the pipe axis may be approximated by the double circular segment defined by the heavy curves on Figure 6.6(b). The area of that double circular segment is given by $a = 2(a_1 - a_2)$, in which a_1 and a_2 are as denoted on Figure 6.6(b) and given by:

$$a_1 = \frac{\pi}{4} d_v^2 \bullet \frac{2\alpha}{2\pi} = \frac{\alpha}{4} d_v^2 \tag{6.32a}$$

$$a_2 = \frac{1}{2} 2xh = xh = (r\cos\alpha)(r\sin\alpha) = \frac{d_v^2}{4} \sin\alpha\cos\alpha \tag{6.32b}$$

Thus:

$$a = \frac{\alpha}{2} d_v^2 - \frac{d_v^2}{2} \sin\alpha\cos\alpha = \frac{d_v^2}{4}(2\alpha - 2\sin\alpha\cos\alpha) = (2\alpha - \sin 2\alpha)\frac{d_v^2}{4} \tag{6.33}$$

in which d_v is the open diameter of the valve and α is the half angle of the opening as shown on Figure 6.6(b).

For a gate valve [Figure 6.6(c)], the upper circle of Figure 6.6(b) can be considered to represent the gate with the lower circle representing the full opening. The open area is then given by the full opening minus the area of the double circular segment formed by the intersection of the two circles. Since the latter is given by Equation 6.33, the area is given by:

$$a = \frac{\pi}{4} d_v^2 - (2\alpha - \sin 2\alpha)\frac{d_v^2}{4} = (\pi - 2\alpha + \sin 2\alpha)\frac{d_v^2}{4} \tag{6.34}$$

The rate of closure (or opening) is important when considering hydraulic transients. If the rotary valve of Figure 6.6(a) rotates at a constant rate, the width of the opening, denoted by y on Figure 6.6(b), decreases linearly with time; thus:

$$y = 2r(1 - t/T_c) \tag{6.35}$$

in which T_c is the valve closure time. Then from Figure 6.6(b) we have $y = 2(r - h)$ and $h = r\cos\alpha$ so that:

$$y = 2r(1 - \cos\alpha) \tag{6.36}$$

From the above two equations, we have:

$$t/T_c = \cos\alpha \tag{6.37}$$

The open area at a given time expressed as a fraction of the closure time may be obtained by eliminating α between Equations 6.36 and 6.37. The rate of change of open area will be of interest in our consideration of hydraulic transients (Sections 6.7 and 8.5). For the rotary valve, this rate is obtained from Equations 6.33 and 6.37 as:

$$\frac{da}{dt} = -\frac{d^2}{T_c}\sin\alpha \tag{6.38}$$

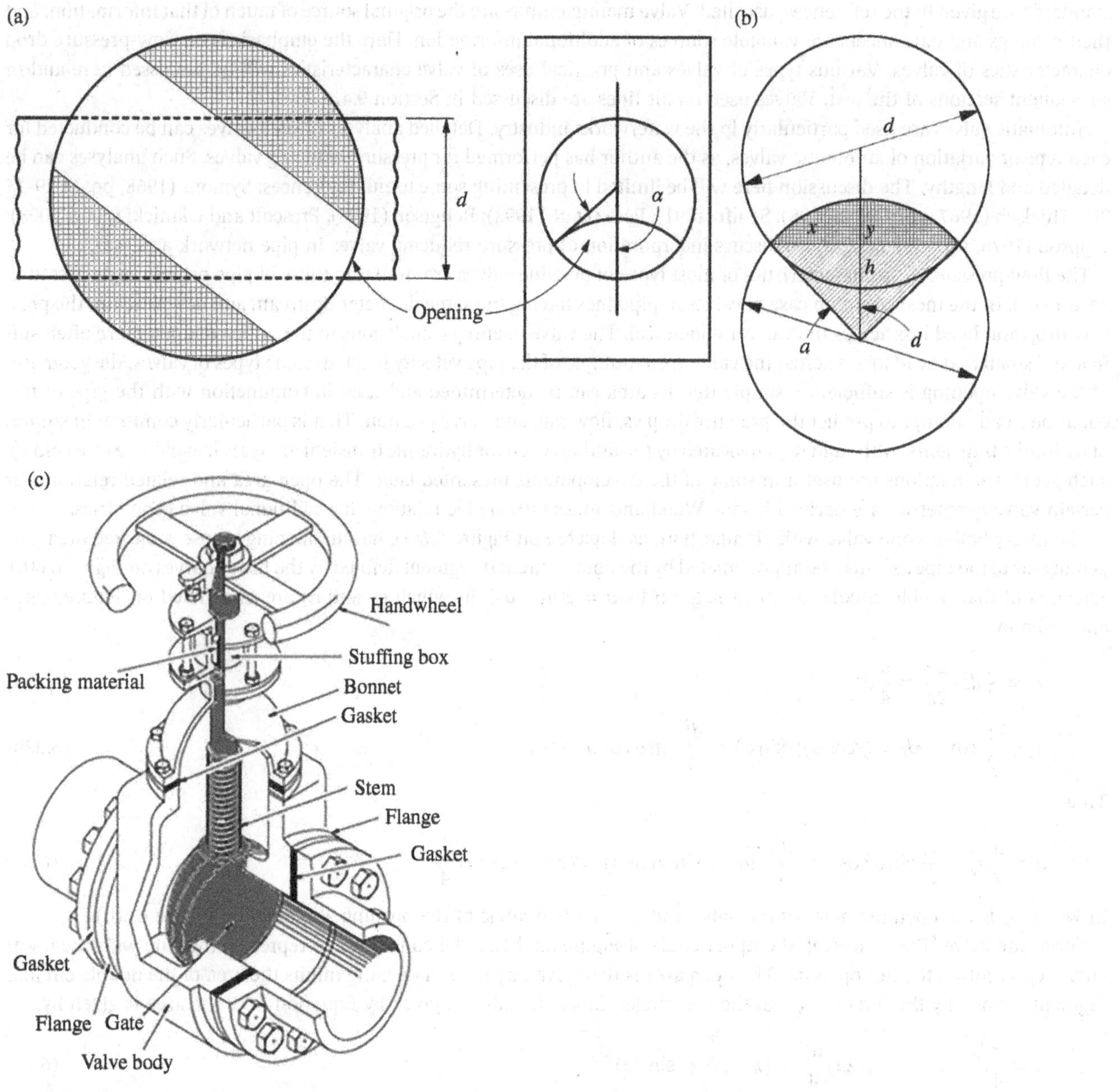

Figure 6.6 (a) Rotary Ball or Cone Valve with Circular Port (After McNown 1950, p. 458); (b) Rotary/Cone Valve Geometry; (c) Gate Valve Geometry (Adapted from Gate Valve – Working, Parts, Diagram, Types, Symbol, Advantages, Applications, 2022)

For the gate valve, an analysis similar to the above for uniform rate of closure gives:

$$y = 2rt/T_c \tag{6.39}$$

in place of Equation 6.36, while Equation 6.37 remains valid. Thus:

$$t/T_c = 1 - \cos\alpha \tag{6.40}$$

The rate of change of open area is obtained from Equations 6.34 and 6.40 as:

$$\frac{da}{dt} = -\frac{d^2}{T_c}\sin\alpha \tag{6.41}$$

which is identical to Equation 6.38.

A final relationship which we shall discuss here is one that has been employed in oft-cited analyses of hydraulic transients. That is for "uniform gate closure," in which the open area of the gate is assumed to decrease linearly with time. In that case, we have simply:

$$a = \frac{\pi}{4}d^2\left(1 - \frac{t}{T_c}\right) \tag{6.42}$$

for which:

$$\frac{da}{dt} = -\frac{\pi}{4}\frac{d^2}{T_c} \tag{6.43}$$

Whether such relationships are approached in practice is questionable, but they are important to understanding some of the classical studies of hydraulic transients.

The substitution of the valve open area in the pipe orifice equation (Equation 6.25) provides an idealization enabling calculation of the pressure drop-flow relationship. This is commonly (and often unnecessarily) done in sophisticated water-hammer analyses, in which the discharge coefficient is assumed constant. However, valve open area relations such as the above represent geometric idealizations to varying degrees. Furthermore, the flow pattern through valves can be much more complex than such an idealization would imply. This type of approach is of limited accuracy and represents a poor substitute for actual data.

6.5.1.1 Pressure Drop-Flow Data

Pressure drop-flow data for valves are basically of two types. The first is for valves in their fully-open positions; such data are used primarily to calculate valve losses along with other form losses along pipelines, usually for valves providing open-shut service. Such data are usually given in terms of K values for use in the equation $h_\ell = K[V^2/(2g)]$ with the terms as given for Equation 6.15c.

The second type of pressure drop-flow data pertains to valves used for control (throttling) purposes in partially open positions. When used in conjunction with a flow meter as part of an automatic flow control system, the precise pressure drop-flow relations are usually not of direct concern to the user. For nonmetered flow control settings, cavitation studies, and analyses of hydraulic transients, the pressure drop-flow relations are of direct concern. Such relations are expressed in a variety of ways, usually involving an equation in conjunction with a chart or table. The equations may be nondimensional or dimensional; in the latter case, derivation of the equation can reduce the risk of error in use (including the acceptance of someone else's erroneous equation). The equations are usually in the following form:

$$Q = C_D a\sqrt{2g\Delta h} \tag{6.44a}$$

$$\Delta h = KV^2/(2g) \tag{6.44b}$$

Note that the discharge coefficient C_D and K are related by:

$$\frac{1}{C_D} = \frac{a}{A}\sqrt{K} \tag{6.45}$$

in which a and A are the two relevant areas.

It is important to note that values of C_D for valve types in general should be taken from experimental data as provided by the manufacturer or otherwise. Attempts to calculate those values by analytical methods can give results that differ substantially from experimental results.

6.5.2 Appurtances

The term "form loss" has been used above, and is preferred to the sometimes used term "minor losses", because the losses are not necessarily minor in the commonly-used meaning of the word. Form losses can be particularly significant in short pipe lengths. Such losses are expressed in terms of K values which are multiplied by velocity head in turbulent flows to give the form head loss. K values can usually be taken as 0.9 for 90° bends and 0.45 for 45° bends. For check valves, a K of 2.5 can usually be used (obviously just for forward flow). Fully-open butterfly valves have K values that vary with valve size, manufacturer, and model; significant variations are not uncommon. An example of butterfly valve characteristics is given under CAVITATION below.

Smaller K values can often be neglected. That includes open gate valves (K typically 0.2) and 1/8 curves (K about 0.2). Under worst-case conditions (largest velocity, C-value, and diameter), the equivalent length of a $K = 0.2$ fitting is on the order of 15 pipe diameters. The head loss of such fittings will often be negligible compared to friction losses in the much greater length of piping adjacent to such appurtenances or to head losses of other adjacent appurtenances.

Additional information regarding form losses in pressure conduits may be found in Graber (2017a, §7.2).

6.5.3 Cavitation

Cavitation is a phenomenon which can occur in a variety of situations, but in environmental engineering applications is mostly associated with valves, pumps, and waterhammer. We will introduce the phenomenon here and discuss it further in later sections (particularly in Section 8.2).

Cavitation refers to the formation of vapor cavities in a liquid flow when the pressure falls below the vapor pressure of the liquid. The concept of vapor pressure is discussed in Section 9.2. We will briefly note here that just as a liquid will turn to gas at a given pressure when the temperature rises sufficiently (e.g., water at 1 atm and 212°F), so will the liquid turn to gas at a given temperature when the pressure falls sufficiently. The pressure at which this occurs is called the *vapor pressure*; the vapor pressure of water as a function of temperature is tabulated in Table 9.1. Thus, for example, if the liquid flow through the venturi tube becomes high enough for a given downstream pressure, the throat pressure will drop to the vapor pressure of the liquid.

In *steady* pipe flows, it is at the small flow restrictions created at valves that drops to vapor pressure are most common. Valves used for flow throttling are particularly susceptible due to the small flow passages necessary for effective flow control. Any brief cavitation occurring in valves used for open-and-shut service is generally of no practical consequence.

The damaging consequences of cavitation are associated with the collapse or implosion of the vapor pockets downstream of the restriction when the pressure again rises above the vapor pressure. The collapse can create extremely high local pressures which, if they impinge upon solid boundaries, can cause severe localized pitting and erosion, often referred to as cavitation corrosion. The corrosion is thought to be caused not only by the direct effects of high localized pressures, but also by the liberation of air in the water, presumably in the region of low pressures, which persists in the region of high localized pressures. Prandtl's lucid description of the corrosion process follows:

> "The air liberated from the water is richer in oxygen than ordinary air and this oxygenated air is forced into the pores of the metal at high pressure and then sucked out again as soon as the pressure falls so that there is a rapid alternation of stresses reaching into the finest pores. The result of these processes ... is that the surface of the metal becomes roughened, and if the cavitation process persists for a long time quite large hollows are eventually formed (Prandtl 1952, p. 318)."

That some empiricism would be required in predicting cavitation conditions can be readily appreciated in light of the difficulty of analytically predicting the minimum pressure at a valve given the complexity of the flow pattern. If we let p_{min} represent this minimum pressure, dimensional analysis (see Chapter 5) suggests that for a given valve-pipe geometry, one of the following relationships would apply:

$$\frac{p_d - p_{\text{min}}}{\rho V^2} = \phi(\mathbf{R}) \tag{6.46}$$

or noting $\rho V^2 \tilde{\propto} (p_u - p_d)$:

$$\frac{p_d - p_{\text{min}}}{p_u - p_d} = \phi(\mathbf{R}) \tag{6.47}$$

in which p_u and p_d are, respectively, the upstream and downstream pressures at consistently defined locations and $\mathbf{R}$ is a conveniently-defined Reynolds Number. Then, cavitation conditions can be defined as occurring when p_{min} equals the vapor pressure p_g. Such reasoning led to development of a cavitation parameter defined by:

$$\sigma = \frac{p_d - p_g}{p_u - p_d} \tag{6.48}$$

In equations such as Equation 6.48 and subsequent equations involving subtraction of the vapor pressure, absolute pressures, or gage pressures must be consistently employed. Since vapor pressure is expressed in absolute pressure herein (psia – Table 9.1), we shall employ absolute pressures for this purpose.

In theory, the parameter σ could be established for a given geometry as a function of Reynolds Number and scaled to other sizes of geometrically similar valves. Historically, this has been attempted, and cavitation data based on this method may still be found in valve catalogs, etc. Such dimensional reasoning is, unfortunately, oversimplified and sounder methods should be applied as discussed below.

Due partly to the impracticality of measuring the minimum pressure (finding its location, measuring it without disturbing the flow), the onset of cavitation is experimentally determined by sensitive noise or vibration sensors affixed to the pipe or valve body. As noted by Tullis (1973), attempts to correlate the cavitation parameter with Reynolds Number have not been successful. Tullis demonstrates this by plotting the relationship between the critical cavitation parameter σ_c (defined below) and Reynolds Number (based on pipe diameter and velocity) for geometrically similar ball valves; the two valves plot on two widely separated curves. Tullis (1973) concluded from this and other data that the complexity of the cavitation process precludes the correlation of scale effects with standard dimensional analysis techniques. One possible explanation (this author's) is that the common determination of vapor pressure (such as given in Table 9.1) is based on thermodynamic equilibrium, which may not be accurately approached in the fluid as it is subjected to abrupt pressure changes. Another (also this author) is the lack of geometric similarity between vapor bubbles and other characteristic dimensions (akin to the problem of modeling two-phase flows discussed in Section 13.3).

Tullis (1973) developed the following equation to account for scale effects:

$$V_i = C_1 V_{io}\left(\frac{p_u - p_g}{p_{uo} - p_{go}}\right)^N \left[1 - \frac{\log_{10}(d_o/d_{oo})}{10^{(0.40-0.52\log_{10}C_d)}}\right] \tag{6.49a}$$

or:

$$V_i = C_1 V_{io}\left(\frac{p_d - p_g}{p_{do} - p_{go}}\right)^M \left[1 - \frac{\log_{10}(d_o/d_{oo})}{10^{(0.40-0.52\log_{10}C_d)}}\right] \tag{6.49b}$$

in which V_{io} is an experimentally determined pipe velocity at which incipient cavitation occurs under test conditions of upstream pressure p_{uo} (or downstream pressure p_{do}) and vapor pressure p_{go} in a test valve of diameter d_{oo}; the corresponding terms without the added subscript "o" refer to the prototype conditions of interest; C_1 is a scale-effects multiplier; M and N are pressure-scale-effects exponents; and C_d is the valve discharge coefficient defined by:

$$C_d = \frac{V}{\sqrt{2(p_u - p_d)/\rho + V^2}} \tag{6.50}$$

Equation 6.49a is used if the upstream pressure p_u is known, and Equation 6.49b is used if the downstream pressure p_d is known. The multiplier C_1 has the following values:

$$C_1 = \begin{cases} 0.94 \; , & do/doo > 1 \\ 1 \; , & do/doo = 1 \\ 1.06 \; , & do/doo < 1 \end{cases} \tag{6.51a}$$

The exponent M or N must be experimentally determined for each valve geometry. Tullis gives values of M and N obtained by him and the Metropolitan Water District of Southern California for butterfly, ball, cone, globe, and Pelton needle valves; for those valves, M ranged from 0.36 to 0.46 and N ranged from 0.34 to 0.43.

The butterfly valve is the primary liquid-throttling valve in our applications and will be considered more thoroughly. Before doing so, however, we will note that the above-mentioned results of Tullis' work (1973) may be used for the other types of valves mentioned if data from the manufacturer or other source provides cavitation data for one set of conditions (pressures and valve size). If the test conditions (pressure and valve size) are not specified, one should be very cautious about accepting cavitation data expressed in terms of the cavitation parameter and conventional dimensionless variables.

Approach conditions for butterfly valves are important. Upstream disturbances caused by pumps, other valves, and elbows can influence the dynamic torque of the butterfly valve and reduce the valve's total flow capacity. Kurkjian (1978) and Hoff and Libke (1995a, 1995b) discuss ideal approach conditions, then give practical guidelines recommended by most valve manufacturers of between 6 and 8 unobstructed pipe diameters upstream of the valve. Upstream straightening vanes can also be installed. Other important considerations are also discussed by Hoff and Libke.

Ball and Tullis (1973) tested butterfly valves which represented wide variations in body and leaf design and with sizes ranging from 4 in. (100 mm) to 20 in. (510 mm). They considered three cavitation limits, all of which may be identified quantitatively from a plot of accelerometer (vibration sensor) output vs. σ:

1) *Incipient cavitation* is the stage at which cavitation becomes audible as light intermittent pops.
2) *Critical cavitation* is the stage at which cavitation becomes audibly steady (while remaining light).
3) *Choking cavitation (or supercavitation)* is the stage of heavy cavitation where the discharge becomes and remains constant for a given upstream pressure irrespective of further decreases in downstream pressure. (Choking is a phenomenon of compressible gas flow, discussed in Section 9.4). Cavitation intensity at the valve is a maximum at this stage.

Between critical and choking cavitation, the degree of cavitation changes from mild (sounding like "pebbles in the line") to severe (sounding like "heavy hammer blows or a loud roar").

Guins (1968) derived a relationship for the discharge coefficient of butterfly valves by the following progression: $Q = C_d A_o V_d \Rightarrow V_u = Q/A_o = C_d V_d$, $V_u^2/(2g) + p_u/\gamma = V_d^2/(2g) + p_d/\gamma \Rightarrow V_d = \sqrt{V_u^2 + (2g/\gamma)(p_u - p_d)}$, $Q/A_o = C_d\sqrt{2(p_u - p_d)/\rho + V_u^2} \Rightarrow C_d = (Q/A_o)/\sqrt{2(p_u - p_d)/\rho + V_u^2}$; the subscripts "$u$" and "$d$" denote conditions 1 pipe diameter upstream of the valve and about 10 diameters downstream of the valve respectively. Ball and Tullis plotted the pipe velocity at incipient cavitation vs. this C_d for the different valve types and sizes.

Ball and Tullis' (1973) plot of pipe velocity at incipient cavitation V_{io} vs. discharge coefficient C_d for the different valve types and sizes is given in their Figure 10. The author has fitted the plotted curve with accuracy sufficient for practical purposes by the straight line given by:

$$V_{io}(\text{fps}) = 23.9C_d \tag{6.51b}$$

Ball and Tullis (1973) also plotted the pipe velocity at critical cavitation V_{co} in a similar manner, which the author has similarly fitted by:

$$V_{co}(\text{fps}) = 27.4C_d \tag{6.51c}$$

Ball and Tullis stated that the difference between V_{io} and V_{co} is about 13%, which is close to the difference given by the author's fitted equations. Noting that the variation in velocity between V_i and V_c only changes the cavitation level from light intermittent to light steady, Ball and Tullis suggested that a 10% error in estimating V_i and V_c would not seriously alter the cavitation intensity.

Example 6.3

A 30-inch butterfly valve is proposed for use as a gravity filter backwash control valve in an industrial wastewater treatment plant (American Cyanamid, Bound Brook, New Jersey). The backwash water is to be taken from a large "clear well" as shown on Figure 6.7. System elevations are as shown on the figure; the 4.05 ft total head loss is approximately constant (independent of flow) and includes 2.5 ft of head loss through the fluidized filter media. The backwash flow in the absence of the valve has been calculated to be 10,120 gpm. The only significant head loss upstream of the butterfly valve is the pipe entrance loss ($K = 0.5$). The proposed valve has characteristics of valve opening and valve flow coefficient C_v as given, respectively, in Columns (1) and (2) of Table 6.3, but no useful cavitation data. The tabulated value of $C_v = Q\sqrt{SG/\Delta p}$ in which Q is flow in gpm, SG = specific gravity = 1 for water, and Δp is pressure drop in psi.

It is desired to determine the valve setting which will throttle the backwash flow to 50% of the unthrottled flow and check for cavitation. Assume a water temperature of 68°F and barometric pressure of 14.7 psia.

Solution

We will first apply the Bernoulli equation (from Section 2.6) between points "1" and "2" (see Figure 6.7) to obtain $z_1 = z_2 + h_{\ell,1-2}$ or $h_{\ell,1-2} = z_1 - z_2 = 4.05$ ft, in which $h_{\ell,1-2}$ is the head loss between points "1" and "2", comprised of the head losses between points "1" and "u" (just upstream of the valve) $h_{\ell,1-u}$, across the valve $h_{\ell,v}$, and $h_{\ell,d-2}$ between points "d" (just downstream of the valve) and "2." Recalling $h_{\ell,1-u} = 0.5V^2/(2g)$ where V is the pipe velocity, we have:

$$0.5\frac{V^2}{2g} + h_{\ell,v} + h_{\ell,d-2} = 4.05$$

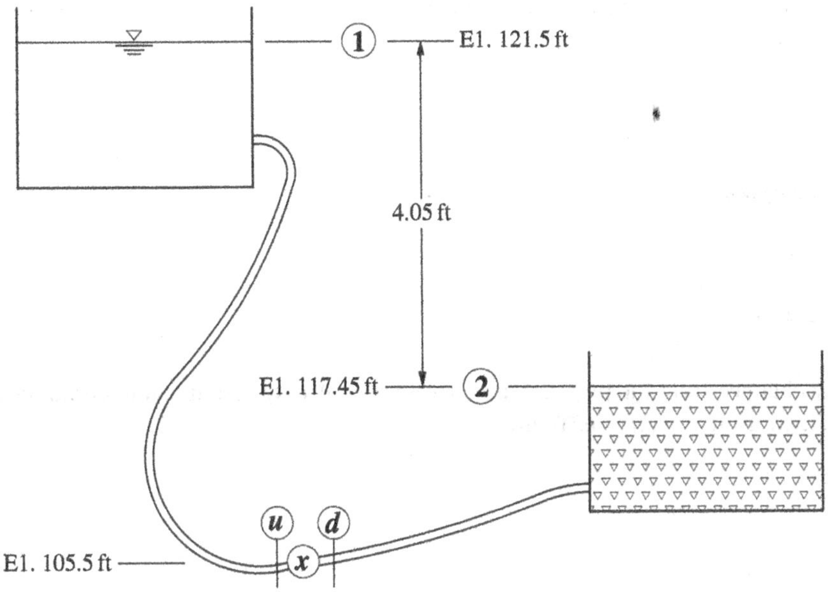

Figure 6.7 Configuration of Gravity Filter Backwash Control Example (NTS)

Table 6.3 Computation of Parameters for Example 6.3

Valve Opening (in degrees) (1)	C_v (gpm/$\sqrt{\text{psi}}$) (2)	C_d (3)	V (ft/sec) (4)	Q (gpm) (5)	p_u (psig) (6)	V_{io} (ft/sec) (7)	V_i (ft/sec) (8)
No valve	∞	1	4.59	10,114	—	—	—
90	37,200	0.811	4.36	9600	6.737	19.38	9.64
80	30,900	0.755	4.26	9394	6.745	18.04	9.03
70	23,400	0.657	4.06	8947	6.762	15.70	7.95
60	17,500	0.546	3.75	8267	6.786	13.05	6.70
50	11,500	0.394	3.13	6898	6.829	9.41	4.93
40	7440	0.267	2.37	5224	6.871	6.38	3.42
30	4090	0.151	1.45	3184	6.907	3.60	1.97
20	2230	0.0830	0.819	1800	6.921	1.98	1.101
10	744	0.0277	0.276	609	6.927	0.66	0.376

Since the head loss across a fluidized bed is independent of flow (see Section 14.8), the head loss $h_{\ell,d-2}$ is given by:

$$h_{\ell,d-2} = 2.5 + K_{d-2}\frac{V^2}{2g}$$

Eliminating $h_{\ell,d-2}$ between the above two equations, and solving for K_{d-2} gives:

$$K_{d-2} = \frac{4.05 - 0.5V^2/(2g) - h_{\ell,v} - 2.5}{V^2/(2g)}$$

When the butterfly valve is fully open, the head loss across the valve $h_{\ell,v}$ is negligibly small, and the pipe velocity, velocity head, and K_{d-2} may be determined as follows:

$$A = \frac{\pi}{4}\left(\frac{30}{12}\right)^2 = 4.91\text{ ft}^2$$

$$V = Q/A = (10{,}120/449)/4.91 = 4.59\text{ ft/sec}$$

$$V^2/(2g) = 0.327\text{ ft}$$

$$K_{d-2} \cong \frac{4.05 - 0.5(0.327) - 0 - 2.5}{0.327} = 4.24$$

The head loss across the valve is directly obtained from the pressure drop across the valve by first applying Bernoulli's equation between points "u" and "d" and then using Equation 6.50 thus:

$$h_{\ell,v} = \frac{p_u - p_d}{\rho g} = \left(\frac{1}{C_d^2} - 1\right)\frac{V^2}{2g}$$

The above equations result in the following:

$$0.5\frac{V^2}{2g} + \left(\frac{1}{C_d^2} - 1\right)\frac{V^2}{2g} + 2.5 + (4.24)\frac{V^2}{2g} = 4.05$$

or, rearranging:

$$\left(3.74 + \frac{1}{C_d^2}\right)\frac{V^2}{2g} = 1.55 \qquad (1)$$

The pressure p_u will be useful below and is determined by application of the Bernoulli equation between points "1" and "u," giving:

$$z_1 = p_u/(\rho g) + z_u + V^2/(2g) + h_{\ell,1-u}$$

From above, $h_{\ell,1-u} = 0.5V^2/(2g)$, so that:

$$p_u/(\rho g) = z_1 - z_u - V^2/(2g) - 0.5V^2/(2g)$$

$$p_u = (62.4/144)\left[121.5 - 105.5 - 1.5V^2/(2g)\right]$$

$$p_u = 0.433\left[16 - 1.5V^2/(2g)\right] \qquad (2)$$

in which $V^2/(2g)$ and p_u are in units of ft and psig, respectively.

The incipient cavitation velocity may be determined from Equation 6.49a. For that purpose, we must determine C_d for various valve angles. The value of C_d can be related to C_v as follows. From Equation 6.50, taking $\Delta p = p_u - p_d$ and expressing the flow in terms of pipe area A and velocity V (using $Q = AV$), we obtain:

$$\sqrt{\Delta p} = \sqrt{p_u - p_d} = Q/C_v = AV/C_v$$

Substituting the above into Equation 6.50 and rearranging gives:

$$C_d^2 = \frac{1}{2A^2/(\rho C_v^2) + 1}$$

For the 30-in. pipe and taking dimensional units of C_v into account (gpm/$\sqrt{\text{psi}}$ per above) the above equation becomes:

$$C_d^2 = \frac{1}{7.21(10)^8/C_v^2 + 1} \qquad (3)$$

Required numbers may now be tabulated as in Columns (3) to (8) of Table 6.3. The value of C_d [Column (3)] is calculated from the Column (2) value of C_v using Equation (3). The pipe velocity [Column (4)] is then obtained using Equation (1). The flow in gpm [Column (5)] is calculated from the velocity in Column (4) by:

$$Q(\text{gpm}) = AV = \frac{\pi}{4}\left(\frac{30}{12}\right)^2 V \times 449 = 2204\ V(\text{ft/ sec})$$

The pressure upstream of the valve [Column (6)] is then obtained using Equation (2). The unadjusted value of pipe velocity at incipient cavitation is then obtained using Equation 6.51b with the value of C_d in Column (3) and is given in Column (7). That is then adjusted to the actual pressures and valve size of interest. The adjustment is accomplished by means of Equation 6.49a, in which $d_o/d_{oo} = 30/12 = 2.5$, $p_{uo} - p_{go} = 71.5$ psi, giving:

$$V_i = 0.94 V_{io}\left(\frac{p_u - p_g}{71.5}\right)^N \left[1 - \frac{\log_{10}(30/12)}{10^{(0.40 - 0.52\log_{10} C_d)}}\right]$$

The value of N is taken as 0.39, which is the average value for butterfly valves given by Tullis (1973). The vapor pressure at 68°F is obtained from Table 9.1 as $p_g = 0.690$ psia. Taking p_u as the *gage* pressure upstream of the valve, we have $p_u - p_g = p_u(\text{psig}) + 14.7 - 0.690 = p_u(\text{psig}) + 14.0$. The above equation then becomes:

$$V_i = 0.94 V_{io}\left[\frac{p_u(\textbf{psig}) + 14.0}{71.5}\right]^{0.39} \left[1 - \frac{0.398}{10^{(0.40 - 0.52\log_{10} C_d)}}\right] \tag{5}$$

The above equation is then used with the values of V_{io} [Column (7)], p_u [Column (6)], and C_d [Column (3)] to obtain the adjusted incipient cavitation velocities in Column (8).

The pipe velocities [Column (4)] are all less than the corresponding incipient cavitation velocities [Column (8)], so cavitation is not expected with valve openings down to 10°.

If cavitation is a potential problem ($V > V_i$ or V_c), several measures are available. The acceptable velocity can be increased by increasing C_d for a given degree of control ($h_{\ell,v}$). Such an increase in C_d can be achieved by employing two or more valves in series. The head loss across a single valve and that across n valves in series are given, respectively, by:

$$h_{\ell,v} = \left(\frac{1}{C_d^2} - 1\right)\frac{V^2}{2g} \tag{6.52a}$$

and:

$$h_{\ell,v} = n\left(\frac{1}{C_d'^2} - 1\right)\frac{V^2}{2g} \tag{6.52b}$$

in which C_d refers to the drag coefficient for a single valve configuration and C_d' refers to the drag coefficient for each individual valve in a n-valves-in-series configuration. For the same $h_{\ell,v}$, Q, and A (hence same $V = Q/A$), we have:

$$n\left(\frac{1}{C_d'^2} - 1\right) = \left(\frac{1}{C_d^2} - 1\right) \tag{6.53}$$

Rearrangement of the above equation gives:

$$C_d' = \sqrt{\frac{1}{(1/n)(1/C_d^2 - 1) + 1}} \tag{6.54}$$

Values of C_d' vs. C_d are tabulated below for $n = 2$ and 3. The values of C_d' are higher, as desired, and represent the drag coefficients of the individual valves in the series if the valves are adequately spaced.

	C_d'	
C_d	$n = 2$	$n = 3$
0.1	0.141	0.171
0.2	0.277	0.333
0.3	0.406	0.478
0.4	0.525	0.603
0.5	0.632	0.707
0.6	0.728	0.792
0.7	0.811	0.862
0.8	0.883	0.918
0.9	0.946	0.963
1.0	1.0	1.0

Other means of cavitation prevention include increasing the valve size (thus reducing V), depressing the pipeline (thus increasing the minimum pressure), the use of cavitation-damage-resistant materials or coatings (Kohler and Ball 1969, pp. 22–67), and air injection. The latter takes advantage of the fact that vapor pressures are subatmospheric at the temperatures of interest (less than 212°F – see Table 9.1) so that bleeding of atmospheric air into such low-pressure regions can raise the minimum pressure above vapor pressure. The air may be introduced into a hollow valve disc via a hollowed valve shaft and distributed to the low-pressure region via disc injection holes (see Figure 6.8).

6.5.4 Forces at Bends and Other Appurtenances

Bends and other appurtenances often require anchorage design to resist imposed forces. Pipe anchorages should be considered early in the progression of a design project to allow timely consideration of the requirements as part of the integrated design process. It should not be treated as a structural design problem independent of the determination of pipe alignment and grades and other factors affecting pipe anchorage requirements. Elaborate, complex, and "in need of further work" procedures for the design of thrust blocks are given by Shumaker *et al.* (2011, 2017). However, the low cost of thrust blocks justifies using simplified design procedures that may not give the smallest size of anchorage for the particular situation. Of the simplified procedures that have been published, the author finds one that has long been available by Everett J. Prescott Company (EJP 2022, most recently), to be the most comprehensive. Those procedures are detailed in Figure 6.9. Shumaker *et al.* (2011) note the importance of not casting the joints into the thrust block.

Pipe hanger design is an important, related topic for which excellent manufacturer's information is available (e.g., ITT Grinnell Corporation 2008). Section 11.5 discusses forces due to drag on chemical feed pipes, Kármán vortex streets, and an associated example of the full range of forces on a distribution pipe spanning the inside of a large conduit. Such forces also occur on submerged pipe stream crossings.

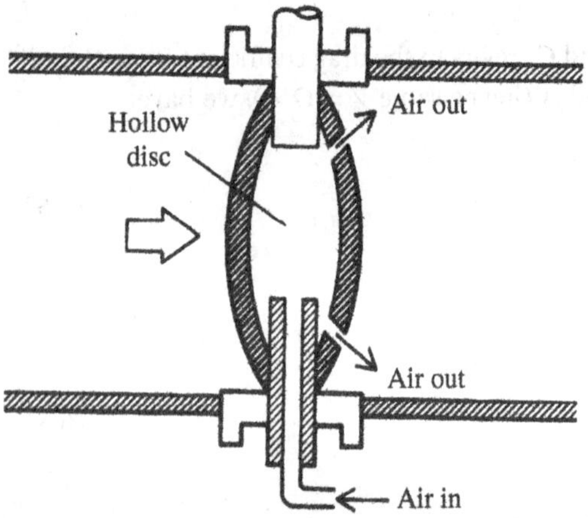

Figure 6.8 Hollow Disc Valve

Example 6.4a

Derive the equation for the thrust required to restrain a 90° bend on a pipe of circular cross section with outer diameter d.

Solution

The geometry is shown below. The thrust T_a is derived as follows:

$$T_a = 2\frac{\pi d^2}{4} p \cos\left(\frac{\pi - \alpha}{2}\right) = \frac{1}{2}\pi d^2 p \cos\left(\frac{\pi}{2} - \frac{\alpha}{2}\right) = \frac{1}{2}\pi d^2 p \sin\left(\frac{\alpha}{2}\right)$$

For $\alpha = 90^\circ$, $T_a = (\pi/4)pd^2/\sqrt{2}$.

Engineering & Technical Data R-16

Thrust Blocking at Fittings & Valves

Thrust Blocking

Before the development of Mechanical Joint Restraint Systems, (See Section B) the only established method for preventing joint separation at bends, tees, crosses and valves was by "Thrust Blocking" the fitting.

Although seldom used today we include this information on Thrust Blocking design as one of several proven methods of resisting thrust in varying soil conditions.

Determining Size and Type of Thrust Blocking

Size and type of the thrust block depends on maximum pressure, pipe size, kinds of soil and types of fittings.

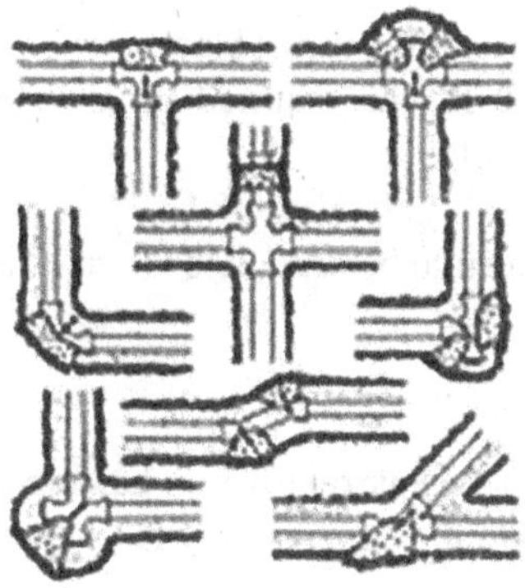

If thrusts due to high pressure are expected, anchor the valves as shown below. At vertical bends anchor to resist upward thrusts.

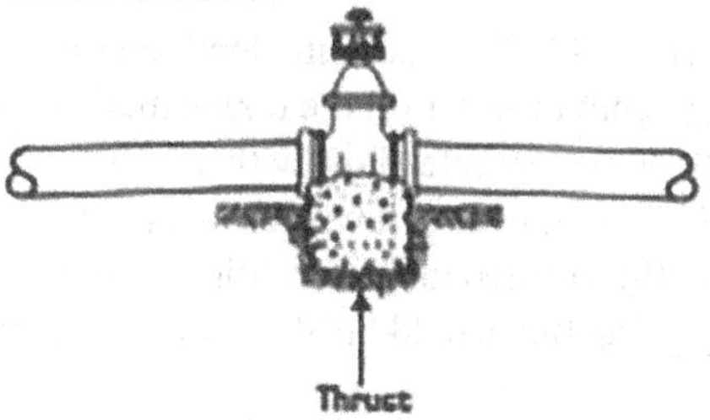

The following table shows the approximate thrust generated at fittings for each 100 psi of water pressure.

THRUST AT FITTINGS IN POUNDS AT 100 LBS. PER SQUARE INCH WATER PRESSURE.

PIPE SIZE (INCHES)	90° BEND	45° BEND	VALVES, TEES AND DEAD ENDS
4"	2,600	1,420	1,850
6"	5,400	2,900	3,800
8"	9,300	5,000	6,500
10"	13,900	7,550	10,850
12"	19,700	10,800	13,900
16"	34,000	18,600	24,200
20"	53,200	28,400	37,000
24"	74,300	40,400	52,700
30"	114,100	62,000	80,800
36"	163,400	88,700	115,600

If size of thrust block has not been specified by engineer, the following example shows steps required to determine the bearing area.

Assume thrust block is resisting horizontal thrust at an 8", 90° bend; pipeline to be tested at 200 psi and the soil is sand.

- Check the above table and you will find that the thrust developed on an 8", 90° bend is 9300 lbs. for each 100 lbs. of water pressure. Since the pipeline is to be tested at 200 psi, the total thrust is 2 x 9300 or 18,600 lbs.
- In the table below, you will find that the bearing power of sand is 2000 lbs. per square foot. Dividing the total force of 18,600 lbs. by 2000 lbs., gives a total required thrust backing of 9.3 square feet or an area slightly over 3 ft. x 3 ft.

SAFE BEARING LOAD	
TYPE OF SOIL	LBS. PER SQ. FT.
Muck, Peat, etc.	0
Soft Clay	1,000
Sand	2,000
Sand and Gravel	3,000
Sand and Gravel Cemented with Clay	4,000
Hard Pan	10,000

Caution: While often used in construction, EJP assumes no responsibility for the above bearing load data. The engineer is responsible for determining safe bearing loads and when doubt exists, soil bearing tests should be specified. The bearing loads given are for horizontal thrusts when depth of cover exceeds 2 ft.

R-16 Phone: 800-EJP-24HR (357-2447) • Fax: 207-582-5637

Figure 6.9 Thrust and Anchorage Fittings and Valves (Source indicated on figure)

Example 6.4b

Yard piping at a wastewater treatment plant (Massachusetts Correctional Institute, Bridgewater) includes the following, along with 250 psi design pressures: 2-inch tees and hydrant ends (pipe O.D. 2.50 in.); 3-in. tee and 90° bend (pipe O.D. 3.96 in.); and 3-in. × 2-in. reducer (same corresponding pipe O.D.s). Based on proximate borings, the corresponding soil types are soft clay, sand, and soft clay. Design pressure is 250 psi. Determine the size and configuration of thrust blocking.

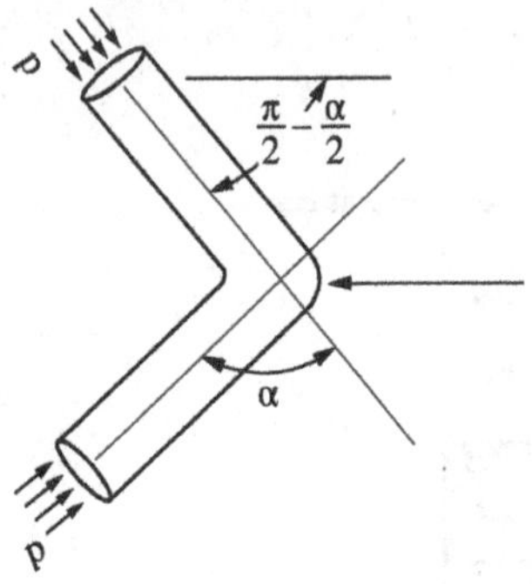

Solution

From Figure 6.9 the soil types correspond to safe bearing loads of 1000, 2000, and 2000 psf. For the 2-inch tee, with T_a being thrust, p being pressure, and d being pipe O.D., we have $T_a = (\pi/4)pd^2 = (\pi/4)(250)(2.50)^2 = 1230$ lb; the required thrust block area is then 1230/1000 = 1.23 sq ft. For the 3-inch tee, we have $T_a = (\pi/4)(250)(3.96)^2 = 3090$ lb; the required thrust block area is then 3090/2000 = 1.55 sq ft. The same size thrust block will be used for the 3-in. × 2-in. reducing tees. For the 3-inch 90° bend, $T_a = (\pi/4)pd^2/\sqrt{2} = (\pi/4)(250)(3.96)^2/\sqrt{2} = 4350$ lb; the required thrust block area is 4350/2000 = 2.2 sq ft. For the 3-in. × 2-in. reducer, $T_a = (\pi/4)(250)[(3.96)^2 - (2.50)^2] = 1840$ lb; the required thrust block area is 1840/1000 = 2.0 sq ft. The corresponding design drawings are shown on Figure 6.10.

If the pipe velocity is included in the above derivation of T_a, it can be shown to be negligible for velocities such as in the above example – an exercise left to the reader.

Example 6.5

A raceway consisting of a 9.34-ft square concrete pipe encasement is to cross the Spicket River in Lawrence, Massachusetts. The channel width is 30.5 ft, maximum design flow is 584 cfs, and the bottom of the raceway is 6.3 ft above the channel bottom. Determine the forces acting on the raceway.

Solution

It is first necessary to determine the worst-case condition. That would occur with the maximum depth on the upstream face of the raceway and separation of flow resulting in no counterbalancing depth of water on the downstream face. That condition would require that supercritical flow occur at the minimum depth of the separated flow with a hydraulic jump downstream. Without presenting details here we just state that the greatest Froude number of the separated flow was found to be well below unity, disallowing the occurrence of this condition. Attention then turned to the condition in which there would be water on downstream side of the raceway and the upstream energy grade line would be at the top of the raceway. The configuration and notation are depicted below.

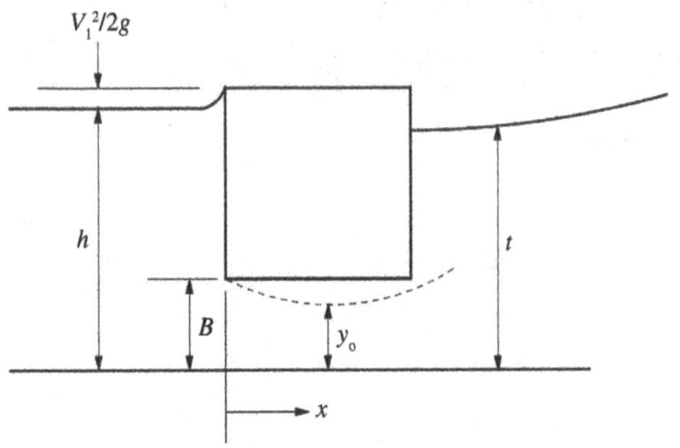

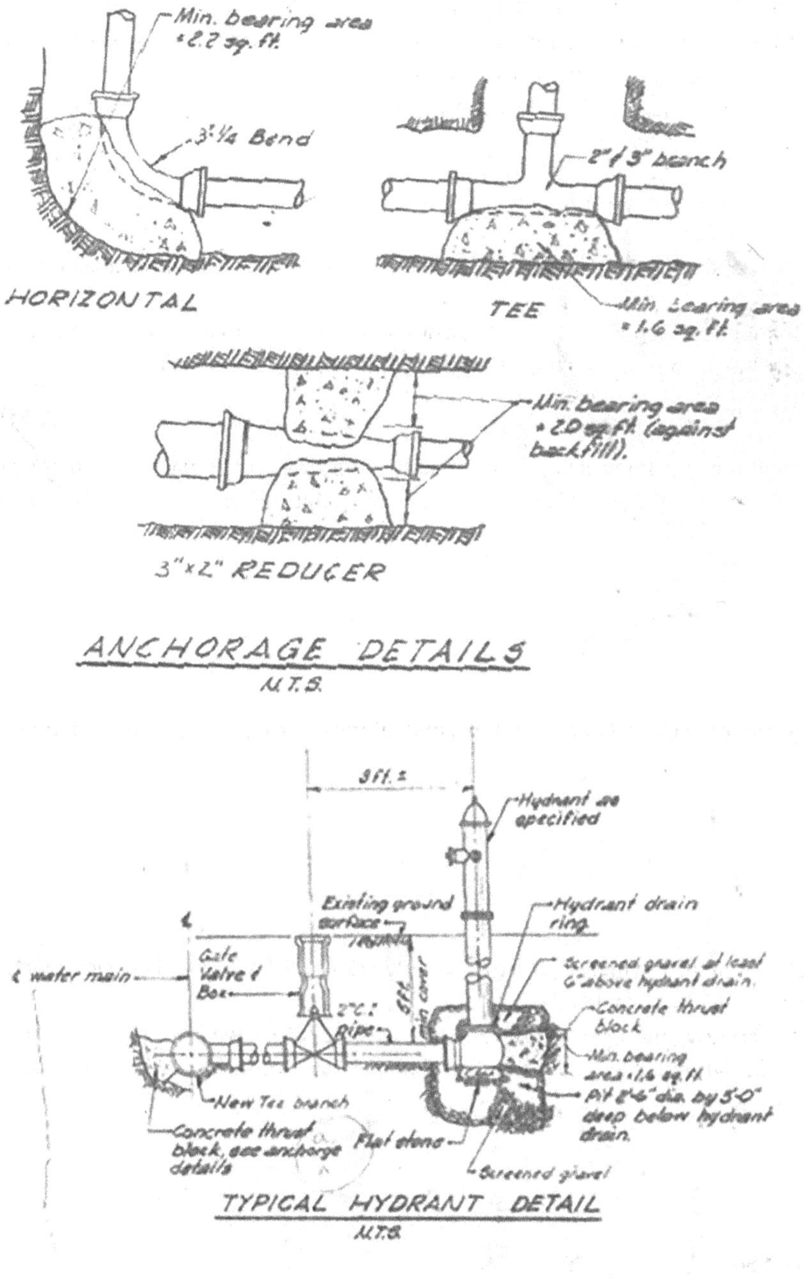

Figure 6.10 Anchorage and Hydrant Details for Bridgewater MCI

A closely applicable analogy is that of a submerged sluice gate, as studied by Rajaratnam and Subramanya (1967). The applicable equation is given by:

$$q = BC_c\sqrt{2g(h-t)}$$

in which q and C_c are, respectively, flow rate per unit width of channel and discharge coefficient, and other terms are as depicted above. From the given values, we have $q = 584/30.5 = 19.15$ cfs/ft. Using a conservative value of $C_c = 0.6$, the above equation gives:

$$h - t = \frac{1}{2g}\left(\frac{q}{BC_c}\right)^2 = \frac{1}{2(32.2)}\left[\frac{19.15}{6.3(0.6)}\right]^2 = 0.399 \text{ ft}$$

Adding run-up due to the upstream velocity head with h assumed to be 15.5 ft (slightly less than 6.3 + 9.34 = 15.64 ft), we obtain $V_1 = 19.15/15.5 = 1.235$ ft/sec and $V_1^2/(2g) = 0.0237$ ft. The differential water surface elevation difference across the raceway is approximately 0.399 + 0.0237 = 0.423 ft, requiring no further iterations.

We consider first the hydrostatic pressure force on the beam, which is highest when the upstream water surface is at the top of the raceway. The force/unit length of raceway, denoted f, is:

$$f = \frac{1}{2}\gamma yA = \frac{1}{2}(62.4)\left[(9.34)^2 - (9.34 - 0.42)^2\right] = 24.3 \text{ lbf/ft}$$

With Rajaratnam and Subramanya (1967), we can demonstrate that the width of the raceway has very little effect on the above calculation.

Next, we evaluate the force as drag to include submerged conditions. The horizontal force F is due to form drag which may be determined from the following equation attributed to Isaac Newton (Camp 1963, p. 36, Newton 1999):

$$F = C_D A \rho V^2/2 \tag{6.55}$$

in which C_D is drag coefficient, A is area on which the drag force acts, ρ is water density, and we set V equal to the approach velocity V_∞. Denoting by D the square encasement side dimension, the force per unit length of raceway for this case is given by:

$$f = \begin{cases} C_D D \dfrac{\rho V_\infty^2}{2}, h > B + D \\ C_D (h - B) \dfrac{\rho V_\infty^2}{2}, h < B + D \end{cases}$$

For $h < B + D$ we determine the maximum f assuming C_D is independent of h and taking $q \cong hV_\infty$ or $V_\infty \cong q/h$ as follows:

$$f = C_D(h - B)\frac{\rho q^2}{2h^2}$$

$$\frac{df}{dh} = C_D \frac{\rho q^2}{2} \frac{h^2 - (h - B)2h}{h^4} = 0$$

which gives the maximum f at $h = 2B = 2(6.3) = 12.6$ ft, at which $V_\infty = q/h = 19.15/12.6 = 1.54 \text{ ft/sec}$. The corresponding Reynolds Number is $\mathbf{R} = V_\infty D/\nu = 1.54(9.34)/10^{-5} = 1.44(10)^6$.

From several references (Rouse 1950, §30; Rohsenow and Choi 1961,§4.9; Streeter 1962, §5.6), we have $C_D = 2$ for flat plates and a square cylinder at turbulent Reynolds Numbers less than that calculated above. We expect C_D to be the same or smaller at a higher Reynolds Number so we have:

$$f = 2(6.3)\frac{62.4(1.54)^2}{2(32.2)} = 28.8 \text{ lbf/ft}$$

which is greater than the 24.3 lbf/ft calculated above.

We now calculate the maximum torsion T as follows:

$$T = fr$$

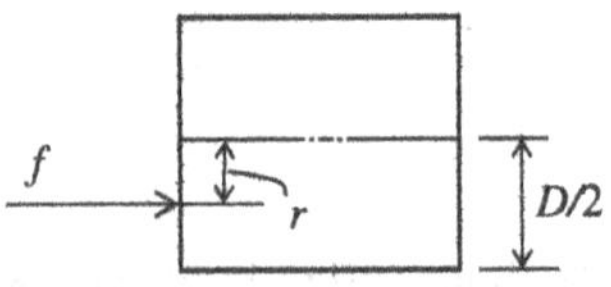

$$r = \frac{D}{2} - \frac{h - B}{2}$$

$$T = C_D(h-B)\frac{\rho q^2}{2h^2}\left(\frac{D}{2} - \frac{h-B}{2}\right) = C_D\frac{\rho q^2}{4}\left[\frac{D(h-B)-(h-B)^2}{h^2}\right]$$

$$\frac{dT}{dh} = C_D\frac{\rho q^2}{4}\left[\frac{h^2D - D(h-B)2h}{h^4} - \frac{h^2 2(h-B)-(h-B)^2 2h}{h^4}\right] = 0$$

Solving the above equation for h and substituting in values yields:

$$h = \frac{2B(B+D)}{2B+D} = \frac{2(6.3)(6.3+9.34)}{2(6.3)+9.34} = 9.0$$

which gives for other pertinent variables using equations given above:

$$r = \frac{9.34}{2} - \frac{9.0-6.3}{2} = 3.32 \text{ ft}$$

$$f = 2(9.0-6.3)\frac{62.4}{2(32.2)}\left(\frac{19.2}{9.0}\right)^2 = 23.8 \text{ lbf/ft}$$

$$T = 23.8(3.32) = 79.2 \text{ ft} - \text{lbf/ft}$$

The final step in the analysis consisted of evaluating vertical alternating forces due to Kármán vortex shedding. A similar analysis is presented for another Lawrence, Massachusetts river crossing in Example 11.5. Similar details are not presented here for the raceway crossing considered here; we simply state that the alternating forces on the raceway were found to be small compared to the weight or buoyant forces. Also, the frequency of the alternating forces is considerably less than the natural frequency of the beam so no resonant condition is anticipated.

6.5.5 Other Pipe Appurtenances

Positive-displacement flow meters are discussed here, including an important shortcoming that the author has encountered. A common example is the nutating-disk meter, consisting of a nutating primary element through which the flow traverses, and a secondary element which receives information from the primary element and converts it to a readout that may be locally displayed and/or telemetered to a passing vehicle or a central location. The author has encountered several cases in which water bill disputes have arisen where such meters have been used to measure water usage at businesses or homes. Related information given below is from Graber (2024a).

Metering is clearly important in reducing water consumption (e.g., Tanverakul and Juneseok 2013) and financing water-supply systems including recovering lost revenue (e.g., Tippin 1978). However, it must be recognized that metering can be imperfect, and common sense needs to be applied to avoid egregious charges. After successfully challenging a water bill, and subsequent posting of the result on the internet, the author has been contacted by others who were questionably billed and has otherwise learned of additional examples of obviously faulty water bills. Such examples are discussed herein, with the hope that some reduction in questionable billing can result. On the other hand, water theft can be an issue and an example of that is presented.

6.5.5.1 Initial Example

The author was engaged by a forensic engineering firm regarding a water bill dispute involving an automobile dealership on Cape Cod. Unfortunately, the owner of the dealership had to hire an engineer and a lawyer to win this case. The owner had a positive-displacement flow meter of the common nutating-disk type, consisting of a nutating primary element through which the flow traverses, and a secondary element which receives information from the primary element and converts it to a readout that may be locally displayed and/or telemetered to a passing vehicle or a central location.

In this particularly egregious example, in which the author served as an expert witness, the dealership was charged by the municipal water company for a nearly 50-fold increase in municipal water usage (and water-usage-based sewer bill) (Stagg *et al.* 2006). The author presented the following arguments: (1) with any reasonable street water pressure and any conjectured leak location in the dealership's piping, it would be impossible to hydraulically convey the claimed rate of flow through the system; (2) the author debunked the water company's "leaky-toilet" theory because even 2500 gallons per day of additional water consumption caused by five perpetually leaking toilets would explain only a small fraction of the water bill; (3) the misreading by the water meter was readily explained. From the record: The author "theorized that the water meter in question incorrectly read the water usage at the Subject Property during the Period at Issue because the meter 'jumped and dragged' an adjacent dial or dials along with it. This jump caused at least the seventh- or millionth-place digit of the meter's mechanical-counting system to over-rotate and erroneously display a higher number in that place. Mr. Graber confirmed the possibility of just such a mechanical-counting-system failure with the manufacturer of this particular Neptune meter and model, which has not been manufactured since 1981. Mr. Graber further testified that the test of the subject water meter conducted by Regan Testing Service did not use a sufficient quantity of water to test for the mechanical-counting-system anomaly, which, in his view, caused the excessive water-usage reading."

The absurdity of the water company's claims was recognized by the Appellate Tax Board, which found in favor of the author's client. That Board abated the water usage charges such that the client was only required to pay 1/50th of the bill. Additional details may be found in Stagg Chevrolet, Inc. vs. Board of Assessors of the Town of Harwich, 2006. The decision of the Appellate Tax Board was affirmed on appeal (Stagg Chevrolet, Inc. vs. Water Commissioners of Harwich 2007).

6.5.5.2 Some Additional Examples Worked by the Author

The posting on the internet of the results of the Stagg case mentioned above resulted in additional requests for the author to investigate similar cases. A sampling of those is discussed here.

Somerville, Massachusetts. The author served as a consultant to owners of a one-family townhouse in Somerville, Massachusetts. The author prepared an affidavit in this case, which included the following. The water bill was normally on the order of \$150–\$200 for two-, three-, or four-month periods over five years, and they then received a four-month bill in 2007 for \$10,568 which was increasing at a 14% annual interest rate. That water bill was 35 times the long-term average for comparable periods. The water department suggested a toilet leak; the homeowners hired a plumber to look for leaks and none were found. There were no exterior taps to blame, nor could toilet leakage under worst-possible conditions account for the quantity of water claimed according to the plumber. The water department also confirmed that there was no leak, and in June 2007 replaced the water meter with a new one. The original meter was tested by the water department four times and was reported to be close to 100% accurate. A separate set of three tests gave similar results. A total of about 30 gallons was used for the first set of tests and a total of about 90 gallons for the second set.

The question naturally arises as to how a meter which when tested exhibits a nearly 100% accuracy can in fact be inaccurate. The meter consists of two parts. The first is the primary portion of the meter (nutating disc), a positive-displacement device which, as it ages, can develop leaks and if anything would measure less than the full quantity of water going through it. The other part of the meter is the secondary portion, which acts in a way similar to an odometer. It is connected magnetically and mechanically to the primary portion of the meter and, in essence, counts the number of revolutions of the primary device and converts that number to a useful measure of water volume. As this component ages, it can perform erratically, and jump as the dials turn, thus measuring more, and in some cases much more, than the actual quantity of water going through the meter. The author confirmed that older meters can malfunction in this manner ("number drag") with a technician at Neptune Company, manufacturer of the meter. A short-term test is generally incapable of determining the type of accuracy reported by the water department. The 120 gallons of total water metered during the city's two tests is less than one-hundredth of one percent of the amount of water metered during the period in question, not nearly enough to test the accuracy of the secondary device. The meter company indicated, based on the serial number, that the meter was about 21 years old at the time of the water bill in question.

The Massachusetts Department of Environmental Protection issues permits that require repair, replacement, or recalibration of all individual service meters over ten years of age. During the period of the questionable bill, the homeowners would have had to use an average of 3.9 gallons/min. The author was able to demonstrate that such a quantity could not flow without being very noticeable or causing considerable damage; the homeowners could testify that there was no such damage, and had a statement from their insurance company confirming that they had no water damage during the period in question. The author demonstrated that worst-case assumptions regarding fixture leakage could not possibly account for the amount of water usage claimed by Somerville. After the meter was replaced, the readings returned to normal. During testimony when asked to explain the questioned water usage, the City's engineer claimed that there could have been an unnoticeable "pinhole leak," but could not explain how such a leak could account for the claimed quantity of water. That engineer further stated that if the homeowner agreed there had been leakage (which they did not) they would have received a 50% rebatement. Where is the logic? The case was settled by a compromise between the homeowner and water department, the homeowner reasoning that the compromise would avoid further engineering and/or legal costs.

Sewer Connection Incentive. It is common in communities for some or all of the water customers to have individual wastewater disposal systems (e.g., septic tanks and leaching fields). Such customers are most often charged for water service only, although some communities charge those not on municipal sewers but with sewer betterments (for which they have been charged) for sewerage service anyway for the revenue and as an incentive for them to connect. The pros and cons of the latter practice are not addressed here, except to say that individual septic systems can be less polluting and usefully augment groundwater. Another situation that arises is discussed next with reference to a client of the author.

Medfield, Massachusetts. A more general issue concerned payment of sewer charges based on a fixed percentage of the water bill. In the case at hand, a sophisticated system recycles car wash water, only wasting that when Total Dissolved Solids exceed a prescribed limit based on automatic conductivity metering to avoid water spotting of the vehicles. That wasted water is then replenished by clean water from the Town system. The total amount of clean water taken in makes up the water wasted as just described plus makes up for the water lost as evaporation and carry-out in the car-washing operation. The net result is that about one-fourth of the total water taken from the Town does not discharge to the sewer. That estimate is most likely on the low side, but without a sewage meter one can only estimate. It is also noted that the number of cars washed during the disputed billing period was less than the periods before and after. This all takes into account the fact that sanitary wastes from public and private facilities and a restaurant (totaling 4 low-flush toilets and 5 washbasins) were also discharged to the municipal sewer system.

With the author's assistance, the owner proposed to install a meter to measure the sewage going to the Town's system, to enable billing based on the metered amount (as is done elsewhere – see below). The Town denied that request. That seems unreasonable since the owner is being billed for services that are not actually received. This obviously creates a disincentive to reduce sewage discharge.

Wastewater metering devices and billing for separate water and sewer services are allowed by regulations of the Massachusetts Water Resources Authority (MWRA 2009), which applies to many of the communities in its large eastern- and central-Massachusetts service area (but not Medfield). Some non-MWRA Massachusetts communities also provide for such monitoring and billing. For example, the City of Attleboro, Massachusetts (2012) provides that "Any user of the city's sewer system may install and maintain, at his own expense, an individual sewer meter or other measuring device acceptable to the Superintendent. Sewage so measured shall be charged for at the ... [sewer use rate]." A car wash in Attleboro was cited by the Attleboro Superintendent that used so much water that it was well worth it to them to install their own sewage meter.

A problem that occurs that requires diligence on the part of water departments is water theft. While working on the Medfield case, the author learned of a case in which a homeowner installed a well from which he asserted he would draw all his water. The homeowner then installed a bypass around his municipal water meter and proceeded to draw all of his water, including for watering his large lawn, from the municipal system. He was caught by an inspector.

Sewer bills are commonly based on a fraction or multiple of the metered water usage (75% in the case of Medfield). An issue that arises concerns requests for subtractive metering of water used for lawn watering to adjust the sewer bill for municipal water that is not returned to the sewer. Two residents who came before the Medfield water and sewer board argued heatedly for the installation of a sewer meter to enable an exemption from sewer charges for water used in their outdoor sprinkler system; their request was denied (Domeshek 2009). ("Sewer meters" are standard water meters that measure some water not returned to the sewer.) Such metering would be expensive and is not generally done in Massachusetts. Furthermore, Medfield concurs with the Massachusetts Department of Environmental Protection's implementation of policies that discourage outdoor water use in the interest of water conservation.

6.5.5.3 Some Other Examples

Brockton, Massachusetts. Another batch of egregious examples was found in Brockton, Massachusetts (Littlefield, 2010, Bolton, 2011). One of the most notorious ones was that of a homeowner who was told he owed the City $17,000. That particular homeowner had a dysfunctional outdoor meter (which the City acknowledged) and disputed the City's claim that they had entered his house to read a meter. Thousands of other residents had similar complaints, and the City acknowledged that they had used estimated billing rather than meter readings for six years.

Randolph, Massachusetts. In (Reardon 2008) a homeowner's combined water and sewer bill (with the sewer portion based on a multiple of the water bill) jumped from $349 to $5222.60 for a comparable period. The Town claimed the charge was legitimate, despite a plumber inspecting the property and, among other things, reporting the meter to be spinning and making "loud rapid clicking noises" when the water was turned off.

Dayton, Ohio. Tobias (2012) cites numerous examples of obviously inaccurate or questionable water bills that customers were required to pay in Dayton, Ohio. He also cites the difficulties in contesting those bills. Most of the problems were due to old meters that were past their assumed 20-year lifespan. Tobias mentions a Dayton attorney who has practiced there for nearly 50 years and said he has taken many calls over his career from people who want to challenge their bills in court. But he turns them down because they are not winning cases. The attorney was quoted as saying "There's no question about the fact that [water meters] can become faulty and generate erroneous data ... But how in the world are you going to prove all of that?" The author has demonstrated that in the above.

6.6 Pipe Networks

A pipe network is a system of interconnected pipes in which the flows entering and leaving the system may be known but the flows within the individual pipes are unknowns to be obtained as part of the problem solution. Such networks generally contain combining- and dividing-flow junctions and often include various of the types of hydraulic components discussed earlier in this chapter.

The most common type of networks encountered in environmental engineering practice are those associated with water transmission and distribution systems. Pipe network analysis methods may also be usefully applied to surcharged sewer networks (such as interconnected parallel interceptors). Conduits in which a prescribed flow distribution is desired are also amenable to network analysis, but are more conveniently treated as *distribution conduits* (Chapter 11).

Methods for the analysis of such systems are introduced in this section. It is not the intention here to delve into the more technical aspects of pipe network analysis. Rather, some perspective will be provided along with guidance to the literature which amply covers the subject. We will begin by discussing the simplification of parallel and series pipes by equivalent pipe representations, with consideration of representations which are simultaneously appropriate to the analysis of transients. Then algorithms for network analysis will be introduced.

6.6.1 Equivalent Pipe Representations

Any number of pipes in parallel or series could be represented as a single equivalent pipe by successive combination of pairs of actual and equivalent pipes. The practical purpose of doing so for network analysis relates to the value of simplifying networks for subsequent analysis whether by computer methods or otherwise in order to reduce the computational effort (including computer time). In the next section we will see that network simplification is also beneficial for analyses of fluid transients. Some care must be taken in the choice of pipes to be combined so that pipes for which pressure or velocity information is specifically desired are not "lost" in equivalent pipes. Depending on the purpose, there are different ways of determining equivalent pipes.

6.6.1.1 Parallel Pipes

From the standpoint of pipe friction, a single equivalent pipe between two junctions is one giving the same total flow between the two junctions at the same differential head, and conversely. Thus:

$$h_{Le} = h_{L1} \tag{6.56a}$$

$$h_{Le} = h_{L2} \tag{6.56b}$$

$$Q_e = Q_1 + Q_2 \tag{6.56c}$$

in which the subscripts "*e*," "*1*," and "*2*" denote equivalent pipe, pipe 1, and pipe 2, respectively. Each of the h_L terms can be expressed in terms of the respective pipe length, diameter, and friction factor. Using the Hazen–Williams equation (Darcy–Weisbach or others could be used in a similar fashion):

$$S_f = \frac{h_f}{L} = \frac{4.73LQ^{1.852}}{C^{1.852}D^{4.87}} \tag{6.57}$$

in which C is the Hazen–Williams coefficient, S_f is friction slope in ft/ft, h_f is head loss in ft, L is pipe length in ft, Q is volumetric flow rate in cfs, and D is pipe internal diameter in inches.

Substituting Equation 6.57 with the appropriate subscripts into Equations 6.56a and 6.56b, we may solve for Q_1 and Q_2 in terms of Q_e, substitute the relations for Q_1 and Q_2 into Equation 6.56c, cancel Q_e, and rearrange to obtain:

$$\frac{C_eD_e^{2.63}}{L_e^{0.54}} = \frac{C_1D_1^{2.63}}{L_1^{0.54}} + \frac{C_2D_2^{2.63}}{L_2^{0.54}} \tag{6.58}$$

The terms on the right-hand side of the above equation are constants. Based on equivalent pipe friction alone, convenient values of L_e and C_e may be selected and the equivalent diameter calculated, or L_e or C_e can be the calculated variable based on selected values for the other equivalent pipe variables.

An alternative method can be derived based on Darcy–Weisbach (Equation 5.1). Setting $V = Q/A$ in that equation, for two parallel pipes of the same length and friction head loss and each of diameter D, we have $Q^2/D^5 = 2^2/D_{eq}^5 = 1^2/D^5 =$ constant in which D_{eq} is the equivalent diameter. Thus $D_{eq} = (4^{1/5})D$.

Example 6.6

Two parallel lines, each of 48-inch diameter and 40,000 ft long, convey a total of 60 mgd (Salem/Beverly Wenham Lake Pumping Station). As part of a network analysis, it is desired to find an equivalent single 40,000 ft pipe. Find the equivalent diameter.

Solution

Using the relationship immediately above, we have $D_{eq} = (4^{1/5})(48) = 63$ inch

The arbitrariness of these choices for equivalent diameter is reduced if inertial characteristics are to be considered. In the next section, we will show that the pipes connecting two nodes have equivalent inertial (but not frictional) characteristics if they have the same $\sum LV = \sum LQ/A \propto \sum LQ/D^2$. For parallel pipes, the single pipe of equivalent inertial characteristics is given by:

$$L_e \frac{Q_e}{D_e^2} = L_1 \frac{Q_1}{D_1^2} + L_2 \frac{Q_2}{D_2^2} \tag{6.59}$$

Substituting the same relations for Q_1 and Q_2 as used above, we may derive an expression for an equivalent pipe based on inertial considerations:

$$C_e D_e^{0.63} L_e^{0.46} = C_1 D_1^{0.63} L_1^{0.46} + C_2 D_2^{0.63} L_2^{0.46} \tag{6.60}$$

We can now eliminate L_e between Equations 6.59 and 6.60 to obtain an expression for the equivalent diameter based on both frictional and inertial considerations:

$$D_e = \left[\frac{C_1}{C_e} D_1^{0.63} L_1^{0.46} + \frac{C_2}{C_e} D_2^{0.63} L_2^{0.46}\right]^{0.348} \bullet \left[\frac{C_1}{C_e}\frac{D_1^{2.63}}{L_1^{0.54}} + \frac{C_2}{C_e}\frac{D_2^{2.63}}{L_2^{0.54}}\right]^{0.30} \tag{6.61}$$

Solving Equation 6.60 for L_e, then substituting for D_e from the above equation results in the corresponding expression for the equivalent length:

$$L_e = \left[\frac{C_1}{C_e} D_1^{0.63} L_1^{0.46} + \frac{C_2}{C_e} D_2^{0.63} L_2^{0.46}\right]^{2.17} / D_e^{1.37} \tag{6.62a}$$

or:

$$L_e = \left[\frac{C_1}{C_e} D_1^{0.63} L_1^{0.46} + \frac{C_2}{C_e} D_2^{0.63} L_2^{0.46}\right]^{1.69} \Big/ \left[\frac{C_1}{C_e}\frac{D_1^{2.63}}{L_1^{0.54}} + \frac{C_2}{C_e}\frac{D_2^{2.63}}{L_2^{0.54}}\right]^{0.41} \tag{6.62b}$$

6.6.1.2 Series Pipes

From the standpoint of pipe friction, a single equivalent pipe between two junctions is one giving the same total flow between the two junctions at the same differential head, and conversely. Thus:

$$h_{Le} = h_{L1} + h_{L2} \tag{6.63a}$$

$$Q_e = Q_1 \tag{6.63b}$$

$$Q_e = Q_2 \tag{6.63c}$$

in which the subscripts "*e*," "*1*," and "*2*" denote equivalent pipe, pipe 1, and pipe 2, respectively.

Substituting the Hazen–Williams equation (Equation 6.2) with the appropriate subscripts into Equation 6.63a, we may use Equations 6.63b and 6.63c to eliminate the flow terms and rearrange to obtain:

$$\frac{L_e}{C_e^{1.852}D_e^{4.87}} = \frac{L_1}{C_1^{1.852}D_1^{4.87}} + \frac{L_2}{C_2^{1.852}D_2^{4.87}} \tag{6.64}$$

The terms on the right-hand side of the above equation are constants. Based on equivalent pipe friction alone, convenient values of L_e and C_e may be selected and the equivalent diameter calculated, or L_e or C_e can be the calculated variable based on selected values for the other equivalent pipe variables.

As in the case of parallel pipes, the arbitrariness of these choices is reduced if inertial characteristics are to be considered. For series pipes, the single pipe of equivalent inertial characteristics is obtained in a manner similar to that for parallel pipes, and given by:

$$\frac{L_e}{D_e^2} = \frac{L_1}{D_1^2} + \frac{L_2}{D_2^2} \tag{6.65}$$

We can now eliminate L_e from Equations 6.64 and 6.65 to obtain an expression for the equivalent diameter based on both frictional and inertial considerations.

$$D_e = \left(\frac{L_1}{D_1^2} + \frac{L_2}{D_2^2}\right)^{0.348} \Big/ \left[\left(\frac{C_e}{C_1}\right)^{1.852}\frac{L_1}{D_1^{4.87}} + \left(\frac{C_e}{C_2}\right)^{1.852}\frac{L_2}{D_2^{4.87}}\right]^{0.348} \tag{6.66}$$

Solving Equation 6.65 for L_e, then substituting for D_e from the above equation results in the corresponding expression for the equivalent length:

$$L_e = D_e^2\left(\frac{L_1}{D_1^2} + \frac{L_2}{D_2^2}\right) \tag{6.67a}$$

or:

$$L_e = \left(\frac{L_1}{D_1^2} + \frac{L_2}{D_2^2}\right)^{1.696} \Big/ \left[\left(\frac{C_e}{C_1}\right)^{1.852}\frac{L_1}{D_1^{4.87}} + \left(\frac{C_e}{C_2}\right)^{1.852}\frac{L_2}{D_2^{4.87}}\right]^{0.696} \tag{6.67b}$$

Using Equations 6.67a and 6.67b, a network containing parallel and series pipes can be reduced to a simpler network of equivalent frictional characteristics. Values of two of the three variables C_e, D_e, and L_e are selected and the value of the remaining variable then calculated. Alternatively, if inertial characteristics are also of concern (and even if they are not), the equivalent diameters and lengths may be readily calculated from Equations 6.65, 6.66, 6.67a, and 6.67b for selected C_e's to obtain a reduced network of equivalent frictional and inertial characteristics. The latter four relationships may be programmed to obtain a convenient tool for network reduction.

Example 6.7

Simplify the transmission network of Figure 6.11 by reducing the series of pipe to one of equivalent inertial characteristics (Massachusetts Correctional Institute, Bridgewater).

Solution

Equation 6.67a is generalized for multiple pipe lengths and applied as follows assuming a pipe diameter of 10 in. for the equivalent pipe:

$$L = 350 + 1390\left(\frac{10}{8}\right)^2 + 540\left(\frac{10}{10}\right)^2 + 1300\left(\frac{10}{12}\right)^2 + 90\left(\frac{10}{8}\right)^2 = 350 + 2180 + 540 + 900 + 140 = 4110 \text{ ft}$$

For more complicated cases in which both frictional and inertial characteristics are considered, computer programs facilitate efficient computations.

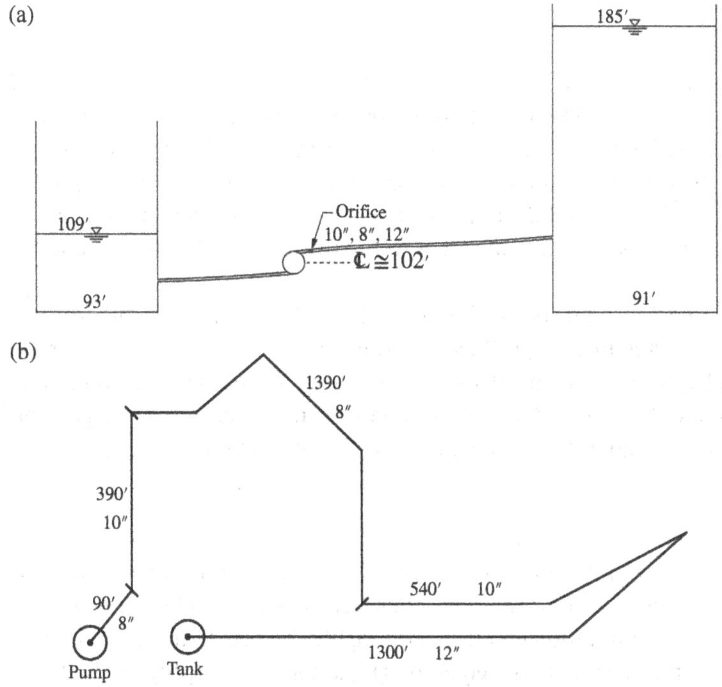

Figure 6.11 (a) Section View of Example Series Transmission Network; (b) Plan View of Example Series Transmission Network

6.6.2 Algorithms for Network Analysis

The knowns and unknowns in the network just mentioned are represented on the figure. Historically, the first method developed for the solution of such problems was that of Hardy Cross, developed at the Clarkson College of Technology during 1970–1971 (Fair and Geyer 1954, §13-6; Washington State Dept. of Social and Health Services 1975). Michalos (1972) has provided an interesting account of the development of this method. The Hardy Cross method provides a readily understood introduction to network analysis.

The Hardy Cross method has been replaced by more recent methods which provide improved convergence and computational efficiency. These methods are generally implemented on digital computers. Wood and Rayes (1981) have published an excellent overview discussion of the various methods. That reference, plus *Analysis of Flow in Pipe Networks* (Jeppson 1976) and *Modeling, Analysis, and Design of Water Distribution Systems* (Cesario 1995) are especially recommended.

The more successful techniques currently in use [e.g., those based on the methods of Epp and Fowler (1973) and Wood and Charles (1972)] are fairly sophisticated and not always as well understood as they should be by users of acquired programs based on those methods. Familiarity with the techniques and programming algorithms confers a decided advantage to those willing to make the educational investment. The author has made use of water system network analyses using KYPIPE2 (Wood 1991), e.g., in Winthrop, Massachusetts, a system of more than 2500 pipes. Related software for preliminary steady-state analyses in the transient analysis of water distribution systems is discussed in Section 8.5. For the Greater Lawrence Sanitary District along the Spicket River in Massachusetts, the author modified Wood's Extended Period Simulation (EPS) Model (Wood 1982), which was possible because at that time the BASIC source code was provided (which unfortunately is no longer done for proprietary reasons). The network was a combined wastewater-stormwater system, modeled under surcharged conditions, but with the surcharge all below manhole elevations. The BASIC model was compiled for fast execution (and error checking). Multiple runs were made for the system of up to 96 pipes and 73 junction nodes, including a proposed interceptor along the river. Special consideration was given to high-velocity junction losses. With the aid of these runs, a six-foot diameter interceptor was recommended.

Numerous methods have been put forth for optimizing water distribution networks. Swamee and Khanna (1974) provided an important contribution which rigorously demonstrates that numerous existing theories for optimizing water networks are erroneous. Caution must be exercised in optimization attempts, and common-sense approaches are often preferable.

6.7 Time-Varying Flow

The preceding sections of this chapter have dealt with flows which are temporally constant. In many environmental engineering applications, the flows are in fact varying with time. The variations are often sufficiently slow that the flow may be regarded as quasi-steady and analyzed accordingly. For example, if the flow changes by only 1% over a time period on the order of the time of travel through the system, the assumption of temporally constant flow is obviously acceptable. If the changes are abrupt, then inertial effects become significant, and the analysis becomes more complex. If the changes are sufficiently abrupt, then compressibility (elasticity) effects become significant as well, further increasing the complexity of the analysis.

Quasi-steady and inertial flow conditions are discussed below with specific reference to problems of practical interest for which compressibility effects may be neglected. This is referred to as *Rigid Water Column Theory.* Discussion of compressibility effects in liquid flow will be deferred until the Chapter 8 section on *Waterhammer* (Section 8.6), by which point further background for consideration of those effects will have been provided. That section will address criteria for neglecting compressibility effects, that is, for the validity of the types of analyses presented in the present section.

6.7.1 Quasi-Steady Flow

The quasi-steady flow assumption will be demonstrated by reference to a practical example. Consider the tank of Figure 6.12, with a cylindrical top and conical bottom, draining into the atmosphere as shown (a sludge storage tank discharging to a barge, for instance). The quasi-steady flow assumption entails neglecting the inertial terms in the energy equation (Section 2.6), which then reduces to the one-dimensional energy equation (Equation 2.71). Applying the latter between points a and b of Figure 6.12 with the elevation at the hose valve taken at datum ($z_4 = 0$) results in:

$$z + \frac{V_a^2}{2g} + \frac{p_{atm}}{\rho g} = 0 + \frac{V_b^2}{2g} + \frac{p_{atm}}{\rho g} + h_L \tag{6.68}$$

We will neglect the friction head loss in the tank. Representing h_L by $FV_b^2/(2g)$, employing continuity ($V_aA_a = V_bA_b$), and denoting $R = A_b/A_a$, we obtain:

$$z = \left(F + 1 - R^2\right)\frac{V_b^2}{2g} \tag{6.69}$$

Since $V_b = A/A_b$, the flow is related to the head z by:

$$Q = C_1 z^{1/2} \tag{6.70a}$$

$$C_1 = \left(\frac{2gA_b^2}{F + 1 - R^2}\right)^{1/2} \tag{6.70b}$$

We will assume $R^2 \ll 1$ so that the term R^2 can be deleted in the following.

The flow and tank volume $\forall$ are related by $Q = -d\forall/dt$, and the tank cross-sectional area A is related to tank volume by $A = d\forall/dz$. Thus:

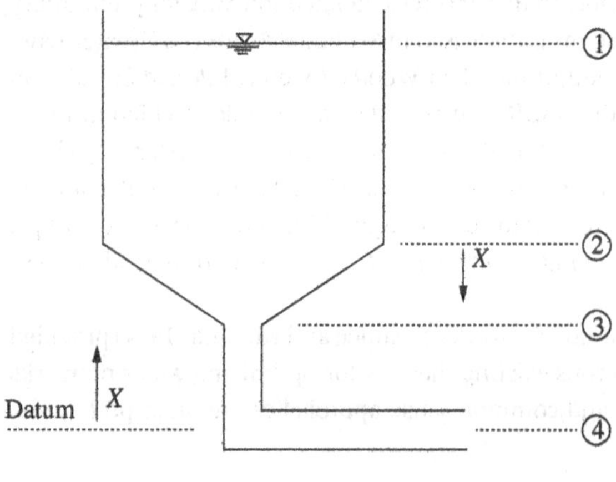

Figure 6.12 Storage Tank Example

$$Adz = -Qdt \tag{6.71}$$

Eliminating Q between Equations 6.70a and 6.71 results in:

$$dt = -\frac{Adz}{Q} = -\frac{Adz}{C_1 z^{1/2}} \tag{6.72}$$

Integrating,

$$\Delta t = -\frac{1}{C_1}\int \frac{Adz}{z^{1/2}} \tag{6.73}$$

in which A is a function of z.

For the particular tank shown in Figure 6.12, the time to drain the tank from elevation 1 to elevation 2 is obtained by setting $A = A_2 = \text{const}$ in Equation 6.73, giving:

$$t_2 - t_1 = -\frac{A_2}{C_1}\int \frac{dz}{z^{1/2}} = \frac{2A_2}{C_1}\left(z_1^{1/2} - z_2^{1/2}\right) \tag{6.74}$$

The tank area A between elevations 2 and 3 is given by:

$$A = A_2[1 - n(z_2 - z)]^2 \tag{6.75a}$$

$$n = \frac{D_2 - D_3}{D_2(z_2 - z_3)} \tag{6.75b}$$

which is substituted into Equation 6.74 to give:

$$t_3 - t_2 = -\frac{A_2}{C_1}\int \frac{[1 - n(z_2 - z)]^2 dz}{z^{1/2}} = \frac{A_2}{C_1}\left[2\left(z_2^{1/2} - z_3^{1/2}\right)(1 - nz_2)^2 + \frac{4}{3}n\left(z_2^{3/2} - z_3^{3/2}\right)(1 - nz_2) + \frac{2}{5}n^2\left(z_2^{5/2} - z_3^{5/2}\right)\right] \tag{6.76}$$

The total time to drain the tank is given by $(t_2 - t_1) - (t_3 - t_2)$ as given by the above equations.

Application to most other tank geometries should be apparent. The following example uses the equations above.

Example 6.8

A homogeneous, digested sludge (10% maximum suspended solids) is to be barge loaded from an existing tank of the configuration shown in Figure 6.12 (Tallman Island WWTP, New York City). The cylindrical portion of the tank is 35 ft in diameter by 30 ft high, while the conical portion has a bottom diameter equal to the diameter of the discharge pipe and is 3.75 ft high. The bottom of the conical portion is 6.75 ft above the invert of the discharge pipe, which is assumed horizontal. The discharge pipe, including the length along the pier and barge, is 800 ft long, and preliminarily assumed to be new 20-inch diameter steel. Valves and horizontal bends will be neglected at this stage. Determine the range of pipe velocities and estimate the time to drain the tank.

Solution

Referring to Figure 6.12, the pertinent elevations are as follows: $z_3 = 6.75$ ft, $z_2 = 6.75 + 3.75 = 10.5$ ft, $z_1 = 10.5 + 30 = 40.5$ ft. The initial and final pipe velocities may be obtained using Equation 6.69, expressed as follows:

$$V_b = \sqrt{\frac{2gz}{K+1}}$$

The value of K is dependent on the velocity V_b, so a trial-and-error procedure is necessary. For steel pipe, we have $\varepsilon = 0.00015$ ft, giving a roughness ratio of $\varepsilon/D = 0.00015/(20/12) = 0.00009$. From the Moody diagram (Figure 5.1) or Equation 5.1, assuming "complete turbulence, rough pipes" for the first trial, we have $f \cong 0.012$. We will increase f by 1% for each percent solids in the sludge (see Section 13.4) giving $f = 1.1(0.012) \cong 0.013$ and $fL/D = 0.013(800)/(20/12) = 6.24$. Including the entrance loss, exit loss, and one vertical bend, we have $\sum K = 0.5 + 1.0 + 0.9 = 2.4$. Therefore $K = (fL/D + \sum K) = 6.24 + 2.4 = 8.64$. The two pipe velocities are then calculated based on the assumed friction factor: $V_b = \sqrt{2(32.2)(40.5)/(8.64 + 1)} = 16.4$ ft/sec initially, and $V_b = \sqrt{2(32.2)(6.75)/(8.64 + 1)} = 6.7$ ft/sec finally.

The kinematic viscosity of 10% digested sludge is in the vicinity of $5(10)^{-4}$ ft^2/sec and the resulting Reynolds Number range is given by:

$$\mathbf{R} = \frac{VD}{\upsilon} = \frac{6.7(16)/12}{5(10)^{-4}} = 2.24(10)^4 \text{ to } \frac{16.4(16)/12}{5(10)^{-4}} = 5.48(10)^4$$

The Moody Diagram gives a corresponding range of f values of 0.021 to 0.025. The higher of these values is selected as the next trial value and increased as before, giving $f = 1.1(0.025) = 0.028$.

Using the new trial value of f, we obtain: $fL/D = 0.028(800)/(20/12) = 13.4$, $K = (fL/D + \sum K) = 13.4 + 2.4 = 15.8$, $V_b = \sqrt{2(32.2)(40.5)/(15.8+1)} = 12.5$ ft/sec initially, and $V_b = \sqrt{2(32.2)(6.75)/(15.8+1)} = 5.1$ ft/sec finally.

The revised Reynolds Number range is:

$$\mathbf{R} = \frac{5.1(16)/12}{5(10)^{-4}} = 1.36(10)^4 \text{ to } \frac{12.5(16)/12}{5(10)^{-4}} = 3.33(10)^4$$

A revised conservative f value of 0.028 is given by the Moody Diagram, which is adjusted for solids concentration to 1.1 (0.028) = 0.030. No further iteration is necessary.

Using the final value of f, we obtain: $fL/D = 0.030(800)/(20/12) = 14.4$, $K = (fL/D + \sum K) = 14.4 + 2.4 = 16.8$, $V_b = \sqrt{2(32.2)(40.5)/(16.8+1)} = 12.1$ ft/sec initially, and $V_b = \sqrt{2(32.2)(6.75)/(16.8+1)} = 4.94$ ft/sec finally.

Velocities between 5 and 8 ft/sec are considered satisfactory to prevent deposition of solids (see Section 13.2), and the above is considered acceptable (a backup plunger pump was provided to clean out the line and handle occasional clogging).

The value of C_1 may now be determined. The pipe cross-sectional area is $A_b = \pi(20/12)^2/4 = 2.18$ ft^2. Then, from Equation 6.70b, $C_1 = \left[2(32.2)(2.18)^2/(16.8+1)\right]^{1/2} = 4.15$ ft$^{5/2}$/sec.

The time to drain the cylindrical portion of the tank is given by Equation 6.74, as follows: $A_2 = (\pi/4)(35)^2 = 962$ ft^2, $t_2 - t_1 = [2(962)/4.15]\left[(40.5)^{1/2} - (10.5)^{1/2}\right] = 1448$ sec $\div$ 60 = 24.1 minutes

The time to drain the conical portion is determined from Equation 6.75b and Equation 6.76, as follows: $n = (35-1.67)/[35(10.5-1.67)] = 0.108$ ft^{-1}, $t_3-t_2 = (962/4.15)\{2[(10.5)^{1/2}-(6.75)^{1/2}][1-0.108(10.5)]^2 + (4/3)(0.108)[(10.5)^{3/2}-(6.75)^{3/2}][1-0.108(10.5)] + (2/5)(0.108)^2[(10.5)^{5/2}-(6.75)^{5/2}]\} = 190$ sec $\div$ 60 = 3.2 minutes.

The total time to drain the tank is estimated to be 24.1 + 3.2 = 27.3 minutes.

A similar problem is that of determining the time to drain a pipeline such as that shown in Figure 6.13. The figure (based on an actual situation analyzed by the author in New Bedford, Massachusetts) represents a water transmission line from a

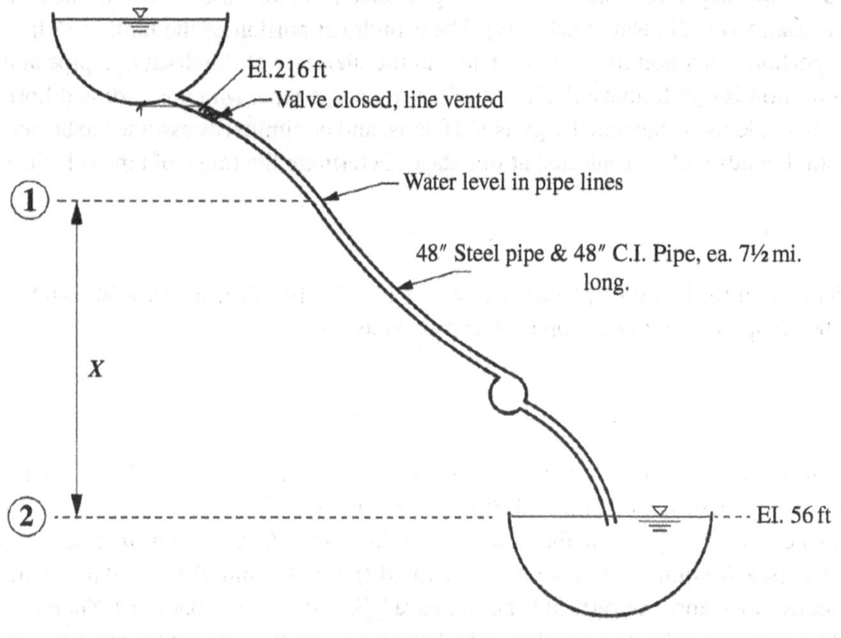

Figure 6.13 Tank Draining Through Pipe to Atmosphere

pumping station to an elevated reservoir. The valve at the elevated reservoir can be closed to avoid reservoir water loss during draining of the line, and the line immediately below that valve can be opened for venting during draining. The impellers would be removed from the pumps during draining, and pump resistance to backflow is then negligible. We will assume quasi-steady flow, and apply the one-dimensional energy equation between section "1" at the water level in the pipe at any time and section "2" at the pump reservoir:

$$\frac{p_1}{\rho g} + \frac{V_1^2}{2g} + z = \frac{p_2}{\rho g} + h_L \tag{6.77}$$

The pipe is assumed adequately vented, so $p_1 = p_2 = p_{atm}$. We will further assume $V_1^2/(2g) \ll h_L$. The above equation then reduces to:

$$z \cong h_L \tag{6.78}$$

The head loss is given by:

$$h_L = \sum \frac{fL}{d_o}\frac{V^2}{2g} + \sum K\frac{V^2}{2g} \tag{6.79}$$

Using continuity ($V = Q/A$), the above two equations can be combined and the result is expressed as follows:

$$z = \frac{Q^2}{2g}\left(\sum \frac{fL}{d_o A^2} + \sum \frac{K}{A^2}\right) \tag{6.80}$$

The transmission line can be divided into segments of constant diameter and slope with fittings at the ends of segments. We will number the segments from $j = 1$ to n starting at the downstream end of the line. When the water surface is in the ith segment, the flow is given by the following rearrangement of the above equation:

$$Q_i = \left\{2gz_i \Big/ \left[\sum_{j=1}^{i}\left(\frac{f\ell}{d_o A^2}\right)_j + \sum_{j=1}^{i}\left(\frac{K}{A^2}\right)_j\right]\right\} \tag{6.81}$$

where the second term in the denominator refers to fittings at the downstream end of the jth segment. The time to drain the line is given by:

$$T = \int_0^L \frac{A d\ell}{Q} \cong \sum_{i=n}^{1} \frac{A_i \ell_i}{Q_i} \tag{6.82}$$

For the simple case of a line of constant slope and diameter and negligible form losses, Equation 6.81 reduces to:

$$Q = A\sqrt{\frac{2gd_o^2}{f\ell}} \tag{6.83}$$

in which ℓ is the length of line downstream of elevation z. Since for constant slope z/ℓ is constant, if f may also be approximated as constant then Q will be constant. The time to drain is then given by substituting Q from the above equation into the following simplification of Equation 6.82:

$$T = \frac{AL}{Q} \tag{6.84}$$

Combining the above two equations and noting $z/\ell = \sin\theta$, we have:

$$T = \left(\frac{f}{2gd_o \sin\theta}\right)^{1/2} L \tag{6.85}$$

In the more general case, a convenient procedure entails first plotting the quasi-steady instantaneous flow versus location along the pipe using Equation 6.81. The pipe area divided by instantaneous flow can then be plotted vs. pipe length.

According to Equation 6.82, the time to drain the line is then given by the area under the resulting curve. Tabular methods can be substituted for graphical methods as appropriate. Simpler methods such as those exemplified below, coupled with field observation, will usually suffice.

Example 6.9
It was desired to estimate the time required to drain the line shown on Figure 6.13 (New Bedford, Massachusetts). The impellers were removed from the pump, making pump resistance negligible. Assume the pipe is adequately ventilated, the slope is constant, and the flow between Sections 1 and 2 is quasi-steady. Datum is taken at Section 2.

Solution
With the exit velocity in the receiving tank fully dissipated and using gage pressures, Bernoulli's equation reduces to: $V_1^2/(2g) + z = h_L$. The velocity head is much less than z over most of the time period and is neglected, and, neglecting form losses, head loss h_L is taken as approximately equal to pipe friction loss h_f. Taken together, this gives $h_f \cong z$. Friction loss is determined from the Darcy–Weisbach equation: $h_f = f(\ell/D)V^2/(2g)$ where ℓ is the length of the full portion of the pipe. With constant pipe slope, we have $z/\ell = z_{max}/L = k$ and the above equations combined to give:

$$\frac{fz}{kD}\frac{Q^2}{2gA^2} = z$$

The z terms cancel and, therefore, Q is constant (independent of z). Using Manning's equation to determine the flow through each pipe and taking $n = 0.016$, we have: $A = (\pi/4)(48/12)^2 = 12.57\ \text{ft}^2$, $R = (48/12)/4 = 1\ \text{ft}$, $L = 7.5(5280) = 39,600\ \text{ft}$, $S = (216 - 56)/39,600 = 0.00404$, and $Q = \frac{1.486}{0.016}(12.57)(1)^{2/3}(0.00404)^{1/2} = 74.2\ \text{cfs}$

The pipe volume is $\text{V} = (\pi/4)(4)^2(39{,}600) = 497{,}000$ cu ft, and the time to drain is $\text{V}/Q = 497{,}000/74.2 = 111\ \text{min} \cong 2$ hours.

Before leaving this subject, brief discussion of proper disposal of chlorinated tank or pipe drainage is warranted. Because chlorine residuals greatly exceed water quality criteria for protection of aquatic life, direct discharge to receiving waters or wetlands should be avoided. Among the methods employed in the author's practice is discharge to groundwater via a constructed basin a sufficient distance from receiving waters and dechlorination with sodium bisulfite.

Prior to construction by the Massachusetts Water Resources Authority (MWRA) of a water tank on Deer Island, the sole water tank in Winthrop supplied Winthrop and Deer Island where a regional wastewater treatment plant was located. In 1992, a new water storage tank had been constructed on Deer Island and, as part of a mitigation agreement (Graber 2024b, 2024c), MWRA prepared to drain and paint the interior of the Winthrop tank. MWRA's draft specifications called for draining the tank with the discharge to a drainage swale and thence directly to nearby coastal marine waters. The author noted the applicable criterion per *EPA Quality Criteria for Water, 1986*: Saltwater criterion, not to be exceeded for more than an average of one hour every three years, for chlorine is 13 μg/L (0.013 mg/L). It was pointed out that with little adsorption before discharge, 0.3 mg/L would exceed the criterion by a factor on the order of 20. Since dechlorination to a much lower limit is not feasible for this purpose, such discharges should be limited to the sanitary sewers. MWRA concurred and that was reflected in the final specifications

Subsequently, in 1995 MWRA was preparing an Operating & Maintenance Manual for their water system, including the new Deer Island water tank, in which options were given for disposal of chlorinated/dechlorinated water. The draft stated that discharge could be to storm drains if chlorine residual did not exceed 0.3 mg/L; with higher chlorine residuals discharge was to be to the sanitary sewer. Since these storm drains discharge to Boston Harbor, the author noted again the applicable criterion stated above. It was pointed out that with the minimal dilution expected from a stormwater outfall, an exceedance similar to that mentioned in the previous paragraph would result. MWRA concurred and that was reflected in the final Manual.

6.7.2 Inertial Effects

To introduce inertial effects, we will reconsider the case considered in the previous section of the draining of a line of constant diameter and slope. Consider the *closed system* defined at a particular instant by the fluid between sections "1" and "2" at the upstream and downstream ends of the column of water, respectively. The momentum equation (from Section 2.4) may be applied to that system to give:

$$p_1A - p_2A + \rho gA\ell \sin\theta - \frac{f\ell}{d_o}\frac{\rho V^2}{2}A = \rho\ell A\frac{dV}{dt} \tag{6.86}$$

in which ℓ is the length of pipe in which water remains.

Setting $p_1 = p_2 = p_{atm}$, the above equation may be rearranged to give:

$$\frac{dV}{dt} + \frac{f}{2d_o}V^2 - g\sin\theta = 0 \tag{6.87}$$

This equation may be separated:

$$dt = \frac{dV}{g\sin\theta - fV^2/(2d_o)} \tag{6.88}$$

then integrated (Dwight 1961, §140.1):

$$t = \frac{2d_o}{f}\int \frac{dV}{\frac{2gd_o}{f}\sin\theta - V^2} = \frac{2d_o}{f}\left[\frac{1}{\left(\frac{2gd_o}{f}\sin\theta\right)^{1/2}}\tanh^{-1}\frac{V}{\left(\frac{2gd_o}{f}\sin\theta\right)^{1/2}}\right] + C_1' \tag{6.89}$$

then solved for V:

$$V = \left(\frac{2gd_o}{f}\sin\theta\right)^{1/2}\tanh\left[\frac{f}{2do}\left(\frac{2gd_o}{f}\sin\theta\right)^{1/2}t\right] + C_1 \tag{6.90}$$

The initial condition $V = 0$ at $t = 0$ gives the constant of integration $C_1 = 0$.

Let x be the displacement of the water surface measured from its initial location. Then $V = dx/dt$ and we have:

$$x = \int\left(\frac{2gd_o}{f}\sin\theta\right)^{1/2}\tanh\left[\frac{f}{2d_o}\left(\frac{2gd_o}{f}\sin\theta\right)^{1/2}t\right]dt \tag{6.91}$$

This equation can be integrated (Dwight 1961, §71.01) to give:

$$x = \frac{2d_o}{f}\ln\cosh\left[\frac{f}{2d_o}\left(\frac{2gd_o}{f}\sin\theta\right)^{1/2}t\right] + C_2 \tag{6.92}$$

The initial condition $x = 0$ at $t = 0$ gives the constant of integration $C_2 = 0$. The above equation can be solved for the total drainage time T for $x = L$:

$$T = \left(\frac{2d_o}{fg\sin\theta}\right)^{1/2}\cosh^{-1}\left[\exp\left(\frac{fL}{2d_o}\right)\right] \tag{6.93}$$

Several points may be made regarding the above solution. First note that the inertial term in the original equation is the right-hand side of Equation 6.86. That term represents the tendency of the fluid mass to resist changes in velocity. Noting that the derivative dV/dt has order of magnitude V/T, the ratio of the frictional to inertial terms in Equation 6.86 is given by:

$$\frac{(f\ell/d_o)(\rho V^2/2)A}{\rho\ell A dV/dt} = \frac{fV^2/(2d_o)}{dV/dt} \sim \frac{fV^2/(2d_o)}{V/T} = \frac{fVT}{2d_o} \sim \frac{fL}{2d_o} \tag{6.94}$$

Thus, we would expect inertial effects to become less significant as $fL/(2d_o)$ becomes large.

The quantitative significance of the term $fL/(2d_o)$ can be ascertained by investigating the ratio of the inertial solution for T as given by Equation 6.93 to the noninertial solution for T as given by Equation 6.85. That ratio becomes:

$$\frac{T_{inertial}}{T_{noninertial}} = \frac{\cosh^{-1}\{\exp[fL/(2d_o)]\}}{fL/(2d_o)} \tag{6.95}$$

Since $\cosh^{-1}x = \ln\left(\sqrt{x^2-1} + x\right)$, we have $\cosh^{-1}x \cong \ln(2x)$ for $x^2 >> 1$. Thus:

$$\frac{T_{inertial}}{T_{noninertial}} \cong \frac{\ln\{2\exp[fL/(2d_o)]\}}{fL/(2d_o)} \tag{6.96}$$

The above relation simplifies to:

$$\frac{T_{inertial}}{T_{noninertial}} \cong 1 + \frac{\ln 2}{fL/(2d_o)} \tag{6.97}$$

The inertial and noninertial solutions are equal to within 1%, for example, when $fL/(2d_o) \geq (\ln 2)/0.01 = 69$.

The above result can be used to conservatively test for the significance of inertial effects in pipeline draining. This is demonstrated by using Example 6.9, for which $L = 7.5(5280) = 39,600$ ft, $n = 0.016$, $d_o = 48$ in. $/12 = 4$ ft, $R = d_o/4 = 4/4 = 1$ ft, $f = 117n^2/R^{1/3} = 117(0.016)^2/1 = 0.030$, and $fL/(2D) = 0.030(39,600)/[2(4)] = 148 > 69$. The author has also demonstrated this by deriving the full equation including inertial effects and applying it to this example, although that is clearly not necessary in this case and that derivation is not presented.

The analytical approach employed above may be extended to more complex situations. For example, the draining of a pipe of variable slope and diameter could be handled if necessary by applying the above approach to each section of constant slope and diameter and relating conditions at adjoining sections by Bernoulli's equation and continuity. Burgreen (1960) used such an approach in his analysis of tank draining with inertial effects. Both the analysis and its results are instructive and will be reviewed here.

The configuration analyzed by Burgreen is basically as shown on Figure 6.14 (Burgreen included tank pressurization, which we shall ignore). The momentum equation (Section 2.4) is applied separately to the closed system consisting of the liquid in the tank, the liquid in the vertical length of drain pipe, and the liquid in the horizontal length of drain pipe. Neglecting friction forces in the tank, and employing the notation of Figure 6.14, the following equations result:

$$p_{atm}A_1 - p_1A_1 + \rho g(H-y)A_1 = \rho(H-y)A_1\frac{d^2y}{dt^2} \tag{6.98a}$$

$$p_2A_2 - p_{22}A_2 + \rho gA_2L_1 - F_1\frac{\rho A_2}{2}\left(\frac{dx}{dt}\right)^2 = \rho L_1A_2\frac{d^2x}{dt^2} \tag{6.98b}$$

$$p_{22}A_2 - p_{atm}A_2 - F_2\frac{\rho A_2}{2}\left(\frac{dx}{dt}\right)^2 = \rho L_2A_2\frac{d^2x}{dt^2} \tag{6.98c}$$

in which $F = fL/d + \sum K$ is the sum of friction and form losses in the respective sections, and y and x represent the displacement of fluid in the tank and drain pipe, respectively. These displacements are related by continuity as follows:

$$A_1y = A_2x \tag{6.99}$$

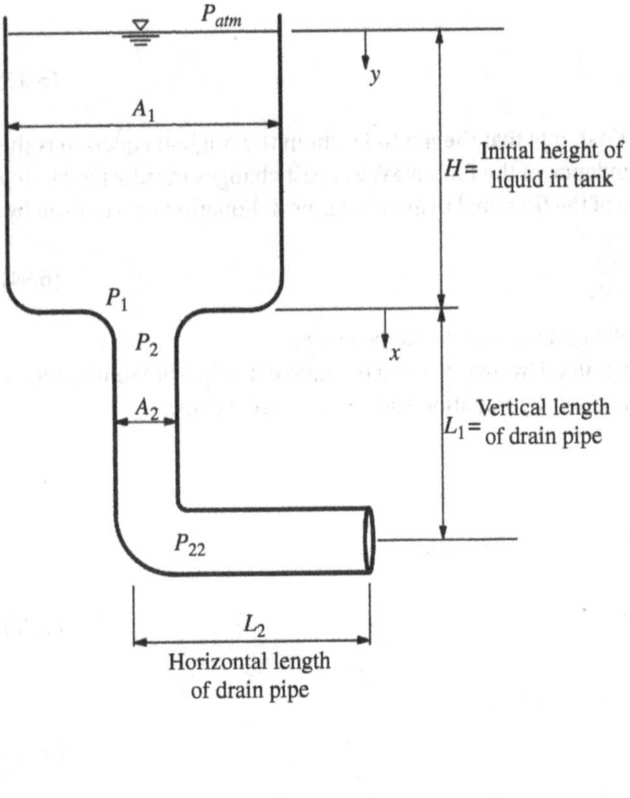

Figure 6.14 Tank and Drain Pipe Arrangement (McNown 1954/ American Society of Civil Engineers).

Adding Equations 6.98b and 6.98c yields:

$$-p_{atm} + p_2 + \rho g L_1 - \rho L \frac{d^2x}{dt^2} - \frac{\rho}{2} F \left(\frac{dx}{dt}\right)^2 = 0 \tag{6.100}$$

in which $L = L_1 + L_2$ and $F = fL_1/d_1 + fL_2/d_2 + \sum K$ represents the number of velocity heads lost through pipe friction and form losses. At the junction of the pipe and tank, Bernoulli's steady-flow energy equation (Section 2.6) applies:

$$p_1 - p_2 = \frac{\rho}{2}\left(\frac{dx}{dt}\right)^2 \left[1 - \left(\frac{A_2}{A_1}\right)^2\right] \tag{6.101}$$

Substituting the above equation into Equation 6.99, substituting Equation 6.98a into that result, employing Equation 6.99 (and the related relations $A_2 dx/dt = A_1 dy/dt$ and $A_2 d^2x/dt^2 = A_1 d^2y/dt^2$), and rearranging results in:

$$\left(\frac{A_1}{A_2} L + H - y\right)\frac{d^2y}{dt^2} + \frac{1}{2}\left[1 + F - \left(\frac{A_2}{A_1}\right)^2\right]\frac{A_1}{A_2}\left(\frac{dy}{dt}\right)^2 - g(H + L_1 - y) = 0 \tag{6.102}$$

Although Equation 6.102 is a second-order, nonlinear differential equation with variable coefficients, the variable *t* is not present explicitly and the equation may be solved by available methods (Hildebrand 1962, pp. 38–40) to give (Burgreen 1960):

$$\frac{dy}{dt} = \left\{\frac{2gR^2}{1 + F - R^2}\left[H + L_1 + \frac{R(L + RH)}{1 + F - 2R^2}\right]\left[1 - \left(1 - \frac{Ry}{L + RH}\right)^{(1 + F - R^2)/R^2}\right] - \frac{2gR^2 y}{1 + F - 2R^2}\right\}^{1/2} \tag{6.103}$$

If further integration is necessary to determine the time of draining by the above equation, a numerical solution would be required. However, the author has yet to find that necessary, as a comparison of the above to the corresponding noninertial solution shows inertial effects to be commonly negligible. The comparable noninertial solution is obtained from Equation 6.103, noting $A_1 dy/dt = A_2 V_b$ and $z = L_1 + H - y$:

$$\frac{dy}{dt} = R\left[\frac{2g(L_1 + h - y)}{1 + F - R^2}\right]^{1/2} \tag{6.104}$$

The ratio of the inertial to noninertial solutions is obtained from Equations 6.103 and 6.104 with some rearrangement, as follows:

$$\frac{(dy/dt)_{inertial}}{(dy/dt)_{noninertial}} = \frac{\left\{\left[1 + \frac{L_1}{H} + \frac{R(L/H + R)}{1 + F - 2R^2}\right]\left[1 - \left(1 - \frac{Ry/H}{L/H + R}\right)^{\frac{(1 + F - R^2)}{R^2}}\right] - \frac{1 + F - R^2}{1 + F - 2R^2}\frac{y}{H}\right\}^{1/2}}{(1 + L_1/H - y/H)^{1/2}} \tag{6.105}$$

Investigation of the above equation shows the conditions under which inertial effects may be neglected (i.e., the above ratio is approximately one) to be as follows:

$$2R^2 \ll 1 \tag{6.106a}$$

$$R\frac{L}{H} \ll (1 + F) \tag{6.106b}$$

Thus inertial effects may be disregarded when the ratio of drain pipe area to tank cross-sectional area is small *and* the length of the drain pipe is a small fraction of the initial height of fluid in the tank. If either of these criteria is not satisfied, then the conventional manner of computing the rate of draining (i.e., neglecting fluid inertia) may not estimate the rate of draining and draining time with acceptable accuracy.

Example 6.10

Check the validity of the drainage time estimated in Example 6.8 (Tallman Island – in which fluid inertia was neglected) by determining the relative importance of fluid inertia on the rate of draining.

Solution

The criteria for negligible inertial effects are given by Equations 6.106a and 6.106b. Considering first the cylindrical portion of the tank, we have:

$$R = \frac{A_2}{A_1} = \frac{(20)^2}{[35(12)]^2} = 0.00227$$

$$2R^2 = 0.0000103$$

$$R\frac{L}{H} = 0.00227\frac{800}{30} = 0.0605$$

$$1 + F = 1 + 8.64 = 9.64$$

The criteria for neglecting fluid inertia (Equations 6.106) are readily satisfied.

The conical portion of the tank has variable R, for which the analysis of inertial effects is not strictly correct. As a practical matter, however, the volume and drainage time of the conical portion are relatively small. This, coupled with the fact that once terminal (noninertial) flow has been established for the cylindrical portion, inertial effects for the conical portion should be relatively small (and would in fact *increase* the rate of drainage), indicates the validity of the approximation for the tank as a whole.

We consider now a *gate valve* with a linear rate of closure [see Section 6.5, including Figure 6.6(b)] using an analysis and solution procedure which are a variation and extension of that given by McNown (1950, pp. 458–459). This is for the case of a valve at the end of a horizontal (zero slope) pipe at which the piezometric head is h, supplied by a tank with piezometric head at the upstream end of the pipe of h_o. Our analysis begins with the equation derived below as Equation 6.131 adapted to the present problem:

$$h - h_o = -\frac{L}{g}\frac{dV}{dt} - \frac{fL}{d_o}\frac{V^2}{2g} \tag{6.107}$$

During the final portion of closure, when the maximum h (denoted by h_m) occurs, the orifice equation may be used to relate the flow and pressure in the pipe:

$$V = \frac{C_c a}{A}\sqrt{2gh_m} \tag{6.108}$$

in which A and a are the cross-sectional areas of the pipe and valve opening, respectively, and C_c is the valve discharge coefficient. With the variables being a and h_m, differentiating the above equation with respect to time gives:

$$\frac{dV}{dt} = \frac{C_c\sqrt{2gh_m}}{A}\left(\frac{a}{2h_m}\frac{dh_m}{dt} + \frac{da}{dt}\right) \tag{6.109}$$

The open area a approaches zero as t approaches t_c, and since dh_m/dt becomes large but does not tend toward infinity, we can neglect the first term in the parentheses above. Equation 6.109 thus modified is substituted into Equation 6.107 giving:

$$h_m - h_o = -\frac{C_c L}{A}\sqrt{\frac{2h_m}{g}}\frac{da}{dt} - \frac{fL}{d_o}\frac{V^2}{2g} \tag{6.110}$$

And substituting Equation 6.108 into Equation 6.110:

$$h_m - h_o = -\frac{C_c L}{A}\sqrt{\frac{2h_m}{g}}\frac{da}{dt} - \frac{fL}{d_o}\left(\frac{C_c a}{A}\right)^2 h_m \tag{6.111}$$

McNown considered the period just prior to valve closure for which, referring to Figure 6.6(b), $\alpha \cong \pi/2$ in Equation 6.41, which gives:

$$da/dt = -d_v^2/T_c \tag{6.112}$$

in which d_v is the diameter of the valve opening. [As indicated in Section 6.5, such predictive relations are limited in their accuracy. They are useful for the exemplifications given here, but may not be suitable for design use for which purpose actual valve data (see below) should be used.]

Substituting Equation 6.112 into Equation 6.111 and rearranging gives:

$$\frac{h_m}{h_o}\left[1+\frac{fL}{d_o}\left(\frac{C_c a}{A}\right)^2\right]-\frac{C_c L}{A}\sqrt{\frac{2}{gh_o}}\left(\frac{d_v^2}{T_c}\right)\sqrt{\frac{h_m}{h_o}}-1=0 \tag{6.113}$$

Defining the following term:

$$K_v \doteq \left(\frac{A}{aC_c}\right)^2 \tag{6.114}$$

Equation (6.113) is rewritten as:

$$\left(1+\frac{fL}{d_o K_v}\right)\frac{h_m}{h_o}-\frac{2Ld_v^2 C_c}{T_c A\sqrt{2gh_o}}\sqrt{\frac{h_m}{h_o}}-1=0 \tag{6.115}$$

In McNown's example, we have $L = 2000$ ft, $d_o = 1$ ft, $f = 0.02$, $T_c = 10$ sec, $h_o = 100$ ft, $C_c = 0.6$, $d_v = 10$ in. $= 0.8\overline{33}$, and $A = (\pi/4)(12/12)^2 = 0.785$ ft^2. Substituting those values into Equation 6.115 gives:

$$\left(1+\frac{40}{K_v}\right)\frac{h_m}{h_o}-2.64\sqrt{\frac{h_m}{h_o}}-1=0$$

The solution is given by:

$$\frac{h_m}{h_o}=\frac{+\,2.64+\sqrt{(2.64)^2+4(1+40/K_v)}}{2(1+40/K_v)}$$

The value of h_m/h_o is largest when K_v is largest, i.e., when K_v approaches infinity which occurs as the valve approaches fully closed. That maximum is given by:

$$\frac{h_m}{h_o}=\frac{2.64+\sqrt{(2.64)^2+4}}{2}=2.98 \Rightarrow h_m=2.98(100)=298\text{ ft}$$

With the maximum piezometric head at station 2 under steady-state conditions being $h_o = 100$ ft the effects of fluid inertia are seen to be very important in this case. (The above corrects McNown's numerical result.)

We consider now uniform gate valve closure on a pipe with negligible friction. The gate is assumed to have an open diameter equal to that of the pipe. For this case, Equation 6.110 applies with $f = 0$. Equation 6.35 also applies and da/dt is given by Equation 6.43. The resulting equation is:

$$\frac{h_m}{h_o}-2\frac{CL}{T_c\sqrt{2gh_o}}\sqrt{\frac{h_m}{h_o}}\frac{\pi}{4}\frac{d_v^2}{A}-1=0 \tag{6.116}$$

Since $\pi d_v^2/(4A) = 1$, and substituting from above $C = V_o/\sqrt{2gh_o}$ gives:

$$\frac{h_m}{h_o}-\frac{V_o L}{gh_o T_c}\sqrt{\frac{h_m}{h_o}}-1=0 \tag{6.117}$$

The solution to this quadratic equation is given by:

$$\frac{h_m}{h_o}=\frac{V_o L}{2gh_o T_c}+\sqrt{\left(\frac{V_o L}{2gh_o T_c}\right)^2+1} \tag{6.118}$$

A rearranged version of this equation is as follows:

$$\frac{h_m}{h_o} = \frac{K_1}{2} + \sqrt{1 + \frac{K_1^2}{4}}, \quad K_1 = \frac{V_oL}{gh_oT_c} \tag{6.119}$$

This form differs from the equation given by Parmakian (1963, Chapter I) for a problem that is similar but tacitly assumes less than complete closure. Equation 6.119 gives a result for h_m/h_o applied to his example substantially higher than his. Specifically, for $V_oL/(gh_oT_c) = 0.20$ Equation 6.119 gives $K_1 = 0.20$ and $h_m/h_o = 1.105$ compared to Parmakian's value of $h_m/h_o = 0.22$.

The above procedure can be extended to other valve configurations which provide more effective control (including lower pressure rises for comparable closure times) than the gate valve considered above, but entail more complex mathematical relationships. The mathematical complexities stem from the nature of the K_v and dK_v/dt vs. t relationships, and the occurrence of h_m at an unknown intermediate $t/T_c < 1$. A convenient computational approach developed by the author will be described below, then extended and applied in Section 8.6.

If continuous mathematical relationships for K_v and dK_v/dt vs. t can be obtained for the valve, then one can obtain h_m/h_o by considering successive values of t/T_c and corresponding values of K_v and dK_v/dt until the maximum of h_m/h_o is obtained. A programmable calculator or computer is a considerable aid in employing the procedure. The basic computational procedure will be described first with reference to another part of McNown's example (1950), after which some general guidance for other types of valves will be given.

McNown considered the rotary valve having the geometry shown on Figure 6.6(a) and (b). The projected area of the valve opening perpendicular to the pipe axis is given by Equation 6.33. The valve is assumed to rotate at a constant rate so that Equation 6.35 applies. Then also using Equations 6.36 and 6.45 we obtain:

$$\sqrt{K_v} = \frac{2A/(d_v^2C_v)}{\cos^{-1}\frac{t}{T_c} - \frac{t}{T_c}\sqrt{1-(t/T_c)^2}} \tag{6.120}$$

Substituting Equation 6.38 into Equation 6.107:

$$\frac{dK_v}{dt} = \left(\frac{A}{C_v}\right)^2 \frac{2}{a^3}\frac{d^2}{T_c}\sin\alpha \tag{6.121}$$

(See the discussion earlier in this chapter regarding the inaccuracy of such relations.) Substituting Equation 6.120 into Equation 6.121, and expressing in terms of t/T_c from Equation 6.42, we obtain:

$$\frac{dK_v}{dt/T_c} = \frac{2d_v^2C_v}{A}K_v^{3/2}\sqrt{1-\left(\frac{t}{T_c}\right)^2} \tag{6.122}$$

Equations 6.115 and 6.122 provide the relationships to be used in conjunction with Equation 6.107 to solve for h_m/h_o.

Programmable calculator output results are $h_m/h_o = 3.89$, occurring at $t/T_c = 0.76$. This result is identical to that reported by McNown (1950, p. 459). Note that the overpressure of $h_m = 3.89(100) = 389$ ft is only 44% of that resulting with the gate valve in the same closure time, demonstrating the advantages for pressure control of the rotary valve over the gate valve.

As a practical matter, the mathematical derivation of valve characteristics is seldom desirable. This has been stated previously and merits further emphasis. If the valve geometry is known with sufficient accuracy, a mathematical relation for the open area may be attainable [such relations for several common types of valves are given in Wood and Jones (1973)]. However, the assumption of a constant or assumed variable relation for C_v (such as the "orifice" value of 0.6 used above) is highly questionable as discussed previously. Furthermore, the valve control mechanism's cams and linkages do not necessarily (or desirably) result in closing characteristics of simple mathematical description. Any valve and operator combination worthy of consideration for pressure control should have an experimentally determined curve of K_v vs. t/T_c and, from the slope of that curve, $dK_v/(dt/T_c)$ vs. t/T_c. The use of such curves in the overall analysis is discussed below.

Actual K_v vs. percent stroke curves for the same rotary cone valve for which Figure 6.15 applies are given on Figure 6.15. (The reader is cautioned not to use these curves for applications without assuring their accuracy for the particular valves in question.) A linear drive will result in the stroke equaling t/T_c in which case the ordinate can also be regarded as $100t/T_c$. Each curve represents a different valve mechanism, as available from the valve manufacturer. We will refer specifically to the curve labeled "R-mechanism". That curve could be "fit" in a variety of ways, including: (1) a series of piecewise

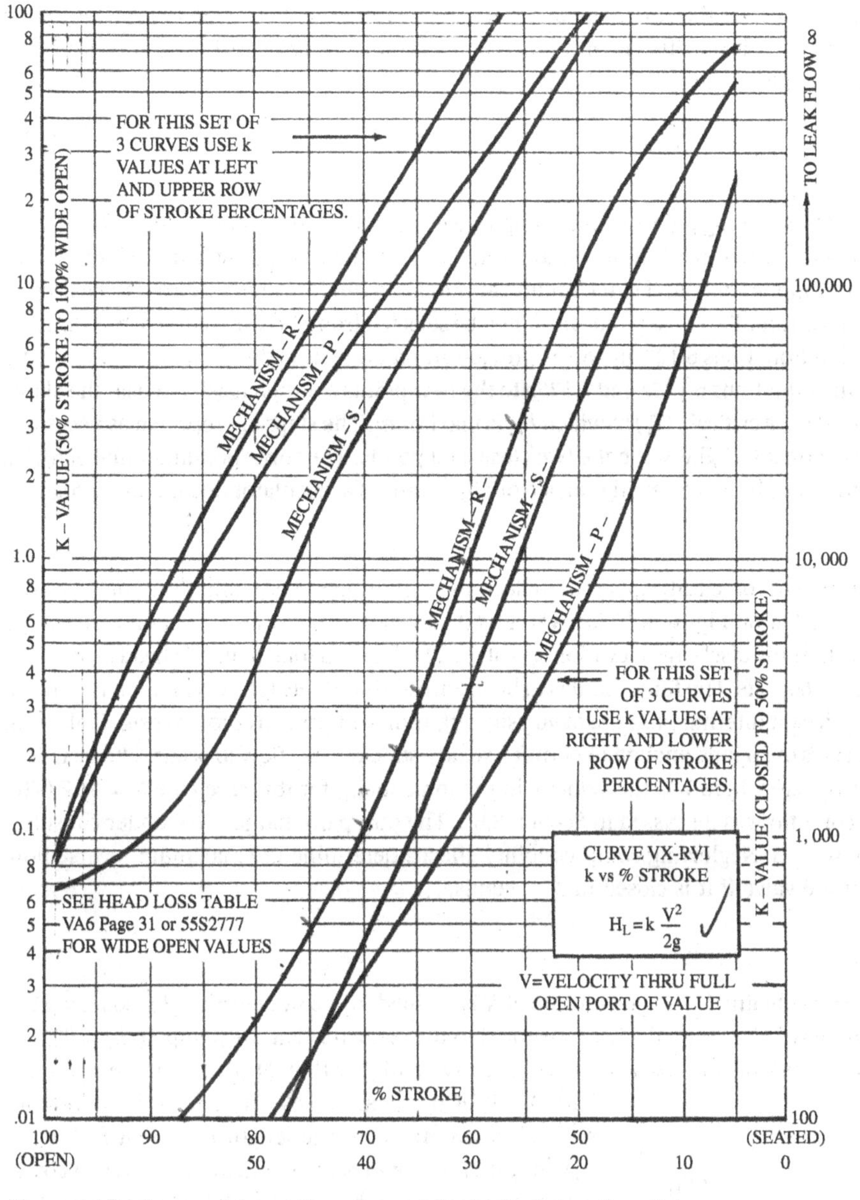

Figure 6.15 Curves of Actual K_v vs. Percent Stroke for Rotary Cone Valve

continuous log-parabolic or log-linear regression relations, (2) an nth-order polynomial or log-polynomial relation, or (3) a single continuous regression equation. The author has had only limited success with the first two of these, due largely to the discontinuous and/or rapidly fluctuating values of dK_v/dt. Considerable "smoothing" of the dK_v/dt vs. t relationship is necessary for these methods to give acceptable results. The third method would be fine if an accurate regression equation could be found; however, the nature of the curve is such that this is often not feasible. Nevertheless, a variation of this method has been found to work well and is described below.

The best simple fit to segments of valve curves such as shown on Figure 6.15 is provided by a log-linear equation of the following form:

$$K_v = b\exp(m^t/T_c) \tag{6.123}$$

Differentiating, we have:

$$\frac{dK_v}{d(t/t_c)} = mK_v \tag{6.124}$$

Although such a fit is not sufficiently accurate over the entire valve curve, it is accurately applied to short segments of the curve. This suggests a useful iterative procedure. The coefficient and exponent of Equation 6.123 may be obtained from two points on the valve curve, denoted by subscripts "1" and "2" as follows:

$$m = (\ln K_{v2} - \ln K_{v1})/(t_2/T_c - t_1/T_c) \tag{6.125a}$$

$$b = \ln K_{v2} - mt_2/T_c \tag{6.125b}$$

Initial values of m and b may be obtained by selecting points near the opposite ends of the curve. Those values may then be substituted into Equations 6.123 and 6.124 to give expressions for K_v and $dK_v/(dt/T_c)$ which may be substituted into Equation 6.115. Recognizing that Equation 6.115 applies only to the maximum value of h_m, a numerical procedure can be used to solve for the value of t/T_c at which h_m is maximum. Two data points may then be selected from the valve curve which bracket this initial t/T_c and are sufficiently close that Equations 6.123 and 6.124 provide an accurate fit within that range of t/T_c. The new values of m and b obtained by applying Equations 6.125a and 6.125b to the new points are then used to repeat the above steps, leading to new values of t/T_c (referred to as critical t/T_c) at which h_m is maximum. The steps are repeated as necessary until convergence is obtained on a value of critical t/T_c between the two close data points. The corresponding value of h_m/h_o is the desired solution. The entire procedure may be conveniently carried out by means of a calculator or computer program.

Example 6.11

The practical significance of this example will become fully apparent in Section 8.6 (Spring Street Pumping Station, Arlington, Massachusetts). The pump system depicted on Figure 6.16 has a wet well elevation (z_3) of El.−9.5 ft, pump discharge and control valve elevation (z_2) of El. +9.5 ft, static discharge elevation (z_1) of El. 22.5 ft, and a total dynamic head of 40.7 ft at the maximum pumping station discharge of 60.5 cfs. The force main is to be prestressed concrete (steel cylinder type), 42-in. diameter, and 3160 ft long, specified for a design internal pressure of 40 psig with transient pressures to 100 psig. Following power failure, the pump discharge valve stays open long enough to permit a steady-state reverse flow to occur. The head loss at the back-spinning pump is given by $KV^2/(2g)$ where V is the velocity in the force main; for this case take $K = 81.7$ (The resistance of pumps under reverse flow conditions is discussed in Section 8.6.). The pump discharge valve under consideration has the valve curve shown on Figure 6.15. Neglecting compressibility effects, determine the maximum reverse flow and the maximum pressure occurring at the valve if it is closed in 25 seconds.

Solution

The values of K_o, h_o, and $f'\ L/d_o$ must first be obtained. The given value of K is defined in the same way as K_o, so $K_o = 81.7$. The value of $h_o = z_1 - z_3$ (see Figure 6.16) may be conservatively taken equal to the static head at the pump, so $h_o = 22.5 - (-9.5) = 32$ ft (note that the wet well elevation during reverse flow will usually be higher than at maximum pumped discharge). All of the head losses external to the pump are assumed distributed over the length L, so that $f'\ L/d_o \cong h_f/[V^2/(2g)]$ in which h_f is the pump system loss at a corresponding pipe velocity head of $V^2/(2g)$. The pump system loss at the maximum pumped flow is 40.7–32 = 8.7 ft, corresponding to a force main velocity head determined by the following succession: $A = (\pi/4)(42/12)^2 = 9.62\ \text{ft}^2$, $V = Q/A = 60.5/9.62 = 6.29$ ft/sec, $V^2/(2g) = (6.29)^2/[2(32.2)] = 0.614$ ft. Thus $f'\ L/d_o \cong 8.7/0.614 = 14.2$.

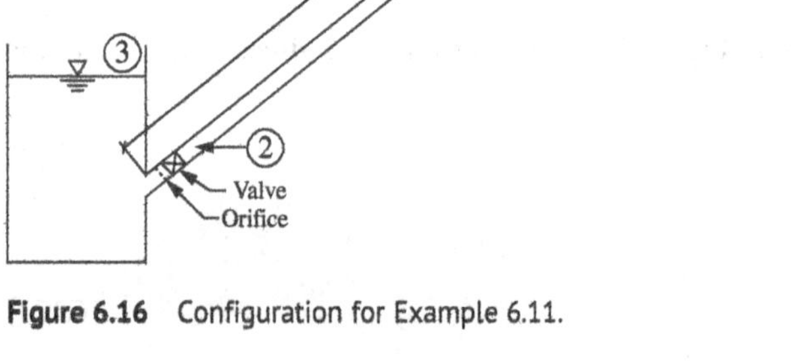

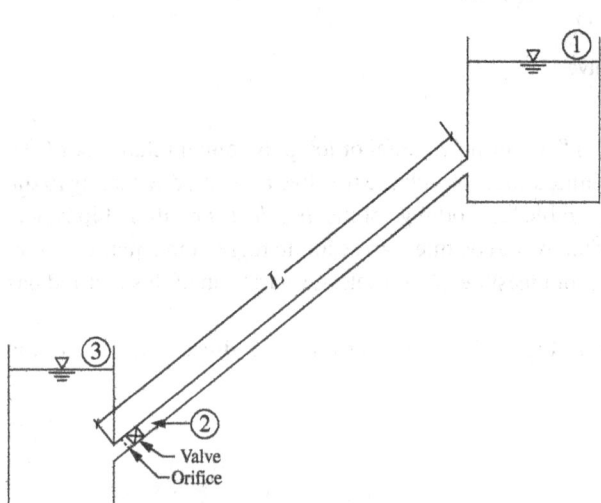

Figure 6.16 Configuration for Example 6.11.

The magnitude of the steady-state reverse flow (prior to valve closure) may be determined by applying the energy equation between stations "3" and "1" of Figure 6.16 as follows:

$z_1 = z_3 + (K_o + f'\ L/d_o)V^2/(2g) \Rightarrow V^2/(2g) = h_o/(K_o + f'\ L/d_o) = 32/(81.7 + 14.2) = 0.334$ ft, $V = \sqrt{0.334(2)(32.2)} = 4.64$ ft/ sec, $Q = VA = 4.64(9.62) = 44.6$ cfs. The assumption is made that $f'\ L/d_o$ at this reverse flow is the same as that at the maximum pumped flow; this assumption neglects the effect of Reynolds Number for certain types of fittings such as tees and wyes). A more precise estimate of $f'\ L/d_o$ could be made if deemed appropriate.

Given the valve curve and values of K_o, $f'\ L/d_o$, and Z, the ratio h_m/h_o can be determined. The values of K_o and $f'\ L/d_o$ were calculated above. The value of Z is calculated by Equation 6.137b, giving: $Z = 25\sqrt{2(32.2)(32)}/3160 = 0.359$. The corresponding value of h_m/h_o can be calculated by means of the procedure discussed above (and aided by a programmable calculator or computer). The value of h_m/h_o thus determined is 3.66, giving $h_m = 3.66(32) = 117$ ft. Recalling that the elevation at the wet well is datum, the pump discharge pressure head is 117-(−13.5) + (−9.5) = 121 ft gage = 52.5 psig. This is well within the permissible surge pressure for the pipe of 100 psig.

Another situation of practical importance in which fluid inertia plays a major role is for valve closure in the system represented on Figure 6.16. The system can be considered to represent a practical idealization of a simple pump system following power failure. The idealization will be explained briefly here, and discussed in greater detail in Section 8.6.

Consider a pump to be located in the position of the orifice on Figure 6.16. (We will show in Section 8.6 that a pump can be idealized as a *fixed* orifice during backflow under certain circumstances.) The pump takes suction from the lower elevation (such as a wet well) and pumps through the line of length L to a higher elevation. The valve depicted on the figure represents the pump discharge valve. In the event of pump power failure, the flow through the pump and line from the lower to higher elevation will slow down and eventually stop. The forward pump spin will similarly decrease and eventually stop. Unless prevented, reverse flow and reverse spin of the pump will commence and increase until a steady-state reverse flow condition is reached. We will assume that such a steady-state condition has been reached, and will analyze the system to determine the effects of slow closure of the pump discharge valve.

We shall employ the one-dimensional differential energy equation in the analysis, and will derive the appropriate differential equation using principles presented in Section 2.6. Equation 2.57 will be applied to the assumed one-dimensional flow in a control volume encompassing a differential length of pipe as shown on Figure 6.17. The flow is *assumed incompressible*, so that the internal energy term u may be dropped from Equation 2.56. Equation 2.57 can then be written in terms of the total head H as follows:

$$-\dot{Q}_f = \oint\!\!\oint_{c.s.} gH\rho\bar{v}\cdot d\bar{A} + \frac{\partial}{\partial t}\iiint_{c.v.} \rho\left(\frac{v^2}{2} + gz\right)d\forall \tag{6.126}$$

in which $\dot{Q}_f$ represents the rate of energy loss due to friction. Application of the above equation to the control volume shown results in:

$$-d\dot{Q}_f = \rho g V A\frac{\partial H}{\partial x}dx + \frac{\rho A dx}{2}\frac{\partial V^2}{\partial t} \tag{6.127}$$

or

$$-\frac{\bullet dQ_f}{\rho g A V dx} = \frac{\partial H}{\partial x} + \frac{1}{g}\frac{\partial V}{\partial t} \tag{6.128}$$

Noting by continuity that $\dot{Q}_f/(\rho g A V) = \dot{Q}_f/(\rho g Q)$ which, in turn, equals h_ℓ, the above equation can be rewritten as follows:

$$\frac{1}{g}\frac{\partial V}{\partial t} = -\frac{\partial(H + h_\ell)}{\partial x} \tag{6.129}$$

The above expression of the energy equation for incompressible flow is useful for analyses of certain time-varying flows in pipelines including the one depicted on Figure 6.17.

We will first investigate the term $\partial(H + h_\ell)/\partial x$ over a finite distance L on Figure 6.17. We will make the assumption (common to analyses of this type) that form losses are either negligible or uniformly distributed over the distance L (and that the pipe is uniform), so that $\partial h_\ell/\partial x = \text{constant} = (f'/D)[V^2/(2g)]$, in which f' is increased over f to account for any distributed form losses.

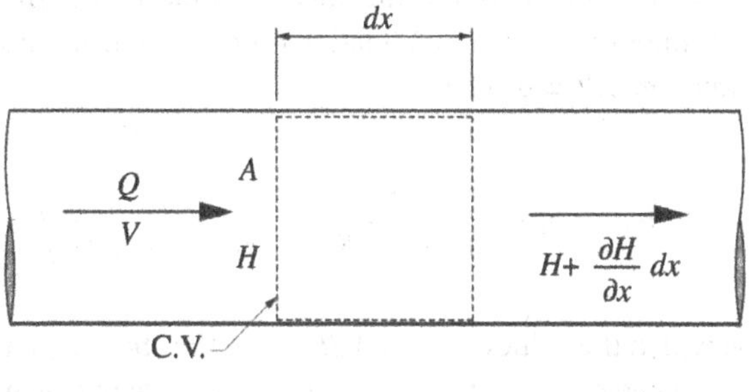

Figure 6.17 Control Volume Encompassing a Differential Length of Pipe.

For incompressible flow, the velocity along the length L is also constant at any instant, and $H = V^2/(2g) + p/(\rho g) + z = V^2/(2g) + h_p$, so we then also have $\partial H/\partial x = \text{constant} = (H_2 - H_1)/L = (h_{p2} - h_{p1})/L$. Thus:

$$\frac{\partial(H + h_\ell)}{\partial x} = \frac{f'}{d_o}\frac{V^2}{2g} + \frac{h_{p2} - h_{p1}}{L} = \frac{1}{L}\left(h_{p2} - z_1 + \frac{f'L}{d_o}\frac{V^2}{2g}\right) \tag{6.130}$$

Substituting the above equation into Equation 6.129 using the total derivative, we obtain:

$$h_{p2} - z_1 = -\frac{L}{g}\frac{dV}{dt} - \frac{f'L}{d_o}\frac{V^2}{2g} \tag{6.131}$$

Fluid inertia can be disregarded in considering the flow across the valve and orifice. Thus the modified Bernoulli equation can be applied between stations 2 and 3, to give:

$$h_{p2} - z_3 = K\frac{V^2}{2g} = (K_v + K_o)\frac{V^2}{2g} \tag{6.132}$$

in which K_v is the form loss factor of the valve and K_o is the form factor loss of the "orifice" (including any other form losses between stations 2 and 3 deemed significant).

We wish to solve for the pressure increase at station 2 and thus seek to eliminate the velocity terms and solve for h_{p2}. Note that the velocity terms occur both as $V^2/(2g)$ and dV/dt. To eliminate dV/dt, we first differentiate Equation 6.132 with respect to t, to obtain:

$$\frac{dh_{p2}}{dt} = \frac{dK_v}{dt}\frac{V^2}{2g} + (K_v + K_o)\frac{V}{g}\frac{dV}{dt} \tag{6.133}$$

We can then eliminate dV/dt between Equations 6.131 and 6.133, giving:

$$h_{p2} = z_1 - \frac{f'L}{d_o}\frac{V^2}{2g} + \frac{L}{(K_v + K_o)V}\left[\frac{dK_v}{dt}\frac{V^2}{2g} - \frac{dh_{p2}}{dt}\right] \tag{6.134}$$

We can now use Equation 6.132 to eliminate V from Equation 6.134. Doing this, letting $h_o = z_1 - z_3$, and selecting the elevation at z_3 as datum yields:

$$h_{p2} = h_o - \frac{f'L}{d_o}\frac{h_{p2}}{K_v + K_o} + \frac{L}{(K_v + K_o)\sqrt{2gh_{p2}/(K_v + K_o)}}\left[\frac{dK_v}{dt}\frac{h_{p2}}{K_v + K_o} - \frac{dh_{p2}}{dt}\right] \tag{6.135}$$

Note that the first two terms on the right-hand side of the above equation constitute the steady-state solution; the subsequent terms represent the effects of fluid inertia in the line of length L.

Rather than integrate Equation 6.135, we will attempt only to obtain the maximum value of h_{p2}, denoted by h_{pm}, at which time $dh_{p2}/dt = 0$. Thus:

$$h_{pm} = h_o - \frac{f'L}{d_o}\frac{h_{pm}}{K_v + K_o} + \frac{L}{(K_v + K_o)\sqrt{2g}}\sqrt{\frac{h_{pm}}{K_v + K_o}}\frac{dK_v}{dt} \tag{6.136}$$

in which the values of K_v and dK_v/dt in the above equation are understood to be values at the time when $h_{p2} = h_{pm}$. To solve the above equation for a given valve and rate of closure (corresponding to a particular curve of K_v vs. t), successive *simultaneous* values of K_v and dK_v/dt may be substituted until the maximum of h_{pm} is determined.

To facilitate the solution, we can nondimensionalize Equation 6.136 by introducing the time to complete valve closure T_c, and rearrange to give:

$$\left[1+\frac{f'L}{d_o(K_v+K_o)}\right]\frac{h_{pm}}{h_o}-\frac{1}{Z(K_v+K_o)^{3/2}}\frac{dK_v}{dt/T_c}\sqrt{\frac{h_{pm}}{h_o}}-1=0 \tag{6.137a}$$

$$Z_c=\frac{T_c}{L/\sqrt{2gh_o}} \tag{6.137b}$$

in which Z_c will be called the *valve closure parameter*. For a given valve curve (K_v vs. t and, hence, dK_v/dt), Equation 6.137a (which applies to the maximum value of h_{pm}) implies the following functional relationship:

$$\frac{h_{pm}}{h_o}=\phi\left(K_o,\frac{f'L}{d_o},Z_c\right) \tag{6.138}$$

7

Flow in Open Channels

7.1 Introduction

The emphasis in the present chapter is on the man-made open channels associated primarily with environmental engineering systems. (The reader may find it helpful to review the preface at this point insofar as it discusses the orientation of this text). Rivers (and streams) and estuaries are addressed in Chapter 15. Asymmetric flows in symmetric expansions are discussed for rivers in Section 15.2, even though such flows also occur in man-made open channels for which related discussion is also included in Chapter 15.

The present chapter is limited to spatially-constant flows, which (as in the previous chapter) are considered to include branching flows. Distribution channels and collection channels are considered in Chapters 11 and 12, respectively. The initial sections of the present chapter consider only steady flows – flows which are (or may be assumed to be) temporally constant. A later section introduces the subject of time-varying flows in channels.

The relatively large characteristic dimensions of the channels of concern generally results in turbulent flow conditions. The dominant analytical complexity attending open-channel flows is the addition of cross-sectional area as an unknown. Unlike pressure conduits, care must be taken in the selection of a starting position and the direction (upstream or downstream) in which computations proceed. The starting conditions are most often conditions of *critical flow* or *uniform flow*, important concepts which are addressed in the two subsections which follow. Subsequent sections deal with flows undergoing transition from one control condition to (or toward) another. This will all take on meaning as practical applications are interwoven into the conceptual scheme.

7.2 Critical Flow Conditions

Consider a partially-full conduit with a shallow slope discharging water freely into the atmosphere. (Shallow here refers to a *subcritical* slope, as defined below.) The specific energy of the flow at the free discharge is given by:

$$E = y + \frac{Q^2}{2gA^2} \tag{7.1}$$

Since Q is constant and A is an increasing function of y, the above equation indicates that the specific energy is a function only of y.

It has been well established experimentally that, of the various hypothetical depths and corresponding energy levels that may occur at such a free discharge, the one that will occur will be that which minimizes the specific energy. This condition may be determined by differentiating Equation 7.1 with respect to y, and setting $dE/dy = 0$, as follows:

$$\frac{dE}{dy} = 1 - \frac{Q^2}{gA^3}\frac{dA}{dy} = 1 - \frac{V^2}{gA}\frac{dA}{dy} = 0 \tag{7.2}$$

The differential water area dA is at the free surface and is equal to Tdy, where T is the *top width* of the flow. Thus:

$$\frac{dA}{dy} = T \tag{7.3}$$

Hydraulics and Pneumatics in Environmental Engineering, First Edition. S. David Graber.

The *hydraulic depth D* is now introduced, defined by:

$$D = \frac{A}{T} \tag{7.4}$$

The hydraulic depth may be thought of as an *average* depth. This may be seen by recalling that the average of the function y is given by $\bar{y} = \int_a^b y dx / \int_a^b dx$, where x is the cross-sectional direction perpendicular to y and a and b are the limits of the function y along the x-axis. Since $\int_a^b y dx = A$ and $\int_a^b dx = T$, then $y = A/T \doteq D$. Substituting Equations 7.3 and 7.4 into Equation 7.2 gives the following as the requirement of minimum energy:

$$\frac{V^2}{gD} = 1 \tag{7.5}$$

The square root of the dimensionless term in Equation 7.5 is known as the *Froude Number* denoted by **F**:

$$\mathbf{F} = \frac{V}{\sqrt{gD}} \tag{7.6}$$

For a channel of *large* slope ($S_o = \tan\theta > 0.15$ as defined in Chapter 2) and energy coefficient (α as defined in Chapter 2) significantly different than unity, the Froude Number may be approximated by $\mathbf{F} = V/\sqrt{gD\cos\theta/\alpha}$ (Chow 1959, §3-3).

A Froude Number of 1 corresponds to *critical* flow conditions, so-called because it divides the region of *subcritical* **F** < 1) and *supercritical* flow conditions, about which more will be said later.

Returning to Figure 7.1, the conditions *near* (see below) the free discharge can now be examined. Since $V = Q/A$, a Froude Number of 1 implies:

$$\frac{Q}{A\sqrt{gD}} = 1 \tag{7.7}$$

which can be rewritten as:

$$A\sqrt{D} = Q/\sqrt{g} \tag{7.8}$$

Knowing $Q/\sqrt{g}$ gives $A\sqrt{D}$ which, for a particular cross section, is a function of y only. Knowing the functional relation, y can be determined. The mechanics of doing so will be dealt with below.

Before going further, it is noted that critical flow conditions occur not only at free channel discharges but also under conditions of free or partially free discharge at weirs, spillways, and critical flow flumes. Each of these critical flow conditions is explored in practical terms below.

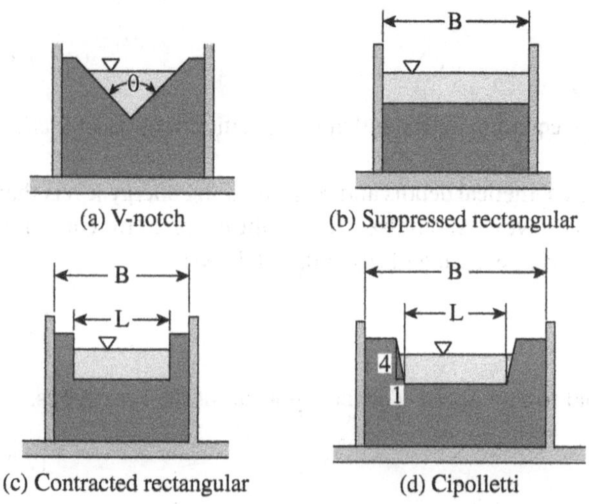

Figure 7.1 Common Sharp-crested Weir Shapes (Bengtson 2019 / Continuing Education & Development, Inc). (a) V-notch; (b) suppressed rectangular; (c) contracted rectangular; (d) cipolleti.

7.2.1 Channel Sections

In open channels, the determination of critical flow conditions requires the development of a relation between $A\sqrt{D}$ and y for use with Equation 7.8. For natural channel cross sections, a mathematical relation is usually not practical, but a plot can be developed of A vs. y (e.g., by planimetering for A) and T vs. y, from which $D = A/T$ vs. y and $A\sqrt{D}$ vs. y can be determined. With such a plot, together with Equation 7.8, the critical flow depth y_c can be determined from $Q\sqrt{g}$ or conversely. For computer calculations, the cross section is usually represented by a series of connected dots, each dot being a surveyed bottom elevation and transverse station. The area, top width, etc. can then be determined by summing triangular and rectangular subareas and/or appropriate interpolations.

Most man-made channels (and some natural channels) are amenable to mathematical expression of the relations between the dependent variables A, T, and D and the independent variable y. For the simple rectangular channel of width b, $A = by$, $T = b$, and $D = y$, from which $A\sqrt{D} = by^{3/2}$ and Equation 7.8 gives the following for the critical depth:

$$y_c = \left(\frac{Q}{b\sqrt{g}}\right)^{2/3} \tag{7.9}$$

For a triangular channel of side slope $1/z$, $A = zy^2$, $T = 2zy$, and $D = y/2$, from which $A\sqrt{D} = zy^{2.5}/\sqrt{2}$ and Equation 7.8 gives the following for the critical depth:

$$y_c = \left(\frac{\sqrt{2}Q}{z\sqrt{g}}\right)^{2/5} \tag{7.10}$$

Tables of geometric elements giving the relations between the dependent variables A, T, D, and $A\sqrt{D}$ and the independent variable y for cross sections encountered in environmental engineering practice are given by, for example, Chow (1959, Table 2-1) and on various web sites. For most of the channel shapes, the relation between $A\sqrt{D}$ and y is too complex to allow an explicit solution for y such as was given above for rectangular and triangular channels. In such cases, a trial-and-error solution for y is required and may be readily accomplished with a programmable calculator.

Substituting Equation 7.7 into Equation 7.1 enables the specific energy to be written in terms of the critical depth and critical hydraulic depth as follows:

$$E = y_c + \frac{D_c}{2} \tag{7.11}$$

For a rectangular channel, $D_c = y_c$, so that $E = (3/2)y_c$. In this case, the depth constitutes 2/3 of the specific energy whereas the velocity head $V^2/(2g) = E - y_c$ constitutes 1/3 of the specific energy. Using the relationship $E = (3/2)y_c$ and Equation 7.10, allows the flow to be expressed in terms of the specific energy in the rectangular channel:

$$Q = \left(\frac{2}{3}\right)^{3/2} b\sqrt{g}E^{3/2} \tag{7.12a}$$

Substituting $g = 32.2$ ft/sec^2 into the above gives the following dimensional relation:

$$Q = 3.09bE^{3/2} \tag{7.12b}$$

in which b, E, and Q have units of ft, ft, and cfs, respectively, and the constant 3.09 has units of ft$^{1/2}$/sec. For the triangular channel ($D_c = y_c/2$), Equation 7.11 gives $E = (5/4)y_c$ which with Equation 7.10 gives:

$$Q = \left(\frac{4}{5}\right)^{5/2} \sqrt{\frac{g}{2}}zE^{5/2} \tag{7.13a}$$

or

$$Q = 2.30zE^{5/2} \tag{7.13b}$$

In Equation 7.13b, E and Q have units of ft and cfs, respectively. Further use will be made of Equations 7.12b and 7.13b in subsequent sections of this chapter.

The predominant occurrence of critical depth controls in channel sections for the applications of interest is near the free fall end of channels on shallow (subcritical) slopes. Examples include skimming troughs, washwater troughs, and other types of collection channels (see Chapter 12).

The critical depth will actually occur a short distance upstream of the free fall section. This is due to curvature effects which, as noted in Chapter 2, are assumed negligible in the derivation of the energy equation. For virtually all practical applications encountered, the assumption that the critical depth actually occurs at the free fall section is adequate.

Note also that downstream submergence will have no effect on the critical depth control as long as the tailwater depth y_t is less than or equal to y_c. If y_t exceeds y_c, the critical depth control will be submerged.

Finally, it is noted that the depth at the free overall may be less than the critical y_c if the channel slope is steep, as discussed in Section 7.5 – Supercritical Flow.

7.2.2 Weirs

Here we discuss sharp-crested, broad-crested, side, and leaping weirs, and the ogee spillway. The Sutro weir is discussed under channel contractions below.

7.2.2.1 Sharp-Crested Weir

Sharp-crested weirs include V-notch (triangular), rectangular (contracted or suppressed), and cipolletti; these are depicted below. The circular weir is a variation of the suppressed rectangular weir as discussed below.

The author specified a V-notch weir for the Tallman Island, New York WWTP for metering waste secondary sludge at a maximum flow rate of 3.5 mgd, with float-operated transmitters. Also for that facility, the author specified two rectangular weirs for storm-overflow metering of primary effluent, each of 10-foot length and 0–20 mgd flow range, with bubble tube sets and gate position sensors. For the Massachusetts Correctional Institute, Bridgewater, the author specified a sludge flow metering system consisting of two V-notch weir plates at the air lift pump boxes adjacent to the sedimentation basins; with float-operated transmitters, each has an operating flow range of 0.1–1.0 mgd. Another example of that facility is given below.

Example 7.1

Determine the notch spacing of a 90° peripheral V-notch weir on a circular clarifier required to limit the head h on the weir to not greater than 2 in. (Massachusetts Correctional Institute, Bridgewater). The clarifier diameter is 20 ft and maximum flow is 0.72 mgd.

Solution

Howe (1950) gives Cone's equation for flow over a triangular weir:

$$Q = 2.49h^{2.48}$$

in which Q is flow rate in cfs and h is head on the weir in ft. The flow per unit length of weir q is $0.72 \times 1.547/(20\pi) = 0.0177$ cfs/ft. The maximum flow per V-notch q_N is determined from the above equation as follows: $h = 2/12 = 0.167$ ft, $h^{2.48} = 0.0118$, $q_N = 2.49(0.0118) = 0.0293$ cfs/notch. The maximum notch spacing is then given by $q_N/q = 0.0293/0.0177 = 1.65$ ft. A manufacturer's standard 12-in. center-to-center notch spacing was selected in this case.

Example 7.2

Discussion of the *circular weir* will be provided in the context of a practical example (Medfield, Massachusetts). [By "circular weir" is meant a suppressed rectangular weir bounded by the circular ends of an encasing pipe.] A boot opening on a sophisticated system for recycling of car wash water discharges over such a weir. The weir spans a horizontal pipe of 8-inch diameter and is 2.22 in. above the invert. The approach "channel" is a tank of width much greater than the circular weir. Determine if this configuration will allow a flow of 20 gpm to pass over the weir without filling the pipe.

Solution

The relevant equations are presented with attributions by Ghobadian and Meratifashi (2012) [the empirical results of Ghobadian and Meratifashi which attempt to improve an earlier attribution are erroneous]:

$$Q = \Omega S\sqrt{2gh}$$

where S is the flow area between the free surface corresponding to the water head h relative to the pipe bottom and the weir crest, and Ω is given by:

$$\Omega = [0.350 + 0.002(D/h)]\left[1 + (S/S')^2\right]$$

where S' is the approach channel flow area, which is much greater than S so the $(S/S')^2$ term can be ignored such that:

$$\Omega \cong [0.350 + 0.002(D/h)]$$

Geometric elements give:

$$\frac{S}{D^2} = \frac{1}{8}(\theta_h - \sin\theta_h) - \frac{1}{8}(\theta_w - \sin\theta_w)$$

$$\theta = 2\sin^{-1}\left[2\sqrt{\frac{h}{D}\left(1 - \frac{h}{D}\right)}\right]$$

in which the subscript "h" denotes the water surface and the subscript "w" denotes the weir crest.

Expressing the first equation above in the following dimensionless form:

$$\frac{Q}{D^2\sqrt{2gh}} = \Omega\frac{S}{D^2}$$

The above four equations have the following four dimensionless variables:

$$\frac{Q}{D^2\sqrt{2gh}},\ \Omega,\ \frac{S}{D^2},\ \frac{h_h}{D_h}$$

yielding a solvable set of equations. An iterative solution to these equations for the given values of the variables yields $h_h/D_h = 0.244$ giving $h = 0.355(8) = 2.84$ in., which is 0.62 in. above the weir and within the boot opening.

Submergence of sharp-crested weirs may occur at high tailwater conditions. The Villemonte (1947) equation applies and is given by:

$$Q/Q_1 = \left[1 - (d/h)^{3/2}\right]^{0.385} \tag{7.14a}$$

$$Q_1 = CLh^{3/2} \tag{7.14b}$$

in which Q is actual flow, Q_1 is unsubmerged flow (for the same h), L is weir length, h is head relative to the weir crest, and d is downstream submergence relative to the weir crest.

Example 7.3

An inverted siphon crossing of the Spicket River, Lawrence, Massachusetts is planned. The upstream structure incorporates a sharp-crested weir which is submerged under design conditions with a flow rate of 21.1 cfs. Pertinent elevations are channel Inv. El. 41.81, weir crest El. 42.86, downstream water surface elevation 43.19, allowable upstream water surface elevation 43.87. Determine the required weir length.

Solution

Pertinent terms in Equation 7.14 include $d = 43.19–42.86 = 0.33$ ft, $h = 43.87–42.86 = 1.01$ ft, and $(d/h)^{3/2} = (0.33/1.01)^{3/2} = 0.1867$. The unsubmerged weir flow is determined using Equation 7.14b. We take $C = 3.3$ so Equation 7.14b gives $Q_1 = 3.3L(1.01)^{3/2}$ cfs. Substituted into Equation 7.14a, we obtain $21.1 = 3.3L(1.01)^{3/2}[1 - 0.1867]^{0.385}$ which is solved for L to obtain a weir length of 6.82 ft.

The sharp-crested weir with end contractions was first analyzed by James B. Francis at Lowell, Massachusetts (Rouse and Ince 1957) and is referred to as the Francis weir. Howe (1950) reported the Francis formula as $Q = 3.3(L - nh/10)h^{3/2}$ with flow rate Q in cfs, n being the number of end contractions, and h being the head on the weir in ft with $h/L < 1/3$.

Example 7.4
The author analyzed a Francis weir at the outlet of Nagog Pond in Acton, Massachusetts as part of a project to determine pond releases and the intermittence of the downstream Nagog Brook (as also mentioned in Section 15.3 and in Graber 2017b, §10.10). The weir is of steel, 10 ft long, with 18½ in. (1.542 ft) from the top of the weir to the top of the concrete end walls. Determine the flow-head relationship at four equally-spaced values of head up to the maximum head within the weir boundaries, and the applicability of the Francis formula.

Solution

The pertinent formula is $Q = 3.3(10 - 2h/10)h^{3/2}$, from which the following values are calculated:

h (ft)	*Q* (cfs)
0.386	7.84
0.771	22.0
1.157	40.1
1.542	61.2

The maximum value of h/L is 1.542/10 = 0.1542, which is less than 1/3 so the formula is applicable over this range.

7.2.2.2 Broad-Crested Weir

Broad-crested weirs may be designed specifically to serve as a weir on, for example, dyke overflows for detention basins, or function fortuitously as broad-crested weirs as in the case of roadway overflows. In the present context, the latter may be encountered immediately downstream of detention basins and are conjunctive in their performance. Ippen (1950) and Chow (1959, Section 3–6) give relevant formulae.

Example 7.5
An emergency dyke overflow is to be planned for a detention basin (Westborough, Massachusetts). The dyke crest is to be at El. 68.5 and the top of the dyke is to be at El. 70.0. Design is to be for a 1 ft head, with a 100-year flow Q of 33.3 cfs. Determine the crest length and whether hand-placed riprap would be suitable.

Solution

Chow gives the following equation for flow over a broad-crested weir, expressed here in a slightly modified form for convenience:

$$q = C_{bc}H^{3/2}$$

$$C_{bc} = 0.433\sqrt{2g}\left(\frac{y_1}{y_1 + h}\right)^{1/2}$$

in which q is volumetric rate of flow per unit length of weir crest, y_1 is the depth of the approach channel, h is the height of the weir above the approach channel, and H is the total head on the weir equal to $y_1 - h$. Units are in ft and sec. Chow notes that the practical range of the coefficient C_{bc} obtained by actual observations is from 3.05 to 2.67. For conservatism (smaller C_{bc} gives larger H for a given q), the coefficient 2.67 is selected, giving the following:

$$L = Q/q = \frac{33.3}{2.67(1)^{3/2}} = 12.47 \text{ ft}$$

A 12-ft weir crest is selected, together with 6-ft horizontal distances of sloped riprap extending to El. 70 on each side of the crest. Considering just the 12-ft crest for conservatism, the velocity on the crest is 33.3/[12(1)] = 2.78 ft/sec, for which hand-placed riprap is suitable.

Submergence of broad-crested weirs may occur at high tailwater conditions. In one instance analyzed by the author (Gaslight Detention Basin, Framingham, Massachusetts), a broad-crested weir was employed in conjunction with culverts for detention of floodwaters. The hydraulics transitioned successively with increasing flow between culvert flow Types 6, 4, and 1 (discussed in Section 7.7) and overflowed a broad-crested weir under high headwater conditions together with Type 1 culvert flow. The weir was submerged under a combination of high headwater and high tailwater conditions. The Villemonte equation (Equation 7.14) was used as an approximation in that case.

7.2.2.3 Side Weirs

Single-side weirs find use as a means of diverting a portion of the flow in an open channel. An example is for diversion of storm flows from a WWTP influent sewer to a stormwater detention basin. Multiple side weirs along the sides of open channels provide an effective means of distributing flow under certain circumstances. In the former usage generally and sometimes in the latter usage, the channel velocities parallel to the weir and the variation in piezometric head along the length of the weir preclude the use of the simple sharp-crested weir equations given above.

The traditional DeMarchi (Chow 1959, §12-3) solution is for a side weir along a rectangular channel. This configuration occurs both in cast-in-place box sewers and in treatment plant influent channels. However, one of the more common configurations, constructed in manhole structures has a semicircular bottom with vertical sides above the spring line. Other channel cross sections may also be desirable. Accordingly, a more generalized solution is presented here. The initial focus is on the single-side weir configuration used for flow diversion. Multiple side weirs used in process influent channels are discussed in Chapter 11.

The length of the individual side weirs is usually sufficiently small that channel friction and slope effects can be neglected, in which case the DeMarchi solution gives the following:

$$\frac{dy}{dx} = \frac{4C_M D}{3A}\frac{\sqrt{(E-y)(y-S)^3}}{D-2E+2y} \tag{7.15}$$

Substituting $A = TD$ and $D = y - K_1T$ for the case of constant top width, the integral of the above equation can be expressed as follows:

$$x = \frac{3T}{4}\int \frac{3y - K_1T - 2E}{C_M\sqrt{(E-y)(y-S)^3}}dy \tag{7.16}$$

Assuming C_M to be independent of x, the above equation is integrated to give:

$$x = \frac{3T}{2C_M}\psi(y, E, S, K_1T) + \text{const} \tag{7.17}$$

in which:

$$\psi = \frac{2E - 3S + K_1T}{E-S}\sqrt{\frac{E-y}{y-S}} - 3\sin^{-1}\sqrt{\frac{E-y}{E-S}} \tag{7.18}$$

For a channel of constant top width and semicircular bottom, $K_1 = (1 - \pi/4)/2$. For a rectangular channel, $K_1 = 0$, the above equation reduces to the varied-flow function of DeMarchi (Chow 1959, §12-3). [The reader is cautioned that the argument of $\sin^{-1}$ in the DeMarchi function is given erroneously as $\sqrt{(E-y)/(y-S)}$ in some publications.] Designating the channel sections at the upstream and downstream ends of the side weir of length L by suffixes 1 and 2, respectively,

$$L = \frac{3T}{2C_M}(\psi_2 - \psi_1) \tag{7.19}$$

Subramanya and Awasthy (1972) derived and experimentally substantiated the following relationship for C_M vs. $\mathbf{F}_1$ for subcritical flow:

$$C_M = 0.611\sqrt{1-\left(\frac{3\mathbf{F}_1^2}{\mathbf{F}_1^2+2}\right)} \tag{7.20}$$

When used as a passive regulating device, the design objectives of a side-weir regulator are generally: (1) to prevent the loss of sanitary flow during dry-weather periods, and (2) to bypass the requisite capacity to avoid excessive flows in the downstream channel during storm periods. The first objective may be accomplished by providing a side weir at an elevation equal to the elevation of the peak sanitary flow. The above equations provide the means of designing the overflow weir to achieve the second objective.

An investigation of the functional relations of the above equations suggests a convenient means of solution, presented here for subcritical flow. For a channel of constant top width, $\mathbf{F} = V/\sqrt{gD} = V/\sqrt{g(y-K_1T)}$, and since $V^2/(2g) = \sqrt{(E-y)}$ the Froude Number approaching the weir can be written as:

$$\mathbf{F}_1 = \sqrt{2\left(\frac{E-y_1}{y_1-K_1T}\right)} \tag{7.21}$$

Substituting the above into Equation 7.20 and rearranging gives:

$$C_M = 0.611\sqrt{\frac{3y_1/E-2-K_1T/E}{1-K_1T/E}} \tag{7.22}$$

Equations 7.17, 7.18, 7.19, and 7.22 can then be combined into a single equation of the following functional form:

$$\frac{3}{2(0.611)\sqrt{\dfrac{3y_1/E-2-K_1T/E}{1-K_1T/E}}}\left[\psi_2\left(\frac{y_2}{E},\frac{S}{E},\frac{K_1T}{E}\right)-\psi_1\left(\frac{y_1}{E},\frac{S}{E},\frac{K_1T}{E}\right)\right]-\frac{L}{T}=0 \tag{7.23}$$

Knowing values of L, S, T, and K_1, and knowing the desired continuing flow Q_2 from which y_2 and E can be determined (e.g., for uniform flow or backwater conditions in the interceptor), the above equation becomes a function of the one unknown y_1/E. The "zero" or root of that function can be readily obtained, such as on a programmable calculator. (The graphical bisection method works well; the nature of the function does not seem as amenable to a Newton–Raphson type of iterative solution.) For subcritical flow, y/E will be in the range $2/3 + K_1T/(3E)$ to 1, providing useful lower and upper limits for the numerical procedure. The above thus provides the value of y_1/E from which $V_1 = \sqrt{2g(E-y_1)}$, $Q_1 = (y_1T - K_1T^2)V_1$ and $Q_s = Q_1 - Q_2$ can be calculated. Alternatively, if the desired Q_1 is known, y_1/E can be calculated and the above equation used to determine values of S and L necessary to divert the required flow. The procedure in either case is easily facilitated by a programmable calculator.

In some cases, it may be possible to set the side weir such that loss of dry-weather sanitary flow is prevented (i.e., if the maximum dry-weather sewer depth is less than the weir height), while bypassing the requisite flow under storm conditions. However, an evident disadvantage of the passive (non-adjustable) side weir is its inability to bypass the required flow while maximizing utilization of plant treatment capacity. An adjustable side weir becomes necessary if this is to be accomplished, as demonstrated in the following example.

Example 7.6

A twin 96 in. sewer is estimated to carry a peak storm flow of 240 mgd and has an existing stormwater diversion consisting of a fixed 22.5 ft long side weir at an elevation 4.75 ft above the approximately horizontal sewer invert (Tallman Island, New York, Linden Street regulator). Determine whether the side weir is sufficient to prevent the flow continuing to a downstream WWTP from exceeding 135 mgd. Backwater calculations indicate a water depth of 7.18 ft in the sewer just downstream of the side weir at 135 mgd flow.

Solution

Initially known values are as follows: $L = 22.5$ ft, $S = 4.75$ ft, $T = 2 \times 8 = 16$ ft, $K_1 = (1/2)(1/2)(1 - \pi/4) = 0.05365$, $Q_2 = 135$ mgd = 209 cfs, $y_2 = 7.18$ ft. The energy head E is determined as follows:

$$V_2 = \frac{Q_2}{(y_2T - K_1T^2)} = \frac{209}{[7.18(16) - 0.05365(16)^2]} = 2.07 \text{ ft/sec}$$

$$\frac{V_2^2}{2g} = 0.0663 \text{ ft}$$

$$E = y_2 + \frac{V_2^2}{2g} = 7.18 + 0.0663 = 7.25 \text{ ft}$$

Using a trial-and-error procedure with an equation given above gives $y_1/E = 0.951$ or $y_1 = 0.951(7.25) = 6.89$ ft. The corresponding upstream flow is determined as follows:

$$V_1 = \sqrt{2g(E - y_1)} = \sqrt{2(32.2)(7.25 - 6.89)} = 4.78 \text{ ft/sec}$$

$$Q_1 = (y_1T - K_1T^2)V_1 = [6.89(16) - 0.05365(16)^2](4.78) = 461 \text{ cfs} = 298 \text{ mgd}$$

Therefore, the existing side weir will divert 298–135 = 163 mgd, which is more than the 240–135 = 105 mgd required.

Example 7.7
It is desired to modify the design of the side weir in the previous example to eliminate diversion of flows below 135 mgd. Find the dimensions and vertical travel of a side weir that will accomplish this.

Solution

An adjustable side weir is necessary, to be set at least 6.18 ft above the sewer invert and be lowered by a level-sensing control system to maintain a water depth of 7.18 ft in the diversion structure when diversion is necessary. Initially known values are as follows: $T = 16$ ft, $K_1 = 0.05365$, $Q_2 = 135$ mgd = 209 cfs, $y_2 = 7.18$ ft, $Q_1 = 240$ mgd = 371 cfs, $E = 7.25$ ft. The upstream depth can be determined by eliminating V_1 from the equations $V_1 = \sqrt{2g(E - y_1)}$ and $Q_1 = (y_1T - K_1T^2)V_1$ and rearranging to obtain: $(y_1T - K_1T^2)^2(2g)(E - y_1) - Q_1^2 = 0$.

Inserting known values, this equation can be solved for the single unknown y_1 by trial-and-error to obtain $y_1 = 7.03$ ft. The remaining unknowns in Equation 7.23 are L and S, different combinations of which will satisfy the above equation. By assuming S, we can solve explicitly for corresponding values of L and a selection made based on structural, mechanical, and cost considerations. In this case, $S = 4.75$ ft was selected to minimize structural modifications, for which the resulting value of L is 14.0 ft and the required side-weir travel is 6.18–4.75 = 1.43 ft.

It should be mentioned before leaving the subject of side weirs that numerous empirical relations are available for side overflow weirs, such as those of Babbitt and Baumann (1958, Chapter 6); and Engels and Forcheimer as cited in Fair and Geyer (1968) and ASCE/WPCF (1960, 1969).

7.2.2.4 Leaping Weir

Although combined sewer overflows are to be avoided, in some cases, they are necessary, at least temporarily. In one case reviewed and modified by the author, to accommodate space constraints a consultant proposed to install an opening on the side of a trunk sewer conveying flow to the treatment plant. The consultant's proposed opening would convey flow to a bypass with an invert matching that of the trunk sewer invert, with capacity to bypass the combined sewage exceeding the capacity of the treatment plant. The author was asked to design the bypass. In another case where a new interceptor sewer was to be installed, the author designed a similar bypass in an intercepting chamber to bypass flows exceeding the downstream treatment plant capacity. In both cases, such designs would allow dry-weather sewage flows to be discharged to the receiving river, with additional complications discussed below in the second case. In both cases, the author proposed and successfully designed variations on leaping weirs [the concept in different forms is described by Fair and Geyer (1954, pp. 72–73) and Greeley *et al.* (1969b, pp. 40–73 to 40–75)] that would prevent bypassing dry-weather flows to the rivers. The second of these examples is described below.

Example 7.8

In this example (Lewiston/Auburn, Maine) the 48-inch Hart Brook sewer follows a brook of the same name which discharges into the Androscoggin River. The sewer is laid on a steep slope (2.5%) and flow will be supercritical under all conditions. A standard overflow weir intercepting chamber on this line would create cross waves (waves perpendicular to the main channel axis, discussed in Sections 7.5 and 7.8) greater than twice the uniform flow depth in the sewer under peak dry-weather flow conditions. To properly design such a structure would require that the sewer discharge into a stilling basin which would create a predictable hydraulic jump and allow interception to occur under subcritical (tranquil) flow conditions. Unless a sizable stilling basin was provided, the flow would "jump" to a depth requiring an overflow weir elevation which would present a considerable restriction to flow under storm conditions.

Solution

A leaping weir regulator circumvents these problems by maintaining supercritical flow in the sewer under dry-weather flow conditions and presenting no restriction in the line. The weir opening is proportioned to intercept the maximum design dry-weather flow (10 mgd) and 18 mgd total under storm flow conditions. The author's design is depicted on Figure 7.2(a) and (b). It is basically a curved, horizontal, Sutro weir, a more conventional form of which is discussed below. The Sutro weir formulae discussed below were applied as a suitable approximation in this case. The flow is from left to right in the 48-inch sewer. The flow drops down to the 24-inch sewer that continues to the wastewater treatment plant. Higher flows of combined stormwater/sanitary sewage continue in part beyond the drop to the storm sewer outlet to the river.

7.2.2.5 Ogee Spillway

The ogee spillway is discussed here because of its important use in self-cleaning trench-type wet wells for wastewater as depicted on Figure 7.3 and discussed further in Section 8.7. For that application, a primary purpose is to create supercritical flow at the base of the ogee, from which a hydraulic jump may be formed. Although there are elaborate relationships for the trajectory of the nappe after springing clear of the weir crest and the corresponding design of the spillway (e.g., Linsley and Franzini 1964, §9-1, USBR 1965, §189), it is sufficient for the present application to use the basic projectile trajectory relationship down to the point of inflexion, given by:

$$x = \sqrt{2(v_o^2/g)y} \tag{7.24}$$

with y, v_o, g, and x being, respectively, the vertical distance down from the weir crest, efflux velocity, gravitational acceleration, and horizontal distance from the weir crest. The shape of the spillway below the point of inflexion is exemplified below.

Example 7.9

An ogee spillway for a self-cleaning trench-type wet well (Washington Highway Pumping Station, Lincoln, Rhode Island) is planned to have a maximum flow rate of 3500 gpm (7.80 cfs), trench width of 32 in. = 32/12 = 2.67 ft, and horizontal distance of 4.0 ft to the point of inflexion. Determine the dimensions of the ogee and whether supercritical flow is achieved at the base of the ogee. Assume critical flow conditions occur at the top of the spillway. If supercritical flow is achieved, determine the sequent depth of the jump.

Solution

Critical-flow conditions correspond to $\mathbf{F} = V/\sqrt{gy_c} = 1 = Q/\left(by_c\sqrt{gy_c}\right)$ from which we obtain the critical depth as

$$y_c = \sqrt[3]{Q^2/(gb^2)} = \sqrt[3]{(7.80)^2/[32.2(2.67)^2]} = 0.643 \text{ ft}$$

The flow area at the top of the spillway is 0.643(2.67) = 1.715 sq ft, and the efflux velocity is $v_o = 7.80/1.715 = 4.55$ ft/sec. Equation 7.24 then gives $x = \sqrt{2[(4.55)^2/32.2]y} = 1.134\sqrt{y}$ or $y = 0.778x^2$ with x and y in units of ft. The drop of the

(a)

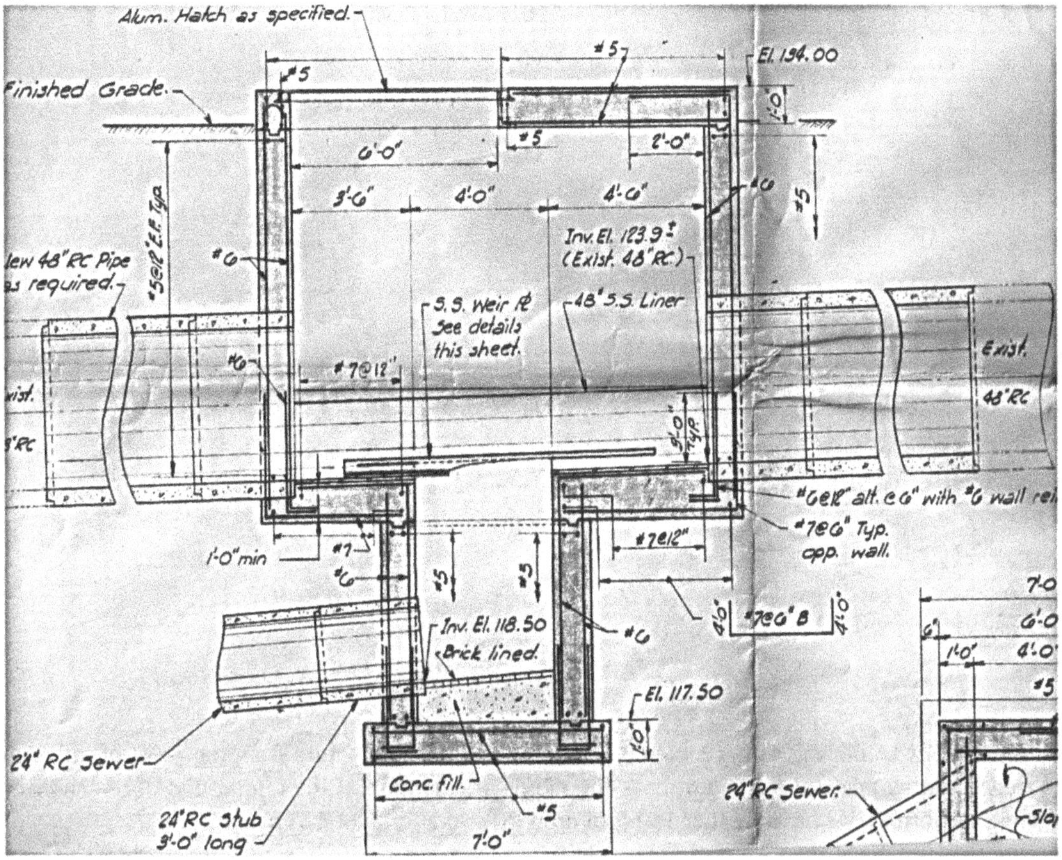

(b)

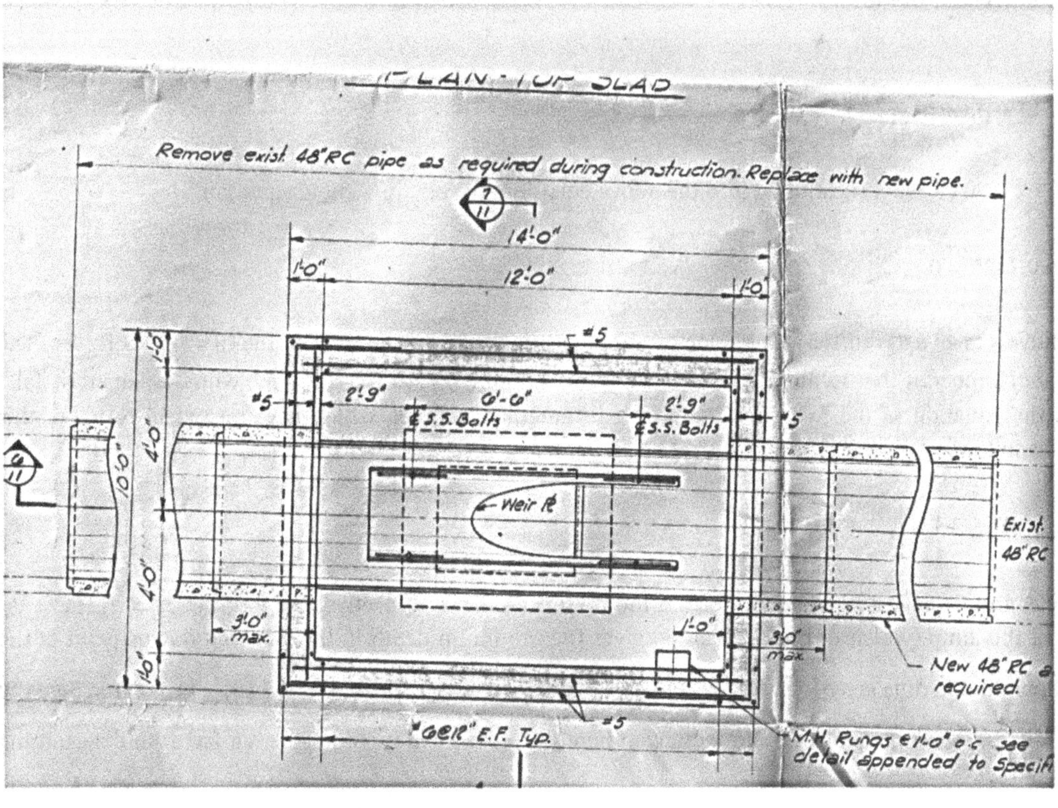

Figure 7.2 (a) Leaping Weir Elevation; (b) Leaping Weir Plan

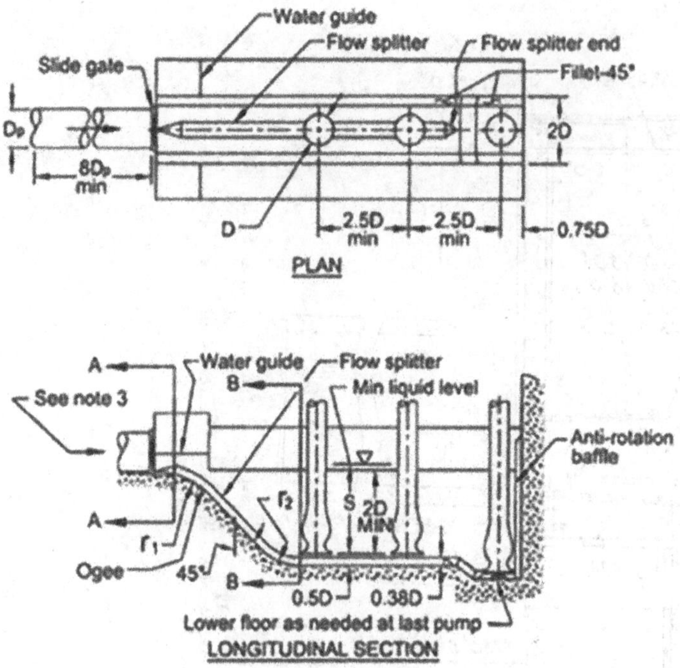

Figure 7.3 Self-Cleaning Trench-Type Wet Well (Elberti 2021 / Water Environment Federation)

ogee to the point of inflexion is $y = (0.778)^2(4.0) = 2.42$ ft. The reverse radius of curvature from the point of inflexion to the bottom of the trench can be taken at 0.5 to 1.0 times the radius of curvature of the top portion of the spillway (ANSI/HI 1998). Selecting the former, we have another 1.21 ft of drop for a total drop of 3.63 ft.

Applying Bernoulli's equation (Equation 4.16) from the top of the spillway to the bottom (denoted station "1" and "2," respectively):

$$\frac{(4.55)^2}{2(32.2)} + 0.643 + 3.63 = \frac{V_2^2}{2(32.2)} + d_2$$

With $V_2 = Q/(bd_2) = 7.80/(2.67d_2)$ substituted into the above equation, we obtain a cubic equation:

$$d_2^3 - 4.59d_2^2 + 0.1325 = 0$$

which is solved iteratively to give $d_2 = 0.1664$. The velocity at that location is $V_2 = 7.80/[2.67(0.1664)] = 17.56$ ft/ sec. The corresponding Froude Number at that location is $\mathbf{F}_2 = V_2/\sqrt{gd_2} = 17.56/\sqrt{32.2(0.1664)} = 7.59$, which is supercritical.

Section 7.5 gives the equation of the hydraulic jump in a rectangular channel, which is expressed in terms of this example's notation as follows:

$$\frac{y_2}{y_1} = \frac{1}{2}\left(\sqrt{1 + 8\mathbf{F}_1^2} - 1\right)$$

in which the variables in this equation relate to those in the present example respectively as follows: $y_1 = d_2$, $\mathbf{F}_1 = \mathbf{F}_2$. The sequent depth of the jump (which in this application gives the minimum depth in the wet well downstream of the jump at the given flow rate) is thus given by $y_2/y_1 = (1/2)\left(\sqrt{1 + 8\mathbf{F}_1^2} - 1\right) = (1/2)\left(\sqrt{1 + 8(7.59)^2} - 1\right) = 10.25$ giving $y_2 = 10.25(0.1664) = 1.706$ ft. Although details are not given here, the calculated jump length is a factor in determining the wet-well trench length.

7.2.3 Channel Contractions

Critical flow channel contractions are used for *flow metering* and as components in *velocity-control systems*. Devices used to create channel contractions include the Parshall flume, Camp regulator, cutthroat flume, and Sutro weir. Hydraulic characteristics of each of those devices and pertinent applications are discussed below. Additional details on Parshall flumes and weir configurations discussed below are given by Babcock (1966, Part 3). Kennison (aka parabolic) nozzles are discussed in WPCF (1978). For level conditions and smooth channels, the simple flumes described by Samani (2017) merit consideration for trapezoidal, rectangular, and circular channels.

7.2.3.1 Parshall Flume

The Parshall flume (Figure 7.4 and Table 7.1), developed by R. L. Parshall (1926) (also see WPCF 1978), is widely used in water and wastewater treatment plants for flow measurement and is occasionally applied for the additional purposes of velocity control, chemical mixing, and postaeration, the latter two taking advantage of the hydraulic jump that forms near the downstream end of the flume. The Parshall flume's flow-measuring properties stem from its ability to create a critical flow control section having a definitive relation between depth and discharge, with a relatively high degree of independence of upstream and *downstream* conditions. It has been extensively calibrated and is available in the form of convenient channel inserts covering a wide range of flows.

The flume contraction creates the critical flow control, and the configuration downstream of the contraction creates conditions (see below) which extend the range of downstream submergences to which the flume may be subjected while still maintaining a one-to-one relationship between flow and gage depth. Extensive calibration data obtained from 1926 to 1957 have resulted in seven empirical formulas relating flow, gage depth, and throat width for throat widths ranging from 1 in. to 50 ft (Davis 1961). The formulas are all of the general form:

$$Q = K_e b y_1^n \tag{7.25a}$$

in which Q is the rate of flow in cfs, b is the throat width in ft, y_1 is depth in feet measured near the upstream end of the flume crest, and the values of K_e and n given in Table 7.2. Although Davis (1961) did not explicitly state the units just mentioned, they can be deduced by comparing the values in the following table with representative formulae given by Chow (1959, §4–6) for which the units are stated.

The above tabulation applies provided that the flume is not submerged by downstream conditions. Davis (1961) gave the criteria that the elevation of the flume above the bottom of the channel must be set such that there is a submergence of less than 60% for the "inch" flumes and 70% for the "foot" flumes in order to maintain free flow. Skogerboe *et al.* (1967a, 1967b) provide theory and data for submerged Parshall flumes and variations thereof.

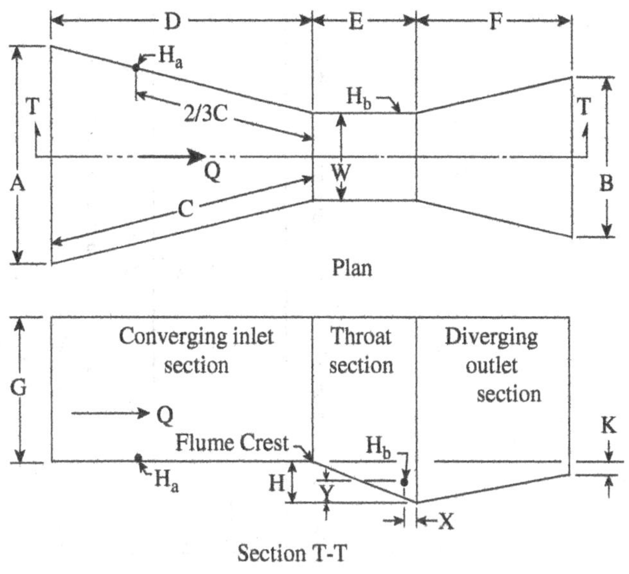

Figure 7.4 Parshall Flume Characteristics (after Skogerboe *et al.* 1966 / with permission of Elsevier)

Table 7.1 Parshall Flume Dimensions and Capacities

Throat Width *W*	Dimensione in Feet and Inches												Free Flow Capacities	
ft. in.	A	B	C	2/3 C	D	E	F	G	H	K	X	Y	Min. cfs	Max. cfs
0′ 3″	0′ 10 3/16″	0′ 7″	1′ 6 3/8″	1′ 0 1/4″	1′ 6″	0′ 6″	1′ 0″	1′ 3″	0′ 2 1/4″	0′ 1″	1″	1 1/2″	0.03	.6
0′ 6″	1′ 3 1/2″	1′ 3 1/2″	2′ 0 7/16″	1′ 4 5/16″	2′ 0″	1′ 0″	2′ 0″	1′ 6″	0′ 4 1/2″	0′ 3″	2″	3″	0.05	2.9
0′ 9″	1′ 10 5/8″	1′ 3″	2′ 10 5/8″	1′ 11 1/8″	2′ 10″	1′ 0″	1′ 6″	2′ 0″	0′ 4 1/2″	0′ 3″	2″	3″	0.1	5.1
12″	2′ 9 1/4″	2′ 0″	4′ 6″	3′ 0″	4′ 4 7/8″	2′ 0″	3′ 0″	3′ 0″	0′ 9″	0′ 3″	2″	3″	0.4	16.0
18″	3′ 4 3/8″	2′ 6″	4′ 9″	3′ 2″	4′ 7 7/8″	2′ 0″	3′ 0″	3′ 0″	0′ 9″	0′ 3″	2″	3″	0.5	24.0
24″	3′ 11 1/2″	3′ 0″	5′ 0″	3′ 4″	4′ 10 7/8″	2′ 0″	3′ 0″	3′ 0″	0′ 9″	0′ 3″	2″	3″	0.7	33.0
30″	4′ 6 3/4″	3′ 6″	5′ 4 1/4″	3′ 6 3/4″	5′ 3″	2′ 0″	3′ 0″	3′ 0″	0′ 9″	0′ 3″	2″	3″	0.8	41.0
3′ 0″	5′ 1 7/8″	4′ 0″	5′ 6″	3′ 8″	5′ 4 3/4″	2′ 0″	3′ 0″	3′ 0″	0′ 9″	0′ 3″	2″	3″	1.0	50.0
4′ 0″	6′ 4 1/4″	5′ 0″	6′ 0″	4′ 0″	5′ 10 5/8″	2′ 0″	3′ 0″	3′ 0″	0′ 9″	0′ 3″	2″	3″	1.3	68.0
5′ 0″	7′ 6 5/8″	6′ 0″	6′ 6″	4′ 4″	6′ 4 1/2″	2′ 0″	3′ 0″	3′ 0″	0′ 9″	0′ 3″	2″	3″	2.2	86.0
6′ 0″	8′ 9″	7′ 0″	7′ 0″	4′ 8″	6′ 10 3/8″	2′ 0″	3′ 0″	3′ 0″	0′ 9″	0′ 3″	2″	3″	2.6	104.0
7′ 0″	9′ 11 3/8″	8′ 0″	7′ 6″	5′ 0″	7′ 4 1/4″	2′ 0″	3′ 0″	3′ 0″	0′ 9″	0′ 3″	2″	3″	**4.1**	121.0
8′ 0″	11′ 1 3/4″	9′ 0″	8′ 0″	5′ 4″	7′ 10 1/8″	21 0″	3′ 0″	3′ 0″	0′ 9″	0′ 3″	2″	3″	4.6	140.0

Table 7.2 Values of K_e and n in Equation 7.25a

Throat Width, b	K_e	n
1 in.	4.06	1.55
2 in.	4.06	1.55
3 in.	3.97	1.547
6 in.	4.12	1.580
9 in.	4.09	1.530
1–8 ft	4	$1.522b^{0.026}$
10–50 ft	$3.6875 + 2.5/b$	1.6

Source: Davis 1961 / American Society of Civil Engineers.

A Parshall flume of molded fiberglass-reinforced polyester with a 6-in. throat width was applied by the author to the screen channel of the Massachusetts Correctional Institute Wastewater Treatment Plant, Bridgewater, with an operating flow range of 0.1–1.0 mgd. Also, two Parshall flumes of fiberglass-reinforced polyester with 2-ft throat widths were applied by the author to meter return secondary sludge at the Tallman Island WWTP, New York City, with operating flow range of 2.0–7.0 mgd. Another of the same material was specified by the author for that facility's Grit Building, with a 30-in. throat width for thickener influent with an operating flow range of 300–450 gpm.

An underutilized application is constant-velocity grit chambers or screen channels which have a Parshall flume (or Camp regulator as discussed below) operating *in conjunction with a channel drop* (Lee and Babbitt 1946; Fair and Geyer 1954, §22-17; Babbitt and Baumann 1958, §18-9; Greeley *et al.* 1969a, Chapter 41, §36). Babbitt and Baumann's configuration has not received the application that it merits, perhaps because the significance of the drop has not been widely appreciated. For their purpose, Babbitt *et al.* suggested a somewhat simplified form of the Parshall flume equation (in which units are again in cfs and ft):

$$Q = 4.1by_1^{3/2} \tag{7.25b}$$

We will consider a rectangular channel with a drop Δz in the channel bottom just upstream of a Parshall flume. Denoting the sections just upstream and downstream of the drop by the subscripts "u" and "d," respectively, the depths of flow y just upstream and just downstream of the drop are related by:

$$y_u + \Delta z \cong y_d \tag{7.26}$$

Since the velocity in the rectangular channel of width w is given by $V = Q/(wy)$, for equal velocities at two different flows Q and Q' we have:

$$\frac{Q}{Q'} = \frac{y_u}{y'_u} = \frac{y_d - \Delta z}{y'_d - \Delta z} \tag{7.27}$$

Equation 7.25b is expressed in the somewhat more general form:

$$Q = Cby_d^{3/2} \tag{7.28}$$

Combining the above equation with Equation 7.27, and rearranging yields:

$$b = \frac{Q}{C}\left[\frac{1-(Q/Q')^{1/3}}{(1-Q/Q')\Delta z}\right]^{3/2} \tag{7.29}$$

The depth of flow upstream of the drop is given by $y_u \cong y_d - \Delta z$ and the velocity at that location is given by:

$$V_u = \frac{Q}{wy_u} = \frac{Q}{w\left[\left(\frac{Q}{Cb}\right)^{2/3} - \Delta z\right]} \tag{7.30}$$

The maximum deviation from the desired velocity (i.e., the extreme value of V_u) is obtained as follows:

$$\frac{dV_u}{dQ} = \frac{w\left[\left(\frac{Q}{Cb}\right)^{2/3} - \Delta z\right] - Q\frac{w}{(cb)^{2/3}}\frac{2}{3}Q^{-1/3}}{\left\{w\left[\left(\frac{Q}{Cb}\right)^{2/3} - \Delta z\right]\right\}^2} = 0 \tag{7.31}$$

which, combined with Equation 7.28, reduces to the following at the extreme value of V_u:

$$\left(\frac{Q}{Cb}\right)^{2/3} = 3\Delta z = y_d \tag{7.32}$$

Taking the derivative again of Equation 7.31 gives:

$$\frac{d^2V_u}{dQ^2} = \frac{w\left[\left(\frac{Q}{Cb}\right)^{2/3} - \Delta z\right] - Q\frac{w}{(cb)^{2/3}}\frac{2}{3}Q^{-1/3}}{\left\{w\left[\left(\frac{Q}{Cb}\right)^{2/3} - \Delta z\right]\right\}^2} \tag{7.33}$$

Substituting the first portion of Equation 7.32 into Equation 7.33 gives:

$$\frac{d^2V_u}{dQ^2} = \frac{1}{6wQ\Delta z} > 0 \tag{7.34}$$

indicating that the extreme value given by Equation 7.32 is a minimum.

7.2.3.2 Camp Regulator

The Camp regulator, named after its developer (Camp 1935, 1942), has been used to control velocities in grit chambers and through screens. It has the advantage over the Parshall flume for that purpose of operating successfully with larger tailwater depths. The channel immediately upstream and downstream of the rectangular regulator is itself rectangular and horizontal. As long as the tailwater depth y_t remains below the critical depth y_c, a critical depth control will occur at the regulator. Submergence effects are discussed in greater detail later in this section. Since $y_c = (2/3)H$ another way of expressing this requirement is $y_t < (2/3)H$.

If contraction of the flow at the regulator and friction losses are neglected, Equation 7.12b applies with b taken as the open width of the regulator. A coefficient of 3.5 has been suggested (Camp 1935, 1942) to conservatively account for these factors, giving:

$$Q = 3.5bE^{3/2} \tag{7.35}$$

in which b, E, and Q have units of ft, ft, and cfs, respectively. The Camp regulator has the advantage over the other contraction devices of being manually adjustable and can be used as a gate to isolate a channel for maintenance purposes. It has not been calibrated for accurate flow measurement. The Camp regulator is depicted below in Figure 7.5.

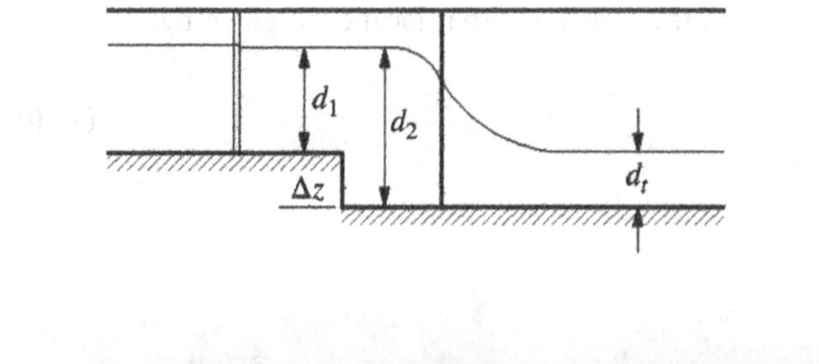

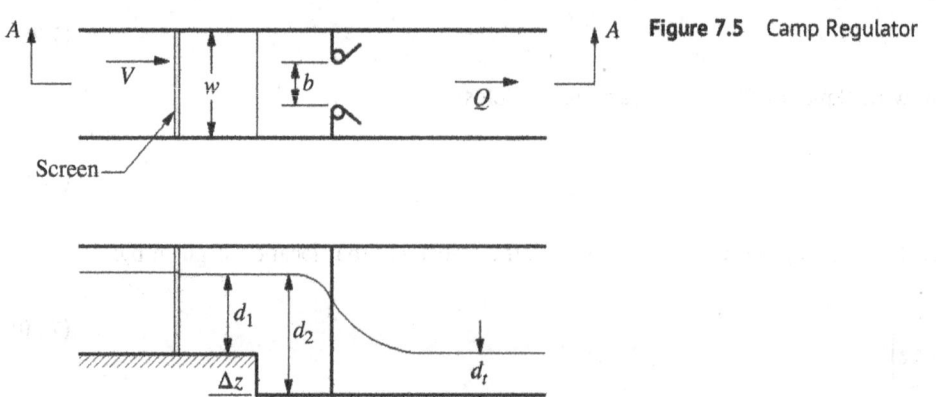

Figure 7.5 Camp Regulator

Example 7.10

Four parallel grit channels are to be converted to screen channels (vortex degritters will be installed further downstream), with velocities through the screens to be controlled by regulating devices installed downstream of the screens (Tallman Island WWTP, New York City). Proportional weirs are presently installed in the channels but would be submerged by the planned downstream wet well elevations. If compatible with equipment to be installed, the channels will be modified to have rectangular cross sections of 6.0-foot width. Design flows are 80 mgd (31.0 cfs) with four channels operating corresponding to [80 (1.547 cfs/mgd)/4] = 30.9 cfs per channel and 160 mgd (82.6 cfs) with three channels operating corresponding to [160(1.547 cfs/mgd)/3] = 82.5 cfs per channel. The channels must be capable of conveying the peak flow of 160 mgd with one channel out of service and a minimum flow of 36 mgd. Channel velocities are to remain within the range of 2–3 fps over the prescribed range of flows, the former to prevent deposition of grit and the latter to prevent carrying screenings through the screens.

Based on downstream hydraulics, the downstream channel invert is at El. −10.5 ft and the maximum water surface immediately downstream of the screens does not exceed El. −3.0 ft. The channels will flow into a wet well having a maximum water surface elevation of −9.25 ft at the maximum channel flow of 82.6 cfs. Referring to Figure 7.5 for notation, maintaining the existing channel invert the maximum permissible value at the downstream end of the grit channels d_2 is −3.0–(−10.5) = 7.5 ft, corresponding to that maximum channel flow. Determine a means of regulating the flow to maintain the desired velocity range in the screen channels.

Solution

The maximum and minimum design flows through each screen channel are $Q_{max} = 160/3 = 53.3$ mgd × 1.547 = 82.5 cfs and $Q_{min} = 36/4 = 9$ mgd × 1.547 = 13.9 cfs. A conventional downstream critical depth control (without drop in channel floor) would result in an approximate velocity variation determined as follows. From Equation 7.5 with $D = y$, $\mathbf{F}^2 = V^2/(gy) = 1$, and noting $V = Q/(by)$, we have $Q^2/(gb^2y^3) = 1 \Rightarrow y^3 = Q^2/(gb^2) \Rightarrow (Q_1/Q_2)^2 = (y_1/y_2)^3$ for constant b. Since $Q_2 = V_2y_2b$ and $Q_1 = V_1y_1b$ we can write $y_1/y_2 = (Q_1/Q_2)(V_2/V_1)$ and then $(Q_1/Q_2)^2 = [(Q_1/Q_2)(V_2/V_1)]^3 \Rightarrow V_1/V_2 = (Q_1/Q_2)^{1/3}$. In terms of minimum and maximum values, we have $V_{min}/V_{max} = (Q_{min}/Q_{max})^{1/3} = (13.9/82.5)^{1/3} = 0.55$, which would not maintain velocities within the desired range.

We will then consider the improved velocity-control concept presented above using Camp regulators, with a step to be installed to lower the invert between the screens and the regulator. Such regulators also have the advantage of enabling isolation of the screen channels. Using the notation in Figure 7.5 and Equation 7.35, neglecting velocity heads, and assuming $d_t < (2/3)d_2$ we have: $Q = Cbd_2^{3/2}$ with $C = 3.5$ for the Camp regulator provided $d_t < (2/3)d_2$, $d_1 + \Delta z \cong d_2$, and $V = Q/(wd_1)$. For equal velocities at two different flows denoted Q and Q', we have:

$$\frac{Q}{Q'} = \frac{d_1}{d_1'} = \frac{d_2 - \Delta z}{d_2' - \Delta z}$$

$$Q(d_2' - \Delta z) = Q'(d_2 - \Delta z)$$

$$Q\left[\left(\frac{Q'}{Cb}\right)^{2/3} - \Delta z\right] = Q'\left[\left(\frac{Q}{Cb}\right)^{2/3} - \Delta z\right]$$

Through several steps, the above equation is solved for b to obtain:

$$b = \frac{Q}{C}\left[\frac{1 - (Q/Q')^{1/3}}{(1 - Q/Q')\Delta z}\right]^{3/2}$$

We now want to determine the maximum deviation from the desired velocity. From relationships above, we have:

$$V = \frac{Q}{w\left[\left(\frac{Q}{Cb}\right)^{2/3} - \Delta z\right]}$$

Differentiating and equating to zero gives:

$$\frac{dV}{dQ} = \frac{w\left[\left(\frac{Q}{Cb}\right)^{2/3} - \Delta z\right] - Q\frac{w}{(cb)^{2/3}}\frac{2}{3}Q^{-1/3}}{\left\{w\left[\left(\frac{Q}{Cb}\right)^{2/3} - \Delta z\right]\right\}^2} = 0$$

which reduces and equates to:

$$\left(\frac{Q}{Cb}\right)^{2/3} = 3\Delta z = d_2$$

at the extreme value of V which we will denote as V_o given by: $V_o = \frac{Q}{w(d_2 - \Delta z)} = \frac{Q}{2w\Delta z} = \frac{Cb(3\Delta z)^{3/2}}{2w\Delta z} = \frac{Cb(3)^{3/2}}{2w}\Delta z^{1/2} = 2.60C\frac{b}{w}\Delta z^{1/2}$ and taking $C = 3.5\ \text{ft}^{1/2}/\text{sec}$ from above we have:

$$V_o = 9.1\frac{b}{w}\Delta z^{1/2}$$

in which V_o, b, w, and Δz have units of ft/sec, ft, ft, and ft, respectively, and the constant 9.1 has units of $\text{ft}^{1/2}$/sec. We will see below that this corresponds to a minimum value of V_o.

Using the equation for b given above, and maximum and minimum flows given above, we have $Q/Q' = 30.9/82.5 = 0.375$. Using the above equation for b with $C = 3.5$:

$$b = \frac{Q}{C}\left[\frac{1-(0.375)^{1/3}}{(1-0.375)}\right]^{3/2}\Delta z^{-3/2} = 0.298\frac{Q}{C}\Delta z^{-3/2} = 0.298\frac{Q}{3.5}\Delta z^{-3/2} = 0.0851Q\Delta z^{-3/2}$$

We next start with the lower flow rate of $Q = 30.9$ cfs, for which the above equation gives:

$$b = 0.0851Q\Delta z^{-3/2} = 0.0851(30.9)\Delta z^{-3/2} = 2.63\Delta z^{-3/2}$$

The process now proceeds iteratively in which for a selected mid-range value of $V = 2.5$ fps we assume Δz and seek to obtain the desired width of $w = 6$ ft and other acceptable parameters. So assuming $\Delta z = 1.0$ ft, we calculate:

$$b = 2.63\Delta z^{-3/2} = 2.63(1)^{-3/2} = 2.63\ \text{ft}$$

$$d_2 = \left(\frac{Q}{cb}\right)^{2/3} = \left[\frac{30.9}{3.5(2.63)}\right]^{2/3} = 2.24\ \text{ft}$$

$$d_1 = d_2 - \Delta z = 2.24 - 1.0 = 1.24\ \text{ft}$$

$$w = \frac{Q}{d_1 V} = \frac{30.9}{1.24(2.5)} = 9.97\ \text{ft}$$

Skipping actual intermediate steps, we assume $\Delta z = 1.65$ ft, for which:

$$b = 2.63(1.65)^{-3/2} = 1.241\ \text{ft}$$

$$d_2 = \left[\frac{30.9}{3.5(1.241)}\right]^{2/3} = 3.69\ \text{ft}$$

$$d_1 = d_2 - \Delta z = 3.69 - 1.65 = 2.04\ \text{ft}$$

$$w = \frac{30.9}{2.04(2.5)} = 6.05\ \text{ft}$$

This is close to the desired 6 ft.

At the higher flow of 160 mgd with all four channels operating corresponding to [160(1.547 cfs/mgd)/4] = 61.9 cfs per channel, we have:

$$d_2 = \left[\frac{61.9}{3.5(1.241)}\right]^{2/3} = 5.88\ \text{ft}$$

$$d_1 = d_2 - \Delta z = 5.88 - 1.65 = 4.23\ \text{ft}$$

$$V = \frac{Q}{d_1 w} = \frac{61.9}{4.23(6.0)} = 2.43\ \text{ft/sec}$$

which is well within the desired range. This value of V is close to the minimum value given by the equation given above: $V_o = 9.1(1.241/6)(1.65)^2 = 2.42$ ft/sec.

Over the range of conditions determined above, the velocity head at d_1 on Figure 7.5 ranges from 1% to 5% of d_1, justifying the assumption of neglecting that velocity head.

At the higher flow of 82.5 cfs per channel when one channel is out of service, we have

$$d_2 = \left[\frac{82.5}{3.5(1.241)}\right]^{2/3} = 7.12 \text{ ft}$$

$$d_1 = d_2 - \Delta z = 7.12 - 1.65 = 4.32 \text{ ft}$$

$$V = \frac{Q}{d_1 w} = \frac{82.5}{4.32(6.0)} = 3.18 \text{ ft/ sec}$$

This is slightly above the desired 3.0 ft/sec but acceptably so for an out-of-service condition.

Finally, we check for satisfaction of the assumption $d_t < (2/3)d_2$ at the minimum d_2 determined above of 2.24 ft. The maximum downstream tailwater depth is $-9.25 - (-10.5) = 1.25$ ft which is less than $(2/3)(2.24) = 1.49$ ft. Thus the regulator will remain unsubmerged and function properly.

The discharge slot of the Camp regulator is manually adjustable from 0 to 24 in., one-half of this travel being obtainable on each side of the regulator by means of handwheels and gear mechanisms. This adjustability can enable accommodation of future conditions.

In the application to the Tallman Island WWTP mentioned above, grit is removed in chambers with scouring velocities controlled to allow grit to settle and remain on the chamber floor but with lighter putrescible materials that settle being scoured back into the flow. Grit can also be removed by means of other devices, such as aerated grit chambers and detritors. The detritor is a continuous flow tank in which the grit settles due to gravity and the water overflows through the outlet weirs on the opposite side; the settled grit is scraped by means of a scraper mechanism toward the openings in the bottom of the sidewall.

Aerated grit chambers were designed by the author for the 2.3 mgd secondary wastewater treatment plant in Hanover, New Hampshire.

For the Rochester, New Hampshire 3.93 mgd advanced wastewater treatment plant, the author specified a 10-ft diameter pista grit chamber [a type of detritor]. The process has two concentric cylinders, the lower one being smaller than the upper one. Paddles in the upper cylinder keep organic material in suspension, while grit settles into the lower chamber. A cycle of air and water backwash is used to remove organic material that may have settled into the lower chamber. That cycle is followed by an airlift cycle to discharge the grit to a storage basin or classification chamber. The specified manufacturer is Ecodyne Limited, now Marmon Industrial Water Limited, Burlington, Ontario, Canada.

Detritors for grit removal were designed by the author for the 23 mgd Cranston, Rhode Island Wastewater Treatment Plant. A specified manufacturer is FMC Link-Belt Division (now WSG & Solutions, Inc., Montgomeryville, Pennsylvania). Detritors (Babbitt and Baumann 1958, §18-6) are also manufactured by Dorr-Oliver Eimco, Salt Lake City, Utah.

The author recommended improvement to two 55′ × 55′ Detritor tanks for grit removal for the 65 mgd Providence, Rhode Island Wastewater Treatment Plant. Improvements included a timing system to eliminate continuous operation at higher flow rates and prevent plugging problems.

Also noteworthy is the WEMCO Hydrogritter Separator which removes and washes grit by cyclonic action, manufactured by Trillium Flow Technologies, Houston, Texas; the HeadCell stacked tray grit separators manufactured by Hydro International Ltd., Portland, Maine; and the vortex Grit Remover by Infilco Degremont, Richmond, Virginia. Particularly effective grit removal with a hydrocyclone at Vancouver, British Columbia, Canada has been reported by Scott (1972); one manufacturer otherwise mentioned here is Smith & Loveless, Inc., Lenexa, Kansas.

Rolling Grit Equipment from Walker Process Equipment, Aurora, Illinois provides a total system for grit removal and dewatering. In the grit basin, a centrally located eductor tube induces a rolling action that maintains organics in suspension while causing settled grit to be carried to a center hopper where an airlift pump transfers the grit to an inclined screw grit washer outside of the basin. Preaeration is provided and there are no moving parts in the grit basin.

An interesting grit-removal concept is the Bendy Channel process, which utilizes the secondary currents (see Section 4.7) created by a channel bend in conjunction with a grit collection hopper and control flume (Kratch 1998, Patel *et al.* 2011).

A good discussion of the relative performance of different types of grit removal systems is provided by McNamara *et al.* (2013).

7.2.3.3 Cutthroat Flume

The cutthroat flume is similar to the Parshall flume but is simpler in being flat-bottomed and with zero throat length. Its design flexibility, in terms of dimensional variations, and simple geometry provides significant advantages. The flumes side slopes have a 3:1 upstream converging section and a 6:1 downstream diverging section (Skogerboe *et al.* 1967, 1972; Walker *et al.* 1973). A comprehensive report on the cutthroat flume is given by Temeepattanapongsal *et al.* (2013). For the example given below, the simplicity of the cutthroat flume allowed it to be fabricated of wood at the site in a matter of hours.

Example 7.11

Design a cutthroat flume for flows ranging from 1/2 mgd = 347 gpm = 0.774 cfs to 2100 gpm = 3.03 mgd = 4.69 cfs in an 8-foot sluiceway with a 1 ft downstream depth and maximum allowable change in water surface level of 15 in. (Homestead Woolen Mills, Swanzey, New Hampshire).

Solution

Assume a 1-ft flume width W and length $L = 9$ ft per Skogerboe *et al.* (1967). From that reference, the upstream flume width is given by $B = W + L/4.5 = 1 + 9/4.5 = 3$ ft, which comports with the 3:1 ratio mentioned above. Also from that reference, the flume width at the gage station is given by $W_g = W + 4L/27 = 1 + 4(9)/27 = 2.33$ ft. The relationship between gage reading h_a in feet and flow rate Q in mgd is given by $h_a = (Q/C)^{1/n_1}$. The coefficient C is given by $C = KW^{1.025} = K(1)^{1.025} = K$. The coefficient K is obtained from Figure 6 of Walker *et al.* (1973) which gives $K = 2.25$. With the exponent $n_1 \cong 1.57$ (from Skogerboe *et al.* 1972) we have the depth at the measurement station $h_a = (3.03/2.25)^{1/1.57} = 1.21$ ft [which is very close to the value predicted in Table 2 of Skogerboe *et al.* (1967)]. The velocity at the gage section is $V = Q/(h_a W_g) = 4.69/[1.21(2.33)] = 1.66$ ft/sec. The corresponding velocity head is $V^2/(2g) = 0.04$ ft. The change in water surface elevation, assuming the flume crest is placed at the elevation of the tailwater, is 1.21 + 0.04 = 1.25 ft = 15 in., within the allowable change.

7.2.3.4 Sutro Weir

The configuration of the Sutro or proportional weir was shown and equations and references were presented in Section 5.3, the application there being for rooftop detention basins; they can also be used to control outflow from ground-level stormwater detention basins. Sutro weirs are also used to maintain constant overflow rates in grit chambers and to maintain constant velocity in an upstream channel, such as one with a screen to remove larger materials ahead of a pump or primary wastewater treatment. In addition to their symmetric form, as depicted in Figure 5.5(a), a Sutro weir may also have an asymmetric half shape.

We note here that overflow rate and longitudinal velocity are the two principal design parameters of grit chambers. The overflow rate is related to the settling of grit-sized particles, whereas the longitudinal velocity is provided to scour and resuspend smaller (and presumably putrescible) particles which fortuitously settle in the grit chamber. Since bar screens commonly precede grit chambers, the hydraulic design of the pretreatment area has commonly been such that the velocity-control devices concurrently maintain the velocity in the screen channel within an acceptable ratio (typically 2–3 fps).

The Howe (1950) equation for the shape of the Sutro weir is given on Figure 5.5(a). Section 5.4 presented the discharge equation of the Sutro weir as:

$$Q = K(h - a/3) \tag{7.36}$$

in which h is water depth measured from the bottom of the weir (above $y = a/3$) and in which:

$$K = Ca^{1/2}b\sqrt{2g} \tag{7.37}$$

The coefficient C varies slightly with water depth (Howe 1950) and can be accurately taken as 0.6 for $a > 0.30$ ft. Even at smaller values of a, when the Sutro weir is used for velocity control rather than flow measurement (for which there are better devices), $C = 0.6$ is adequate.

Example 7.12

A Sutro weir is to be used to control velocity through a screen upstream of a stormwater pumping station (Kennebec Sanitary Treatment District, Maine). Given design values are: $Q = 16.60$ cfs, channel width = 4-4 in., weir base $b = 4$ ft, total weir height $h = 2.5$ ft.

Solution

Equations 7.36 and 7.37 give, respectively:

$$16.60 = K(2.5 - a/3)$$

$$K = 0.6a^{1/2}4.0\sqrt{2(32.2)} = 19.26\sqrt{a}$$

from which K can be eliminated, giving:

$$16.60 = 19.26\sqrt{a}(2.5 - a/3)$$

The above equation can be solved for a as a cubic equation or as a transcendental equation which can be solved by trial-and-error or using, for example, Microsoft Excel Solver. The trial-and-error result is $a = 0.125$ ft = 1½ inch. Substituting that value of a into the above equation gives 16.74 ft^2/sec for the right-hand side, which is within 1% of the desired value of 16.60. Using the above values of a and b, the weir shape is specified according to the equation of Figure 5.5(a).

7.2.3.5 Wet-Well Control of Screen Channels

Velocity control in screen channels is commonly provided by means of a critical-depth control at the downstream end of the channels. This control is often provided by a Parshall flume so that it will also serve as a flow meter. Camp regulators are also used; although they are not calibrated flow meters, they provide adjustability for velocity control and a means of isolating the channel. The variation of upstream velocity associated with either the Parshall flume or Camp regulator can be further reduced by placing a properly-sized step in the channel floor between the screens and hydraulic control. For very precise velocity control a Sutro (proportional) weir can be used. These critical-depth controls all require limiting downstream submergence for their proper operation and thus result in head loss. Based on their usual placement, the Sutro weir has the largest head loss, followed by the Parshall flume and Camp regulator.

Although only feasible under certain circumstances, control of water surface elevations in a common downstream wet well can also be used to control velocities in screen channels as an alternative to other methods mentioned above. This will be demonstrated by the example given below.

Example 7.13

In a treatment plant upgrading (North Charleston Sewer District, South Carolina), three 4-ft wide channels operating in parallel are to be reconfigured as screen channels discharging to a common wet well. From the wet well pumps will lift the wastewater to the elevation of aerated grit chambers. The screen channels have invert elevations of 92.39 ft relative to project datum and a depth of 7″–10″. Using wet-well control, it is desired to design for a screen-channel velocity of 2 ft/sec with all channels operating and to limit the channel velocity to not more than $3/2 \times 2 = 3$ ft/sec with two channels operating. Peak and minimum wastewater flows are 66 and 11 mgd, respectively. Determine whether the screen channels can accommodate these conditions and corresponding wet well elevations for purposes of pump control.

Solution

With three channels operating, a maximum flow of 66/3 = 22 mgd per channel at 2 ft/sec corresponds to a channel depth of 22(1.547)/[4(2)] mgd [(ft^3/sec)/mgd]/[ft(ft/sec)] = 4.25 ft. Exit head loss into the wet well is negligible (demonstration is left to the reader), so this channel depth corresponds to a wet well elevation of 92.39 + 4.25 = 96.64. At the minimum flow of 11/3 = 3.67 mgd per channel with three channels operating at 2 ft/sec, the desired channel depth is 3.67(1.547)/[4(2)] = 0.709 ft, corresponding to a wet well elevation of 92.39 + 0.709 = 93.10. With two screen channels operating, the corresponding elevations at maximum and minimum flow are 33(1.547)/[4(2)] = 6.38 ft and 5.5(1.547)/[4(2)] = 1.06 ft, respectively. The maximum depth of 6.38 ft is well within the screen channel depth. Adjustable pump controls were set for the flow-elevation relationships thus determining the number of channels operating.

Screening is commonly employed as part of the preliminary treatment in wastewater treatment plants for the purpose of removing rags and other materials to prevent problems such materials can cause in downstream treatment components. In one bizarre incident, the writer was inspecting the Bridgewater, Massachusetts Correctional Facility (MCI) preparatory to designing an upgrade and expansion of the facility. The plant operator said the screen was particularly important because the inmates at times coordinated in flushing bed sheets down the toilets which were then intercepted by the screen to prevent downstream damage. [Such actions by inmates are not uncommon (Brzozowski 2016).] During the writer's inspection, the writer observed that the screen was missing and asked why. The operator asked the writer to follow him toward an area in nearby woods where smoke could be seen rising from a fire. The writer was introduced there to inmates who had temporarily removed the screen to cook their lunches on it (hopefully after exposing it to the flames for a suitable time).

For screens, which usually have automatic rakes at larger facilities, it is necessary to control velocities in the screen channel. Velocities that are too low will allow deposition of grit and the material intended to be screened ahead of the screen and grit removal. Velocities that are too high will force the material through the screens without being removed. Velocities in the range of 1.25–3.0 fps are recommended (Camp 1951; Fair and Geyer 1954, §15-11; Linsley and Franzini 1964, §19-17). In some facilities, comminutors are used to macerate rag-like materials either instead of screens or in conjunction with them.

The following from WPCF/ASCE (1959) is important: "The velocity distribution in the approach channel is very important to good screen operation. A straight channel ahead of the screen insures good velocity distribution across the screen and maximum effectiveness of the device. Use of other than a straight approach channel has often resulted in large parts of the flow passing through one side of the screen."

For the 3.93 mgd advanced wastewater treatment plant for Rochester, New Hampshire the author specified one 36-inch comminutor of 10 mgd peak capacity with a coarse screen and bypass screen. Current manufacturers of such comminutors include Franklin Miller, Livingston, New Jersey; Grind Hog Comminutors by G.E.T. Industries, Caledon, Ontario, Canada; and Smith & Loveless, Inc., Lenexa, Kansas.

A climber-type mechanically cleaned screen was specified by the author for the North Charleston, South Carolina 41 mgd Wastewater Treatment Facility. A specified manufacturer still providing such equipment was Envirex/Evoqua Water Technologies, Pittsburgh, Pennsylvania. Other current manufacturers are Huber Technologies, Denver, North Carolina; Infilco Degremont, Richmond, Virginia; and Vulcan Industries, Missouri Valley, Iowa.

For the 23 mgd Cranston, Rhode Island Wastewater Treatment Plant, the author specified two catenary screens. Current manufacturers of such screens include E & I Corporation, Westerville, Ohio; ESMIL Group, Mogadore, Ohio; and Fairfield Service Company, Michigan City, Indiana.

Four mechanical bar screens were in use at the 65 mgd Providence, Rhode Island Wastewater Treatment Plant. Due to operational problems in the winter, the author recommended enclosure of the bar screens and a means of conveying the screenings to a centralized collection point.

For the 380,000 gpd Bridgewater MCI facility, the author specified two mechanically cleaned catenary bar screens.

Also, meriting consideration for wastewater screening is the Hycor Rotoshear self-cleaning wedge wire rotary drum screen, Fort Lauderdale, Florida (Lenhart and McGarry 1984). Also, rotary-gravity-type screening machines (Christman 1962) merit consideration for certain applications. A good discussion of different types of screens and considerations in screen selection is provided by Forstner (2007).

Screens are also used to pretreat river water before conveyance for potable water treatment (Shelley *et al.* 1981) and to remove trash from stormwater runoff before discharge to receiving waters (Sobelman *et al.* 2005).

7.2.3.6 Point-Velocity Discharge Measurements

Parr *et al.* (1981) addressed a point-velocity discharge measurement method for circular sewers, which gives an accurate, economical method for monitoring sewer flow. A depth measurement and a single velocity measurement, together with the given discharge equations provide the discharge values for sewers flowing full or partially full. Section 13.3 discusses the author's use of a Price current meter for measuring velocities in a wastewater treatment plant channel. As discussed in Section 15.4, the author also used a Price current meter to measure streamflow in a river in Panamá at cross-sectional locations complying with U.S. Geological Survey procedures (Chow 1959, §2-6).

7.3 Uniform-Flow Conditions

Let us consider an open channel (e.g., a pipe flowing partially full) with a mild slope (as defined below), discharging freely. As discussed in the previous section, critical depth occurs close to the free discharge. The channel is assumed prismatic, and the following equation applies:

$$\frac{dy}{dx} = \frac{S_o - S_f}{1 - \mathbf{F}^2} \tag{7.38}$$

If the channel slope is less than the friction slope ($S_o < S_f$) at any location and the flow is subcritical throughout ($\mathbf{F} < 1$), then the above equation indicates that dy/dx is negative. A negative dy/dx corresponds to a water surface slope which increases with distance *upstream*. Note from Equation 7.39 that, at constant flow, as the depth increases the friction slope decreases. It must eventually decrease to the value of S_o, at which location (according to the above equation) $dy/dx = 0$. The condition $dy/dx = 0$, corresponding to a water surface parallel to the channel bottom, is called *uniform flow*. The corresponding depth is called the *uniform flow depth* or *normal depth* (although there is nothing particularly "normal" about it).

Under uniform-flow conditions, the rate of decrease of potential energy as the flow goes down the slope S_o is exactly equal to the rate of energy loss due to channel friction ($S_o = S_f$). Uniform-flow conditions will persist upstream of the point at which dy/dx is first close to zero, as long as the channel cross section, slope, roughness characteristics, and rate of flow remain constant. [Uniform flow conditions are actually approached asymptotically, but are considered to exist from a practical standpoint if the depth is within a small percentage (say 1%) of the uniform-flow depth.] In essence, uniform-flow conditions are attained far enough upstream in any open channel of *mild slope* in which the channel characteristics just mentioned remain constant. Under uniform-flow conditions, the slope of the energy grade line, water surface, and channel are all equal.

Note that the downstream tailwater condition of the flow could be a value of y_t intermediate between the critical and uniform flow depths. The related channel configuration has the tailwater depth greater than the uniform flow depth, and the depth decreases in the upstream direction toward the uniform flow depth (in this case $S_o < S_f$ at the downstream end, and reasoning similar to the above may be employed to deduce the profile shape).

If the channel slope is steep (see below), the flow may be supercritical ($\mathbf{F} > 1$), in which case the uniform or normal flow depth is approached in the downstream direction provided the channel characteristics are constant for a sufficient distance. (If the slope is too steep, air entrainment and wave instabilities can cause deviations from uniform flow – see Section 7.5.) The possible profiles in this case may be deduced using similar reasoning to that given above. A freely discharging channel on a steep slope will discharge at a depth less than the critical depth and may discharge at a depth close to the normal depth if the channel characteristics are constant for a sufficient distance upstream of the free discharge. This can be envisioned by considering the profile in a channel changing from a mild to a steep slope, or a steep culvert. The depth near the brink of the drop will be between y_c and y_n if the drop occurs upstream of the location where the normal depth is approached, and will be close to y_n if the drop is downstream of that location.

The terms *mild slope* and *steep slope* as used above will now be quantified. [These terms should not be confused with the term *small slope* as defined in Chapter 2. A *steep* slope may be (and commonly is) a *small* slope.] A channel with a *mild slope*, also called a *subcritical slope*, is one having the normal depth greater than the critical depth. A channel having a *steep slope*, also called a *supercritical slope*, is one having the normal depth less than the critical depth. The determination of subcritical vs. supercritical slope may be made by calculating the Froude Number at uniform flow conditions and comparing it with 1, or by calculating and comparing the critical and normal depths in the channel reach of interest. The determination of critical flow depth was discussed in Section 7.2. The determination of uniform flow depth is discussed below. The reader may wish to demonstrate after reading a little further that the critical slope is not independent of flow; the terms *subcritical slope* and *supercritical slope* are convenient ones but should be understood to be dependent on flow.

7.3.1 Determination of Uniform Flow Parameters

The following equation was proposed by Manning in 1889 for open-channel flows (Yen 1992a, 1992b):

$$Q = \frac{K_n A R^{2/3} S_f^{1/2}}{n} \tag{7.39}$$

in which $K_n = 1.486$ for customary units or 1 for metric units; Q = flow rate (cubic feet per second or cubic meters per second); A = cross-sectional area of the water stream (square feet or square meters); R = hydraulic radius $R = A/P$ (feet or meters); S_f = friction slope or slope of the energy grade line (feet per foot or meters per meter) (which, as discussed above, equals the slope of the drain only for steady uniform flow); WP = wetted perimeter, which is the length of the periphery of the cross-sectional shape of the liquid in contact with the pipe wall surface (feet or meters); and n = Manning's roughness coefficient for the drain product.

As noted above, uniform flow depth has friction slope S_f equal to the channel slope S_o. In terms of the Manning equation, we have (setting $S_f = S_o$ in Equation 7.39):

$$Q = AV = \frac{K_n}{n} A R^{2/3} S_o^{1/2} \tag{7.40a}$$

$$V = \frac{K_n}{n} R^{2/3} S_o^{1/2} \tag{7.40b}$$

Values of Manning's n for closed conduits and open channels are given in Table 6.1. A nomograph for the solution of Equation 7.40b is given in Appendix C of Chow (1959).

Equation 7.40a may be written in the following alternate form:

$$\frac{nQ}{K_n S_o^{1/2}} = \frac{AR^{2/3}}{\sigma^{8/3}} \tag{7.41}$$

in which σ is a characteristic length of the cross section (e.g., diameter of a circular channel, bottom width of a trapezoidal channel, etc.). The term on the right-hand side of the above equation is a function of y/σ (in which y is depth of flow) for most of the man-made channels of interest. Chow (1959) plots the relationships between $AR^{2/3}/\sigma^{8/3}$ and y/σ for circular and trapezoidal (including rectangular) channels. Chow also presents in tabular form (Chow 1969, Appendix A) the relation between y/d_o and $AR^{2/3}/d_o^{8/3}$ for circular channel sections. The applicable relationships can be readily programmed on a programmable calculator or computer.

Knowing y/σ for the channels discussed in the preceding paragraph, we can determine $AR^{2/3}/\sigma^{8/3}$ and hence $nQ/(K_n S_o^{1/2})$. Knowing n and S_o, we can then determine the flow Q under uniform flow conditions. Conversely, knowing $nQ/(K_n S_o^{1/2})$, we can determine $AR^{2/3}/\sigma^{8/3}$ and hence y/σ.

For natural channels, $AR^{2/3}$ in Equation 7.40a may be plotted against flow depth to aid in the calculations, or the channel may be represented as a series of line segments connecting coordinate points, from which A, R, and $AR^{2/3}$ may be determined by means of a simple computational algorithm. These and other aspects of uniform flow in natural channels (in which uniform flow is more the exception than the rule) are discussed in Section 15.2. In keeping with the thrust of the present chapter, the remainder of the present section will focus on uniform flow in man-made channels.

The assumption of uniform flow is a convenient design device for various types of channels. Topography commonly establishes the channel slope, the flow is known and, for a given channel roughness and cross-sectional shape, the channel dimensions (including an allowance for freeboard, which is the vertical distance from the maximum water surface to the top of the channel) can be determined. Backwater calculations (see Section 7.4) are thus avoided, and it is only necessary to ensure that uniform flow conditions (or a lower flow depth) are permitted to occur.

In environmental engineering practice, uniform flow design is mostly limited to the design of sewers (both wastewater and storm). Although street gutters and stormwater grates have traditionally been designed based on the assumption of uniform flow, they are more accurately designed for spatially-increasing flow as addressed in Section 12.7.

Example 7.14

A main stormwater pumping station (Kennebec Sanitary Treatment District, Maine) has an overflow pipe which is a 66-inch diameter concrete sewer ($n = 0.013$) with a design flow of 97.0 mgd. The pipe has a terminal slope of 0.0050 ft/ft, with lesser slopes upstream. The outlet is at a sufficient elevation that it discharges freely into the river. Determine the conditions at the pipe discharge.

Solution

If the terminal section of pipe has a subcritical slope, the pipe should discharge at critical flow conditions (the reader should confirm this reasoning in light of the stated upstream and downstream conditions and with reference to Sections 7.4 and 7.5).

The full flow and corresponding velocity are obtained from Equation 6.3 as $Q_f = 153$ mgd and $V_f = Q_f/A_f = 153(1.547)/[(\pi/4)(66/12)^2] = 10.0$ ft/sec. The ratio of design flow:full flow is equal to $Q_d/Q_f = 97/153 = 0.634$, From a figure of flow characteristics of circular sections for variable n (see Section 5.3), we then have $y/d_o = 0.65$ and $V/V_f = 0.92$. From a table or figure of geometric elements [e.g., Chow (1959) Table 2-1 or Appendix A, or related websites], for $y/d_o = 0.65$ we have $D/d_o = 0.567$. The uniform flow Froude Number can then be calculated as follows: $V = 0.92(10.0) = 9.2$ ft/sec, $d_o = 66/12 = 5.5$ ft, $D = 0.567(5.5) = 3.11$ ft, $\mathbf{F} = V/\sqrt{gD} = 9.2/\sqrt{32.2(3.11)} = 0.92$. Therefore, the slope is subcritical and critical flow conditions occur at the discharge. The critical flow section factor (discussed in Section 7.4) is given by $Z = Q/\sqrt{g} = 97(1.547)/\sqrt{32.2} = 26.4\ \text{ft}^{5/2}$. Thus $Z/d_o^{2.5} = 26.4/(5.5)^{2.5} = 0.372$. From a table or figure of geometric elements, we obtain $y_c/d_o = 0.62$ and $A_c/d_o^2 = 0.512$. Thus: $A_c = 0.512(5.5)^2 = 15.49$ ft^2, $V_c = Q_c/A_c = 97(1.547)/15.49 = 9.69$ ft/sec. This velocity is high enough to justify scour protection as discussed in Section 7.8 and Example 7.23.

If the slope had been determined to be supercritical with the uniform flow velocity higher than the critical flow velocity, then the discharge velocity would be dependent on upstream conditions as discussed in Section 7.5.

The Chezy formula will be used for certain purposes below and is given as $V = C\sqrt{Ri}$ in which new terms C and i are the Chezy coefficient and hydraulic gradient (or friction slope S_f), respectively. For normal depth of flow, i equals the channel bottom slope. Units are metric, with V, C, R, and i being in units of m/s, m$^{1/2}$/s, m, and m/m. The Chezy coefficient is related to the Manning roughness coefficient by $C = R^{1/6}/n$.

7.4 Subcritical Flow

7.4.1 Speed of Surface Waves; Subcritical Flow

Consider the flow in a channel as it approaches a local channel variation or disturbance. The "information" regarding the disturbance is propagated along the channel via *surface waves*, in a way analogous to the propagation of circular ripples outward from the point at which a rock is thrown into a pond.

The surface waves move upstream and downstream of the disturbance at absolute velocity V_w as shown on Figure 7.6(a). Let us consider only the waves moving upstream. In order to make the problem a steady-flow one, we will view such a wave and the oncoming flow from the perspective of an observer moving with the wave, i.e., the observer moves to the left at velocity V_w relative to Figure 7.6(a). The flow as seen by that observer is shown on Figure 7.6(b), in which the wave appears stationary and the approaching flow has velocity $V + V_w$.

Frictional effects are neglected; parallel flow, small (negligible) slope, and $\alpha = 1$ are assumed; and the steady-flow energy equation applied between sections "1" and "2" of Figure 7.6(b) to give:

$$y + \frac{(V + V_w)^2}{2g} = y + h + \frac{V'^2}{2g} \tag{7.42}$$

in which h is the height of the wave and V' is the velocity of flow through the wave location relative to the wave (or relative to the observer moving with the wave). By continuity:

$$(V + V_w)A = V'(A + hT) \tag{7.43}$$

in which A is the undisturbed channel flow area (at section "1"), T is the top width of the undisturbed flow, and hT is the additional area at the cross section of the wave for h assumed small.

Substituting V' from Equation 7.43 and $A = TD$ into Equation 7.40 and solving for $V + V_w$ gives:

$$V + V_w = \sqrt{\frac{2g(D + h)^2}{2D + h}} \tag{7.44}$$

For small waves ($h \ll D$), the above equation reduces to the following approximation:

$$V + V_w \cong \sqrt{gD} \tag{7.45}$$

or:

$$V = \sqrt{gD} - V_w \tag{7.46a}$$

$$V_w = \sqrt{gD} - V \tag{7.46b}$$

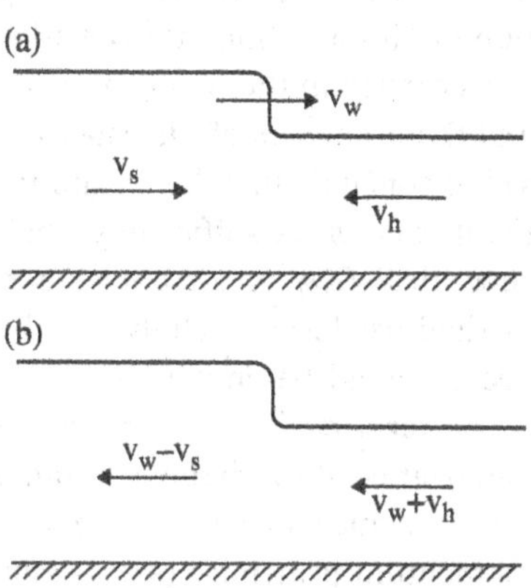

Figure 7.6 Wave Moving into Ebb Current: (a) Relative to Fixed Observer; (b) Relative to Observer Moving with Wave (Graber 2012a / ASME)

The term $V + V_w$ represents the velocity of the wave relative to the velocity of the flow; this relative velocity is referred to as the celerity c of the wave, so that $c = \sqrt{gD}$.

For $V < \sqrt{gD}$, V_w is greater than zero, that is, the surface wave is capable of moving into the oncoming flow, and "informing" *the oncoming flow of the presence of the disturbance*. This is referred to as *subcritical* flow. The oncoming flow is "aware" of the existence of the disturbance and is capable of adjusting gradually to it. For $V = \sqrt{gD}$, the upstream wave is stationary ($V_w = 0$) relative to the fixed observer or the disturbance [Figure 7.6(a)] and the flow is termed *critical* (this is the physical significance of the term "critical" as used in Section 7.2). If $V > \sqrt{gD}$, V_w is negative, the surface wave can only move downstream of the disturbance, and the flow is referred to as *supercritical* flow. For either critical or supercritical flow, the oncoming flow is not "aware" of the existence of the disturbance until it is right upon it, and must adjust abruptly.

The above criteria can be expressed in terms of the Froude Number defined by $\mathbf{F} = V/\sqrt{gD}$. Thus, $\mathbf{F} < 1$ corresponds to subcritical flow, $\mathbf{F} = 1$ corresponds to critical flow, and $\mathbf{F} > 1$ corresponds to supercritical flow. There are very distinct differences in the behavior of subcritical and supercritical flows. The differences are directly analogous to those in subsonic and supersonic gas flows as discussed in aerodynamics texts. The surface wave celerity is directly analogous to the sonic speed or speed of sound in gases (Chapter 9). The present section will focus on subcritical flows, while Section 7.5 will deal with pertinent aspects of supercritical flows.

The critical-flow section factor Z is a useful parameter, defined as the product of the water area and the square root of the hydraulic depth under critical-flow conditions. Denoting critical-flow conditions with the subscript "c," we have the following progression: $Z = A_c\sqrt{D_c} = A_c\sqrt{A_c/T_c}, D_c/2 = V_c^2/(2g), Z = A_c\sqrt{V_c^2/g} = A_c\sqrt{Q_c^2/(A_c^2 g)} \Rightarrow Z = Q_c/\sqrt{g}$. Values of $Z/d_o^{2.5}$ are tabulated by Chow (1959, Appendix A).

7.4.2 Backwater Calculations

A problem encountered in environmental engineering applications is that of determining the depth and other flow characteristics at one or more locations along a channel upstream of a known *critical or subcritical* "tailwater" or control condition. In most such situations, subcritical flow conditions will occur in at least a portion of the channel upstream of the known condition. Representative examples include a sewer discharging to an inverted siphon river crossing (see Chapter 6) for which we wish to know the upstream profile and the distance within which uniform flow conditions (on which sewer design is generally based) occur; an overflow regulator (such as discussed in Section 7.2) upstream of a wastewater treatment plant for which the relation between water surface elevation and flow rate is known at the treatment plant and we wish to know this relation back at the regulator; and treatment plant channels in which the downstream water surface elevations are known and upstream water surface elevations are to be determined.

The above are representative backwater problems. [Chow (1959) states that the term "backwater" was originally used in reference to the "longitudinal surface curve of the water backed up above a dam or into a tributary by flood in the main stream." He suggests the use of "backwater curve" if the flow depth increases in the direction of flow (as behind a dam) and the use of "drawdown curve" if the depth decreases in the direction of flow. He notes that many authors have extended the term "backwater" to include all types of flow profiles, as does the present author (at least when dealing with subcritical flows).] The computations usually proceed backward or upstream in subcritical flow situations (and downstream in supercritical flow situations). These directions of computation correspond to the direction of "information" transmission by surface waves in channels as discussed above. Chow states that "... computations carried in the wrong direction tend inevitably to make the result diverge from the correct profile." Prasad (1970) and the discussers of his paper generated some interesting discussion on the direction of computation which led Prasad to conclude that, by certain methods (including his), if the computational steps, velocity, or tolerance used to stop the iterations are sufficiently small, the computations can proceed in either direction (e.g., proceeding downstream in subcritical flow is sometimes convenient when the depth is known at an upstream location). However, unless a method has been carefully tested for directional sensitivity and the limitations established, the usual directions noted above should be employed.

The term *backwater curve* is used herein to refer to the longitudinal water-surface profile in a subcritical, gradually-varied flow channel. *Gradually varied* implies that the channel retains a constant cross section and slope for some distance. (That distance should be at least a small multiple of a characteristic depth of flow.) Either energy or momentum principles (see Chapter 2) can be applied to the flow in such a channel, and both yield the same basic equation:

$$\frac{dy}{dx} = \frac{S_o - S_f}{1 - Q^2/(gA^2 D)} \tag{7.47}$$

Equation 7.47 assumes various similar forms depending on the expression for friction slope. For S_f expressed by the Chezy formula given above, we can write $V = Q/A = C\sqrt{RS_f}$ or $S_f = Q^2/(C^2A^2R)$ giving:

$$\frac{dy}{dx} = \frac{S_o - Q^2/(C^2A^2R)}{1 - Q^2/(gA^2D)} \tag{7.48}$$

Noting that A, R, and D are functions of y, it can be readily appreciated that exact, closed-form solutions of Equation 7.48 have not been obtained, except for the simplest geometric case of a wide rectangular channel (of width b and with the Chezy C assumed constant). In that case, $A = by$, $R = y$, and $D = y$, and Equation 7.48 reduces to a form for which solutions have been obtained. However, that simple case is of little practical value. [For $S_o \neq 0$, the solution was obtained by Breese, as described in Chow (1959) and Posey (1950). A solution for $S_o = 0$ is also feasible (Chow 1959).]

Practical solution techniques for backwater problems encountered in the large majority of environmental engineering applications can be categorized under (1) approximate solutions, (2) standard-step numerical solutions, and (3) direct-step solutions. The distinction between the first two categories is minor. Generalized numerical solutions are included under the second category. The three categories are discussed below. In keeping with the purposes stated in Section 7.1, the computational methods discussed are geared primarily to man-made channels, backwater in natural channels being addressed in Section 15.2. [Additional methods of computation are given by Chow (1959, Chapter 10). His categorization of *graphical integration* and *direct-integration* methods may be included under the category of direct-step solutions.]

7.4.2.1 Approximate Solutions

Application of the steady, parallel flow, small slope energy equation between stations "1" and "2" at the upstream and downstream ends, respectively, of an open channel of length L results in the following:

$$y_1 + \frac{V_1^2}{2g} + z_1 = y_2 + \frac{V_2^2}{2g} + z_2 + h_f \tag{7.49}$$

In the absence of form losses, the head loss term can be calculated from:

$$h_f = \overline{S}_f L \tag{7.50}$$

in which $\overline{S}_f$ is the average friction slope, $(S_{f_1} + S_{f_2})/2$.

Knowing or assuming y, Q, z, and the channel cross section at a particular station, the specific energy $[E = y + V_1^2/(2g) + z]$ and S_f can be determined. Thus, knowing y_2, Q, z_2, z_1, the cross section at stations "1" and "2", the value of y_1 can be determined by trial-and-error. The *approximation* stems from the application of Equation 7.49 in one step to the entire channel length and the resulting inaccuracy in the estimate of $\overline{S}_f$. This approximation is adequate in most constant-flow treatment plant channels connecting process units, in which $\Delta WSL = (y_1 + z_1) - (y_2 + z_2)$ is small (intentionally so), so that h_f is small compared to E. In many such cases, $\overline{S}_f$ can be approximated by S_{f2} (since $S_{f1} \cong S_{f2}$), simplifying the trial-and-error process.

7.4.2.2 Standard-Step Numerical Solutions

In cases where the change in water surface level is large, Equation 7.49 can be used with good accuracy if applied successively to short segments Δx of the channel length. By selecting Δx sufficiently small, the procedure is accurate for relatively large changes in WSL. Substituting $z_1 - z_2 = S_o\Delta x$ (for S_o less than 0.15) and $h_f = \overline{S}_f\Delta x$, Equation 7.49 can be rewritten as follows:

$$y_1 - y_2 = \frac{V_2^2 - V_1^2}{2g} + \overline{S}_f\Delta x - S_o\Delta x \tag{7.51}$$

Chapter 5 presented a dimensional analysis which led to useful deductions regarding generalized solutions of the basic equation governing constant flow in open channels. That analysis, coupled with the feasibility of obtaining standard-step numerical solutions to virtually any type of subcritical backwater problem, suggests that generalized or "once-and-for-all" numerical solutions can be obtained for one particular category of problems. Such solutions enable the use of a design chart or other convenient device to obtain a ready solution to a problem that would otherwise require a special-case (individual) numerical solution for comparable accuracy.

The category of backwater problems for which generalized solutions are possible is horizontal ($S_o = 0$), prismatic channels having a characteristic cross-sectional dimension σ and free discharge (i.e., downstream critical-flow conditions). The nondimensional, functional form of the solution is given by Graber (2004b):

$$\frac{y_o}{\sigma} = \phi\left(\frac{fL}{\sigma}, \frac{y_c}{\sigma}\right) \text{ for } S_o = 0, \mathbf{F}_L = 1, y_L = y_c \tag{7.52a}$$

Most practical cross-sectional shapes are theoretically included, such as rectangular (σ = channel width, b), circular (σ = diameter d), etc. (Triangular channels are theoretically excluded, but can be included by the simple artifice of considering a trapezoidal channel of small bottom width).

The author has obtained a generalized numerical solution for a horizontal, circular channel with a free discharge. Using a numerical method, the relation between the variables of Equation 7.52a was established and plotted over a typical range of values (Graber 2004b). These results are of practical value for the design of skimmer channels (see Chapter 12). A similar chart could be prepared for other cross sections; the reader should be aware of this possibility should the need arise. Yao (1971) has obtained a generalized numerical solution for a horizontal rectangular channel with a free discharge, which he presented in nomographic form. His solution would be useful in cases (relatively few) for which the approximate solution technique presented above is inadequate.

In cases of constant flow, a simple device may be used to extend the generalized solutions to horizontal channels with submerged discharges ($\mathbf{F}_L < 1$). That consists of considering an additional imaginary downstream channel length L' to a free discharge such that the depth y_o' at the upstream end of the imaginary channel equals the downstream depth in the channel of length L. Functionally:

$$\frac{y_o}{\sigma} = \phi\left[\frac{f(L + L')}{\sigma}, \frac{y_c}{\sigma}\right] \text{ for } S_o = 0, \mathbf{F}_L = 1, y_L = y_c \tag{7.52b}$$

$$\frac{y_o'}{\sigma} = \phi\left[\frac{fL'}{\sigma}, \frac{y_c}{\sigma}\right] \text{ for } S_o = 0, \mathbf{F}_L = 1, y_L = y_c \tag{7.52c}$$

The length L' is obtained from the relationship represented by Equation 7.52c and used together with L in the relationship represented by Equation 7.52b. For circular channels, this has practical application, for example, to the design of the constant-flow portions of skimmer channels. The author has also previously obtained a generalized solution for circular channels with uniform inflow corresponding to Equation 7.52a; it has application, for example, to terminal skimmer channels.

7.4.2.3 Direct-step Solutions

Equation 7.51 can be rewritten as follows:

$$\Delta x = \frac{(y_1 - y_2) + (V_1^2 - V_2^2)/(2g)}{\bar{S}_f - S_o} \tag{7.53}$$

Knowing y_2, Q, S_o, and the cross section at stations 1 and 2, an assumption for the value of y_1 will enable the calculation of Δx. This procedure, known as the direct-step method, avoids the trial-and-error computations at each step that are associated with the standard-step method. It has the disadvantages of: (1) giving increments Δx which may not be particularly convenient and (2) applying only to prismatic channels. The latter limitation is due to the dependence of the geometric elements on channel station in nonprismatic channels.

The practical value of the direct-step method in environmental engineering applications is related primarily to its use in solutions for channels of complex geometric elements. Then the disadvantages of the method may be outweighed by the advantage of avoiding trial-and-error computations. The method developed by Keifer and Chu (1955) is a particularly useful one of the direct-step methods, which the author has employed on numerous projects, including ones exemplified later in this chapter. The method is discussed below (and further explained by Graber 2004b) and applied in Section 7.7.

Equation 7.48 can be expressed in a form analogous to Equation 7.53 as follows:

$$dx = \left[\frac{1 - Q^2/(gA^2D)}{S_o - Q^2/(C^2A^2R)}\right] dy \tag{7.54}$$

Substituting Equation 7.4 into the above equation and rearranging gives:

$$dx = \frac{1}{S_o}\left[\frac{1-Q^2T/(gA^3)}{1-Q^2/(C^2A^2RS_o)}\right]dy \tag{7.55}$$

Keifer and Chu cast the terms $Q^2T/(gA^3)$ and $Q^2/(C^2A^2RS_o)$ into a much more convenient form by proceeding essentially as follows.

Note that the following functional relationship applies:

$$\frac{gA^3}{T\sigma^5} = f_2\left(\frac{y}{\sigma}\right) \tag{7.56}$$

in which σ is the characteristic cross-sectional dimension introduced above. Using the Chezy formula given above, we have $Q_* = CA_*\sqrt{R_*S_o}$ in which the subscript "$*$" denotes uniform flow conditions at a reference depth. In the case of a circular cross section, the reference depth can be conveniently taken as the full depth such that $y_*/\sigma = y_*/d = 1$, in which case Q_* is the flowing-full capacity Q_f. For other channels with gradually closing crowns, y_* and Q_* can be similarly taken as the full depth and capacity, respectively. For channels with open tops, (e.g., rectangular or trapezoidal), y_* can be selected, for example, at $y_*/\sigma = 1$. The following relationship then applies for a prescribed y_*/σ:

$$\frac{Q^2}{C^2A^2RS_o} = \frac{Q^2}{Q_*^2}\frac{A_*^2}{A^2}\frac{R_*}{R} = \left(\frac{Q}{Q_f}\right)^2 f_1\left(\frac{y}{\sigma}\right) \tag{7.57}$$

Note that Q/Q_* is a constant for a prismatic channel segment of constant flow and slope. The Keifer–Chu method is not applicable to channels of zero slope (although it is for negative slope) because, as noted by Chow (1959, p. 267), if $S_o = 0$ then $Q_* = 0$, $Q/Q_* = \infty$, and the varied-flow functions given below become meaningless.

Combining Equations 7.55, 7.56, and 7.57 and integrating, we can write:

$$L = \frac{\sigma}{S_o}(\Phi_2 - \Phi_1) - \frac{Q^2}{S_o\sigma^4}(\psi_2 - \psi_1) \tag{7.58a}$$

$$\Phi\left(\frac{y}{\sigma}, \frac{Q}{Q_f}\right) = \int_0^{y/\sigma} \frac{d(y/\sigma)}{(Q/Q_f)^2 f_1(y/\sigma) - 1} \tag{7.58b}$$

$$\psi\left(\frac{y}{\sigma}, \frac{Q}{Q_f}\right) = \int_0^{y/\sigma} \frac{d(y/\sigma)}{f_2(y/\sigma)\left[(Q/Q_f)^2 f_1(y/\sigma) - 1\right]} \tag{7.58c}$$

in which the integration is begun at a convenient starting point of y/σ, not necessarily zero (see below). Keifer and Chu (1955) called Φ and ψ *backwater functions*, and noted that for a particular cross-sectional shape they are functions only of y/σ and Q/Q_f. For *any* prismatic cross-sectional shape having mathematically describable (and continuous) functions f_1 and f_2, the backwater functions can be evaluated numerically to give a "once-and-for-all" graphical and/or tabular representation. Keifer and Chu developed such representations for circular channels; those graphs and tables are cited below but not presented as the author uses a method that is more widely applicable as exemplified below.

Note that with Keifer and Chu's method L can be as large as desired, subject only to the assumption of constant S_o and C (or other friction factor used in the determination of Q^*). The computations can be repeated for different values of y_1 and corresponding value of L until the desired L is obtained; each intermediate solution conveniently gives an actual point on the backwater curve. The use of Keifer and Chu's backwater functions must be restricted to conduits having positive bottom slopes, i.e., downward in the direction of flow.

The Keifer–Chu solution is useful, as an example, for the backwater calculations that are occasionally necessary for the analysis of circular storm and sanitary sewers when there are departures from the usual uniform-flow assumption. Keifer and Chu (1955) calculated values of their functions for circular conduits by numerical integration using Simpson's rule and presented the results in tabular and graphical form. Chow (1959, p. 262, Appendix E) has presented more detailed tables furnished by Keifer and Chu in which the interval for values of y/d and α are 0.01 and 0.05, respectively, and the Waterways Experiment Station (USACE 1963) has published tables that are even more detailed with α ranging from 0.05 to 1.60 by intervals of 0.05. Escoffier (1955) presented a graphical construction that can facilitate the determination of L. Nalluri and Tomlinson (1978) suggested that when highly accurate results are required it is important to accurately determine Φ and ψ by graphical interpolation.

Two problems can arise when using the Keifer–Chu relationship. The first, mentioned above, is the need to interpolate between tabulated values; second is that problem variables can be outside the range of tabulated values (Keifer and Chu 1955, Escoffier 1955). The author avoids these problems by using his own software employing Romberg integration for determining Φ and ψ, as employed in subsequent examples. One caution: be very careful about crossing critical depth with Keifer–Chu.

Referring to Equations 7.58 with $\sigma = d_o$ and noting the sign conventions, we have:

$$\begin{aligned} L - 5d_o = &-\frac{d_o}{S_o}\left[\{\Phi(y_c/d_o, Q/Q_o) - \Phi(y_e/d_o, Q/Q_o)\} - \frac{Q}{d_o^{5/2}}\{\psi(y_c/d_o, Q/Q_o) - \psi(y_e/d_o, Q/Q_o)\}\right] \\ &+ \frac{d_o}{S_o}\left[\{\Phi(1, Q/Q_o) - \Phi(y_c/d_o, Q/Q_o)\} - \frac{Q}{d_o^{5/2}}\{\psi(1, Q/Q_o) - \psi(y_c/d_o, Q/Q_o)\}\right] \end{aligned} \tag{7.59}$$

in which L is the culvert control length, and Φ and ψ are the Keifer–Chu functions given in Section 7.4.

The parenthetic values associated with Φ and ψ denote the values of y/d_o and Q/Q_f at which the functions are to be evaluated, Q_f being the flowing-full capacity of the pipe. Rearranging the above equation, we have:

$$\begin{aligned} \frac{L}{d_o} = &5 + \frac{1}{S_o}[\Phi(1, Q/Q_o) - 2\Phi(y_c/d_o, Q/Q_o) + \Phi(y_e/d_o, Q/Q_o)] \\ &- \frac{1}{S_o}\left(\frac{Q}{d_o^{5/2}}\right)[\psi(1, Q/Q_o) - 2\psi(y_c/d_o, Q/Q_o) + \psi(y_e/d_o, Q/Q_o)] \end{aligned} \tag{7.60}$$

These equations will be used in examples given below.

7.5 Supercritical Flow

Supercritical flows occur less commonly than subcritical flows in environmental engineering applications. However, it is very important to be able to determine when supercritical conditions will occur, to determine the possible consequences of such conditions, and make appropriate design decisions.

The basic distinction between subcritical and supercritical flows has been discussed earlier in this chapter. Supercritical flows have bulk mean velocities greater than the celerity of surface waves ($\mathbf{F} = V/\sqrt{gD} > 1$). In environmental engineering applications, such flows most often occur in steep sewers (particularly outfall sewers). [The term "outfall" is used herein to refer to a wastewater (including treatment plant effluent) or storm sewer pipe or other conduit which discharges into a receiving water body (river, ocean, etc.). This differs from the usage, found in some parts of the U.S., of the term "outfall" for an interceptor sewer.]

Further explanation of supercritical flow can be accomplished with reference to a commonly-observed phenomenon. Consider the flow of water past a small object, such as rainwater runoff past a pebble in a street gutter. Applying reasoning analogous to that of Section 7.4, the "information" regarding the presence of the pebble is propagated via surface waves moving outward from the pebble. If the approaching flow velocity is very small, the waves form concentric circles moving outward at velocity c about the pebble [Figure 7.7(a)] similar to those observed when a pebble is thrown into a pond. As the flow velocity increases, the wave pattern is displaced in the direction of flow, as in Figure 7.7(b) in which the flow remains subcritical. When the flow velocity reaches the celerity c, the pattern shown on Figure 7.7(c) results, in which the waves can no longer travel upstream, corresponding to critical flow conditions. A further increase in flow velocity results in the supercritical wave pattern shown on Figure 7.7(d). An abrupt wave front is formed as represented by the line tangent to the wave circles. The angle β is called the wave angle and, with reference to Figure 7.7(d), is seen to have a magnitude given by:

$$\sin\beta = \frac{c}{V} \cong \frac{\sqrt{gD}}{V} = \frac{1}{\mathbf{F}} \tag{7.61}$$

in which $\mathbf{F}$ is the Froude Number. (The relation $c = \sqrt{gD}$ applies strictly to a small wave disturbance as discussed in Section 7.5; Equation 7.61 is of a more qualitative nature for abrupt wave fronts.) The wave front depicted on Figure 7.7

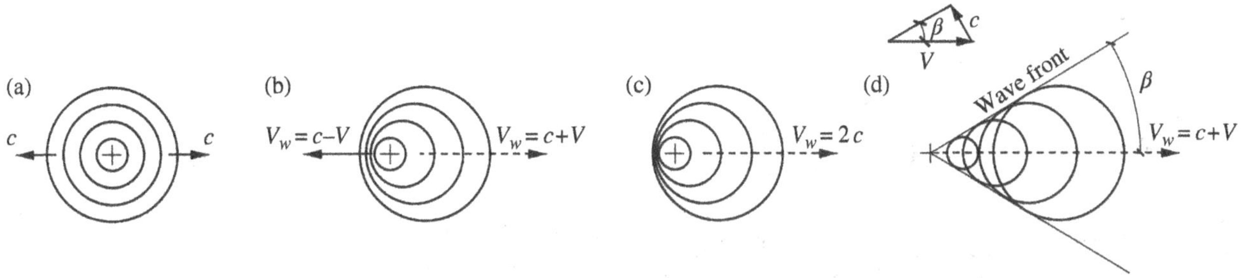

Figure 7.7 Wave Patterns Created by Disturbances: (a) Still Water, $V = 0$; (b) Subcritical Flow, $V < c$; (c) Critical Flow, $V = c$; (d) Supercritical Flow, $V > c$

(d) is easily observed in the vicinity of pebbles in gutters carrying runoff from steep streets. Such a wave front is directly analogous to the wave created on the bow of a ship or the shock wave created on the nose of a supersonic aircraft, and discussions similar to the above are found in marine engineering and aerodynamics texts. The reader may find it easier to visualize Figure 7.7 (after Chow 1959, p. 539) in terms of a boat or plane moving through their respective media than in terms of free-surface flow moving past a stationary object.

A number of adverse effects may be associated with supercritical flows. These include erosive velocities in sewers (discussed in Section 6.2), scouring velocities at outfalls (discussed in the next section), cross waves (discussed later in this section and in Section 7.8), air entrainment, and roll-wave instabilities. The latter three are discussed in this section. The hydraulic jump, on the other hand, may be beneficially utilized for mixing, postaeration, and energy dissipation. Supercritical flow should not necessarily be avoided. Many sewers and other hydraulic structures have been designed successfully for supercritical flows where conditions such as steep slopes make this economically desirable. However, supercritical flows can present surprises, and it is important that potential problems be anticipated. The Froude Number should be calculated and compared to 1 as a matter of course in the design and analysis of sewers and drainage channels, and supercritical conditions (Froude Number greater than 1) flagged for careful assessment.

7.5.1 Flow Profiles

Supercritical-flow profiles occasionally require determination in the analysis and design of sewers and drainage channels (all referred to as "channels" below). Insofar as supercritical *gradually-varied* flow profiles are concerned, by far the most common need is to determine the *drawdown curve* on the steep-slope side when the channel changes from a shallow (subcritical) slope to a steep (supercritical) slope or from steep to steeper. [Such a drawdown curve is the so-called S2 profile; additional details regarding gradually-varied supercritical flow profiles may be found in *Open-Channel Hydraulics* (Chow 1959).] When the channel upstream of the steep channel has a shallow slope, critical depth starting conditions may be assumed at the change in slope (see Figure 7.8). For subsequent changes to steeper slopes, the steeper section will have starting conditions determined by the drawn-down depth in the upstream channel. The other element of supercritical flow profiles with which we shall be concerned is the hydraulic jump, discussed in the next section.

If the channel slope and cross section remain constant, the supercritical drawdown curve is usually a relatively short transition ending in uniform flow conditions. Due to the abruptness of this transition, it is sometimes computationally convenient to assume that uniform flow conditions will be reached in the channel length of interest, and employ a direct-step method to determine the distance within which uniform-flow conditions are approached (recall that uniform-flow conditions are approached asymptotically). For circular sewers, the Keifer–Chu method (Keifer and Chu 1955), discussed above and used in Section 7.7, may be a useful direct-step method; the first and second stations must be taken as the upstream and downstream supercritical flow stations respectively. (Interpolation problems may limit the applicability or accuracy of the Keifer–Chu method, particularly at low Q/Q_f and/or when uniform-flow conditions are approached.)

A standard-step method may also be used, with the computations proceeding in the downstream direction. The procedure is readily programmed for automatic computation. The next example demonstrates a supercritical drawdown curve computation. The methodology (and the Keifer–Chu method) applies to all types of supercritical gradually varied flow profiles in circular sewers. A similar program may be used for trapezoidal channels.

The primary reason for the abruptness of the drawdown curve is the fact that the water surface curve is extremely steep in the vicinity of critical depth conditions. Note that at $y = y_c$, we have $dy/dx = \infty$; that is, the flow profile would, in theory, be

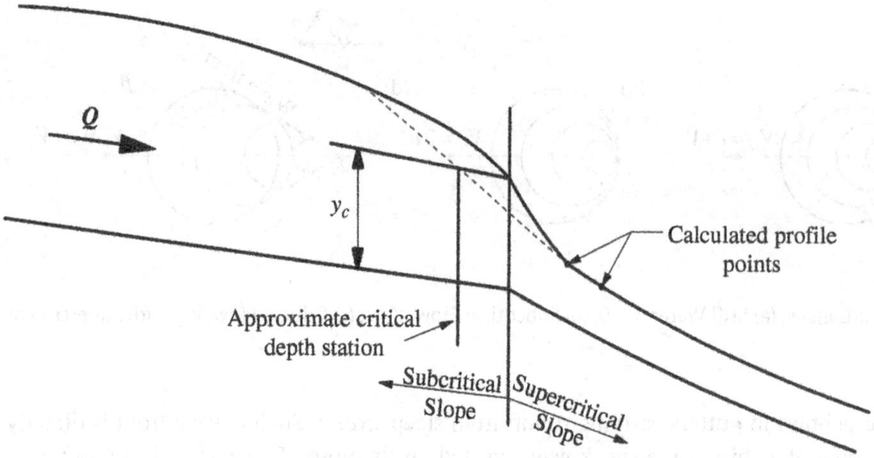

Figure 7.8 Transitional Flow Profile from Subcritical to Supercritical Slope

vertical in crossing the critical depth line. If the depth *increases* in the direction of flow across the critical depth line, a hydraulic jump occurs (see next subsection); if the depth *decreases* across the critical depth line, the abrupt decrease is termed a *hydraulic drop*. There are two important ramifications of the singularity at $y = y_c$. The first is that the curvature of the flow profile near $y = y_c$ is so great that the assumption of parallel flow on which Equation 7.49 is based becomes inaccurate (thus at $y = y_c$ the water surface slope is steep but not actually vertical). The second is that the singularity at $y = y_c$ often plays havoc with numerical methods used to calculate profiles when near $y = y_c$. A convenient resolution of both of these problems, which will often be conservative, is to start the drawdown computations at a point where the Froude Number is about 1.15. This will usually avoid computational problems, and result in an initial and subsequent point on the computed profile. A straight-line extrapolation may then be carried back to locate the approximate upstream distance to the critical depth station; this station will be conservatively upstream of the actual critical depth point as shown on the figure, with the difference usually being small relative to the overall length of interest. This procedure is demonstrated in the following example.

Example 7.15a

An inverted siphon river crossing has an approach sewer profile which conforms approximately with the ground surface profile with a steep slope of 5% (Springfield, Massachusetts). The energy and hydraulic grade lines intersect the approach sewer, which is proposed to be 66-inch concrete ($n_f = 0.013$), at the initial year average daily flow of 20.6 mgd and the design year (40 years in the future) peak flow of 160.1 mgd. The sewer will convey a combined flow of wastewater and stormwater. Determine whether the maximum velocities in the approach sewer would be acceptable.

Solution

The uniform and critical flow depths will first be calculated at the above-stated flows. The flowing-full area, capacity of the pipe Q_f using Manning's Equation 7.39, and corresponding velocity V_f are determined as follows:

$$A_f = \left[\left(\frac{\pi}{4}\right)\left(\frac{66}{12}\right)^2\right] = 23.8\ \text{ft}^2$$

$$Q_f = \frac{1.486}{n} A_f R^{2/3} S^{1/2} = \frac{1.486}{n}(23.8)\left(\frac{66/12}{4}\right)^{2/3}(0.05)^{1/2} = 751\ \text{cfs}/1.547 = 485\ \text{mgd}$$

$$V_f = Q_f/A = 751/23.8 = 31.6\ \text{ft/sec}$$

For the initial year average daily flow, we have $Q/Q_f = 20.6/485 = 0.0425$ for which we obtain for uniform-flow conditions (see Section 7.3) $y_n/d_o = 0.16$, $y_n = 0.16(5.5) = 0.88$ ft, $V/V_f = 0.42$, and $V = 0.42(32.0) = 13.4$ ft/sec. The corresponding values for the design year peak flow are $Q/Q_f = 0.330$, $y_n/d_o = 0.45$, $V/V_f = 0.75$, $y_n = 0.45(5.5) = 2.48$ ft, and $V = 0.75(32.0) = 24.0$ ft/sec.

The critical flow section factor is given by $20.6(1.547)/\sqrt{32.2} = 5.62\ \text{ft}^{5/2}$ at the average daily flow and $160.1(1.547)/\sqrt{32.2} = 43.6\ \text{ft}^{5/2}$ at the design year peak flow. Thus $Z/d_o^{2.5} = 5.62/(5.5)^{2.5} = 0.0792$ at the average daily flow and $43.6/(5.5)^{2.5} = 0.615$ at the design year peak flow. From a table or figure of geometric elements, we obtain, respectively, $y_c/d_o = 0.278$, $A_c/d_o^2 = 0.1779$, giving $y_c = 1.529$ ft and $A_c = 0.1779(5.5^2) = 5.38\ \text{ft}^2$; and $y_c/d_o = 0.798$, $A_c/d_o^2 = 0.6723$, giving $y_c = 4.39$ ft and $A_c = 0.6723(5.5^2) = 20.3\ \text{ft}^2$.

The slope is thus supercritical ($y_n < y_c$) over the range of flows, and the uniform flow velocities are considered excessive (see Section 6.2). The question now becomes whether the uniform-flow conditions are actually approached in the steep sewer; if so the preliminary design will require modifications.

Immediately upstream of the points where the HGL (Hydraulic Grade Line) intersects the sewer crown, subcritical flow backwater conditions would occur. The subcritical backwater may persist over the entire upstream length of the steep sewer or may become conjugate to the lower conjugate depth at a particular point, in which case a hydraulic jump will occur. Substantial further analysis showed that unacceptably high velocities could not be avoided with the proposed design. A wide rectangular channel was considered but found to be impracticably wide. The author proposed that drop structures or multiple pipes were the only realistic alternatives for further consideration.

Example 7.15b

The steep concrete approach sewer of Example 7.15a ($n = 0.13$, 5% slope) is preceded by a subcritical approach sewer (Springfield, Massachusetts). From Example 7.15a, that slope is known to be supercritical at a flow of 20.6 mgd (31.9 cfs). Determine the distance within which uniform flow conditions occur in the steep sewer.

Solution

A critical-depth control will occur near the upstream end of the steep sewer, from which a drawdown toward uniform flow conditions will occur. The critical flow depth may be obtained by calculating $Q/\left(d_o^{2.5}\sqrt{g}\right) = 31.9/\left[(66/12)^{2.5}\sqrt{32.2}\right] = 0.0792$, which gives $y_c/d_o = 0.2773$ or $y_c = 0.2773(5.5) = 1.53$ ft. The flowing full capacity of the sewer is $Q_f = 485$ mgd from the previous example. Thus $Q/Q_f = 20.6/485 = 0.0425$ giving the uniform flow condition (from Figure 5.2 for variable n) as $y_n/d_o = 0.17$ or $y_n = 0.17(5) = 0.94$ ft.

Distance Downstream from Starting Condition (ft)	Depth of Flow (ft)
0	1.430
10	1.091
20	0.991
30	0.933
40	0.895
50	0.867
60	0.847
70	0.832
80	0.820
90	0.810
100	0.803
110	0.797
120	0.793
130	0.789
140	0.786
150	0.783

For this application, a standard-step method (discussed in Section 7.4) was used. A starting depth of $y = 1.43$ ft was selected, which is 0.1 ft less than y_c and corresponds to $\mathbf{F} = 1.135$ (calculated from $\mathbf{F} = V/\sqrt{gD}$ with $V = Q/A$ and A and D obtained with the aid of a table or figure of geometric elements). The calculated profile is tabulated above.

A straight-line extrapolation from the starting point of $y = 1.43$ ft and the calculated profile point of $y = 1.091$ ft a distance of 10 ft downstream, gives the approximate distance from the critical depth control downstream to the start at 2.9 ft (the actual distance would be slightly less). The uniform-flow depth of $y_n = 0.94$ ft (variable n) is within 30 ft of the starting location or within $30 + 2.9 \cong 33$ ft of the upstream end of the steep sewer. Note the slight inconsistency involved in basing the drawdown computation on constant n (a practical matter) and the uniform flow depth on variable n.

Example 7.16

A stormwater pumping station (Fairfield, Maine) has an overflow pipe which is a 36-inch nominal diameter reinforced concrete sewer (35.49-inch actual diameter) with a design flow of 22.3 mgd. The pipe has a slope of 0.0100 ft/ft. The outlet is at a sufficient elevation that it discharges freely into the river. Determine the conditions at the pipe discharge.

Solution

If the pipe has a subcritical slope, the pipe should discharge at critical flow conditions. The full flow and corresponding velocity are obtained from Equation 6.3 as $Q_f = 64.2$ cfs $\div 1.547 = 41.5$ mgd (an exercise left to the reader) and $V_f = Q_f/A_f = 64.2/\left[(\pi/4)(35.49/12)^2\right] = 9.35$ ft/sec. The ratio of design flow:full flow is $Q_d/Q_f = 22.3/41.5 = 0.537$. From a figure of flow characteristics of circular sections for variable n (see Section 5.3), we then have for uniform flow $y/d_o = 0.59$ and $V/V_f = 0.875$. From a table or figure of geometric elements [e.g., Chow (1959) Table 2-1 or Appendix A, or related websites], for $y/d_o = 0.59$ we have $D/D_o = 0.553$. The uniform flow Froude Number can then be calculated as follows: $V = 0.875$ $(9.35) = 8.18$ ft/sec, $d_o = 35.49/12 = 2.96$ ft, $D = 0.553(2.96) = 1.635$ ft, $\mathbf{F} = V/\sqrt{gD} = 9.35/\sqrt{32.2(1.635)} = 1.289$. Therefore, the flow is supercritical and supercritical flow conditions occur at the discharge. The discharge velocity of 8.18 ft/sec is high enough to justify scour protection as discussed in Section 7.8.

Example 7.17

The outlet sewer from a siphon chamber is of 9.0 ft diameter, slope of 0.0021, and Manning's $n = 0.013$ (Nashua, New Hampshire). Additional flows are to be brought in due to upstream expansion of the system. Those flows are dry weather and wet weather of $Q_D = 32.2$ mgd and $Q_p = 343$ mgd, respectively. Determine whether the peak flow is within the capacity of the sewer, the flow depths under the two flow conditions, and whether flow will be subcritical or supercritical under those conditions.

Solution

The flowing-full area, wetted perimeter, and hydraulic radius are calculated, respectively, as $A = (\pi/4)(9^2) = 63.6$ ft^2, $P = 9\pi = 28.3$ ft, $R = 63.6/28.3 = 2.25$ ft. Equation 6.3 gives the capacity of the sewer: $Q_f = \left[1.486(63.6)(2.25^{2/3})\right.$ $\left.(0.0021)^{1/2}\right]/(0.013) = 572$ cfs $\div 1.547 = 370$ mgd. The full flow velocity $V_f = Q_f/A_f = 572/63.6 = 8.99$ ft/sec. The wet weather flow of 343 mgd is within the 370 mgd capacity of the sewer. For peak flow conditions, $Q_p/Q_f = 343/370 = 0.927$. From a figure of flow characteristics of circular sections for variable n [e.g., from Camp 1946a or Chow's (1959) Figure 6-5 from Camp] we then have $y/d_o = 0.84$ and $y = 0.84(9.0) = 7.56$ ft. From a table or figure of geometric elements [e.g., Chow (1959) Table 2-1 or Appendix A, or related websites], for $y/d_o = 0.84$ we have $A/d_o^2 = 0.7043$ and $D/d_o = 0.9606$. The uniform flow Froude Number can then be calculated for peak flow as follows: $A = 0.7043$ $(9)^2 = 57.0$ sq ft, $V = Q/A = 343(1.547)/57.0 = 9.31$ ft/sec, $= 0.9606(9) = 8.65$ ft, $\mathbf{F} = V/\sqrt{gD} = 9.31/\sqrt{32.2(8.65)} = 0.558$ < 1 so the flow is subcritical. A similar exercise for dry-weather flow (left up to the reader) gives $y = 2.07$ ft and a subcritical Froude Number of 0.655.

7.5.2 Hydraulic Jump

7.5.2.1 General Characteristics

We will introduce the phenomenon of the hydraulic jump by considering a steep sewer with a subcritical tailwater submergence as shown on Figure 7.9. Such a condition may occur, for example, at the inlet to an inverted siphon river crossing. The normal depth and critical depth lines are shown on Figure 7.9, as is the downstream backwater curve. The upstream flow is assumed to be uniform. How does the transition from the upstream condition to the downstream condition occur?

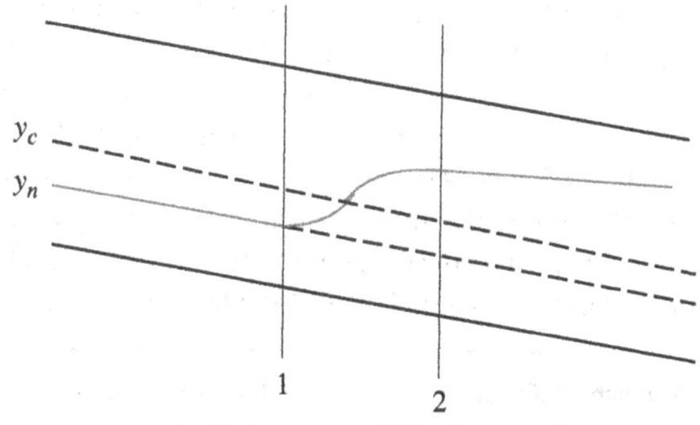

Figure 7.9 Steep Sewer with Subcritical Tailwater Submergence

Because the supercritical approach flow cannot "communicate" with downstream conditions, it must respond abruptly to the downstream subcritical conditions at some point along the backwater curve. The abrupt transition occurs by means of a *hydraulic jump* as shown on Figure 7.9. To analyze the hydraulic jump, we will apply the steady-flow momentum equation to the control volume bounded by the lateral flow boundaries and stations "1" and "2" immediately upstream and immediately downstream of the jump, respectively. Equation 2.35 applied to the direction of the conduit gives (with references to Section 2.7):

$$\rho g\tilde{z}_1 A_1 - \rho g\tilde{z}_2 A_2 - F_D + W'\cos\theta = \beta_2\rho V_2 Q - \beta_1\rho V_1 Q \tag{7.62}$$

in which F_D is the drag force exerted by the flow boundaries and $W'\cos\theta$is the component in the direction of the conduit (of slope angle θ) of the weight of water between stations "1" and "2." The above relationship applies to channels of small slope ($S_o < 0.15$) as discussed in Section 2.7. Setting $\beta_1 = \beta_2 = 1$, and noting $V = Q/A$, the above equation can be rewritten as follows:

$$\tilde{z}_1 A_1 + \frac{Q^2}{gA_1} + \frac{W'\cos\theta}{\rho g} = \tilde{z}_2 A_2 + \frac{Q^2}{gA_2} + \frac{F_D}{\rho g} \tag{7.63}$$

The sum $\tilde{z}A_1 + Q^2/(gA)$ is called the *specific force* at a section.

We will neglect the weight and drag terms for the moment. For a rectangular cross section of width b ($\tilde{z} = y/2, A = by$), Equation 7.63 then results in a quadratic equation, the solution to which is given by:

$$\frac{y_2}{y_1} = \frac{1}{2}\left(\sqrt{1 + 8\mathbf{F}_1^2} - 1\right) \tag{7.64}$$

The depth y_2 is referred to as the *sequent depth* or *conjugate depth* of y_1 for a rectangular channel and may be thought of as the downstream depth which will force the hydraulic jump to occur. (The term *alternate depth* is used by hydraulicians to refer to the depth of flow at which the *specific energy* is equal to that at a higher or lower depth. There is, however, some intermixing of the terms *alternate depth* and *sequent depth* in the environmental engineering literature; that should be avoided.). The depths y_2 and y_1 are referred to as the upper and lower sequent (or conjugate) depths, respectively. Many experimental observations have confirmed the accuracy of Equation 7.64 over a wide range of Froude Numbers (Chow 1959, §15-2, §15-4).

We will focus on the rectangular channel to discuss some general characteristics of the hydraulic jump before proceeding to more complex cross sections. The hydraulic jump satisfies the momentum equation, and an energy loss attends the jump. Neglecting channel slope effects, the energy loss is given by:

$$\Delta E \cong \left(y_1 + \frac{V_1^2}{2g}\right) - \left(y_2 + \frac{V_2^2}{2g}\right) \tag{7.65}$$

Employing continuity ($V_2 y_2 = V_1 y_1$) to eliminate V_2 from the above equation, expressing V_1 in terms of the Froude Number $\left[\mathbf{F}_1^2 = V_1^2/(gy_1)\right]$, and using Equation 7.64 to eliminate $\mathbf{F}_1$ from the resulting equation, enables us to express the energy loss in terms of y_2/y_1 as follows:

$$\frac{\Delta E}{y_1} = \frac{(y_2/y_1 - 1)^3}{4y_2/y_1} \tag{7.66}$$

By substituting y_2/y_1 from Equation 7.64 into the above equation, we may alternately express $\Delta E/y_1$ in terms of the approach Froude Number $\mathbf{F}_1$. The accuracy of Equation 7.66 has been well-established for $\mathbf{F}_1 > 2$. For $1 < \mathbf{F}_1 < 2$ the observed energy loss is less than that predicted by Equation 7.66 (Chow 1959, §15-4).

Referring to Equation 7.63, note that the weight or slope effect adds to the specific force at section "1." This suggests that the effect of slope is to increase the sequent depth y_2 or increase y_2/y_1. This is in fact the case as indicated by Chow's (1959, §15–16) compilation of data by various investigators. The ratio y_2/y_1 is dependent on channel slope and Froude Number. The ratio of y_2/y_1 in a sloped channel to that in a horizontal channel is fairly insensitive to Froude Number, and the average value for rectangular channels with Froude Numbers of about 2–20 are approximately as follows:

S_o (Channel Slope)	$(y_2/y_1)/(y_2/y_1)_{S_o=0}$
0.05	1.18
0.10	1.42
0.15	1.67

The effect of slope is to move the jump downstream if the downstream backwater curve is of the type shown on Figure 7.9. Neglecting the slope effect will cause the jump location to be estimated as being upstream of its actual location in such a case.

The drag term in Equation 7.63 is seen to add to the specific force at section "2," opposite to the effect of slope. The jump length (the distance between sections "1" and "2") is relatively small and the drag term may be safely neglected if there are no obstructions in the channel. However, such obstructions in the form of baffles and sills may be used advantageously to compensate for the lack of tailwater and create jumps in stilling basins as discussed below.

The hydraulic jump in a horizontal, rectangular channel has a length which is roughly 4–6 times the downstream depth y_2, depending on the upstream Froude Number. Chow (1959, §15-5) presents an empirically-based figure giving that length L in terms of L/y_2 as a function of the upstream Froude Number $\mathbf{F}_1 = V_1/\sqrt{gy_1}$. That figure is employed in the following example.

Example 7.18

Example 7.9 (Washington Highway Pumping Station, Lincoln, Rhode Island) determined the supercritical flow conditions at the base of an ogee spillway as follows: flow depth at the upstream end of the jump = 0.1664 ft, velocity = 17.56 ft/sec, Froude Number = 7.59, flow depth at the downstream depth of the jump = 1.706 ft. Determine the jump length, which for this case is important to the sizing of the trench-type wet well.

Solution

In terms of the variables pertinent to the present problem, we have $\mathbf{F}_1 = 7.59$ and $y_2 = 1.706$ ft. Chow's figure indicates that this Froude Number is in the "Steady jump," "Best performance" range, and gives $L/y_2 = 6.15$. The jump length is thus $L = 1.706(6.15) = 10.49$ ft.

The reader is referred to Chapter 15 of *Open Channel Hydraulics* (Chow 1959) for details regarding other characteristics of hydraulic jumps. The moving hydraulic jump in the form of the tidal bore is discussed in Section 15.4.

7.5.2.2 Environmental Engineering Applications

One of the most common occurrences of hydraulic jumps in environmental engineering applications is a transition from supercritical to subcritical conditions in sewers such as shown on Figure 7.9, or near the change in channel slope from supercritical to subcritical as discussed earlier. Similar circumstances are found in drainage channels, such as at the discharge of steep drainage channels into flat wetland areas. Another common occurrence of the hydraulic jump is in Parshall flumes as discussed in Section 7.2. The strong eddying and turbulence in the hydraulic jump can make it effective for chemical mixing, aeration, and postaeration, as discussed in Sections 14.4, 14.7, and 14.7, respectively. The hydraulic jump may also be used to remove air pockets from water-supply lines and thus prevent air locking. The hydraulic jump is widely used as an energy dissipator at outlets from dams and other hydraulic structures, but it is not usually the means of choice for outfalls used in environmental engineering applications. This is primarily because the tailwater range within which the jump will be confined to the stilling basin is fairly small compared to the wide variations of river levels which usually comprise the outfall tailwater conditions. Types of outlet structures which may be more suitable are discussed in Section 7.7

along with hydraulic jump stilling basins. The latter should be kept in mind, and the reader should be aware of the existence of well-tested, generalized designs (Chow 1959, Chapter 15; Peterka 1964). These designs often make use of baffles and sills, the purpose of which may be, at least partially, to increase the drag force opposing the jump along the lines discussed earlier.

7.5.2.3 Hydraulic Jumps in Sewers

The common occurrences of hydraulic jumps in sewers are in the vicinity of breaks in slope from supercritical to subcritical and in steep sewers discharging to high tailwaters. In considering the break in slope, we will focus on the case where normal depth conditions have been achieved (closely achieved, to be precise) upstream and downstream of the break by virtue of the upstream drawdown and downstream backwater profiles (extension to other cases should be apparent to the reader after reading a little further). If the downstream normal depth is greater than the upstream sequent depth ($y_{n2} > y'_{n1}$), then the hydraulic jump must occur upstream of the break. The location of the jump could, in theory (however, see below), be determined by calculating the backwater profile (using methods discussed in Section 7.4) from y_{n2} at the break upstream to the station where the backwater depth equals the upper sequent depth y'_{n1}; the jump will be located at this station. If the downstream normal depth is less than the upstream sequent depth ($y_{n2} < y'_{n1}$), the hydraulic jump will occur downstream of the break. Then the jump location could, in theory (however, see below), be determined by calculating the gradually varied flow profile (using methods discussed earlier in this section) from y_{n1} at the break downstream to the station where the flow depth equals the lower sequent depth y'_{n2}; the jump will be located at this station.

For jumps in horizontal sewers, computations may be carried out along the lines discussed above. Sequent depths for horizontal rectangular sewers are calculated by Equation 7.64; sequent depth calculations for more general cross sections are discussed below. For sloped, rectangular sewers, the sequent depths may be adjusted for slope as discussed above. If jump length is deemed important in establishing jump locations, methods given by Chow (1959, §15-7, §15-16) may be used. For sloped sewers of more general cross section, data for determination of the sequent depth may not be available. In such a case, the relationships for a horizontal channel may be used with the realization that an upstream limit of the jump location is being established. [Textbook methods which purport to accurately establish jump location in sewers (sometimes going to the extent of accounting for jump length) without considering slope are plainly erroneous.] Fortunately, it is not necessary to locate the hydraulic jump in every case; it is often enough to understand that it will occur. If, for instance, a sewer is designed on the uniform-flow assumption (see Section 7.3), in which uniform-flow conditions are acceptable in the upstream and downstream sewers (in terms of depth and velocity), then it may be sufficient to understand that the hydraulic jump is part of the transition between the upstream and downstream uniform depth conditions. Before attempting to establish the jump location, a qualitative understanding should be gained and the need for going beyond that ascertained. Situations do occasionally occur in which it becomes necessary to consider jump location. For example, if the velocity at supercritical uniform-flow conditions would be excessive, we may wish to know if the hydraulic jump could occur far enough upstream in the pipe to prevent excessive velocities from occurring. Other such situations are occasionally encountered.

For rectangular channels, Equation 7.64 is used to calculate sequent depths, y_1 giving the lower sequent depth and y_2 giving the upper sequent depth. For more complex channel shapes, we return to Equation 7.63 with $F_D = W' \cos\theta = 0$:

$$\tilde{z}_1 A_1 + \frac{Q^2}{gA_1} = \tilde{z}_2 A_2 + \frac{Q^2}{gA_2} \tag{7.67}$$

Knowing Q, y_1 (or y_2), and the cross-sectional shape, we can obtain the left-hand side (or right-hand side) of the above equation. The right-hand side (or left-hand side) is then a function of only the unknown y_2 (or y_1) which can be found by trial-and-error.

Equation 7.67 can be usefully specialized for different channel shapes to facilitate calculations. Denoting the specific force by F_s, we have:

$$F_s = \tilde{z}A + \frac{Q^2}{gA} \tag{7.68}$$

For the rectangular channel:

$$F_s = \frac{by^2}{2} + \frac{Q^2}{gby} \tag{7.69}$$

or:

$$F_s/b = y^2/2 + q^2/(gy) \tag{7.70}$$

where $q = Q/b$. Nondimensionalizing F_s with respect to the critical depth gives the following form:

$$\frac{F_s}{by_c^2} = \frac{1}{2}\left(\frac{y}{y_c}\right)^2 + \frac{y_c}{y} \tag{7.71}$$

For the circular cross section, and other cross sections with a characteristic dimension σ (Graber 2004b), we may nondimensionalize Equation 7.68 and equate that form at two sections as follows:

$$\frac{F_s}{\sigma^3} = \frac{\tilde{z}_1}{\sigma}\frac{A_1}{\sigma^2} + \frac{Q^2/(g\sigma^5)}{A_1/\sigma^2} = \frac{\tilde{z}_2}{\sigma}\frac{A_2}{\sigma^2} + \frac{Q^2/(g\sigma^5)}{A_2/\sigma^2} \tag{7.72}$$

The terms z/σ and A/σ^2 are functions only of y/σ. For the circular cross section, $\sigma = d_o$ and $Q^2/(g\sigma^5) = Q^2/(gd_o^5) = (A_c/d_o)^2 D_c/d_o = \phi(y_c/d_o)$. Thus F_s/d_o^3 can be expressed as a function of y/d_o and y_c/d_o.

The situation sometimes arises in which a closed conduit (such as circular pipe) on a supercritical slope is surcharged at its downstream end. This upper sequent depth can be greater than the depth of the conduit, that is, the jump can go from free surface to surcharged conditions. In this case, the downstream specific force is given by:

$$F_{s2} = \frac{p_c A_f}{\rho g} + \frac{Q^2}{gA_f} \tag{7.73}$$

in which p_c is the pressure at the centroid of the conduit cross section and A_f is the full cross-sectional area. For parallel flow (and neglecting the effect of air entrainment on density), the centroidal pressure is given by $p_c = \rho g \tilde{z}_c$, where $\tilde{z}_c$ is the vertical distance from the hydraulic grade line down to the centroid. Thus:

$$F_{s2} = \tilde{z}_c A_f + Q^2/\left(gA_f\right) \tag{7.74}$$

or

$$\frac{F_{s2}}{\sigma^3} = \frac{\tilde{z}_c}{\sigma}\frac{A_f}{\sigma^2} + \frac{Q^2/(g\sigma^5)}{A_f/\sigma^2} \tag{7.75}$$

which is readily evaluated. For a circular pipe, $\tilde{z}_c = y' - (1/2)d_o$ in which y' is the elevation of the hydraulic grade line above the pipe invert. In this case:

$$\frac{F_{s2}}{d_o^3} = \left(\frac{y'}{d_o} - \frac{1}{2}\right)\frac{A_f}{d_o^2} + \frac{Q^2/(gd_o^5)}{A_f/d_o^2} = \frac{\pi}{4}\left(\frac{y'}{d_o} - \frac{1}{2}\right) + \frac{4}{\pi}\frac{Q^2}{gd_o^5} \tag{7.76}$$

Kalinske and Robertson (1943) (and earlier workers whom they cite) experimentally confirmed the applicability of the momentum equation as above to the hydraulic jump for circular conduits. (Kalinske and Robertson computed the term $W' \cos\theta$ in their sloped circular conduits from observations of the jump length and profile.)

Stahl and Hager (1999) investigated experimentally and theoretically the hydraulic jump in circular pipes. For Froude Numbers ranging from 1.5 to 6.5 and upstream depth:diameter ratios from 0.20 to 0.70, they fitted the following equation for the sequent depth ratio:

$$\frac{h_2}{h_1} = 1.00\mathbf{F}_1^{0.90} \tag{7.77a}$$

and for the jump length:

$$\frac{L_a}{h_2} = 4\mathbf{F}_1^{1/2} \tag{7.77b}$$

These equations apply to the case in which the jump remains below the pipe crown. Equation 7.77b gives the "aeration length," which is the length over which air clouds have left the flow as distinct from the shorter "roller length." The aeration length is an index of the overall length of the jump.

Example 7.19

A circular sewer flowing to a self-cleaning trench-type wet well (Washington Highway Pumping Station, Lincoln, Rhode Island) has a 29.44 in. (2.45 ft) internal diameter approach pipe on a 2% slope with Manning's roughness coefficient $n =$ 0.010. For 3.89 cfs flow rate, determine uniform-flow conditions in the pipe and corresponding Froude Number, and whether and under what downstream conditions a hydraulic jump will occur.

Solution

The uniform flow parameter is calculated first: $nQ/\left(1.486d_o^{8/3}S^{1/2}\right) = 0.010(3.89)/\left[1.486(2.45)^{8/3}(0.02)^{1/2}\right] = 0.0170$. The corresponding uniform-flow depth ratio is 0.1582 (which the reader should verify), giving the uniform-flow depth = 0.1582 (2.45) = 0.388 ft. The flow area A, hydraulic depth D, and velocity V are, respectively, 0.479 ft^2, 0.268 ft, and $V = Q/A = 3.89/0.479 = 8.12$ ft/sec. The corresponding Froude Number $\mathbf{F} = V/\sqrt{gD} = 8.12/\sqrt{32.2(0.268)} = 2.77$. The flow is supercritical, and Equations 7.77a and 7.77b give $h_2 = 0.388(2.77)^{0.90} = 0.971$ ft and $L_a = 0.969(4)(2.77)^{1/2} = 6.45$ ft.

7.5.2.4 Air Entrainment

In some cases, it may be necessary to take air entrainment into account insofar as it effects both the density ρ and volumetric rate of flow Q in Equation 7.73 (Kalinske and Robertson 1943; Kennison 1943; Kohler and Ball 1969, pp. 22–75 to 22–77). High-velocity flows have a tendency to entrain air, a phenomenon known as *bulking*. The phenomenon is distinct from atmospheric reaeration (see Section 14.7), which is a diffusion phenomenon. Bulking is characterized by white water and an increase in depth. It is most commonly associated with high-velocity flows on steep chutes and spillways (Hall 1943; Ippen 1950, pp. 546–547; Straub and Lamb 1956), but can also occur in steep sewers. From DeLapp's (1943) discussion of Hall, it may be deduced that bulking in rectangular channels commences at a Froude Number of about 5, although additional variables are certainly involved. Chow (1959, p. 33) suggested that air becomes entrained at velocities of about 20 fps and higher, influenced by such other factors as entrance conditions, channel roughness, distance traveled, channel cross section, volume of discharge, etc. Bulking would require design concern in steep sewers if freeboard is important or if depth is being measured for flow metering purposes.

7.5.2.5 Gas Transfer at Hydraulic Jumps

Studies of gas transfer at hydraulic jumps include those of Wilhelms *et al.* (1981) and Chanson (1995). Chanson's experiments used air as the gas, whereas those of Wilhelms *et al.* used Krypton-85 as a dissolved gas as a tracer for dissolved oxygen. Of the experiments performed and compared by Chanson, one of the empirical formulae of Wilhelms *et al.* seems most useful for estimating dissolved oxygen increase across the jump. That formula is as follows:

$$\frac{D_o}{D} - 1 = 4.924(10)^{-8}\mathbf{F}^{2.106}\mathbf{R}^{1.034} \tag{7.78}$$

in which D_o is the upstream saturation deficit, D is the downstream saturation deficit, $\mathbf{F}$ and $\mathbf{R}$ are, respectively the Froude Number and Reynolds Number of flow upstream of the jump. Wilhelms *et al.* plotted experimental results over a range of Froude Numbers from 2 to 10, which their experiments nearly included; and over a range of Reynolds Numbers over a range of 10^3–10^6. Their experiments covered Reynolds Numbers ranging from $0.5(10)^4$ to $7(10)^4$, but Chanson's comparisons suggest that results of Wilhelm *et al.* comport reasonably well with those of Avery and Novak (1978) over Froude Numbers from 2 to 9 with Reynolds Numbers from $1.5(10)^4$ to $4.4(10)^4$.

Denoting the saturation concentration by C_s and the downstream concentration by C_d, we have:

$$C_d = C_s - \frac{D_o}{1 + (D_o/D - 1)} \tag{7.79}$$

As an example, take $C_s = 9$ mg/L, upstream concentration 3 mg/L giving an upstream deficit $D_o = 9 - 3 = 6$ mg/L, upstream Froude Number $\mathbf{F} = 4$, and upstream Reynolds Number $\mathbf{R} = 1(10)^5$. Equation 7.78 gives:

$$\frac{D_o}{D} - 1 = 4.924(10)^{-8}(4)^{2.106}\left[1(10)^5\right]^{1.034} = 0.135$$

$$C_d = 9 - \frac{6}{1 + 0.135} = 3.71 \text{ mg/L}$$

7.5.3 Roll-Wave Instabilities, Oblique Waves, Design Adjustment

Steep, rock-lined channels are used on the side slopes of some landfills and are thus part of the scope of this book. Such channels are subject to three unique hydraulic concerns: (1) avoiding roll-wave instabilities, (2) avoiding oblique waves (waves oblique to the main axis of the channel) or providing sufficient freeboard for such waves at channel bends, and (3) recognizing that supercritical flow may not actually occur in such channels and that the design needs to be adjusted accordingly.

7.5.3.1 Roll-Wave Instabilities

The tendency for roll waves to form in steep channels was discovered around 1940 and is discussed by Keulegan (1950) and Chow (1959, §7-5, §8-8, §19-9). Roll waves combine to form wave heights which can overtop channels in unpredictable ways, particularly if the channels are curved. They are a free-surface instability occurring at high velocities and have the appearance of a succession of surface waves. They could occur in steep sewers, but require design concern only if freeboard is of importance or if depth is being measured for flow metering purposes.

Roll-wave instability (Vedernikov Number < 1 – see below) can usually be avoided. When necessary, wave heights can be considered in design where suitable predictive methods exist, such as for wide, straight rectangular channels with Froude Numbers > 2, for which Brock's (1970) method considering shock weight reasonably predicts shock profiles.

The criterion for roll waves to form is a Vedernikov Number **V** greater than or equal to one:

$$\mathbf{V} = x\gamma\mathbf{F} \geq 1 \tag{7.80}$$

in which x is the exponent of the hydraulic radius in the uniform-flow equation being used (2/3 for turbulent flow if the Manning equation is used or 1/2 for the Chezy formula), **F** is the Froude number, and γ is a channel section shape factor defined as:

$$\gamma = 1 - R\frac{dP}{dA} \tag{7.81}$$

in which A is the flow area, P is the wetted perimeter, and R is the hydraulic radius.

For a trapezoidal cross section, the shape factor becomes:

$$\gamma = 1 - \frac{2\sqrt{1+z^2}(1+zy/b)y/b}{\left[(1+2y/b)\sqrt{1+z^2}\right](1+2zy/b)} \tag{7.82}$$

in which y is the flow depth, b is the bottom width, and z is the channel side slope (run over rise). For a rectangular channel ($z = 0$), the above equation reduces to:

$$\gamma = 1 - \frac{2y/b}{(1+2y/b)} \tag{7.83}$$

For very wide rectangular channels the above equation gives $\gamma = 1$, whereas for very narrow rectangular channels $\gamma = 0$. For a circular cross section, we have $\gamma = (\sin\phi/\phi - \cos\phi)/(1 - \cos\phi)$ in which ϕ is the total included angle to the top width $T = [\sin(\phi/2)]d_o$ with d_o being diameter.

In calculating the Froude number for steep channels (in which roll waves are more likely to occur), it can be important to account for slope effects (on the energy equation from which the Froude number is derived) and nonuniformities in the velocity profile. The equation for Froude number including those factors is:

$$\mathbf{F} = \frac{V}{\sqrt{(gD\cos\theta)/\alpha}} \tag{7.84}$$

in which V is average velocity, D is hydraulic depth (flow area divided by top width), θ is the slope angle of the channel bottom, and α is the energy coefficient (also known as the velocity-head coefficient or Coriolis coefficient).

The following equations are given by Jarrett (1992) for the energy coefficient in steep rock-lined channels:

$$\alpha = 1.43(R/D_{84})^{-0.77} \tag{7.85a}$$

$$\alpha = 1.97S^{0.08} \tag{7.85b}$$

$$\alpha = 2.51n^{0.22} \tag{7.85c}$$

in which D_{84} is the intermediate particle diameter (riprap) that equals or exceeds that of 84% of the particle diameters (same units as R), S is the channel bottom slope, n is Manning's roughness coefficient, and other terms are as defined previously. For conservatism, the author would use the equation giving the highest value of α (and would only use the first equation if the D_{84} can be derived from a known gradation).

7.5.3.2 Oblique Waves and Superelevation

Oblique waves occur with supercritical flow at abrupt or gradual channel bends. Requiring additional freeboard, they are predictable and can be incorporated in the design in the case of concrete or concrete-lined channels. Methods for calculating the heights of such waves are given by Ippen (1950, Section D.17 – Channel Curves) and Chow (1959, §16-5). For rock swales, however, oblique wave heights would be difficult to predict.

The Quarry Hills landfill in Quincy and Milton, Massachusetts was capped with excavated material from the Central Artery/Tunnel project in downtown Boston (aka "Big Dig"). As part of that work, quarries were filled and a 27-hole golf course and other recreational facilities were constructed on the site. The author reviewed and advised on the technical details.

Approximately 36 downcomers were proposed for the several landfill areas to convey drainage down the landfill slopes. Two of the downcomers as initially proposed included substantial bends. As further discussed below, unless there are very compelling reasons to include bends, it is preferable to avoid bends in such channels. The bends were eliminated based on the author's recommendation.

Superelevation occurs in subcritical flow and is discussed here because subcritical flow is more likely to occur in some or all of the Quarry Hills downcomers. Methods for calculating the height of superelevation at bends are given by Ippen (1950, Section B.8) and Chow (1959, §16-4).

7.5.3.3 Supercritical/Subcritical Flow

There is increasing evidence that supercritical flow is not stable, and may not actually occur in steep, natural streams (or occurs only infrequently and for short channel reaches); and by implication that it would not occur or should not be relied upon in steep, stone-lined channels. The earliest publications which called attention to the problem include those by Bathurst (1985), Thorne (1985), Trieste (1994), Jarrett (1984, 1992), and Wahl (1994). The flows in such channels are observed to alternate between supercritical and subcritical, with the supercritical flow only occurring for short distances. Explanations for this include the following:

- Published Manning's n values (e.g., in the previous chapter's Table 6.1, and in Chow 1959) are generally derived for lowland rivers with shallow slopes, and have not been verified for steeper rivers.
- An increase in energy losses (which are lumped into n-values), as discharge increases, with the resulting effective n-values being substantially larger than those given in the literature for subcritical flows.
- Additional energy losses due to extreme agitation, eddies, and hydraulic jumps.
- Form drag of rocks, free surface distortion and hydraulic jumps associated with locally accelerated flow.

Jarrett (1992) gives one of the more convenient equations for estimating the roughness coefficient for mountainous streams (analogous to steep, rock-lined channels):

$$n = 0.32S^{0.38}R^{-0.16} \quad R \text{ in meters} \tag{7.86a}$$

which in English units becomes:

$$n = 0.387S^{0.38}R^{-0.16} \quad R \text{ in feet} \tag{7.86b}$$

Substituting the above equation into Manning's equation and simplifying:

$$Q = 3.840AR^{0.827}S^{0.12} \tag{7.87}$$

For a trapezoidal channel, this becomes:

$$Q = 3.840(b + zy)y\left[\frac{(b + zy)y}{b + 2y\sqrt{1 + z^2}}\right]^{0.827} S^{0.12} \tag{7.88}$$

For a rectangular cross section ($z = 0$), this reduces to:

$$Q = 3.840\, by\left(\frac{by}{b + 2y}\right)^{0.827} S^{0.12} \tag{7.89}$$

For a given Q, b, z (if the latter is applicable), and S, the above equations can be solved numerically (such as by graphical bisection) for the uniform-flow depth y.

The following summarizes the author's application to a trapezoidal channel proposed for a large technology park in Westborough, Massachusetts. The channel bottom width is 10 ft with a side slope of 1:3, bottom slope of 0.075, and depth of 2.5 ft. The Manning's n value was based on FHwA (2005, Table 5-4) for rock riprap giving $n = 0.104$. The normal depth is calculated by trial-and-error from the values preceding it. Subsequent terms then follow.

Discharge = 115 cfs
Side slope $z = 0.33$
Manning's $n = 0.104$
Bottom Slope = 0.075
Normal Depth = 2.06 ft
Velocity = 5.21 ft/sec
Top Width =11.38 ft
Hyd. Depth = 1.939 ft (calculated using Manning's equation)
$\cos\theta$ = 0.949
$\alpha = 1.601$ (based on slope – Equation 7.85b)
F = 0.68
Shape Factor = 0.725 (from Equation 7.82)
V = 0.328 (from Equation 7.80)

Although the design was based on the above, it is interesting in retrospect (and for reasons discussed below) to apply Jarrett's equation to the example given above, which results in the following:

Discharge = 115 cfs
Side slope $z = 0.33$
Bottom Slope = 0.075
Normal Depth = 2.41 ft
Velocity = 4.42 ft/sec
Top Width = 11.61 ft
Hyd. Depth = 2.24 ft
$\cos\theta = 0.949$
$\alpha = 1.601$ (based on slope – Equation 7.85b)
F = 0.53
Manning's $n = 0.133$ (calculated from Equation 7.86b)
Shape Factor = 0.697 (from Equation 7.82)
V = 0.248

The calculated depth is only 1.17 times higher than the value calculated previously and the flow is subcritical in both cases (with Vedernikov Numbers less than 1 in both cases). In both cases, the flow remains within the 2.5-ft channel depth.

The following summarizes the author's application to a proposed 16 ft wide rectangular channel and bottom slope of 0.40 at the Quarry Hills MDC North landfill in Quincy, Massachusetts. The Manning's n value is first based on FHwA (1988, Table 3) for rock riprap with 6-inch D_{50} and 0-0.5 ft depth range. The normal depth is calculated by trial-and-error from the values preceding it.

Discharge = 15.55 cfs
Manning's $n = 0.104$
Bottom Slope = 0.40
Normal Depth = 0.266 ft
Velocity = 3.66 ft/sec

Hyd. Depth = 0.266 ft (calculated using Manning's equation)
$\cos \theta = 0.928$
$\alpha = 1.831$ (based on slope – Equation 7.85b)
F = 1.30
Shape Factor = 0.968 (from Equation 7.83)
V = 0.837 (from Equation 7.80)

Now applying Jarrett's equation to the example given above results in the following. The normal depth is calculated by trial-and-error from the values preceding and following it.

Discharge = 15.55 cfs
Bottom Slope = 0.40
Normal Depth = 0.515 ft
Velocity = 1.887 ft/sec
Hyd. Depth = 0.515 ft
$\cos \theta = 0.928$
$\alpha = 1.831$ (based on slope)
F = 0.48
Hyd. Radius = 0.484
Manning's $n = 0.307$ (calculated from Equation 7.86b)
Shape Factor = 0.940 (from Equation 7.83)
V = 0.301 (from Equation 7.80)

The calculated depth is nearly 2 times higher than the value calculated previously and the flow is now subcritical (both with a Vedernikov Numbers less than 1).

The question in this case was whether to base design on Jarrett which predicted a subcritical flow or FHwA which predicted a supercritical flow. The latter would require a stilling basin on each channel at the toe of the slope. The design proceeded incrementally (a few channels first) without stilling basins, which proved to be successful with subcritical flow at the base.

Giving further thought to the causes of subcritical flow in steep, rock-lined channels, the phenomena involved appear related to the effect of depth reduction on relative roughness (as defined, e.g., for the Moody Diagram), which overwhelms the effect of reduced hydraulic radius and the relative importance of form drag (as distinct from friction drag), and hence on the effective Manning's n.

It should be noted that Jarrett's equations are regression equations derived or checked for slopes less than 0.1, and for natural mountain rivers for which the river bed may bear a relation to other hydraulic variables which makes the regression better for such rivers than it would be for a man-made channel. There are a number of other references that deal with the problem of steep, rocky channels, which may allow or lead to the derivation of more appropriate equations for the variables of interest. More recent such references are Aguirre-Pe and Fuentes (1990), Afzalimehr and Anctil (1998), and Ferro (1999). The publication "Hydraulic Engineering '94'" is the proceedings of a 1994 conference that contains numerous papers on the "Hydraulics of mountain rivers"; the specific papers are listed in the Database under the key words of "Mountain streams" and "Flow resistance."

7.5.3.4 Overall Considerations

Putting all of the above together, it would be prudent to:

1) Use Jarrett's relationships as given above or other appropriate relationships to design the channel as a steep, rock-lined channel. Include ample freeboard (Chow 1959, §7-5).
2) Check for roll-wave stability by calculating the Vedernikov number **V** under the assumption of supercritical flow (where applicable) based on a conventional n-value (i.e., n-value applicable to a small slope) and accounting for α and $\cos \theta$ in calculating Froude number; and modify the design (e.g., reduce flow by increasing the number of channels) if necessary to make $\mathbf{V} < 1$.
3) Avoid channel bends, or if not avoidable include sufficient freeboard (Chow 1959, §7-5) for cross waves assuming supercritical flow and superelevation assuming subcritical flow (the latter by assuming an n-value applicable to a steep slope).

Freeboard for cross waves in rock-lined channel should be designed very conservatively in view of the large uncertainty of predicting cross waves in such channels. (Also see Section 7.8.)

4) Riprap stability should be checked for the most critical hydraulic conditions of depth and velocity.

Elaborating on this latter point, the matter of riprap stability forcefully gained attention when a seat-of-the-pants design of slope drains was implemented at the Quarry Hills MDC North landfill slope. On July 18, 2000, subjected to an intense 0.5-inch rainfall, a number of the stone-lined slope drains (downcomers) failed severely, with the stones displaced downhill to the foot of the slope. Sediments from the failed slope wound up on a downstream roadway and posed a threat to nearby wetlands. A similar failure occurred subsequently at the Milton portion of the site leading to wetland deposits. These incidents led to the author's involvement in the design of the slope drains, eventually numbering 36 as mentioned above. The design included appropriate numbers and spacing of slope drains and liner design.

Riprap design is based on maximum velocities. For the downcomers discussed above, the 4.86 ft/sec velocity was selected for conservatism. The designer selected a six-inch stone in some areas and eight-inch stone in others (representative of approximately the 50% size), which has permissible velocities well in excess of 4.86 ft/sec by all of the permissible velocity procedures considered (including USBR 1963b, USBPR 1967, FHwA 1990a). Keyed-in HDPE liners were provided below the stones.

7.6 Sewer Transitions, Junctions, and Drops

7.6.1 Sewer Transitions

The quoted portion of the following discussion of sewer transitions is from a paper by Thomas R. Camp entitled "Hydraulics of Sewers" (Camp 1952).

> In sewer design it is usually assumed for simplicity that the flow is steady and uniform throughout a stretch of sewer between increments of flow, provided of course there is no change in the size or grade within the stretch. To simplify the design of sanitary sewers it is usually assumed that all increments of flow from house sewers within a stretch enter at the upper end of the stretch. For small sewers it is desirable to provide manholes at each end of a stretch and at intervals of 200 to 600 ft to facilitate inspection and cleansing and to make the stretch straight in line and grade between manholes. For small sewers, transitions, curves and junctions may be confined within the manholes. For larger sewers special structures are required except for curves where the sewer itself may be designed to conform to a definite curve.
>
> If the transitions between stretches are properly designed the assumption of uniform steady flow within the stretch for the short period during which peak flow takes place is a close approximation and fully warranted. If the transitions are not properly designed, however, and sufficient allowance is not made for invert drops at transitions, the transitions may reduce the expected capacity of the sewer and cause surcharge at the peak design flow. It should be pointed out that the practice of designing transitions so that the crowns (or the inverts) of abutting sewers are placed at the same elevation without regard to the energy conditions is faulty.
>
> It is convenient in transition design to assume first that the energy loss, invert drop and change in water surface due to the transition are concentrated at the center of the transition and then to distribute these changes smoothly throughout the length of transition when it is detailed. The method is illustrated in Fig. [7.5] where H_T represents the head loss due to the transition and y represents the required drop in the invert due to the transition. This procedure permits the sewer design to progress before the transition design is undertaken. It is necessary only that adequate allowances be made during sewer design for the invert drops.
>
> It is to be noted particularly that the transition head loss must be applied to the energy grade line and not to the hydraulic grade line (i.e., the water surface). The errors involved are insignificant in sewers where the velocity is less than 2.5 fps because the velocity head itself is small (0.1 ft.), but for higher velocities grave errors may result from improper design of transitions. Even with low-velocity sewers, consideration must always be given to invert drops where increments of flow are accepted in order to avoid backwater curves with attendant low velocities in the upstream stretches of sewer. Except for some special and infrequent cases involving junctions, the energy grade line

falls through a transition. The hydraulic grade line however may rise and usually does in transitions involving substantial decrease in velocity.

For convenience, the sewer inverts may be taken as datum and the total energy heads H_u and H_l in the upper and lower stretches of sewer where uniform flow is assumed to take place, may be measured upward from the inverts. In each case, the energy head is the depth plus the velocity head, or:

$$H = d + \frac{V^2}{2g}$$

and the energy grade line is parallel with the invert and water surface at a height equal to the velocity head above the water surface.

From Fig. [7.5] it is obvious that the required invert drop y is readily obtainable from the relation:

$$y = H_\ell + H_T + H_u$$

In the design of a transition, y may have a value from zero to several feet in magnitude. If a negative value of y is obtained from the equation, it should not be used in design because it will result in placing the invert of the lower pipe at a higher elevation than that of the upstream pipe, creating a sump for the collection of grit and other suspended matter. If y is negative from the equation, y should be taken at zero whereupon the head loss h_T becomes greater than the minimum computed and there will be a drawdown curve in the upper stretch of sewer. This is the case which frequently occurs when a small steep sewer discharges into a larger sewer with a flat slope.

Data on the magnitude of the transition loss H_T for sewers are almost completely lacking. No reliable measurements of such losses in sewers have ever come to the author's attention. There have been experimental studies in recent years on both contracting and expanding transitions but usually, the cross section has been rectangular. The most reliable data available for use are measurements of head losses in flume and siphon transitions for irrigation purposes made by the U.S. Bureau of Reclamation. These data are analyzed in a paper by Julian Hinds (1928, p. 1423) who recommends that the head loss for a well-designed transition may be estimated by the equation $H_T = k\Delta[V^2/(2g)]$ where $\Delta[V^2/(2g)]$ is the change in velocity head through the transition and k is a factor which may have a value as low as 0.1 for increasing velocity transitions and 0.2 for decreasing velocity transitions. Since the velocity heads are quite small in most sanitary sewers, it is the opinion of the author that larger allowances should be made for transition losses with a minimum value of 0.02 ft. in any case.

The transition data studied by Hinds involved flow in the upper alternate stage [subcritical flow] in all cases and the values of k which he recommends are strictly applicable only to upper stage flow. There are many sewers in which the flow must take place in the 'lower alternate stage'; hence sewer transitions may involve flow from upper to lower, from lower to upper or from lower to lower stage. Where a lower stage is involved the value of k will probably be greater than stated above and if flow is from lower to upper stage a hydraulic jump will take place with considerable energy loss.

Because of the above considerations, it is advisable as part of design procedure to determine at what stage flow will take place in all sewers.

It should be emphasized that the above equation for H_T with the suggested values of k applies to *well-designed* transitions. Hinds suggested that these values be applied to transitions having an angle of divergence between the center line and each wall of less than 12.5° and for Froude Numbers less than 0.5. For ordinary sewer conditions, Davis and Sorensen (1969, pp. 40–49) suggested that $k = 0.2$ be employed for increasing velocity transitions and $k = 0.3$ for decreasing velocity transitions. There are important cases in practice for which a "well-designed" transition is not feasible. Drop structures, required to reduce velocities (as discussed below) or to avoid interference with existing utilities, are an example. For drop structures, the head loss should be taken as follows, which gives a value of H_T considerably higher than determined by the equation given above.

$$H_T = 1.0\frac{V_u^2}{2g} + 0.5\frac{V_\ell^2}{2g} \tag{7.90}$$

Failure to properly apply the basic principles presented above may result in serious surcharging of the sewers. In high-velocity storm sewers surcharges as high as five or six feet due to transition energy losses have been encountered. In this connection, it should be mentioned that surcharging of deep storm sewers, (i.e., design for hydraulic grade line above the sewer crown) may be desirable and economical (Jones 1971). However, this must be done purposefully and by employing the principles discussed herein.

7.6.2 Sewer Junctions

The quoted portion of the following discussion of sewer junctions is from Camp (1952).

> Most sewer transitions are also junctions where a branch sewer enters a main sewer. In this case there are two flow paths, the lower stretch of main sewer being common to both. In at least one of the streams there is curvature and the value of H_T for this path should include the effect of curvature. An additional allowance in head loss should also be provided for both paths to care for the impact loss due to the converging streams. The hydraulic design of a junction is, in effect, the design of two transitions, one for each flow path.
>
> The impact loss at a junction is, theoretically, subject to computation by the momentum principle. This principle has been successfully applied to the computation of head loss in a hydraulic jump occurring in rectangular open channels. For a junction of two or more streams meeting at an angle, the pressure plus the linear momentum below the junction point must equal the sum of the components of the pressure and momentum of each tributary stream in the direction of flow of the stream below the junction. The energy loss due to impact in the junction is the difference between the total energy in the stream below and the sum of the energies of the tributary streams. There will always be an energy loss of appreciable magnitude due to impact at the junction. It is conceivable, however, that in some cases, such as with a large stream at high velocity into which a small low-velocity stream is discharged, there may be a gain in energy along one path at the expense of a greater loss along the other path. In sewer design it is not safe to assume that an impact loss may be negative, for if the assumption be in error the upper stretch of sewer will be subjected to backwater and possible surcharge.
>
> Unfortunately, it is not practicable to apply the momentum theory to the actual design of junctions. In making the application, it is necessary to choose cross-sections just above and below the junction for which pressure and momentum computations are made. To the pressures thus computed must be added the pressure components in the direction of flow along the walls and floors of the channels. The pressure components on the walls and floor are a considerable part of the total pressure, but they cannot be accurately determined because their magnitudes are influenced by the impact of the streams and the curvature.

Since the publication of the above in 1952, empirical work has been completed (McNown 1954, Sangster *et al.* 1961, Blaisdell and Manson 1963, Pinkayan 1972) which allows estimates to be made of junction losses. Where junctions have high-velocity outflows, head losses may be conservatively determined by Equation 7.90. This assumes that the inflows are completely stilled in the junction and the flow re-accelerated in the outlet sewer. This will predict excessive losses in some cases, and, where this appears to be the case, careful analysis should be performed by means of the above-mentioned techniques. For rectangular channel junctions such as those found at wastewater treatment plants, Shabayek *et al.* (2002) and references given therein are useful.

Also relevant to this section are Pedersen and Mark (1990) who presented theoretical and experimental results for straight-through flow in manholes with fully submerged inlet and outlet pipes of equal diameter. And also, Law and Reynolds (1966); Ramamurthy *et al.* (1988); and Ramamurthy and Satish (1988); who collectively presented theoretical and experimental results for combining and dividing open channel flows at right-angled junctions.

7.6.3 Sewer Drops

Sewer drops in ordinary manholes can result in erosive velocities at the bottom of the drop and release of sewage gas, particularly odorous and corrosive hydrogen sulfide. The erosive velocities are sometimes dealt with by placing a granite block at the bottom of the drop. A particularly suitable way of dealing with the erosion problem while substantially diminishing the odor and corrosion potential is with the vortex insert assembly or drop shaft, details of which are given by Pisano *et al.* (1990), Switalski *et al.* (1992), Pisano and Brombach (1995), Banister *et al.* (1999), Natarius (2001), and Winkler and Guswa

(2002). Similar devices that trap stormwater debris are the swirl concentrator and the like (Field and Masters 1977, Helm *et al.* 1994, Wolf and Phillips 1994, Field *et al.* 1995) and the continuous deflection technology unit (WEF 1998) now provided by Contech Engineered Solutions LLC, West Chester, Ohio. Lager and Smith (1974) provide additional discussion of vortex regulators for combined sewer control.

7.7 Culverts and Related Outlet Structures

Culverts are used to convey the flow of a natural or man-made channel through an embankment, such as provided for a roadway. Culvert-like structures are also used as detention-basin outlet controls (see Section 5.4), and as "culvert spillways" for dams (USBR 1965, pp. 265–267, 326–334). Siphon spillways (USBR 1965, pp. 267–269, 334–335) and drop inlet (shaft or morning glory) spillways (USBR 1965, pp. 264–265, 311–326) in certain control modes are hydraulically related to culvert spillways. Culverts and such related structures can be quite complex hydraulically and embody many of the types of flows discussed in previous sections of this chapter and Chapter 6. Additional aspects of culverts in connection with a tidal pond network analysis are discussed in Section 15.3. Additional details regarding culverts are found in PCA (1962), FHwA (1965), USDA (1972) and Graber (2017a, §7.3.2). Of particular importance is the design or correction of culverts for aquatic organism passage (FHwA 2010, Singler *et al.* 2012, Sowby 2015).

But, before going further: In commenting on a published paper on culverts, Mavis *et al.* (1966) provided an interesting perspective on culverts and more which is worthy of quoting here: "Is the 'flow (of water) in culverts' really difficult to analyze? In climbing the stairs from the hydraulic laboratory ... to the minarets of 'philosophies' has not the author tripped (as many others have tripped) over the pebbles in an otherwise useful monolithic structure? If one becomes so concerned in the minutiae of science or philosophy that he loses perspective, he can become as frustrated as an anonymous author relates lightly in the following lines:

> A centipede was happy quite
> Until a frog in fun
> Said 'Pray, which foot comes after which?'
> This wrought her mind to such a pitch,
> She lay distracted in the ditch,
> Considering how to run.

The major types of flows occurring in culverts are depicted on Figure 7.10 (a larger number, including various subtypes and transitional states, are sometimes listed, e.g., USBR 1965). The flow type (and flow-head relation) is dependent on headwater depth, culvert slope, culvert length, entrance shape, shape of the conduit, conduit size, tailwater depth, and the roughness of the conduit material (PCA 1964). The predictive aspects of flow type and corresponding flow-head relations have been studied quite thoroughly, most notably by the U.S. Bureau of Standards and the Federal Highway Administration. For purposes of the design of culverts *per se*, the Bureau of Public Roads design charts (PCA 1964, Chapters 3–5; FHwA 2012) incorporate those predictive aspects in a practical manner suitable for many drainage design situations. We shall limit detailed discussion here to the hydraulic details of those culvert-type outlet structures suitable for detention-basin controls.

For ordinary culvert design, we are normally concerned with the hydraulic capacity of the culvert under maximum flow conditions. For detention-basin outlet control, however, as a practical matter, we will usually want a single flow-head ($Q-H$) relation throughout the effective storage range of the basin. For flow commencing as Types IV, V, or VI (which are essentially open-channel flows), the flow type will usually change to Type I, II, or III as H increases. This results in a change in the H-datum and $Q-H$ relation. The $Q-H$ relations for each of these flow types are fairly difficult to predict, as is the transition between flow types. The overall $Q-H$ relation is not only difficult to predict with accuracy, but the $Q-H$ relation is not of a type allowing the use of convenient design aids such as those of Section 5.4. To a lesser extent, these same drawbacks may be ascribed to Type III flow (as discussed below). For most detention basins with culvert-type outlets, it is convenient to design for Type I or Type II flow characteristics. Type II is usually the most convenient basis for the design of detention basins with culvert-type outlets, as discussed below, and requires that effective storage be considered as that for which the headwater depth HW (equal to the depth above the upstream invert – see Figure 7.10) exceeds approximately $1.5d_o$ (see below). Type I may occur as a result of an artificial or natural increase in tailwater and may be a useful basis for detention-basin design in some cases (also discussed below). Occasionally a design for Type II conditions must be checked for Type I conditions.

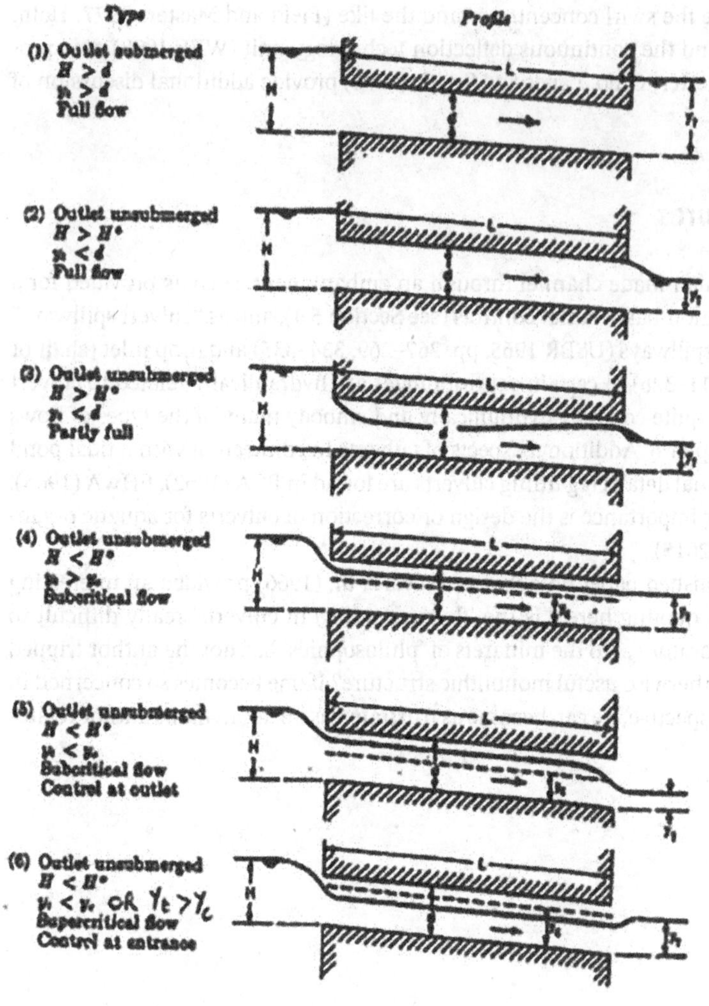

Figure 7.10 Culvert Flow Types (Chow 1959 / McGraw-Hill Education)

It should be understood before proceeding to the further discussion of Type I and II flows in detention-basin design that the complexity of the other flow types should not preclude their use for major structures or other special circumstances for which the hydraulic complexities (including flow routing studies) may justifiably be considered. Types I and II are emphasized here because they are suitable and convenient for the large majority of detention basins with culvert-type outlets which are likely to be encountered. An understanding of Type III conditions is necessary to understand when Type II conditions will occur.

Since the progression with increasing flow is often from one of the Types IV, V, or VI to one of the Types I, II, or III, we shall discuss these two groupings in that order. In keeping with the above, the discussion of Types IV, V, and VI will be qualitative in nature, while Types I, II, and III will be discussed in quantitative detail.

7.7.1 Flow Types IV, V, and VI

When flow first commences through a culvert, the flow is usually open-channel along the full length of the culvert (exceptions may occur if the culvert has an adverse slope or a downstream control submerges the culvert outlet). For low tailwater ($y_t < y_c$), the resulting flow profile will then depend on whether the culvert slope is subcritical or supercritical. For a subcritical slope, the flow (Type V) will be essentially as discussed in Section 7.4, with a critical flow control immediately upstream of the outlet and backwater profile in the culvert. The headwater will be above the depth immediately inside the pipe entrance by an amount depending on the entrance loss. For a supercritical culvert slope, the control will be at the entrance (Type VI) where the depth will be close to the critical depth for the given flow and channel section. The headwater will again be above this depth by an amount depending on the entrance loss. For $y_c < y_t < d_o$ and HW sufficiently small to prevent sealing of the entrance, the flow will be Type IV. Such flow may be subcritical throughout or may be subcritical at the downstream end and supercritical upstream if the culvert slope and length are sufficiently large. If detailed analyses of flows of the above types are required, they can be carried out by the methods of Sections 7.4 and 7.5.

For Type V and subcritical Type VI, additional variations are possible as the flow increases, which may be regarded as transitional to Type II or Type III. For all of the types just discussed (IV, V, and VI), as the flow increases the headwater will increase leading eventually (for y_t remaining less than d_o) to a transition to Type II or Type III conditions. The nature of these transitions will be elaborated upon below.

7.7.2 Flow Types I, II, and III

Flow Types I, II, and III all require that the inlet be sealed. The transition from the initial impingement of the flow on the upstream crown to a condition of full seal is complex and may entail vortexing and other phenomena. For Types II and III, sealing of inlets of most circular types occurs when the headwater depth HW exceeds about 1.2 to $1.5d_o$, d_o being the pipe internal diameter. We will take $HW > 1.5d_o$ as a conservative criterion for our purposes. For $HW > 1.5d_o$, the distinction between Types I, II, and III depends on tailwater conditions and/or the length of the pipe. When the tailwater depth y_t (see Figure 7.10) is less than the pipe diameter, the flow will be of Type II or Type III, depending on whether the culvert is "hydraulically long" or "hydraulically short", as discussed below. Under certain circumstances, the progression with increasing flow is from Type III to Type II to Type I, and it is convenient to discuss the three types in that order.

7.7.2.1 Type III

Consider the short culvert with $y_t < d_o$, as depicted on Figure 7.10. With $HW > 1.5d_o$, the flow entering the pipe will separate from the pipe periphery forming a contracted free surface and water eddies below the free surface. Studies by Bossy (PCA 1964) and others have shown that the contracted depth is well below critical depth and that a good approximation of the depth within a distance of $5d_o$ of the entrance may be obtained from the one-dimensional energy equation (see Chapter 2). Denoting by subscript "e" conditions at $5d_o$, we have:

$$HW + z_i \cong H_e' + z_e + K_e \frac{V_o^2}{2g} \tag{7.91}$$

in which H_e' denotes the supercritical specific energy $\left[y_e + V_e^2/(2g)\right]$ at $5d_o$ and K_e is the entrance loss coefficient applied to the full-pipe velocity head. Since $z_i - z_e = 5d_oS_o$ where S_o is culvert slope, we have:

$$\frac{H_e'}{d_o} = \frac{HW}{d_o} + 5S_o - K_e \frac{V_o^2}{2gd_o} \tag{7.92}$$

The depth and other conditions corresponding to H_e' can be readily established by trial-and-error using $H_e' = \left[y_e + Q_e^2/(2gA_e^2)\right]$ and hydraulic elements.

Since the flow is supercritical downstream of the inlet, tailwater conditions cannot be communicated to the headwater, and hydraulic control in Type III flow is at the inlet. The culvert inlet then acts as an orifice, and the headwater depth of a culvert with Type III inlet control is given by:

$$\frac{HW}{d_o} = -0.5S_o + \frac{h_1}{d_o} + k_1 \left(\frac{Q}{d_o^{5/2}}\right)^2 \tag{7.93}$$

for values of $Q/d_o^{5/2}$ exceeding an experimentally determined minimum given in Table 7.3 (from PCA 1964, Tables 8 and 9). The factors h_1/d_o and k_1 are also empirical and are given in Table 7.3 for various entrance shapes.

Downstream of the free-surface contraction depicted on Figure 7.10, the depth will increase toward the critical and/or uniform-flow depths as discussed below. If the culvert length is too short to allow expansion of the free surface to the crown of the pipe, the culvert is said to be "hydraulically short" and the control will remain at the inlet (Type III flow). Otherwise, Type II flow may occur as discussed below. Type III examples are given below; those examples will be built upon in culvert examples given below.

Table 7.3 Inlet Control Performance Coefficients and Entrance Loss Coefficients

	Submerged Inlet Flow			Nonsubmerged Inlet Flow			Entrance Head Loss Coefficient
Entrance Shape	h_1/D	k_1	$Q/D^{5/2*}$	k	m	H_e/D^{**}	K_e
With headwall							
Groove edge, .05*D*x.07*D*	0.74	0.0468	3.3	0.0018	2.5	0.035	0.19
Rounded edge, .15*D* radius	0.74	0.0419	2.58	0.00065	2.67	0.016	0.15
Square edge	0.67	0.0645	2.58	0.0098	2.0	0.105	0.43
Headwall and 45° wingwalls							
Groove edge, .05*D*x.07*D*	0.73	0.0472	3.0	0.0018	2.50	0.035	0.20
Square edge	0.70	0.0594	3.5	0.0030	2.67	0.072	0.35
Headwall and parallel wingwalls							
Groove edge, .05*D*x.07*D*	0.74	0.0528	4.0	0.0020	2.67	0.048	0.30
Miter (square edge)							
2:1 embankment slope	0.74	0.0750	4.0	0.0210	1.33	0.091	0.62
Projecting entrance							
Groove edge, .05*D*x.07*D*	0.70	0.0514	2.58	0.0045	2.0	0.049	0.25
Square edge (thick wall)	0.64	0.0668	3.5	0.0145	1.75	0.116	0.46
Thin edge	0.53	0.0924	4.0	0.0420	1.33	0.205	0.92

* The equation for inlet control with submerged inlet only applies when $Q/D^{5/2}$ is larger than the listed values.
** The equation for inlet control with nonsubmerged inlet only applies when $Q/D^{5/2}$ is less than the listed values.
Source: Portland Cement Association (PCA) 1964 / Portland Cement Association.

7.7.2.2 Type II

In order to obtain a Type II culvert flow, two conditions (in addition to a submerged inlet) must be satisfied. First, the flow must exceed the maximum free-surface flow capacity of the pipe. Referring to Section 5.3 and Figure 5.2 (hydraulic elements), based on constant n for conservatism, this will be the case for $AR^{2/3}/d_o^{8/3}$ greater than its value at $y/d_o = 0.94$, or (in English units):

$$\frac{nQ}{1.486 d_o^{8/3} S_o^{1/2}} = \frac{AR^{2/3}}{d_o^{8/3}} > 0.3353 \tag{7.94}$$

This requirement may be expressed in the following rearranged form:

$$Q > 0.4983 \frac{d_o^{8/3} S_o^{1/2}}{n} \tag{7.95}$$

If the above inequality is satisfied, the normal depth may be considered to be at or above the crown of the pipe. The depth of flow downstream of the contracted free surface then must tend toward the crown of the pipe as depicted on Figure 7.10. Noting that y_c must be less than d_o [see, e.g., Chow's (1959) Appendix A tabulation], the flow, initially supercritical, must pass through critical depth en route to depth d_o. The change from supercritical to subcritical flow must occur via a hydraulic jump across the critical depth line. The flow profile upstream and downstream of the jump will be gradually varied. In addition to the flow (or normal depth) condition stated above, the second condition for Type II flow is that the culvert be long enough ("hydraulically long") for the flow depth to reach the culvert crown. When the depth of flow reaches the crown, the air in the culvert above the free surface is isolated and entrained by the high-velocity flow. The air pocket is removed fairly quickly (the culvert is said to "prime" itself) and replaced by a water eddy, submerging the inlet control. The pipe will now flow full for a distance at the upstream end, and the free surface will break away from the pipe crown and decrease to approximately y_c at the downstream end. With $HW > 1.5d_o$, this will generally be a steady-state condition without vortexing or pulsation.

The length of pipe required for Type II flow conditions, other requirements being satisfied, is called the "culvert control length". A culvert longer than the culvert control length is said to be "hydraulically long". An approximate, conservative determination of the culvert control length may be made by application of the direct-step method of Keifer and Chu discussed above. [The overall method used here is a variation on a procedure given by PCA (1964). The accuracy of the PCA method is limited by the accuracy with which the friction term can be estimated when the one-dimensional energy equation is applied in one step from y_e to d_o.]

The input values for the Keifer–Chu method are obtained by first solving Equation 7.93 for the headwater depth ratio HW/d_o for the culvert entrance (h_1/d_o and k_1 in that equation are obtained from Table 7.3). Knowing the value of HW/d_o, Equation 7.92 can be solved for the supercritical specific energy at section "e". For that specific energy H'_e/d_o, the value of y_e/d_o may be obtained. Note that y_e/d_o is a function only of S_o and $Q_f/d_o^{5/2}$ for a particular culvert entrance. The value of Q_f is obtained from Equation 6.3. Equation 7.61 is then used to calculate the control length or length for which a culvert satisfying the other necessary conditions [$HW/d_o > 1.5$ and $Q > (K_n/n)AR_f^{2/3}S_o^{1/2}$] will flow fully.

Recalling the functional dependencies of y_e/d_o, y_c/d_o, and Q_o noted above, we find that the control length L is a function of d_o, S_o, Q, and n. [Dimensional analysis may be employed to show that a minimum of four dimensionless variables are involved in the relations for culvert control length. The three-variable plot of PCA (1964) is functionally incorrect. Calculations with the relations used herein also show it to be an inadequate approximation in some cases.] The culvert control length will generally decrease with increasing Q and n and decreasing S_o and d_o. Note that the discharge depth must be less than d_o, although it can be very close to d_o if the flow is large enough to make y_e/d_o close to 1. Examples of the use of the above are given below.

We shall now discuss the head-discharge relation for Type II flow. Two cases are of interest here, the first being atmospheric discharge and the second "channel" or "supported" discharge (examples of each are shown on Figure 7.10.) The largest headwater (for $y_t < d_o$) results when $y_c < d_o$ is close to 1, in which case the pipe flows full through almost all of its length. Application of the Bernoulli equation (Section 2.6) between the headwater (assumed to have negligible velocity head) and discharge plane then gives the following relationship:

$$HW + LS_o - \left(\frac{d_o}{2} + \frac{p_2}{\rho g}\right) = \left(K_e + 1 + \frac{fL}{d_o}\right)\frac{V_o^2}{2g} \tag{7.96}$$

in which p_2 is the pressure at the discharge centerline. For atmospheric discharge, we can take $p_2 \cong 0$, giving:

$$HW + LS_o - \frac{d_o}{2} = \left(K_e + 1 + \frac{fL}{d_o}\right)\frac{V_o^2}{2g} \tag{7.97}$$

For channel discharge, the pressure at the discharge centerline is greater than atmospheric due to the pressure distribution at the discharge plane. This pressure distribution differs from hydrostatic due to the curvature of the streamlines as the water surface drops abruptly to the downstream channel water surface. Values of the term $h_2 = d_o/2 + p_2/(\rho g)$ are dependent on the pipe slope, material, and discharge factor $Q/d_o^{5/2}$. [The value of h_2 falls approximately halfway between y_c and the crown of the conduit; i.e., $h_2 \cong (y_c + d_o)/2$ (PCA 1964, p. 244)]. The maximum values of h_2 are generally less than d_o (the pressure p_2 is less than hydrostatic), and h_2 can conservatively be taken as d_o for many applications (such as detention basins).

The further below 1 that y_c/d_o is, the more conservative an estimate of HW the above approach gives. For purposes of culvert design, a conservative estimate of HW results in a conservative overall design. For detention-basin design, however, a conservative estimate of HW, while contributing to a conservative estimate of required freeboard, may have the effect of overestimating the flow attenuation. The effect is usually overcome, however, by the other conservatisms inherent in the design process (most notably that resulting from the finite increments in available pipe size).

The following are not all the cases in which the author has used the Keifer–Chu method, but are presented because the values were outside the range of tabulated values of Keifer–Chu given in tables cited above. The author thus used the Romberg integration method mentioned above to determine Φ and ψ values.

Example 7.20
A concrete culvert of 45-ft length and 1-ft diameter is to serve as the outlet for a detention pond (Rockport, Massachusetts). Design flow is 4.043 cfs. The slope and estimated Manning's n are 0.0028 and 0.015 respectively. Determine the Type III flow capacity with the headwall having 45° wingwalls and groove edge. Then calculate the culvert control length to enable determination of the type of culvert flow.

Solution
From Table 7.3, we have $h_1/D = 0.73$, $k_1 = 0.0472$, and $K_e = 0.2$. From Equation 7.93:

$$k_1\left(Q/d_o^{5/2}\right)^2 = HW/d_o + 0.5S_o - h_1/d_o = 1.5 + 0.5(0.0028) - 0.73 = 0.7714$$

from which $Q/d_o^{5/2} = \sqrt{0.7714/0.0472} = 4.043$ and $Q = 4.043(1)^{5/2} = 4.043$ cfs. Using Manning's equation (an exercise left to the reader), the flowing-full capacity is $Q_o = 1.634$ cfs(using the notation for use in Keifer–Chu) giving $Q/Q_o = 4.043/1.634 = 2.474$ as the ratio of actual flow to culvert capacity. The critical flow section factor is given by $Z = Q/\sqrt{g} = 4.043/\sqrt{32.2} = 0.7125\ \text{ft}^{5/2}$. Thus $Z/d_o^{2.5} = 0.712/(1)^{2.5} = 0.712$. From a table or figure of geometric elements, we then obtain $y_c/d_o = 0.8512$. The full entrance velocity equals $4.043/\left[(\pi/4)(1)^2\right] = 5.148$ ft/sec. Equation 7.92 then gives the specific energy $H_e'/d_o = 1.5 + 5(0.0028) - 0.2(5.148)^2/[2(32.2)(1)] = 1.432$. The equation $H_e' = \left[y_e + Q_e^2/(2gA_e^2)\right]$ given above is solved for y_e by trial-and-error and geometric elements with $H_e' = 1.432(1) = 1.432$ ft and $Q_e = 4.043$ cfs giving $y_e = 0.70216$ ft and $y_e/d_o = 0.70216$.

The Romberg integration method is used to determine Φ and ψ values thus:

$$\Phi = (1, Q/Q_o) = 0.021422$$

$$\psi = (1, Q/Q_o) = 0.005313$$

and similarly for y_c/d_o:

$$\Phi = (y_c/d_o, Q/Q_o) = 0.013128$$

$$\psi = (y_c/d_o, Q/Q_o) = 0.004991$$

and for $y_e/d_o = 0.70216$:

$$\Phi = (y_e/d_o, Q/Q_o) = 0.006151$$

$$\psi = (y_e/d_o, Q/Q_o) = 0.004353$$

The culvert control length is calculated using Equation 7.60 as follows:

$$\frac{L}{d_o} = 5 + \frac{1}{0.0028}[0.021422 - 2^*0.013128 + 0.006151]$$
$$- \frac{1}{0.0028}\left[\frac{4.043}{(1)^{5/2}}\right][0.005313 - 2^*0.004991 + 0.004353] = 5.93$$

The culvert length of 45 ft is greater than the calculated control length of 5.93(1) = 5.93 ft. This information enabled the culvert head-flow characteristics to be calculated at various flows. In this case the culvert flow type transitioned with increasing flow from Type V to Type V with upstream full to Type II.

Example 7.21
A concrete pipe of 8-ft length and 0.5-ft diameter is to serve as the outlet for a detention pond (Jean–Cor, Topsfield, Massachusetts). Design flow is 0.715 cfs. The slope and estimated Manning's n are 0.0025 and 0.013 respectively. Determine the Type III flow capacity with the headwall having 45° wingwalls and groove edge. Then calculate the culvert control length to enable determination of the type of culvert flow.

Solution

From Table 7.3, we have $h_1/D = 0.73$, $k_1 = 0.0472$, and $K_e = 0.2$. From Equation 7.93:

$$k_1\left(Q/d_o^{5/2}\right)^2 = HW/d_o + 0.5S_o - h_1/d_o = 1.5 + 0.5(0.0025) - 0.73 = 0.7713$$

from which $Q/d_o^{5/2} = \sqrt{0.7713/0.0472} = 4.042$ and $Q = 4.042(0.5)^{5/2} = 0.715$ cfs. Using Manning's equation (an exercise left to the reader), the flowing-full capacity is $Q_o = 0.2806$ cfs (using the notation for use in Keifer–Chu) giving $Q/Q_o = 0.715/0.2806 = 2.548$ as the ratio of actual flow to culvert capacity. The critical flow section factor is given by $Z = Q/\sqrt{g} = 0.715/\sqrt{32.2} = 0.1260\ \text{ft}^{5/2}$. Thus $Z/d_o^{2.5} = 0.1260/(0.5)^{2.5} = 0.71278$. From a table or figure of geometric elements, we then obtain $y_c/d_o = 0.8516$. The full entrance velocity equals $0.715/\left[(\pi/4)(0.5)^2\right] = 3.641$ ft/sec. Equation 7.92 then gives the specific energy $H_e'/d_o = 1.5 + 5(0.0025) - 0.2(3.641)^2/[2(32.2)(0.5)] = 1.430$. The equation $H_e' = \left[y_e + Q_e^2/(2gA_e^2)\right]$ given above is solved for y_e by trial-and-error and geometric elements with $H_e' = 1.430(0.5) = 0.715$ ft and $Q_e = 0.715$ cfs giving $y_e = 0.35108$ ft and $y_e/d_o = 0.70216$.

The Q/Q_o of 2.548 is again outside the tabulated values of the Keifer–Chu functions given in tables cited above, so the Romberg integration method mentioned above is used to determine Φ and ψ values thus:

$$\Phi = (1, Q/Q_o) = 0.072957$$
$$\psi = (1, Q/Q_o) = 0.016978$$

Proceeding as in the previous example, $Z/d_o^{2.5} = 0.71278$ and $y_c/d_o = 0.85174$, giving:

$$\Phi = (y_c/d_o, Q/Q_o) = 0.043854$$
$$\psi = (y_c/d_o, Q/Q_o) = 0.015866$$

and similarly for $y_e/d_o = 0.70216$:

$$\Phi = (y_e/d_o, Q/Q_o) = 0.019868$$
$$\psi = (y_e/d_o, Q/Q_o) = 0.013667$$

The culvert control length is calculated using Equation 7.60 as follows:

$$\frac{L}{d_o} = 5 + \frac{1}{0.0025}[0.072957 - 2^*0.043854 + 0.019868]$$
$$- \frac{1}{0.0025}\left[\frac{0.715}{(0.5)^{5/2}}\right][0.016978 - 2^*0.015866 + 0.013667] = 8.81$$

The culvert length of 8 ft is greater than the calculated control length of 8.81(0.5) = 4.41 ft. This information enabled the culvert head-flow characteristics to be calculated at various flows. In this case the culvert flow type transitioned with increasing flow from Type V to Type V with upstream full to Type II.

7.7.2.3 Type I

Flow Type I requires $y_t > d_o$ and $HW > d_o$. If $y_t > d_o$ but the hydraulic grade line intersects the crown of the culvert, then free surface flow will occur in the upstream portion of the culvert.

Applying Bernoulli's equation between the headwater and tailwater (assumed to have negligible velocity head) gives:

$$HW + LS_o - h = (K_e + K_o + fL/d_o)\frac{V_o^2}{2g} \tag{7.98}$$

in which h is the tailwater ($h > d_o$) measured from the downstream invert. The tailwater h is generally a function of flow, and this dependency must be incorporated in the flow-head relation.

In the above example, the outlet pipe was found to be hydraulically long and satisfy the other requirements of Type II culvert flow. This will not always be the case, and adjustments may be necessary. Among the more practical of such adjustments are: (1) increasing the minimum head (and flow) at which effective storage (Type II flow) is considered to begin and making any corresponding changes required in basin size or shape, and (2) employing multiple outlet pipes.

7.8 Outfall Structures and Scour Protection

Outfalls from wastewater treatment plants and storm sewers discharge to receiving waters directly or indirectly via lateral channels (defined below). In the case of stormwater outfalls, the usual design considerations are: (1) scour protection and (2) aesthetics. For wastewater outfalls, additional considerations include (3) need for postaeration, (4) initial dilution requirements, and (5) zones of passage. Initial dilution requirements may establish the need for a submerged diffuser type of outfall, which may be designed by principles given in Chapter 11 – Distribution Conduits. In that case, zones of passage must be addressed (USEPA 2014). The present section will be limited to those types of outfalls which terminate at the edge of the receiving body of water, for which the first three of the above considerations pertain. Postaeration will be addressed in Section 14.7, and the reader should be aware of the possibility of providing postaeration at the outfall structures. The focus of the present section will be on scour protection and related aesthetic considerations.

Most of the discussion of outfalls here pertains to those discharging at the side bank of a stream, perpendicular to the streamflow or nearly so. In such cases, the outfall tailwater is dependent upon the streamflow, and a wide range of tailwater conditions can be expected. In some cases, the outfall will discharge to the head of a natural or man-made channel which flows (usually a short distance) into a larger stream or river. Such channels will be referred to as *lateral channels*. Outfalls to lateral channels are very similar to storm drain *outlets* (culverts or otherwise) to drainage channels; subsequent discussion of outfalls to lateral channels is equally applicable to drain outlets. In lateral channels, the outfall tailwater is determined at least in part by the outfall discharge, and a known minimum tailwater generally exists.

Depending on the magnitude of the discharge velocity, the configuration of the outfall, and the nature of the river or channel bottom, measures may be necessary to prevent excessive scour. Outfalls from wastewater treatment plants and storm sewer systems must occasionally be laid on steep slopes due to topographic and economic considerations, and this creates a particularly high scour potential. The purposes of scour protection are prevention of the undermining of the outfall and destabilization of the receiving channel. Destabilization can be associated with functional and aesthetic damage to the channel and increased downstream sediment load. The visual appearance of the outfall may also be important in a broader aesthetic context.

In this section, we will discuss outfall scour protection with both subcritical and supercritical outfall approach flows, and scour protection by riprapping. The reader is referred to Section 13.2 for a discussion of the mechanism of scour.

7.8.1 Outfall Structures

The culvert-type end wall is the most common type of outfall structure (PCA 1964, Chapter 6). The end wall protects the embankment and outfall conduit from the scouring action of eddies created adjacent to the issuing jet. Angled end walls, called "wing walls," concentrate the eddies away from the adjacent streambank.

A primary purpose of scour protection is to prevent undermining of the outfall pipe or structure. If topography and other considerations are such that the outlet may be placed well above the receiving water surface, then the flow will strike the water surface some distance out from the channel bank. A roller will form downstream from the outfall, which may tend to move scoured material toward the bank and prevent serious scour at the base of the outfall. This is akin to the principle of the spillway bucket employed in conjunction with dam spillways (USBR 1965, §'s 200, 201). Such an arrangement may not be aesthetically desirable due to the visibility of the outfall and discharge, and any scour pool created and visible at low water levels may be unsightly. When the water level rises and submerges the outfall, the submerged jet may deflect toward the channel bottom (the Coanda effect discussed in Section 15.2) and increase the scour potential.

Topography, aesthetics, and other considerations often dictate bringing the outfall invert close to the bottom of the receiving channel. The first step in the determination of scour protection requirements in this case is a comparison of the outfall discharge velocity with the erosive velocity of the channel bottom material (see, e.g., PCA 1964, Tables 10 and 11; and Peterka 1964, Section 11). In the case of direct discharge to a natural stream bed, the latter will be indirectly determined by the hydrologic, hydraulic, and geologic characteristics of the stream.

If the outfall discharge velocity is less than the erosive velocity of the channel bottom, then little or no scour protection is required other than possibly than that required to protect the outfall from the velocity of the stream into which it discharges in the case of a transverse discharge. When the outfall discharge velocity exceeds the erosive velocity of the channel bottom, two measures may be taken individually or in combination to prevent scour: (1) decrease the outfall discharge velocity, (2) add a protective lining to the channel bottom. These measures will be discussed below in the contexts of various types of velocity-reducing outfall structures.

7.8.1.1 Subcritical Approach Flow

When the approach flow is subcritical, a transition may be readily designed which will reduce the velocity to the desired value prior to discharge to the river. Under the common conditions of low river level, this may be accomplished by ending the transition with a critical depth control ($\mathbf{F} = V_c/\sqrt{gD_c} = 1$) having $D_c < V_d^2/g$ in which V_d is the maximum desired discharge velocity. For a rectangular control ($D_c = y_c$), for example, this requirement can be translated to one for the control width, as follows:

$$b^2 = \frac{Q^2}{gy_c^3} > \frac{Q^2}{gV_c^6/g^3} = \frac{g^2Q^2}{V_c^6} \Rightarrow b > \frac{gQ}{V_c^3} \tag{7.99}$$

in which Q is the maximum design flow and V_c is the maximum desired discharge velocity.

The width of the terminal section of the culvert-type outlet may be proportioned using the above relationship. Downstream of the terminal control, the flow will spread and the velocity will decrease from that at the control. The transition from the approach sewer to the terminal control section must adequately spread the flow over the terminal section and do so in a reasonable length. The discussion of subcritical-flow transitions in Section 7.6 is helpful in this regard; the baffled outlet structure of Smith and Yu (1966) seems particularly suitable.

The above approach is used most commonly for discharges to lateral ditches (or drainage ditches). The velocity to which the flow may be reduced is often limited by the boundaries of the channel.

7.8.1.2 Bureau of Reclamation Basin IX

Basin IX, depicted on Figure 7.11 (Peterka 1964), consists of a baffled apron which maintains moderate velocities as the flow drops to the lower elevation. The basin requires no tailwater to be effective and performs well with high tailwater. The multiple rows of baffled piers on the apron (also called the "chute") prevent excessive acceleration of the flow and provide a reasonable terminal velocity regardless of the height of drop (USBR 1965).

The chute is constructed on an excavated slope 2:1 or flatter extending to below the channel bottom. Backfill is placed over one or more rows of baffles to restore the original streambed elevation. When scour occurs, successive rows of baffle piers are

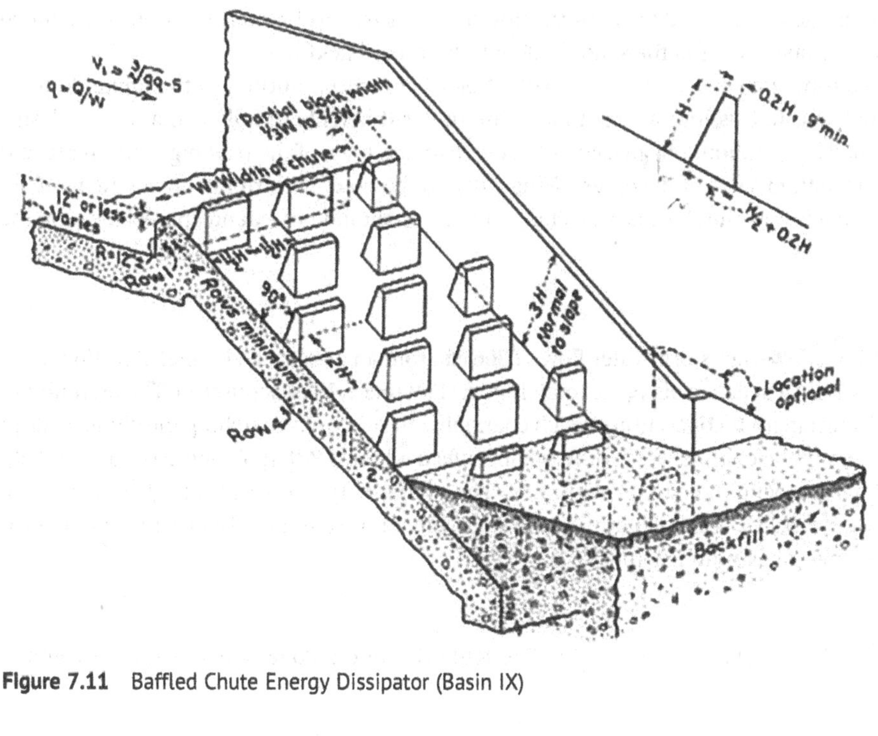

Figure 7.11 Baffled Chute Energy Dissipator (Basin IX)

exposed to prevent excessive acceleration of the flow entering the channel. In effect, a stable scour pool is created (equilibrium scour conditions are discussed below).

Design criteria have been established by field experience and hydraulic model tests. A summary of these criteria based on Peterka (1964, Section 9) and the author's modifications follows (see Figure 7.11 for dimensional notation):

1) The approach channel must have subcritical flow and should distribute the flow laterally across the width of the chute. [The chute entrance conditions that have been suggested elsewhere (USBR 1965) are not attainable in the author's opinion. The best that can be done is to provide a uniformly distributed subcritical approach flow, in which the geometry at the top of the chute (including baffles) will control and determine the entrance velocity. A supercritical approach flow would result in a higher velocity (the top of the chute would not necessarily control), and should be avoided; a short energy dissipating pool upstream of the chute (USBR 1965) may be provided in such cases. The energy dissipating pool should create a transition from supercritical to subcritical conditions via a hydraulic jump. A vertical offset between the approach channel floor and the chute may be used to create the hydraulic jump (see USBR 1965). The offset may be reticulated to reduce sediment accumulation.]
2) The chute slope should be in the range 2:1 (steep) to 4:1 (flat).
3) The chute width W is selected to provide a unit discharge $q = Q/W$ of less than 60 cfs/ft based on the design flow Q. The 60 cfs/ft is the maximum value, and a value on the order of 20 cfs/ft is usually preferable from the standpoints of performance (particularly downstream channel protection) and overall economics (USBR 1965).
4) The baffle pier height, H, should be about $0.8D_c$ where D_c is the critical depth on the rectangular chute given by $D_c = \sqrt[3]{q^2/g}$. The height is not critical and may be increased to about $0.9D_c$ if desired.
5) Baffle pier widths and spaces should be equal, preferably about $(3/2)H$, but not less than H. Other baffle pier dimensions are not critical. Partial blocks, width $(1/3)H$ to $(2/3)H$, should be placed against the training walls in Rows 1, 3, 5, 7, etc., alternating with spaces of the same width in Rows 2, 4, 6, etc.
6) For a 2:1 chute slope, the slope distance between rows of baffle piers should be $2H$, twice the baffle height H. When the baffle height is less than 3 ft, the row spacing may be greater than $2H$ but should not exceed 6 ft. For slopes flatter than 2:1, the row spacing may be increased to provide the same vertical differential between rows as expressed by the spacing for a 2:1 slope.
7) The baffle piers are usually constructed with their upstream faces normal to the chute surface; however, piers with vertical faces may be used. Vertical face piers tend to produce more splash and less bed scour, but differences are not significant.
8) Four rows of baffle piers are required to establish full control of the flow, although fewer rows have operated successfully. Additional rows beyond the fourth maintain the control established upstream, and as many rows may be constructed as is necessary. The chute should be extended to below the normal downstream channel elevation as explained in the text of this section, and at least one row of baffles should be buried in the backfill.
9) The chute training walls should be three times as high as the baffled piers (measured normal to the chute floor) to contain the main flow of water and splash. It is impractical to increase the wall heights to fully contain the splash.
10) Riprap consisting of 6-inch to 12-inch stones should be placed at the downstream ends of the training walls to prevent eddies from eroding the bank at the bottom sides of the chute. Wingwalls can be used, with riprap downstream of the wingwalls (and upstream as well if that is how the slope is protected). The riprap should not extend appreciably into the flow area.

Example 7.22

A type IX Basin is being considered to lower a design stormwater flow of 306 cfs from an approach channel elevation of El. 13.30 to a river bed elevation of El. 7.00 within a distance of approximately 40 ft (Lawrence, Massachusetts). The preliminary layout calls for a 17-ft chute width and a rectangular concrete approach channel with side walls matching the chute walls in alignment and 1-ft wide training walls along the channel centerline (giving effectively two 8-ft approach channels) to help maintain a good flow distribution following a bend in channel alignment. The preliminary bottom profile gives a tentative apron slope of 21.375/(12.22–7.00) = 21.375/5.22 = 4.1:1. Determine the suitability of the tentative layout, make modifications as necessary, and proportion remaining basin dimensions.

Solution

The tentative nominal unit flow is $q = Q/W = 306/17 = 18$ cfs/ft. The tentative apron slope is flatter than 2:1 and is acceptable.

A subcritical approach flow will be provided if the slope of the approach channel is subcritical. Whether that slope is subcritical can be ascertained by calculating the Froude Number corresponding to uniform flow conditions. From Section 7.3, we have:

$$\frac{AR^{2/3}}{b^{8/3}} = \frac{nQ}{1.486 b^{8/3} S^{1/2}}$$

$$Q = 306/2 = 153 \text{ cfs}$$

$$b = 8 \text{ ft}$$

$$b^{8/3} = (8)^{8/3} = 256$$

$$S = (13.30 - 12.22)/19.0 = 1.08/19.0 = 0.0568$$

$$\frac{AR^{2/3}}{b^{8/3}} = \frac{0.013(153)}{1.486(256)0.238} = 0.0220$$

$$y_n/b = 0.11$$

$$y_n = 0.11(8) = 0.88 \text{ ft}$$

$$V_n = Q/(y_n) = Q/(y_n b) = 153/[0.88(8)] = 21.7 \text{ ft/sec}$$

$$\mathbf{F}_n = V_n/\sqrt{g y_n} = 21.7/\sqrt{32.2(0.88)} = 4.1$$

The uniform-flow velocity is clearly excessive, and the slope is in fact supercritical ($\mathbf{F}_n > 1$).

In this case, either a stilling basin must be employed to force a jump to subcritical approach conditions or the approach channel slope must be decreased to subcritical. The latter approach was selected in this case, and a change was made in the preliminary layout to a horizontal approach channel (with no vertical offset). The new chute slope is 21.375/(13.30–7.00) = 21.375/6.30 = 3.39, which is acceptable.

The critical depth on the rectangular chute is given by $D_c = \sqrt[3]{q^2/g} = \sqrt[3]{(18)^2/32.2} = 2.16$ ft. A 2 ft baffle height is selected and gives $H/D_c = 2/2.16 = 0.925$, which is acceptable. Good baffle pier widths and spacings are $(3/2)H = 3$ ft, with partial blocks as suggested by Item 5 above. The figure below aids in the determination of block spacing. The triangle on the right represents the 2:1 slope condition (see Item 6 above), in which "a" is the vertical differential between rows spaced $2H$ apart. The left triangle represents the actual slope (3.39:1) for which the spacing x must be determined for the same vertical differential "a." The unknown x is solved from the trigonometric relationships as follows:

$$(2H)^2 = a^2 + (2a)^2 = 5a^2$$

$$x^2 = (3.39a)^2 + a^2 = \left[(3.39a)^2 + 1\right](2H)^2/5 = 12.5[2(2)]^2/5 = 40.0$$

$x = 6.32$ ft, say 6 ft-4-in.

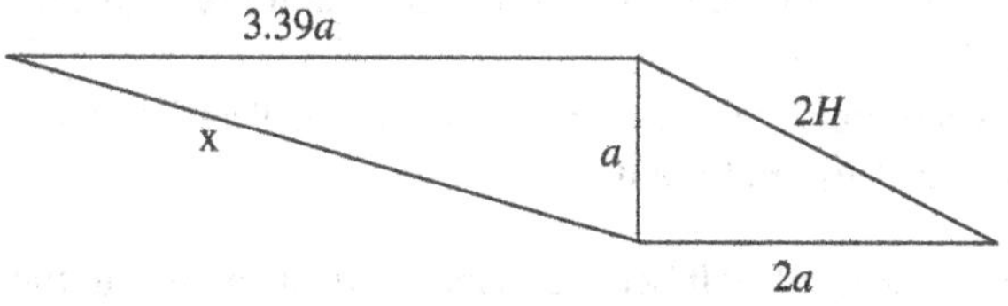

The exposed slope length is $\sqrt{(21.375)^2 + (6.30)^2} = 22.3$ ft, along which four exposed rows of baffles can be placed starting at the top of the chute. Two rows of baffles will be buried at the bottom of the chute.

The retaining walls at the sides of the chute (chute training walls) will be $3H = 3(2) = 6$ ft ft along their entire length (in this case the fill sloped downward from the top of the walls to the adjacent stream bank, which was easily accomplished in an attractive manner). Wingwalls and riprap were placed in accordance with Item 10 above.

7.8.1.3 Bureau of Reclamation Basin VI

Basin VI, depicted on Figure 7.12 (Peterka 1964), is an impact-type energy dissipator contained in a relatively small box-like structure. It is used primarily with pipe outlets, but may also be used with an open channel entrance. No tailwater is required for successful performance.

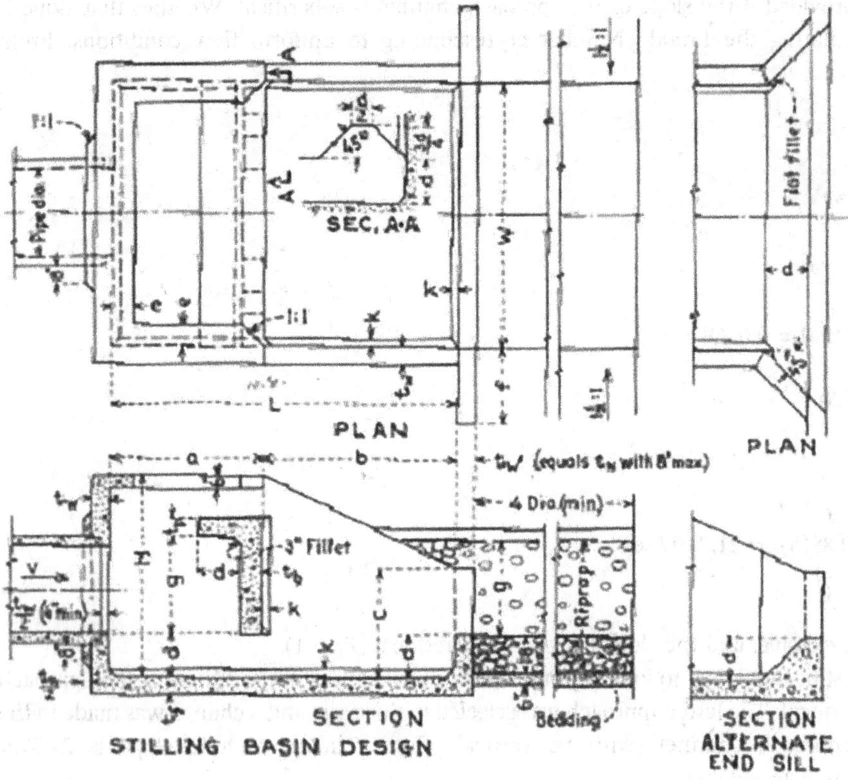

Figure 7.12 Impact-type Energy Dissipator (Basin VI)

Energy dissipation occurs when the flow strikes the vertical hanging baffle and is turned upstream by the horizontal portion of the baffle and the floor, creating energy-dissipating eddying upstream of the baffle. The energy loss is considered to be greater than would occur in a hydraulic jump with the same inlet Froude Number. The notches in the baffle are provided to aid in cleaning sediment out of the structure; sediment deposition could occur, for example, as a result of very low stormwater flows discharging into the basin. The notches provide concentrated jets of water to clean the basin. If such cleaning is not considered necessary, the notches may be omitted. The basin is designed to pass the full discharge over the top of the baffle should the space beneath the baffle become clogged. Performance is not as good under such conditions but is regarded as acceptable. High tailwater may also cause some flow to pass over the top of the baffle and may be unavoidable in many applications. [Smith and Korolischuk (1973) proposed a modified design of the impact basin, which they found gave improved performance (most particularly reduced downstream scour potential) for cases where definite tailwater conditions are known to exist.]

Design criteria have been established by hydraulic model tests. A summary of these criteria based on Peterka (1964, Section 6) and the author's modifications follows (see Figure 7.12 for dimensional notation):

1) The velocity at the entrance to the stilling basin must not be much in excess of 30 ft/sec. For conservatism and due to the uncertainties of doing otherwise, the inlet velocity should be determined as that occurring in the absence of the stilling basin. Thus, an approach pipe on a subcritical slope would usually be assumed to discharge at critical flow conditions, whereas one on a supercritical slope would be assumed to discharge at uniform flow conditions (or some depth intermediate between uniform and critical flow depths based on a drawdown calculation). Velocities should normally not be allowed to approach 30 ft/sec anyway, because of the likelihood of pipe erosion.
2) Based on the maximum design discharge, determine the stilling basin dimensions from Table 7.4, Columns 3 to 13. (Multiple units side by side may be used for flows larger than 339 cfs and may be economical for smaller flows in some cases). Oversizing should be avoided (use the minimum dimensions accommodating the maximum design flow), as performance may worsen at lower flows. The alternative end sill and 45° wall design shown on Figure 7.12 are recommended for best performance, including reduced potential for downstream erosion.
3) Suggested thicknesses of various parts of the basin are given in Table 7.4, Columns 14–18.

Table 7.4 Stilling Basin Dimensions for Impact-type Energy Dissipator

Suggested Pipe Size*			Feet and Inches										Inches					
Dia in. (1)	Area (sq ft) (2)	Max Dis-charge Q (3)	W (4)	H (5)	L (6)	a (7)	b (8)	c (9)	d (10)	e (11)	f (12)	g (13)	t_w (14)	t_f (15)	t_b (16)	t_p (17)	K (18)	Suggested Riprap Size (19)***
18	1. 77	21**	5–6	4–3	7–4	3–3	4–1	2–4	0–11	0–6	1–6	2–1	6	6½	6	6	3	4.0
24	3. 14	38	6–9	5–3	9–0	3–11	5–1	2–10	1–2	0–6	2–0	2–6	6	6½	6	6	3	7.0
30	4. 91	59	8–0	6–3	10–8	4–7	6–1	3–4	1–4	0–8	2–6	3–0	6	6½	7	7	3	8.5
36	7. 07	85	9–3	7–3	12–4	5–3	7–1	3–10	1–7	0–8	3–0	3–6	7	7½	8	8	3	9.0
42	9. 62	115	10–6	3–0	14–0	6–0	8–0	4–5	1–9	0–10	3–0	3–11	8	8½	9	8	4	9.5
48	12. 57	151	11–9	9–0	15–8	6–9	8–11	4–11	2–0	0–10	3–0	4–5	9	9½	10	8	4	10.5
54	15. 90	191	13–0	9–9	17–4	7–4	10–0	5–5	2–2	1–0	3–0	4–11	10	10½	10	8	4	12.0
60	19. 63	236	14–3	10–9	19–0	8–0	11–0	5–11	2–5	1–0	3–0	5–4	11	11½	11	8	6	13.0
72	28. 27	339	16–6	12–3	22–0	9–3	12–9	6–11	2–9	1–3	3–0	6–2	12	12½	12	8	6	14.0

* Although Columns 1 and 2 are repeated from Peterka (1964), the suggested pipe sizes may generally be disregarded. Pipe sizes are determined by other factors for most of our applications, and are not critical to basin operation.

** For discharges less than 21 second-feet, obtain basin width from curve of Figure 42 in Peterka (1964). Other dimensions proportional to W; H = 4W/3, L = 4W/3, d = W/6, etc.

*** Determining riprap size explained in text.

Source: Peterka 1964 / United States Department of the Interior-Bureau of Reclamation / Public Domain.

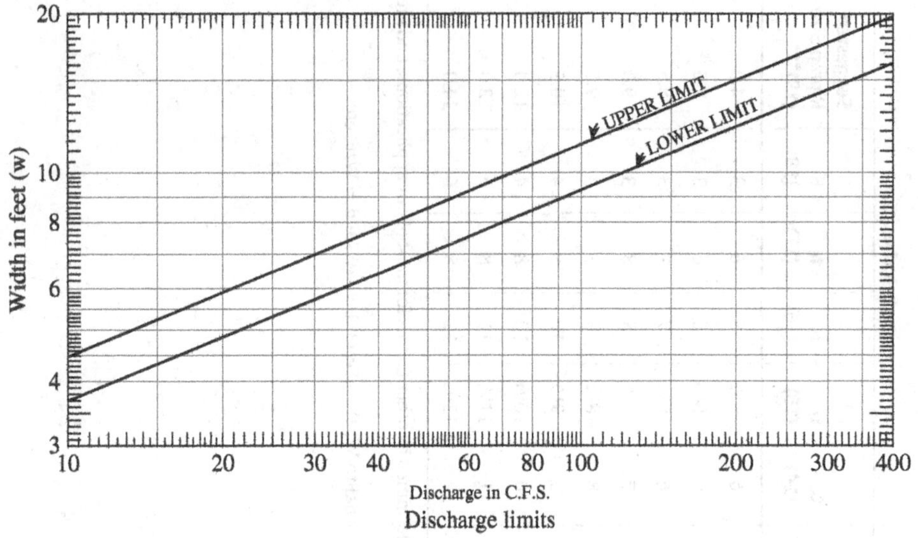

Figure 7.13 Discharge Limits for Impact-type Stilling Basin (Basin VI)

4) The relative invert elevation of the pipe or channel at the basin entrance must be as shown on Figure 7.12. The maximum width of the channel flow (or pipe diameter) must be less than the basin width. The entrance pipe or channel should have a slope not greater than 15°. If the outfall pipe or channel has a greater slope, the slope should be decreased to less than 15° for a distance upstream of the basin sufficient to redirect the flow (two or more diameters are suggested for a pipe).
5) When the approach flow is supercritical, a hydraulic jump may occur in the downstream end of the pipe. If the pipe is not well vented, a vent at least one-sixth the pipe diameter should be installed at a convenient location upstream of the jump. Otherwise, deleterious pressure fluctuations may occur in the system.
6) Riprap having minimum median stone sizes as given in Table 7.4, Column 19 should be placed downstream of the end sill. Smith and Korolischuk (1973) suggested that the riprap extend about $4W$ (see Figure 7.12) downstream from the end sill if the receiving stream has a sand bed. The riprap should extend laterally to the width of the end walls and may be placed upstream of the end walls as well. Discharge limits are given on Figure 7.13.

Example 7.23

A concrete storm sewer outfall (Fitchburg, Massachusetts) with a maximum design flow of 275 cfm has been tentatively selected to have a 66-inch diameter on a slope of 0.0077 ft/ft and with Manning's $n = 0.013$. Assess the suitability of this outfall for an impact-type stilling basin (Basin VI) and determine the basin dimensions.

Solution

The critical and uniform flow depths will be determined first. The critical flow section factor is given by $Z = Q/\sqrt{g} = 275/\sqrt{32.2} = 48.4\ \text{ft}^{5/2}$. Thus $Z/d_o^{2.5} = 48.4/(5.5)^{2.5} = 0.682$. From a table or figure of geometric elements, we obtain $y_c/d_o = 0.84$. The uniform flow depth is determined as follows: $A_f = (\pi/4)(66/12)^2 = 23.8\ \text{ft}^2$, $R = D/4 = 66/[4(12)] = 1.375\ \text{ft}$, $Q = (1.486/0.013)(23.8)(1.375)^{2/3}(0.0077)^{1/2} = 295\ \text{cfs}$. Then $Q/Q_f = 275/295 = 0.933$, from which Figure 5.2 gives (for variable n) $y_n/d_o = 0.75$.

The uniform flow depth is less than the critical flow depth, from which we conclude that the slope is supercritical and that the maximum velocity that could occur at the outlet corresponds to critical depth conditions. From a table or figure of geometric elements, for $y_n/d_o = 0.75$ we obtain $A/d_o^2 = 0.6318$. Thus: $A = 0.6318(5.5)^2 = 19.1\ \text{ft}^2$ and $V = 275/19.1 = 14.4\ \text{ft/sec}$. The velocity is acceptable (<30 ft/sec) according to Item 1 above. The outfall slope of $S = 0.0077$ corresponds to a slope angle of $\tan^{-1}(0.0077) = 0.44°$, which is well below the 15° maximum allowed by Item 4. The outfall is thus suitable for an impact-type stilling basin (Basin VI).

Referring to Figure 7.12 and Table 7.4, the basin dimensions are determined for the design discharge of 275 cfs to be as follows: $W = 16$ ft-6 in., $H = 12$ ft-3 in., $L = 22$ ft, $a = 9$ ft-3 in., $b = 12$ ft-9 in., $c = 6$ ft-11 in., $d = 2$ ft-9 in., $e = 1$ ft-3 in., $f = 3$ ft, and $g = 6$ ft-2 in. The alternate end sill and 45° wall angle (Figure 7.12) are selected (the end wall of length f is provided at the end of the wall angle). The thicknesses suggested by Columns 14 to 18 of Table 7.4 and Figure 7.12 ($t_w = 12$ in., $t_f = 12½$ in., $t_b = 12$ in., $t_p = 8$ in., $k = 6$ in.) were checked by a structural engineer.

In accordance with Figure 7.6 and Table 7.4, riprap of 14-inch median stone diameter was extended 30 ft from the end walls along the basin centerline, closed in along the river bank for a distance 30 ft along the river downstream of the basin, and brought around liberally in the upstream river direction.

The two types of stilling basins discussed above (Basins VI and IX) are somewhat unique among stilling basins in requiring no tailwater depth.

7.8.1.4 Riprap Scour Protection

A common method of scour protection is by placing stones of equivalent diameter greater than the d_s for which scour would occur at the prevailing velocity. Such stones are referred to as *riprap*, which may be defined as a "relatively thin layer of specially selected and graded rock fragments used for protecting earth slopes from erosion by water currents and waves" (USBR 1963a). Basic riprap design concepts will be presented first, followed by discussion of application of those and additional concepts to lateral channels and outfalls.

7.8.1.5 Riprap Design Concepts

The specific riprap design methods presented here are those presented for *dumped riprap* by the U.S. Bureau of Public Roads (1967). Those methods are based largely on a 1948 report by the American Society of Civil Engineers (Subcommittee on Slope Protection 1948) and a report by the Corps of Engineers Waterways Experiment Station (Campbell 1966).

Figure 7.14 relates velocity to *median* (see below) stone size, expressed as the equivalent spherical diameter or weight of the stone. The top curve is for stone on a flat channel bottom and is of the form $V \propto \sqrt{d}$. The lower curves (together with the top curve) pertain to the side slopes of trapezoidal channels with side slopes from 12:1 (shallow) to 1:1 (steep).

The velocity at which the stones would be displaced decreases as the side slope increases because the weight components which tend to cause the stone to "roll" down the side slope add to the destabilizing shear and lift forces due to the flow. The curves give the size k of stone with a unit weight of 165 pounds per cubic foot; for stones of other unit weights w, Creager's equation (Subcommittee on Slope Protection 1948) gives adjusted stone sizes k_w:

$$k_w = \frac{102.5k}{w - 62.5} \tag{7.100}$$

The velocity of Figure 7.14 is regarded as the average velocity against the stone. With certain exceptions noted below, the latter variable is related to the mean velocity of flow by means of Figure 7.15. We will not attempt to assign undue physical significance to this largely empirical adjustment but will state simply that the adjustment compensates for the vertical velocity profile and projection of stone dimensions into the overlying flow.

With reference to Figure 7.14, note that for channels having high velocity (hence requiring large k) and small depth (a common combination for channels on steep slope), k/d could actually exceed 1. In such a case, V_s/V of 1 should be used. The Bureau of Public Roads recommended (USBPR 1967) that 0.4 of the total depth be used for computing d in Figure 7.15 when the depth of flow exceeds about 10 ft. The reason stated for this is that the total depth would result in a stone size which would be adequate at the channel bottom but which might be too light to provide protection near the water surface. When the stone is subjected to more direct attack from the current than would occur in a straight channel, then V_s/V should not be taken less than 1, and should in effect be increased above 1 in some cases. The following pertains (FHwA 1990a, p.V-19):

> "The size of stone required to resist displacement from direct impingement of the current as might occur with a sharp change in stream alignment is greater than the value obtained from [Figure 7.14], although research data is lacking on just how much larger the stone should be. The California Division of Highways ... recommends doubling the velocity against the stone as determined for straight alignment before entering [Figure 7.14] for stone size. Lane (1955, p. 1248) recommends reducing the allowable velocity by 22% for very sinuous channels; for determining stone size by [Figure 7.14], the velocity (V_s) would be increased by 22%. Until data are available for determining the stone size at the point of impingement, a factor which would vary from 1 to 2 depending upon the severity of the attack by the current, should be applied to the velocity V_s before entering [Figure 7.14]."

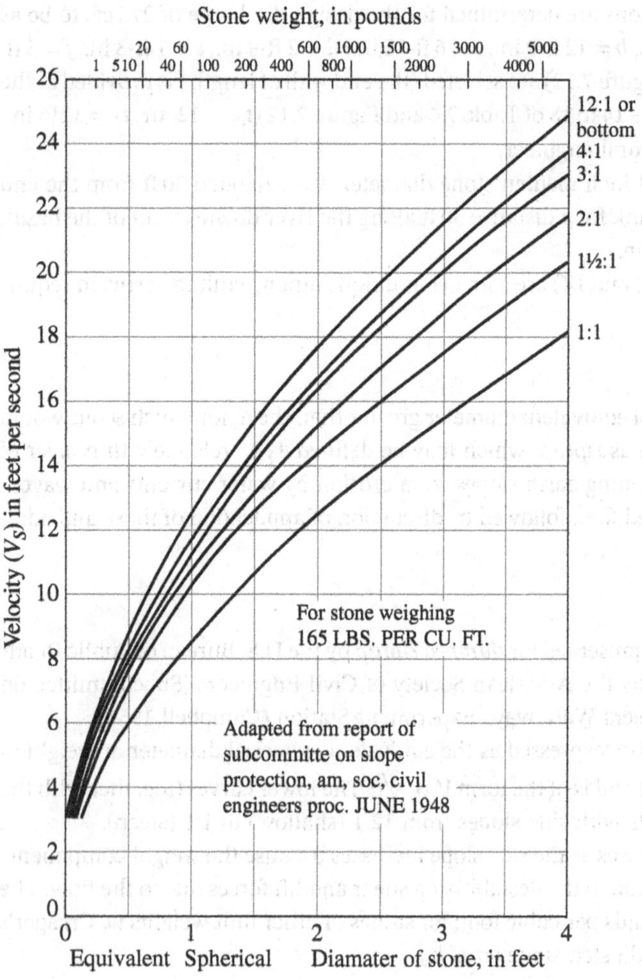

Figure 7.14 Size of Stone That Will Resist Displacement for Various Velocities and Side Slopes (USBPR 1967 / U.S. Department of Transportation / Public Domain)

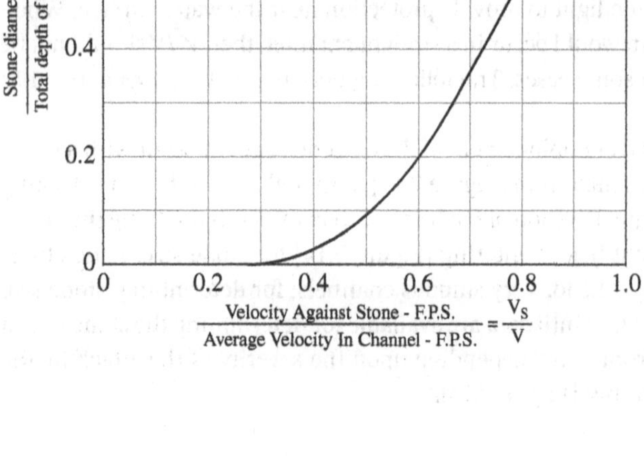

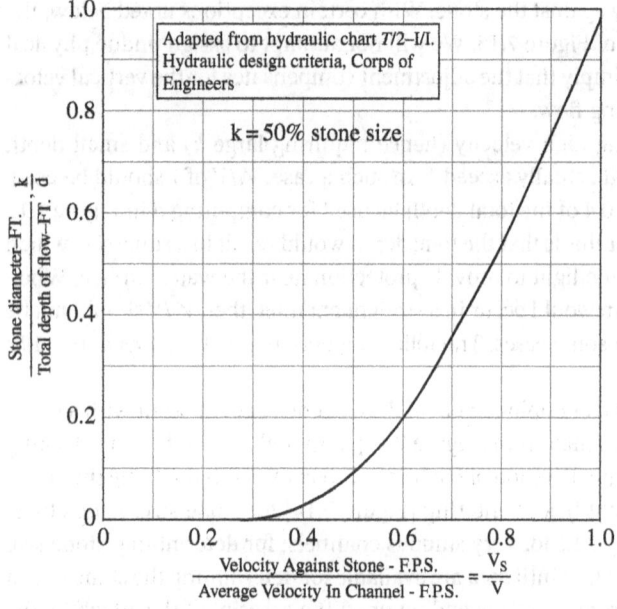

Figure 7.15 Velocity against Stone on Channel Bottom (USBPR 1967 / U.S. Department of Transportation / Public Domain)

7.8.1.6 Lateral Channels

Lateral channels (defined above) and related drainage channels may require scour protection not only in the vicinity of the outfall, but also along the downstream channel length if scouring velocities persist downstream of the transition from the outfall to the channel. The discussion here is limited to what Chow (1959) refers to as nonerodible channels, which are channels that are lined or built up to withstand erosion. The dimensions of such channels are decided on the "basis of hydraulic efficiency, or empirical rules of best section, practicability, and economy" (Chow 1959). Unlined channels may be designed in a similar fashion if the resulting velocities would be below those eroding the natural channel material (exposed or grassed); such cases are limited to channels on shallow slopes and/or having stable foundation materials. Otherwise, channel dimensions may be proportioned for stability rather than so-called hydraulic efficiency. [Chow (1959, Chapter 7) provides introductory discussion of such channels.]

The trapezoidal cross section is most common for nonerodible channels. Topography generally determines the bottom profile and may suggest the section dimensions, particularly if a natural drainage swale is being followed. Bureau of Reclamation experience curves (Chow 1959, §7-7; USBR 1963b) provide guidance in the selection of bottom width based on channel flow capacity. Side slopes depend partially on the channel material; they cannot exceed the stable slope for that material. For sandy loam or porous clay, a 3:1 side slope is usually safe, with steeper slopes acceptable for firmer materials (Chow 1959, §7-4). The steeper the slope, the smaller the acceptable riprap on the slope (see below). For a trapezoidal channel section, a side slope of $\sqrt{3}/3$ (approximately ½:1 = 1:2) is most efficient hydraulically in the sense that it has the least wetted perimeter for a given area. To demonstrate this we will elaborate on and make more understandable the derivation given by Streeter (1962, §11.2). The derivation will also be presented more rigorously by using the calculus of variations and other modifications (Hildebrand 1952, §2.1).

The water area and wetted perimeter of the trapezoidal section are given respectively by:

$$A = (b + zy)y \tag{7.101a}$$

$$P = b + 2y\sqrt{1 + z^2} \tag{7.101b}$$

in which b, y, and z are, respectively, the bottom width, water depth, and side slope. Manning's equation (Equation 7.39) with Q, n, and S known gives $AR^{2/3}$ = **constant**. Progressing from this relationship, we have $AR^{2/3} = A(A/P)^{2/3} = A^{5/3}/P^{2/3}$ giving $A^{5/3} \propto P^{2/3}$ or $A \propto (P^{2/3})^{3/5} = P^{2/5}$ or $A = CP^{2/5}$ in which C is a constant. Using the latter relationship together with Equations 7.101a and 7.101b, gives the following progression:

$$b = P - 2y\sqrt{1 + z^2} \tag{7.102a}$$

$$A = \left(P - 2y\sqrt{1 + z^2} = zy\right)y \tag{7.102b}$$

$$A = P - 2y^2\sqrt{1 + z^2} + zy^2 = CP^{2/5} \tag{7.102c}$$

We now have two equations relating the dependent variable A to the independent variables y and z. We can minimize A (and hence minimize P) by taking $A(y, z)$ as our objective function and finding conditions at which $\partial A/\partial y = \partial A/\partial z = 0$.

Taking the partial differential of Equation 7.102c with respect to y gives:

$$\frac{\partial A}{\partial y} = y\frac{\partial P}{\partial y} + P - 2(2y)\sqrt{1 + z^2} + 2yz = C\frac{2}{5}P^{-3/5}\frac{\partial P}{\partial y} = 0 \tag{7.103}$$

Because $\partial A/\partial y = 0$ implies $\partial P/\partial y = 0$, the above equation reduces to:

$$P = 4y\sqrt{1 + z^2} - 2yz \tag{7.104}$$

Now taking the partial differential of Equation 7.102c with respect to z:

$$\frac{\partial A}{\partial z} = P\frac{\partial y}{\partial z} + y\frac{\partial P}{\partial z} - 2\sqrt{1 + z^2}2y\frac{\partial y}{\partial z} - 2y^2\frac{\partial}{\partial z}\left(\sqrt{1 + z^2}\right) + y^2 + z2y\frac{\partial y}{\partial z} = 0 \tag{7.105}$$

Since $\partial A/\partial z = 0$ implies $\partial P/\partial z = 0$ and we have $\partial/\partial z(\sqrt{1 + z^2}) = (1/2)(1 + z^2)^{-1/2}2z = z(1 + z^2)^{-1/2}$, the above equation gives:

$$P\frac{\partial y}{\partial z} - 2\sqrt{1 + z^2}2y\frac{\partial y}{\partial z} - 2y^2z(1 + z^2)^{-1/2} + y^2 + 2yz\frac{\partial y}{\partial z} = 0 \tag{7.106}$$

Substituting Equation 7.104 for P and canceling terms in the above equation we reduce to:

$$2z/\sqrt{1+z^2} = 1 \tag{7.107}$$

from which we obtain:

$$z = \sqrt{3}/3 \tag{7.108}$$

Substituting z from Equation 7.108 into Equation 7.104 and reducing terms we obtain:

$$P = 4y\sqrt{1+\left(\sqrt{3}/3\right)^2} - 2y\sqrt{3}/3 = 2\sqrt{3}y \tag{7.109}$$

Using Equations 7.102a, 7.108, and 7.109 and reducing terms we obtain:

$$b = 2\frac{\sqrt{3}}{3}y \tag{7.110}$$

From Equations 7.107, 7.101a, and 7.109 we obtain $A = \sqrt{3}y^2$. From Equations 7.109 and 7.110, $b/P = \left[2(\sqrt{3}/3)y\right]/(2\sqrt{3}y) = 1/3$ or $b = P/3$ so $P = 3b$. Therefore the side lengths are equal to the bottom width for the least wetted perimeter. Using Equation 7.109, we have $\tan\theta = z = \sqrt{3}/3$ where θ is the channel with the least perimeter.

The following discussion of riprap is presented in the context of a project in Groton, Massachusetts, for which the author designed a stormwater drainage channel with side slopes of 1:2. With a bottom width of 4 ft, the design called for 6 inches of 1″ to 3″ traprock on the bottom and grassed side slopes over 6 in. of topsoil where the channel slope exceeded 10% and all loam and seed for lesser channel slopes.

The size of riprap, height of riprap above the water surface, and channel freeboard (height of bank above the water surface) are the final things to be determined. We will concern ourselves here with that portion of the lateral channel over which the flow has been well distributed downstream of the outfall. The channel hydraulics can be established by methods discussed in earlier sections of this chapter once the channel roughness has been characterized. For unlined channels or channels with a known lining (such as concrete), the determination of Manning's n is straightforward. For riprapped channels, it is less straightforward because the roughness is dependent on the size of stone, and the stone size is a variable to be determined.

Determining the size of channel riprap usually involves trial-and-error computations in which the riprap size (or Manning's n) is assumed, hydraulic computations are carried out to determine the velocities and corresponding riprap size, Manning's n is determined for that size riprap, and the steps are repeated as necessary. The hydraulic computations are often based on the assumption of uniform flow. In the author's opinion, gradually-varied flow computations (backwater and drawdown) should be used more frequently for this purpose for reasons of both function and economy in riprap sizing.

The necessary height of the riprap above the water surface and the freeboard (taken here to mean the height of the channel bank above the water surface) depends primarily on the expected extent of superelevation, cross waves, and roll waves. Determining factors are size of channel, velocity, alignment, debris, and wind. The height of riprap above the water surface is often less than the freeboard (i.e., the riprap need not extend to the top of the bank), because the velocities of water above the riprap due to localized waves, etc., would be substantially less than velocities at the design depth. The U.S. Bureau of Reclamation (USBR 1965, p. 291) gives a recommendation for freeboard in the form of the following empirical expression:

$$\text{Freeboard (ft)} = 2.0 + 0.025Vd^{1/3} \tag{7.111}$$

in which V is mean velocity in ft/sec and d is depth of flow in feet.

For supercritical flow, Equation 7.111 will generally provide suitable freeboards. For subcritical flows, superelevation Δh on the outside of a smooth curves of width b, radius of curvature r_c, and hydraulic depth D may be approximated by the following variation of an equation given by Chow (1959, §16-4):

$$\frac{\Delta h}{D} = \text{F}^2\frac{b}{r_c} \tag{7.112}$$

in which the Froude Number is based on mean velocity.

The following additional information is from Graber (2017a). "When flow velocities exceed the capacity of the natural channel bottom soils and vegetation, a channel lining should be provided. Channel lining can consist of concrete-filled fabric mats, rock riprap, precast concrete erosion revetments, cables and interlocking block revetments, soil cement, bagged concrete, gabions, soil bioengineered bank protection. ASCE/WEF (1992) provides additional information on design criteria and procedures for channel erosion protection measures. The utilization of flexible liner protection materials such as riprap and gabions is generally limited to subcritical flow conditions, supercritical flow conditions with Froude numbers generally less than 1.7, and streamflow velocities that do not exceed the practical limitations of these materials to withstand the associated tractive forces. Flexible linings are not recommended in flow situations where hydraulic jumps are possible."

"'Hard armor,' consisting of select stone or other materials, may be more suitable. In these conditions, the stone may require separation from the site soil by either a filter blanket of graded sand or gravel or a site-suitable geocomposite in order to prevent loss of soil particles and subsiding of the armor. The filter blanket or geocomposite must be of such type to minimize blinding or clogging from soil fines to prevent the development of hydrostatic groundwater forces beneath the structure. Determination of the need for a filter blanket and design methods are given in the Bureau of Public Roads, *Use of Riprap for Bank Protection*, (USBPR 1967). The need for and design of geotextile separators are discussed by Koerner (2005, Section 9.1). Geosynthetic (e.g., expanding geocells) or recycled plastic products are also available which can be used as a replacement for riprap."

"Channels lined with riprap or gabions are also used at roadsides and other stormwater drainage applications, and also as lateral collector channels at the top of slopes, such as at landfills, connected to multiple riprap- or gabion-lined channels continuing down the slope to a collector channel at the foot of the slope. For channels on shallow slopes, Manning's n values of 0.020–0.035 generally apply. However, for steep channels careful consideration is necessary in the selection of Manning's n values, design for channel stability, and awareness of the possibility of roll waves and cross waves. On steep slopes with shallow flow, substantially larger n values than those corresponding to channel depictions in, e.g., Chow (1959) will apply. Charts given in *Design of Roadside Channels with Flexible Linings* (FHwA 2005) are given for slopes up to 25%. However those charts or the formulas given for use with the associated iterative procedure are based on experimental work for a maximum slope of 8%. At larger slopes, the iterative procedure may not converge. When convergence is obtained, the procedure will enable determination of flow depth and channel stability. This procedure, based on Bathurst *et al.* (1981), will commonly reveal a subcritical flow and flow depth higher (with implications for channel stability) than with supercritical flow that would occur at lower n values. The same effect on flow regime was noted by Trieste (1992, 1994). For slopes beyond the range of FHwA (1990a) and Bathurst *et al.* (1981), n values as a function of slope, hydraulic radius or hydraulic depth, and riprap size can be obtained from other references; a good review is given by Bettess (1999); with the flow depth thus determined, the stability design method in FHwA continues to apply. For n values, a more recent comprehensive review, new developments, and recommendations (applicable to man-made triangular, parabolic, and trapezoidal cross-sections) are given by Froehlich (2012)."

"When supercritical flow does occur, cross waves occurring at channel bends may cause overtopping of the channel and should be considered, see, e.g., Ippen (1950) or Chow (1959). Roll-wave instability should be either avoided [Vedernikov number < 1 (Chow 1959)] or resulting wave heights be considered, where suitable predictive information exists [e.g., wide rectangular channels with Froude numbers > 2 (Brock 1970)], to avoid overtopping."

7.8.1.7 The Flow Pattern Below an Outfall

The flow issuing from an outfall at the bank of a river commonly has the river bottom as its lower boundary. As the flow moves out into the river, it tends to remain "attached" to the river bottom due to the Coanda effect discussed in Section 15.2. To an extent dependent upon their relative momentums, the outflow flow will bend in the downstream direction of the river flow.

In a paper on "Three Dimensional Turbulent Wall Jets", Rajaratnam and Pani (1974) reported on their experimental investigations of straight (unbent) flows having some of the essential characteristics of the flows of interest. The flows which they studied originated with square, rectangular, circular, elliptic, and triangular cross sections, all of which were submerged. They depict a representative horizontal velocity distribution in the center plane. They also depict the potential core region typical of issuing jet flows, in which the velocity distribution near the core remains constant (in this case also uniform) while the boundary layers grow inward. The jet is fully developed at or shortly downstream of the point where the boundary layer encroachment on the potential core is complete.

Using a Preston tube, Rajaratnam and Pani measured the bed shear stress along the center plane (at which the shear stress is a maximum). Their results are plotted in terms of the dimensionless shear stress coefficient which incorporate the bed shear τ_o and mean nozzle exit velocity U_o, and a dimensionless distance ratio which incorporates the longitudinal distance from the nozzle x' and the cross-sectional area of the nozzle A. Conveniently, a single curve fits the data points for all of the

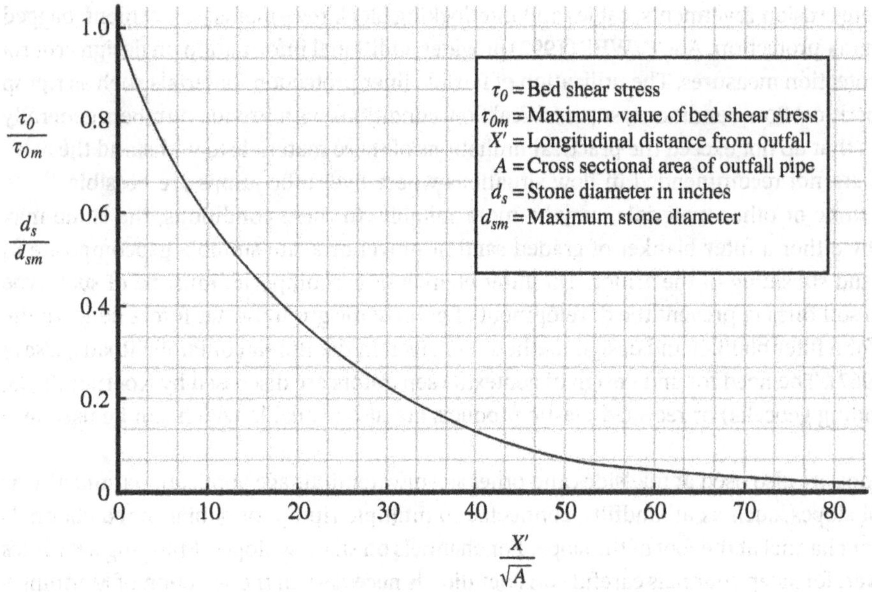

Figure 7.16 Bed Shear and Stone Diameter Relationships

tested nozzle cross sections. Noting that the bed shear is a maximum at $x' = 0$, we can replot their data in terms of τ_o/τ_{om} as shown on Figure 7.16, in which τ_{om} is that maximum of the bed shear (at $x' = 0$) and τ_o is the reduced bed shear at $x' > 0$. Using the relationship from the Shields equation that the bed shear τ is proportional to the required stone diameter d (see Section 13.2), we can alternately express the ordinate on Figure 7.16 as d_s/d_{sm} in which d_{sm} is the maximum stone diameter required right at the outfall (such as determined for the outfall velocity from Figure 7.14) and d_s is the stone diameter required at downstream location x'. The above, in conjunction with practical details presented below, can provide useful guidance for the design of riprap for some outfalls in the absence of cross flows which would bend the wall jet. Such a situation is rare, however, and all factors considered, the above together with consideration of cross flows will most often demonstrate the infeasibility of relying *solely* on riprapping downstream of high-velocity outfalls.

7.8.1.8 Influence of Crossflows

The conservative determination of x' and the extent of riprapping must be cognizant of the lateral extent and bending of the wall jet. Rajaratnam and Pani (1974) give data with which the lateral extent of an unbent jet may be estimated. If the jet is in fact unbent, that knowledge of the jet's lateral extent could enable riprap design. However, discharge to a stream bank results in an unbent jet only when there is no appreciable stream velocity.

The investigations of Pratte and Baines (1967) of profiles of round turbulent jets in a cross flow, although not for wall jets, give some indication of the extent of jet curvature and downstream translation. Their work demonstrates that riprapping might have to be carried a considerable distance downstream and provided over a large area to properly protect the stream bed. A more recent paper by Rajaratnam (1980) gives actual erosion patterns by jets in a crossflow.

7.8.1.9 Equilibrium Scour Conditions

An important notion is that of equilibrium scour conditions, such as might occur in the erodible channel below an outfall. If the velocity of flow discharging from an outfall near a stream bed exceeds the scouring velocity of the stream bed particles, scour will occur and a scour hole will be created. Although difficult to quantify, it is reasonable to expect that an equilibrium scour configuration will be reached. If the bed material is fairly homogeneous, this will occur at such time as the velocity at each section in and downstream of the scour hole has been reduced to the scour velocity of the material at that section. The mean velocity is reduced by the increase in the vertical flow dimension in the scour hole (and possibly upward expansion of the flow if the discharge is submerged) and lateral spreading of the flow. At the downstream limit of scour, the lateral spreading would conceptually have reduced the velocity of the flow over the unscoured bed to the scour velocity. The scour depth may be further limited by less erodible materials below the surface of the original streambed. The equilibrium scour concept is implicit in the design of certain types of outlet structures (see, e.g., Basin IX discussed above).

Chow (1959, §7-9) gives charts for the permissible (scouring) velocities (i.e., velocities that will not cause erosion) for non-cohesive and cohesive soils, and a curve of correction factors to be applied to the permissible velocities to account for channel depth. Those curves include the pertinent particle characteristics and provide an unambiguous relation to soil types.

Example 7.24

Following up on Example 7.23 (Fitchburg, Massachusetts), estimate (a) the extent of riprapping that would be required to prevent scour of the river bed in the absence of the stilling basin and (b) whether the riprap determined for use with the stilling basin would completely protect the river bed from scour. The river bed material is estimated to scour at a velocity of 4 ft/sec. First neglect and then consider the effect of river cross flow.

Solution

From Example 7.23, the outfall has an outlet velocity of 13.2 ft/sec at a flow of 275 cfs with a critical depth flow area of 20.8 ft^2. The effective "scouring" velocity can be considered to be related to bed shear according to $\tau_o \propto V_b^2$. Thus a decrease in scouring velocity from 13.2 to 4 ft/sec corresponds to a bed shear of $(4/13.2)^2 = 0.092$ times the initial bed shear. From Figure 12 of Rajaratnam and Pani (1974) for an unbent jet (i.e., neglecting cross flow), this bed shear reduction would occur at $x'/\sqrt{A} = 34$, or $x' = 34\sqrt{20.8} = 155$ ft. This distance was considered excessive in this case (the river is only about 80 ft wide).

The effect of cross flow cannot be quantified without resorting to model studies, but the discussion of such effects in the text suggests that the high-velocity bent jet could persist and scour the river bed for a considerable distance downstream. If all scour is to be prevented, even more extensive riprapping would be required for the bent jet.

The velocity of flow over the stilling basin end sill can be estimated from the following, developed with the aid of Equation 7.9 (in which $b = W$, the basin width):

$$V_d = \frac{Q}{by_c} = \left(\frac{gQ}{b}\right)^{1/3} = \left[\frac{32.2(275)}{16.5}\right]^{1/3} = 8.13\ \text{ft/sec}$$

From Figure 7.17 (Peterka 1964, Figure 165), a 10-inch median stone size corresponds to this velocity, providing approximate verification of the 14-inch median stone size selected in Example 7.23. The depth over the end sill is approximately the critical depth given by:

$$y_c = \frac{V_d^2}{g} = \frac{(8.13)^2}{32.2} = 2.05\ \text{ft}$$

Thus $A = 16.5(2.05) = 33.9$ ft^2 and $\sqrt{A} = 5.82$ ft. The 25 ft length of riprap corresponds to $x'/\sqrt{A} = 25/5.82 = 4.3$, which from Figure 7.16 corresponds to $\tau_o/\tau_{om} = 0.75$. This corresponds to a scouring velocity of $\sqrt{0.75} = 0.87$ times the discharge velocity at the sill, or 0.87(8.13) = 7.0 ft/sec. Thus, some erosion of the river bed outside of the riprap zone is likely. This may be made more severe by the cross flow.

The outfall and outfall structure would themselves be well protected by the extent of riprap provided, but some river bed scour is expected. The above does, however, exemplify the need for the outfall discharge velocity to usually be reduced to the desired velocity rather than relying on moderate riprapping to provide significant velocity reduction. In this particular application, the compromise represented by the stilling basin and moderate riprapping was considered satisfactory.

The above demonstrates the impracticality (and provide means for quantifying such demonstrations) of designing scour protection for wastewater and stormwater outfalls by methods relying *solely* on riprapping when crossflows are involved. Secondarily, the above is of some quantitative significance for minor structures when cross flows are absent (akin to culvert outlets) and when used in conjunction with subcritical outfalls and stilling basins as discussed below.

Albertson *et al.* (1950) described the lateral diffusion, deceleration, and entrainment of turbulent submerged jets. Both circular and slot jets were studied. A jet issuing from an orifice or slot has an initial zone of flow establishment with potential flow (defined in Chapter 4). For a three-dimensional (i.e., circular) jet, that zone extends to 6.2 times the initial jet diameter and the maximum velocity decreases according to $(v_{max}/v_o)(x/D_o) = 6.2$ in which the terms represent the maximum (centerline) velocity, initial velocity, distance from the orifice, and orifice diameter. For the slot, the potential-flow zone extends to 5.2 times the initial jet width and the maximum velocity decreases according to $(v_{max}/v_o)\sqrt{x/B_o} = 2.28$ in which the terms represent the maximum velocity, initial velocity, distance from the slot, and slot width.

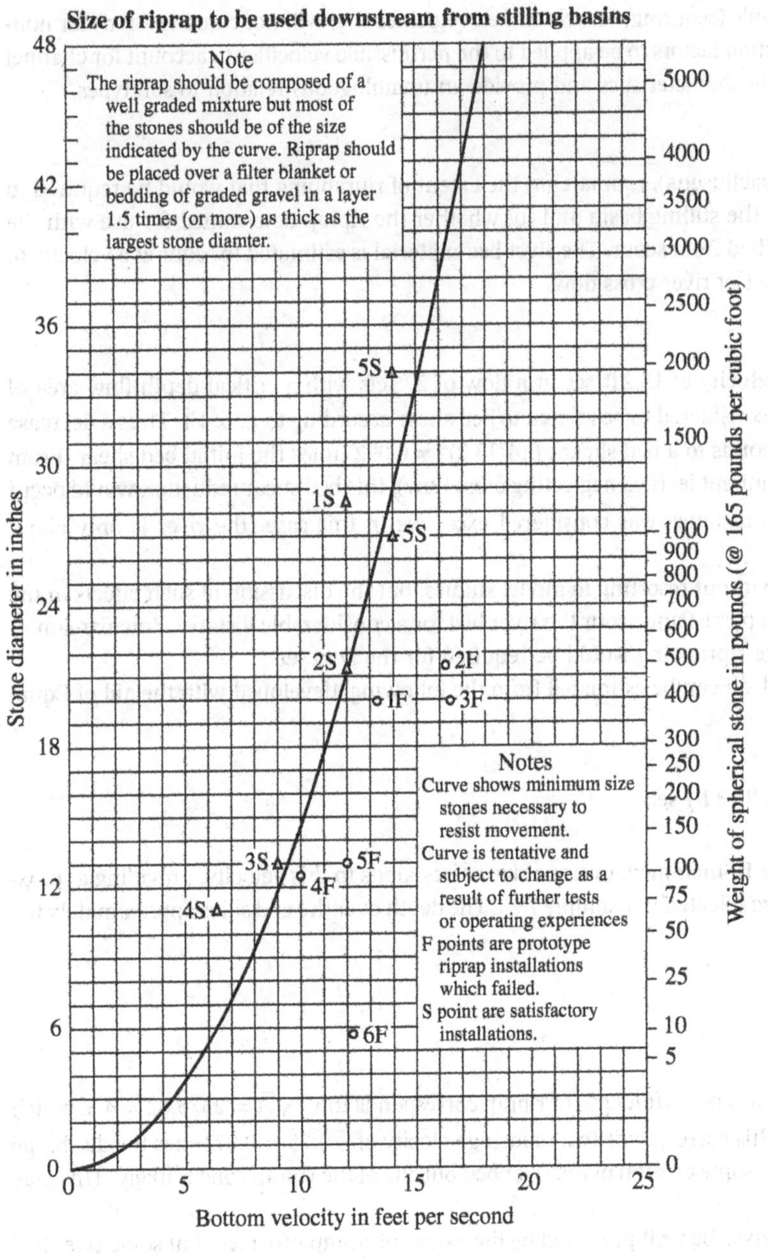

Size of riprap to be used downstream from stilling basins
Note
The riprap should be composed of a well graded mixture but most of the stones should be of the size indicated by the curve. Riprap should be placed over a filter blanket or bedding of graded gravel in a layer 1.5 times (or more) as thick as the largest stone diamter.
Notes
Curve shows minimum size stones necessary to resist movement.
Curve is tentative and subject to change as a result of further tests or operating experiences
F points are prototype riprap installations which failed.
S point are satisfactory installations.
Stone diameter in inches
Bottom velocity in feet per second
Weight of spherical stone in pounds (@ 165 pounds per cubic foot)
1S
2S
3S
4S
5S
1F
2F
3F
4F
5F
6F

Figure 7.17 Curve to Determine Maximum Stone Size in Riprap

Example 7.25

Example 7.14 (Kennebec Sanitary Treatment District, Maine) gave an outfall discharge velocity to the river of 9.6 ft/sec, and a discharge area in the partially full pipe of 15.5 ft^2. Determine the requirement for riprap to prevent erosion of the river bottom.

Solution

Estimate the initial velocity of the jet assuming it is the same as that of a circular jet of equivalent area. Then that diameter $D_o = \sqrt{4A/\pi} = \sqrt{4(15.5)/\pi} = 4.44$ ft. The potential core length $\cong 7 \times 4.44 = 31$ ft, in which distance the centerline velocity does not decrease. Beyond the potential core distance, the maximum velocity decreases according to $v_{max} = 6.2v_oD_o/x = 6.2(9.6)(4.44)/x = 264/x$ ft/sec in which x is distance in ft beyond the potential core length. The natural river velocity is about 3 ft/sec, so the riprap will be terminated at a total distance from the outfall of $264/v_{max} + 31 = 264/3 + 31 \cong 120$ ft.

The riprap is installed in a semicircular pattern to account for spreading of the discharge and downstream deflection. The riprap size is based throughout on the 9.6 ft/sec velocity for conservatism, which, per Section 11 of "Hydraulic Design of Stilling Basins and Energy Dissipators" (Peterka 1964), calls for a 21-inch thick graded mixture of gravel bedding of 14-inch (or 140 lb) average stone size with most of the larger stone on top.

7.9 Time-Varying Flow

A useful perspective is established by dividing open-channel time-varying (unsteady) flows into two categories. First are those in which the volumetric changes occur almost entirely in basins with horizontal water surfaces. The detention basins discussed in Section 5.4 and equalization basins discussed in Section 8.5 fall in this category. The volumetric changes associated with the flow control components (pipes, weirs, etc.) are either nonexistent or comparatively very small, and the hydraulic relationships (e.g., outflow-elevation) may be regarded as quasi-steady. The flow routing that is involved may be regarded as *hydrologic routing*, as opposed to the *hydrodynamic routing* discussed below. The basic concepts involved in hydrologic routing have been addressed in sufficient detail, insofar as environmental engineering applications are concerned, in the aforementioned sections.

Hydrodynamic routing problems in man-made channels encountered in environmental engineering applications are confined to sewers (sanitary, storm, and combined), the flow in other channels being adequately handled by quasi-steady approximations. Chapter 15 addresses unsteady free-surface flows in estuaries and the ocean.

The time-varying flows occurring in sewers subject to normal wastewater and stormwater flow variations fall under the category of "gradually varied unsteady flow." As defined by Chow (1959), such flow is characterized by mild curvature of the wave profile, gradual change in depth, and negligible vertical component of acceleration of water particles. The assumption of parallel flow (see Section 2.6) is reasonable in this case. The general differential equation for gradually varied unsteady flow may be derived by the application to an appropriately defined control volume, subject to the assumption of parallel flow, of the Law of Conservation of Mass expressed for a control volume as given in Section 2.3, and the energy equation for a control volume given in Section 2.6. Good details may be found in Chow (1959), Harris (1970), and Pinkayan (1972). Methods of flow routing in storm sewers are discussed in Graber (2017a, §7.5). Taking storage in sewers into account can allow more economical sewer design.

On the other hand, the time-varying flows resulting from surges of various kinds and quick operation of controlling structures have characteristics which are opposite those described above and are categorized as "rapidly varied unsteady flow" (Chow 1959, p. 523). One example of that is the tidal bore discussed in Section 15.4. Another relevant example of rapidly varied unsteady flow occurs in sewer flushing, for which the parallel flow assumption would not be accurate. A thorough discussion of sewer flushing, including the use of flush tanks, is provided by Babbitt and Baumann (1958, §'s 4-26, 6-11, 13-10). The first of these sections discusses the "hydraulic bore or traveling wave" created by the sudden release of water for sewer flushing; the formulae presented are akin to those presented in Section 15.4.

The reader interested in an overview of unsteady flow in open channels for applications beyond the scope of this book (such as flood routing, wave propagation, surges in canals, etc.) is referred to Gilcrest (1950) and Part V of *Open-Channel Hydraulics* by Chow (1959).

8

Pumping, Equalization, and Waterhammer

8.1 Introduction

This chapter addresses pumps of various types, pump cavitation, equalization, waterhammer, wet wells, and stormwater pumping station design. However, it is emphasized at the outset that equalization and waterhammer are covered here because they are most often *but not always* associated with pumped systems and for organizational convenience. Both topics can be important in gravity systems, as will be expanded upon in the sections that follow.

The material on pumps in this chapter is not intended to be a standard discourse on material readily found elsewhere (e.g., numerous handbooks and manuals of practice). Rather, it is intended to provide a concise overview and presentation of less available information, oriented toward the engineer in charge of selecting and specifying the hydraulic characteristics of the pump (as opposed to the manufacturing design engineer). Pump theory, for instance, is presented only to the extent needed for application by such pump users. Pumps may be categorized according to type as centrifugal pumps, positive-displacement pumps, and miscellaneous types. These types are defined and discussed in three separate sections.

For system synthesis, upper and lower estimates of the system curves should be used to reflect changing pipe friction with system aging and uncertainties or variable data about fitting losses. The actual coefficient of pipe friction may vary with the age of the force main, and it is desirable to study the hydraulics over an "envelope" of friction values. Form losses often have a degree of uncertainty, and a range of values can be found in the literature. It is often prudent to estimate minimum as well as maximum system-head curves during pump selection, both to guard against surge (in this case, meaning separation of flow over the impeller blades) and since pump power requirements may actually be greater under minimum head (and, hence, maximum flow conditions).

8.2 Centrifugal Pumps

Centrifugal pumps include radial (volute centrifugal and diffuser centrifugal), axial flow, and mixed-flow types (Symons 1966, Karassik *et al.* 1976). Performance test codes for centrifugal pumps are given by ASME (1973).

8.2.1 Turbomachine Theory

The turbomachine theory of pumps presented in Section 2.5 provides a useful point of departure for developing an understanding of pump characteristics. For a given pump and speed, the total head H theoretically varies linearly with discharge Q, as shown in Figure 8.1. Centrifugal pumps are usually designed with $\beta_2 < 90°$, which gives decreasing head with increasing discharge. A variety of factors cause the actual head–discharge curve to be lower than the theoretical curve shown in Figure 8.1. A detailed understanding of those factors is of great importance to pump researchers and designers, who have developed a very impressive empirical and theoretical body of knowledge. The intricacy with which dimensions of well-designed pumps are proportioned and the complexity of actual pump design merits the respect of the user.

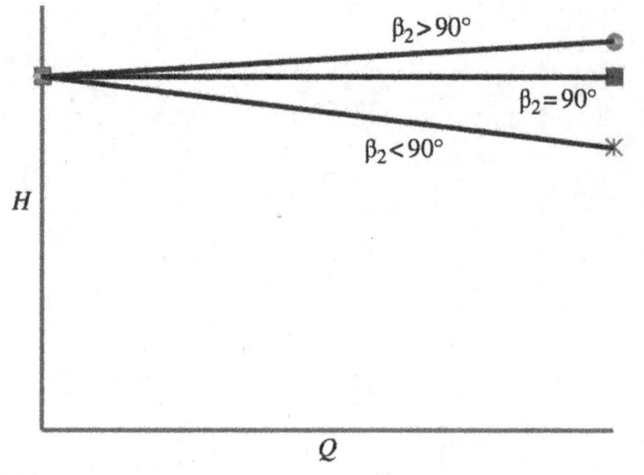

Figure 8.1 Theoretical Head–Discharge Curves (Streeter 1962 / McGraw Hill)

8.2.2 Dimensional Analysis

A dimensional analysis, following the principles set forth in Chapter 5, can be carried out to deduce important relations between parameters of actual pumps. The *independent* variables of importance can be listed in the initial functional relation as follows:

$$\Phi(Q, \Delta p, D, N, \rho, \mu) = 0 \tag{8.1}$$

in which Q is rate of flow, Δp is pressure increase across the pump, N is impeller rotational speed, ρ is fluid density, and μ is absolute viscosity, D is impeller diameter, and is the only dimension necessary to characterize *geometrically similar* pumps. Applying the Buckingham–π Theorem to Equation 8.1 gives the number of independent variables $m = 6$ and number of fundamental units $r = 3$, from which the number of dimensionless π- parameters is $n = m - r = 6 - 3 = 3$.

Kinematic similarity requires that the ratio of the impeller tip velocity to a characteristic fluid velocity be equal. The former is proportional to ND, while the latter is proportional to Q/D^2. This suggests the first π-parameter, as follows:

$$\pi_1 = \frac{ND}{Q/D^2} = \frac{ND^3}{Q} \tag{8.2a}$$

Two more independent π-parameters can be derived (an exercise left to the reader), and convenient choices are:

$$\pi_2 = \frac{\Delta p}{\rho N^2 D^2} \tag{8.2b}$$

$$\pi_3 = \frac{\rho V D}{\mu} \tag{8.2c}$$

Noting that π_3 is a Reynolds Number $\mathbf{R}$, we can express the overall functional relation as follows:

$$\Phi\left(\frac{ND^3}{Q}, \frac{\Delta p}{\rho N^2 D^2}, \mathbf{R}\right) = 0 \tag{8.3}$$

The change in total head across the pump is given by $H = \Delta p/\gamma + \Delta v^2/(2g) + \Delta y$ which, since Δy is negligible, is approximated by $H \cong \Delta p/\gamma + \Delta v^2/(2g)$. We may thus write:

$$\frac{H}{N^2 D^2} \cong \frac{\Delta p}{\gamma N^2 D^2} + \frac{\Delta v^2}{2gN^2 D^2} \tag{8.4}$$

The notation v is used for velocity in the above expression to avoid confusion with the characteristic fluid velocity denoted by V. If v_a and v_b denote the fluid velocities at the pump inlet and outlet, respectively, then $\Delta v^2 = v_b^2 - v_a^2$. By kinematic similarity within a given machine, we have $v_b \propto V$ and $v_a/v_b = constant$, so that $\Delta v^2 = v_b^2(1 - v_a^2/v_b^2) \propto V^2$. Therefore, the second term on the right-hand side of the above equation is proportional to $V^2/(2N^2D^2)$. Since $V \propto Q/D^2$, the following succession of proportionalities holds true:

$$\frac{\Delta v^2}{2gN^2D^2} \propto \frac{V^2}{2gN^2D^2} \propto \frac{Q^2}{2gN^2D^6} \propto \frac{Q}{ND^3} \tag{8.5}$$

Since the final term in the above succession must be constant for kinematic similarity (see the first π- parameter, π_1 above), then the first term in the succession is also constant. Then, from Equation 8.4, the variables $H/(N^2D^2)$ and $\Delta p/(\gamma N^2 D^2)$ are mutually dependent in dynamically similar machines, so that Equation 8.3 can be expressed in the following more general form:

$$\Phi\left(\frac{ND^3}{Q}, \frac{Hg}{N^2 D^2}, \mathbf{R}\right) = 0 \tag{8.6}$$

The reader should demonstrate that the first two dimensionless variables in the above equation may be obtained by non-dimensionalizing and regrouping variables. By eliminating D between the first two variables of the above relationship, we obtain another (alternative) form:

$$\Phi\left[\frac{N\sqrt{Q}}{(gH)^{3/4}}, \frac{Hg}{N^2D^2}, \mathbf{R}\right] = 0 \tag{8.7}$$

The above two functional relations have important uses in characterization of types of pumps, in pump modeling or testing, and in extrapolating measured performance characteristics of a particular pump to other conditions. The first variable in Equation 8.7, with the parameter g dropped, gives the important *dimensional* variable *specific speed*, defined by:

$$N_s = \frac{N\sqrt{Q}}{H^{3/4}} \tag{8.8}$$

Different types of pumps fall into fairly distinct ranges of specific speed, and this is of occasional value in the preliminary selection of pumps. The specific speed increases in going from centrifugal (radial) pumps to mixed-flow pumps to axial-flow pumps. The above equation for specific speed will be used below in a more definitive way.

In most applications of moderate-to-large-capacity pumps, the Reynolds Number is sufficiently large that the flow is fully turbulent and $\mathbf{R}$ can be dropped from the above functional relations. Then, if we are considering various operating conditions for the same pump, D is constant and Equation 8.7 reduces to the following:

$$\Phi\left(\frac{N}{Q}, \frac{H}{N^2}\right) = 0 \tag{8.9}$$

(The similarity relations given by Equations 8.6 and 8.7 are expressly not applicable for changes in impeller diameter, as may be seen from the inclusion of D in Equations 8.6 and 8.7. The reader should be aware of incorrect claims to the contrary.) The above equation says that dynamic similarity will be maintained if Q and H are simultaneously changed in accordance with $Q \propto N$ and $H \propto N^2$. For a change between operating conditions 1 and 2, this can be expressed by:

$$\frac{H_1}{H_2} = \frac{Q_1^2}{Q_2^2} = \frac{N_1^2}{N_2^2} \tag{8.10a}$$

This relationship allows, for example, a pump curve provided for one speed to be used to develop a pump curve for another speed, or the determination of the speed of a pump at a particular operating point (Q and H) given pump curves at another speed.

Simpson and Marchi (2013) note that pump efficiency does not scale in accordance with the affinity relationships of dimensional analysis. They provide the basis for the following relationship, in which η_1 and η_2 are the efficiencies corresponding, respectively, to speeds N_1 and N_2.

$$\eta_2 = 1 - (1 - \eta_1)(N_1/N_2)^{0.1} \tag{8.10b}$$

Example 8.1

Head–discharge and pump efficiency curves for an existing pump having a 45-in. diameter impeller are given for a speed of 300 rpm by the curve marked "45 DIA." in Figure 8.2. The unit is an engine-driven raw sewage pump in a wastewater treatment plant (Tallman Island WWTP, New York City). It is desired to determine the speed and efficiency at which the pump can discharge the minimum plant flow of 20 mgd at a head of 25.5 ft.

Solution

The dynamic similarity relations of Equation 8.10a apply, in which we will have the subscript "2" denote conditions at

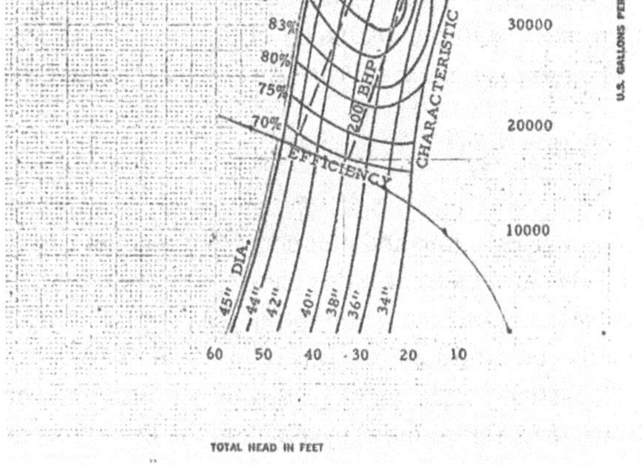

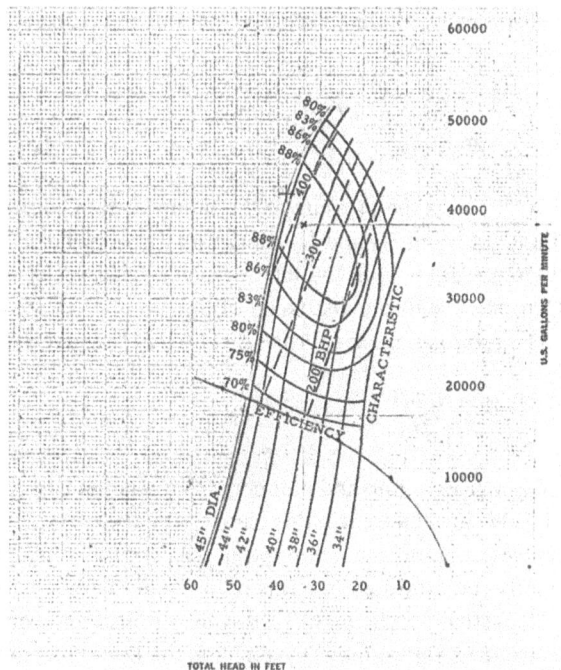

Figure 8.2 Pump Performance Curve and Superimposed System Curve at 300 rpm

20 mgd × 695 = 13,900 gpm and subscript "1" denote conditions at 300 rpm. Thus:

$$\frac{H_1}{25.5} = \frac{Q_1^2}{(13,900)^2} = \frac{(300)^2}{N_2^2}$$

The relationship between H_1 and Q_1 is plotted as a parabola on the pump performance chart at 300 rpm (Figure 8.2). The parabola intersects the pump curve for the 45-inch impeller at a point corresponding to $H_1 = 46.5$ ft, $Q_1 = 9300$ gpm, and an efficiency of slightly less than 70%. The pump speed at 20 mgd is then given by:

$$N_2 = 300\frac{13,900}{19,300} = 216 \text{ rpm}$$

Application of Equation 8.10b gives a decrease in efficiency at 216 rpm of about 0.1% (an exercise left to the reader). In other cases, examples of which are given by Simpson and Marchi (2013), the change in efficiency can be more significant.

8.2.3 Pump Cavitation and Surge

Just as the high velocities through valves can create pressures sufficiently low for cavitation to occur (see Section 6.5), so can the high velocities created by the pump impeller promote cavitation conditions in the pump casing. As with valves, it is the combination of low initial pressures (in the case of pumps, at the pump inlet) and high subsequent velocities that creates the potential for cavitation. The damaging effects of pump cavitation are similar to those discussed for valves (Section 6.5). For reasons similar to those discussed in connection with valves, cavitation limits in pumps must be established experimentally (Wislicenus 1942).

Cavitation is due to reduction of pressure at some point to less than vapor pressure of water at the prevailing temperature. It is a result of a combination of a high velocity at some point within a hydraulic device (meter, valve, pump, etc.) and sufficiently low upstream pressure, which causes (according to Bernoulli's theorem or conservation of energy) a sufficiently low pressure at the high-velocity location. A subsequent downstream increase in pressure above vapor pressure causes the sudden, implosive elimination of the vapor bubbles, which results in cavitation damage and attendant noise.

Historically, a pump cavitation parameter analogous to Equation 6.48 has been employed, defined by:

$$\sigma = \frac{\left(p_s - p_g\right)/(\rho g)}{(H_d - H_s)} \tag{8.11}$$

in which p_s is the pressure at the pump suction connection, p_g is the vapor pressure at the prevailing fluid temperature (Table 9.1), and H_d and H_s are the total energy heads at the pump discharge and suction connections, respectively. The difference $H_d - H_s$ is the total dynamic head (TDH) defined previously, which is conventionally denoted by $H_d = H$. (Consistent with the discussion of Equation 6.48, we shall employ absolute pressure rather than gage pressures here.)

The term *net positive suction head* (*NPSH*) is given to the numerator in Equation 8.11. Thus:

$$\sigma = \frac{NPSH}{H} \tag{8.12}$$

For a given pump and pump speed, a relation can be experimentally established between pump discharge rate and the critical value of the cavitation parameter σ_c below which cavitation will occur. Since H is a function only of pump discharge for a given pump and pump speed, the corresponding *NPSH*, referred to as the *minimum required NPSH* and equal to $\sigma_c H$, can also be expressed in terms of pump discharge. An example of this is given in Figure 8.2.

The actual *NPSH* can be readily determined and compared to the minimum required *NPSH*. The actual *NPSH* may be determined, for example, by applying the Bernoulli equation between the wet-well surface and the pump suction connection in the case of Figure 8.3 to obtain:

$$z_1 = \frac{p_s}{\rho g} + z_s + \frac{V_s^2}{2g} + h_{L,1-s} \tag{8.13}$$

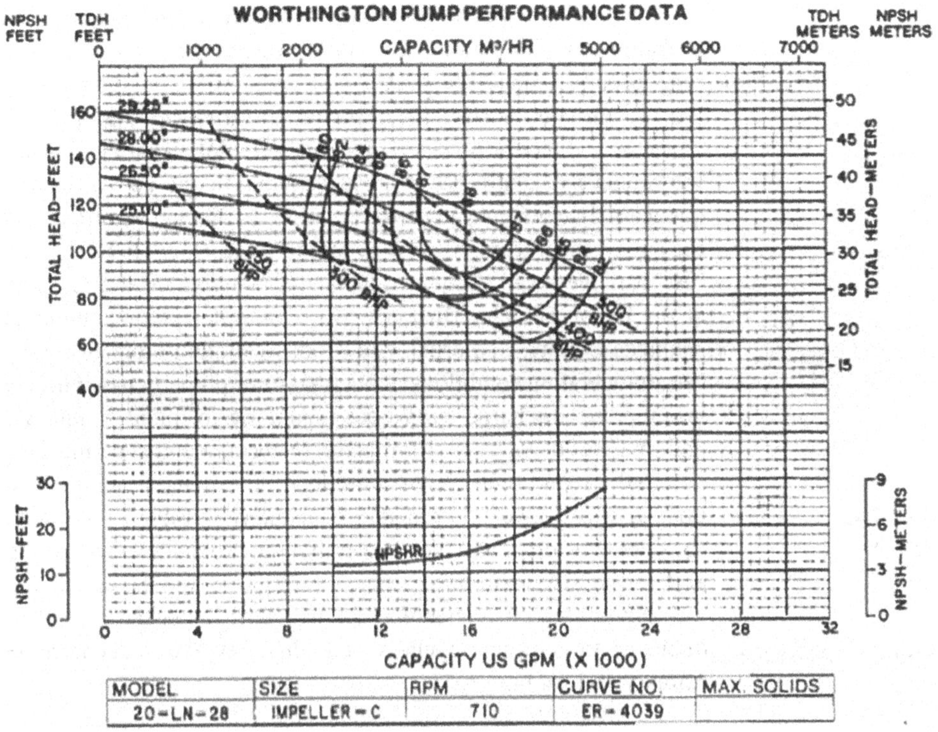

Figure 8.3 Pump Curves for Worthington Corporation Horizontal Split Case Pump Type LN

in which p_s is the gage pressure at the pump suction connection, $h_{L,1-s}$ is the head loss in the suction line, and other terms are as shown in Figure 8.3. The actual *NPSH* is then given by:

$$NPSH = \frac{p_{s,abs} - p_v}{\rho g} = \frac{p_s + p_{atm} - p_v}{\rho g} = (z_1 - z_2) + \frac{p_{atm} - p_v}{\rho g} - \frac{V_s^2}{2g} - h_{L,1-2} \tag{8.14}$$

The minimum required *NPSH* should be available from the pump manufacturer. However, it is sometimes necessary to extrapolate to other conditions (of pump speed, for example) or to estimate the minimum *NPSH* in the absence of manufacturer's data.

Surge limit is sometimes confused with *NPSH*. Surge is a completely separate phenomenon from cavitation. When pumps operate beyond their design range, the condition known as "surge" can result. The flow on the impeller blades separates from the blades much as happens with the wings of an airplane during stall. Then the pumping efficiency can fall off and the pump discharge can decrease abruptly until the impellers can recover. The process repeats itself, resulting in a surging action as the flow increases and decreases. In addition to ineffective pumping, the associated vibration can reduce the life of the pump.

When the pump head–flow curve is plotted on the same graph as the *NPSH* required-flow curve (see, e.g., Figure 8.3), the head and *NPSH* curves point toward each other at the high-flow end, and the two curves end at about the same maximum flow. Because of this, it is sometimes mistakenly deduced that the end of the head–flow curve has something to do with NPSH and thus cavitation. That is not so. The vertical scales of the two curves are completely different. The head–flow curve is independent of the suction head. One could have, for example, a 50-foot TDH with 1000 ft of static head above the pump on the suction side and still have the same end point on the head–flow curve, clearly with no possibility of cavitation occurring because of the very high *NPSH* available.

The phenomenon that determines the high-flow limit of the head–discharge curve can be explained with reference to Figure 8.4. That figure shows the absolute and relative velocities of an impeller vane and approaching flow. At moderate flows, the vane acts similar to an airfoil (an airplane wing), with the approach flow at a modest "angle of attack" and the flow going smoothly over the top surface of the vane. As the flow increases, the angle of attack also increases. As a sufficiently high flow, the flow separates from the top surface of the vane in a fashion similar to "stall."

When that occurs, the pumping efficiency falls drastically, then recovers and cycles as described above. In compressors with large numbers of vanes (unlike most pumps), rotating surge cells can result, which cause a portion of the compressor vanes to function improperly at a given time.

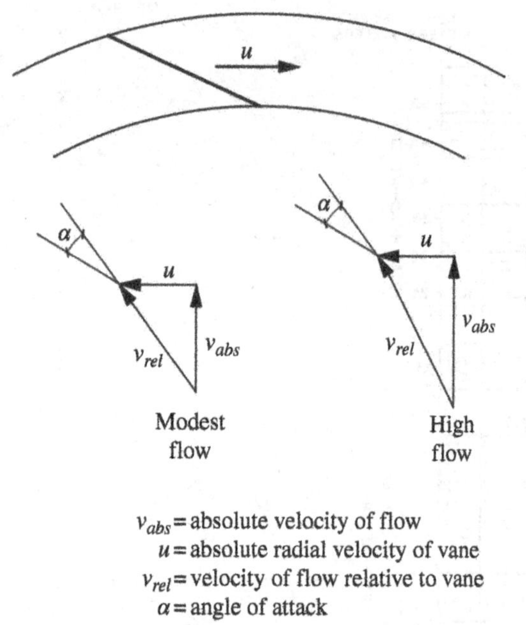

Figure 8.4 Absolute and Relative Velocities of Impeller Vane and Approaching Flow

The surge phenomenon is totally independent of the cavitation phenomenon, although if the suction pressure is low enough when surge occurs, it can be accompanied by cavitation due to the resulting low pressure on the stall side of the pump vane (Kittredge 1976).

The importance of considering the pipe friction variation with age was noted in Section 8.1. A good example of the adverse consequences of such failure to consider pump discharge hydraulics over the age of the system is given by Whiton and Ham (1993). In that case, variable-speed pumps of 150 hp and 1180 rpm maximum speed at the Union Sanitary District's two wastewater pump stations in Alameda County, California, had serious vibration problems under certain operating conditions. Design was not properly based on the lower pump discharge curves corresponding to initial conditions, and thus was beyond the manufacturer's recommended pump curve and to the right of maximum conditions. An extensive pump vibration analysis was conducted in which many possible causes were considered before it was recognized that the problems were attributable to the aforementioned mismatch of pump and system curves. The solution entailed changing the two-vane impellers to smaller-diameter three-vane impellers and changing the sequencing of primary and lag pumps.

Example 8.2

A Myers WGL20F submersible grinder pump was installed in the sump at a commercial installation in Medfield, Massachusetts. The pump required frequent repair and made a rattling sound when operating. The pump head–discharge curve was obtained and is shown in Figure 8.5. Determine the problem.

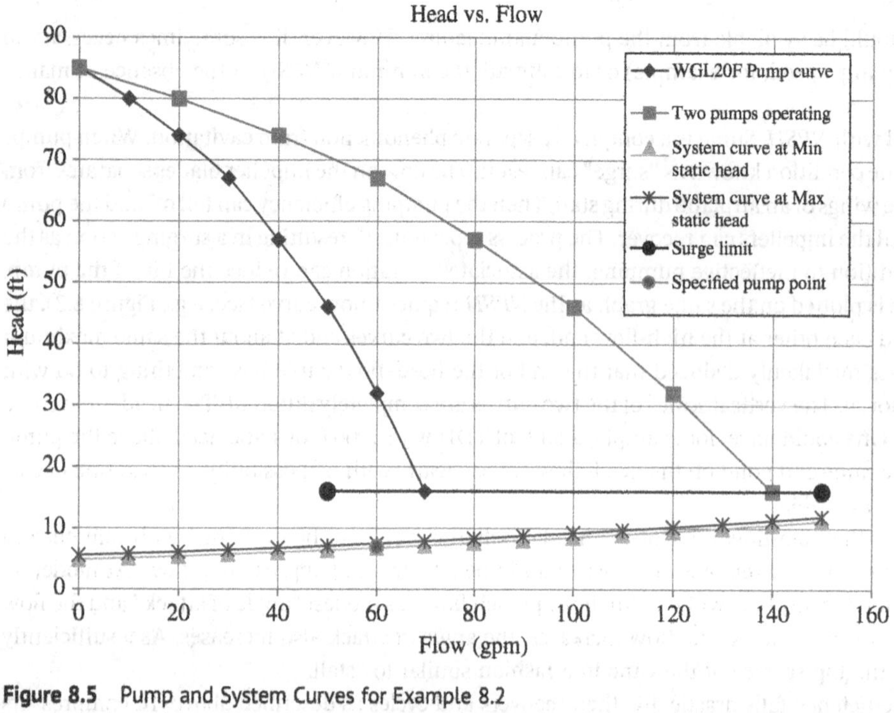

Figure 8.5 Pump and System Curves for Example 8.2

Solution

The system curve was calculated and added to Figure 8.5. The surge limit is also added based on the manufacturer's data. Points along the system-head curve can be seen to be beyond the surge limit, explaining the poor operation of the pump. Other grinder pumps were considered, but none would operate in the required range. A smaller discharge line was recommended by the author (the existing one was oversized) to bring the system curve up into the operating range of the pump.

The author has specified a large number of centrifugal pumps for numerous facilities. Such pumps can be categorized according to orientation (horizontal radial, vertical radial, axial aka propeller pumps, and mixed-flow propeller pumps), stages (single stage, two stage, and multistage with three or more impellers), and impeller designs (single suction aka end suction, double suction, semi-closed aka semi-open, enclosed impeller). In the following, examples in each category are discussed. All of the pumps discussed below were specified by the author.

8.2.3.1 Horizontal Radial Pumps

For New York City's Tallman Island WWTP, the author specified two horizontal, end suction, horizontal discharge, non-clog, mixed flow, centrifugal pumps to pump digested sludge from storage tanks to the sludge vessel. The pumps at maximum speed of 720 rpm are rated at 8300 gpm, TDH 45 ft of water, 79% minimum efficiency at rated conditions, 20-inch suction and discharge diameters, and 125 hp two-speed squirrel cage induction motors.

For the Cranston, Rhode Island Wastewater Treatment Plant, the author specified five return-activated sludge pumps. The pumps at maximum speed of 720 rpm are rated at 3200 gpm, TDH 64 ft of water, 74% minimum efficiency at rated conditions, 10-inch suction diameter, 8-inch discharge diameter, and 75 hp totally enclosed overhead mounted with V-belt drive.

Also for the Cranston WWTP, the author specified two horizontal centrifugal pumps for froth spray with process effluent. The pumps at maximum speed of 1750 rpm are rated at 325 gpm, TDH 60 ft of water, 78% minimum efficiency at rated conditions, 4-inch suction diameter, 3-inch discharge diameter, and 10 hp totally enclosed motor.

Also for the Cranston WWTP, the author specified a horizontal centrifugal chlorine water booster pump. The pump at 3500 rpm is rated at 48 gpm, TDH 115 ft of water, 52% minimum efficiency at rated conditions, 1-inch suction diameter, 1-1/4-inch discharge diameter, and 3 hp totally enclosed motor.

Also for the Cranston WWTP, the author specified two horizontal centrifugal effluent sample pumps. The pumps at 1750 rpm are rated at 20 gpm, TDH 19 ft of water, 40% minimum efficiency at rated conditions, 1-inch suction diameter, 1-1/4-inch discharge diameter, and 1/3 hp totally enclosed motor.

Also for the Cranston WWTP, the author specified two horizontal centrifugal gravity settler supernatant pumps. The pumps at 1160 rpm are rated at 700 gpm, TDH 28 ft of water, 85% minimum efficiency at rated conditions, 6-inch suction and discharge diameters, and 7-1/2 hp totally enclosed motor.

8.2.3.2 Vertical Radial Pumps

For the Powell's Cove Pumping Station (sewage) in New York City, the author specified two vertical, extended-shaft, bottom-suction, non-clog centrifugal pumps. The pumps at a maximum speed of 585 rpm are rated at 4200 gpm, TDH 30 ft of water, 80% minimum efficiency at rated conditions, 10-inch suction and discharge nozzles, and 40 hp four-speed, wound rotor motors.

The author specified two vertical centrifugal units for raw wastewater at the Pontiac Avenue Wastewater Pumping Station in Cranston, Rhode Island. The pumps at 875 rpm are rated at 5500 gpm, TDH 73 ft of water, 85% minimum efficiency at rated conditions, 12-inch suction and discharge diameters, 125 hp variable speed, and wound rotor motors.

For the Cranston, Rhode Island Wastewater Treatment Plant, the author specified three vertical centrifugal process wastewater pumps of 1000 gpm capacity at rated head, TDH at rated head of 50 ft of water, 80% minimum efficiency at rated conditions, 6-inch suction and discharge diameters, and 20 hp totally enclosed motors.

For Cranston's Mayflower Drive Wastewater Pumping Station, the author specified two vertical centrifugal pumps of 4200 gpm capacity at rated head, TDH at rated head of 45 ft of water, 76% minimum efficiency at rated conditions, 12-inch suction and discharge diameters, and 75 hp wound rotor motors.

For the Rochester, New Hampshire, 3.9 mgd AWTP, the author specified four vertical centrifugal non-clog, variable-speed raw sewage pumps. The pumps consist of one pair of one capacity and one pair of another capacity, with one of each capacity normally serving as standby. For use in conjunction with the equalization system described in Section 8.5, one of the larger variable-speed pumps operating at full speed is sufficient to pump a peak equalized flow of 4.68 mgd. Two variable-speed pumps one of the smaller capacity and one of the larger capacity, operating at full speed are sufficient to pump a peak flow of 6.93 mgd.

Vertical centrifugal sump pumps also fall into this category. Unlike the submersible sump pump, they utilize a shaft and bearing support configuration that allows the volute to hang in the sump while the motor and bearings are outside the sump. For New York City's Powell's Cove Pumping Station and Tallman Island WWTP, the author specified five such pumps, with capacities ranging from 75 to 250 gpm, rated head TDH of 15 to 25 ft, 67% efficiencies, discharge sizes of 1-1/2 to 4 in., and totally enclosed motors of ½ to 3 hp.

Submerged centrifugal sump pumps were specified for the Cranston, Rhode Island Wastewater Treatment Plant. Six duplex and one single unit are single stage, vertical shaft, bottom inlet, side discharge, enclosed impeller, with 50 and 100 gpm capacities at rated head, rated head TDH ranging from 19 to 42 ft, 1750 rpm, 2-inch and 4-inch discharge, and totally enclosed motors of 0.6–2 hp. The unit of single design was specified for the same facility with 50 gpm capacity at rated head, rated head TDH of 45 ft, 3450 rpm, 3-inch discharge, and totally enclosed 1 hp motor.

Also specified for the same facility were two portable submersible sump pumps for the 1st-stage aeration tanks, with 700 gpm capacity at rated head TDH of 22 ft, speed of 1150 rpm, 37 ft shutoff head, 4-inch discharge, and 7-1/2 hp totally enclosed submersible motor.

Also, there are two submersible screenless sewage ejectors and two submersible screenless drainage ejectors, the former rated for 75 gpm at 13 ft TDH with 2 hp motors; the latter rated for 150 gpm at 27 ft TDH with 3/4 hp motors.

One duplex centrifugal sump pump was specified for the Cranston, Rhode Island Wastewater Treatment Plant, having 50 gpm capacity at rated head, the latter being 22 ft TDH, 2-inch discharge, and 0.5 hp totally enclosed motor.

At Cranston's Pontiac Avenue Waste Water Pumping Station, one simplex centrifugal sump pump was specified, having 50 gpm capacity at rated head, the latter being 29 ft TDH, 2-inch discharge, and 0.5 hp totally enclosed motor.

Also for the Cranston WWTP, submersible screenless sewage and drainage pumps were specified for various locations, the former type rated for 75 gpm at 36 ft TDH with 2 hp motors; the latter type rated for 100 gpm at 22 ft TDH with 1-1/2 hp motors.

8.2.3.3 Mixed-Flow Propeller Pumps

The author specified four vertical, mixed-flow (diagonal impeller, between radial and axial) propeller pumps at the Tallman Island WWTP for location in primary effluent channels to pump back to the primary settling tank influent channels for use as agitation water (see Section 13.3). The pumps are two-stage, with maximum speed of 1800 rpm, 2450 gpm rated capacity, rated TDH of 48 ft of water, 87% minimum efficiency at rated conditions, 12-inch column diameter, 12-inch discharge nozzle diameter, and 40 hp totally enclosed motor.

8.2.3.4 Double Suction Pumps

For the Tallman Island WWTP, the author specified two single-stage, vertical shaft, split case, double suction pumps. The pumps served to circulate hot water from the main heat exchanges located in the Pump and Blower Building to the sludge heaters located in the Digester Complex. The pumps have 600 gpm rated capacity, rated 67 ft of water, 77% minimum efficiency at rated conditions, 82 ft shutoff head at rated speed, 5-inch suction diameter, 4-inch discharge diameter, and are directly driven by two-speed squirrel cage induction motors.

8.2.3.5 Axial Pumps

An axial-flow pump has a propeller-type impeller running in a casing. Specified four vertical, mixed-flow propeller pumps at New York City's Tallman Island WWTP for location in primary effluent channels to pump back to the primary settling tank influent channels for use as agitation water (see Section 13.3). The pumps are two-stage, with maximum speed of 1800 rpm, 2450 gpm rated capacity, rated TDH of 48 ft of water, 87% minimum efficiency at rated condition, 12-inch column diameter, 12-inch discharge nozzle diameter, and 40 hp totally enclosed motor.

8.2.3.6 Turbine Pumps

Regenerative turbine pumps fill a need between centrifugal and positive-displacement designs. They are a low-capacity, high-head type that combines the high discharge pressure of displacement types with the flexible operation of centrifugal pumps. Regenerative turbine pumps are known by several names, such as vortex (but see also below), peripheral, and regenerative. The chief difference between centrifugal and regenerative turbine pumps is in the impeller. Compared to most centrifugal pumps, turbine pumps have smaller-diameter impellers with rows of numerous small vanes. In the regenerative turbine pump, a double row of vanes is cut in the impeller's rim. These vanes rotate in a channel, with liquid flowing in at the suction and picked up by the impeller's vanes. After making nearly one revolution in the annular channel, the fluid has a high velocity that sends it out of the discharge. Liquid entering a centrifugal pump's impeller passes between its vanes only once and has energy added to it only while going from the impeller's eye to its rim. In a regenerative turbine pump, the vanes recirculate the fluid as it travels from the suction end to the outlet. Because of this action, the fluid flows in a path like a screw thread (helical) as it is carried forward. Consequently, energy is added to the fluid in a regenerative motion by the impeller's vanes as it travels from suction to discharge. This regenerative action has the same effect as multistaging in a centrifugal pump. In a multistage centrifugal, the fluid's pressure is the result of energy added in the different stages. In a regenerative turbine pump, pressure at its discharge is the result of energy added to the fluid by a number of impeller vanes.

For the Tallman Island WWTP, the author specified eight horizontal turbine pumps, two each in four locations, to pump city water from open float tanks to seal water connections on pumps and to flushing connections on meters. Pump speeds were all 1750 rpm, capacities at rated head ranged from 5 to 60 gpm, rated head TDH ranged from 75 to 175 ft, and motors totally enclosed ranging from 0.5 to 7.5 hp.

For the 23 mgd Cranston, Rhode Island Wastewater Treatment Plant specified five horizontal turbine seal water pumps to pump chlorinated plant effluent at two outlying pumping stations. Pump speeds are 1750 rpm, capacity at rated head 4 gpm, rated head TDH 75 or 125 ft; motors totally enclosed, squirrel cage induction, 1/3 or 3/4 hp.

Also for the Cranston, Rhode Island Wastewater Treatment Plant, the author specified four vertical turbine variable-speed utility water pumps to pump chlorinated plant effluent for various purposes. Pump speeds are 770 rpm, capacity at rated head 800 gpm, rated head TDH 130 ft, 8-in. discharge; motors totally enclosed, weatherproof, fan cooled, 40 hp.

8.2.3.7 Torque-Flow Pumps

Torque-flow pumps (semi-open) have a recessed impeller, which provides excellent solids handling capability. They were pioneered by the Western Equipment Manufacturing Company and are now provided by Trillium Flow Technologies, Houston, Texas. In our applications, they are used for pumping primary and digested sludge, for which the recessed impeller facilitates solids' passage. Specified six primary sludge and grit pumps and four digested sludge pumps for the Tallman Island WWTP, New York City. Specifications for the primary sludge/grit pumps included 1/2% solids concentration, 6 × 6 pump size, pump speed of 1200 rpm, rated head of 127/126/124 ft of water at corresponding capacities of 300/375/450 gpm; motors were totally enclosed 60 hp, belt drive with maximum motor speed of 1800 rpm. Specifications for the digested sludge pumps included 6% solids concentration, 4 × 4 pump size, pump speed of 1030 rpm, rated head of 52/46/40 ft of water at corresponding capacities of 400/600/800 gpm; motors were explosion proof 25 hp, belt drive with maximum motor speed of 1200 rpm. Backup plunger pumps were provided, as described below.

Two vertical centrifugal, torque-flow pumps were specified for pumping septage at the headworks of the Cranston, Rhode Island Wastewater Treatment Plant. Specifications included solids content of 4%, 4 × 4 pump size, 1160 rpm, one at 250 gpm capacity at rated head, 35 ft TDH rated head, and the other 150 gpm capacity at rated head, 38 ft TDH rated head, both with 10 hp explosion proof motors.

8.3 Positive-Displacement Pumps

Positive-displacement pumps are of the reciprocating and rotary types. In the reciprocating type, water is drawn into a cylinder on one stroke and forced out on the next. Pumps in this category include the piston pump, plunger pump, and diaphragm pump; the latter two are the most common and are discussed below. Rotary pumps have cams, gears, vanes, or screws (progressive cavities) whose action relative to each other creates a space that is filled with liquid and moves continuously from the inlet to the discharge sides of the pump (Reethof 1960; Linsley and Franzini 1964, §'s 12-12, 12-13, 12-17; Little 1976).

8.3.1 Plunger Pumps

Plunger pumps (Buse 1976) are used as the primary or backup pumps for pumping primary sludge and grease/scum. The pumps are classified as simplex, duplex, triplex, or quadraplex depending on the number of plungers. As backup pumps, they can be operated to unclog pipelines, such as those specified for the Tallman Island WWTP, New York City, to back up torque-flow primary sludge pumps. Hydraulic specifications include liquid type and concentration, rated flow capacity, TDH at rated flow capacity and suction lift, strokes per minute, number of cylinders, and suction and discharge pipe size.

Specified 31 plunger pumps for all the abovementioned purposes at the 23 mgd Cranston, Rhode Island Wastewater Treatment Plant. For the sludge pumping station at the Bridgewater Massachusetts Correctional Facility, the author specified a duplex plunger pump having a volumetric capacity of 117 gpm at 80 ft TDH including a maximum suction lift of 10 ft at 49 rpm maximum or at an RPM selected so that mean linear piston speed does not exceed 500 in./min. For the Tallman Island WWTP, New York City, specified plunger pumps as follows: one for primary sludge pumping, eight for thickened sludge pumping, and one for pumping digested sludge. All were duplex pumps, with sludge concentration of 6–7%, capacities of 120–200 gpm, TDH of 240 ft at rated capacity, 10 ft suction lift, 48 strokes per minute, and 6-in. suction and discharge side. Motors are 1200 rpm horizontal, single-speed, squirrel cage induction motors of 5, 7-1/2, and 20 minimum horsepower. Manufacturers include Carter Pump, Peapack, New Jersey; Komline–Sanderson, Peapack, New Jersey; and ITT Marlow (now Wastecorp Pumps, Grand Island, New York).

Overload protection devices are important for plunger pumps to protect against dead heading, closed valves in the system, and unexpected objects in the suction line. For example, in the Cranston facility mentioned above each pump was provided with two overload protection devices. The primary device is a pressure switch designed to shut off the pump motor at a discharge pressure equal to 125–150% of the specified dynamic head. The second device was a shear pin protection device designed to shear at 150–175% of equivalent normal torque.

8.3.2 Diaphragm Pumps

A diaphragm pump uses a combination of the reciprocating action of a rubber, thermoplastic, or teflon diaphragm and suitable valves on either side of the diaphragm (check valves, butterfly valves, flap valves, or any other form of shutoff valves) to pump a fluid. Such pumps are used in our applications for pumping sodium hypochlorite. The type of diaphragm pump used for that purpose has the diaphragm sealed with one side in the fluid to be pumped and the other in hydraulic fluid. The diaphragm is flexed, causing the volume of the pump chamber to increase and decrease. A pair of nonreturn check valves prevents reverse flow of the fluid. Bartlett (1999) provides important recommendations for maximizing the performance of diaphragm metering pumps.

For the Bridgewater Massachusetts Correctional Facility, the author specified two diaphragm pumps to handle 12-1/2% sodium hypochlorite, each with a capacity of 1% against a suction lift of 16 ft and discharge pressure established by a back pressure valve to maintain pressure at the pump discharge sufficient for accurate delivery. The motors were 1/3 hp for operation on 120 volt, 60 hertz, single-phase circuit. Single ball check valves were specified for suction and discharge. The hydraulic (water) reservoir and gear reduction lubrication system are totally separate.

8.3.3 Progressive Cavity Pumps

In 1930, René Moineau, a pioneer of aviation, while inventing a compressor for jet engines, discovered that this principle could also work as a pumping system. The University of Paris awarded René Moineau a doctorate of science for his thesis on "A new capsulism." His pioneering dissertation laid the groundwork for the progressive cavity pump. The name of the inventor and licensing led to the name of the American pump company, Moyno. The progressive cavity pump normally consists of a helical rotor and a twin helix, twice the wavelength helical hole in a stator. The rotor seals tightly against the stator as it rotates, forming a set of fixed-size cavities in between. The cavities move when the rotor is rotated, but their shape or volume does not change. The pumped material is moved inside the cavities, and the design results in a nearly constant flow.

Progressive cavity pumps are installed in pipelines; a single screw rotates in a cylindrical cavity, thereby moving the material along the screw's spindle. Whereas the Archimedean Screw lifts the liquid, the progressive cavity pump provides pressure increase in the pipeline. Covering a wide range of capacities, manufacturers include Moyno Pumps, Dayton, Ohio; NETZSCH Pumps & Systems Ltd., Dorset, United Kingdom; Roper Pumps, Commerce, Georgia; Robbins & Myers Inc., Springfield, Ohio; and Sulzer Engineering Company, Winterthur, Switzerland.

For the Cranston, Rhode Island Wastewater Treatment Plant, specified progressive cavity pumps for various purposes: two thickened sludge pumps, three sludge transfer pumps with macerators, three gravity settler sludge pumps, two polymer metering pumps, two lime slurry feed pumps, two pumps for grease concentration, and one polymer solution metering pump for liquid automatic polymer make-up.

8.4 Miscellaneous Types of Pumps

Miscellaneous types of pumps include air-lift pumps, screw pumps, hydraulic ram pumps, jet pumps, and other classifications (Symons 1966, Karassik *et al.* 1976). Such pumps will be discussed here and in subsequent sections of the text.

8.4.1 Air-Lift Pumps

In air-lift pumps, air is forced down a vertical pipe that elbows up into a riser and entrains the pumped liquid (Linsley and Franzini 1964, §12–14). The liquid can be solid-bearing, such as activated sludge. A minimum pump diameter (and connecting pipe diameter) is recommended to avoid clogging for waste-activated sludge service. The City of New York and the Chicago Sanitary District have used air-lift pumps successfully for this purpose for many years. New York's Tallman Island and Jamaica Plains wastewater treatment plants have used such pumps for return of waste-activated sludge since the late 1930s and early 1940s, respectively, and have had air-lift pumps installed in subsequent expansions of these plants. Hundreds of air-lift pumps have been installed in wastewater treatment plants. They are also used in dry wells where the compressor is on the surface and there are no mechanical parts in the well. For the Bridgewater Massachusetts Correctional Facility, the author specified two air-lift pumps for pumping activated sludge and skimmings for return to the oxidation ditches and for conveyance to holding and waste pumping facilities. Each air-lift pump has a capacity of 264 gpm at 70% submergence when supplied with 52 cfm of free air. For the Tallman Island Wastewater Treatment Plant, the author specified two 18-inch air-lift pumps for pumping activated sludge for return to aeration tanks. Each air-lift pump has a capacity of 2800 gpm with 75% submergence with air required not to exceed 475 scfm. References on water wells also note the use of air-lift pumps for well pumping (e.g., Johnson 1966, pp. 303–305, 388–389).

8.4.2 Screw Pumps

Screw pumps are rotary positive-displacement pumps that produce an axial-flow through the pumping elements (Czarnecki and Lippincott 1976). In the third century B.C.E., the Greek mathematician and physicist, Archimedes, invented the original screw pump and gave it his name, calling it the "Archimedean Screw." The Archimedes screw pump (Benjes and Foster 1976) is particularly useful in wastewater applications because it allows small wet wells and thus has the advantage over centrifugal pumps of substantially reducing grit and scum accumulations in the wet well. It also allows the non-clog conveyance of solids. Structural failures of large screw pumps have led to technical innovations in their structural design (Leitch 1988). Manufacturers include Lakeside Equipment Corp., Bartlett, Illinois, and Externalift by Evoqua Water Technologies, Pittsburgh, Pennsylvania.

Specified four screw pumps for the 65 mgd Providence, Rhode Island Wastewater Treatment Plant. The pumps were specified to have 30° angles of inclination and lifts of 12.75 ft. Three of the pumps have 96-inch screw diameters, 28 rpm pump speed, and 200 hp motors. The fourth pump has a 66-inch screw diameter, 32 rpm pump speed, and 60 hp motor. For the 23 mgd Cranston, Rhode Island Wastewater Treatment Plant, the author specified three screw pumps. The pumps were specified to have 38° angles of inclination, lifts of 23 ft, 54-inch screw diameters, 43 rpm maximum operating speed, and 50 hp motors. Also specified are screw pumps for the Rochester, New Hampshire, 3.9 mgd advanced wastewater treatment plant (AWTP).

8.4.3 Hydraulic Ram Pumps

A hydraulic ram pump is a cyclic water pump powered by hydropower. The hydraulic ram has a particularly interesting history, starting in the year 1740 (Gardner-Thorpe *et al.* 2007). The pump takes in water at one hydraulic head and flow rate and discharges water at a higher hydraulic head and lower flow rate. The device uses an air chamber and the waterhammer effect to develop pressure that allows a portion of the input water that powers the pump to be lifted to a point higher than where the water originally started. The hydraulic ram is sometimes used in remote areas where there is no electric power but where there is a source of low-head hydropower and a need for pumping water to a destination higher in elevation than the source (Fair and Geyer 1954, §30.4; Wagner and Lanoix 1959, pp. 149–151; Linsley and Franzini 1964, §12–16). This pump is particularly suitable for remote areas where electric power is not available. Two of the author's projects with hydraulic ram pumps are discussed below.

Example 8.3

The author designed a system utilizing a hydraulic ram pump in the town of Nuevo Chagres on the Caribbean coast of Panamá. The existing situation entailed a single open dug well serving a population of about 250. The community had a high incidence of water-borne diseases. A perennial stream flowed through the town through a breached dam. The dam had been used to divert water by pumping it to the small encampment of an American antiaircraft battery on a nearby hill during World War II. The author arranged for the restoration of the dam and conducted a topographic survey, which demonstrated that the hill had a suitable elevation to gravity-feed water to the town. Further study led the author to select and acquire a hydraulic ram pump to lift water from behind the dam to a tank on the aforementioned hill. The author located a surplus tank and arranged for the tank to have suitable exterior and interior coatings and be transported by Army landing craft to the shore of the town. The author designed and oversaw construction of a concrete cradle to which the tank was hauled and mounted. Piping was installed from the hydraulic ram pump to the tank and from the tank to sanitary spigots in the town. The pump specified by the author is a RIFE Ram with a capacity of 12–35 gpm, 2½-inch intake, 1-inch discharge, and minimum required vertical fall of 2½ ft.

Example 8.4

At Cerro Azul, Panamá, the source of water supply is a small pool immediately upstream from a waterfall on a stream north of a campsite to which a one-inch diameter pipe transmits water from the pool to a hydraulic ram pump. The author was asked to determine the capacity of the system to assist in planning a camp expansion. The "fall" or difference in elevation between the drive pipe inlet and its connection to the ram pump is approximately 33 ft. Approximately 500 ft of ½-inch galvanized pipe carry the pumped flow uphill to storage drums. The base on which the drums rest is 171 ft above the source of supply. The delivery pipe discharges at an elevation of 5 ft above the tank base. This results in a net static head on the system of 176 ft. The author calculated that the 200+ ft of 1" drive pipe with a 33-foot fall has a capacity of 6 gpm. The pump in use was designed to operate with a 4 gpm drive flow. The existing piping system thus has adequate capacity for the existing pump. That information can be used to aid in further planning.

8.4.4 Jet Pumps

Jet pumps function by directing a high-pressure stream of fluid through a nozzle to entrain the suction fluid (Linsley and Franzini 1964, §12–15; Sanger 1969; Jumpeter 1976). Jet pumps are used as a part of systems for reducing radon and other gaseous contaminants from groundwater by the venturi-aeration process (O'Brien 1995). The author specified such a system to treat private on-site well waters for homes in Dedham, Massachusetts, and Lakeville, Massachusetts. Specifications include Alraider Model 433, 115 volt, 1/2 hp regenerative blower, 1/2 hp Goulds Submersible Pump J+ Model C48A93A06, up to 16 gpm at 50 psi, and 44 gallon tank capacity.

8.5 Equalization

Distribution storage in water supply systems (Section 6.6) permits a variable distribution system demand to be met with a relatively constant supply to distribution storage. A relatively constant supply is preferable in terms of reduced transmission costs. The fluctuating demands on the water system are converted to roughly similar fluctuations in the wastewater collection system. The provision of storage to smooth those fluctuations can significantly reduce downstream transmission and treatment costs and improve treatment efficiency (more about these benefits will be said below). Such storage is called *equalization storage*. Equalization storage is similar in function to detention basin storage for stormwater flow attenuation, as discussed in Section 5.4. Here, the term *equalization storage* refers specifically to storage provided to convert a cyclic (usually diurnal) inflow to an outflow having a peak:average flow ratio smaller than that of the inflow and to otherwise modify the flow-time characteristics.

Pump station wet wells are occasionally used to equalize time-varying inflow. An increasingly common example is in septage receiving units in which it is desired to convert an intermittent inflow to a more constant pumped outflow. Long wet-well detention times may be required, so aeration and/or chemical treatment (e.g., chlorination) may be employed to minimize sedimentation and septic odors. The sizing of the wet well for equalization is straightforward, particularly

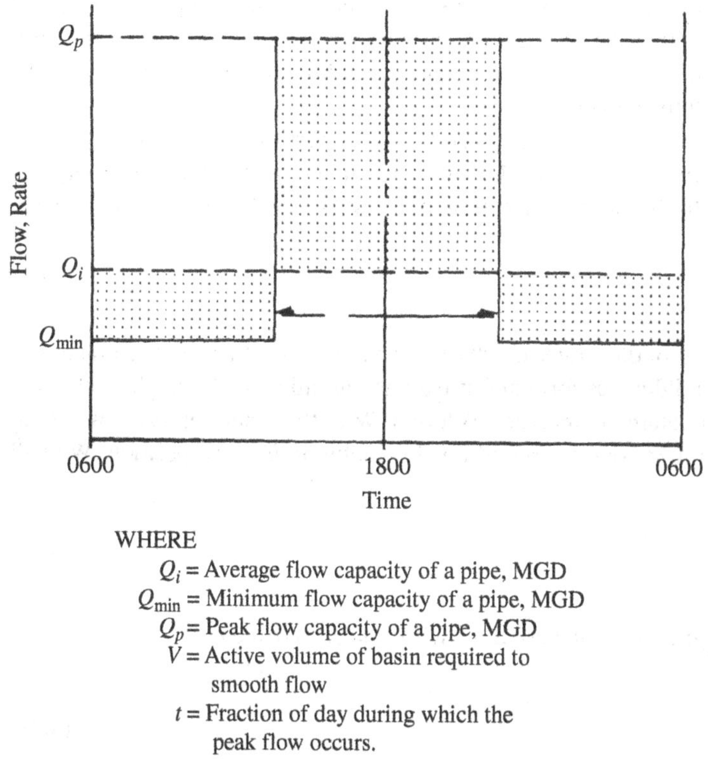

Figure 8.6 Square-Wave Hydrograph Assumed for Inflow to Equalization Basin

considering that alarm control can be exercised over the inflow (the rate of which will commonly exceed pumping capacity). Equalization of the more continuous wastewater inflows is generally accomplished in a separate equalization basin.

We shall commence our discussion of the analytical principles of equalization hydraulics by considering the somewhat idealized inflow pattern of Click and Mixon (1974) shown in Figure 8.6. In that pattern, it is assumed that the average-to-minimum flow ratio is identical to the peak-to-average ratio. Thus:

$$X = \frac{Q_p}{Q_i} = \frac{Q_i}{Q_{min}} \tag{8.15}$$

in which Q_{min} is the minimum wastewater flow, Q_i is the average flow, and Q_p the peak wastewater flow.

Letting t_f equal the fraction of the day during which the peak flow occurs, the average flow Q_i is related by definition to Q_{min} and Q_p as follows:

$$Q_i = Q_{min} + \left(Q_p - Q_{min}\right)t \tag{8.16}$$

Solving for t_f gives:

$$t = \frac{Q_i - Q_{min}}{Q_p - Q_{min}} \tag{8.17}$$

Noting from Equation 8.15 that X^2 equals $(Q_p/Q_i) \bullet (Q_i/Q_{min}) = Q_p/Q_{min}$, Equation 8.17 can be rewritten as:

$$t_f = \frac{X - 1}{X^2 - 1} \tag{8.18}$$

The *active* volume of the basin required to completely smooth the flow is:

$$\mathcal{V}_{active} = \left(Q_p - Q_i\right)t_f T = Q_i T \frac{(X-1)^2}{X^2 - 1} \tag{8.19}$$

in which T is the number of time units in a day. Thus, the active storage volume can be calculated from the average flow and the peak-to-average flow ratio.

Complete smoothing would generally require a pumped outflow, although a gravity outflow with a control valve could theoretically be provided. Click and Mixon (1974) found that the square-wave inflow pattern of Figure 8.6 provided uniformly conservative estimates of equalization basin capacities for pumped outflows. The author has used this inflow pattern to assess the economics of equalization in wastewater transmission systems.

Equalization can also be advantageous for wastewater treatment plants (USEPA 1974, Ongerth 1979). Overall economics can be achieved by equalization preceding the treatment processes, thus reducing the treatment process design capacity and achieving total economic benefit. Our analysis starts with the basin conservation of mass relationship given in Section 5.4:

$$\frac{dS}{dt} = I - O \tag{5.4}$$

in which I is the volumetric inflow rate, O is the volumetric outflow rate, and dS/dt is the differential increase in storage with respect to time. One can express the above equation in difference form and use an actual inflow hydrograph to obtain a numerical solution, or assume a sinusoidal inflow and obtain an analytical solution. With the latter approach, we could later express the actual inflow hydrograph in terms of a Fourier series and obtain a solution by superposition. We will assume a sinusoidal inflow for preliminary purposes:

$$I = \overline{Q} + (Q_p - \overline{Q}) \sin \frac{2\pi t}{T} \tag{8.20}$$

in which $\overline{Q}$ is average flow, Q_p is peak flow, t is time, and T is cycle period, which we take as one day.

For a basin with vertical sides:

$$S = AH \tag{8.21}$$

where A is surface area and H is depth measured from the bottom of the outflow control band (i.e., at $O = 0$). We assume a linear relationship between outflow O and depth H. This can be obtained by: (1) gravity discharge through a Sutro (proportional) weir; or (2) pumping controlled by a depth-paced proportional controller. By the first method, pumping (flow matched) must be from a sump to which the basin weir discharges freely; the static head in the sump must be below the weir crest at all times, requiring greater pumping head. By the second method, the equalization basin is used directly as the sump, and the pumping head is substantially less. The latter method is preferable and is used for the example given below.

For either method, letting C be a proportionality constant, we have:

$$O = CH \tag{8.22}$$

and therefore $H = O/C$, and with Equation 8.21 $S = (A/C)O$. Substituting the latter into Equations 5.4 and 8.20 and rearranging:

$$\frac{A}{C}\frac{dO}{dt} + O = \overline{Q} + (Q_p - \overline{Q}) \sin \frac{2\pi t}{T} \tag{8.23}$$

Equation 8.23 is a linear, first-order differential equation that is solved by the method of integrating factor and rearranging to obtain:

$$O = \overline{Q} + \frac{C}{A}(Q_p - \overline{Q})\frac{(C/A)\sin(2\pi t/T) - (2\pi/T)\cos(2\pi t/T)}{(C/A)^2 + (2\pi/T)^2} + \frac{C_1}{\exp(Ct/A)} \tag{8.24}$$

The last term will be dropped because it represents a nonsteady effect that will damp out to zero at large enough t.

We want to determine O_{max} which occurs at $dO/dt = 0$. Carrying out that operation with the above equation with the last term dropped gives:

$$\tan\left(\frac{2\pi t}{T}\right) = -\frac{C}{A}\frac{T}{2\pi} \tag{8.25}$$

The above equation gives the phase lag at which O goes through a maximum and minimum.

Combining Equations 8.24 and 8.25 and proceeding with further derivation gives maximum and minimum values of O as follows:

$$O_{\substack{\max \\ \min}} = \overline{Q} \pm (Q_p - \overline{Q}) \left(\frac{\dfrac{CT}{A2\pi}}{\sqrt{1 + \left(\dfrac{CT}{A2\pi}\right)^2}} \right) \tag{8.26}$$

For $O_{\max}$ Equation 8.26 can be rearranged to give the following solution for $CT/(A2\pi)$ with $H_{\max}$:

$$\left(\frac{CT}{A2\pi}\right)^2 = \frac{1}{\left[(Q_p - \overline{Q})^2/(O_{\max} - \overline{Q})^2\right] - 1} \tag{8.27}$$

For fully equalized flow ($C = 0$), we must obtain a separate solution for the case of $O = \overline{Q}$. For this case, from Equations 5.4, 8.20, and 8.21 we obtain:

$$\frac{dH}{dt} = \frac{Q_p - \overline{Q}}{A} \sin \frac{2\pi t}{T} \tag{8.28}$$

Integrating this equation gives:

$$H = -\frac{Q_p - \overline{Q}}{A} \frac{T}{2\pi t} \cos \frac{2\pi t}{T} + C_1 \tag{8.29}$$

with C_1 being a constant of integration. The minimum of H occurs at $t/T = 0$, giving:

$$C_1 = \frac{Q_p - \overline{Q}}{A} \frac{T}{2\pi} \tag{8.30}$$

and

$$H = -\frac{Q_p - \overline{Q}}{A} \frac{T}{2\pi} \left(1 - \cos \frac{2\pi t}{T}\right) \tag{8.31}$$

Differentiating the above equation gives stationary values of H, which are minima at $t/T = 0, 1, \ldots$ and maxima at $t/T = 1/2, 3/2, \ldots$. The effective maximum depth is:

$$H_{\max} = \frac{Q_p - \overline{Q}}{A} \frac{T}{2\pi}[1 - (-1)] = H = \frac{Q_p - \overline{Q}}{A} \frac{T}{\pi} \tag{8.32}$$

A somewhat similar derivation and result (but dimensionally incorrect) are presented by Ongerth (1979, Section 3). Equation 8.32 gives the surface area as follows:

$$A = \frac{Q_p - \overline{Q}}{H_{\max}} \frac{T}{\pi} \tag{8.33}$$

For more inflow variations that are not adequately represented by sine wave variations, numerical procedures can be developed by generalizing methods presented above. The author did so, for example, for the Rochester, New Hampshire, AWTP after preliminary sizing by the method given above, as exemplified below.

Example 8.5

It is desired to design an equalization basin at the head of a wastewater treatment plant (Rochester, New Hampshire, AWTP) with $\overline{Q} = 3.93$ mgd, $Q_p = 6.93$ mgd, $Q_{\min} = 1.60$ mgd, and one day cycle period. The equalization basin is to be designed for linear outflow with maximum outflow $O_{\max} = 4.68$ mgd. Determine the required volume for a basin with an effective depth of 10 ft.

Solution

Terms in Equation 8.27 have values as follows:

$$(Q_p - \overline{Q})^2 = (6.93 - 3.93)^2 = 9\ (\text{mgd})^2$$

$$(O_{\max} - \overline{Q})^2 = (4.68 - 3.93)^2 = 0.5625\ (\text{mgd})^2$$

$$\left(\frac{CT}{A2\pi}\right) = \sqrt{\frac{1}{(9/0.5625) - 1}} = 0.258$$

Equation 8.22 gives $C = O_{\max}/H_{\max} = 4.68/10 = 0.468$ mgd/ft, yielding:

$$\begin{aligned} A &= [CT/(2\pi)][A2\pi/(CT)] = [(0.468\ \text{mgd/ft})(1\ \text{day})/2]]/0.258 = 0.288\ \text{million gallons/ft} \\ &= 288,000\ \text{gallons/ft} \div 7.48\ \text{cu ft/gallon} = 38,600\ \text{sq ft} \end{aligned}$$

The configuration as designed has a width of 60 ft with three filet walls per tank and two tanks each of 325-ft length, giving a total surface area of approximately (because of filet wall widths) 2(60)(325) = 39,000 sq ft.

For use of a more specific inflow hydrograph, one can express Equation 5.4 in difference form, use the actual inflow hydrograph, and obtain a numerical solution. Alternatively, one can obtain an analytical solution for a sinusoidal inflow (such as derived above), use that solution in conjunction with a Fourier series analysis, and obtain the final solution using superposition. For either of these two methods, one should decide whether such precision is necessary.

Boksiner *et al.* (2014) reported on a creative, multifaceted peak-flow storage system at the Trinity River Wastewater Treatment Facility in Texas. The system was designed for a 715,500 m^3/d (189 mgd) average daily flow and a 2.36 million m^3/d (623 mgd) 2-hour peak flow. Wet-weather flows in excess of plant capacity can be diverted to two inline storage basins, an underground tunnel, and an offline storage basin of area of 18,600 m^2 (200,000 ft^3).

An alternative to equalization of the type discussed above has been used successfully in San Antonio, Texas (Macias 2006). At several wastewater facilities there with a combined capacity of 225.5 mgd, taking certain units off line during low-flow periods was found advantageous. By successively taking primary and secondary clarifiers off line at night during low-flow periods, nuisance algae that had required costly manual removal were deprived of sunlight, water, and nutrients; the algae died, and when flow increased again, the dead algae sloughed off at weirs. Taking one clarifier off line also reduced gasification in the online units, enabled additional time to lower sludge blankets, and enabled better weir overflow rates. By taking chorine contact times offline, adequate detention times were maintained during low flows, which enabled reduction in chemical and electricity costs. Other methods and benefits of using low-flow periods and excess capacity in the early years of the design period are noted by Macias. Motorized valves were essential in implementing all of these methods so that valves could be opened and closed quickly enough.

8.6 Waterhammer

In Section 6.7, time-varying (transient) flows such as those that result from the *slow* opening and closing of valves on pipelines were discussed. Pump startup, shutdown, and power failure, often in conjunction with valve operation, are also examples of time-varying flows. In Section 6.7, we considered two levels of simplification in the analysis of time-varying flows: (1) the neglect of fluid inertia and compressibility and (2) the neglect of compressibility only. Criteria for ascertaining the significance of fluid inertia were discussed in that section. In the present section, we shall consider the effects of compressibility in flowing liquids, develop criteria for ascertaining the significance of compressibility, and discuss methods for considering compressibility effects in certain simple, commonly encountered cases. We will see that compressibility effects commonly assume their greatest importance in pump systems and can most often be neglected in the applications considered in Section 6.7. The significance of the title of this section will become apparent below.

8.6.1 Speed of Pressure Waves

In Section 7.4, we discussed the transmission of "information" via surface waves in open-channel flow. An analogous but less conspicuous mode of "information" transmission occurs in pressure-conduit flows. In pressure conduits, the

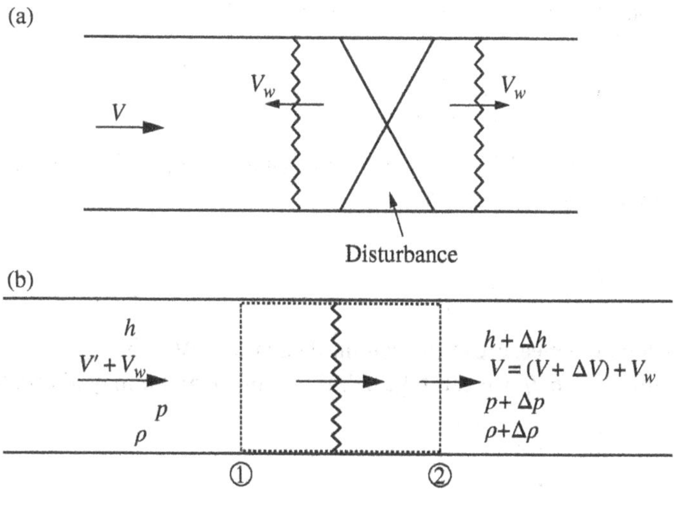

Figure 8.7 (a) Pressure Wave in Closed Conduit Relative to Stationary Observer; (b) Pressure Wave in Closed Conduit Relative to Moving Observer

"information" regarding the existence of a local conduit variation or disturbance is propagated through the conduit via *pressure waves.* We shall first determine the speed of such waves by means analogous to those employed for surface waves in Section 7.4.

Consider the flow in a closed conduit moving at velocity V as it approaches a local conduit variation or disturbance (such as a valve). As shown in Figure 8.7(a), pressure waves convey the "information" regarding the disturbance upstream and downstream of the disturbance at absolute velocity V_w. Let us consider only the waves moving upstream. In order to make the problem a steady-flow one, we will view such a wave and the oncoming flow from the perspective of an observer moving with the wave, i.e., the observer moves to the left at velocity V_w relative to Figure 8.7(a). The flow as seen by that observer is shown in Figure 8.7(b), in which the pressure wave appears stationary and the approaching flow has velocity $V + V_w$.

Frictional effects are neglected, one-dimensional flow is assumed, *and the conduit walls are assumed rigid.* The one-dimensional energy equation (Equation 2.64b) is then applied between sections "1" and "2" of Figure 8.7(b) to give:

$$\tilde{H} + \frac{(V+V_w)^2}{2} = (\tilde{H} + \Delta\tilde{H}) + \frac{V'^2}{2} \tag{8.34}$$

in which $\Delta\tilde{H}$ is the enthalpy change across the wave and $V' = V + V_w + \Delta V$ is the velocity of flow downstream of the wave location relative to the wave (or relative to the observer moving with the wave). By continuity:

$$\rho(V+V_w)A = (\rho+\Delta\rho)V'A \tag{8.35}$$

Substituting V' from Equation 8.35 into Equation 8.34 and solving for $V + V_w$ gives:

$$V + V_w = \sqrt{\frac{2\Delta\tilde{H}}{1-\left(\frac{\rho}{\rho+\Delta\rho}\right)^2}} \tag{8.36}$$

For small pressure waves, we may pass to differentials and drop higher-order terms to obtain:

$$V + V_w = \sqrt{\rho\frac{d\tilde{H}}{d\rho}} \tag{8.37}$$

The terms reversible, adiabatic, and isentropic will be discussed in Chapter 9, which deals with compressible gas flows. Here we shall state simply that the frictionless flow across a small pressure wave may be regarded as reversible and adiabatic, which is called isentropic (constant entropy). From thermodynamics (Mooney 1959, p. 152) for an isentropic process, the differential enthalpy is related to the differential pressure by:

$$d\tilde{H} = \frac{1}{\rho}dp \tag{8.38}$$

Substituting the above relationship into Equation 8.37 gives:

$$V + V_w = \sqrt{\left(\frac{\partial p}{\partial \rho}\right)_s} \tag{8.39}$$

in which the subscript "s" denotes constant entropy. The compressibility of liquid is expressed in terms of the *bulk modulus of elasticity*, defined by:

$$\beta = -\frac{dp}{dV/V} \tag{8.40}$$

which is a measure of the volume decrease $-dV$ resulting from a pressure increase dp. β has units of p, such as psi or kPa. Since $V = m/\rho$ we can take $dV/V = -d\rho/\rho$, and β may alternately be written:

$$\beta = \frac{dp}{d\rho/\rho} \tag{8.41}$$

Equation 8.39 may thus be written in terms of the bulk modulus as follows:

$$V + V_w = \sqrt{\beta/\rho} \tag{8.42}$$

The bulk modulus of water has a value of 300,000 psi × 144 = 43.2(10)^6 lbf/ft^2 [2.07(10)^6 kPa], which, with $\rho = \gamma/g = 62.4$ lbm/ft^3/32.2 = 1.938 (lbf/ft^3)/(ft/sec^2) or slugs/ft^3 (1 gm/cm^3), gives $V + V_w = 4720$ ft/sec (1440 m/sec).

The term $V + V_w$ represents the velocity of the pressure wave relative to the velocity of the flow, referred to as the celerity a of the wave, in this case for a rigid conduit (elastic conduits are considered below). [Following common convention, we shall use the notation a to denote the wave celerity in liquids and c (in Chapter 9) to denote the wave celerity in gases.] Water velocities could not possibly approach the value of the wave celerity without the water turning to steam, except under impractically high pressures (the reader may verify this by referring to Bernoulli's equation and the discussion of cavitation in Section 6.5). Thus, an interpretation analogous to that for open-channel surface waves (supercritical vs. subcritical, etc.) has no meaning. (We shall see in Chapter 9, however, that such an analogous interpretation for *gas* flows is quite meaningful.) Furthermore, the value of the celerity is so high compared to practical flow velocities that the celerity can be interpreted for practical purposes as the absolute velocity of the pressure wave (i.e., $V_w >> V$ so $a = V + V_w \cong V_w$).

A rigid conduit wall was assumed in the above, whereas actual conduits have elastic walls which influence the pressure wave celerity. The celerity a of pressure waves in an elastic pipeline filled with liquid is given by (Parmakian 1963, Chapters II and III):

$$a = \sqrt{\frac{1}{\rho\left(\frac{1}{\beta} + \frac{Dc_1}{Ee}\right)}} \tag{8.43}$$

in which D is the pipe's internal diameter, E is the modulus of elasticity (Young's modulus) of the pipe wall material, e is the pipe wall thickness, and c_1 is a pipe support coefficient.

For thin-walled ($D/e > 25$) steel, cast iron, and asbestos-cement pipes anchored against longitudinal movement throughout their length, $c_1 = 1 - \mu^2$ and $\mu = 0.3$ (Parmakian 1963, Chapter III), giving $c_1 = 0.91$. The nonconstant variables are then D/e and E. For water (with ρ and β as given above) and $c_1 = 0.91$, Equation 8.43 can be used to prepare curves for given pipe materials (E); the value of E used should be that for the specific pipe material (alloy and manufacturing method) if possible. For reinforced concrete pipe, the wave velocity may be determined based on an equivalent thickness of steel pipe; Parmakian (1963, Chapter III) suggests that this equivalent thickness be taken as the average circumferential thickness of the steel (reinforcing bars or cylinder) plus one-twentieth of the concrete thickness (see Example 8.7). [A more complex procedure, which takes into account the variations in effective modulus of elasticity as the concrete cracks, has been suggested by Kennison (1956) and may merit consideration in some cases.] A similar procedure may be used for steel-banded wood-stave pipes (Parmakian 1963, Chapter II). Parmakian gives recommendations and curves for other support conditions (pipe anchored at one end only; pipe with expansion joints), circular tunnels, and steel-lined circular tunnels. Additional pipe conditions are considered by Wylie and Streeter (1983, §2–3).

The gas content of the liquid being pumped is an important consideration in estimating wave celerities. Wylie and Streeter (1983, §1–3) note the substantial reductions in celerity that can result from very slight gas content. The effect of such reductions and related recommendations are given later.

Example 8.6

Find the waterhammer wave speed for Class 52 ductile iron pipe of 16-inch nominal diameter anchored against longitudinal movement (Admiralty Hill, Chelsea, Massachusetts).

Solution

For 16-inch Class 52 ductile iron, ANSI A21.50 gives $E = 34.6(10)^8$ psf, O.D. = 17.40 in., and thickness $e = 0.40$ in. Then I.D. = 17.40 − 2 × 0.40 = 16.6 in. So $Dc_1/(Ee)$ = 16.6 in.(0.91)/[34.6(10)^8 lbf/ft^2(0.40 in.)] = 4.36(10)$^{-7}$ ft^2/lbf. Then by Equation 8.43:

$$a = \sqrt{\frac{1}{\frac{\gamma}{g}\left(\frac{1}{\beta} + \frac{Dc_1}{Ee}\right)}} = \sqrt{\frac{1}{\frac{62.4\ \mathrm{lbf/ft^3}}{32.2\ \mathrm{ft/sec^2}}\left\{1/\left[43.2(10)^6\ \mathrm{lbf/ft^2}\right] + 4.36(10)^{-7}\ \mathrm{ft^2/lbf}\right\}}} = 3,890\ \mathrm{ft/sec}$$

Example 8.7

Find the waterhammer wave speed for 48-inch diameter AWWA C301-58 Class 250 Prestressed Concrete Steel-Cylinder Type pipe anchored against longitudinal movement (Arlington, Massachusetts).

Solution

For this piping, the AWWA standard gives $E = 30(10)^6$ psf for the steel cylinder. Using the same formula and dimensional units as in the previous example: O.D. = 48 in., and thickness e = 0.103+(48/16+5/8)/20 = 0.28425 in. Then I.D. = 48 − 2 × 0.28425 = 47.4 in. So $Dc_1/(Ee)$ = 47.4 in. (0.91)/[30(10)^6 lbf/ft^2(0.288425 in.)] = 3.56(10)$^{-8}$ ft^2/lbf. Then by Equation 8.43:

$$a = \sqrt{\frac{1}{\frac{62.4}{32.2}\left\{1/\left[43.2(10)^6\right] + 3.56(10)^{-8}\right\}}} = 2,964\ \mathrm{ft/sec}$$

8.6.2 Wave Motion and Waterhammer

The approaching flow in a conduit is continuously "informed" of conduit variations and disturbances by the emanating pressure waves. Under steady-state conditions, the information content and resulting flow adjustments vary spatially but not with time. A change in boundary conditions, such as the gradual opening or closing of a valve on a conduit of length L, creates a succession of small pressure waves that travel from the valve and cause the flow to adjust to the changed conditions. The succession of pressure waves will travel through the pipe length, reflect from the opposite end (the nature of the reflected wave will be discussed later), and return to the valve having traveled a total distance $2L$. The total travel time of each wave from and back to the valve is $\tau = 2L/a$, with a being the wave celerity, as given above.

When boundary changes are abrupt, such as when valves are opened or closed rapidly, the strength, reflection, and interaction of waves can cause large variations in pressure. The extreme case of sudden closure of a valve in a line is often cited to demonstrate the overpressures that can potentially result. Applying the momentum equation (Equation 2.40) to the control volume of Figure 8.7(b), we have:

$$pA - (p + \Delta p)A = (\rho + \Delta\rho)(V')^2 - \rho(V + V_w)^2 \tag{8.44a}$$

Substituting $\rho + \Delta\rho$ from the continuity equation (Equation 8.35) into the above, recalling $V' = V + V_w + \Delta V$, and simplifying yields:

$$-\Delta p = \rho(V + V_w)\Delta V = \rho A V \tag{8.44b}$$

If a valve is closed abruptly, then $V + \Delta V = 0$ or $\Delta V = -V$, for which the above equation gives:

$$\Delta p = \rho A V \tag{8.44c}$$

Such a pressure increase could occur, for instance, if a pump check valve was to hang open and, instead of closing upon the beginning of reverse flow, close after reverse flow reached velocity V.

As an example, with water at a flow velocity of $V = 5$ ft/sec in a steel pipe with $a = 3000$ ft/sec, sudden valve closure would result in a pressure increase of $\Delta p = \rho AV = 1.938(3000)(5)/144 \cong 200$ psi. If the valve closure is more gradual, a continuous succession of pressure waves will emanate from the valve as the pressure at the valve continuously increases. The positive pressure waves emanating in the upstream direction will be reflected as a negative wave at the open end of the pipe (e.g., at a reservoir). The negative waves will reduce the pressure toward the pre-closure value as they move past a point in the pipe. If the valve closure is completed in a time period less than the time ($2L/a$) it takes for the initial waves to return to the valve, then there will be no cancellation of the pressure increase at the valve until after valve closure is completed. Thus, if the valve closure period $T_c < 2L/a$, the pressure increase at the valve will be approximately that due to abrupt closure.

On the downstream side of a closing valve, negative waves cause decreasing pressures. Such pressure decreases are not directly detrimental in most pipes (the exception being some thin-walled pipes, which are not designed to sustain negative internal pressures in addition to external loadings). However, if the pressure reduces to the vapor pressure at a point along the pipe, a vapor cavity forms and the water column separates. Similar to valve and pump cavitation discussed previously (Sections 6.5 and 8.2, respectively), any detrimental effects are due to the pressure increase that occurs when the vapor cavity subsequently collapses. In this case, the collapse occurs when the negative waves are reflected as positive waves, which, when they reach the cavity, cause the water column to rejoin. The collapse of the cavity creates a hard blow, which is aptly described as *waterhammer*.

A related natural phenomenon is quite interesting. The two-inch-long snapping shrimp (aka pistol shrimp) has an outsize claw that it cocks open and snaps shut at high speed. This generates tiny bubbles that implode, sending a shock wave that stuns small crabs, worms, other prey, and combative snapping shrimp (Zuckerman 2011, Callier 2022, Thompson 2024).

Waterhammer analyses may be carried out in a variety of ways. The graphical method of waterhammer analysis (Parmakian 1963) is probably the best intuitive aid for understanding waterhammer phenomena. However, such an understanding is not necessary to use the methods discussed herein. Most modern analyses of *complex* systems are computer-based and employ the Wave Characteristic Method (WCM, a.k.a. Wave Plan Method) or the Method of Characteristics. However, computer-based methods are sometimes used in cases where the graphical method or other methods (discussed below) would result in a more rapid and intuitive solution. The emphasis here will be on relatively simple systems to which generalized waterhammer solutions or other approximations may be applied. Such solutions often provide a good first cut in more complex situations, allowing the problem or potential problem to be identified and the analytical effort and control methods to be scoped. An understanding of the detailed methods of waterhammer analysis is not essential to use the methods discussed below. However, the methods will have more meaning to the reader who has gained some understanding of the analytical methods (particularly the graphical method). Alternatively, the material presented below should provide a good point of departure for the reader desiring to go further with this subject.

We shall renew our consideration of the system depicted in Figure 6.16, specifically, the transient response of that system due to valve closure. Quick's waterhammer chart for uniform gate closure (i.e., gate open area decreasing uniformly with time from fully open to closed with constant discharge coefficient) with negligible pipe friction, as given by Parmakian (1963, Chapter X), is presented in Figure 8.8. The maximum head rise at the valve ΔH is given by:

$$\Delta H = 2\zeta K H_o \tag{8.45}$$

in which ζ is called the pipeline constant given by:

$$\zeta = \frac{cV_o}{2gH_o} \tag{8.46}$$

with V_o being the steady-state pipe velocity prior to valve closure and H_o being the static head (see Figure 8.8). The term K in Equation 8.45 is read from Figure 8.8 for given values of ρ' (ρ on the figure) and N, the latter being the time of gate closure T_c nondimensionalized with respect to the wave travel time:

$$N = \frac{T_c}{2L} \tag{8.47}$$

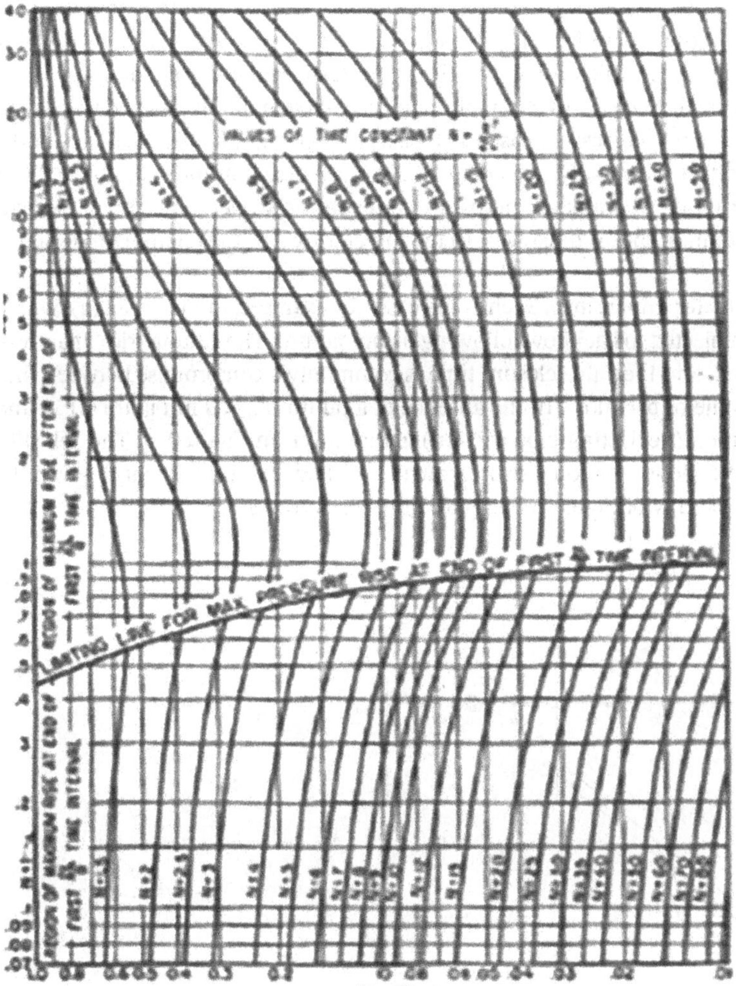

Figure 8.8 Quick's Waterhammer Chart for Uniform Gate Operation

Note from Figure 8.8 that K (and hence ΔH) decreases as N increases, slower gate closures giving smaller head rises. A major valve manufacturer suggests an approximation that can be derived by assuming $K \cong 1/N$ (this is the author's derivation, the manufacturer's basis being unknown), with which the above three equations give:

$$\Delta H \cong \frac{aV_o}{gN} = \frac{2LV_o}{gT_c} \tag{8.48}$$

or

$$\Delta p = \rho g \Delta H = \frac{\rho' a V_o}{N} = \frac{2\rho' L V_o}{T_c} \tag{8.49}$$

The approximation $K \cong 1/N$ can be seen by inspection of Figure 8.8 to give conservative values of K (and hence conservative ΔH) for all N when $\rho' < 2$ and for larger N at larger values of ρ'. For small N and large ρ', the approximation may underpredict ΔH. Further discussion of the use of this approximation is presented below.

8.6.3 The Incompressible Flow Approximation

Intuitively, it seems reasonable that if the valve closure or opening time is a sufficient multiple of τ, then compressibility effects will be negligible. This has in fact been demonstrated to be the case, as noted below. For the tank-draining applications of Section 6.7, T_c (valve closure time) $\ll \tau$ imposes little practical limitation. Note, for example, that the τ for the drain pipe of Tallman Examples 6.4 and 6.6 is on the order of $2L/a = 2(800)/4720 = 0.34$ sec. Clearly, a valve opening time in such a case that is a large multiple (say 100 times) of τ will be of no practical significance relative to draining time. As a practical matter, valves in such cases should be opened slowly, with elimination or acceptable minimization of waterhammer noise at the valve (and concurrently acceptably low overpressure) used as a basis for determining the adequacy of the rate of opening. With manual operators, the traditional (and usually conservative) approach is to gear the operator so that a sufficiently large number of turns of the handwheel, nut, or crank are necessary to insure slow operation. When electric, hydraulic, or pneumatic operators are used, the operating speeds must be selected judiciously and the equipment designed accordingly.

For valve closure, the rate of closure may be of greater importance. Such is the case, for example, in Figure 8.9(a), which can be considered to represent a pumping system subjected to backflow following power failure. The volumetric capacity of pump station wet wells to accept backflow is limited, and the valve closure time is commonly a compromise between limiting waterhammer pressures and limiting the volume of backflow. In the absence of a pump ($K_o = 0$ in Figure 6.15), and with uniform gate closure, the waterhammer solution (neglecting pipe friction) is that given in Figure 8.8. The "elastic" solution may be compared to the incompressible flow solution (also known as "rigid column" theory) given by Equation 6.112. In terms of the variables ρ (our ρ') and N of Figure 8.8, Equation 6.112 becomes:

$$\frac{h_m}{h_o} - 1 = \frac{1}{2}\left(\frac{\rho'}{N}\right)^2 + \sqrt{\left(\frac{\rho'}{N}\right)^2 + \frac{1}{4}\left(\frac{\rho'}{N}\right)^4} \tag{8.50}$$

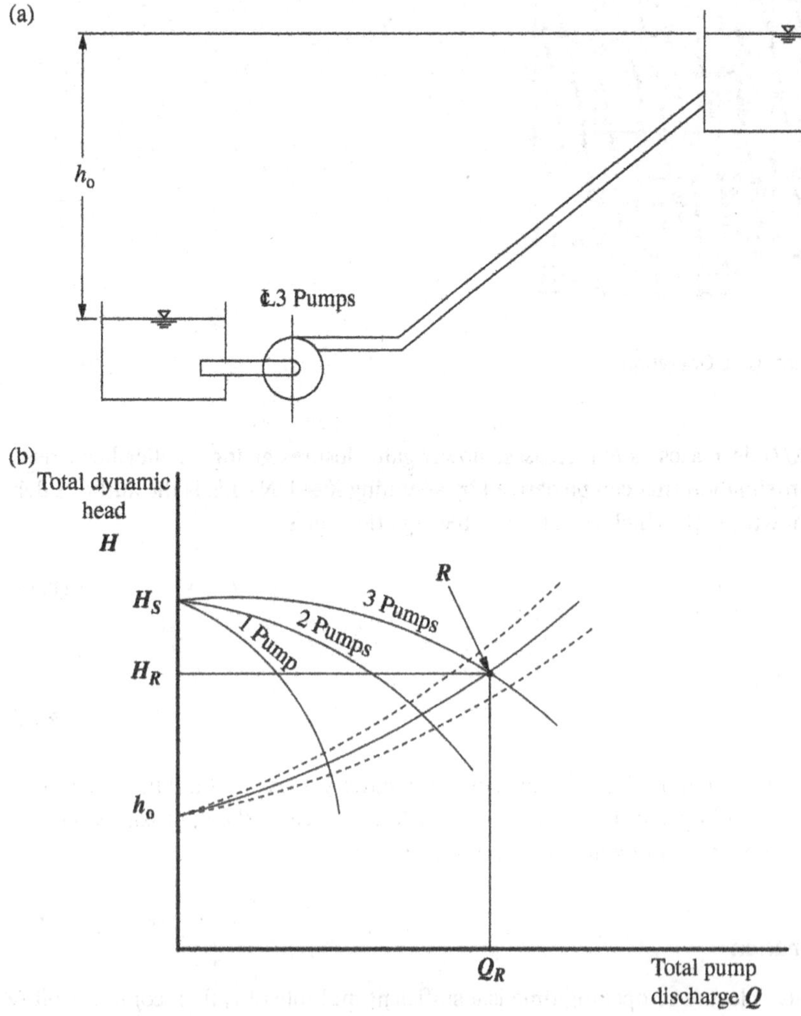

Figure 8.9 (a) Simple Centrifugal Pump System; (b) Pump-System Curves

The McNown example of Section 6.7 with d_v changed to 12 inches and a = 4000 ft/sec gives $\rho' = aV_o/2gh_o = aC/\sqrt{2gh_o} = 4000(0.6)/\sqrt{2(32.2)(100)} = 30$ and $N = T_c/(2L/a) = 10/[2(2000)/4000)] = 10$, for which Equation 8.50 gives $h_m/h_o = 10.9$. Figure 8.8 gives $K \cong 0.175$ for which the equation at the bottom of Figure 8.8 (with use of our notation) gives $h_m/h_o = 1 + 2(30)(0.175) = 11.5$, in reasonable agreement.

Vallentine (1965) carried out a formal comparison of the rigid water column theory and "elastic" theory for the configuration of Figure 6.6 (with $K_o = 0$) under conditions of uniform gate closure (the gate valve problem of Section 6.5 and above), ignoring pipeline friction. He used the Allievi charts for the elastic solution; the Allievi charts agree essentially with Quick's results (Figure 8.8) but cover a wider range of conditions. Valentine found close agreement for times of closure $T_c > 8(2L/a)$ provided $h_o/V_o < 50$ sec. Just outside this range, the rigid column theory underestimates the waterhammer pressures. Stephenson (1966) considered the same system but included line friction; he found good agreement between the rigid column and elastic theories for $T_c > 10(2L/a)$. These criteria are strictly applicable to the assumed uniform closure but provide useful guidance for actual valve closures considered later in this section. The rigid column results have the practical advantage of involving fewer variables, the relations among which may be presented in simple equation or graphical form for a particular type of valve.

8.6.4 Nature of Waterhammer in Centrifugal Pump Systems

The subject of waterhammer is a specialized one about which entire books have been written (e.g., Parmakian 1963, Rich 1963, Wylie and Streeter 1983). The treatment here begins with simplified methods of waterhammer analysis based on an intuitive understanding and generalized results of analyses presented by others and above. The approach is hierarchic (and generally conservative), whereby increasingly sophisticated methods are used as appropriate. It is anticipated that the approximate techniques described herein will be fully applicable to some of the waterhammer problems encountered. The material presented here should help establish a proper perspective as to when the more sophisticated methods presented below are required.

Figure 8.9(a) depicts a simple centrifugal pump system. The corresponding pump-system curves are presented in Figure 8.9(b). The system characteristics are represented by a *region* bounded by maximum friction and minimum friction curves. The "friction" losses include both pipe friction and "form" or "minor" losses due to valves and fittings. The value of estimating minimum as well as maximum system-head curves for use in pump selection was discussed at the beginning of this chapter. As explained below, maximum and minimum system-head curves can also be helpful in waterhammer studies.

Waterhammer in pumping stations is usually most severe during pump power failure. When power failure occurs, a continuous succession of negative pressure waves emanates from the pump as it decelerates causing reduced pressures or "downsurge" to occur in the pipe system. These negative pressure waves are reflected as positive pressure waves at the open end of the pipe (e.g., a water tank or open channel), resulting in "upsurge" in the piping system.

For the moment, let H_R and Q_R (point "R" in Figure 8.9(b)) be the head and flow at which the pump is operating just prior to power failure. Waterhammer characteristics are commonly nondimensionalized with respect to H_R and Q_R, such as in Figure 8.10, which gives $h = H/H_R$ and $q = Q/Q_R$ at the pump discharge. The initial conditions are represented by point 1 ($h = 1$ and $q = 1$) in Figure 8.10. Immediately after power failure, the head and flow will decrease, as shown in Figure 8.10. When point 2 is reached, flow reversal occurs, i.e., water starts flowing backward through the pump. If there is no valve control and the open end of the pipe discharges to a sufficient reservoir of water, the flow and pressure will eventually stabilize at point 5, with the pump acting as a turbine.

Check valves of one type or another are generally provided, which ultimately stop the flow and cause the system to stabilize at point 6. Of primary concern are the minimum and maximum pressure conditions occurring at points 3 and 4, respectively. Generalized solutions have been developed, by means of which critical points such as 3 and 4 can sometimes be predicted without the need for applying complete methods to each individual problem. Such generalized solutions are the basis for some of the procedures discussed below.

Waterhammer can also occur during pump startup (Joseph and Hamill 1972). If the piping system has been designed to withstand the pump shutoff head (occurring at zero Q_R) and the check valves are in close proximity to the pump, then this is usually not of concern (Joseph and Hamill 1972). An exception to this occurs when throttling valves are located downstream in the piping system, such as in plant process-water pumping systems with swing-check valves located at the pump. Pressure-relief valves are useful in such situations, as discussed below.

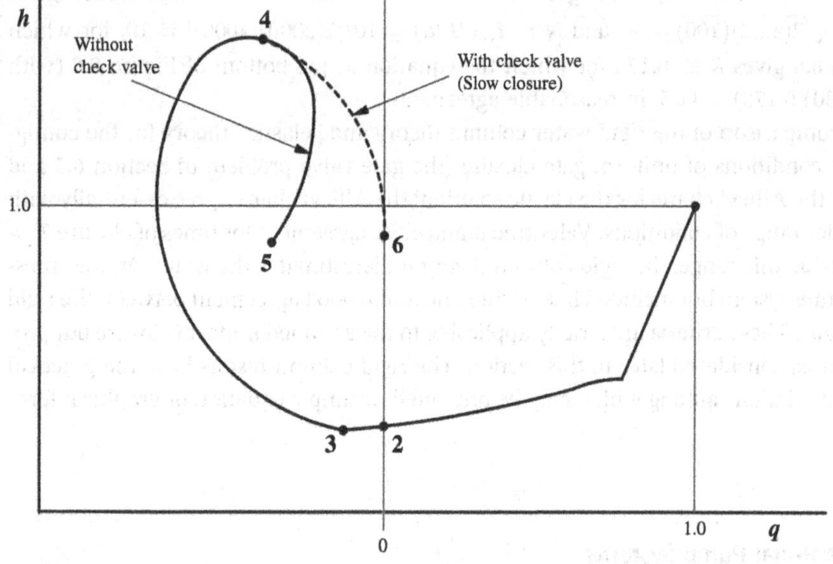

Figure 8.10 Head and Flow Variations during Transient

Waterhammer can also occur in pipe systems without pumps. Rapid opening and closing of valves on piping systems supplied by reservoirs is an example. Also, abrupt opening of gate valves to drain long water transmission lines (such as for repair work) can generate large underpressures, which can buckle thin-walled pipes.

Valve cavitation has been discussed in Section 6.5. A relationship between valve cavitation and waterhammer merits discussion here. A company manufacturing waterhammer controls claimed to be able to provide complete services, obviating the need for design engineering analysis. Their founder (Frank Parks of Parco Hydraulic Control) touted a method, which the author criticized in discussions with him, entailing the use of a stethoscope to listen for cavitation at the pipe just downstream of the valve and adjust the valve control speed to eliminate or minimize the cavitation noise at that location. However, the critical location is the location on the discharge line of greatest cavitation potential and associated negative pressures, as discussed elsewhere in this section. An example is given below of a case in which the resulting negative waves traveled back to the pump, where they were reflected as excessive positive waves, which resulted in catastrophic pipe failure. A degree of audible cavitation at the pump valve is of much less importance; replacing a pump valve as a routine matter is far less important than a catastrophic failure.

It is important to use the correct inertia for the pump, as discussed later in this chapter.

8.6.4.1 Generalized Waterhammer Charts

For centrifugal pump systems, generalized waterhammer charts have been prepared by Parmakian (1963, Chapter XI), Schnyder (Rich 1963, p. 180), and Kinno and Kennedy (1965). All of those charts are based on the assumption that there is negligible valve control. The Kinno and Kennedy (referred to hereafter as "K&K") charts are the most comprehensive and were obtained by the graphical method, including pipe friction for a pump of specific speed (defined by Equation 8.65) equal to 2700. The K&K charts apply with sufficient accuracy to pumps of specific speed of less than 2700 (Kinno and Kennedy 1965). Donsky's (1961) work suggests that characteristics for radial-flow pump similar to those assumed by K&K will provide conservative estimates of waterhammer effects for pumps having higher specific speeds.

The following parameters may be obtained from the K&K charts:

1) Minimum head at pump.
2) Minimum head at mid-length of pipe.
3) Maximum head at pump.
4) Maximum head at mid-length of pipe.
5) Time of flow reversal at pump.
6) Maximum head at pump if reverse rotation is prevented.

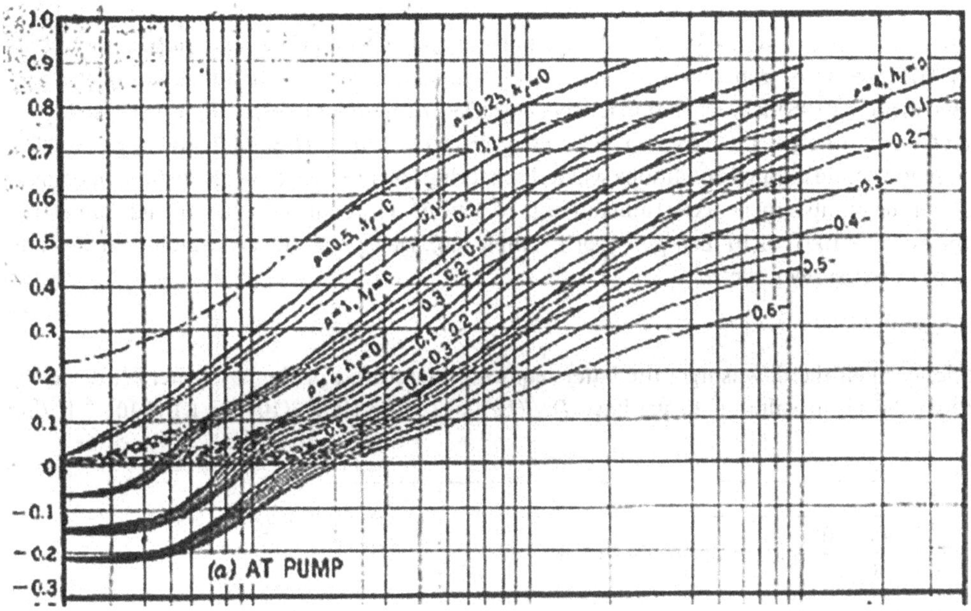

Figure 8.11 Minimum Head at the Pump after Power Failure

The Parmakian and Schnyder charts are similar to some of the K&K charts but exclude friction and with one exception (see below), do not provide as many parameters as the K&K charts. For frictionless flow, the three sets of generalized charts show good agreement with each other. The only advantage of the other charts is that the Parmakian chart for time of flow reversal (frictionless case) is easier to read than the K&K chart. Also, only Parmakian gives results for magnitude and time of maximum reverse pump speed and time of zero pump speed.

The reader is referred to Kinno and Kennedy (1965) for their charts, only one of which [their Figure 10(a)] is presented here as Figure 8.11 for dimensionless minimum head at the pump after power failure (defined below), which is used in an example presented below. Supplementary charts given by Parmakian are also discussed below without inclusion here [note that Parmakian's K equals K&K's K_1 divided by two so that Parmakian's $K(2L/a) = 1/(2\tau)$]. The K&K charts characterize the transient behavior of a pump-pipe system in terms of the following parameters:

1) The pipeline constant (here using a different notation than used in Equation 8.46 and Figure 8.11):

$$\zeta = \frac{aV_R}{2gH_R}$$

in which a is the celerity of pressure waves in the pipe (determined as described above), V_R is the pipe velocity at the rated pump discharge Q_R, H_R is the rated pumping head, and g is the acceleration of gravity. The terms Q_R and H_R are further explained below. (The term ζ here is identical to K&K's ρ.)

2) The pump inertia time constant, given by:

$$T_m = \frac{WR^2\eta_R(2\pi N_R)^2}{\gamma g H_R Q_R}$$

in which WR^2 is the moment of inertia of the rotating parts of the pump (including fluid and motor, as appropriate), η_R is the rated pump efficiency, and N_R is the rated speed in revolutions per unit time. [The term T_m is the reciprocal of K&K's pump constant K_1. The notation T_m and T_e (see next item) and the interpretation here is that of Paynter and Free (1966). The terms as given here are less prone to dimensional error when working with various dimensional systems.]

3) The round-trip wave time $T_e = 2L/a$, in which L is the length of the pipeline. This is the time required for pressure waves emanating at the pump to travel down the pipeline, reflect, and return to the pump.
4) The ratio (dimensionless) of the pump inertia time constant to the round-trip wave time: $\tau = T_m/T_e$
5) The ratio, λ_f, of the friction head in the discharge line, H_f, to the rated pump head: $\lambda_f = H_f/H_R$
6) The ratio, h_d, of the minimum head at the pump during transient, H_d, to the rated pump head: $h_d = H_d/H_R$

Example 8.8
A single pump takes flow from a wet well and discharges to a storage tank, as shown in Figure 6.11 (Bridgewater Massachusetts Correctional Facility). The pipe is cast iron anchored against longitudinal movement with a modulus of elasticity $E = 1.73(10)^9$ lbf/ft^2 and wall thickness $e = 0.45$ inch. The pump is driven by a 75-hp motor with respective rated speed, flow, rated head, friction head, and efficiency of $N_R = 1750$ rpm, $Q_R = 1000$ gpm, $H_R = 170$ ft, $H_f = 94$ ft, and $\eta_R = 83\%$. The orifice is sized to cause the pump to operate at rated conditions when discharging solely to the tank against the maximum static head. The combined WR^2 for the pump impeller (wet) and motor is 10.5 lbf-ft^2. The equivalent length and diameter of piping from the reservoir to the tank is 4100 ft of 10-inch pipe (per Example 6.6). Determine the acceptability of a swing-check valve as the sole means of waterhammer control.

Solution

We first estimate the wave celerity, then the downsurge, the latter requiring calculation of the parameters ζ, τ, and h_f. For $D/e = 10/0.45 = 22.2$ and the given value of E, we have $Dc_1/(Ee) = 22.2(0.91)/1.73(10)^9 = 1.168(10)^{-8}$ ft^2/lbf. Then Equation 8.43 gives:

$$a = \sqrt{\frac{1}{\frac{\gamma}{g}\left(\frac{1}{\beta} + \frac{Dc_1}{Ee}\right)}} = \sqrt{\frac{1}{\frac{62.4\ \text{lbf/ft}^3}{32.2\ \text{ft/sec}^2}\{1/[43.2(10)^6\ \text{lbf/ft}^2] + 1.168(10)^{-8}\ \text{ft}^2/\text{lbf}\}}} = 3,850\ \text{ft/sec}$$

The value of ζ is determined by the following progression:

$Q_R = 1000\ \text{gpm}/449 = 2.23$ cfs
Area $A = (\pi/4)(10/12)^2 = 0.545$ ft^2
Velocity $V_R = Q_R/A = 2.23/0.545 = 4.08$ ft/sec

From the equation given in Item 1 above:

$$\zeta = \frac{aV_R}{2gH_R} = \frac{3850(4.08)}{2(32.2)(170)} = 1.435$$

For use below, we calculate $2L/a = 2(4100)/3850 = 2.13$ sec, and from the reciprocal of the equation for pump inertia time constant given in Item 2 above, we calculate K&K's pump constant K_1, which is double that of Parmakian and given by:

$$K_1 = 1/T_m = \frac{2\gamma g H_R Q_R}{WR^2 \eta_R (2\pi N_R)^2} = \frac{2(62.4)(32.2)(170)(2.23)(60)^2}{10.5(0.83)[2\pi(1750)]^2} = 5.21\ \text{sec}^{-1}$$

From which:

$$\tau = \frac{1}{K_1(2L/a)} = \frac{1}{5.21(2.13)} = 0.0901$$

The ratio of friction head to rated head is:

$$\lambda_f = H_f/H_R = 94/170 = 0.553$$

From Figure 8.11, using the above values of ζ, τ, and λ_f (ζ being Figure 8.11's h_f), we obtain $h_d \cong -0.12$ at the pump. Using H_R and h_d from above, the maximum suction water elevation of 109 ft from Figure 6.11 and the pump ℄elevation from Figure 6.11, we obtain the minimum head at the pump = 109 + 170(−0.12) − 92 = −3.4 ft. The minimum absolute pressure head at the pump is given by 32.86 − 3.4 = 29.46 ft, which is well above vapor pressure at any possible temperature (see Table 9.1). Therefore, column separation is not a concern.

The maximum overpressure due to abrupt check valve closure equals $[2(1 + 0.11)]\ H_R = 2.22(170) = 378$ ft. With the maximum suction well water level from Figure 6.11(a) of 109 and pump ℄Elev. of 92 ft, we have head at the pump = 378 + 109 − 92 = 395 ft, which is multiplied by 0.434 to give 171 psig. The pump manufacturer confirmed the acceptability of this pressure at the pump.

Elaboration on some of the parameters used above follows.

Pressure Wave Celerity

The celerity (velocity) of propagation of pressure waves in a pipeline was discussed above. It was noted that very slight gas content can cause substantial reductions in wave celerity. Reductions in celerity will generally increase downsurge and may increase upsurge (such as if swing-check valves are used) in pump systems. The author's practice, as discussed above, is to base waterhammer computations on wave celerities equal to the celerity for gas-free liquid and that for one-half of the celerity determined for gas-free liquid when the liquid is raw sewage, or water expected to have entrained gases. In cases where gas concentration is determinable (e.g., for air injection into a force main), the relationships provided by Wylie and Streeter (1983, §1.3) can be used to estimate a.

WR^2 for Motor and Pump

Special mention should be made of a couple of potential pitfalls in establishing WR^2 values for the motor and pump. In obtaining the electric motor WR^2 from catalogs and/or manufacturers, one should be certain that the impeller WR^2 and *not the load inertia* WR^2 is used. The latter is commonly given and is the load WR^2 which the motor can accelerate without injurious temperature rise. The load inertia is typically one or two orders of magnitude larger than the motor/shaft $WR^2$and is of no significance in waterhammer studies.

The pump WR^2 should preferably be the "wet impeller" WR^2 to account for fluid in the pump casing. The reader is advised to check with manufacturers and use values for the particular type of motor and pump to be specified. The motor typically provides about 90% of the total WR^2 of the pump and motor (Parmakian 1976), so the total WR^2 can be conservatively assumed to be approximately that of the motor in many cases. This approximation is particularly useful if information on the pump WR^2 is difficult to obtain.

Rated Pump Characteristics

The terms H_R, Q_R, η_R, and N_R are all based on "rated" pump characteristics. The rated pump speed N_R is the speed at which the pump will be operating when waterhammer conditions are most critical (usually maximum speed if a variable-speed drive) and for which the pump curve has been established. The rated efficiency is the *maximum* efficiency of the pump at the rated speed. H_R is the head corresponding to N_R and η_R, which can be obtained from manufacturer's curves. If the operating pumps are all identical, and if the actual pump head is approximately equal to H_R, then Q_R equals the discharge of a single pump (operating at N_R and H_R) multiplied by the number of pumps operating. If the pumps are not identical, then judgment may suggest conservative approximations, whereby the generalized charts may still be employed. If the actual pump head differs significantly from H_R, then the generalized charts can be used as described below to provide acceptable approximations, provided that the pump discharge is within 20% of Q_R (Kinno and Kennedy 1965, p. 263). More specifically, computer-based methods may be required if the abovementioned approximations are not adequate.

Downsurge Analysis

It is conceptually convenient to divide the waterhammer problem into two time periods: (1) the period prior to significant valve control, and (2) the period commencing when valve control becomes significant. It is assumed here that, regardless of the type of check valve employed, valve control does not become significant during the time prior to flow reversal. This is usually a desirable objective since significant valve control during this time will increase the magnitude of the downsurge. It is sometimes necessary, however, to begin valve closure during downsurge so that the valve will be in position to dissipate the energy of the flow when it reverses, and thereby prevent large reverse flow and large reverse acceleration of the pump. Some increase in downsurge may be an acceptable compromise.

The initial purpose of downsurge analysis is to determine whether or not the minimum waterhammer pressures in the pipeline reach the vapor pressure of water. If so, a vapor pocket will form and column separation may occur. The major potential problem with column separation is that very large overpressures can be created when the water columns subsequently rejoin. The reduced pressure itself is usually not of concern from the standpoint of pipeline buckling, since most pipelines are capable of withstanding an internal vacuum with full external loading. The possible exceptions are thin-walled steel pipe, which may be encountered under certain circumstances, and large-diameter pipes with small thickness-to-diameter ratios.

In performing the downsurge analysis, first determine the following pump characteristics for each individual pump: N_R, E_R, Q_R, H_R, n_S. Next, determine the actual pump head H_u and discharge Q_u based on the maximum system-head curve. Since the downsurge is increased by pipe friction, this will provide conservative estimates. [Upper and lower estimates of friction losses should be made in determining the envelope of system-head curves for pump section. Similar reasoning regarding the effects of friction should be employed when interpolating on the K&K curves.] Check whether or not Q_u is within 20% of the sum of the Q_u's for the operating pumps. Next, determine the following pipe characteristics: flow area A, flow velocity $V_u = Q_u/A$, pipe profile, and length L. If the pipe diameter varies, a conservative approximation is obtained if V_u is based on the minimum area, and an equivalent length is determined by methods presented by Parmakian (1963). Then, determine the wave celerity a as discussed above, and use one-half a as a lower estimate if gas content is anticipated.

Compute the quantities ρ_u, τ_u, and $h_{fu} = H_{fu}/H_u$, using V_u, H_u, Q_u, and H_{fu}. Do so for both a and one-half a. In determining WR^2 for use in K, use the sum of WR^2 values for all operating pumps. Next, determine H_{mu} and h_{du} from Figure 8.11, using K&K's Figure 11 as an aid. The minimum heads are then computed as $H_d = h_{du}H_R$ and $H_m = h_{mu}H_R$. The minimum absolute pressure during downsurge should be determined at the pump and pipe mid-length as follows: Absolute pressure (feet of water) = H_m (or H_u) + 32.86 – pipe ℄elevation + wet well elevation.

The point where the absolute pressure is minimum may be estimated graphically by plotting the absolute pressure on a convenient scale above the pipe ℄at the two points on the pipe profile. One straight line should be drawn through the two points. A second straight line should be drawn between the point at the pipe's mid-length and the water surface level of the downstream reservoir. The minimum pressure should be scaled at the point where the straight line comes closest to the pipe ℄(this approximation should be used judiciously). Unless the pipe has some particularly high points, the minimum pressure will usually occur at the pump. This absolute minimum pressure should be compared with the vapor pressure to determine if water column separation could occur. The vapor pressure of water increases with increasing temperature, ranging from 0.698 psia = 1.56 ft of water at 90°F up to 14.7 psia (atmospheric pressure) at 212°F. The vapor pressure of water as a function of temperature, as tabulated in Table 9.1, will generally be less than 1 psia for temperatures of interest. The concept of vapor pressure is fully discussed in Section 9.2, and its role in valve cavitation is discussed in Section 6.5.

It will be assumed for the moment that vapor pressure is not reached, and the analysis may proceed to the next step (upsurge analysis). A subsequent section on downsurge control discusses steps to take if the downsurge is excessive.

Upsurge Analysis – Swing-Check Valve Control

A properly functioning swing-check valve will close abruptly at the instant of flow reversal. When this occurs, the effect of swing-check valve closure may be conservatively estimated by assuming that the pressure will increase above $h = 1$ (datum at elevation of the wet well) by an amount equal to the downsurge relative to $h = 1$ (along path 2-3-4 in Figure 8.12). This procedure provides a conservative estimate for two reasons: (1) it assumes that minimum pressure occurs at the pump at the instant of flow reversal, whereas it may have occurred prior to that and the pressure would have increased slightly; and (2) point 3 is actually below $h = 1$ due to pipe friction (Donsky 1961, Parmakian 1963, Kinno 1968), the upsurge actually following the dashed line 3′–4′ in Figure 8.12.

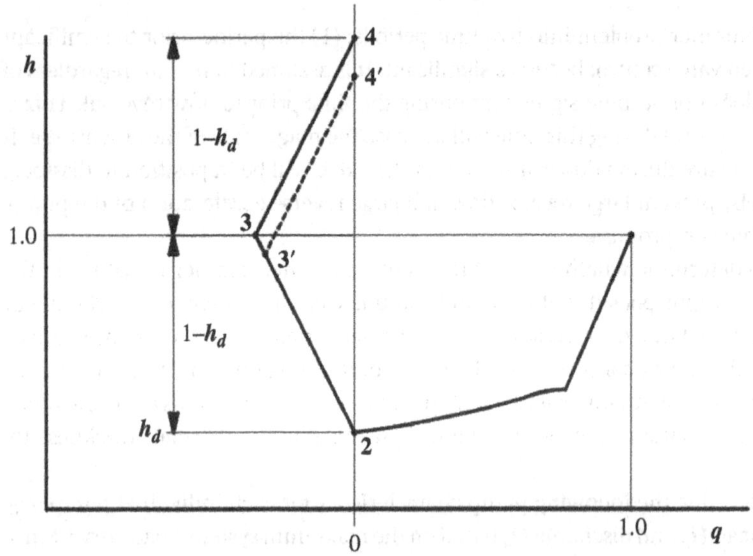

Figure 8.12 Transient Characteristics with Swing-Check Valve

To state the above mathematically, the maximum gage pressure at the pump due to check valve closure is given by:

$$\text{Gage pressure(feet of water)} = (1 - h_{du})H_R - \text{pipe } \Phi \text{ elevation at pump} + \text{wet well elevation}$$

Other Methods of Upsurge Control

The upsurge resulting from swing-check valves is sometimes excessive, and other methods of upsurge control must be considered. Even when swing-check valves adequately limit upsurge and when downsurge is not a problem, other factors may also dictate alternate types of check valves. Such factors include reliability, lower head loss, reduced maintenance, and more positive valve control for both pump startup and shutdown (including power failure). Cone check valves are commonly used for this purpose for sewage service, with butterfly valves being common for water service. The methods of upsurge control discussed below are inertial control, anti-reverse control, power-operated valve closure, pressure relief, and miscellaneous methods. Pipeline profile changes and route changes may also merit consideration. The methods of downsurge control discussed later are also applicable to upsurge control.

Inertial Control

Inertial control involves increasing the WR^2 of the rotating pump-motor parts. This increases the parameter τ which in turn decreases both the downsurge and upsurge. Motors with lower speeds usually have higher WR^2 and this might be taken advantage of in some cases. The construction of synchronous motors is usually such that they have larger WR^2 than induction motors of similar ratings. Similarly, totally-enclosed, fan-cooled motors generally have higher WR^2 than drip-proof motors. Although not a common method of waterhammer control, it may be feasible to add a flywheel (Yang 2001) or specify a pump or motor with higher-than-normal WR^2. Although motor WR^2 values are fairly standard between manufacturers for smaller motors (up to about 500 hp), larger motors have a wider range of WR^2.

A more common application of inertial control is employed when magnetic drives are used. In the event of power failure, the magnetic coupling is normally deactivated, so that only the pump and magnetic members of the drive contribute their WR^2. A backup battery system can be used to energize the magnetic coupling so that the motor and ring member of the drive contribute their WR^2, which is typically 50% of the total. If very slow valve closure is employed, the K&K charts can be used to determine the resulting upsurge (Figures 13 and 14) as well as downsurge. Since upsurge is increased by pipe friction, the minimum system-head curve will provide conservative estimates for upsurge analysis. The feasibility of very slow valve closure depends largely on the capacity of the wet well and upstream conduits to receive backflow. The reliability of the battery-activated system can be enhanced by providing a low charge alarm. Cost and maintenance should be considered. When fast valve closure is desired, the methods discussed below under "power-operated valve closure" may be employed in an analysis including inertial control.

Pump Anti-Reverse Control

Unless it is prevented from occurring, the pump and connected parts will begin to spin backward shortly after flow reversal. A pump spinning in reverse provides a higher resistance to reverse flow than would be the case if reverse spin was prevented. Consequently, reverse spin will increase the upsurge pressures. Furthermore, if power comes on during reverse spin, the pump shaft may snap; the controls should assure that this cannot happen.

An anti-reverse ratchet may sometimes be employed to prevent backspin. Kinno and Kennedy (1965) presented a chart with which the maximum head at the pump may be determined if reverse rotation of the pump is prevented. The feasibility of using this method depends in part on clearances, which may limit the space available for the brake or ratchet mechanism. If clearances prevent the use of an anti-reverse ratchet on the pump, then the use of an anti-reverse device on the motor can be considered. If magnetic drives are used, motor-anti-reverse control could necessitate battery energizing of the magnetic drive during power failure. Slow valve closure is required for the K&K to apply. The same backflow limitations apply as for inertial control. When fast valve closure is desired, the methods discussed immediately below may be employed in an analysis, including anti-reverse control.

Power-Operated Valve Closure

Power-operated valves usually should not begin to control the flow significantly until the time of flow reversal in order to prevent worsening of the downsurge. Thus, the valve opening prior to flow reversal should be such that the head loss across the valve at any time is negligible compared to the total head across the pump at that time. Then the valve should close at a rate that will limit upsurge while not allowing excessive backflow. Valve control systems for this purpose are commonly used in conjunction with cone valves or butterfly valves.

When backflow is of little concern, and very slow valve closure is acceptable, the K&K charts may be used to predict upsurge. Such an approach relies on the assumption that valve closure will be so slow that it cannot increase the upsurge. (How slow is "slow" depends, among other things, on the type of valve – either judgment or approximate analysis of the type discussed below can be employed.) This method is marginally practical for short pipelines, such as on raw sewage pumping stations at treatment plants.

Most often, it will be desirable to close the valve within a shorter period of time. The solution to waterhammer problems involving controlled valve closure on a pumping system with appreciable pipeline friction is somewhat formidable. There are basically four approaches to the design of a power-operated waterhammer control valve: (1) complete solution by the graphical method; (2) practical experience coupled with field adjustment of valve closure; (3) complete solution by computer-based methods using the method of characteristics; and (4) approximate analysis. The first method is very time-consuming and is not recommended. The second method is commonly employed but can be hazardous unless one can be reasonably certain that conditions are similar enough to merit extrapolation.

Solution by more-complete computer-based methods has the primary advantage of accuracy and is exemplified below for the Spring Street Pumping Station. An approximate computer-based method developed by the author is particularly convenient in that generalized charts can be presented for each type of valve. An example chart (for Allis-Chalmers Rotovalves, now VAG USA) is presented in Figure 8.13. The basic variables are given by the following functional equation:

$$\frac{h_m}{H_o} = \phi\left(Z, \frac{fL}{D}, K_o\right) \tag{8.51}$$

in which:

$Z = \dfrac{T_c}{L/\sqrt{2gh_o}}$, valve closure parameter

h = pressure head relative to wet well

h_m = maximum head at pump

H_o = static head on system

T_c = time of valve closure

L = length of pipe

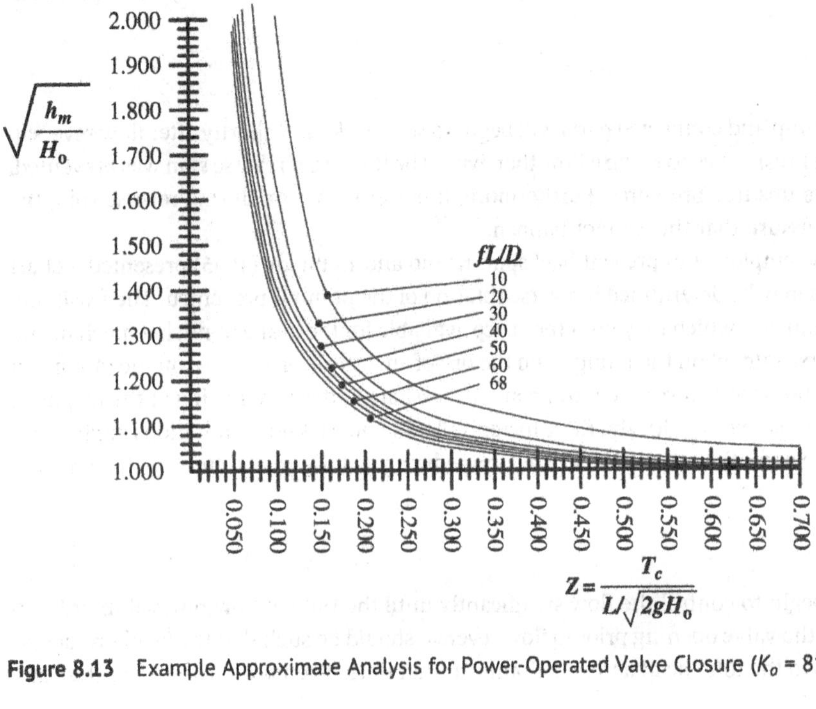

Figure 8.13 Example Approximate Analysis for Power-Operated Valve Closure (K_o = 81.7)

f = pipe friction factor (Darcy–Weisbach)
D = pipe diameter
K_o = pump loss coefficient (defined above)

The basic assumptions in the author's approximate method are as follows: (1) elastic effects can be neglected so that rigid column theory applies; (2) valve control commences under conditions of maximum reverse flow resulting from the static head of the system; and (3) the resistance of the pump remains constant.

For an assumed constant value of K_o, one chart such as Figure 8.13 (for which $K_o = 81.7$) can be prepared for each type of valve. There are at most only five or six different valve types or characteristics of practical interest. [The importance of considering the actual valve characteristics was emphasized by Wood and Jones (1973).] Referring to Figure 8.13, all terms but h_m and T_c are given for a particular system. Thus, the chart relates time of valve closure to the maximum head at the pump.

Although the above method may provide a "worse case" approximation by virtue of the second assumption mentioned above, this assumption may be desirable in view of: (1) recognition that the calculated valve closure setting should be viewed only as the initial setting for field adjustment; (2) the possibility of improper adjustments of valve-closure settings; (3) the variation of wave speeds due to gas entrainment referred to previously; and (4) the variability of pipe friction losses. The variation of wave speeds and the variability of pipe friction losses can cause a variation in the actual time of flow reversal. Thus, if flow reversal occurs in less than the assumed time, the valve could actually close during the occurrence of significant backflow.

The K&K figures are useful aids when programming valve closures for waterhammer control. So-called "delayed fast closure" should be considered; this is particularly important if the valve (e.g., rotovalve) closes to a small percentage of open area early in the valve stroke. The pump specification should include the requirement for the reverse speed capability.

Pressure Relief

Pressure-relief valves [e.g., Golden Anderson (now VAG USA)] are adjusted to open automatically at a predetermined pressure. As a sole method of waterhammer control, relief valves are most applicable to small pipelines where the escape of a relatively small amount of water will alleviate waterhammer pressures.

Another application of pressure-relief valves is as backup to check-valve control. Pressure-relief valves have been employed in conjunction with both swing-check and power-operated control valves. If the check valve closes improperly (such as when a swing-check valve suddenly slams after hanging open too long), the pressure-relief valve will limit the resulting pressure increase. If power-operated check valves are controlled by a fail-safe accumulator system, backup by pressure relief would be somewhat redundant.

Miscellaneous Methods

Several miscellaneous methods of waterhammer control merit mention because they may be applicable to special situations or be found on existing equipment. These methods are described in manufacturer's literature and in hydraulic reference texts (e.g., McNown 1950). They include:

1) Surge arrestors such as the Greer SurgeKushon (Parker Hannifin Corporation, Rockford, Illinois) and Smith Hydrotrol (Jay R. Smith Mfg. Co., Montgomery, Alabama). These devices employ gas chambers in closed vessels in which nitrogen gas is often used because of its inert nature. They are installed on low-pressure piping, particularly on plumbing systems.
2) Surge suppressors, which have been used for upsurge control.
3) Pump bypass lines provide a direct connection between pump suction and discharge lines during surge conditions (Kinno 1968).
4) "Smart" motor controllers that provide controlled acceleration and deceleration of the pump motor without feedback devices, e.g., provided by Allen-Bradley Co. LLC, Roseville, Montana (WE&M March 1990).

Also worthy of mention is the novel vortex chamber proposed by Kolf and Zielinski (1959). This device is located upstream of the pump and restricts backflow. It may have applicability under certain circumstances where very slow valve closure is desired. The use of siphons in large lift stations serves a similar purpose. The siphon action is disrupted by power failure so that backflow cannot occur.

Downsurge Control

If the downsurge analysis indicates that vapor pressure would otherwise be reached in the pipeline, there are several possible methods of controlling the magnitude of the negative pressure and/or subsequent overpressure:

1) Valve control
2) Air-inlet valves
3) Air chambers (hydropneumatic surge arresters)
4) Surge tanks

In conjunction with air-inlet valves, or with no downsurge control, the water column separation phenomenon can be analyzed to estimate the ensuing overpressures.

Air chambers offer a particularly practical solution for sewage pumping stations where open surge tanks and air-inlet valves should be avoided. A preliminary evaluation of air-chamber control may be made using the generalized charts of Evans and Crawford (1954; also Parmakian 1963, Chapter XVIII; Wood 1970; and Stephenson 2002). The charts give upsurge and downsurge pressures at the pump and mid-length of the pipeline. Derivations and formulae by Gardner and Gummer (1973a, 1973b) are also informative. It is necessary to evaluate the downsurge at the air chamber to assure that the volumes of air and water in the air chamber are such that air will not enter the force main during downsurge.

Once the feasibility of air-chamber control has been established, the final design should be left to the manufacturers because of device-specific internal details. Fluid Kinetics (Fluid Kinetics, Inc., Orchard Park, New York) and PULSCO (Santa Ana, California) have performed analyses of the system, recommended air chamber equipment, and guaranteed performance under waterhammer conditions.

Surge tanks are installed on large water (rather than sewage) pipelines to relieve excess pressure caused by waterhammer and to provide a supply of water to reduce negative pressures. Their function is thus similar to that of the air chamber. The simplest surge tanks are little more than a vertical standpipe connected to the pipeline. Surge tanks are usually an economical alternative on larger pipelines for which the heads are not extremely high (because of the great height of surge tank required). The design of surge tank systems is complex and is best handled by computer-based methods such as those employed by Eriksen (1972).

Air-inlet valves open automatically when the pressure in a pipe drops below some predetermined value; they allow air to enter the pipe and thus limit the negative pressures. A combination air-release and air-inlet valve is commonly employed so that the entrapped air can be subsequently released. Air-inlet valves do have a tendency to become inoperable.

Computer programs are available for analyzing the overpressures occurring after water column separation. In considering such an analysis, one should critically assess the uncertainty of the hydraulic conditions resulting when water columns do separate (Kalkwijk and Kranenburg 1971, Streeter 1971, Safwat 1972, Safwat and Van Den Polder 1972).

Summary of Hierarchal Approach

To start the hierarchic approach discussed herein, first check downsurge using the generalized charts presented. It is important to conduct the downsurge analysis in the layout phase because if downsurge control devices are required, they require significant space.

If the downsurge analysis shows that column separation is not a problem, proceed to an analysis of upsurge. Upsurge control by swing-check valves, inertial control, and pump anti-reverse control can be analyzed using generalized charts discussed herein. Upsurge control by power-operated valve closure or pressure relief can be analyzed by the author's approximate method presented above or by means of the appropriate computer-based methods discussed herein.

If the downsurge analysis shows that column separation is a potential problem, then the downsurge control methods discussed herein should be considered for both upsurge and downsurge control. Air chambers may be evaluated by means of generalized charts referred to above; the final design should usually be done by the manufacturers. Surge tanks and air-inlet valves should be analyzed by appropriate computer-based methods.

The conditions that must be met in order to properly use approximate techniques were discussed above. Branching piping systems, such as distribution systems, require special consideration. Computer-based solutions are often required for distribution systems.

8.6.4.2 Pump Characteristics

Pump characteristics are needed for purposes of the transient analysis. It is often the case that no information exists, from the pump manufacturer or otherwise, for pump curves beyond the region of the curves for normal operation nor are data always available for pump inertia. This is especially so for older pumps, making it necessary to derive the extended curves and make assumptions regarding pump inertia. The extended pump curves will be derived first, after which assumptions for pump inertia will be discussed. The derivation is presented in somewhat rigorous detail because the author knows of no single reference source in which this is done.

The derivation of extended pump curves relies on principles of pump similitude. Similitude refers to the relationships between the variables of a process at different scales or other conditions, having certain similarities. For complete similitude, the similarity must be both geometric and dynamic. Geometric similarity refers to the need to have complete similarity between all dimensions in order to properly relate process variables at one scale to those at another. Similitude is closely tied to *dimensional analysis* because to achieve dynamic similarity (defined in Section 5.2) in geometrically similar processes, at least all but one of the dimensionless groups yielded by dimensional analysis must be equal.

8.6.4.3 Dimensional Analysis Extended

Dimensional analysis for centrifugal pumps was introduced in Section 8.2. Proceeding more generally than in Section 8.2, the change in total head across the pump is given by $H = \Delta p/(\rho g) + \Delta v^2/(2g) + \Delta y$, in which the terms on the right-hand side are the change in pressure head (pressure energy), velocity head (kinetic energy), and elevation (potential energy), successively. Thus, we can express the total head in nondimensional form as follows:

$$\frac{\Delta p}{\rho N^2 D^2} = \frac{Hg}{N^2 D^2} = \frac{\Delta p}{\rho N^2 D^2} + \frac{\Delta v^2}{2N^2 D^2} + \frac{g\Delta y}{N^2 D^2} \tag{8.52}$$

$$\frac{\Delta v^2}{2N^2 D^2} \propto \frac{V^2}{2N^2 D^2} \propto \frac{Q^2}{2N^2 D^6} \propto \left(\frac{Q}{N^2 D^3}\right)^2 \tag{8.53}$$

Continuing to proceed more generally than in Section 8.2, we have the following general form:

$$\Phi\left(\frac{ND^3}{Q}, \frac{Hg}{N^2 D^2}, \frac{g\Delta y}{N^2 D^2}, \mathbf{R}\right) = 0 \tag{8.54}$$

The other variables in the π-parameter containing Δy are already present in the other π-parameters so that parameter can be simplified giving the following:

$$\Phi\left(\frac{ND^3}{Q}, \frac{Hg}{N^2 D^2}, \frac{\Delta y}{D}, \mathbf{R}\right) = 0 \tag{8.55}$$

Since $\Delta y/D$ must be constant for geometric similarity, that term may be dropped. Furthermore, in applications of moderate-to-large water pumps, the Reynolds Number is sufficiently large that the flow is fully turbulent and **R** can be dropped from the above functional relationship. Thus:

$$\Phi\left(\frac{ND^3}{Q}, \frac{Hg}{N^2 D^2}\right) = 0 \tag{8.56}$$

Since the first of these parameters must be constant for kinematic similarity (see above), and since the second parameter is a function of only the first, the second parameter must also be constant, giving more specifically for dynamic similarity:

$$\frac{Q}{ND^3} = \text{const} \tag{8.57a}$$

$$\frac{H}{N^2 D^2} = \text{const} \tag{8.57b}$$

For two geometrically similar pumps of different sizes D_1 and D_2, this relationship allows the pump curves for one to be used to derive the pump curves for the other:

$$\frac{Q_1}{Q_2} = \frac{N_1}{N_2}\left(\frac{D_1}{D_2}\right)^3 \tag{8.58a}$$

$$\frac{H_1}{H_2} = \left(\frac{N_1}{N_2}\right)^2\left(\frac{D_1}{D_2}\right)^2 \tag{8.58b}$$

When dealing with one particular pump, the size variable drops out of the above equations, and they reduce to the following useful forms:

$$\frac{Q_1}{Q_2} = \frac{N_1}{N_2} \tag{8.59a}$$

$$\frac{H_1}{H_2} = \left(\frac{N_1}{N_2}\right)^2 \tag{8.59b}$$

These equations say that dynamic similarity will be maintained if Q and H are simultaneously changed in accordance with $Q \propto N$ and $H \propto N^2$. This allows, for example, a pump curve provided for one speed to be used to develop a pump curve for another speed, or the determination of the speed and efficiency of a pump at a particular operating point (Q and H) given pump efficiency curves at another speed.

We will now consider the question of how to deal with the size variable more generally, particularly for waterhammer analyses. For this purpose, it is desirable to define a reference operating condition for every pump, one for which none of the variables of interest (such as flow) is zero. The only such point that generally exists is the maximum-efficiency point, which we will denote by the subscript "R" (for reference). Applying Equations 8.58 to that point gives:

$$\frac{Q_{R1}}{Q_{R2}} = \frac{N_{R1}}{N_{R2}}\left(\frac{D_1}{D_2}\right)^3 \tag{8.60a}$$

$$\frac{H_{R1}}{H_{R2}} = \left(\frac{N_{R1}}{N_{R2}}\right)^2\left(\frac{D_1}{D_2}\right)^2 \tag{8.60b}$$

We can now eliminate the size terms (D_1 and D_2) between Equation 8.58a and Equation 8.60a and between Equation 8.58b and Equation 8.60b. If we also introduce the ratios $q = Q/Q_R$, $h = H/H_R$, and $\alpha = N/N_R$, the following relationships are obtained:

$$\frac{q_1}{q_2} = \frac{\alpha_1}{\alpha_2} \tag{8.61a}$$

$$\frac{h_1}{h_2} = \frac{\alpha_1^2}{\alpha_2^2} \tag{8.61b}$$

or, more generally:

$$\frac{q}{\alpha} = \text{const} \tag{8.62a}$$

$$\frac{h}{\alpha^2} = \text{const} \tag{8.62b}$$

Returning to functional form (see analogous Equations 2.108 and 2.109), we have the following general relationship:

$$\Phi\left(\frac{h}{\alpha^2}, \frac{\alpha}{q}\right) = 0 \tag{8.63a}$$

Because we are interested in values of α (and q) going through zero, we add the following variation of the above functional relation:

$$\Phi\left(\frac{h}{q^2}, \frac{q}{\alpha}\right) = 0 \tag{8.63b}$$

By eliminating D between the first two variables of Equation 8.56, we obtain the following alternative functional relationship:

$$\Phi\left[\frac{N\sqrt{Q}}{(gH)^{3/4}}, \frac{Hg}{N^2D^2}\right] = 0 \tag{8.64}$$

Equations 8.56 and 8.64 have important uses in the characterization of types of pumps, in pump modeling or testing, and in extrapolating measured performance characteristics of a particular pump to other conditions. The first variable in the above relationship, applied to the maximum-efficiency reference point and with the variable g dropped, gives the important *dimensional* variable *specific speed*, defined by:

$$N_s = \frac{N_R\sqrt{Q_R}}{H_R^{3/4}} \tag{8.65}$$

As noted above, different types of pumps fall into fairly distinct ranges of specific speed, with specific speed increasing as they go from radial-flow pumps to mixed-flow pumps to axial-flow pumps.

Equations 8.56, 8.63, or 8.64 fully characterize the pump curves of dynamically similar (and geometrically similar) pumps. These relationships hold true at the reference points and all other points on the pump curves of such pumps. Therefore, if the relationships are true for geometrically similar pumps at the reference points, then they are true at all other corresponding points on the pump curves. Furthermore, since one of the parameters in any one of Equations 8.56, 8.63, or 8.64 is uniquely dependent on the only other parameter in its equation, then only one of those parameters has to be equal at the reference points of two geometrically similar pumps for those pumps to be dynamically similar. Referring to the first parameter in Equation 8.64 and the definition in Equation 8.65, such equality at the reference point can be stated in a form not involving pump size by $N_{S1} = N_{S2}$. In other words, two geometrically similar pumps are dynamically similar and will have identical pump curves if their specific speeds are identical.

By this reasoning and with reference to Equation 8.63, geometrically similar pumps with identical specific speeds will have a unique curve of h/α^2 vs. q/α^2. This is not only true in the zone of normal pump operation (forward or positive flow, positive speed, positive TDH) but also in the other zones of potential interest in transient analyses. In the most complete classification, there are seven additional zones, involving various sign combinations of flow, speed, head, and torque. Those other zones are discussed later in this chapter.

Torque is an additional important pump variable in waterhammer analyses. Pump power is equal to torque times rotational speed and is also equal to the specific gravity of the pumped liquid times rate of flow times head divided by shaft to water efficiency. Homologous pump theory assumes that efficiency does not change with pump size. On that assumption, for water (or any given liquid of constant specific gravity) torque is proportional to QH/N. Thus, torque is not an independent variable, rather it is dependent on other variables included in, e.g., Equation 8.56. To determine its functional relation, it can be used instead of one of the other variables, and (for reasons that will become apparent), head is the variable that it best replaces for purposes of characterizing pump torque in waterhammer analyses.

The functional relationship for torque can be quickly derived at this point by starting with the following expression of the relationship for two different pumps based on the proportionality discussed in the preceding paragraph:

$$\frac{T_1}{T_2} = \frac{Q_2H_2/N_2}{Q_1H_1/N_1} \tag{8.66}$$

Using Equation 8.58b to eliminate head from the above, we obtain:

$$\frac{T_1}{T_2} = \frac{Q_2N_2}{Q_1N_1}\left(\frac{D_2}{D_1}\right)^2 \tag{8.67}$$

In terms of the maximum-efficiency reference point, the above equation gives:

$$\frac{T_{R1}}{T_{R2}} = \frac{Q_{R2}N_{R2}}{Q_{R1}N_{R1}}\left(\frac{D_2}{D_1}\right)^2 \tag{8.68}$$

Eliminating the size terms between Equations 8.67 and 8.68, using the ratios q and α defined earlier, and introducing the ratio $\beta = T/T_R$, we have:

$$\frac{\beta_1}{\beta_2} = \frac{q_1\alpha_1}{q_2\alpha_2} \tag{8.69}$$

or, more generally:

$$\frac{\beta}{q\alpha} = \text{const} \tag{8.70}$$

Equation 8.70 can be combined with Equation 8.62a to eliminate the variable q and yield the following parameter:

$$\frac{\beta}{\alpha^2} = \text{const} \tag{8.71}$$

Coupling this parameter with the parameter q/α (see Equations 8.62 and 8.63) gives the following form for the torque relationship:

$$\Phi\left(\frac{\beta}{\alpha^2}, \frac{q}{\alpha}\right) = 0 \tag{8.72a}$$

As before (Equation 8.63b), we add the following variation:

$$\Phi\left(\frac{\beta}{q^2}, \frac{\alpha}{q}\right) = 0 \tag{8.72b}$$

Equations 8.63 and 8.72 are the important functional relationships for characterizing pumps in waterhammer analyses.

8.6.4.4 Plotting and Use of Pump Data

The conventional plot of pump data for normal pump operation has TDH as the y-axis and flow (discharge) as the x-axis. For each speed, a different flow–head curve can be plotted, and all of the data falls in the first quadrant (positive flow, positive head). Such a plot could be extended into the other three quadrants to represent the head–flow–speed relationship in other zones. However, such a plot does not usefully distinguish between the different zones of interest in waterhammer.

A more useful form is the Karman–Knapp plot, which is presented as a polar-type diagram in the general form shown below. On a Karman–Knapp plot, multiple curves of constant head and constant torque are plotted vs. the flow and speed ratios. The corresponding functional relationship represented is in a less general form than Equations 8.63 and 8.72:

$$h = \Phi(\alpha, q) \tag{8.73a}$$
$$\beta = \Phi(\alpha, q) \tag{8.73b}$$

Such a plot does not represent a form that lends itself to digital computation because we are interested in continuous changes in the head and torque variables.

That can be overcome by taking full advantage of the functional relationships in Equations 8.63 and 8.72. Multiple plots can be presented corresponding to these relationships in the different zones, with the full range of α and q represented by separate plots for $0 \le \alpha/q \le 1$, $-1 \le \alpha/q \le 0$, $0 \le q/\alpha \le 1$, and $-1 \le q/\alpha \le 0$. Such plots are presented by Thomas (1972). However, those relationships are difficult to handle from a computational standpoint because h, β, q, and α may all go through zero and change signs during a transient. Those relationships also do not facilitate a single "four-quadrant plot."

Marchal *et al.* (1965) developed a particularly useful solution to this problem. They, in effect, brought together the Karman–Knapp plot (depicted schematically below and represented by Equations 8.73) and the relationships expressed by Equations 8.63 and 8.72 by transforming the Karman–Knapp plot to polar coordinates. Letting r and θ represent polar coordinates on the Karman–Knapp plot, and considering the x-axis and y-axis to be the $+\alpha$ and $+q$ Karman–Knapp axes, we have the following relationships:

$$h = \Phi_p(r, \theta) \tag{8.74a}$$
$$\beta = \Phi_p(r, \theta) \tag{8.74b}$$
$$r = \sqrt{\alpha^2 + q^2} \tag{8.75a}$$
$$\theta = \tan^{-1}\frac{q}{\alpha} \tag{8.75b}$$

which combine to give:

$$h = \Phi_p'\left(\alpha^2 + q^2, \tan^{-1}\frac{q}{\alpha}\right) \tag{8.76a}$$
$$\beta = \Phi_p'\left(\alpha^2 + q^2, \tan^{-1}\frac{q}{\alpha}\right) \tag{8.76b}$$

Table 8.1 Zones of Possible Pump Operation

Zone	Quadrant	Flow (q)	Speed (α)	Head (h)	Torque (β)	Definition
A	I	+	+	+	+	Normal pumping
B	I	+	+	−	+	Energy dissipation
C	I	+	+	−	−	Reverse turbine
D	IV	+	−	−	−	Energy dissipation
E	IV	+	−	+	−	Reverse rotational pumping
F	III	−	−	+	−	Energy dissipation
G	III	−	−	+	+	Normal turbine
H	II	−	+	+	+	Energy dissipation

Comparing Equations 8.76 with Equations 8.72 and noting that:

$$\frac{h}{\alpha^2 + q^2} = \frac{h/\alpha^2}{1 + q^2/\alpha^2} \tag{8.77}$$

and similarly, for other like parameters, we can conclude that dynamic similarity will be maintained by the following more general variations on Equations 8.76:

$$\frac{h}{\alpha^2 + q^2} = \Phi\left(\tan^{-1}\frac{q}{\alpha}\right) \tag{8.78a}$$

$$\frac{\beta}{\alpha^2 + q^2} = \Phi\left(\tan^{-1}\frac{q}{\alpha}\right) \tag{8.78b}$$

This form has the advantages of allowing the variables to go through zero and change signs during transients (without values becoming infinite) and facilitates a very simple "four quadrant" plot. In that plot, there will be one curve for the head term (represented by Equation 8.78a) and one curve for the torque term (represented by Equation 8.78b). That can be presented in polar form but is generally found in the literature (Wylie and Streeter 1983, Wood and Funk 1988) in rectangular-coordinate form in which the variables on the left-hand side of Equations 8.78a and 8.78b are plotted on the y-axis and the x-axis gives values of the angle $\pi + \tan^{-1}(q/\alpha)$.

Table 8.1 below gives a complete listing of all possible zones in which the curves can fall (including some of theoretical interest only) in relation to the zone designations and quadrants depicted in Figure 8.14 below. It should be noted that the zone designations, quadrants, and definition terminology differ between various literature sources. The zones and definitions given here coincide with those given in the *SURGE5 User's Manual* (Wood and Funk 1988), except for the quadrants, which conform to Figure 8.14.

8.6.4.5 Development of Specific Pump Data

In the absence of test data for a particular pump, pump curves of the type described immediately above can be developed by interpolating based on the pump's specific speed between curves for other pumps whose specific speeds bracket that of the pump of interest. Another alternative is to use data for another pump whose specific speed is closest to that of interest. Yet another approach is to combine both of these approaches by deriving pump curves by interpolation and then conducting sensitivity analyses based on data for pumps of higher and lower specific speeds. As discussed later, there are limitations to any of these procedures, and judgment must be applied in interpreting the results.

Thomas (cited above) developed equations for interpolating between curves of different specific speeds. For each of his pump zones, equations are given that have the functional form of Equations 8.63 and 8.72 generalized by the addition of the specific speed variable, commensurate with the discussion presented above:

$$\frac{h}{\alpha^2} = \Psi\left(\frac{q}{\alpha}, N_s\right) \tag{8.79a}$$

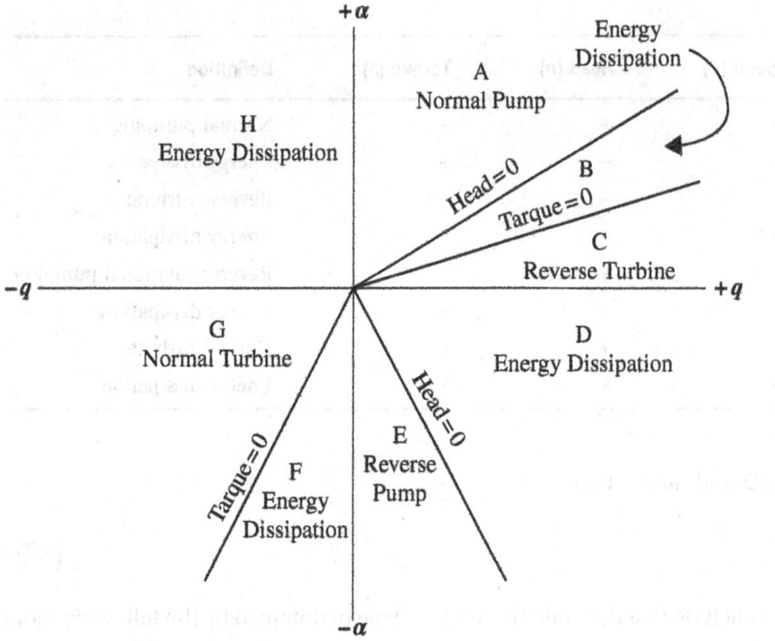

Figure 8.14 General Form of Karman–Knapp Plot

$$\frac{h}{q^2} = \Psi\left(\frac{\alpha}{q}, N_s\right) \tag{8.79b}$$

$$\frac{\beta}{\alpha^2} = \Psi\left(\frac{q}{\alpha}, N_s\right) \tag{8.79c}$$

$$\frac{\beta}{q^2} = \Psi\left(\frac{\alpha}{q}, N_s\right) \tag{8.79d}$$

For each of his pump zones, Thomas gives second-degree polynomial equations in the appropriate one of the above forms for each of seven points. (Thomas uses different definitions of pump zones than those given above, but he in effect includes all but Zones D and E, which generally have no relevance to waterhammer.) Having calculated the values at each of those seven points for the seven values of q/α or α/q, a sixth-degree polynomial can be fitted (exactly) to the seven points.

The author has developed methods for automating this procedure with the aid of spreadsheet functions. For a given specific speed, any of the dependent variables (left-hand side) in Equation 8.79 can be expressed as the following matrix product:

$$\mathbf{D} = \mathbf{Q} \cdot \mathbf{C} \tag{8.80}$$

D is a column matrix of dimensions 7 × 1, representing the values of the *dependent* variables at each of the seven points:

$$\mathbf{D} = \begin{Bmatrix} d_1 \\ d_2 \\ d_3 \\ d_4 \\ d_5 \\ d_6 \\ d_7 \end{Bmatrix} \tag{8.81}$$

C is also a column matrix of dimension 7 × 1, representing the coefficients of the sixth-degree polynomial:

$$\mathbf{C} = \begin{Bmatrix} A \\ B \\ C \\ D \\ E \\ F \\ G \end{Bmatrix} \tag{8.82}$$

Q is a square matrix of dimension 7 × 7, representing the values of the *independent* variable α/q or q/α, denoted by i:

$$\mathbf{Q} = \begin{pmatrix} i_1^6 & i_1^5 & i_1^4 & i_1^3 & i_1^2 & i_1^1 & i_1^0 \\ i_2^6 & \bullet & \bullet & \bullet & \bullet & \bullet & \bullet \\ i_3^6 & \bullet & \bullet & \bullet & \bullet & \bullet & \bullet \\ i_4^6 & \bullet & \bullet & \bullet & \bullet & \bullet & \bullet \\ i_5^6 & \bullet & \bullet & \bullet & \bullet & \bullet & \bullet \\ i_6^6 & \bullet & \bullet & \bullet & \bullet & \bullet & \bullet \\ i_7^6 & \bullet & \bullet & \bullet & \bullet & \bullet & \bullet \end{pmatrix} \tag{8.83}$$

Given values (from Thomas) for the elements of matrices **D** and **Q**, we can solve for the necessary polynomial coefficients by matrix inversion and multiplication:

$$\mathbf{C} = \mathbf{Q}^{-1} \bullet \mathbf{D} \tag{8.84}$$

Having obtained those coefficients, the value of the dependent variable can be calculated using the polynomial equation at any desired value of the dependent variable, for example:

$$h/q^2 = A(\alpha/q)^6 + B(\alpha/q)^5 + C(\alpha/q)^4 + D(\alpha/q)^3 + E(\alpha/q)^2 + F \tag{8.85}$$

All of the calculations can be performed and the resulting curves plotted using appropriate spreadsheet software.

8.6.4.6 Moment of Inertia

The pump unit's moment of inertia is predominantly due to the driving motor. Since information on the pump impeller inertia is usually not available, motor inertia is used as an approximation to the total inertia (in the absence of flywheels or magnetic drives, which can be usefully employed for waterhammer control). The motor moment of inertia for 3-phase motors depends on the number of poles, denoted by p and given by $p = 3fV/N$ in which f is alternating-current frequency in cycles per second (cps), V is voltage, and N is synchronous motor speed in rpm. Motor manufacturers provide charts of moments of inertia, denoted WR^2 for the rotor and standard shaft combined. For example, for the 30 hp motors at the Admiralty Hill project mentioned above, we have $f = 60$ cps and $V = 120$ volt, and we calculate $p = 3(60)(120)/N = 7200/N$. For $N = 900$ rpm, we calculate $p = 7200/900 = 8$. With these values of hp and N, the manufacturer's chart gives for the particular motor $WR^2 = 11.2$ lb-ft^2. The use of moment of inertia is discussed in an example given below.

Example 8.9

This example reports on portions of the author's work on a pumping station (Spring Street Pumping Station, Arlington, Massachusetts), which draws water directly from a 56-inch diameter main via 1400 ft of 36-inch diameter reinforced concrete pipe. The station is a component in a system operated by the Massachusetts Water Resources Authority (MWRA – the client). The station has three 300-hp pumps and one 150-hp pump. Station discharge is to a 36-inch diameter reinforced concrete main that feeds the Northern Extra High (NEH) Service Area. The discharge line from each pump is equipped with a control valve. Cone valves were originally installed for this purpose. On April 12, 1994, at approximately 7:15 p.m., there was a "catastrophic" break of the discharge piping in the basement of the Spring Street Pumping Station. A substantial length of interconnected water distribution piping ruptured and several homes were flooded. As part of the repairs to the station, the cone valves were replaced with butterfly valves and new control mechanisms. Although this alteration changes somewhat the performance of the system during hydraulic transients, comparative analyses showed that the transient results would be similar for either set of valves. Therefore, only analyses with the new butterfly valves are reported here.

The author was tasked with determining the cause of the pipe failure and recommending protective measures. That entailed modeling hydraulic transients associated with various operating conditions at the pumping station. Transients occurring during pump startup, normal pump shutdown, and power failure were considered. Recommendations were made for pumping station control valve timing and system operating procedures.

The system was modeled from a suction node of specified pressure at one wall of the pumping station, through the station, and into the transmission system. The minimum elevation in the system occurs in the discharge line where it exits the pumping station (Elevation 86.8). (Pipe elevations are centerlines; a consistent datum was used.) From that location, the 36-inch transmission main continues into an elaborate pipe network that services the towns of Arlington, Belmont, Lexington, and Waltham. The system extends out to 15,600 ft from the pumping station.

Two storage facilities are in the network: the Arlington Heights Standpipe and the Walnut Hill Elevated Water Tank in Lexington.

The two-million gallon Walnut Hill Elevated Water Tank has a bottom elevation of 410.0, normal operating range of 433–440, and overflow at Elevation 445.0.

Pump Characteristics

For purposes of transient analyses, pumps are characterized by the "rated" conditions (denoted by subscript "R") occurring at the maximum-efficiency point. Based on those conditions, the *specific speed* N_S (gpm units) can be derived for each pump from Equation 8.8 expressed as follows:

$$N_S = \frac{N_R\sqrt{Q_R}}{H_R^{3/4}} \tag{8.86}$$

in which N_R, Q_R, and H_R are the rated impeller rotational speed (rpm), discharge (gpm), and TDH (ft), respectively. These values plus maximum efficiencies (η_R) and pump inertias WR^2 (provided by the client) are tabulated below for the two pump sizes.

Pump Size (hp)	N_R (rpm)	Q_R (gpm)	H_R (ft)	Specific Speed (gpm units)	Efficiency	Inertia (lb-ft^2)
300	1750	4100	236	1861	88%	121
150	1750	2300	210	1521	83%	48

For purposes of transient analyses, pump characteristics are needed beyond the region of the curves for normal operation. As is commonly the case for older pumps, no such information exists for these particular pumps nor were data available for pump inertia. It was thus necessary to extrapolate the extended curves and make assumptions regarding pump inertia. The specific speeds tabulated above fall within the range of radial-flow centrifugal pumps, for which transient analyses are fairly insensitive to specific speeds. For purposes of transient modeling, the SURGE5 software's (University of Kentucky) standard pump profile having the closest specific speed was utilized. As discussed above, that includes characteristics for operation in various zones as described earlier in this chapter.

For pumps of 1960s vintage, the motor typically provides about 90% of the total WR^2 of the pump and motor, so the total WR^2 can be conservatively assumed with good accuracy to be approximately that of the motor alone. That common assumption was made. It is also necessary to know the allowable reverse speed of the pump and driver. The client advised that the pump drive motors have reverse overspeed capacities of 125% for both the 150-hp and 300-hp units.

Steady-State Operation

An understanding of steady-state forward-flow operation, the normal operating condition, is necessary to define initial conditions for transient modeling. The SURGE5 transient network model was used for both steady-state and transient modeling. Data necessary for the steady-state modeling were gathered, including pipe plan stations and centerline elevations, nominal and actual pipe diameters, pipe type (material/class), plan sheet and source, descriptive location, roughness coefficient, form losses (tees, bends, check valves, etc.) for forward flow and reverse flow. Based on discussions with the client during a visit to the pumping station, a constant pressure (fixed-head node) of 66.09 psi was assumed in the 24-inch cast iron suction main where it enters the station. Also based on the visit, the pump combinations considered are two 300-hp pumps, one 300-hp and one 50-hp pump, one 300-hp pump alone, and one 150-hp pump alone.

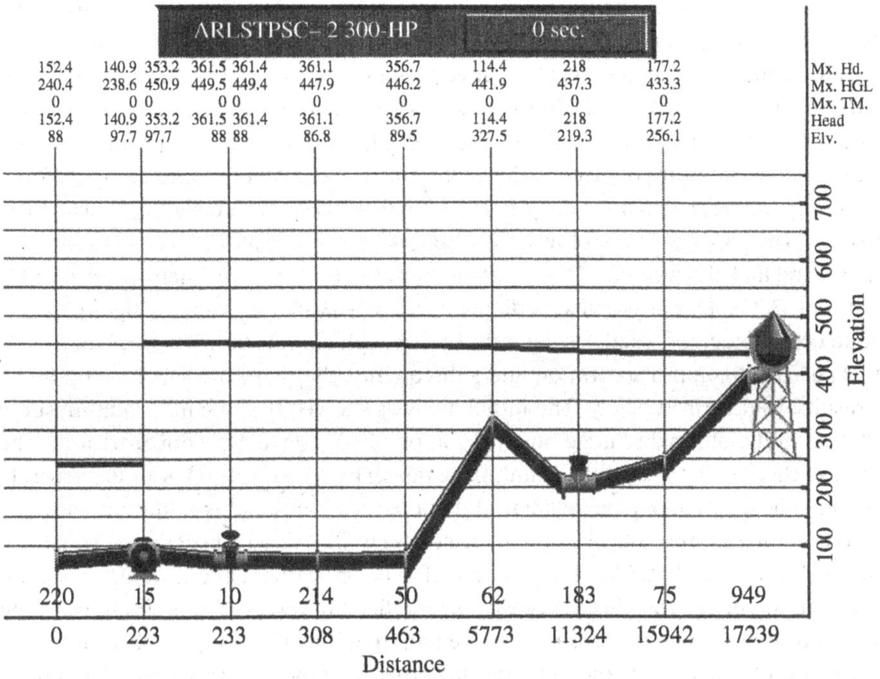

Figure 8.15 GEMS Initial Condition Steady-State Run

Various steady-state runs were made with various combinations of numbers of pumps running and water demands and a minimum Walnut Hill Tank operating elevation. A combination that gave larger flows over most of the length of the discharge main (due to larger total demands but lower demand at an intermediate metered takeoff) was selected as the initial condition for transient modeling. A FEMS (Graphical Engineering Module/Surge) profile plot of that steady-state run is shown in Figure 8.15. The scale on that figure is distorted so that all important elements are distinguishable.

The initial-condition steady-state run revealed that the highest pressures occur within the pumping station in the pump discharge lines at their lowest elevations (Elevation 88.0) immediately upstream of the control valves. Even though the discharge main as it exits the pumping station is at a slightly lower elevation (Elevation 86.8), the head loss within the discharge piping between these two locations causes the pressure to be slightly higher at the upstream location. For the initial condition of two 300-hp pumps operating with the maximum demands, that maximum pressure is 156.9 psi (362.0 ft). With the Walnut Hill Tank at its minimum water elevation (to maximize flows), this is the maximum normal operating pressure to which waterhammer pressures can be compared.

System Transient Performance

Wave speeds were calculated using Equation 8.43, and values discussed with that equation. An example of the use of that equation for a pipe segment in the Arlington project is given above as Example 8.7. The nonconstant variables D/e and E were calculated or estimated for each pipe length. Tabulated wave speeds are regarded as *upper* wave speeds. Where there was uncertainty regarding the precise value of a variable (such as modulus of elasticity), an assumption was made giving a conservatively high wave speed. It was also assumed, by using Equation 8.43, that there is no entrained air. Very slight air content can substantially reduce wave speeds. As part of the sensitivity analysis discussed below, runs were also made with the wave speeds reduced to 50% of the tabulated values to account for entrained air and to compensate for the assumptions used in deriving upper-wave speeds.

For the transient analyses, demands were modeled as "side-discharge orifices." This overcomes the shortcomings of some models that assume demands remain constant during transients. In fact, the pressure variation in the mains during the transients causes the demands to fluctuate. The model accounts for this by allowing demands to be modeled as square-root resistance elements between the main being modeled and the actual demand discharge location (of which there are actually many due to the multitude of consumer services). By inputting an exit head along with the steady-state demand, the model calculates the square-root resistance. The demand is assumed to vary instantaneously through the orifice in response to changes in the transient pressures in the main. It is therefore desirable to locate the demand orifices as close to the demand location as is practical.

An initial run was made for the case of the two largest pumps tripping due to power failure without any valve control. That case was selected to determine whether downsurge was a possible problem in the absence of valve control (see further discussion below). When power failure occurs, a continuous succession of negative pressure waves emanates from the pump as it decelerates, causing reduced pressures of "downsurge" to occur in the pipe system. These negative pressure waves are reflected as positive pressure waves at the open end of the pipe (i.e., the Walnut Hill Elevated Tank), resulting in "upsurge" in the piping system. If pressures during the downsurge are reduced to the vapor pressure of water, a vapor cavity (column separation) can form, which can lead to a large pressure increase when the vapor cavity subsequently collapses. This is most likely to occur at high points in the system. [The vapor pressure of water is 0.306 psi absolute at 65°F, decreasing as water temperature decreases (see Table 9.1 and related discussion). The vapor pressure at 65°F is equivalent to 0.306(144/62.4) = 0.706 ft of water (absolute) or 0.706–33.90 (standard atmospheric pressure) = −33.19 ft of water (gauge).]

Column separation was in fact found to occur under this modeled condition at the high point (Elevation 327.5) in the 36-inch transmission main near Station 54+65. That column separation and subsequent collapse of the vapor cavity had a profound effect on the waterhammer results, as described below. The initial analysis showed that the maximum upsurge pressures occur within the pumping station in the pump discharge lines (Elevation 88.0) immediately downstream of the control valves. The modeled system was also checked for stability by running the model for a short period prior to purposely initiating the transient. Because reasonable consolidation of pipe segments did not reduce computational time and because of the desire to treat column separation as accurately as possible, the full piping system was retained for subsequent analysis.

For all the initial analyses, the Walnut Hill Elevated Water Tank was assumed to be operational and in communication with the system. The storage facility plays an important role during transients by reflecting waves of opposite signs. If the valve at the tank was closed during, e.g., pump startup, then the positive waves emanating from the pump would be returned as reinforcing positive waves by the closed tank valve, with the potential for creating large overpressures. The author explained that to the client, with the result that important additional analyses were done as explained below under *Transient Control*.

The characteristics of the new butterfly valves and controllers were derived from valve and controller characteristics provided by the valve manufacturer. The effective area of the valve as a fraction of full port vs. time to open or close was determined. This effective area, as required for SURGE5, is the valve opening area having the same orifice characteristics as if the orifice coefficient (standard orifice formula) remained constant. It properly simulates the valve hydraulic characteristics, compensating for the SURGE5 assumption of constant orifice coefficient. Points on the curves thus derived were used as input to the SURGE5 model for cases with the new butterfly valves. Actual valve opening and closing times will be field set, based largely on the analyses given below.

For future projects, the author recommended that the information necessary to perform or confirm transient analyses be required by the specifications to be submitted during the shop drawing reviews. Specifically, it was recommended that the specifications require submittal of data giving: (1) valve coefficient (K or C_v) vs. valve (plug or disc) angle; and (2) the relationship between valve angle and percent time to fully open and fully close.

Transient Analyses – Butterfly Valves

After comparing the valve curves and considering the results given above, a valve closure time of 40 seconds was selected for the SURGE5 runs with the butterfly valves. The transient analysis results for the case of two 300-hp pumps tripping (losing power) simultaneously together with 40-second closure of the control (butterfly) valves gave the highest waterhammer pressures compared to other scenarios for the same closure time. The next four plots show graphic output for this run. The pumps were tripped and valve closure began at time equal to one second. Figure 8.16 is the GEMS (Graphical Engineering Module/SURGE) plot of the run after 60 seconds, including the envelope of maximum and minimum pressure heads. Figure 8.17 is a plot of pressure head immediately downstream of the control valve. Figure 8.18 is a plot of flow rate at the location where the discharge pipe exits the station. Figure 8.19 depicts the pump speed vs. time, showing the maximum reverse pump speed and duration.

As had been found for the cone valves, column separation still occurred at the high point in the line. The subsequent collapse of the vapor cavity caused the large, sudden overpressures shown in Figure 8.17. The pressure at the control valve increased abruptly from the normal operating pressure of 157 psi to a maximum waterhammer pressure of 257 psi. This overpressure equals the maximum of 257 psi occurring in the initial run discussed above, in which the pumps trip without valve closure. This indicates that the butterfly control valves with 40-second closure would result in no further reduction in overpressures. The reduced overpressures (butterfly vs. cone valves) are achieved at the expense of an increase in reverse pump speed, from 1860 to 2010 rpm (Figure 8.19 and Table 8.2).

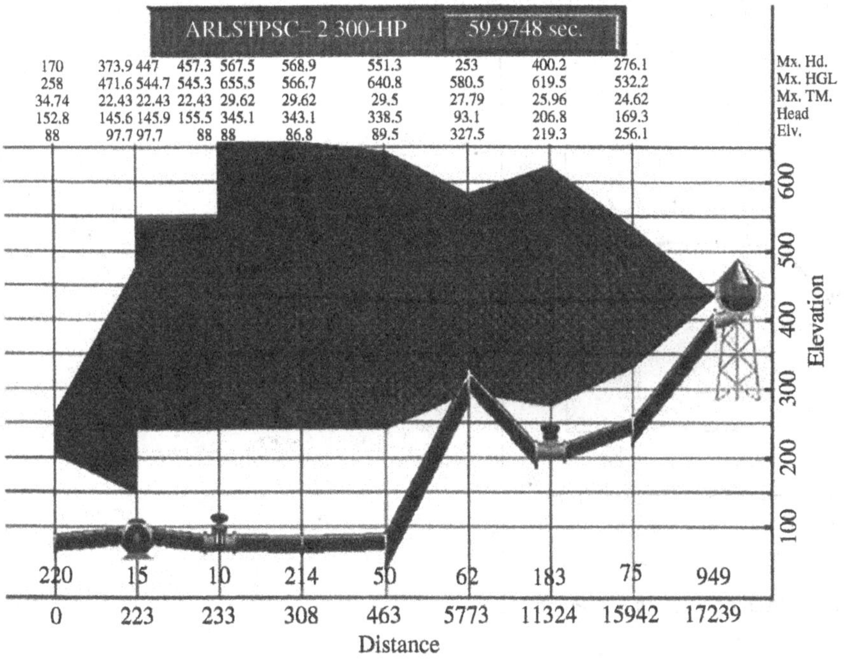

Figure 8.16 Maxim and Minimum Pressure Heads at 60 Seconds with Butterfly Valves

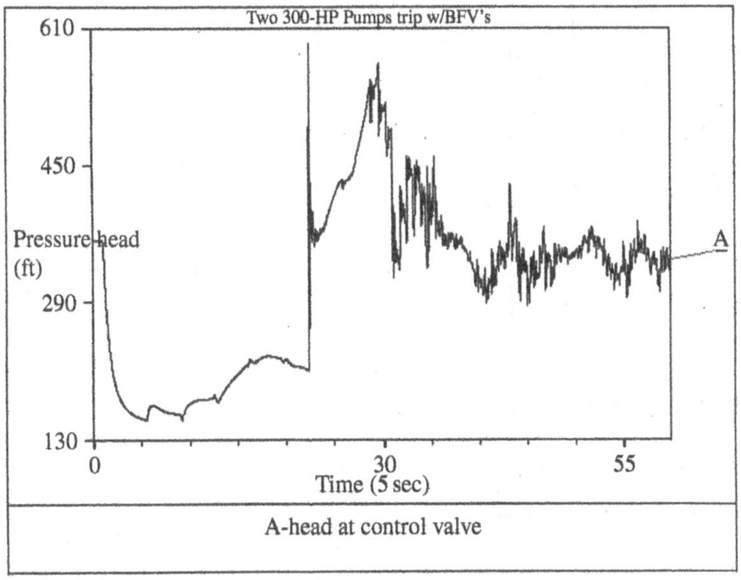

Figure 8.17 Pressure Head Immediately Downstream of Control Valve

Hydraulic transients are also associated with pump startup. Each pump is normally started against a closed discharge valve, which is very close to the pump. Because of the short distance between the pump and the discharge valve, the pressure in the intervening pipe quickly approaches the pump shutoff head as the pump comes up to speed. Depending on the control sequence of operation, the discharge valve begins to open before or after shutoff head is reached. If the discharge valve is slowly opened, it allows gradual attainment of a steady-state head–discharge condition. This condition is reached by the propagation of positive pressure waves from the pump control valve downstream into the distribution system, the subsequent return of waves of opposite signs reflected from the storage tank, reflections at the pump, and the successive

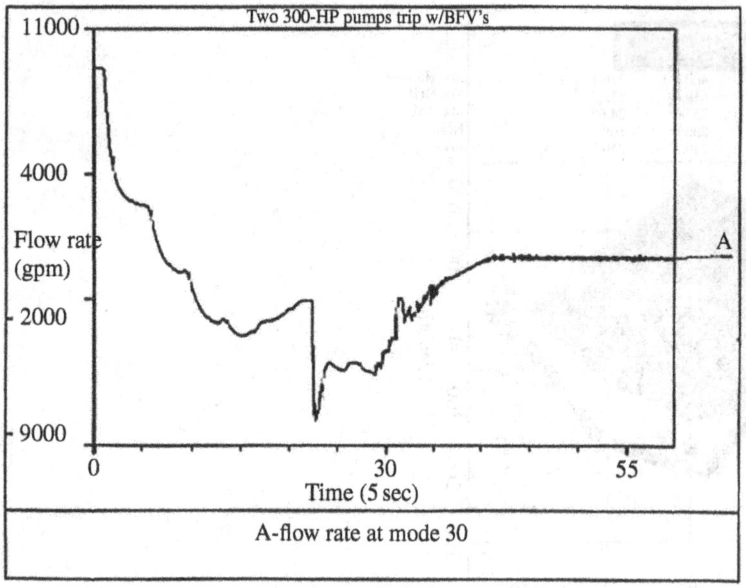

Figure 8.18 Butterfly Valves Flow Rate Where Discharge Valves Exit Pump Station

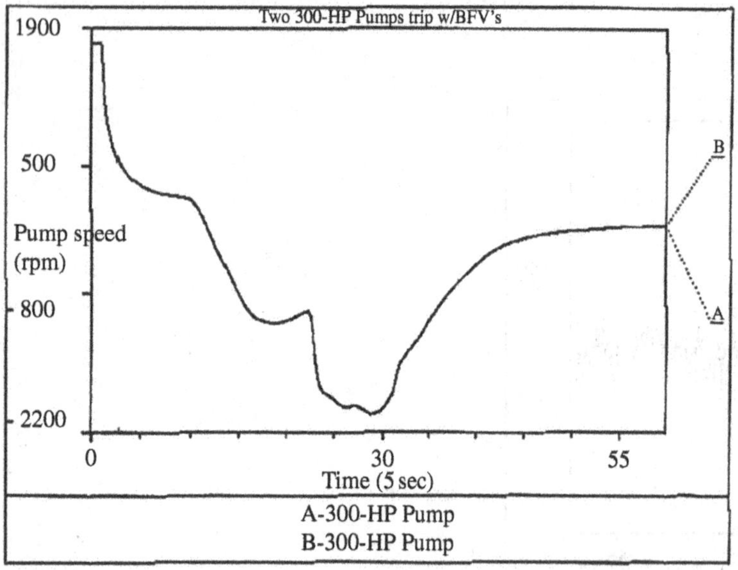

Figure 8.19 Pump Speed vs, Time Showing Maximum Reverse Speed and Duration

occurrence of such reflections until the steady-state condition is reached. If the discharge valve is opened too quickly, damaging transient pressures can occur.

We have made the worst-case assumption that the pump has come up to full speed before the valve is opened. Table 8.2 results for the two worst conditions (300-hp pumps alone and together) indicate that the valve opening is sufficiently slow and the maximum pressures are acceptable. One important qualification should be noted. Much higher pressures than those tabulated can result if air is trapped in the line during startup. If the lines are being filled or if for other reasons air is in the system, particular care should be exercised when starting up the system. The air should be permitted to escape by operation of the air-release valves in the system, and startup should be sufficiently slow. Only one 150-hp pump should be operated until all air has been released. If prior experience during initial startup has so suggested, then reduced-speed manual opening of the control valve should be used to further reduce the speed of initial startup.

Table 8.2 Transient Results with 40-second Butterfly Valve Closure

Condition	Steady State Pressure at Valve (psi)	Maximum Transient Pressure at Valve (psi)	Maximum Transient Pressure at Station 1 +84 (psi)	Maximum Reverse Pump Speed (rpm)	Column Separation
Two 300-hp pumps trip, with control valve	157	257	245	2010	Yes
300-hp and 150-hp pumps trip, with control valve	154	216	214	2280 (300-hp) 2250 (150-hp)	Yes
One 300-hp pump trips, with control valve	152	232	225	2480	Yes
One 150-hp pump trips, with control valve	149	192	190	2400	No
Two 300-hp pumps, normal valve closure on one	157	199	155	0	No
One 300-hp pump, normal valve closure	152	200	161	0	No
One 300-hp pump, normal valve opening	152	198	168	0	No
Two 300-hp pumps, normal valve opening on one	157	198	168	0	No

Selected results for various conditions analyzed are presented in Table 8.2. The two locations at which pressure results are given are the maximum-pressure location at the pump station control valve (upstream or downstream of the valve as appropriate) and a node near the junction between buried cast iron and reinforced concrete pipe at 184 ft from the wall of the station at which near-maximum pressures were found to occur in the piping downstream of the station.

The four possible cases of pump tripping with control valve were analyzed. For the three most severe cases, column separation occurred followed by large and abrupt pressure increases at the locations tabulated. For the least severe case (150 hp pump), column separation did not occur, and pressure increases were moderate and gradual. Again, in the absence of column separation, the overpressures are spread out over time, and the peak is much lower.

For the cases of pump tripping in which column separation occurs, Table 8.2 overpressures are lower than those with the cone valves. There is a tradeoff between reduced overpressures and increased reverse pump speed in some of those cases with the new butterfly valves.

As noted under *Pump Characteristics*, the client gave the maximum allowable reverse speed of the motors as 125% of their normal forward speed (1750 rpm), or $1.25 \times 1750 = 2190$ rpm. Referring to Table 8.2, that speed is exceeded in all but the case of two 300-hp pumps tripping. The allowable reverse speed of the motors discussed above (2190 rpm) is exceeded for three of the four cases given in Table 8.2.

As noted previously, any reverse pump speeds exceed what the manufacturer initially recommended, but the speeds are reversed for only about 30 seconds, and it is possible that the reverse speeds are acceptable for such a short duration. This is given further consideration below.

Normal pump shutdown was assumed to entail 40-second closure of the valve on the discharge of the pump to be de-energized, with the pump energized until complete valve closure. Pressures between the pump and valve increased gradually to shutoff, as is commonly the case when the valve closure time is a large multiple of the wave travel time between the pump and valve. As shown in Table 8.2, the valve closure is sufficiently slow to avoid excessive downsurge and subsequent upsurge in the piping downstream of the valve, and the maximum pressures were found to be acceptable.

Conclusions

Model results for the replacement butterfly valves show that normal pump startup and shutdown are readily controlled. However, power failure will result in large and abrupt overpressures. These overpressures are a consequence of column separation at a high point in the system and the shock waves that follow collapse of the vapor cavity. Column separation is avoided only in the case of one 150-hp pump operating alone, and only in that case would valve control alone suffice. For all other pump combinations, the overpressures cannot be sufficiently ameliorated by the control valves alone. It seems unlikely that the large, unbalanced pressures and thrusts could be tolerated by the piping system for very long.

The reverse pump speeds exceed, in some cases, the maximum allowable reverse speed of the motors as given by the client. The reverse speeds in all cases exceed what was initially recommended by the pump manufacturer. However, the pump manufacturer's recommendations seem conservative. Presumably, there have been power failures with reverse speed previously without damage to the pumps. As mentioned above, the manufacturer was asked to review more specific information in this regard and provided subsequent information, which is considered below. Some preliminary runs with the wave speed reduced to 50% of the values derived above (to account for possible air entrainment) indicate that the effect is to somewhat decrease overpressures but to further increase pump speeds.

The only case for which we would recommend valve control alone under the conditions analyzed above would be for one 150-hp pump operating. However, the downsurge controls that would be required for the other cases would modify the valve setting for the case of one 150-hp pump, necessitating subsequent revisions. Decisions regarding the downsurge controls should be made before proceeding with the determination of valve settings.

There are a number of possible ways of controlling downsurge to eliminate column separation, which generally can be categorized as follows:

1) Change in pipeline profile
2) Inertial control
3) Surge tanks
4) Air chambers (hydropneumatic surge arrestors)
5) Pump bypass
6) Air-inlet valves (air-release/vacuum valves)

Change in pipeline profile is only practical in limited circumstances, and probably not here. Inertial control entails increasing the drive inertia to reduce downsurge. It is commonly used for magnetic variable-speed drives by using battery power to keep the drive components connected to the pump following power failure; it is generally not practical for constant-speed pumps.

Surge tanks and air chambers are excellent methods of downsurge control. They are most effective if located at or near the high point(s) where column separation occurs, but sometimes work if located at another location. That other location is commonly the pumping station, where operator attention can be more conveniently provided.

A pump bypass entails piping around the pumps in which valves can open on power failure to provide a low-resistance path around the pump. This method has a good chance of working in the present circumstances.

Air-inlet valves (Parmakian 1950) located at high points in the line also offer a method that will, in concept, prevent column separation. This method can be as effective as surge tanks and air chambers, and significantly less costly. However, such valves tend to become inoperable and require concerted maintenance attention to remain operational. Redundancy can help offset this limitation.

Controls 3 and 4 and certain other methods, such as surge-anticipation valves, can be used to reduce the overpressures without eliminating downsurge. However, elimination of column separation usually affords the most cost-effective protection. Overpressure-reduction methods are sometimes used as backup methods intended to reduce column separation (particularly as backup to air-inlet valves).

Two additional analyses were made to demonstrate the potential effectiveness of eliminating column separation. For the first, the system was modeled with a large-diameter surge tank (or water tank) at the high point in the line in the case of two 300-hp pumps tripping with 40-second butterfly valve closure. The resulting maximum pressure at the control valves was 192 psi (vs. 257 psi in Table 8.2 with only valve control). The maximum reverse pump speed in this case is 2230 rpm, which is higher than occurs with valve control alone (2010 rpm per Table 8.2). However, the pump speed could be reduced by faster valve closure while still achieving reductions in overpressure compared to Table 8.2.

The second additional analysis was the same as the first described in the previous paragraph with the addition of anti-reverse ratchets to the pumps. This does two things. First, it eliminates the concern with reverse rotation. Second, it reduces upsurge pressures because pumps restrained from reverse rotation (at the low specific speeds of these pumps) offer less resistance to backflow than when rotating in reverse. The resulting maximum pressure at the control valves was only 171 psi. The maximum reverse pump speed is, of course, zero.

It is possible that anti-reverse controls at the pumps would reduce the overpressures with column separation enough that there would be greater willingness to rely on air-inlet valves as a means of preventing column separation. If the air-inlet valves failed, the anti-reverse ratchets might keep the overpressures closer to acceptable limits. These overpressures could be predicted by modeling.

8.6.4.7 Transient Control

Introduction

In accordance with the original Scope of Services, the analyses discussed above were based on the assumption that the branch to the Arlington Heights Standpipe was closed. In subsequent discussions with the client based on familiarity of the system gained from the above, the client was advised of the potentially important role of the Arlington Heights Standpipe as a surge tank, which could mitigate waterhammer.

The two-million gallon Arlington Heights Standpipe has a bottom elevation of 382.0, normal operating range of 432–441, and overflow at an elevation of 442.5. The client advised that the Standpipe has an altitude valve that is set to the overflow elevation of 442.5 and to a lower level which is 3 ft below the overflow. This description is characteristic of a Differential Single Acting Altitude Valve, which closes at maximum level to prevent overflow plus remains closed until the tank level drops a predetermined amount (3 ft in this case) to allow fluctuation and then opens to refill the standpipe.

A run was made for the condition of the two largest pumps (300-hp) tripping due to power failure without any valve control. The initial run for this condition without the connection to the Arlington Heights Standpipe resulted in column separation at the high point in the transmission pipe and high overpressures due to the subsequent collapse of the vapor cavity. With the connection to the Arlington Heights Standpipe now included in the model, the results were dramatically different – there was no column separation and much lower overpressures (e.g., 193 psi at the pump control valve).

The condition of the two 300-hp pumps tripping with 40-second closure of the butterfly valves was run next. This is analogous to the run discussed under *Transient analyses – Butterfly Valves*, except for addition of the Arlington Heights Standpipe connection. The results were much more favorable than for the case with that connection closed. There was no column separation and overpressures were much lower. The pressure at the control valve increases gradually from the normal operating level of 155 psi to a maximum transient pressure of 194 psi vs. the sudden increase to 257 psi predicted with the connection to Arlington Heights closed. However, the maximum reverse pump speed was increased from 2010 rpm for the case with Arlington Heights closed to 2244 rpm with that connection open.

Control Valve Timing

The butterfly valve timing discussed above gave a maximum reverse pump speed for the condition of two 300-hp pumps tripping in excess of that recommended by the pump manufacturer (2100 rpm) and considered acceptable for the motors (2190 rpm), although only slightly greater than that which would have occurred for this condition (subject to the stated assumptions) with the original cone valves. Furthermore, the results in Table 8.2 suggest that other pump trip conditions give higher reverse pump speeds. Results given above for the butterfly valves (without the Arlington Heights connection) indicate that slower valve closure would not result in further reduction in overpressures and suggest that slightly faster valve closure might not significantly increase overpressures while possibly reducing reverse pump speed.

Accordingly, a run was next performed with faster, 30-seconds closure of the butterfly valves. The maximum pressure was not significantly worsened (compared to 40-seconds valve closure), increasing only to 195 psi. However, the reverse pump speed was not significantly reduced either, decreasing from 2244 to 2239 rpm. The pump speed curves were very similar for the 40-second and 30-second closures. The pump only remains near the maximum reverse speed for a few seconds, spending more time in the neighborhood of negative 1700 rpm. Furthermore, the pump has operated this way in the past with the cone valves.

The valves are presently set for 40-seconds closure and little would be gained by changing that to 30 seconds or less. The control valves would have to be closed much faster than 30 seconds to significantly reduce the reverse pump spin. Faster closure would make the system less forgiving of other factors in the system (e.g., a closed valve). Therefore, having consulting with the client, it was decided to continue the analysis with 40-seconds closure.

In the runs described above, the tank elevations were assumed to be 433 for the Walnut Hill Tank and 439.5 for the Arlington Heights Standpipe. In addition, maximum demands (and maximum wave speeds) were assumed as described previously. The flow initially divides, with flow going to both Walnut Hill and Arlington Heights. A few seconds after the pump trip, the flow in piping in the vicinity of the Arlington Heights Standpipe reverses, and the standpipe acts as a surge tank, providing flow into the adjoining area and preventing the column separation at the high point in that area. Another benefit of the division of flow is that the initial flow at the high point in that adjoining area is reduced, lowering the inertia and thus further reducing the tendency for column separation at that high point.

The next step consisted of sensitivity analyses to determine the effects of inflow and outflow along the network in the vicinity of the Arlington Heights Standpipe, and such other permutations as wave speeds and certain demands and inflows. An exploratory run was next made with outflows added to appropriate nodes. The outflows and demands were taken as all-connected average-day 1994 demands listed in the Annual Summary tabulation provided by the client and identified by schematics provided by the client. The demands were conservatively assumed to remain constant throughout the transient. Since average-day demands at major locations gave lower pressures than no demands at those locations, it was reasoned that there was no need to consider higher demands at those locations.

A run was next made without certain demands, and the Arlington Heights Standpipe was reduced to its lower operating water elevation of 432. This gave pressures that were generally a few psi higher than the runs discussed above. With this as a base case, runs were made for additional permutations of wave speeds and demands. As expected, maximum demands (included in the base case) gave the highest pressures, as did maximum wave speeds. Flow reversal occurs at about 5.5 seconds (4.5 seconds after pump trip), and zero pump speed occurs at about 7 seconds. The maximum reverse pump spin of 2270 rpm occurs at about 14 seconds. However, the pump only remains near the maximum reverse speed for a few seconds, spending more time in the neighborhood of negative 1600 rpm. The pump has operated this way in the past with cone valves. Negative 2270 rpm and negative 1600 rpm are, respectively, 130% and 91% of normal forward speed (1750 rpm). Referring to *Pump Characteristics* above and considering the duration of the reverse speeds and past operating experience, the reverse speeds with control valve closure should be acceptable for the motors and pumps. The runaway speed with two 300-hp pumps and no control valve closure of about 2010 rpm (Table 8.2) is 115% of normal forward speed. Table 8.2 shows higher reverse speeds under certain other conditions. Operation under these conditions (such as with a failed control valve) for an extended period of time would be risky for both motors and pumps.

The three remaining conditions for pumps tripping were also analyzed, as shown in Table 8.3, except for the conditions of both 300-hp pumps tripping, which has a maximum reverse pump speed higher than that with Arlington Heights valved off, which results in lower reverse speeds than given in Table 8.2. The condition of the 300-hp and 150-hp pump tripping gives a maximum reverse speed of 2410 rpm for the 150-hp pump, which is the highest reverse speed. The maximum reverse speed again occurs for only a few seconds, with the 150-hp plump spending more time in the neighborhood of negative 1880 rpm. This should also be acceptable.

Normal pump shutdown and startup were analyzed for the 40-second valve operating periods. Selected results for various conditions analyzed are presented in Table 8.3. As for Table 8.2, the two locations at which pressure results are given are the maximum-pressure location at the pump station control valve (upstream or downstream of the valve as appropriate) and a node near the junction between buried cast iron and reinforced concrete pipe at a station 184 ft from the wall of the station, at which near-maximum pressures were found to occur in the piping downstream of the station. Table 8.3 results can be directly compared to those in Table 8.2, the latter being for the new butterfly valves with Arlington Heights valved off.

8.6.4.8 Summary

The transient pressures in Table 8.3 are all less than those in Table 8.2 and demonstrate the considerable benefit of the open interconnection to the Arlington Heights Standpipe. Assuming that the pump station cement-lined ductile iron discharge lines have the usual 250 psi pressure rating and that the pipe supports (including thrust restraints) have been designed for that pressure rating, the pump station internal piping should readily withstand the transient pressures given in Table 8.3. The reasonably gradual manner in which these transient pressures occur is an important factor in making this determination.

Downstream buried piping in the vicinity of the Walnut Hill Tank connection and along an adjoining area consists of cast iron and reinforced concrete. Some of the cast iron and reinforced concrete pipe is rated at 250 psi, which would withstand the transient pressures given in Table 8.3. The rest of the cast iron and reinforced concrete pipe is of unknown strength. In particular, the 184 ft length of buried cast iron pipe between the pump station wall and the junction with reinforced concrete pipe (Station 1+84) has relatively high transient pressures (see Table 8.3) and is of unknown strength.

Should this length of pipe show evidence of leakage in the future, it would probably be more economical to replace it, to provide additional transient controls, or sustain the additional wear and tear on the pumps that would result from slower valve closure. The slower valve closure would only reduce the overpressures by a few psi, such as from 199 psi at the control valve for two 300-hp pumps tripping (Table 8.3) to 193 psi at best (with no control valve closure).

Table 8.3 Transient Results with 40-second Butterfly Valve Closure and Open Connection to Arlington Heights Standpipe

Condition	Steady State Pressure at Valve (psi)	Maximum Transient Pressure at Valve (psi)	Maximum Transient Pressure at Station 1+84 (psi)	Maximum Reverse Pump Speed (rpm)	Column Separation
Two 300-hp pumps trip, with control valve	154	199	199	2270	No
300-hp and 150-hp pumps trip, with control valve	152	196	196	2235 (300-hp) 2410 (150-hp)	No
One 300-hp pump trips, with control valve	151	184	184	2140	No
One 150-hp pump trips, with control valve	149	181	180	2270	No
Two 300-hp pumps, normal valve closure on one	154	192	152	0	No
One 300-hp pump, normal valve closure	151	193	150	0	No
One 300-hp pump, normal valve opening	151	190	157	0	No
Two 300-hp pumps, normal valve opening on one	154	190	158	0	No

It is necessary that the Arlington Heights Standpipe be in open communication with the adjoining area in order for the favorable Table 8.3 predictions to be realized. The interrelation between the standpipe altitude valve controls and the Arlington Spring Street Station's pump controls should be checked and modified as necessary to insure that the altitude valve stays open on power failure and that the system is in all other respects operated so that the altitude valve is open when the Spring Street pumps are in operation. The pumps must be shut down *before* the altitude valve closes so that the associated pumps cannot trip when the altitude valve is closed. Normal shutdown and startup should take place with the altitude valve open, primarily to prevent the possibility of the pump tripping with the altitude valve closed.

If the Arlington Heights Standpipe is out-of-service or, for any other reason, the standpipe temporarily is not in communication with the adjoining area, then the Spring Street Pumping Station operation should be limited to one 150-hp motor-driven pump (see Table 8.2) or the diesels driving the pumps. An alternative that would give more flexibility and would forgive operator error in this regard would be the installation of a vacuum relief valve at an appropriate location. One such possibility, the high point in an adjoining area, was discussed previously. In light of the subsequent results given in this section, a vacuum relief valve could be successfully located on piping in the vicinity of the Arlington Heights Standpipe, between its junction with the adjoining area and a nearby line valve. The runs made without the Arlington Heights connection showed a large negative pressure in the vicinity of the Arlington Heights Standpipe, which suggests that a vacuum valve near that location could quite possibly prevent column separation. The advantage of that nearby location over the high point in the adjoining area is that the former is cast iron pipe, whereas the latter is concrete. A vacuum relief valve could be more readily installed on the cast iron pipe, such as by using a flanged tapping sleeve.

Another advantage of that nearby location is that whereas negative pressures occur at the abovementioned high point even with Arlington Heights on line (albeit well above vapor pressure), the pressure at that nearby location remains positive. Therefore, the vacuum relief valve at the abovementioned location could be left always open (i.e., not valved off by, e.g., a butterfly valve between the flange and the vacuum relief valve) without admitting air that would have to be purged following power failure with Arlington Heights on line. The vacuum relief valve at that location would only activate with Arlington Heights off line.

8.6.4.9 Other Waterhammer Projects

Types of waterhammer controls mentioned above include flywheels and magnetic drives, control of pump discharge valve opening and closing, vacuum relief valves, and maintenance of standpipe hydraulic communication (with the standpipe acting as a surge tank). Some of these same controls have been used on others of the author's projects, plus additional ones discussed below.

Salem/Beverly Wenham Lake Water Pumping Station, Salem, Massachusetts

This 20 mgd station had complex waterhammer problems involving the pumping station, two distribution storage facilities, and approximately 17 miles of transmission lines. Most of the transmission lines date to the 1800s and have limited over-pressure tolerances. A preliminary downsurge analysis was performed, which identified various operational conditions at which water column separation would occur. Several reliable methods for preventing water column separation were evaluated, including (1) bypass line around the pumps with quick opening check valve(s), (2) hydropneumatic surge arrester, (3) surge tank(s), and (4) pipeline changes. The nature of the system precluded the first two options and led to the recommendation of a surge tank and a change in the size of a proposed pipeline. Air-inlet valves were planned for portions of the system that could not feasibly be protected by other methods. Upsurge control measures recommended were prevention of reverse pump rotation and power-operated slow valve closure on the pump discharges. The upsurge analysis included the determination of backflow and characterization of the transmission network for simplified rigid column analysis of the pressure rises accompanying valve closure. The author's valve closure analysis determined the valve closure time for the actual valve characteristics and other system parameters. Torques required to hold the pumps against backspin were determined, and advice was provided on associated equipment. Specifications were prepared for the pump discharge valves and operators. Used Kinno and Kennedy (1965) charts.

Spot Pond Pumping Station, Stoneham, Massachusetts

This station has a complex system of intake lines, wet wells, and multiple discharge lines that terminate at the Bear Hill Standpipe and Fells Reservoir. The station's three deep-well turbine pumps, totaling 1400 hp, have gradually been converted from diesel drives to motors, which presents the challenge of reduced inertia. The primary concern was with transient pressures in the discharge lines associated with pump power failure. Recommended control measures included vacuum-release/pressure air-release valves, for which locations and sizing were determined. It was recommended that certain cleaning and lining be deferred because they would substantially worsen column separation and overpressures.

Belmont Pumping Station, Belmont, Massachusetts

Plans to replace one of three pumps with a modern motor with much lower inertia than the original and to use that pump as the lead pump of three 125-hp centrifugal pumps and control valve modifications were analyzed for various existing and future conditions. With a distance from the pumps to the Arlington Covered Reservoir of 6780 ft, the focus was on power failure and required control valve modifications for the three pumps, and pump control valve settings for a combination of butterfly (one pump) and cone valves (two pumps) timings for normal startup, normal shutdown, and emergency shutdown.

Lake Avenue Wastewater Pumping Station, Worcester, Massachusetts

Air chambers (surge arrestors) were recommended by the author for this pumping station. Used Kinno and Kennedy (1965), Evans and Crawford (1954), and Parmakian (1963, Chapter XVIII).

Wastewater Pumping Station, Haverhill, Massachusetts

Cone valves were specified by the author to have hydraulic cylinder operators and a control system providing timed opening and closure under normal operation and delayed fast closure for a set time period after power failure.

8.7 Wet Wells

Entrainment of air into the pump inlet must be avoided. We will address this subject by starting with the simpler case of the water drain vortex. The formation of such vortices was a mystery for many years; it is thought to have been caused by an instability late into the twentieth century (Levi 1995, p. 351). However, unlike such phenomena as atmospheric tornadoes, dust devils (on Mars as well as on Earth), waterspouts (Levi 1972, Lugt 1983, Moskowitz 2021, Fischetti 2023), and fire tornadoes (Forthofer 2019), the water drain vortex is not caused by an instability. It is, rather, a continuous phenomenon beginning with a surface dimple, which increases in size until the vortex reaches the drain and air entrainment occurs (Odgaard 1986, 1988; Gulliver 1988). Although primarily of theoretical value for pumping stations, good descriptions and a formula for simple drain vortices are provided in the references just cited. Odgaard's (1988) formula for incipient air entrainment is:

$$H^2 = -0.9\frac{\sigma}{\rho g}\sqrt{\frac{\upsilon H}{\nu + \varepsilon}} + 0.0043\frac{\Gamma^2 \upsilon}{g(\nu + \varepsilon)} \tag{8.87}$$

in which H = critical depth of submergence, σ = surface tension, ρ = density, g = gravitational acceleration, ν = kinematic viscosity, υ = axial velocity near drain entrance, Γ = circulation, and ε = eddy viscosity $\approx 6(10)^{-5}\,\Gamma$.

Neglecting the surface tension term (as per Odgaard 1988) and rearranging results in:

$$\frac{H^2 g(\nu + \varepsilon)}{\Gamma^2 \upsilon} = 0.0043 \tag{8.88}$$

Greater circulation and drain velocity favor the drain vortex occurring at greater H and conversely.

Practical advice for preventing air-entraining vortices in conventional wet wells is provided in ANSI (2012); Bauer *et al.* (1996); and Jones and Sanks (2008).

The author provided detailed guidance on the design of a self-cleaning trench-type wet well for the 5.04 mgd (maximum flow) Washington Highway Wastewater Pumping Station, Lincoln, Rhode Island. Partial design guidelines for such wet wells are provided by Sanks *et al.* (1993a, 1993b, 1995), ANSI/HI (1998), Jones (2002), Dunn *et al.* (2002), and Kinshella *et al.* (2005). The author developed methods to improve upon existing published design techniques and put the design of such systems on a more rational footing. The full design is too detailed to discuss here. The design of the ogee weir for this facility is discussed in Section 7.2, and the length of the jump is discussed in Section 7.5.

8.8 Stormwater Pumping Station Design

8.8.1 Introduction

The material in this section is taken from Graber (2010a, 2017a). Consideration is given first to small to medium-sized pumping stations, defined by WEF (1993, pp. 36–39) as those for which a single pump provides design capacity (with one or more others for low flows, alternating sequences, and/or backup); such stations are of less than 200 l/s (3200 gpm) capacity (WEF 1993). Larger submersible pumps have become available (ASCE/WEF 1992, pp. 390–395), which allows larger stormwater pump stations to rely on a single submersible pump. One can use a generalized solution for the hydrologic and hydraulic design of such small to medium-sized stormwater pumping stations (Graber 2010a). The benefits of the solution include (1) the ability to consider the effects of storm duration rather than the fixed hydrology assumed in other methods; (2) a simpler and more intuitive design procedure than other available methods; and (3) enabling the calculation of storage requirements independent of the geometry of the wet well and any inundated upstream area.

The solution can be derived by first specializing four basic relationships on which a generalized solution for detention basin design is based (Graber 2009a). The first is conservation of mass, relating the terms I = volumetric inflow rate, Q = volumetric outflow rate, and $dS\,/\,dt$ = time rate of increase of effective storage volume. Second is the inflow relationship, which is assumed to have the form of the trapezoidal inflow hydrograph depicted in Figure 8.20 with peak inflow I_p, linear rising limb to the time of concentration t_c, duration t_r, and linear receding limb of duration Rt_c; additional terms in Figure 8.20 are defined below. The third and fourth relationships are simplified forms of the storage and outflow terms, expressed, respectively, by Equations 5.5 and 5.6 of Chapter 5.

Equation 5.5 is cited only for comparison with related formulations (Graber 2009a, 2009b). For present purposes, no restrictive assumption is made at this stage regarding the nature of S. Equation 5.6 is specialized for the case of constant pumped outflow (when the pump is on) so that $m = 0$, $H^m = 1$, and $Q = M$. Constant pumped outflow is a useful preliminary design assumption, also made by others (e.g., Burton 1980; FHwA 1982, 2001a; Baumgardner 1983; ASCE/WEF 1992;

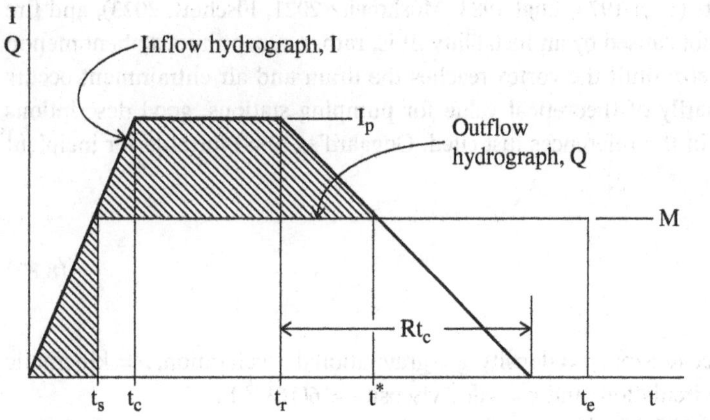

Figure 8.20 Inflow and Outflow Hydrograph Characteristics

Froehlich 1994). The constant pumped flow is assumed to start at t_s = the time of intersection of M with the rising limb of the hydrograph, as also assumed by Burton (1980) and depicted in Figure 8.20.

$$\frac{t_s}{t_c} = \frac{M}{I_p} \tag{8.89}$$

The maximum outflow (constant M) and concurrent maximum storage volume denoted by S^* (depicted by the cross-hatched area between the inflow hydrograph and constant pumped flow in Figure 8.20), occur at time denoted by t^*; nondimensionalized t^* and S^* are given by (Graber 2010a):

$$\frac{t^*}{t_c} = \frac{t_r}{t_c} + R\left(1 - \frac{M}{I_p}\right) \tag{8.90a}$$

$$\frac{S^*}{I_p t_c} = \frac{t_r}{t_c}\left(1 - \frac{M}{I_p}\right) + \frac{1}{2}(R-1) - \frac{M}{I_p}\left[R - \frac{M}{I_p}\left(1 + \frac{1}{2}R\right)\right] \tag{8.90b}$$

The nondimensional inflow volumes are given by (Graber 2010a):

$$\frac{\forall}{I_p t_c} = \begin{cases} \dfrac{1}{2}\left(\dfrac{t}{t_c}\right)^2, & t \le t_c & (8.91a) \\ \dfrac{t}{t_c} - \dfrac{1}{2}, & t_c < t \le t_r & (8.91b) \\ \dfrac{t}{t_c} - \dfrac{1}{2R}\left(\dfrac{t}{t_c} - \dfrac{t_r}{t_c}\right)^2 - \dfrac{1}{2}, & t_r < t \le (t_r + Rt_c) & (8.91c) \end{cases}$$

The total inflow volume $\forall_i$ is given by Equation 8.91c for $t = (t_r + Rt_c)$:

$$\frac{\forall_i}{I_p t_c} = \frac{R+1}{2} + \frac{t_r}{t_c} - 1 \tag{8.92}$$

The ratio of maximum storage volume to total inflow volume is obtained from Equations 8.90b and 8.92:

$$\frac{S^*}{\forall_i} = \frac{\frac{t_r}{t_c}\left(1 - \frac{M}{I_p}\right) + \frac{1}{2}(R-1) - \frac{M}{I_p}\left[R - \frac{M}{I_p}\left(1 + \frac{1}{2}R\right)\right]}{\left(\frac{R+1}{2} + \frac{t_r}{t_c} - 1\right)} \tag{8.93}$$

The following ratio of S^*/V_i for $t_r/t_c \to \infty$ is obtained by taking the limit of Equation 8.93:

$$\lim_{t_r/t_c \to \infty} \frac{S^*}{V_i} = \left(1 - \frac{M}{I_p}\right) \tag{8.94}$$

Equations 8.93 and 8.94 are plotted in Figure 8.21 for $R = 1.67$; the selection of this R value is explained in Graber (2009a). The values of S^*/V_i are equal (independent of t_r/t_c) at $M/I_p = (1+R)/(2+R)$ at which $S^*/V_i = 1/(R+2)$; the numerical values for $R = 1.67$ are $M/I_p = 0.728$ and $S^*/V_i = 0.272$.

8.8.2 Comparison to Other Methods

Also plotted in Figure 8.21 is the relationship for estimating the required storage given in ASCE/WEF (1992) [although attributed therein to the Federal Highway Administration (FHwA 1982), the method does not appear in that publication]. That relationship is based on fitting a triangular hydrograph to the peak-flow portion of an SCS (1986) [Soil Conservation Service, now National Resource Conservation Service (NRCS)] type inflow hydrograph and deriving an equation given, in terms of present notation, by $S^*/V_i = (1 - M/I_p)^2$. The triangular hydrograph corresponds in Figure 8.20 to $t_r/t_c = 1$; the approximation can be seen to follow the curve in Figure 8.21 for $t_r/t_c = 1$ up to $M/I_p \cong 0.15$ and to significantly underpredict the required storage for larger values of M/I_p [if the upper curves are plotted for $R = 1$, their distance above the ASCE/WEF (1992) curve becomes even greater]. Even for $t_r/t_c = 1$ and $R = 1$ as assumed in the ASCE/WEF (1992) method, Equation 8.93 does not reduce to the ASCE/WEF equation; the method fails to consider what Baumgardner (1983) refers to as "Storage Below Pump-On Elevation" and which, for the case considered here, is the storage associated with the area to the left of t_c (see Figure 8.20), which becomes more significant as M/I_p gets larger. This is further addressed in an example presented below.

FHwA (2001a) properly cautions against using long-duration events, such as SCS/NRCS 24-hour hyetographs: "Shorter duration storms that compare with the estimated time of concentration for the drainage area are usually more appropriate for pump station design." A method that accomplishes that and is related to the method discussed in the previous paragraph is given in ASCE/WEF (1992). It entails a triangular approximation to the maximum intensity portion of the SCS/NRCS hyetograph. A comparative example of that method is also given below.

Comparison with other methods, including Burton (1980) and Froehlich (1994), is given in Graber (2010a, 2011a).

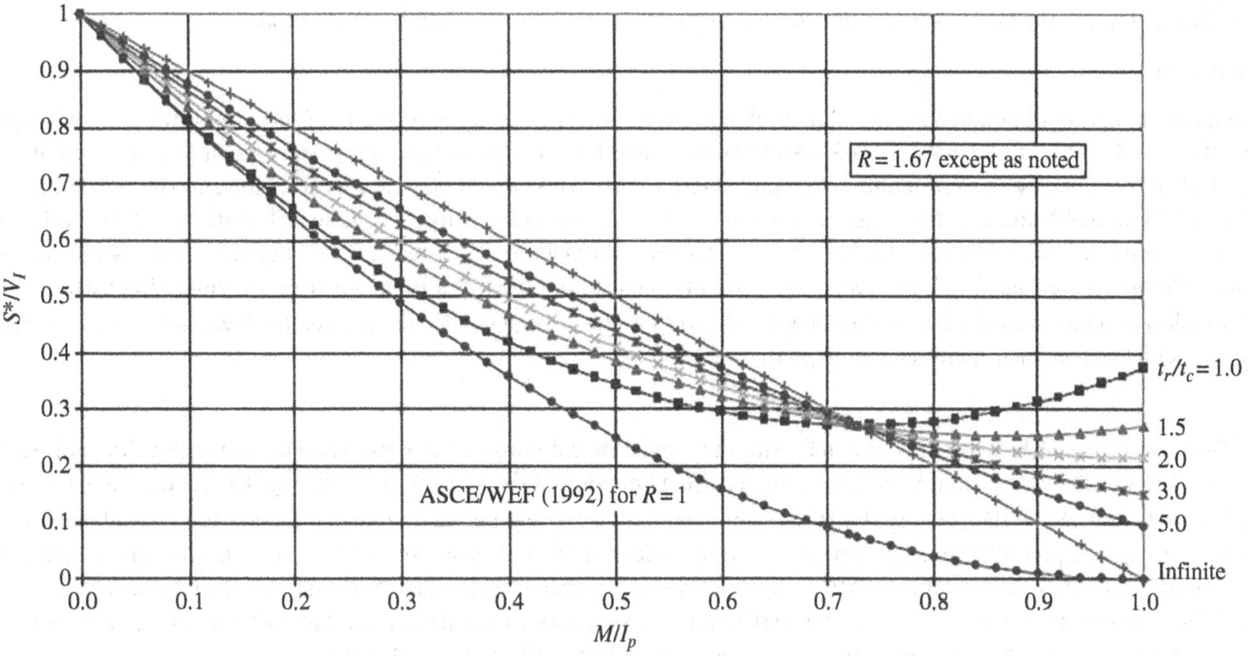

Figure 8.21 S^*/V_i vs. Q^*/I_p and t_r/t_c

8.8.3 Pump Operation

From Figure 8.20 and Equation 8.89, the pump start volume is given by:

$$\forall_{p,start} = (1/2)t_s M = (1/2)M^2 t_c / I_p \tag{8.95}$$

The start volume is measured upward from the dead-storage elevation in the wet well, which may generally be taken to be equal to the pump stop elevation. Water below that elevation must be assumed to remain in the wet well. The dead-storage elevation can be based on the minimum required submergence of submersible pumps or the minimum acceptable submergence of bell inlets. Avoidance of air-core vortices is one important consideration in this regard. For submersible pumps, the minimum submergence is also based on cooling requirements. Some submersible pumps have a cooling jacket with a glycol coolant that allows the pump to operate without the motor being submerged; in such cases, the minimum submergence may be at the elevation of the top of the volute (the manufacturer should always be consulted). For bell inlets and other configurations, minimum submergence guidance is provided by ANSI/HI (1998).

Referring to Figure 8.20, the pump will continue to run until the time that maximum storage has been reached. Beyond that time, the pump will continue to run until the difference between the cumulative inflow volume and the cumulative pump discharge volume equals zero. Referring again to Figure 8.20, this will generally occur sometime after $t = t_r + Rt_c$. The cumulative inflow volume based on Figure 8.20 has reached a maximum and remains unchanged for $t > t_r + Rt_c$, and is given by Equation 8.92.

Denoting the cumulative pump discharge volume by $\forall_P$, the nondimensional cumulative pump discharge volume is given by:

$$\frac{\forall_P}{I_p t_c} = \frac{M}{I_p}\left(\frac{t}{t_c} - \frac{t_s}{t_c}\right) \tag{8.96}$$

Setting $\forall_i = \forall_P$ using Equations 8.92 and 8.96, and solving for t/t_c, gives the pump stop time, denoted by t_e, as follows:

$$\frac{t_e}{t_c} = \frac{M}{I_p} + \frac{t_r/t_c + R/2 - 1/2}{M/I_p} \tag{8.97}$$

Example 8.10

A drainage area containing homes, appurtenant structures, and pervious surfaces is protected from external flooding by a berm (Winthrop, Massachusetts). However, runoff from the drainage area itself requires a permanent solution to address flooding within the area behind the berm during times when the external water level is too high to permit one-way gravity discharge through the berm. A wet well and pumping station are to be designed for that purpose.

Solution

At a time of concentration of 6 minutes, the peak design (50-year) rate of runoff using the SCS Graphical Peak Discharge Method (SCS 1986) is 0.182 m^3/s (6.43 cfs). An essentially identical value of peak runoff (within 1% in this case) can alternatively be obtained by converting the curve number (CN) value to a Rational Method runoff coefficient (Graber 1992) and using the Rational Method. Table 8.4 gives values of t_r, t_r/t_c, and rainfall intensity i. The latter is based on published rainfall data for Topsfield, Massachusetts (Graber 1992), with linear interpolation for intermediate values of i. Successively larger storm durations are considered; the peak rate of runoff decreases in proportion to the decrease in rainfall intensity as the storm duration increases (Graber 2009a, 2009b). Values of $\forall_i/(I_p t_c)$ are calculated using Equation 8.92; values of $\forall_i$ are also given, which shows their increase with increasing t_r.

Different values of M are then considered, with the remaining columns of Table 8.4 based on one such value. For each value of M/I_p, values of t^*/t_c and $S^*/\forall_i$ are calculated using Equations 8.90a and 8.93, respectively. The values of $S^*/(I_p t_c)$ are calculated from $(S^*/\forall_i)[\forall_i/(I_p t_c)]$; Equation 8.90b could alternatively be used. Values of S^* are then calculated from $I_p t_c[S^*/(I_p t_c)]$. The procedure is easily repeated for other values of M in the same spreadsheet (not shown). The maximum values of S^* all correspond to values of M/I_p, t_r/t_c, and $S^*/\forall_i$ that comport with Figure 8.21; that figure could be used instead of Table 8.4 to estimate values of $S^*/\forall_i$ giving maximum S^*. It should be understood that values in Table 8.4 of t^*/t_c, $S^*/\forall_i$, $S^*/(I_p t_c)$, and S^* for other than the maximum S^* are only trial values with no physical significance.

Table 8.4 Stormwater Pumping Station Design Example

									M =	0.0849 m^3/s		=	3 cfs	
t_r (min)	t_r/t_c	50-year i (mm/hr)	50-year i (in/hr)	i_p (m^3/s)	i_p (cfs)	$V_i/(i_pt_c)$	V_i m^3	V_i ft^3	M/i_p	t^*/t_c	S^*/V_i	$S^*/(i_pt_c)$	S^*(m^3)	S^*(ft^3)
5	0.833	195.6	7.700	0.188	6.658	1.143	79.3	2.800	0.451	1.751	0.353	0.413	28.0	990
6 (t_c)	1.000	188.9	7.436	0.182	6.429	1 335	87.5	3.090	0.467	1.891	0.366	0.489	32.0	1.131
7	1.167	182.1	7.172	0.176	6.201	1502	94.9	3.352	0.484	2.029	0.372	0.559	35.3	1.247
3	1.333	175.4	6.907	0.169	5.972	1.668	101.6	3.587	0.502	2.164	0.373	0.623	37.9	1.339
9	1.500	168.7	6.643	0.163	5.744	1.835	107.4	3.794	0.522	2.298	0.371	0.680	39.8	1.406
10	1.667	162.0	6.379	0.156	5.516	2.002	112.5	3.974	0.544	2.428	0.365	0.730	41.0	1.449
11	1.833	157.3	6.193	0.152	5.355	2.168	118.3	4.180	0.560	2.568	0.360	0.782	42.7	1.507
12	2.000	152.6	6.007	0.147	5.194	2.335	123.6	4.366	0.578	2.705	0.354	0.827	43.8	1.547
13	2.167	147.8	5.821	0.142	5.033	2.502	128.3	4.533	0.596	2.841	0.346	0.867	44.5	1.570
14	2.333	143.1	5.635	0.138	4.872	2.668	132.5	4.680	0.616	2.975	0.337	0.899	**44.6**	**1.577**
15	2.500	138.4	5.449	0.133	4.711	2.835	136.1	4.808	0.637	3.107	0.326	0.924	44.4	1.567
16	2.667	134.7	5.303	0.130	4.585	3.002	140.3	4.954	0.654	3.244	0.316	0.950	44.4	1.568
17	2.833	131.0	5.156	0.126	4.458	3.168	144.0	5.085	0.673	3.380	0.306	0.969	44.0	1.555
18	3.000	127.2	5.010	0.123	4.332	3.335	147.2	5.201	0.693	3.513	0.294	0.981	43.3	1.530
19	3.167	123.5	4.863	0.119	4.205	3.502	150.1	5.301	0.713	3.645	0.281	0.985	42.2	1.491
20	3.333	119.8	4.717	0.115	4.079	3.668	152.5	5.386	0.736	3.775	0.267	0.981	40.8	1.440
21	3.500	117.2	4.613	0.113	3.988	3.835	155.9	5.506	0.752	3.914	0.257	0.984	40.0	1.413
22	3.667	114.5	4.509	0.110	3.898	4.002	159.0	5.616	0.770	4.051	0.245	0.982	39.0	1.377
23	3.833	111.9	4.404	0.108	3.808	4.168	161.8	5.715	0.788	4.188	0.233	0.972	37.7	1.332
24	4.000	109.2	4.300	0.105	5.718	4.335	164.3	5.803	0.807	4.323	0.220	0.955	36.2	1.278
25	4.167	106.6	4.196	0.103	3.628	4.502	166.5	5.880	0.827	4.456	0.207	0.930	34.4	1.215
26	4.333	104.6	4.118	0.101	3.561	4.668	169.4	5.984	0.843	4.596	0.196	0.913	33.1	1.170
27	4.500	102.6	4.040	0.099	3.493	4.835	172.1	6.080	0.859	4.736	0.184	0.890	31.7	1.119
28	4.667	100.6	3.962	0.097	3.426	5.002	174.6	6.168	0.876	4.874	0.172	0.860	30.0	1.060
29	4.833	98.6	3.884	0.095	3.358	5.168	176.9	6.248	0.893	5.011	0.159	0.823	28.2	995
30	5.000	96.7	3.806	0.093	3.291	5.335	178.9	6.320	0.912	5.148	0.145	0.779	26.1	923

Note: Bold font corresponds to critical duration.

Figure 8.22 plots the maximum storage volume vs. constant pumped outflow. Based on economic and space considerations, the 0.0849 m^3/s (3.00 cfs) pump capacity and corresponding 44.6 m^3 (1580 ft^3) storage capacity were selected. Pertinent values of the selected design in relation to Figure 8.21 are $M = 0.0849$ m^3/s (3.00 cfs), $I_p = 0.138$ m^3/s (4.87 cfs), $M/I_p = 0.0849/0.138 = 0.616$, $t_s = 0.616(6) = 3.70$ min (from Equation 8.89), $t_c = 6$ min, $t_r = 14$ min, $t^* = 14 + 1.67(6)(1 - 0.616) = 17.8$ min (from Equation 8.90a), $t_r + Rt_c = 14 + 1.67(6) = 24$ min, and $t_e = 0.616(6) + (14/6 + 1.67/2 - 1/2)(6/0.616) = 29.7$ min. (from Equation 8.97). Figure 8.23 plots the mass curves: cumulative inflow volume based on Equations 8.91 and 8.92, and cumulative discharge volume based on Equation 8.96. The maximum difference between the two curves equals the 44.6 m^3 (1580 ft^3) storage capacity, and the two curves can be seen to intersect at t_e where the pump stop volume is zero.

The pump start volume of 9.43 m^3 (333 ft^3) (Equation 8.57), maximum storage volume of 44.6 m^3 (1580 ft^3), and pump stop volume of zero are relative to a datum, which is the minimum submergence. The minimum submergence for the selected submersible, glycol-cooled pumps is 33 cm = 0.33 m (13 in. = 1.083 ft). The selected underground storage chamber has vertical sides and interior plan dimensions of 2.74 m (9.0 ft) × 4.88 m (16.0 ft) = 13.4 m^2 (144 sq ft), for which the corresponding depths from the bottom of the chamber are: 0.33 m (1.083 ft) stop, 0.33 + 9.43/13.4 = 1.03 m (3.39 ft) start, and 0.33 + 44.6/13.4 = 3.66 m (12.0 ft) maximum. These start and stop wet-well water elevations are specified in the design.

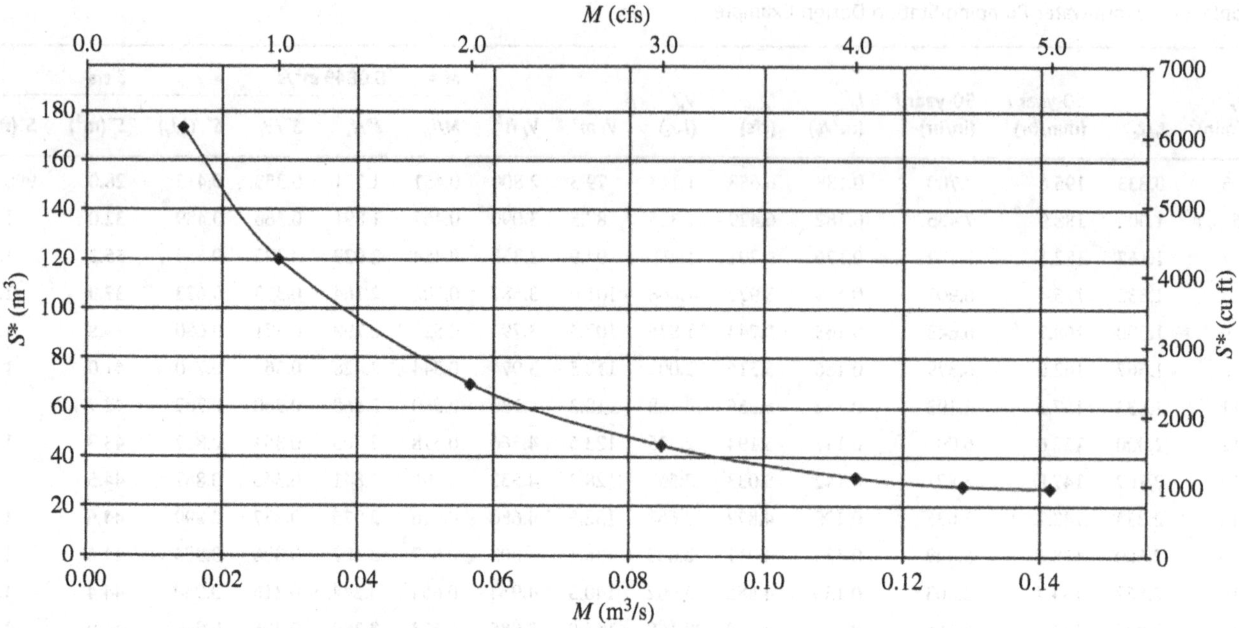

Figure 8.22 Example Maximum Storage Volume vs. Constant Pumped Outflow

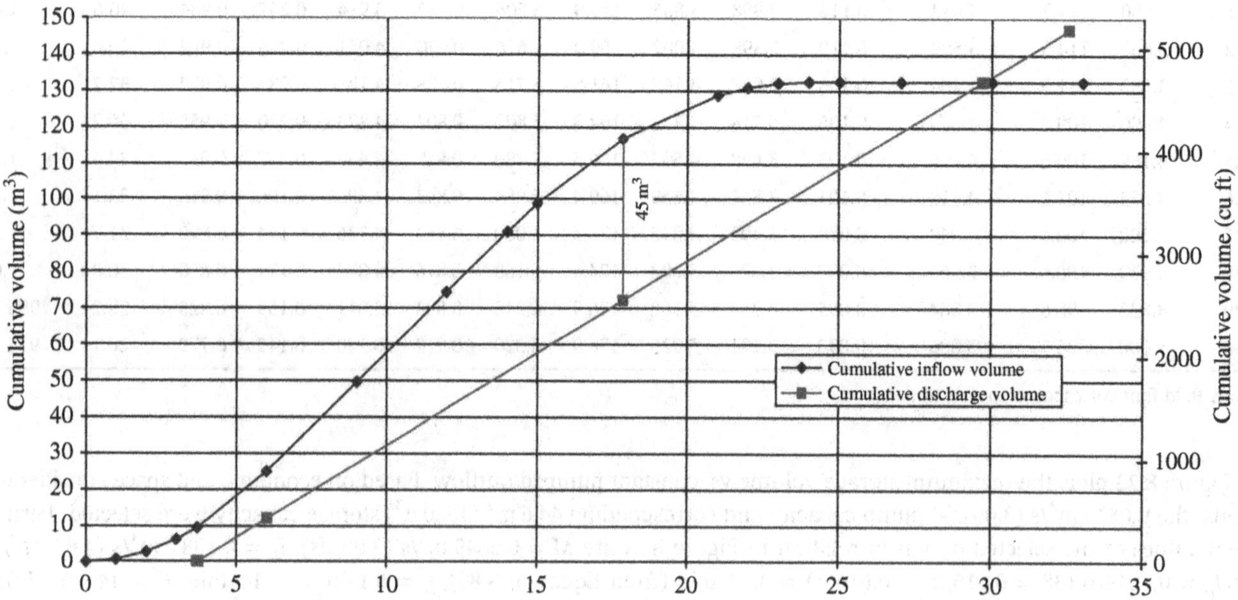

Figure 8.23 Example Mass Curves

To check pump cycling during more frequent storms with lower inflows, advantage can be taken from the demonstration by Pincince (1970) that the minimum cycle time for a single pump occurs when the inflow is one-half of the discharge; that leads to the following equation for minimum cycle time $t_{\min}$:

$$t_{\min} = \frac{4V_{ss'}}{M} \tag{8.98}$$

in which $\forall_{ss'}$ is volume of the wet well between pump start and stop and M is pump discharge as used above. In this case, $\forall_{ss'} = 13.4(1.03 - 0.33) = 9.38\ m^3$, giving $t_{min} = 4(9.38)/[0.0849(60)] = 7.4$ min, or 60/7.4 = 8 starts per hour. This is acceptable for the selected pumps, which are rated for 15 starts per hour. Pincince gives a method of extending his analysis to multiple fixed-speed pump installations.

An additional benefit of the generalized solution given above is that no assumption is made in deriving that solution regarding the wet-well geometry; that assumption did not come in until the chamber geometry was selected in this example. In the case considered in this example, it was desired to know the extent of flooding upstream of the pumping station associated with a 100-year rainstorm. An inlet grate was set at an elevation corresponding to the maximum 50-year chamber water elevation. The same process as demonstrated above for the 50-year storm was used to determine a 100-year storm $S* = 52.4\ m^3$ (1850 ft^3). The difference of 52.4–44.6 = 7.8 m^3 (275 ft^3) can be stored on the site without causing damage.

The example given above was also run using HydroCAD® version 9.00's Modified Rational Method, Dynamic Storage-Indication Method, and pump modeling features. The input matched that used for the above example, except that constant pump discharge was replaced with a pump discharge–TDH relationship having miniscule increases in discharge as the TDH decreased (a necessary workaround). The resulting HydroCAD output values were within 1% of those in this example.

The example given above was run again with one change: an actual pump curve and discharge characteristics were used rather than the 0.0849 m^3/s (3.00 cfs) constant-flow assumption utilized above. That included a pump discharge–TDH relationship for the selected pump and discharge information, which, coupled with wet-well depth variations that affect the static head, enabled calculation of the system flow–head relationship. HydroCAD results showed a pump discharge varying from 0.0813 m^3/s (2.87 cfs) when the pump starts, to a maximum of 0.0892 m^3/s (3.15 cfs), and then decreasing to 0.0796 m^3/s (2.81 cfs) just before the pump stops; this properly reflects the variation in static head. The maximum computed storage volume is 45 m^3 (1580 ft^3), which is identical to the value calculated in the original example (this is good, but possibly somewhat coincidental, and it would suffice for those values to be close to each other). Other assumptions for pump start and start elevations can be tested as necessary, e.g., by using the HydroCAD software discussed above.

8.8.4 Extension of the Method

The method given above can be readily extended to more complex situations if certain assumptions are made. Graber (2010a), does this, for example, for two pumps with constant capacity.

8.8.5 More Generally

For large, multiple-pump installations, inflow hydrographs with varying storm durations should be used to define the critical duration for each condition. FHwA (2001b) discusses the importance of considering different storm durations. Inflow hydrographs developed by the Modified Rational Method, such as shown in Figure 5.3, are useful for that purpose. The analysis will then require an extension of the trial-and-error or iterative procedures discussed in, e.g., Baumgardner (1983), ASCE/WEF (1992), and FHwA (1982, 2001b). Such analyses can be aided greatly by automatic computation.

9

Compressible Fluid Flow

9.1 Introduction

Much of the material in the present chapter is taken from Graber (2023a). The environmental engineer designs and/or specifies various systems in which gases are the fluid medium. Such gases include air, digester gas, molecular oxygen, chlorine, ozone, natural gas, and carbon dioxide. These gases are compressed, conveyed, dispersed, and diffused for a variety of purposes. Some gaseous systems, such as those providing molecular oxygenation, chlorination, and ozonation, are largely manufacturer-designed, with the prime design engineer's role being mainly that of selection and specification. On the other hand, aeration systems, and to a lesser extent, digester gas systems, entail the more detailed involvement of the environmental engineer. In all cases, a basic understanding is desirable.

The compressibility, or density variation, of gases introduces additional considerations, beyond those discussed in previous chapters, that must be taken into account. This chapter addresses those considerations. Gas properties are addressed first, followed by sections on flow through pipes and fittings, certain types of flow meters, dynamic and positive-displacement blowers, and compressors. In addition to addressing those practical applications, the material in this chapter provides background for the section on gas distribution conduits (i.e., spatially decreasing flows) in Chapter 11 and others. The emphasis is on pneumatic (air) systems, although many of the principles are more generally applicable.

9.2 Gas Properties

A *gas* is a molecular compound or mixture of two or more molecular compounds at a temperature greater than the liquid–vapor equilibrium or boiling temperature corresponding to the pressure of the mixture. Air, for example, is a mixture of oxygen, nitrogen, water vapor, and less than 1% of argon and other gases. [The term *vapor* is used here in the conventional sense to mean gas at a state not far from saturation (as defined below), which includes gaseous water in the applications of interest here.] The boiling temperatures of all of the major component compounds of air except water vapor are well below the air temperatures found in the natural environment, so the liquefied states of the compounds are not normally encountered. Water in liquefied form is, of course, commonly encountered, and water vapor will often condense to liquefied water in environmental engineering systems. Other gases of interest, which are also found in liquefied form in environmental engineering applications include chlorine and pure (molecular) oxygen.

A phase equilibrium diagram for water is shown in Figure 9.1. Of specific interest here are the equilibrium lines, and most particularly the liquid–vapor equilibrium or boiling line. The equilibrium line represents *saturation states*, which are states in which two different phases (gas–liquid, gas–solid, solid–liquid) can exist together in equilibrium. If liquid water at 1 atm and 212°F is brought into contact with water vapor at 1 atm and 212°F, there is no apparent disturbance of either phase by the other (actually, the vapor molecules condense on the liquid surface as fast as they evaporate from it). If, on the other hand, all else is the same but the liquid water is at a temperature lower than 212°F, then sufficient vapor must condense to raise the temperature of the liquid to 212°F. Generally, nonequilibrium situations will require energy exchanges between phases as they tend toward an equilibrium state. Note that the solid–liquid equilibrium, or freezing, temperature is little affected by pressure. Further discussion of the various features of Figure 9.1 and comparable diagrams for other substances may be found in thermodynamics texts.

Hydraulics and Pneumatics in Environmental Engineering, First Edition. S. David Graber.

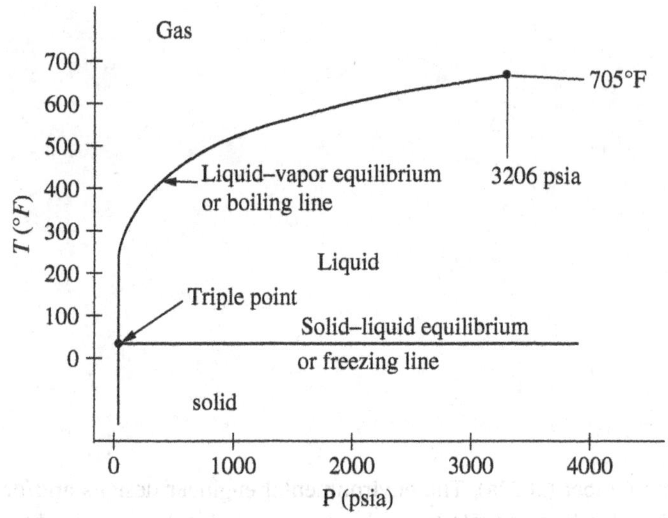

Figure 9.1 Phase Equilibrium Diagram for Water (Mooney 1953/Pearson)

The low-pressure portion of the boiling line is of primary interest here. In that portion, the pressure at the saturation state, known as the *saturation pressure* or *vapor pressure*, is a strong function of the saturation temperature (the seemingly vertical line is actually sloped) and vice versa. The saturation pressure, denoted p_g, is tabulated vs. temperature in Table 9.1 for the range of temperatures expected in environmental engineering air systems. When the pressure is less than p_g for the prevailing temperature, water vapor will form (as in cavitation – see Sections 6.5, 8.2, and 8.6). When the pressure exceeds p_g at the prevailing temperature, water vapor will condense to liquid water. The relation between the saturation pressure and temperature is important in determining certain characteristics of humid air, as discussed later.

Pertinent additional properties of *gases* as defined above are discussed below. The significance of these properties will become apparent in subsequent sections of this chapter.

Table 9.1 Saturation Pressure p_g (psia) of Water Vapor vs. Temperature T (°F)*

T	p_g	*T*	p_g	*T*	p_g	*T*	p_g	*T*	p_g
		75°	0.4298	**120°**	1.6924	**165°**	5.335	**220°**	17.186
		76	0.4443	**121**	1.7400	**166**	5.461	**222**	17.861
32°	0.08854	**77**	0.4593	**122**	1.7888	**167**	5.590	**224**	18.557
33	0.09223	**78**	0.4747	**123**	1.8387	**168**	5.721	**226**	19.275
34	0.09603	**79**	0.4906	**124**	1.8897	**169**	5.855	**228**	20.016
35°	0.09995	**80°**	0.5069	**125°**	1.9420	**170°**	5.992	**230°**	20.780
36	0.10401	**81**	0.5237	**126**	1.9955	**171**	6.131	**232**	21.567
37	0.10821	**82**	0.5410	**127**	2.0503	**172**	6.273	**234**	22.379
38	0.11256	**83**	0.5588	**128**	2.1064	**173**	6.417	**236**	23.217
39	0.11705	**84**	0.5771	**129**	2.1638	**174**	6.565	**238**	24.080
40°	0.12170	**85***	0.5959	**130°**	2.2225	**175°**	6.715	**240°**	24.969
41	0.12652	**86**	0.6152	131	2.2826	**176**	6.868	**242**	25.884
42	0.13150	**87**	0.6351	132	2.3440	**177**	7.024	**344**	26.827
43	0.13665	**88**	0.6556	133	2.4069	**178**	7.183	**946**	27.798
44	0.14199	**89**	0.6766	134	2.4712	**179**	7.345	**248**	28.797
45°	0.14752	**90°**	0.6982	**135°**	2.5370	**180°**	7.510	**250°**	29.825
46	0.15323	**91**	0.7204	**136**	2.6042	**181**	7.678	**252**	30.884
47	0.15914	**92**	0.7432	**137**	2.6729	**182**	7.850	**254**	31.973
48	0.16525	**93**	0.7666	**138**	2.7432	**183**	8.024	**256**	33.093
49	0.17157	**94**	0.7906	**139**	2.8151	**184**	8.202	**258**	34.245
50°	0.17811	**95°**	0.8153	**140°**	2.8886	**185°**	8.383	**260°**	35.429

Table 9.1 (Continued)

T	p_g	T	p_g	T	p_g	T	p_g	T	p_g
51	0.18486	96	0.8407	141	2.9637	186	8.567	262	36.646
52	0.19182	97	0.8668	142	3.0404	187	8.755	264	37.897
53	0.19900	98	0.8935	143	3.1188	188	8.946	266	39.182
54	0.20642	99	0.9210	144	3.1990	189	9.141	268	40.502
55°	0.2141	100°	0.9492	145°	3.281	190°	9.339	270°	41.858
56	0.2220	101	0.9781	146	3.365	191	9.541	272	43.252
57	0.2302	102	1.0078	147	3.450	192	9.746	274	44.682
58	0.2386	103	1.0382	148	3.537	193	9.955	276	46.150
59	0.2473	104	1.0695	149	3.627	194	10.168	278	47.657
60°	0.2563	105°	1.1016	150°	3.718	195°	10.385	280°	49.203
61	0.2655	106	1.1345	151	3.811	196	10.605	282	50.790
62	0.2751	107	1.1683	152	3.906	197	10.830	284	52.418
63	0.2850	108	1.2029	153	4.003	198	11.058	286	54.088
64	0.2951	109	1.2384	154	4.102	199	11.290	288	55.800
65°	0.3056	110°	1.2748	155°	4.203	200°	11.526	290°	57.556
66	0.3164	111	1.3121	156	4.306	202	12.011	292	59.356
67	0.3276	112	1.3504	157	4.411	204	12.512	294	61.201
68	0.3390	113	1.3896	158	4.519	206	13.031	296	63.091
69	0.3509	114	1.4298	159	4.629	208	13.568	298	65.028
70°	0.3631	115°	1.4709	160°	4.741	210°	14.123	300°	67.013
71	0.3756	116	1.5130	161	4.855	212	14.696	302	69.046
72	0.3886	117	1.5563	162	4.971	214	15.289	304	71.127
73	0.4019	118	1.6006	163	5.090	216	15.901	306	73.259
74	0.4156	119	1.6459	164	5.212	218	16.533	308	75.442

* Abridged from Table 1 of *Thermodynamic Properties of Steam* by Joseph H. Keenan and Frederick G. Keyes, John Wiley & Sons, Inc., New York, 1936.
Source: Adapted from Keenan and Keyes (1936).

9.2.1 Density and Perfect Gases

Whereas the density of water and most other liquids is virtually independent of pressure and temperature, the density of gases is a strong function of pressure and temperature. This density variability, or compressibility, of gases adds complexity to the analysis of gas flows. That is analogous to the complexity added in going from the incompressible flows in pressure conduits to the open-channel flow of liquids.

Air at the temperatures and pressures of interest here will generally comply with the equation of state for perfect gases, given by:

$$p\forall = mRT \tag{9.1}$$

in which p is the absolute pressure, m is the mass of the gas in volume $\forall$, T is absolute temperature, and R is an experimental gas constant, which differs for each gas. Since *density* ρ is defined by $\rho = m/\forall$, the above equation says that density varies with absolute temperature and absolute pressure according to:

$$\rho = p/(RT) \tag{9.2}$$

[Some thermodynamicists define a gas as *semiperfect* a gas obeying Equation 9.1, and define a *perfect* gas as a special case of the semiperfect gas for which the specific heats (see below) are constant. This distinction is not made here. A supercompressibility factor $Z = p\forall/mRT$ may be used to quantify departures from Equation 9.1. For air at pressures below 50 psia and temperatures above −30°F (430°R), $Z \cong 1$. For higher pressures, $Z \cong 1$ within a smaller temperature range. Applications considered herein generally fall well within the range of temperatures and pressures for which $Z \cong 1$ and Equation 9.1 is accurate.]

Absolute pressure is gage pressure (the pressure measured by a gage balanced by atmospheric pressure) plus the prevailing atmospheric pressure. Absolute temperature is temperature in degrees Rankine (R) or degrees Kelvin (K), which are related to Fahrenheit (F) or Celsius aka Centigrade (C) temperatures, as follows:

deg R = deg F + 459.7
deg K = deg C + 273
deg R = 1.8°K

Also:

deg F = (9/5)deg C + 32
deg C = (5/9)(deg F − 32)

Values of the gas constant R are given for various gases in Table 9.2.

9.2.2 Humidity

Humidity has a significant effect on the performance variables of air systems in environmental engineering practice. One such example is its effect on density, which in turn influences the volumetric rate of flow and such attendant variables as velocity and pressure drop. The study of mixtures of dry air and water vapor is known as *psychrometrics*, pertinent aspects of which are reviewed here. The term *air* alone will hereafter refer to the mixture of dry air and water vapor.

Table 9.2 Selected Properties of Gases

Gas	Formula	Molecular weight	Specific heats, Btu/lbm °F at 1 atm, ordinary room temperatures C_p	C_v	$k = C_p/C_v$	Gas constant, R, ft lbf/lbm °R $pv/T = R$ 1 atm 32°F
Air*	...	28.97	0.240	0.171	1.40	53.34
Monatomic Gases						
Argon	A	39.94	0.123	0.074	1.67	38.65
Helium	He	4.003	1.25	0.75	1.66	386.3
Diatomic Gases						
Carbon monoxide	CO	28.01	0.249	0.178	1.40	55.13
Hydrogen	H_2	2.016	3.42	2.43	1.41	767.0
Nitrogen	N_2	28.02	0.248	0.177	1.40	55.13
Oxygen	O_2	32.00	0.219	0.156	1.40	48.24
Triatomic Gases						
Carbon dioxide	CO_2	44.01	0.202	0.156	1.30	34.88
Sulfur dioxide	SO_2	64.07	0.154	0.122	1.26	23.55
Water vapor	H_2O	18.016	0.446**	0.336**	1.33	85.58**
Hydrocarbons						
Acetylene	C_2H_2	26.04	0.383	0.303	1.26	58.77
Methane	CH	16.04	0.532	0.403	1.32	96.07
Ethane	C_2H_6	30.07	0.419	0.342	1.22	50.82
Iso-butane	C_4H_{10}	58.12	0.398	0.358	1.11	25.79

* The composition of air, percent by volume, is taken as N_2, 78.03; O_2, 20.99; A, 0.98; following Keenan and Kaye (1948).
** pv/T for water vapor at 1 psia, 300°F, data from Keenan and Keyes (1936); c_p and c_v for water vapor at pressures below 1 psia.
Abridged from Table A1 of Mooney (1953). Data mainly from Rossini *et al.* (1947) and Eshbach (1936).
Source: Data from U.S. Department of Commerce; Ovid Eshbach (1936).

9.2.2.1 Specific Humidity

The *specific humidity* ω is defined as the ratio of the mass of water m_w to the mass of *dry* air m_a in a mixture. Thus:

$$\omega = \frac{m_w}{m_a} \tag{9.3}$$

Another measure of humidity is *relative humidity*, which will be defined below after a brief discussion of partial pressures.

9.2.2.2 Partial Pressures

Consider a mixture of two perfect gases, *A* and *B*, confined in a fixed volume $\mathcal{V}$ as shown in Figure 9.2 (a). (The reasoning can easily be extended to any number of gases.) For concreteness, think of A as dry air and B as water vapor. Denote the pressure and temperature in the fixed volume by *p* and *T*, respectively. If the two gases are separated into separate compartments at temperature *T* and pressure *p*, they would occupy volumes $\mathcal{V}_A$ and $\mathcal{V}_B$, respectively [Figure 9.2(b)], where $\mathcal{V} = \mathcal{V}_A + \mathcal{V}_B$. Then, if each component was allowed to individually occupy the volume $\mathcal{V}$ at temperature *T*, as depicted in Figure 9.2(c), the components would assume pressures p_A and p_B in accordance with Equation 9.1. Thus:

$$p_A\mathcal{V} = m_A R_A T \tag{9.4a}$$

$$p_B\mathcal{V} = m_B R_B T \tag{9.4b}$$

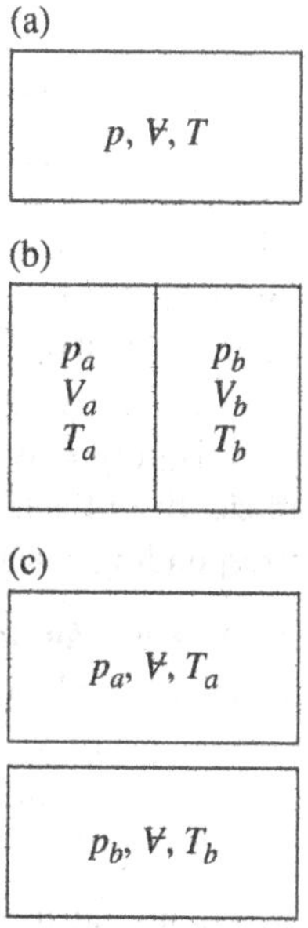

Figure 9.2 (a). Mixture of Two Perfect Gases, A and B (b). Gases Separated into Two Separate Components (c). Each Gas Individually Occupying Volume $\mathcal{V}$

The pressures p_A and p_B are, by definition, the *partial pressures* of the respective components A and B.

Dalton's Law of Partial Pressures, as developed experimentally, states that the pressure of a mixture of gases is equal to the sum of the partial pressures of the individual components, taken each at the temperature and volume of the mixture. Thus:

$$p = p_A + p_B \tag{9.5}$$

9.2.2.3 Relative Humidity

Define now the *relative humidity* ϕ as the ratio of the partial pressure p_w of the water vapor to the saturation pressure p_g for water vapor at the temperature of the mixture. Thus:

$$\phi = \frac{p_w}{p_g} \tag{9.6}$$

Relative humidity data are available from first-order weather stations of the National Oceanic and Atmospheric Administration (NOAA).

We may now derive a useful relation between specific humidity and relative humidity. Denote by subscript "a" *dry* air, and by subscript "w" water vapor. From the definition of specific humidity (Equation 9.3) and Equation 9.4, we have:

$$\omega = \frac{R_a/p_a}{R_w/p_w} \tag{9.7}$$

in which p_a and p_w are the partial pressures of the dry air and water vapor, respectively, and R_a and R_w are their respective gas constants.

From the definition of relative humidity (Equation 9.6), the following can be written:

$$p_w = \phi p_g \tag{9.8}$$

From Dalton's Law of Partial Pressures (Equation 9.5) and Equation 9.8, we have:

$$p_a = p - p_w = p - \phi p_g \tag{9.9}$$

Substituting Equations 9.8 and 9.9 into Equation 9.7 and rearranging results in:

$$\omega = \frac{R_a/R_w}{p/\left(\phi p_g\right) - 1} \tag{9.10}$$

For values of $R_a = 53.34$ ft lbf/(lbm °R) and $R_w = 85.58$ ft lbf/(lbm °R) (from Table 9.2), Equation 9.10 becomes:

$$\omega = \frac{0.623}{p/\left(\phi p_g\right) - 1} \tag{9.11}$$

The value of specific humidity may range from 0 to more than 5% ($\omega = 0$–0.05), the former corresponding to completely dehumidified air and the latter corresponding to a relative humidity of 100% ($\phi = 1$) and atmospheric pressure and temperature of 13.0 psia and 100°F, respectively. Note that ω is largest when p is smallest and ϕ and p_g (hence temperature) are largest. Once the atmospheric air is drawn into an air system, the specific humidity will remain constant in the system unless the air is purposely humidified or dehumidified or condensation occurs. Condensation will occur if the pressure p rises sufficiently or the temperature (and with it p_g) falls sufficiently to cause ϕ in Equation 9.11 to reach unity. The temperature corresponding to a given pressure and specific humidity at which condensation occurs is called the *dew-point* temperature.

9.2.2.4 Moist Air Density

A convenient relationship for determining air density including the effect of humidity may now be derived. The density of the air is given by:

$$\rho = \frac{m}{\forall} = \frac{m_a + m_w}{\forall} = \frac{m_a}{\forall} + \frac{m_w}{\forall} = \rho_a + \rho_w \tag{9.12}$$

Thus, the density of the mixture is equal to the sum of the densities of the components alone, occupying the full volume of the mixture at the pressure p and temperature T of the mixture. Hence, from Equations 9.2, and 9.12 we have:

$$\rho_a = \frac{p_a}{R_a T} \tag{9.13a}$$

$$\rho_w = \frac{p_w}{R_w T} \tag{9.13b}$$

$$\rho = \frac{1}{T}\left(\frac{p_a}{R_a} + \frac{p_w}{R_w}\right) \tag{9.13c}$$

Substituting Equations 9.8 and 9.9 into Equation 9.13 results in:

$$\rho = \frac{1}{T}\left(\frac{p - \phi p_g}{R_a} + \frac{\phi p_g}{R_w}\right) \tag{9.14}$$

or

$$\rho = \frac{1}{T}\left[\frac{p}{R_a} - \phi p_g\left(\frac{1}{R_a} - \frac{1}{R_w}\right)\right] \tag{9.15}$$

For values of R_a and R_w as given above, Equation 9.15 can be expressed in the following dimensionless form:

$$\rho = \frac{p - 0.3767\phi p_g}{R_a T} \tag{9.16}$$

From this equation, it can be seen that the humidity causes a decrease in air density. This is as expected since water vapor is lighter than dry air at the same temperature and pressure.

Equation 9.16 can be written in the following convenient dimensional form:

$$\rho = \frac{2.700}{T}\left(p - 0.3767\phi p_g\right) \tag{9.17}$$

in which T, p, and ρ have units of deg R, psia, and lbm/ft^3, respectively. The American Society of Mechanical Engineers (ASME) and Compressed Air and Gas Institute (CAGI) define *standard air* as air at a temperature of 68°F (528°R), pressure of 14.70 psia, and relative humidity of 36% ($\phi = 0.36$). (An alternative standard used by the gas industry is based on dry air at 60°F. The ASME/CAGI standard is used in this book.) From Table 9.1, we find $p_g = 0.3390$ psia at 68°F. Substituting these parameters into Equation 9.17 gives $\rho_s = 0.0749$ lbm/ft^3, in which the subscript "s" is used here and below to denote *standard conditions.*

9.2.2.5 Specific Gravity

Specific gravity is the ratio of the density of a gas to the density of dry air at the same temperature and standard pressure (14.7 psia). Although this may seem like a somewhat arbitrary parameter at first, it has definite value as a result of its relation to the gas constant R (which, in turn, is important in, e.g., blower and compressor testing). We will demonstrate this value below and will also show that some other definitions of *specific gravity* are improper and can lead to error.

Denoting the specific gravity by G, and taking air as the gas of interest, the above definition and Equation 9.15 allow us to write the following expression:

$$G = \frac{(1/T)\left[p/R_a - \phi p_g(1/R_a - 1/R_w)\right]}{(1/T_{da})(p_{da}/R_a)} \tag{9.18}$$

in which the subscript "*da*" denotes dry air. Canceling the temperatures (which are the same by definition of G) and rearranging gives:

$$G = 1 - \phi \frac{p_g}{p_{da}}\left(1 - \frac{R_a}{R_w}\right) \tag{9.19}$$

Note in the above expression that p_g is a function of the gas temperature (Table 9.1), and that the pressure of the dry air remains. Substituting the values of R_a and R_w from Equation 9.19 results in:

$$G = 1 - 0.3767\phi\frac{p_g}{p} \tag{9.20}$$

A tabulation of G from the above equation, the p_g–temperature relationship of Table 9.1, and pressure of 14.70 psia is given in Table 9.3.

Although the values of G given in Table 9.3 are essentially identical to those tabulated in ASME (1979b) and Rollins (1973) over the range of 40–130°F, the definitions of specific gravity given in those references differ from those given above. The definition in ASME (1979b) adds conditions of standard temperature to the above definition and is clearly erroneous since specific gravity as used therein is a function of temperature. (That reference erroneously defines specific gravity as the ratio of the density of gas at a pressure of 14.7 psia and a temperature of 68°F to the density of dry air at the same pressure and temperature and an apparent molecular weight of 28.970.) The definition in Rollins is technically correct and is consistent with the use of specific gravity therein if it is qualified that the specific gravity is at standard pressure. [That reference defines specific gravity as the ratio of the specific weight (density × g) of air or other gas to that of dry air at the same pressure and temperature.] The definition given above removes the ambiguities and need for qualification.

9.2.2.6 Moist Air Gas Constant

By adding Equations 9.4a and 9.4b and employing Dalton's Law (Equation 9.5), the following is obtained:

$$p\mathcal{V} = (m_a R_a + m_b R_b)T \tag{9.21}$$

Table 9.3 Specific Gravity G of Moist Air at Standard Pressure

	Relative humidity, percent									
Temperature °F	**10**	**20**	**30**	**40**	**50**	**60**	**70**	**80**	**90**	**100**
32	0.9998	0.9995	0.9993	0.9991	0.9989	0.9986	0.9984	0.9982	0.9980	0.9977
40	0.9997	0.9994	0.9991	0.9988	0.9984	0.9981	0.9978	0.9975	0.9972	0.9969
50	0.9995	0.9991	0.9986	0.9982	0.9977	0.9973	0.9968	0.9963	0.9959	0.9954
60	0.9993	0.9987	0.9980	0.9974	0.9967	0.9961	0.9954	0.9947	0.9941	0.9934
70	0.9991	0.9981	0.9972	0.9963	0.9953	0.9944	0.9935	0.9926	0.9916	0.9907
80	0.9987	0.9974	0.9961	0.9948	0.9935	0.9922	0.9909	0.9896	0.9883	0.9870
90	0.9982	0.9964	0.9946	0.9928	0.9911	0.9893	0.9875	0.9857	0.9839	0.9821
100	0.9976	0.9951	0.9927	0.9903	0.9878	0.9854	0.9830	0.9805	0.9781	0.9757
110	0.9967	0.9935	0.9902	0.9869	0.9837	0.9804	0.9771	0.9739	0.9706	0.9673
120	0.9957	0.9913	0.9870	0.9827	0.9783	0.9740	0.9696	0.9653	0.9610	0.9566
130	0.9943	0.9886	0.9829	0.9772	0.9715	0.9658	0.9601	0.9544	0.9487	0.9430

Comparing the above equation to Equation 9.1, we find that the gas constant of the mixture is the mass-weighted mean of the gas constants of the components, i.e.:

$$mR = m_a R_a + m_b R_b \tag{9.22}$$

Since $m = m_a + m_b$, we thus obtain:

$$R = \frac{m_a R_a + m_b R_b}{m_a + m_b} \tag{9.23}$$

Substituting m_a and m_b from Equations 9.4a and b into Equation 9.23 and simplifying, we have:

$$R = \frac{p_a + p_w}{p_a/R_a + p_b/R_b} \tag{9.24}$$

Applying Equation 9.24 to air (i.e., a mixture of dry air and water vapor) and employing Equations 9.5, 9.8, and 9.9, results in the following:

$$R = R_a \frac{1}{1 - \phi\left(p_g/p\right)\left(1 - R_a/R_w\right)} \tag{9.25}$$

Comparing this relationship with Equation 9.19, the following relationship between R and G for moist air is obtained:

$$R = R_a/G \tag{9.26}$$

9.2.2.7 Volumetric Rate of Flow

The *volumetric rate of flow* Q is an important parameter used, for example, in determining pipe velocity ($V = Q/A$) and in specifying blower performance (see Section 9.5). However, great care must be exercised when applying this parameter to avoid ambiguities that can lead to substantial error.

As for incompressible flow, the volumetric rate of flow of a component in a compressible mixture is equal to the mass rate of flow of that component divided by the density of that component. The same relationship applies to the mixture as a whole. *Since the individual components and the mixture occupy the same volume, the volumetric flow rates of the components and mixture are all the same.* Thus, in the case of a mixture of dry air and water vapor, we have:

$$Q = \frac{w_a}{\rho_a} = \frac{w_w}{\rho_w} = \frac{w}{\rho} \tag{9.27}$$

in which the densities are given by Equation 9.13a, b, or c, as appropriate.

Mass rate of flow is an unambiguous quantity, independent of fluid properties. Thus, we may regard an unambiguous specification of volumetric flow rate as being one from which the mass flow rates of all components can be determined. Referring to Equations 9.8, 9.9, 9.13–9.17 and recalling that p_g is dependent on temperature alone, it can be seen that such a specification will result if the pressure, temperature, and relative humidity are specified, along with the volumetric rate of flow. Note that flow, specified in terms of the *standard conditions* given above, satisfies this requirement. The rate of flow specified with reference to standard conditions is often given in units of standard cubic feet per minute, abbreviated scfm, or standard cubic feet per hour, abbreviated scfh. Knowing scfm or scfh, the mass rate of flow of any or all components may be determined, and vice versa.

The process requirements are often first determined in terms of the mass flow rate of a component. In aeration applications, for example, the required oxygenation capacity is determined in terms of the mass flow rate of oxygen, from which the mass flow rate of dry air is readily determined. From this, an unambiguous specification of the process air requirements may be expressed, such as in terms of *standard* volumetric rate of flow.

Given an unambiguous specification of flow rate, it is often necessary to determine the volumetric rate of flow at actual conditions, such as in a pipeline, for determination of velocity and pressure losses. A relation between *standard* flow rate and *actual* flow rate is useful in this connection. Such a relationship may be predicated on the assumption that mass rate of *dry* airflow is the desired product. Thus, by conservation of dry air mass, we have w_a = constant, which, recalling that $Q_a = Q$, gives:

$$\rho_a Q = \rho_{as} Q_s \tag{9.28}$$

With the aid of Equations 9.13a and 9.9, the above relationship gives:

$$Q = \frac{T_i}{T_s}\frac{p_s - \phi_s p_{gs}}{p - \phi p_g} Q_s \tag{9.29}$$

Substituting standard conditions of temperature (and corresponding p_{gs}), pressure, and relative humidity into the above equation gives the following *dimensional* relationship:

$$Q = \frac{0.0276T}{p - \phi p_g} Q_s \tag{9.30}$$

in which T and p are in units of deg R and psia, respectively.

Equation 9.30 represents the relation between volumetric flow rates at actual and standard conditions corresponding to identical mass flow rates of dry air. In a flow stream, one condition cannot generally be reached from the other without humidifying (adding water vapor) or dehumidifying (removing water vapor) from the flow stream.

9.2.3 Specific Heat and Enthalpy

The *specific heat in a constant-volume process* is defined for any substance by:

$$c_v = \left(\frac{\partial u}{\partial T}\right)_v \tag{9.31}$$

in which u is internal energy as defined in Chapter 2. The *specific heat in a constant-pressure process* is defined for any substance by:

$$c_p = \left(\frac{\partial \tilde{H}}{\partial T}\right)_p \tag{9.32}$$

in which $\tilde{H}$ is *enthalpy*, defined for any process by:

$$\tilde{H} = u + p/\rho \tag{9.33}$$

as also given in Chapter 2.

Values of c_p and c_v for various gases are given in Table 9.2. The role of enthalpy in the steady-flow energy equation becomes apparent by relating the above definition to Equation 2.62.

Since internal energy, unlike pressure and density, has no absolute value, an arbitrary base must be set for internal energy or enthalpy so that values of u and h may be fixed. Since the concern here is only with enthalpy *changes* of perfect gases, absolute values and bases for these parameters will not be dealt with here. The interested reader is referred to standard thermodynamics texts for bases in common use.

A perfect gas is, by definition, a gas satisfying the semiperfect gas relationship (Equation 9.1) plus having constant values of c_v and c_p, independent of temperature. For dry air and water vapor, c_p and c_v are actually weak functions of temperature (see, e.g., Table 2 of Keenan and Kaye 1948). However, they may be assumed to be constant with no significant error for the applications considered herein with temperatures below 300°F (760°K). For certain other gases, such as CO_2, the variation of specific heat with temperature is appreciable.}

Thus, by Equations 9.31 and 9.32, a perfect gas has u and $\tilde{H}$ functions of temperature only, and the following can be written:

$$du = c_v dT \tag{9.34a}$$

$$d\tilde{H} = c_p dT \tag{9.35a}$$

or

$$\Delta u = c_v \Delta T \tag{9.34b}$$

$$\Delta \tilde{H} = c_p \Delta T \tag{9.35b}$$

From Equations 9.33, 9.2, 9.34a, and 9.35a, the following relationship between the two specific heats may be derived:

$$c_p - c_v = R \tag{9.36}$$

In using the above relationship it is necessary to use consistent units, keeping in mind that the usual units for the specific heats are BTU/(lbm deg), whereas the usual units for R are ft lbf/(lbm deg).

The First Law of Thermodynamics (see Chapter 2) requires that when gases at equal pressures and temperatures are mixed adiabatically and without work in a vessel (such as in Figure 9.2), the internal energy of the system remains constant. From this, it may be deduced that the internal energy of a mixture of gases is equal to the sum of the internal energies of the individual components, each taken at the temperature and volume of the mixture. Thus:

$$mu = m_A u_A + m_B u_B \tag{9.37}$$

Then, from Equations 9.5, 9.33, and the fact that $m/\rho = \forall = m_A/\rho_A = m_B/\rho_B$, it can be deduced that a similar relationship holds for enthalpy, as follows:

$$m\tilde{H} = m_A\tilde{H}_A + m_B\tilde{H}_B \tag{9.38}$$

Similarly, from Equations 9.37, 9.34(b), and 9.35(b), the following is derived:

$$mc_v = m_A c_{vA} + m_B c_{vB} \tag{9.39}$$

$$mc_p = m_A c_{pA} + m_B c_{pB} \tag{9.40}$$

There will be particular need for the enthalpy change ($\Delta\tilde{H}$) and specific heats of air, i.e., the mixture of dry air and water vapor. Considering any of these properties, we obtain from Equations 9.38–9.40 and the definition of specific humidity (Equation 9.3):

$$\tilde{H} = \frac{\tilde{H}_a + \omega\tilde{H}_w}{1+\omega} \tag{9.41}$$

$$c_v = \frac{c_{va} + \omega c_{vw}}{1+\omega} \tag{9.42}$$

$$c_p = \frac{c_{pa} + \omega c_{pw}}{1+\omega} \tag{9.43}$$

Note that $\tilde{H}$, as given by Equation 9.41, is the enthalpy of the mixture per unit mass of mixture. In psychrometrics, air enthalpy is commonly expressed per lbm of *dry* air. Denoting by $\tilde{H}'$ moist air enthalpy per unit mass of dry air, we have $\tilde{H}' = \tilde{H} \times m/m_a = \tilde{H} \times (m_a + m_w)/m_a = \tilde{H}(1+\omega)$. Using Equation 9.35b, we can write the changes in enthalpy per mixture unit mass or dry air unit mass, respectively, as follows:

$$\Delta\tilde{H} = \frac{(c_{pa} + \omega c_{pw})\Delta T}{1+\omega} \tag{9.44}$$

$$\Delta\tilde{H}' = (c_{pa} + \omega c_{pw})\Delta T \tag{9.45}$$

9.2.4 Specific Heat Ratio and Isentropic Processes

The specific heat ratio k is a particularly important parameter in compressible fluid processes. This ratio is defined as follows:

$$k = c_p/c_v \tag{9.46}$$

in which c_p is specific heat at constant pressure and c_v is specific heat at constant volume.

From Equations 9.36 and 9.46, the following relationships between k, R, and the specific heats may be derived:

$$c_v = \frac{1}{k-1}R \tag{9.47}$$

$$c_p = \frac{k}{k-1}R \tag{9.48}$$

The classical kinetic theory of gases predicts values of k for monatomic (one atom per molecule), diatomic, and polyatomic gases of 5/3, 7/5, and 4/3, respectively. For monatomic and diatomic gases, the values of k observed at 1 atm and ordinary room temperatures are in reasonably good agreement with the classical theory values given in Table 9.2. For polyatomic gases in general (see Table 9.2) and for monatomic and diatomic gases at high temperatures, the deviations from the classical theory are large. The values given in Table 9.2 will suffice for dry air and water vapor over the temperature ranges of concern here. Dry air, for example, has a value of k within one-half of 1% of 1.4 for temperatures ranging from −360°F (100°R) to 300°F (760°R).

The specific heat ratio for air may be written from Equations 9.42 and 9.43 as follows:

$$k = \frac{c_{pa} + \omega c_{pw}}{c_{va} + \omega c_{vw}} \tag{9.49}$$

This expression may be rewritten in terms of c_p's and k's as follows:

$$k = k_a \frac{1 + \omega(c_{pw}/c_{pa})}{1 + \omega(k_a/k_w)(c_{pw}/c_{pa})} \tag{9.50}$$

Substituting the c_p's and k's from Table 9.2 into Equation 9.50, we find that k of moist air remains constant to within less than 1/2 of 1% over a range of specific humidities from 0 to 5%. [The variation is less than 1% if the effect of temperature on the c_p's and k's is taken into account for temperatures below 300°F (760°R).] One may thus, for most purposes, take $k = 1.4$ for air. Furthermore, it may generally be assumed that k is independent of temperature, pressure, and humidity, which is an important fact in blower and compressor test work as discussed in Section 9.5.

The specific heat ratio plays an important role in *isentropic* (constant entropy – reversible and adiabatic) processes involving perfect gases, for which:

$$p/\rho^k = \text{constant} \tag{9.51}$$

[The reader is referred to basic thermodynamics texts for general discussion of isentropic processes. See, for example, Mooney (1959) or Hatsopoulus and Keenan (1965).]

For an isentropic process in a perfect gas undergoing a change from state 1 to state 2, the following relationships may be derived from the above equation and Equation 9.2:

$$p_2\rho_2^{-k} = p_1\rho_1^{-k} \tag{9.52a}$$

$$p_2T_2^{-k(k-1)} = p_1T_1^{-k(k-1)} \tag{9.52b}$$

$$T_2\rho_2^{-(k-1)} = T_1\rho_1^{-(k-1)} \tag{9.52c}$$

9.2.5 Viscosity

As for incompressible flows, the effects of viscosity in compressible flows may generally be expressed in relation to the Reynolds Number. Since kinematic viscosity υ is dependent on density (which varies with pressure, temperature, and humidity), it is more convenient to work with absolute viscosity μ when dealing with gases. Contrary to most liquids, the absolute viscosity of gases increases as temperature increases. The effect of pressure on μ is small. In the case of dry air, for example, the effect of pressure on μ is less than the uncertainty with which μ may be estimated for pressures up to 200 psia (Keenan and Kaye 1948). The absolute viscosity of dry air varies with temperature in accordance with Table 9.4. A good fit to the data points is provided by the following second-degree polynomial:

$$\mu\left(\frac{\text{lbm}}{\text{sec ft}}\right) = (16.4 + 0.2317T - 0.00005667T^2)(10)^{-7}$$

$$400 \le T(^\circ\text{R}) \le 900 \tag{9.53}$$

The absolute viscosity of water vapor alone is less than that of dry air at the same temperature. The effect of water vapor is thus to lower the absolute viscosity of air. By using the viscosity of dry air, the estimated viscosity will be slightly greater than the actual viscosity, and the estimated Reynolds Number will be slightly less than the actual. This is conservative in the determination of the Darcy–Weisbach friction factor f (see Figure 5.1). The conservatism is generally small.

Table 9.4 Absolute Viscosity of Air at Pressures Below 200 psia*

Temperature		
°R	°F	$\mu \times 10^7$ lbm/(sec ft)
400	−59.7	100
450	−9.7	109
500	40.3	118
550	90.3	126
600	140.3	135
650	190.3	143
700	240.3	151
750	290.3	158
800	340.3	166
900	440.3	179

* Abridged from Table 2 of Gas Tables by Keenan and Kaye (1948).
Source: Adapted from Keenan and Kaye (1948).

9.2.6 Speed of Sound and Mach Number

The velocity of propagation of small pressure disturbances, i.e., the *speed of sound*, in a gaseous medium is a key parameter in both the qualitative and quantitative analyses of compressible fluid flow.

The speed of sound in an elastic fluid was derived in Section 8.6, where it is referred to as the speed of pressure waves. For gases, the elasticity of the fluid is much greater than the elasticity of the confining pipes or other vessels, so that the speed of sound is given by:

$$c = \sqrt{dp/d\rho} \tag{9.54}$$

For small pressure disturbances, the wave propagation process is practically reversible and adiabatic (Shapiro 1953, §3.1), i.e., isentropic. For a perfect gas, the relation between pressure and density in an isentropic process is given by Equation 9.51. Taking the logarithmic differential (taking first the logarithm of an expression and then differentiating that result) of Equation 9.51, the following is obtained:

$$\frac{dp}{d\rho} = kp/\rho \tag{9.55a}$$

Using Equation 9.2, the above relationship becomes:

$$\frac{dp}{d\rho} = kRT \tag{9.55b}$$

which may be substituted into Equation 9.54 to give:

$$c = \sqrt{kRT} \tag{9.56}$$

Substituting values of k and R for air from Table 9.2 into the above equation, the following equation for the speed of sound in air at normal temperatures and pressures is obtained:

$$c = 49.02\sqrt{T} \tag{9.57}$$

in which units of c and T are ft/sec and °R, respectively. At standard atmospheric temperature (68°F = 528°R), the speed of sound in air is approximately 1130 ft/sec. This is roughly one-fifth of the speed of sound in water at the same temperature. Whereas the speed of sound in water is much larger than attainable water velocities, the velocity of a gas can be made to reach and exceed the speed of sound.

The ratio of the local velocity of a gas to the local speed of sound at the same point is of great practical importance. This ratio is called the Mach Number and is denoted by **M**. Thus,

$$\mathbf{M} = V/c = V/\sqrt{gkRT} \tag{9.58}$$

9.3 Constant Flow in Pipes

We will commence our analysis of constant flow in pipes by considering the adiabatic flow of a perfect gas in a pipe of constant area. Adiabatic flow is approximated in short pipelines or in well-insulated pipelines in which little heat transfers across the pipe walls. The analysis of adiabatic flow is patterned partly after that of Shapiro (1953, Chapter 6). The density of the gases of interest is sufficiently small that potential energy is miniscule compared to other energy terms and is neglected in the rest of this chapter.

Using the above-mentioned analysis as a point of departure, approximations are made leading to practical methods of design and analysis. Commonly encountered fittings and appurtenances are discussed. The overall approach, in addition to providing practical methods, should provide a clear understanding of the assumptions and accuracy inherent in the methods presented.

9.3.1 Adiabatic, Constant-area Flow of a Perfect Gas

The following derivation applies to the infinitesimal control volume depicted in Figure 9.3. By first taking the logarithmic differential of the perfect gas equation of state (Equation 9.2), the following is obtained:

$$\frac{dp}{p} = \frac{d\rho}{\rho} + \frac{dT}{T} \tag{9.59}$$

The logarithmic differential of the expression for the square of the Mach Number in a perfect gas (Equation 9.58 squared) is obtained next:

$$\frac{dM^2}{M^2} = \frac{dV^2}{V^2} - \frac{dT}{T} \tag{9.60}$$

The conservation of mass, or continuity, relationship (see Chapter 2) gives:

$$\rho AV = \text{constant} \tag{9.61}$$

Taking logarithmic differentials with A assumed constant gives:

$$\frac{d\rho}{\rho} + \frac{dV}{V} = 0 \tag{9.62}$$

As a convenience, the numerator and denominator of the second term are multiplied by V, and the above equation is rewritten as:

$$\frac{d\rho}{\rho} + \frac{1}{2}\frac{dV^2}{V^2} = 0 \tag{9.63}$$

The steady-flow energy equation (see Chapter 2) is applied to the adiabatic flow ($\dot{Q} = 0$) through the control volume of Figure 9.3 with potential energy changes neglected to give:

$$d\tilde{H} + d\left(\frac{V^2}{2}\right) = 0 \tag{9.64}$$

in which $\tilde{H}$ is enthalpy. The differential enthalpy is given for a perfect gas by Equation 9.35a, in which c_p is constant. Substituting Equation 9.35a

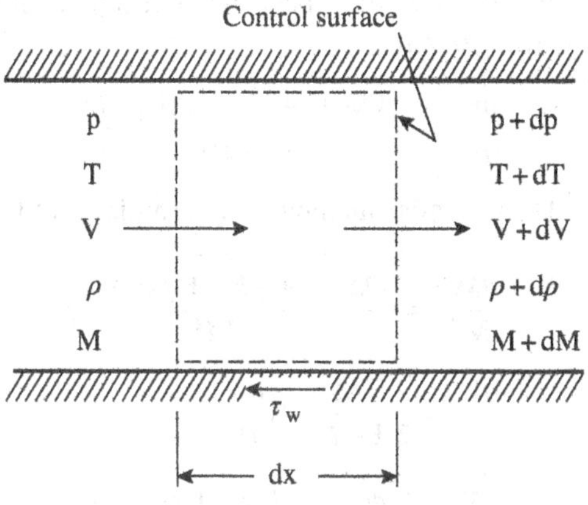

Figure 9.3 Control Surface for Analysis of Adiabatic, Constant-area Flow (Shapiro 1953/The Ronald Press Company)

into Equation 9.64, dividing through by c_pT, and employing Equations 9.58 and 9.48, results in:

$$\frac{dT}{T} + \frac{k-1}{2}\mathbf{M}^2\frac{dV^2}{V^2} = 0 \tag{9.65}$$

The steady-flow momentum equation (Equation 2.37) is applied to the control volume of Figure 9.3, second-order products are dropped, and the continuity relationship (Equation 9.62) is employed to give:

$$-Adp - \tau_w Pdx = \rho AVdV \tag{9.66a}$$

in which τ_w is the shear stress exerted on the flow by the walls, and P is the wetted perimeter over which τ_w acts. The shear stress is related to the Darcy–Weisbach friction factor f:

$$\tau_w = \frac{f}{4}\frac{\gamma V^2}{2g} \tag{9.66b}$$

Substituting the above equation into the momentum relationship (Equation 9.66) results in:

$$-dp - \frac{f}{4}\frac{\gamma V^2}{2g}\frac{P}{A}dx = \rho VdV \tag{9.67}$$

Noting that the hydraulic radius R equals A/P, dividing through by p, and noting from Equations 9.2 and 9.58 that $\rho V^2 = kp\mathbf{M}^2$, the above equation can be rewritten as follows:

$$\frac{dp}{p} + \frac{k\mathbf{M}^2}{2}\frac{f}{4R}dx + \frac{k\mathbf{M}^2}{2}\frac{dV^2}{V^2} = 0 \tag{9.68}$$

For simplicity of notation and since circular pipes are of primary concern, we shall replace $4R$ by pipe diameter D in the following. However, the reader should keep in mind that the more general $4R$ may be substituted.

Equations 9.59, 9.60, 9.63, 9.65, and 9.68 represent five simultaneous, linear, algebraic equations in six differential variables: dp/p, $d\rho/\rho$, dT/T, $d\mathbf{M}^2/\mathbf{M}^2$, dV^2/V^2, and fdx/D. Friction is the physical phenomenon causing changes in the other variables. Therefore, fdx/D is selected as the independent variable, in terms of which the remaining five variables may be found with the aid of the five equations. For example, the term dT/T may be eliminated from Equations 9.59 and 9.65, yielding:

$$\frac{dp}{p} = \frac{d\rho}{\rho} - \frac{k-1}{2}\mathbf{M}^2\frac{dV^2}{V^2} \tag{9.69}$$

Next, $d\rho/\rho$ is eliminated from this expression with the help of Equation 9.63, giving:

$$\frac{dp}{p} = -\frac{1+(k-1)\mathbf{M}^2}{2}\frac{dV^2}{V^2} \tag{9.70}$$

Finally, use the above relationship to eliminate dV^2/V^2 from Equation 9.68 and obtain the following algebraic rearrangement:

$$\frac{dp}{p} = -\frac{k\mathbf{M}^2[1+(k-1)\mathbf{M}^2]}{2(1-\mathbf{M}^2)}f\frac{dx}{D} \tag{9.71}$$

Using similar methods, the formulas that follow are obtained:

$$\frac{d\mathbf{M}^2}{\mathbf{M}^2} = \frac{k\mathbf{M}^2\{1+[(k-1)/2]\mathbf{M}^2\}}{1-\mathbf{M}^2}f\frac{dx}{D} \tag{9.72}$$

$$\frac{dV}{V} = \frac{k\mathbf{M}^2}{2(1-\mathbf{M}^2)}f\frac{dx}{D} \tag{9.73}$$

$$\frac{dT}{T} = \frac{1}{2}\frac{dc}{c} = -\frac{k(k-1)\mathbf{M}^4}{2(1-\mathbf{M}^2)}f\frac{dx}{D} \tag{9.74}$$

$$\frac{d\rho}{\rho} = -\frac{k\mathbf{M}^2}{2(1-\mathbf{M}^2)}f\frac{dx}{D} \tag{9.75}$$

Note that for $f = 0$, no property changes occur in the adiabatic flow. For $f \neq 0$, the direction of change of the various flow properties may be ascertained from Equations 9.71 through 9.75. For subsonic flow ($\mathbf{M} < 1$), which is our almost exclusive concern, the term $(1 - \mathbf{M}^2)$ in the denominator of each of the equations is greater than zero. Thus, frictional effects cause the pressure, temperature, and density to decrease in the direction of flow while the Mach Number and velocity increase.

The differential equations, Equations 9.71 through 9.75, may be readily integrated. For example, Equation 9.72 can be rearranged and integrated as follows:

$$\int_0^{L^*} f\frac{dx}{D} = \int_{\mathbf{M}^2}^{1} \frac{1-\mathbf{M}^2}{k\mathbf{M}^4\{1+[(k-1)/2]\mathbf{M}^2\}}d\mathbf{M}^2 \tag{9.76}$$

where the limits of integration are taken as $x = 0$ at the section where the Mach Number is $\mathbf{M}$ and $x = L^*$ at the section where the Mach Number is unity [for the physical significance of L as the maximum possible pipe length, the interested reader is referred to Shapiro (1953)]. Carrying out the integration, one obtains:

$$\bar{f}\frac{L^*}{D} = \frac{1-\mathbf{M}^2}{k\mathbf{M}^2} + \frac{k+1}{2k}\ln\frac{(k+1)\mathbf{M}^2}{2\{1+[(k-1)/2]\mathbf{M}^2\}} \tag{9.77}$$

where $\bar{f}$ is the mean friction coefficient with respect to length, defined by:

$$\bar{f} = \frac{1}{L^*}\int_0^{L^*} f dx \tag{9.78}$$

Since $\bar{f}L^*/D$ is a function only of $\mathbf{M}$, the length of pipe L required for the flow to pass from a given initial Mach Number $\mathbf{M}_1$ to a given final Mach Number $\mathbf{M}_2$ is found from:

$$\bar{f}\frac{L}{D} = \left(\bar{f}\frac{L^*}{D}\right)_{\mathbf{M}_1} - \left(\bar{f}\frac{L^*}{D}\right)_{\mathbf{M}_2} \tag{9.79}$$

To illustrate how the local flow properties are found in terms of the local Mach Number, we shall take the pressure as an example. Equations 9.71 and 9.72 may be combined to give:

$$\frac{dp/p}{d\mathbf{M}^2/\mathbf{M}^2} = -\frac{1+(k-1)\mathbf{M}^2}{2\{1+[(k-1)/2]\mathbf{M}^2\}} \tag{9.80}$$

or

$$\frac{dp}{p} = -\frac{1+(k-1)\mathbf{M}^2}{2\mathbf{M}^2\{1+[(k-1)/2]\mathbf{M}^2\}}d\mathbf{M}^2 \tag{9.81}$$

Denoting the pressure at $\mathbf{M} = 1$ by the symbol p^*, by integrating between the section where $\mathbf{M} = \mathbf{M}$ and $p = p$, and the section where $\mathbf{M} = 1$ and $p = p^*$, the following is obtained:

$$\frac{p}{p^*} = -\frac{1}{\mathbf{M}}\sqrt{\frac{k+1}{2\{1+[(k-1)/2]\mathbf{M}^2\}}} \tag{9.82}$$

By similar methods, the following relationships are obtained:

$$\frac{V}{V^*} = \mathbf{M}\sqrt{\frac{k+1}{2\{1+[(k-1)/2]\mathbf{M}^2\}}} \tag{9.83}$$

$$\frac{T}{T^*} = \frac{c^2}{c*^2} = \frac{k+1}{2\{1+[(k-1)/2]\mathbf{M}^2\}} \tag{9.84}$$

$$\frac{\rho}{\rho^*} = \frac{V^*}{V} = \frac{1}{\mathbf{M}}\sqrt{\frac{2\{1+[(k-1)/2]\mathbf{M}^2\}}{k+1}} \tag{9.85}$$

The quantities marked with an asterisk in the above expressions represent the values of the flow properties at the section in the pipe where $\mathbf{M} = 1$. Since those quantities are constants for a given adiabatic, constant-area flow, they may be regarded as convenient reference values for normalizing the equations. In order to find the change in some property, say the pressure between the sections where the Mach Numbers are $\mathbf{M}_1$ and $\mathbf{M}_2$, respectively, set $p_2/p_1 = (p/p*)_{\mathbf{M}_2}/(p/p*)_{\mathbf{M}_1}$ where $(p/p*)_{\mathbf{M}_2}$ is the value of the right-hand side of Equation 9.82 corresponding to $\mathbf{M}_2$, and so forth. Shapiro (1953) gives graphs and tabulates the dimensionless ratios of Equations 9.82 through 9.85 using Mach Number as the argument. These equations can also be readily programmed on a programmable calculator.

9.3.2 Relations for Low Mach Numbers

When the costs of compression are balanced against the costs of piping in most environmental engineering systems, economics strongly favors pipe velocities, which are a small fraction of the speed of sound. Thus, relations for low Mach Numbers are of particular interest.

For $[(k-1)/2]\mathbf{M}^2 \ll 1$ or $\mathbf{M}^2 \ll 2/(k-1)$, Equations 9.82 through 9.85 are approximated by:

$$p_2/p_1 \cong \mathbf{M}_1/\mathbf{M}_2 \tag{9.86}$$

$$V_2/V_1 \cong \mathbf{M}_2/\mathbf{M}_1 \tag{9.87}$$

$$T_2/T_1 \cong \text{const} \tag{9.88}$$

$$\rho_2/\rho_1 \cong \mathbf{M}_1/\mathbf{M}_2 \tag{9.89}$$

Note that the adiabatic flow is now approximately isothermal.

If, in addition, we have $\mathbf{M}_2 \ll 1$, Equation 9.77 can be approximated by:

$$\bar{f}\frac{L^*}{D} \cong \frac{1}{k\mathbf{M}^2} + \frac{k+1}{2k}\ln\frac{(k+1)\mathbf{M}^2}{2} \tag{9.90}$$

Equation 9.79 then gives:

$$\bar{f}\frac{L}{D} = \frac{1}{k\mathbf{M}_1^2} - \frac{1}{k\mathbf{M}_2^2} + \frac{k+1}{2k}\ln\frac{\mathbf{M}_1^2}{\mathbf{M}_2^2} \tag{9.91}$$

Employing Equation 9.86, the above equation can be written in terms of the pressure ratio p_2/p_1 as follows:

$$\bar{f}\frac{L}{D} = \frac{1-(p_2/p_1)^2}{k\mathbf{M}_1^2} + \frac{k+1}{k}\ln\left(\frac{p_2}{p_1}\right)^2 \tag{9.92}$$

Note that $\mathbf{M}^2 \ll 1$ is a more stringent criterion than $\mathbf{M}^2 \ll 2/(k-1)$ for $k < 3$, which includes all the gases listed in Table 9.2. Thus, all of Equations 9.86–9.92 are satisfied if $\mathbf{M}^2 \ll 1$. Now consider important simplifications that may be made if restrictions are placed on the magnitude of the pressure drop as well as the Mach Number.

9.3.2.1 Simplifications for Moderate Pressure Drops

Writing the term p_2/p_1 as $1 - \Delta p/p_1$ in which Δp is the pressure drop $(p_1 - p_2)$, the logarithmic term of Equation 9.92 can be written as the following series expansion:

$$\ln\left(\frac{p_2}{p_1}\right)^2 = 2\ln\left(1-\frac{\Delta p}{p_1}\right) = 2\left[\frac{\Delta p}{p_1} - \frac{1}{2}\left(\frac{\Delta p}{p_1}\right)^2 + \left(\frac{1}{3}\right)\left(\frac{\Delta p}{p_1}\right)^3 \mp \ldots\right] \tag{9.93}$$

For $(\Delta p/p_1)^2 \ll 3$, terms of third order and higher may be dropped from the series expansion, and Equation 9.92 may be written as follows:

$$\bar{f}\frac{L}{D} \cong \frac{2\Delta p/p_1 - (\Delta p/p_1)^2}{k\mathbf{M}_1^2} + \frac{k+1}{k}2\left[\frac{\Delta p}{p_1} - \frac{1}{2}\left(\frac{\Delta p}{p_1}\right)^2\right] \tag{9.94}$$

The second term on the right-hand side of the above expression, which corresponds to the logarithmic term of Equation 9.92, may be neglected if, in addition to the above pressure drop criterion, we have $1/(k\mathbf{M}_1^2) \gg (k+1)/k$ or $\mathbf{M}_1^2 \ll 1/(k+1)$. This criterion for $\mathbf{M}_1^2$ is more stringent than the requirement $\mathbf{M}_1^2 \ll 1$ required for Equation 9.92.

Thus, for $\mathbf{M}_1^2 \ll 1/(k+1)$ and $(\Delta p/p_1)^2 \ll 3$, Equation 9.92 may be approximated by:

$$\bar{f}\frac{L}{D} \cong \frac{1-(p_2/p_1)^2}{k\mathbf{M}_1^2} \tag{9.95}$$

Substituting Equation 9.56 into the above equation and employing Equation 9.2 results in:

$$\frac{p_1^2-p_2^2}{2p_1\rho_1} = \frac{\bar{f}L}{D}\frac{V_1^2}{2g} \tag{9.96}$$

Another form of this equation may be obtained by substituting Equation 9.56 into Equation 9.95 and employing Equation 9.26 and $V = Q/(\pi D^2/4)$, resulting in:

$$Q = \frac{\pi}{4}\sqrt{gR_a}\frac{T_1}{p_1}(1/\bar{f})^{0.5}\left[(p_1^2-p_2^2)/(GT_1L)\right]^{0.5}D^{2.5} \tag{9.97}$$

A dimensional version of the above equation, with the first p_1 and the two T_1's taken at appropriate base and average values, is widely used in the gas industry (Hunt 1956, Segeler 1967). {A supercompressibility factor is added in the gas industry equation [and may be omitted for pressures less than 100 psig (Segeler 1967)]. Also, f is commonly taken as $C_f/4$, although the notation f is sometimes used for C_f. Substantial changes in pipeline elevations should and can be taken into account (Segeler 1967).}

The above equation is applicable to isothermal flow (recall Equation 9.88), which is approximated in long pipelines for which there is sufficient pipe area for heat transfer to bring the gas temperature to that of the environment around the pipe. Environmental engineers find it necessary to plan for gas service extensions or modifications for treatment facilities. Although the final design is generally carried out by the gas utility, Equation 9.97 may be of occasional use in preliminary planning and coordination.

9.3.2.2 Simplification for Small Pressure Drops

Equation 9.96 may be rewritten in terms of pressure drop $\Delta p = p_1 - p_2$, as follows:

$$\left(1-\frac{\Delta p}{2p_1}\right)\frac{\Delta p}{\rho_1} = \frac{\bar{f}L}{D}\frac{V_1^2}{2g} \tag{9.98}$$

For $\Delta p/p_1 \ll 2$, the above equation reduces to the pressure drop formula for incompressible flow

$$\frac{\Delta p}{\rho_1} \cong \frac{\bar{f}L}{D}\frac{V_1^2}{2g} \tag{9.99}$$

Equation 9.99 is a good approximation for adiabatic gas flows when $\mathbf{M}^2 \ll 1/(k+1)$ and $\Delta p/p_1 \ll 2$. More specifically, $\mathbf{M}^2 < 0.02/(k+1)$ and $\Delta p/p_1 < 0.04$ will result in an error of less than 2% in the determination of Δp by Equation 9.99 (compared with Equation 9.94). For air ($k = 1.4$), the Mach Number criterion becomes $\mathbf{M}^2 < 0.01$ or $\mathbf{M} < 0.1$. The adiabatic assumption is satisfied only in *infinitesimal* runs of well-insulated pipe [an erroneous application to long pipes is found in a well-known reference (Fair and Geyer 1954, §25-14)]. However, the approximation of Equation 9.99 may usually be used for longer, uninsulated pipe runs with an appropriate selection of temperature and pressure conditions, as discussed next.

9.3.3 Constant Density Approximation for Air Piping

The previous section established the assumptions that must be satisfied for the constant density approximation to apply accurately to gas piping. For air, the assumptions are: (1) Mach Number $\mathbf{M} < 0.1$, (2) adiabatic flow, (3) ratio of pressure drop to inlet pressure $\Delta p/p_1 < 0.04$. These assumptions are discussed below from a practical standpoint.

There are a number of formulae and charts for the flow of air and other gases, among them the formulae of Weymouth, Spitzglass, and Unwin and the charts of McMillan (based on the Fritzshe formula) and Lemke (1952, WPCF 1967, Menon 2005). The assumptions made in their developments are sometimes not clearly stated. All were developed for a limited range of conditions (such as one pipe roughness or fixed blower inlet conditions) or require very cumbersome corrections; none are directly applicable to the full range of conditions encountered in environmental engineering applications. Furthermore, with the availability of programmable calculators and computers, such formulae and charts no longer provide the practical computational advantages they once did, over the more fundamental and widely applicable methods presented below.

9.3.3.1 Mach Number

King (1944) prepared an economic analysis of air piping for activated sludge systems in which he derived formulae for the economical diameter of steel and cast iron pipe. Those formulae were subsequently used to derive an economic range of velocities under a particular set of temperature and pressure conditions, as given in the following table. [Economic velocities from WPCF (1971), based on actual flow at 125°F and 7.5 psig (p_{atm} = 14.7 psia)].

Pipe Diameter	Economic Velocity Range	Mach Number M
(in.)	Range (fpm)	@125°F
1–3	600–1000	0.00843–0.01405
4–10	900–1500	0.01265–0.0211
12–24	1400–2000	0.01967–0.0281
30–60	2000–3200	0.0281–0.0450

King's analysis was published in 1944, and economic velocities would probably be lower for most comparable systems today due to the relatively higher cost of power. However, the above table will provide a general idea of economic velocities. Case-specific analysis of economic pipe sizes is justified for major systems.

The Mach Numbers corresponding to the economic velocity range are given in the third column of the above table (the speed of sound in air @125°F is 1185 ft/sec = 71,100 ft/min). The Mach Numbers are all within the 0.1 limit established above. Although peak flows (on which maximum pressure calculations are based) exceed average flows (on which economic velocities and pipe diameters are often properly based), the peak Mach Numbers are commonly less than 0.1. The author has found this to be the case for almost all systems encountered in environmental engineering applications, although the Mach Number should be checked in each instance. When pipe Mach Numbers greater than 0.1 are encountered, the economy of the pipe system should be checked. (Mach numbers greater than 0.1 are commonly encountered in aeronautical and aerospace applications, where weight is an important factor.)

9.3.3.2 Non-adiabatic Flow

The air piping employed in environmental engineering applications is commonly of substantial length. The piping is occasionally insulated to prevent external condensation and/or head loss but is often uninsulated. Transfer of heat between the internal flow and the environment occurs across the pipe walls. For example, blower inlet piping taking outdoor air will often have heat transferred into it from the warmer building air. The blower discharge will have an elevated temperature, and heat will commonly be transferred from the air in the discharge piping outward into the environment (building air, outdoor air, tank water, etc.). The flow is not truly adiabatic, and the gas density will vary not only due to changes in pressure but also due to changes in temperature within the piping system.

Nevertheless, a conservative evaluation of piping system friction losses, which will be sufficiently accurate in most instances, may be made by assuming a constant density based on the temperature and pressure that give an appropriate *worst-case* condition in a particular pipe segment. The assumptions necessary to insure a conservative estimate will be considered first, after which the conditions under which these assumptions are acceptable will be elaborated upon.

Consider the relationship for friction loss in a pipe, as given in Chapter 6 and in the previous section (Equation 9.99). As noted in Chapter 6, f is a function of the pipe Reynolds Number $\mathbf{R}$ and roughness ratio ε/D. Incorporating this functional relationship for f and setting $V = Q/A$, Equation 9.99 can be rewritten as:

$$\Delta p = f(\mathbf{R}, \varepsilon/D)\frac{L}{D}\frac{\rho Q^2}{2gA^2} \tag{9.100}$$

in which all terms on the right-hand side except $\mathbf{R}$ and Q are essentially independent of temperature.

The meaning of "constant flow" as it applies to a gas system must now be qualified to mean constant *mass rate of flow*. The *mass* rate of flow w is related to the volumetric rate of flow Q by:

$$w = \rho Q \tag{9.101}$$

Since, by conservation of mass, w remains constant along a pipe without the addition or removal of gas, the volumetric rate of flow Q varies inversely with the density ρ.

Substituting Q from Equation 9.101 into Equation 9.100 results in:

$$\Delta p = \frac{1}{\rho} f(\mathbf{R}, \varepsilon/D) \frac{L}{D} \frac{w^2}{2gA^2} \tag{9.102}$$

Substituting ρ from Equation 9.2 into the above gives:

$$\Delta p = \frac{RT}{p} f(\mathbf{R}, \varepsilon/D) \frac{L}{D} \frac{w^2}{2gA^2} \tag{9.103}$$

The Reynolds Number may be expressed in terms of mass rate of flow as follows:

$$\mathbf{R} = \frac{\rho VD}{\mu} = \frac{\rho QD}{\mu A} = \frac{wD}{\mu A} \tag{9.104}$$

Therefore, the Reynolds Number varies inversely as the absolute viscosity μ, the terms w, D, and A being constant.

The maximum temperature and minimum pressure are seen to give the largest pressure drop. That is so not only because of the explicit effect of T and p in Equation 9.103 but also because the one other affected variable, namely f, increases with increasing T. This can be seen by noting that increasing temperature increases gas viscosity μ, which decreases the Reynolds Number $\mathbf{R}$ according to Equation 9.104, which in turn increases f (see the Moody Diagram, Figure 5.1).

Therefore, subject to other applicable limitations, the maximum temperature and minimum pressure in the piping system can be used to obtain a conservative estimate of the pressure drop in the system. This is particularly convenient because the maximum temperature and minimum pressure can often be determined at the outset. As an example, consider a blower and piping system supplying air to a submerged diffuser. The maximum temperature occurs at the blower discharge and can be determined by methods described later in this chapter. The minimum pressure can be taken as the hydrostatic pressure at the submerged diffuser. The use of that temperature and pressure in Equation 9.103 will give a conservative estimate of the pressure drop in the piping system. The conservatism will usually not be excessive, considering the use of the pressure drop in blower selection (as discussed below) and the relatively small changes in temperature and pressure in the piping system.

Shapiro (1953) deals with the more general and considerably more complex cases of fluid flow in conduits with combined friction, heat transfer, and variable area. Consideration of such cases is rarely necessary in environmental engineering applications. It is noteworthy that cooling subcritical pipe flow by itself causes pressure to increase in the direction of flow (Shapiro 1953, p. 199). This suggests that neglecting such cooling results in a conservative estimate of pressure drop in flows with combined friction and cooling, which is in agreement with the above.

9.3.3.3 Pressure Drop

As noted previously, Equation 9.99 may be used for the accurate calculation of air pipe friction losses, as long as the drop in pressure along the line is only a few percent (less than 5% is suggested) of the total pressure; Lemke (1952) provides experimental support of this assertion. Many, if not most, air lines employed in environmental engineering practice will satisfy this requirement. Some will not, however, particularly if fitting losses (see below) are factored in (as they should be). One way of handling cases when $\Delta p/p_1$ exceeds 5% is to divide the pipe into incremental lengths such that $\Delta p/p_1 < 0.05$ in each increment, and the p_1 in each increment is adjusted to account for upstream or downstream pressure losses (depending on the direction in which computations proceed). Thus, if proceeding in the downstream direction, one could take the upstream-most pressure for p_1 in the first increment, neglect $(1 - \Delta p/p_1)$ in Equation 9.98 (i.e., use Equation 9.99), then use a lower p_1, adjusted for the calculated pressure drop, for each subsequent increment. However, by using the minimum downstream pressure as discussed above (and demonstrated below), a conservative calculation of pressure drop is obtained. If a more accurate, less conservative estimate of pressure drop is desired, one could proceed computationally in the upstream direction, raising p_1 for each successive incremental length in which $\Delta p/p_1 < 0.04$.

Example 9.1

This example is for an aeration air header supplying dome diffusers in the east battery-activated sludge tanks at the Tallman Island WWTP, New York City. A 369-ft length of 16-in. diameter cement-lined cast iron pipe (16.53-in. I.D., roughness $\varepsilon = 0.0002$ ft) has a design airflow rate of 8400 scfm. Assume Equation 9.99 is applicable, estimate the pressure drop, and then check the assumption.

Solution

The pipe flow area, velocity, Reynolds Number, and relative roughness are given, respectively, by $A = (\pi/4)(16.53/12)^2 = 1.490$ sq ft, $V = (8,\ 400/60)/1.490 = 94.0$ ft/sec, $\mathbf{R} = 94.0(16.53/12)/10^{-5} = 1.295(10^7)$, and

$\varepsilon/D = 0.0002/(16/12) = 0.0001500$. From the Moody Diagram (Figure 5.1), $f = 0.0130$. The Mach Number $\mathbf{M} = V/c = 94.0/1,130 = 0.0832$, which satisfies the criterion $\mathbf{M} < 0.1$. Then, using Equation 9.99, we have:

$$\frac{\Delta p}{\rho_1} \cong \frac{0.0130(369)}{(16/12)} \frac{(94.0)^2}{2(32.2)} = 494 \text{ ft of air} \times 0.0005604 \text{ psi/(ft of air)} = 0.277 \text{ psi}$$

To use the criterion $\Delta p/p_1 < 0.04$ it is necessary to know upstream pressure p_1, which in this case is 5.65 psi. Then $\Delta p/p_1 = 0.277/5.65 = 0.0490 < 0.05$ and the criterion is satisfied.

9.3.3.4 Fittings and Appurtenances

If the pipe flow can reasonably be assumed to be incompressible and fitting losses are mild (i.e., moderate "*K*" values corresponding to maximum velocities through fitting apertures not greatly in excess of pipe velocities), then the incompressible flow approximation can usually be applied with acceptable accuracy to losses through fittings. In such cases, the "*K*" values and methods of Chapter 6 may be used. There are, however, certain appurtenances that are unique to gas systems and which will be discussed further.

Inlet air filters are used to protect blowers and compressors from damaging particulates and to reduce clogging of any porous-media devices (such as air diffusers). If only the former is of concern, then a simple paper or wire-mesh element filter capable of removing almost all particles 2 μm and above will generally suffice. Filters of this type are available in sizes for capacities of about 15–2200 scfm. Weatherhoods are available for outdoor installation. The author prefers that filter connections be of the same size as the blower inlet piping, without increasers. The reason for this is to reduce pressure drop, the power savings from which should more than offset the cost of "oversized" filters. Longer intervals between changing elements are an additional advantage. Where porous media are to be protected from undue clogging, more elaborate inlet air filtration is usually provided. This may range from glass fiber media filters up to filter systems incorporating automatic renewable media prefilters and electrostatic agglomerators.

In some cases, manufacturers express pressure drop across clean inlet filters in terms of inches WC (water column) at various capacities. [The differential pressures measured by a water manometer are commonly expressed as inches W.C., inches WG (water gage), or inches H_2O. The former is used herein.] This is a clean filter rating, and a multiplier for clogging must be applied. The clean filter loss may be assumed to be a velocity-squared loss, proportional to the square of the flow. An example of the procedure for determination of filter pressure loss in such a case is given below. In other cases, the manufacturer will prescribe the final pressure drop across the filtering system. A pressure gauge should generally be installed between the filter and blower to indicate when the design pressure drop has been reached, at which time the filter element should be changed.

Example 9.2

A 6-in. inlet air filter has been selected for use with a blower of 319 cfm peak design inlet capacity (Rochester, New Hampshire Advanced Wastewater Treatment Facility). The manufacturer's literature indicates a pressure loss for the filter of 2.8 in. W.C. at the filter's rated capacity of 1100 cfm. Determine the clean filter pressure drop in psi and the clogged-filter pressure drop using a clogging multiplier of 3.

Solution

The clean filter pressure drop is given by:

$$\Delta p = (319/1100)^2(2.8) = 0.235\ in.\ W.C. \times 0.433/12 \cong 0.008 \text{ psi}$$

The multiplier 0.433/12 is a conversion factor. The clogged pressure drop is $3 \times 0.008 \cong 0.025$ psi.

Silencers are used on the discharge and/or inlet sides of blowers to reduce noise levels. Their design and selection entail a tradeoff between noise attenuation and pressure drop. Increasing the silencer size to reduce pressure drop increases the exterior surface area of the silencer from which noise is radiated. The internal construction of a chamber-type silencer includes packing (usually hair felt or glass pack), which may be added in combination with chamber-absorptive type silencers or may be used with different internal constructions for absorption silencers. Combination intake filter silencers are also available in smaller sizes. Manufacturers' silencer pressure drop formulae can be reduced to a $KV^2/(2g)$ relationship, as demonstrated by the following example. Silencer pressure drops often constitute the largest single pressure drop in environmental engineering gas systems, and further research and development oriented toward a standard-production low-pressure drop silencer seem warranted. The power savings would often justify a more expensive silencer.

Example 9.3

An 8-in. chamber-type inlet silencer has been selected for use with a blower of 2462 cfm peak design inlet capacity and corresponding inlet density of 0.06107 lbm/ft^3 (Rochester, New Hampshire AWTP). The manufacturer's formula (Burgess-Manning Inc., Orchard Park, New York) for silencer pressure drops with dry air is as follows:

$$\Delta p = C(V/4005)^2(p/14.7)(530/T))$$

in which Δp = pressure drop through the silencer in inches W.C., V = air velocity through the silencer in feet per minute, C = pressure drop coefficient, p = absolute pressure in psia at the silencer (inlet or discharge pressure), and T is absolute temperature in deg R. The pressure drop coefficient is given as 4.2. Determine the corresponding value of K in a pressure drop formula of the form $KV^2/(2g)$ and the pressure drop in psi for the silencer.

Solution

We commence by establishing the basis for the manufacturer's formula and adapting the relationship for use with humid air. Using the relationship $h_L = \Delta p/(\rho g/g_o) = KV^2/(2g)$ and the perfect gas relationship (Equation 9.2), the following is obtained:

$$\Delta p = K\frac{\rho V^2}{2g} = K\frac{p}{R_aT}\frac{V^2}{2g} = K\frac{p}{R_aT}\frac{p_sT_s}{p_sT_s}\frac{V^2}{2g}$$

$$2gR_a = 2(32.2)(53.34)\ \frac{\text{ft}^2}{\text{sec}^{2\circ}R}$$

$$\Delta p = \frac{K}{T}\frac{pV^2}{2(32.2)(53.34)}\frac{530}{530}\frac{14.7}{14.7}\left(\frac{1}{60}\right)^2(2.307)(12) = \left(\frac{V}{4013}\right)^2 K\left(\frac{p}{14.7}\right)\left(\frac{530}{T}\right)$$

in which the units are comparable to those of the manufacturer's formula. The two equations are practically identical (the differences probably are attributable to a small difference in a constant or conversion factor) and indicate that K and C are essentially identical. Thus, K should be taken as 4.2, and the pressure loss can be calculated from $KV^2/(2g)$ with V being the actual velocity based on the 6-in. nominal inlet diameter of the silencer. Thus:

$$V = \frac{2462}{(\pi/4)(8/12)^2} = 7053\ \text{ft/min} \div 60 = 118\ \text{ft/sec}$$

$$\frac{V^2}{2g} = \frac{(118)^2}{2(32.2)} = 215\ \text{ft}$$

$$h_L = K\frac{V^2}{2g} = 4.2(215) = 901\ \text{ft of air} \times \frac{0.06107}{144} = 0.382\ \text{psi}$$

As can be seen from the above examples, it is important to understand the basis for manufacturers' pressure drop formulae and to be able to convert the formulae to a convenient form [the author prefers the $KV^2/(2g)$ form] or to convert results computed from the formulae to the desired dimensional units. Valves, flow meters, and gas distribution systems (e.g., air diffusers) are additional appurtenances for which such conversions are necessary. Two common types of valves, butterfly and check valves, are discussed immediately below. Flow meters are addressed in Section 9.4, while gas distribution systems are taken up in Chapter 11.

Butterfly valves are widely used on main process air lines because of their low-pressure loss when fully open, throttling characteristics, and tight closure. The most common requirement for air service is that of determining the pressure drop across a fully open butterfly valve. A procedure based on one manufacturer's formula is given in the following example. The fully open pressure drop of a properly selected butterfly valve is usually so small as to be almost negligible, and the procedure given herein is an approximate one. When necessary, more precise methods should be readily developed by the reader conversant with the principles presented in this chapter.

Example 9.4

A 6-in. butterfly valve is to be used for open/shut service on the discharge side of a blower discharging 500 scfm (Rochester, New Hampshire AWTP). The manufacturer gives the following dimensional formula for air:

$$Q_s = 963CD^2\sqrt{\frac{\Delta p(p_1 + p_2)}{T}}$$

in which Q_s is standard flow in scfh, D is valve diameter in inches, Δp is pressure drop in psia, p_1 is upstream pressure in psia, p_2 is downstream pressure in psia, T is temperature in deg R, and C is a valve coefficient given the value 71 by the manufacturer for this size valve. The air temperature, barometric pressure, and downstream gage pressure are estimated to be 275°R, 14.0 psia, and 6.139 psig, respectively. The density and velocity head are estimated to be 0.0720 lbm/ft^3 and 33.0 ft, respectively. Determine the pressure drop across the valve.

Solution

The reasonableness of the manufacturer's equation should first be established. This can be done by comparing it with the incompressible orifice equation with ρ given by the perfect gas equation (an exercise left to the reader). The air humidity is not considered in the above equation, so it is regarded as an approximation. Since the pressure drop across an open butterfly valve is normally quite small, one can approximate $(p_1 + p_2) \cong 2p_2$ and rearrange the above equation as follows:

$$\Delta p \cong \frac{T}{2p_2}\frac{Q_s^2}{(963CD^2)^2}$$

Substituting the appropriate values results in:

$$\Delta p = \frac{(275 + 460)(500 + 60)^2}{2(14.0 + 6.139)\left[963(71)(6)^2\right]^2} = 0.00271 \text{ psi}$$

This value can be checked by calculating the corresponding K value as follows:

$$K = \frac{\Delta p/\gamma}{V^2/(2g)} = \frac{0.003(144)}{0.07201(33.0)} = 0.18$$

This value of K is reasonable for an open butterfly valve. The pressure drop is very small, and no further refinement is necessary.

Check valves are used on the discharge side of blowers and occasionally at other locations on air lines. An example of the latter is on drop pipes to large-bubble diffusers in sludge storage tanks, for which the check valves reduce the likelihood of diffuser clogging. A check valve particularly suited to air systems is the double-door spring-loaded type. Manufacturers give head-loss curves for water (see Figure 9.4 for an example), which can be converted for air applications as explained below. Note that $K = \Delta h/[V^2/(2g)] \propto \Delta h/Q^2$ for a particular valve size. Hence:

$$\ln \Delta h \propto \ln(KQ^2) = \ln K + 2\ln Q \tag{9.105}$$

The straight-line portions of the $\ln\Delta h$ vs. $\ln Q$ curves in Figure 9.4 correspond to the fully turbulent regime, over which K in the above relationship is, accordingly, constant. Since the velocity of air is generally at least as high (and often an order of magnitude higher) as that of water through a check valve of the same size, and since the viscosity of air is an order of magnitude lower than that of water, the Reynolds Number of the corresponding airflow will be one to two orders of magnitude higher than that of the water flow. Hence, the value of K obtained for water flow in the fully-turbulent regime can be applied to the corresponding airflow. Thus, any corresponding values of Δh and $V^2/(2g)$ from Figure 9.4 or Figure 9.5 for a particular valve size will give K, which can then be used to determine the pressure loss for a valve of the same size in an airline. The following example demonstrates the procedure.

Example 9.5

A 6-in. double-door spring-loaded check valve is to be used on the discharge side of a blower with an actual peak discharge airflow of 270 cfm (air density = 0.7208 lbm/ft^3) (Rochester, New Hampshire AWTP). Water data for this type of check valve are given in Figures 9.4 and 9.5. Find the K value and air pressure drop across the valve.

Figure 9.4 Double-door Check Valve (APCO/DeZurik, Marietta, Georgia)

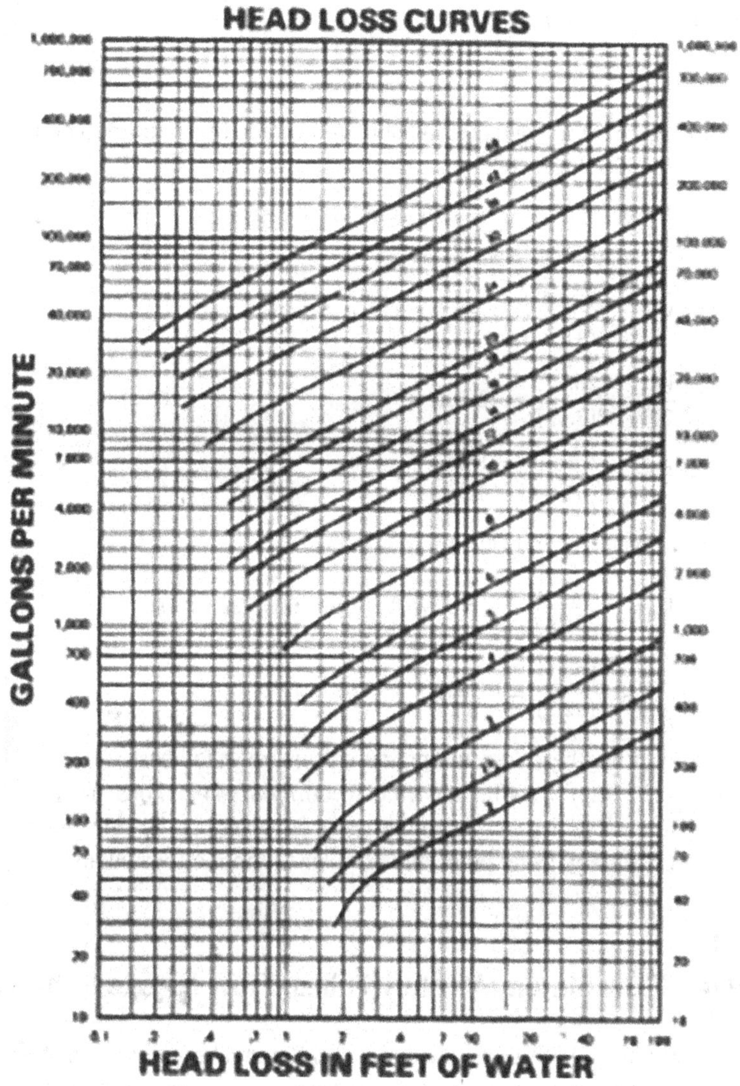

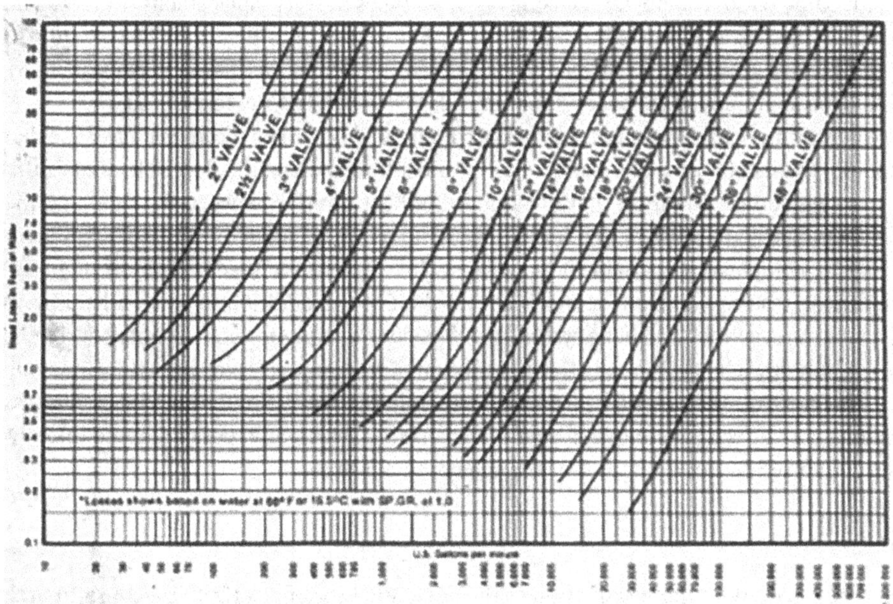

Figure 9.5 Mission Duo-check Check Valve (Crane Valves, The Woodlands, Texas)

Solution

From Figure 9.4 or Figure 9.5, a point on the straight-line portion of the curve for a 6-in. check valve is selected and used to obtain the K value as follows:

$$\Delta h \cong 30 \text{ ft } @ \; Q = 2500 \text{ gpm for water}$$

$$Q = 2500/449 = 5.568 \text{ cfs}$$

$$V = \frac{5.568}{(\pi/4)(6/12)^2} = 28.4 \text{ ft/sec}$$

$$\frac{V^2}{2g} = \frac{(28.4)^2}{2(32.2)} = 12.5 \text{ ft}$$

$$K = \frac{\Delta h}{V^2/(2g)} = \frac{30}{12.5} = 2.4$$

The air velocity and velocity head are:

$$V = \frac{270}{(\pi/4)(6/12)^2} = 1375 \text{ ft/min} \div 60 = 22.9 \text{ ft/sec}$$

$$\frac{V^2}{2g} = \frac{(22.9)^2}{2(32.2)} = 8.16 \text{ ft}$$

The resulting head loss and pressure drop are as follows:
$h_L = KV^2/(2g) = 2.4(8.16) = 19.6 \text{ ft} \times 0.7208/144 = 0.098 \text{ psi}$

Pressure drops across most other types of valves can be determined by using values of K known for those valves (see Section 6.5) or by using procedures similar to those discussed above.

9.4 Flow Meters: Venturi Tubes, Flow Nozzles, Orifices

Chapter 6 dealt with particular types of differential-pressure primary elements used in flow metering, including venturi tubes, flow nozzles, and orifices. The reasons for focusing on those particular types were stated in Chapter 6 in connection with the incompressible flow of liquids. The same types of meters are most commonly applied to air systems in environmental engineering applications, and the same reasons apply for focusing on those types here. (Displacement counters and ammeters are used in metering blower discharges, as discussed in Section 9.5). For digester gas flow monitoring, venturi tubes, positive-displacement diaphragm meters, turbine-type meters, vortex shedding meters (discussed in Section 11.5), propeller meters, and shunt meters have been employed (Babcock 1966, Part 2; WPCF 1978; Crawford 2019).

The preceding section dealt with the constant-area flow of a gas, such as occurs in pipe flow. The *theoretical* basis for that section was derived from the assumption of adiabatic flow of a perfect gas. In most flow meters, the flow velocity is sufficiently high and the lengths of flow channels sufficiently short that the amount of heat transfer between the fluid and the surroundings is low enough to be considered negligible. Thus, the flow can be assumed to be adiabatic without significant error. However, in order to handle the added complexity of variable flow areas, such as are encountered in the flow metering devices considered here, another simplifying assumption is made, namely that the flow, in addition to being adiabatic, is also *reversible* in the thermodynamic sense. A reversible, adiabatic system is one of *constant entropy* and, as noted earlier, is called *isentropic*. The short lengths of smooth flow channels justify the assumption of frictionless (reversible) flow for some well-proportioned venturi meters and flow nozzles. For other cases, particularly orifices, the frictionless assumption is less appropriate. Nevertheless, isentropic flow of a perfect gas provides a useful *theoretical* basis for all types of flowmeters considered here and will be addressed first followed by consideration of the meters themselves.

9.4.1 Isentropic Flow of a Perfect Gas

Equation 9.65 was derived by applying the steady-flow energy equation to the adiabatic flow of a perfect gas; constant area was not assumed. No further assumptions are necessary when integrating Equation 9.65 to obtain Equation 9.84. In dealing with isentropic flows, it becomes convenient to deal with *stagnation properties*, which are the fluid properties at a reference value of $\mathbf{M} = 0$. Equation 9.84 then gives the stagnation temperature as follows:

$$\frac{T_o}{T^*} = \frac{k+1}{2} \tag{9.106}$$

in which the subscript "o" denotes the stagnation property. By eliminating T^* between Equations 9.84 and 9.106, the ratio of the flow temperature to the stagnation temperature is obtained:

$$\frac{T_o}{T} = 1 + \frac{k-1}{2}\mathbf{M}^2 \tag{9.107}$$

Substituting Equations 9.52b and 9.52c into Equation 9.107 gives the stagnation pressure and density ratios as functions of Mach Number:

$$\frac{p_o}{p} = \left(1 + \frac{k-1}{2}\mathbf{M}^2\right)^{k/(k-1)} \tag{9.108}$$

$$\frac{\rho_o}{\rho} = \left(1 + \frac{k-1}{2}\mathbf{M}^2\right)^{1/(k-1)} \tag{9.109}$$

Working charts and tables for the isentropic flow relationships (Equations 9.107, 9.108, and 9.109) are given by Shapiro (1953) and others. These equations can also be readily programmed on a programmable calculator. It is noteworthy that p is within 1% of p_o for $\mathbf{M} < 0.11$. The corresponding values of $\mathbf{M}$ for ρ within 1% of ρ_o and T within 1% of T_o are 0.12 and 0.22, respectively. The typical economic pipe velocities discussed in Section 9.3 are well within these limiting Mach Numbers, suggesting that differences between stagnation properties and actual properties in full-section pipe flow are often negligible.

The values of the temperature, pressure, and density ratios at the critical state [corresponding to minimum flow area (Shapiro 1953)] are found by setting $\mathbf{M} = 1$ in Equations 9.107, 9.108, and 9.109. The results, together with the numerical values for air ($k = 1.4$), are given below:

$$\frac{T^*}{T_o} = \frac{2}{k+1} = 0.8333 \tag{9.110}$$

$$\frac{p^*}{p_o} = \left(\frac{2}{k+1}\right)^{k/(k-1)} = 0.5283 \tag{9.111}$$

$$\frac{\rho^*}{\rho_o} = \left(\frac{2}{k+1}\right)^{1/(k-1)} = 0.6339 \tag{9.112}$$

We now derive a relationship for isentropic mass rate of flow, which will provide a basis for subsequent relationships for flow meters. Consider a flow nozzle or venturi tube, as shown in Figure 6.1. Here we denote state points 1 and 2 with Section 1 being immediately upstream of the nozzle or venturi and Section 2 being at the minimum cross section. The minimum cross section may be either an exit section or a throat section, depending on whether the nozzle is converging or converging–diverging. Application of the continuity relationship (Chapter 2) and the steady-flow energy equation for adiabatic flow (Chapter 2) with zero $\cdot Q$ (the rate of heat flow exchange) between Sections 1 and 2 gives:

$$w = \rho_1 A_1 V_1 = \rho_2 A_2 V_2 \tag{9.113}$$

$$h_1 + V_1^2/(2g) = h_2 + V_2^2/(2g) = h_o \tag{9.114}$$

Equations 9.113 and 9.114 can be rewritten as follows:

$$V_2^2 = 2g(h_1 - h_2) + V_1^2 \tag{9.115}$$

$$V_1 = \frac{\rho_2 A_2 V_2}{\rho_1 A_1} \tag{9.116}$$

Eliminating V_1 from these equations and solving for V_2 gives:

$$V_2 = \frac{\sqrt{2g(h_1 - h_2)}}{\sqrt{1 - (\rho_2/\rho_1)^2 (A_2/A_1)^2}} \tag{9.117}$$

Then substituting the above equation into Equation 9.113, the following is obtained:

$$w = \rho_2 A_2 \frac{\sqrt{2g(h_1 - h_2)}}{\sqrt{1 - (\rho_2/\rho_1)^2 (A_2/A_1)^2}} \tag{9.118}$$

It is desired to express Equation 9.118 in terms of the pressures measured at or near Sections 1 and 2. For isentropic flow, Equation 9.52a gives the density ratio in terms of the pressure ratio as follows:

$$\frac{\rho_2}{\rho_1} = \left(\frac{p_2}{p_1}\right)^{1/k} \tag{9.119}$$

For an isentropic process, the differential enthalpy is related to the differential pressure by (Mooney 1959, §11-4; also note Equation 2.59):

$$d\tilde{H} = \frac{1}{\rho} dp \tag{9.120}$$

the integral of which is given by:

$$\tilde{H}_1 - \tilde{H}_2 = -\int_1^2 d\tilde{H} = -\int_1^2 \frac{dp}{\rho} \tag{9.121}$$

Substituting $\rho = \rho_1 (p/p_1)^{1/k}$ from Equation 9.119 into Equation 9.121 and carrying out the integration results in:

$$\tilde{H}_1 - \tilde{H}_2 = \frac{k}{k-1} \frac{p_1}{\rho_1} \left[1 - \left(\frac{p_2}{p_1}\right)^{(k-1)/k}\right] \tag{9.122}$$

Substituting Equations 9.122 and 9.119 into Equation 9.118 and adding a *discharge coefficient* C gives:

$$w = \frac{CA_2}{\sqrt{1 - (p2/p1)^{2/k} \beta^4}} \sqrt{\frac{2gk}{k-1} p_1 \rho_1 \left(\frac{p_2}{p_1}\right)^{2/k} \left[1 - \left(\frac{p_2}{p_1}\right)^{(k-1)/k}\right]} \tag{9.123}$$

in which $\beta = d_2/d_1$. This rather awkward relationship is often written in terms of the analogous incompressible flow equation:

$$w = \frac{CYA_2}{\sqrt{1 - \beta^4}} \sqrt{2g\rho_1 (p_1 - p_2)} \tag{9.124}$$

in which Y is an *expansion factor* obtained by eliminating w from Equations 9.123 and 9.124, with the results given by:

$$Y = \left\{\frac{k}{k-1} \left(\frac{p_2}{p_1}\right)^{2/k} \left[\frac{1 - (p_2/p_1)^{(k-1)/k}}{1 - p_2/p_1}\right]\right\}^{1/2} \left[\frac{1 - \beta^4}{1 - (p_2/p_1)^{2/k} \beta^4}\right]^{1/2} \tag{9.125}$$

Although formidable in appearance, the above relationship contains only three dependent variables: k, β, and p_2/p_1. This relationship is plotted for air ($k = 1.4$) in Figure 9.6 and is readily programmed on a programmable calculator or computer.

Alternatively, Equation 9.124 is written in terms of the *flow coefficient* K defined as:

$$K = \frac{C}{\sqrt{1 - \beta^4}} \tag{9.126}$$

so that:

$$w = KYA_2 \sqrt{2g\rho_1 (p_1 - p_2)} \tag{9.127}$$

with Y still given by Equation 9.125.

As the pressure drop ratio $\Delta p/p_1 = (p_1 - p_2)p_1$ becomes small, p_2/p_1 approaches 1 and Y approaches 1 (see Figure 9.6), and Equation 9.124 (or Equation 9.127) approaches the relationship for incompressible flow. At the opposite extreme of large

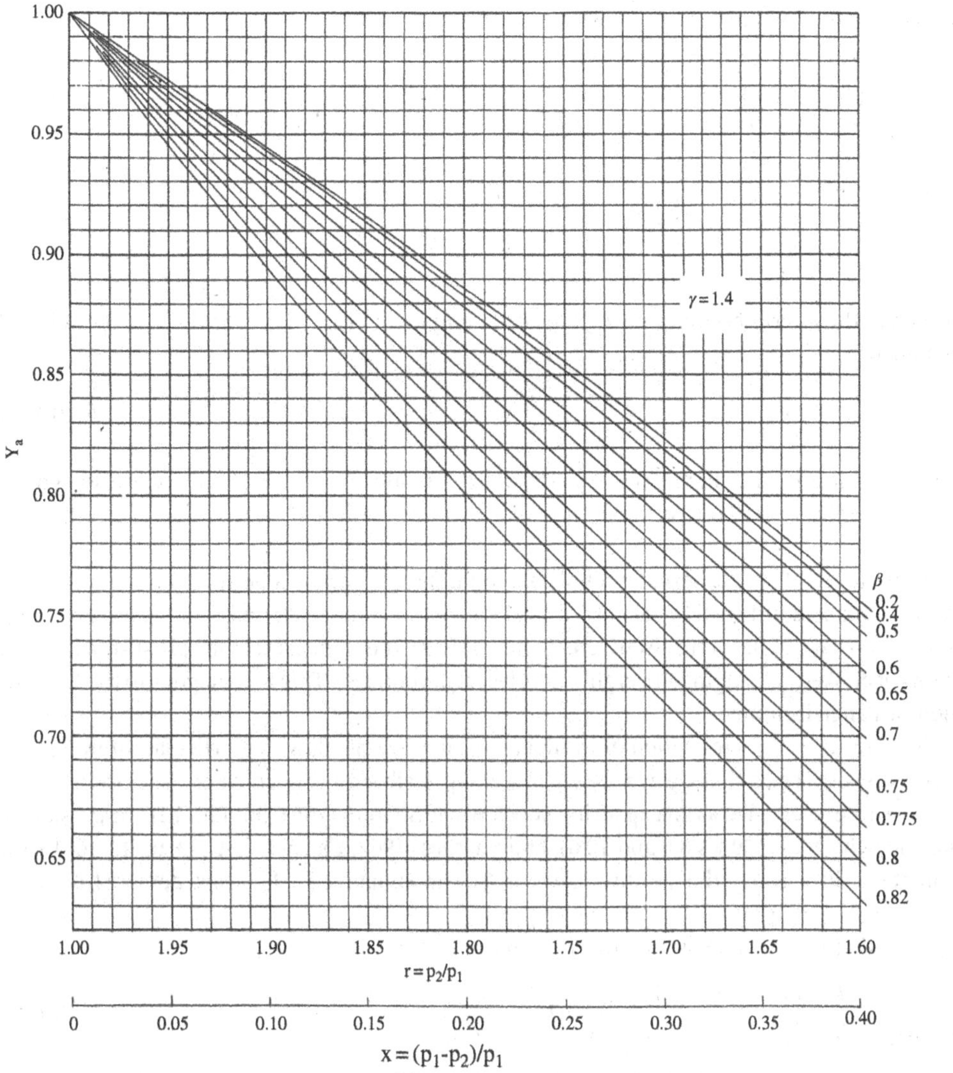

Figure 9.6 Expansion Factors for Flow Nozzles and Venturi Tubes, $\gamma = 1.4$ (ASME 1971/American Society of Mechanical Engineers)

$\Delta p/p_1$, the critical pressure ratio p_2/p_o given by Equation 9.111 is approached. When that critical pressure ratio is reached, i.e., at:

$$\frac{p_2}{p_1} = \frac{p_2/p_o}{p_1/p_o} = \frac{p^*/p_o}{p_1/p_o} = \frac{0.5283}{p_1/p_o} \tag{9.128}$$

the Mach Number at the throat section becomes unity. Further increases in $(p_o - p_2)/p_o$, corresponding to further decreases in p_2/p_o, are not possible (note that p_2 is at the throat or exit section), and the flow is said to be *choked*. Under choked-flow conditions, Equation 9.123 is valid only with p_2/p_1 given by Equation 9.128.

For choked flow, the equation for w takes on the following functional form:

$$w = CA_*\phi(p_1/p_o, \beta, k)\sqrt{gp_1\rho_1} \tag{9.129}$$

where A_* denotes A_2 under choked conditions. Thus, for a choked venturi or flow nozzle, only the properties at station 1 need to be measured to determine the flow. The analogy with open-channel critical-flow meters (Chapter 7) should be apparent. The analogy is continued by noting that the flow and conditions at station 1 in the choked nozzle are independent of downstream pressure (downstream depth in the analogous open channel) as long as the ratio p_2/p_o remains below the critical value p^*/p_o (critical submergence ratio in the open channel).

When $p_1 \cong p_o$ and $p_2/p_o = p^*/p_o$ as given by Equation 9.111, Equation 9.123 reduces to:

$$w = \frac{CA_*}{\sqrt{1-[2/(k+1)]^{2/(k-1)}\beta^4}}\sqrt{k\left(\frac{2}{k+1}\right)^{(k+1)/(k-1)} g p_o \rho_o} \tag{9.130}$$

For air ($k = 1.4$), this equation becomes:

$$w = \frac{3.886CA_*}{\sqrt{1-0.402\beta^4}}\sqrt{p_o \rho_o} \tag{9.131}$$

For a relatively large entrance area, the β term becomes negligible. Then employing Equation 9.2 with $R = 53.34$ ft lbf/(lbm deg R) the following is obtained:

$$w = 0.532CA_*\frac{p_o}{\sqrt{T_o}} \tag{9.132}$$

in which A_*, p_o, and T_o have units of ft^2, lbf/ft^2, and °R, respectively. The above equation (with $C = 1$) is known as Fliegner's formula (Shapiro 1953) after the nineteenth-century investigator who experimentally derived the equation without the benefit of theoretical understanding.

What is emphasized here is the existence of, and criteria for, choked conditions. It is rarely a desired condition in environmental engineering applications, and thus little will be said about the behavior of converging and converging–diverging nozzles under choked-flow conditions. That can be found elsewhere (Shapiro 1953). It is of interest to note that shock waves may occur in the diverging sections of choked converging–diverging nozzles (e.g., venturis). This is analogous to the hydraulic jump in the diverging portion of Parshall flumes.

The shock wave in a compressible flow involves a transition from supersonic to subsonic flow. Of particular interest is the impossibility of rarefaction or reverse shock, i.e., a shock wave arising in a compressible flow in which pressure increases in passing downstream across the wave, also referred to as an expansion shock wave. Shapiro (1953, pp. 120–121: Impossibility of a Rarefaction Shock), Kuethe and Schetzer (1959 §'s 8.6 and 10.6), and Shames (1962, §'s 12-5, 12-6, 13-10, 13-16, 13-18, 15-8) provide proofs of this impossibility, using the Second Law of Thermodynamics. Shapiro's demonstration is recounted here.

Shapiro derives the following equation for the increase in entropy across a shock wave:

$$\frac{s_y - s_x}{R} = \frac{k}{k-1}\ln\left[\frac{2}{(k+1)\mathbf{M}_x^2} + \frac{k-1}{k+1}\right] + \frac{1}{k-1}\ln\left[\frac{2k}{(k+1)}\mathbf{M}_x^2 - \frac{k-1}{k+1}\right] \tag{9.133}$$

in which s is entropy, R is the experimental gas constant which differs for each gas, $\mathbf{M}$ is Mach Number, and subscripts x and y respectively denote conditions upstream and downstream of the shock wave. For typical gases with $1 < k < 1.67$, the entropy change is always positive when $\mathbf{M}_x$ is greater than unity and is always negative when $\mathbf{M}_x$ is less than unity. This can be seen by reversing the numerator on the left-hand side of the above equation and plotting $(s_x - s_y)/R$ vs. $\mathbf{M}_x$ as in Figure 9.7 below. Above the x-axis shocks are possible, whereas below the x-axis shocks are impossible because entropy change must always be positive per the Second Law of Thermodynamics.

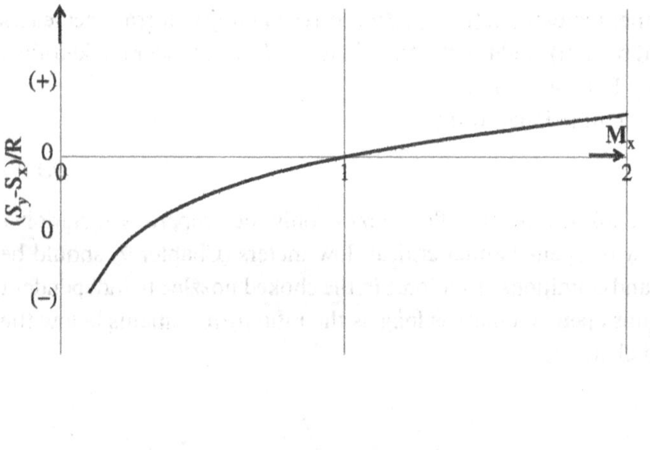

Figure 9.7 Nondimensional Entropy Difference vs. Upstream Mach Number (from Graber 2023a)

9.4.2 Venturis and Flow Nozzles

Figures 6.6(a) and (b) show typical venturis and flow nozzles, which are equally applicable to gas flows. Equation 9.127 is the form of the flow equation commonly used for such meters. Substituting $w = \rho_s Q_s$, $A_2 = (\pi/4)D^2(d/D)^2 = \beta^2(\pi/4)D^2$, and $(p_1 - p_2) = \rho_w h_w$ into Equation 9.127 results in:

$$Q_s = KY\frac{\beta^2 \pi D^2}{4\rho_s}\sqrt{2g\rho_1\rho_w h_w} \tag{9.134}$$

or

$$\beta^2 K = \frac{4\rho_s Q_s}{Y\pi D^2\sqrt{2g\rho_1\rho_w h_w}} \tag{9.135}$$

Substituting $\rho_s = 0.0764$ lbm/ft^3, $\rho_w = 62.4$ lbm/ft^3, and other appropriate constants and conversion factors into the above equation, the following, widely used *dimensional* relationship is obtained:

$$\beta^2 K = \frac{0.0002126 Q_s}{YD^2\sqrt{\rho_1 h_w}} \tag{9.136}$$

in which Q_s is standard volumetric airflow in scfh, D is pipe diameter in inches, ρ_1 is the inlet air density in lbm/ft^3, and h_w is differential pressure in inches W.C.

Frictional effects usually cause the performance of flow nozzles and venturis to differ only slightly from that based on isentropic flow conditions. Thus, the discharge coefficient C, which represents the ratio of the actual nozzle flow to the flow calculated from the isentropic relationship at the actual pressures p_1 and p_2, is often on the order of 0.99. This will be the case when the nozzle or venturi is well proportioned, with straight axes, and has Reynolds Numbers based on the minimum area of 10^6 or higher. For lower Reynolds Numbers, the boundary layers on the wall (to which the frictional effects are primarily confined) may occupy a significant proportion of the flow passages, causing a reduction in discharge coefficient.

Thus, for well-designed and installed nozzles and venturis operated at sufficiently high Reynolds Number, the discharge coefficient C will be constant ($\cong 1$) and the flow coefficient K (see Equation 9.126) will be closely approximated by $K \cong 1/\sqrt{1-\beta^4}$. The term $\beta^2 K$ in Equation 9.135 is then a function of β only, given by $\beta^2 K \cong \beta^2/\sqrt{1-\beta^4}$. This relationship is discussed in Section 6.3.

Equation 9.125 provides the compressibility factor for venturis and flow nozzles. It is found from that equation or Figure 9.6 that, for air, Y is within 1% of unity for $\Delta p/p_1 = (p_1 - p_2)/p_1 = 1 - p_2/p_1$ less than values ranging from 0.01 to 0.02 depending on β. The value of $\Delta p/p_1$ is also equal to $0.036h_w/p_1$ where h_w is the differential pressure measured in inches W.C. and p_1 is in psia. The term $0.036h_w/p_1 < 0.015$ is an approximate criterion for incompressible flow suggested by one manufacturer of flow nozzles and venturis.

The approximately incompressible range ($0.99 < Y < 1$) in Figure 9.6 includes some environmental engineering applications of venturis and flow nozzles. However, it is quite common to have values of $\Delta p/p_1$ on the order of 0.10, with compressibility effects being an important concern for accurate flow measurement. In such cases, it is not only advisable to consider compressibility effects in sizing of the primary element and calculation of pressure loss, but it is also important to consider such effects in specification and operation of the instrumentation. In some cases, the secondary instrumentation (see Section 6.3) can be required to fully compensate (by electronic, electrical, and/or mechanical conditioning) for pressure, temperature, and (as necessary) compressibility factor. In other cases, it may be preferable to require the instrumentation manufacturer to furnish suitable curves or tables for converting meter readings [including differential pressure, static pressure (pressure at a point moving with the fluid), and temperature] to standard flow.

It is to be emphasized that the relationships given above are useful for sizing meters, estimating pressure losses, and communicating with manufacturers, including checking and understanding their formulae. However, when accurate flow measurement is desired with manufacturer-furnished instrumentation (properly specified and checked), the manufacturer's secondary instrumentation or calibration data should provide the basis for flow determination. Pressure losses across venturis and flow nozzles on gas flows are generally approximated as being an empirically determined fraction of the total pressure differential, as for incompressible flow (see Chapter 6).

Example 9.6

A short-form venturi tube is to be installed in a 6-in. pipeline (Rochester, New Hampshire, AWTP). The selected venturi throat diameter is 2.7 in. The maximum airflow rate and corresponding pressure and density are 803 scfm, 20.5 psia, and 0.07753 lbm/ft^3, respectively. Find the corresponding pressure loss.

Solution

The venturi β is 2.7/6 = 0.45, which, from above, gives $\beta^2 K \cong \beta^2/\sqrt{1-\beta^4} = 0.207$. Equation 9.136 then gives the following dimensional relationship:

$$h_w = \left[\frac{0.0002126 Q_s}{\beta^2 KYD^2\sqrt{\rho_1}}\right]^2$$

Assuming $Y = 1$, the above equation yields:

$$h_w = \left[\frac{0.0002126(803)(60)}{0.207(1)(6)^2\sqrt{0.07753}}\right]^2 = 24.4 \text{ inches W.C.}$$

The assumption $Y \cong 1$ may now be checked and adjusted as necessary with $\Delta p/p_1 = 0.036h_w/p_1 = 0.036(24.4)/20.5 = 0.0428$ and $\beta = 0.45$, Equation 9.125 (or Figure 9.6) gives $Y = 0.976$. The adjusted pressure differential is:

$$h_w = 24.4\left(\frac{1}{0.976}\right)^2 = 25.6 \text{ in.W.C.}$$

Further iteration would yield adjustments of little consequence (the reader should check this). For $\beta^2 K = 0.207$, the curve indicates a head loss equal to 15% of the differential pressure. The pressure loss is thus:

$$\Delta p_{loss} = 0.15(25.6) = 3.84 \text{ in.W.C.} \times \frac{1}{2.307(12)} = 0.139 \text{ psi}$$

9.4.3 Orifices

Orifices create a relatively large energy loss and are therefore less frequently used for permanent flow metering. However, they are very convenient for temporary test purposes, as they are inexpensive, easily made in most machine shops, and can often be readily inserted between couplings of existing piping. Figure 9.8 depicts the ASME (1971) thin-plate concentric orifice and required dimensions. Table 9.5 gives minimum recommended thicknesses of the orifice plate based on deflection limitations. Carbon steel is a suitable material for temporary orifices in air lines. The outlet corner of the orifice should be beveled, as shown in Figure 9.8, only when the thickness of the orifice plate exceeds the maximum required orifice width (the smaller of 0.02D or d/8). The minimum orifice width is 0.01D. The purpose of the maximum orifice width and/or beveling requirement is to assure that the jet issuing from the orifice does not reattach to the downstream edge of the orifice. Several configurations of pressure taps may be employed, each of which has strict dimensional specifications. ASME "Fluid Meters" (1971) provides for three pressure tap configurations: taps at 1D and ½D (see Figure 9.8), flange taps, and vena contracta taps. The author has found taps at 1D and ½D to be most convenient for temporary metering. Relations for that type will be discussed here (see "Fluid Meters" for discussion of other types).

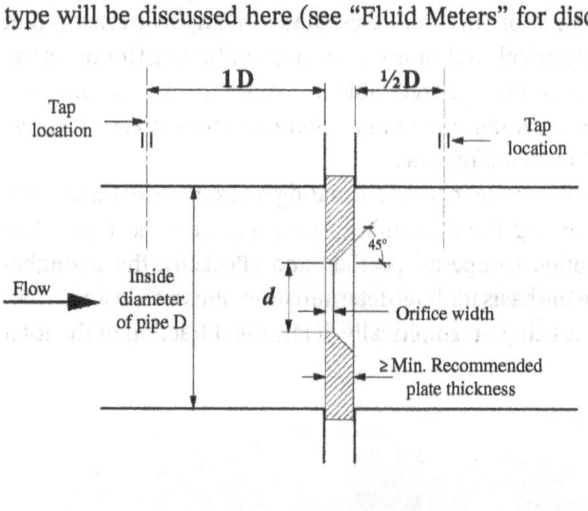

Figure 9.8 Thin-plate Concentric Orifice and Required Dimensions

Table 9.5 Minimum Recommended Thicknesses of Orifice Plates, inches (ASME 1971, Figure II-III-1)

	Internal diameter of pipe (inches)				
Diff'l pressure (in. H_2O)	**3 and less**	**6**	**10**	**20**	**30**
			$\beta < 0.5$		
< 1000	1/8	1/8	3/16	3/8	1/2*
< 200	1/8	1/8	1/8	1/4	3/8
< 100	1/8	1/8	1/8	1/4	3/8
			$\beta > 0.5$		
< 1000	1/8	1/8	3/16	3/8	1/2
< 200	1/8	1/8	1/8	3/16	3/8
< 100	1/8	1/8	1/8	3/16	1/4

* For 1/2-in. plate in 30-in. line, maximum differential = 500 in.
Source: ASME (1971)/American Society of Mechanical Engineers.

(Flange taps are 1 in. from the orifice faces, which usually requires drilling through the pipe coupling, and the downstream vena contracta tap locations must be changed for each size orifice.) Taps at 1*D* and ½*D* have the centers of the inlet and outlet taps 1 and 1/2 pipe diameter from the inlet and outlet faces of the orifice plate, respectively, within prescribed dimensional tolerances (ASME 1971).

Equation 9.125 is commonly applied to concentric orifices, β representing the ratio of orifice diameter to pipe diameter. The discharge coefficient *C* is appreciably less than 1, due to contraction of the flow stream downstream of the orifice (the vena contracta). Its value depends on the location of the pressure taps and is additionally dependent upon *D*, *d* and the Reynolds Number based on orifice diameter *d*, given by:

$$\mathbf{R}_d = \frac{\rho V d}{\mu} = \frac{wd}{\mu A} = \frac{4w}{\pi \mu d} \tag{9.137}$$

The expansion factor *Y* is a function of the same variables as in Equation 9.125, but is given by the following relationship (Lefkowitz 1958, p. 16–20; ASME 1971, §II-III-8):

$$Y = 1 - \left(0.41 + 0.35\beta^4\right)\left(\frac{p_1 - p_2}{p_1 k}\right) \tag{9.138}$$

This relationship applies to all three types of taps. If the static pressure is measured at the downstream tap (p_2) rather than the upstream tap (p_1), a variation of the above equation applies (ASME 1971).

Example 9.7
It is desired to install temporary orifice meters in an airline of Schedule 40 steel (O.D. = 6.625 in., I.D. = 6.065 in.) to measure airflow over a range of 29–716 scfm as part of an investigation of channel agitation (see Chapter 13) (Tallman Island WWTP, New York City). The upstream pressure, temperature, and relative humidity are estimated to be 18.1 psia, 95°F, and 40%, respectively. The orifices will be installed at a pipe joint held by a Victaulic coupling permitting a pipe end separation of 0–1/4 in. Taps at 1*D* and ½*D* and a U-tube manometer with a 36-in. maximum differential will be used. Design a series of orifices suitable for this application.

Solution

From Table 9.5, the minimum recommended orifice plate thickness is 1/8 in. for all β's. Referring to Figure 9.8, the orifice width must be between $0.01D \cong 0.06$ in. and the smaller of 0.12 in. or $d/8$. Available space, availability of sheet steel stock, and ease of machining led to the selection of the plate thickness and tentative orifice width (to be checked later for the $d/8$ criterion) shown in Figure 9.8. Note that a concentric ring is called for at the outer periphery to facilitate centering and securing of the orifice at the pipe end.

For trial sizing of the orifice diameter, we will neglect compressibility ($Y \cong 1$) and the β term in Equation 9.124, which becomes:

$$w \cong CA_2\sqrt{2g\rho_1(p_1 - p_2)}$$

The mass rate of flow is obtained from $\rho_s Q_s$ where the standard density $\rho_s = 0.0749$ lbm/ft^3. Thus, the mass flow range is 0.0749(29) = 2.17 lbm/min to 0.0749(716) = 53.6 lbm/min.

The upstream density is obtained using Equation 9.17 and Table 9.1, as follows: $\rho_1 = \frac{2.700}{(95+460)}[18.1 - 0.3767(0.40)(0.8153)] = 0.0875$ lbm/ft^3

Assuming $C = 0.6$, the orifice area at maximum flow is approximated by:

$$A_2 \cong \frac{w}{C\sqrt{2g\rho_1(p_1 - p_2)}} = \frac{53.6/60}{0.6\sqrt{2(32.2)(0.0875)[(36/12)(62.4)]}} = 0.0458\text{ ft}^2$$

The corresponding orifice diameter is $d = \sqrt{4A_2/\pi} = 0.242$ ft × 12 = 2.90 in..

Using $\beta = 2.90/6.065 = 0.478$ and $(p_1 - p_2)/(p_1 k) = (36/12)(62.4/144)/[18.1(1.4)] = 0.0513$, the results are $\sqrt{1-\beta^4} = 0.9735$ and $Y = 0.9780$ (by Equation 9.138). At 95°F (555°R), the viscosity is $128(10)^{-7}$ lbm/(sec ft) (by Equation 9.53) and $\mathbf{R}_d = 4(53.6)/[\pi(128)(10)^{-7}(60)(2.90/12)] = 368,000$ (by Equation 9.137). Then, from ASME 1971, §II-III-7, $C = 0.6015$ follows. We thus have $CY\sqrt{1-\beta^4} = 0.6043$, which may be used to refine the values of A_2 and d in accordance with Equation 9.124, to obtain:

$$A_2 = 0.0458(0.6/0.6042) = 0.0455\text{ ft}^2$$

$$d = \sqrt{4A_2/\pi} = 0.241(12) = 2.89\text{ inches}$$

This value of d is sufficiently close to the first trial value, and no further iteration is necessary.

The orifice sized for maximum flow will give an imperceptible manometer reading at minimum flow ($h_w \propto w^2$), so the next step consists of sizing an orifice to obtain a reasonable reading (say 4 in.) at minimum flow. Using the same procedure as before, we obtain a trial orifice area and diameter as follows:

$$A_2 \cong \frac{2.17/60}{0.6\sqrt{2(32.2)(0.0875)[(4/12)(62.4)]}} = 0.00557\text{ ft}^2$$

$$d = \sqrt{4A_2/\pi} = 0.0842(12) = 1.010\text{ in.}$$

The terms involved in adjusting this result are: $\beta = 1.010/6.065 = 0.1665$, $(p_1 - p_2)/(p_1 k) = (4/12)(62.4/144)/[18.1(1.4)] = 0.00570$, $\sqrt{1-\beta^4} = 0.9996$, $Y = 0.9977$, $\mathbf{R}_d = 4(2.17)/[\pi(128)(10)^{-7}(60)(1.010/12)] = 42,700$, $C = 0.5988$, and $CY\sqrt{1-\beta^4} = 0.5976$. The refined values of A_2 and d are:

$$A_2 = 0.00557(0.6/0.5976) = 0.00559\text{ ft}^2$$

$$d = \sqrt{4A_2/\pi} = 0.0844(12) = 1.013\text{, say 1.01 inches}$$

which requires no further iteration.

With the 1.01-in. orifice, the full-scale reading corresponds to a flow of approximately $w = 2.17\sqrt{36/4} = 6.51$ lbm/min, which with the 2.89-in. orifice corresponds to a gage reading of approximately $36(6.51/53.6)^2 = 0.53$ in. The overlap is not satisfactory, and at least one more orifice is necessary to adequately cover the range of flows. At 6.51 lbm/min, a 4-in. reading is obtained with an orifice approximately as follows:

$$A_2 \cong \frac{6.51/60}{0.6\sqrt{2(32.2)(0.0875)[(4/12)(62.4)]}} = 0.01670\text{ ft}^2$$

$$d = \sqrt{4A_2/\pi} = 0.1458(12) = 1.750\text{ inches}$$

This diameter is refined as before, giving $d = 1.747$, say 1.75 in.

With the 1.75-in. orifice, the full-scale reading corresponds to a flow of approximately $w = 6.51\sqrt{36/4} = 19.53$ lbm/ min, which with the 2.89-in. orifice corresponds to a gage reading of approximately $36(19.53/53.6)^2 = 4.78$ in. The overlap is acceptable.

The three orifices required to cover the range thus have diameters of 1.01, 1.75, and 2.89 in. Other dimensions comport with Figure 9.8. The dimensions include reasonable machining tolerances. Note that the orifice width of 0.110 in. is less than $d/8$ in all cases.

Example 9.8

Data obtained from tests using the 1.75-in. pipe orifice (diameter verified by micrometry) of the previous example are given in Columns 1 and 2 of Table 9.6 (Tallman Island WWTP, New York City). The static pressure was measured by means of a mercury manometer connected to the upstream 1D tap and added to atmospheric pressure to obtain absolute static pressure. The air temperature, measured by inserting a thermometer into a tap upstream of the orifice, was approximately constant at 29°C (84°F = 544°R). The relative humidity was estimated to be 40%. Determine the standard airflow (scfm) corresponding to each test point.

Solution

Equations 9.124 and 9.125 provide the applicable orifice relationships. For this orifice $\beta = 1.75/6.065 = 0.2885$, giving $\sqrt{1-\beta^4} = 0.9965$. The upstream air density ρ_1 is determined from the upstream pressure and temperature measurements and the humidity estimate using Equation 9.17 and Table 9.1:

$$\rho_1 = (2.700/544)[p_1(\text{in. Hg}) \times 13.6(62.4/12)/144 - 0.3767(0.40)(0.5771)] = [0.4911p_1(\text{in. Hg}) - 0.08696]/201.5$$

The resulting values are given in Column 3 of Table 9.6. [Although the relative humidity was only estimated in this case, it could have been determined by measuring the ambient humidity (i.e., with the aid of a sling psychrometer) and transferring that measurement to the conditions of temperature and pressure upstream of the orifice (by methods discussed in Section 9.5)].

The discharge coefficient C is dependent on $\mathbf{R}_d$, which is not yet determinable, so a value of $C = 0.6$ is initially assumed. The value of $(p_1 - p_2)/p_1k$ is determined from Columns 1 and 2 (with proper conversion of units) and then Y is determined by Equation 9.138 for each test, as given in Column 4 of Table 9.6. Equation 9.124 provides the following relationship for determining values of w for each test run:

$$w = \frac{CY}{0.9965}(0.01670)\sqrt{2(32.2)\rho_1[(h_w/12)(62.4)]} = 18.40CY\sqrt{\rho_1 \Delta h}$$

in which w, ρ_1, and h_w have units of lbm/min, lbm/ft^3, and inches W.C., respectively. Initial (for $C = 0.6$) values of w thus determined are given in Column 5 of Table 9.6.

The viscosity at 544°R is given by Equation 9.53 as $126(10)-^7$ lbm/(sec ft). Using this viscosity and other known parameters in Equation 9.137 gives the following relationship for the orifice Reynolds Number:

$$\mathbf{R}_d = \frac{4w}{\pi(126)(10)^{-7}(60)(1.75/12)} = 1.155(10)^4 w$$

Table 9.6 Tabulation for Example 9.8

(1)	(2)	(3)	(4)	(5)	(6)	(7)	(8)	(9)
Δh (in. W.C.)	p_{1a} (in. Hg)	ρ_1 (lbm/ft^3)	Y	Initial w (lbm/min)	Initial $\mathbf{R}_d$	Refined C	Refined w (lbm/min)	Q (scfm)
38.9	38.4	0.09316	0.9782	20.56	237,500	0.5964	20.43	273
8.0	34.9	0.08463	0.9951	9.039	104,400	0.5977	9.004	120
4.5	34.2	0.08292	0.9972	6.725	77,670	0.5984	6.707	89.5

Initial values of $\mathbf{R}_d$ thus determined are given in Column 6 of Table 9.6. Refined values of C are then determined from ASME 1971, §II-III-7, and are given in Column 7. Then by using the equation for w given above, refined values of w are obtained (Column 8). Using the refined w values from Column 8 to calculate new values of C and w leads to no further refinement in the values of w (the reader should check this).

In Column 9, the flow Q in scfm is given, as determined by dividing the value of w in Column 8 by $\rho_s = 0.0749$ lbm/ft^3.

9.4.4 Coriolis Meters

Although the Coriolis effect is most commonly associated with large-scale phenomena such as hurricanes, there are also small-scale examples of interest. Hurricanes rotate counterclockwise in the Northern Hemisphere and clockwise in the Southern Hemisphere. Ascher Shapiro conducted experiments in large, quiescent tanks in the Northern Hemisphere in which he replicated the counterclockwise rotation (National Committee for Fluids Mechanics Films 1972, pp. 67–68). Similar experiments in Australia demonstrated the clockwise rotation. (Bathtub drains and toilets are not good examples because the drainage rotations are caused by asymmetries in filling of the tub or bowl.) A practical application of the Coriolis effect at small scale is found in the insertion flowmeter, which uses the Coriolis principle (Sietsema *et al.* 2015) to measure the mass flow and density directly. The device can meter liquid, gas, and steam. Manufacturers of insertion flowmeters include EMCO™, Pleasant Prairie, Wisconsin; Omega Engineering Inc., Norwalk, Connecticut; and Khrone America Inc., Beverly, Massachusetts.

9.5 Blowers and Compressors

The term *blower* has historically been applied to a machine in which the gas (air or otherwise) is compressed to a final pressure of up to 40 psig, the term *compressor* being applied to a machine in which the gas is compressed to a final pressure in excess of 40 psig (Rollins 1973).

Modern applications involving a wide range of pressures have reduced the use of this differentiation, with the term *compressor* supplanting the term *blower* and covering the full range of pressures in many industries. However, the term *blower* remains a common one in environmental engineering applications practice, primarily because the distinction remains a useful one. Applications of dynamic compressors and rotary positive-displacement compressors are most often confined to those involving diffusion into water at depths corresponding to pressures of less than 10 psig. Hence, the term *blower* is commonly applied to those machines. Reciprocating compressors are commonly used with receivers for applications requiring intermittent delivery and/or higher pressures, the pressures commonly being at least 40 psig for those machines.

The three broad types of machines mentioned above cover virtually all applications of interest. (Pressure for some applications is supplied by compressed gas storage tanks; in some cases, such as in chlorine systems, the gas is liquefied and must be *evaporated* before use.) Reciprocating compressor selection, from the user's standpoint, involves important decisions about equipment details and features. However, the pneumatic selection considerations are straightforward, involving the simple matching of compressor capabilities with pressure and capacity (flow) requirements. Only limited additional discussion of reciprocating compressors will be included in this chapter. The emphasis here is on rotary positive-displacement blowers and dynamic blowers, with the primary emphasis on pneumatic performance characteristics.

9.5.1 Dynamic Blowers

The mode of operation of dynamic blowers is analogous to that of dynamic pumps, the latter being discussed in Chapter 8. As with dynamic pumps, the flow may leave the impeller in radial or axial directions, the former being termed centrifugal blowers and the latter axial blowers. The emphasis here will be on centrifugal blowers, which are the dominant type of dynamic blower in environmental engineering applications.

For a blower operating with a small pressure rise and hence small density change across the machine, the turbomachine theory and dimensional analysis of pumps presented in Chapter 8 apply. A distinction has been made in mechanical engineering practice and formally approved by key engineering societies between *fans* for low pressures and *compressors* for high pressures (McDonald 1958, p. 14–66). The demarcation is set at 7% increase in air density from the fan inlet to the outlet. For fans operating with less than this density increase, the assumption of incompressibility leads to substantial simplification without significant error. Such fans are suitable, e.g., for heating, ventilating, and air conditioning uses, but have limited use in environmental engineering process applications. Hence, rather than use incompressible turbomachine theory as a basis for discussion (as is done elsewhere), one could derive relationships applicable to centrifugal compressors that give pressure-discharge curves (inlet pressure and temperature become additional variables as discussed below). The basis for developing such a relation is given, for example, by Streeter (1962). Such a development, however, results in curves of little quantitative value (as was also the case for pumps) and adds little to one's physical feeling for the problem. Furthermore, the resulting variables can be derived by dimensional analysis, which provides a more comprehensive basis. Hence, the dimensional analysis approach is taken here.

The independent variables of importance in centrifugal pumps were listed in Equation 8.1, repeated as follows:

$$\Phi(Q, \Delta p.D, N, \rho, \mu) = 0 \tag{8.1}$$

In blowers, compressibility effects are important. This requires not only the addition of the speed of sound c to the above functional relationship but also that recognition be given to the change in the variables Q, ρ, μ, and c between the inlet and outlet of the blower. This suggests the addition of the full complement of inlet and outlet variables to the above functional relationship to give the initial functional relationship for blowers as follows:

$$\Phi(Q_1, \Delta p, D, N, \rho_1, \rho_2, \mu_1, \mu_2, c_1, c_2) = 0 \tag{9.139}$$

Note that Q_2 is not needed because $Q_2 = \rho_1 Q_1/\rho_2$ and is thus not independent of the other parameters.

A problem is encountered in going further with the above functional relationship. In equipment handling compressible gases with an input or output of work, similitude conditions cannot be satisfied at the inlet and outlet simultaneously. This will be elaborated upon below, and the reader is asked to accept this momentarily. We will initially choose to work with inlet conditions and reduce Equation 9.139 to the following approximate functional relation:

$$\Phi(Q_1, \Delta p, D, N, \rho_1, \mu_1, c_1) \cong 0 \tag{9.140}$$

With one more variable than was the case for pumps, we expect one more π-parameter than given by the dimensional variables for pumps. Three of the π-parameters can be selected by analogy with Equation 8.3, and the fourth is the inlet Mach Number $\mathbf{M}_1 = V_1/c_1 \propto (Q_1/D^2)/c_1$. Thus:

$$\Phi\left(\frac{Q_1}{ND^3}, \frac{\Delta p}{\rho_1 N^2 D^2}, \mathbf{R}_1, \mathbf{M}_1\right) \cong 0 \tag{9.141}$$

As for pumps, the Reynolds Number is commonly sufficiently large to make the flow "fully turbulent," rendering the Reynolds Number relatively insignificant. However, the addition of the Mach Number as an additional significant variable precludes the development of speed (N) adjustment relationships analogous to those developed in Chapter 8 for pumps.

Nevertheless, useful possibilities employing the dimensionless variables of Equation 9.141 do remain. One such possibility, employed in modeling work, entails maintaining constant inlet (or outlet, but not both) Mach and Reynolds Numbers and kinematic similarity [$Q_1/(ND^3)$ constant] between model (usually a unit of relatively small size) and prototype, in which case values of $\Delta p/(\rho_1 N^2 D^2)$ developed for the model can be extrapolated to the prototype (see Equation 9.141). Maintaining Mach and Reynolds Number similarity simultaneously requires that a different gas be used for model and prototype (helium model testing is common for air prototypes). It can be demonstrated by means of measurement or thermodynamic calculations that Mach and Reynolds Number similarity will not be simultaneously maintained at the opposite end (outlet or inlet). No further elaboration on this application will be made, but rather focus on another application of greater significance in environmental engineering will be placed.

A common problem is that of predicting the performance of a given machine at constant speed under conditions of varying inlet temperature and pressure. The inlet pressure varies not only with variations in atmospheric pressure but also as a result of inlet throttling when used for flow control. For a perfect gas, the pressure change variable and Mach Number are given by:

$$\frac{\Delta p}{\rho_1 N^2 D^2} = \frac{\Delta p}{p_1}\frac{RT}{N^2 D^2} = \left(1 - \frac{p_2}{p_1}\right)\frac{RT_1}{N^2 D^2} \tag{9.142}$$

$$\mathbf{M}_1 = \frac{V_1}{\sqrt{kRT_1}} \propto \frac{Q_1}{D^2\sqrt{kRT_1}} \tag{9.143}$$

Equation 9.141 can then be rewritten in terms of the following grouping of variables:

$$\Phi\left(\frac{Q_1}{ND^3}, \frac{p_2}{p_1}, \frac{RT_1}{N^2D^2}, k, \mathbf{R}_1\right) = 0 \tag{9.144}$$

The number of variables is now formidable. However, noting that the isentropic work of a blower is given by (see standard thermodynamics texts for the development of the following relationship):

$$W_s = \left(\frac{k}{k-1}\right)RT_1\left[\left(\frac{p_2}{p_1}\right)^{(k-1)/k} - 1\right] \tag{9.145}$$

Carter *et al.* (1960) in effect investigated the relation among the variables resulting from the following combination of variables:

$$\Phi\left(\frac{Q_1}{ND^3}, \mu_s, \mathbf{R}_1\right) = 0 \tag{9.146a}$$

$$\mu_s = \left(\frac{k}{k-1}\right)\frac{RT_1}{N^2D^2}\left[\left(\frac{p_2}{p_1}\right)^{(k-1)/k} - 1\right] \tag{9.146b}$$

The term μ_s is dimensionless and called the *isentropic work coefficient*.

The above relationship is of considerable value in the extrapolation of performance curves during design and for adjustment of test data to specified conditions. For both purposes, we restrict ourselves to one machine (constant D) and the same gas (same k and R), under which conditions the approximate similitude relationships reduce to:

$$\frac{Q_1}{N} = \text{const} \tag{9.147a}$$

$$\mu_s \propto \frac{T_1}{N^2}\left[\left(\frac{p_2}{p_1}\right)^{(k-1)/k} - 1\right] = \text{const} \tag{9.147b}$$

$$\mathbf{R}_1 = \frac{\rho Vd}{\mu} = \frac{Vd}{\nu} \propto \frac{V}{\nu} \propto \frac{Q}{N} \propto \frac{N}{\nu} = \text{const} \tag{9.147c}$$

For sufficiently large Reynolds Numbers (criteria are given below), the Reynolds Number becomes relatively unimportant. Then Reynolds Number equality, which is a definite liability in a lot of test work, may be relaxed. However, to compensate for this, Reynolds Number corrections may be made to the isentropic work coefficient as described below.

For fixed speed and high (thus relatively insignificant) Reynolds Number, Equations 9.147a and 9.147b simplify further to:

$$Q_1 = \text{const} \tag{9.148a}$$

$$T_1\left[\left(\frac{p_2}{p_1}\right)^{(k-1)/k} - 1\right] = \text{const} \tag{9.148b}$$

This last relationship is useful in adjusting blower curve data for different inlet temperatures and pressures.

Performance Specification and Testing. A complete *performance* specification of a centrifugal air blower should include the following information or its equivalent:

1) Rated capacity in scfm (measured at 14.7 pounds per square inch absolute, 68°F, and 36% relative humidity) when operating at specified conditions of the following:
   a) Inlet temperature
   b) Inlet pressure at blower flange
   c) Discharge pressure at blower flange
   d) Relative humidity
   e) Barometric pressure

2) Maximum (not to exceed) brake horsepower at the blower shaft under the above conditions.
3) A characterization of the pressure-discharge curve, such as a requirement that when the volumetric capacity is reduced by a certain minimum percentage under specified inlet conditions, the blower shall develop a certain minimum pressure, in excess of the specified discharge pressure.
4) A brief, general description of the discharge control method (e.g., inlet throttling or speed control), volume and/or speed range, and surge limit is discussed below.
5) Inlet temperature over which the above conditions may occur and over which the blower shall operate satisfactorily.

In addition to the performance specification, a centrifugal blower specification should include requirements pertaining to sound level, construction, accessories (expansion joints, silencers, instrumentation), surge control and other protective devices, drive details and couplings, shop testing, noise testing, shop painting and rust inhibitor, installation, field tests and services, and spare parts.

Surging occurs when the blower or compressor throughput is reduced to a point sufficiently below design conditions that erratic performance results. A good discussion of surge and its control for centrifugal compressors is given by White (1972). Graber (1964) presents experimental work for the detection, by fluid jet amplifiers, of impending stall due to surge in high-performance jet aircraft.

There are several categories of shop testing; the category of particular concern here is that required to demonstrate compliance with a performance specification such as outlined above. One or more blowers of each capacity are commonly required to be tested in accordance with the ASME Power Test Code for Compressors and Exhausters (PTC 10) (ASME 1979b) and the ASME Performance Test Codes for Displacement Compressors, Vacuum Pumps, and Blowers (PTC 9) (ASME 1979a). The former is most pertinent to the present chapter and is the focus here.

The PTC 10 tests should be required to include determination of the surge point and the shape of the characteristic curves at various inlet valve control positions and/or drive speeds. The determination of all data required to evaluate efficiencies and horsepower requirements for the specified conditions should also be required. The tests may be considered representative of duplicate machines.

PTC 10 provides for three classes of tests, designated Class I, II, and III. Class I includes "all tests made on the specified gas...at the speed, inlet pressure, inlet temperature, and conditions of cooling for which the compressor is designed or intended to operate." Class II and Class III tests are "intended for use where the compressor cannot be reliably tested on the specified gas at specified operating conditions." Class III tests are for gases that differ from perfect gas laws beyond prescribed limits and require more complex formulas for test interpretation. Since air testing of air compressors is of concern here, and since air at the temperatures and pressures of interest comply sufficiently with the perfect gas laws, attention will be focused on Class I and Class II tests.

Class I tests are the most accurate and should be used whenever feasible. The allowable departure from specified operating conditions for Class I tests is given in Table 9.7. If inlet throttling is used for capacity control during testing, then the limitations of Class II may apply. For Class II (and Class III) tests, the allowable departures are less stringent, as given in

Table 9.7 Allowable Departure from Specified Operating Conditions for Class I Tests (Boyce *et al.* 1976)

	Variable	Symbol	Unit	Departure (%)	*
(a)	Inlet pressure	P_i	psia	5	**
(b)	Inlet temperature	T_i	R	8	**
(c)	Specific gravity of gas	G	ratio	2	**
(d)	Speed	N	rpm	2	–
(e)	Capacity	q_i	cfm	4	–
(f)	Cooling temperature difference	–	°F	5	***
(g)	Cooling water flow rate	–	gpm	3	–

* Departures are based on the specified value, where pressures and temperatures are absolute.
** The combined effect of items (a), (b), and (c) shall not produce more than 8% departure in inlet gas density.
*** Difference is defined as inlet gas temperature minus inlet cooling water temperature.
Source: Boyce *et al.* (1976)/Texas A&M University.

Table 9.8 Allowable Departure from Specified Design Parameters for Class II and Class III Tests (Boyce *et al.* 1976)

		Range of test values Limits – % of design value	
Variable	**Symbol**	**Min**	**Max**
Volume ratio	q_i/q_d	95	105
Capacity-speed ratio	q_i/N	96	104
Machine Mach Number	M_m		
0–0.8		50	105
Above 0.8		95	105
Machine Reynolds Number	R_e		
where the design value is			
Below 200,000 Centrifugal		90	105
Above 200,000 Centrifugal		10*	200
Below 100,000 Axial Compressor		90	105
Above 100,000 Axial Compressor		10**	200
Mechanical losses shall not exceed 10% of the total shaft power input at test conditions			

* Minimum allowable test Machine Reynolds number is 180,000.
** Minimum allowable test Machine Reynolds number is 90,000.
Source: Boyce *et al.* (1976)/Texas A&M University.

Table 9.8. (Additional factors may call for narrower ranges, as noted in PTC 10.) Since geometric and kinematic similarity are required, the Reynolds and Mach Numbers may be defined with respect to any blower dimension and velocity. PTC 10 defines *Machine Reynolds Number* and *Machine Mach Number* as follows:

"Machine Reynolds Number is defined ... by the equation $\mathbf{Re} = UD/\nu$ where U is the velocity at the outer diameter of the first impeller or of the first-stage rotor leading edge, ν is the kinematic viscosity of the gas at the compressor inlet stagnation conditions, and D is a characteristic length. For *centrifugal compressors*, D shall be taken as the exit width of the first-stage impeller. For *axial compressors*, D shall be taken as the chord at the tip of the first-stage rotor blade. These variables must be expressed in consistent units so as to yield a dimensionless ratio. Machine Mach Number is defined as the ratio of the velocity at the outer diameter of the first impeller or of the leading edge of the first-stage rotor blade to the acoustic velocity in the gas at the inlet conditions of stagnation pressure and stagnation temperature. Note: This is not to be confused with local Mach Number as used in aerodynamics."

Note that **Re** also equals $\rho UD/\mu$ in which ρ and μ are the density and absolute viscosity at the blower inlet conditions and is proportional to $\rho N/\mu$ for a given machine. The requirements for Class I tests (Table 9.7) may be interpreted as follows: (1) requiring that the Machine Reynolds Number must be maintained relatively constant, and (2) insuring the acceptability of the approximations made in the dimensional analysis leading up to the equations given above. Recognizing that Class II tests are of lesser accuracy, the requirements for such tests (Table 9.8) are intended to serve the second purpose given above for Class I test requirements, plus insure that Mach and Reynolds Number similarities are maintained over the ranges in which those parameters are significant. Thus, the Machine Mach Number is less significant as a similitude parameter when its value is less than 0.8, for which a wider range of Mach Number test values is allowed. For Machine Mach Numbers above 0.8, where Mach Number becomes particularly significant, the range of allowable Mach Number test values is substantially reduced. The same reasoning applies to Reynolds Number, and here the analogy to turbulent pipe flow should be apparent.

PTC 10 "provides rules for establishing the following test quantities and predicting these quantities, and to predict these quantities for performance of the same compressor when handling a suitably similar gas under specified and dynamically similar operating conditions:

a) Quantity of gas delivered
b) Pressure rise produced
c) Shaft power required
d) Efficiency of the compressor
e) Surge limit of the compressor."

These rules and their rationale are discussed here. We note at the outset that, for Class I tests, it is the intention of PTC 10 that "the magnitude of the adjustments to be applied will be minimum…." Larger adjustments are required for Class II (and Class III) tests, and the accuracy of those tests is reduced accordingly. The emphasis in the following discussion will be on Class I test adjustments, with some mention of Class II adjustments necessary for a fuller understanding of the limitations of the adjustments for Class I.

Following PTC 10, we will denote by subscripts "*sp*" and "*t*" parameters under specified and test conditions, respectively. Starting with inlet capacity (volumetric flow rate referred to inlet conditions of pressure, temperature, and gas composition) conversion, the following relationship is given for all classes of tests:

$$q_{sp} = \frac{q_t}{N_t} N_{sp} \tag{9.149}$$

This relationship is a direct result of Equation 9.147a.

For Class I tests with air or other perfect gases, the pressure-ratio conversion is given by:

$$\frac{T_{1sp}}{N_{sp}^2}\left[\left(\frac{p_2}{p_1}\right)_{sp}^{(k-1)/k} - 1\right] = \frac{T_{1t}}{N_t^2}\left[\left(\frac{p_2}{p_1}\right)_t^{(k-1)/k} - 1\right] \tag{9.150}$$

This relationship is a direct result of Equation 9.147b. For Class II tests, the Reynolds Number is considered to be significant for accurate test work. For Reynolds Numbers below the limits discussed above (200,000 for centrifugal compressors and 100,000 for axial compressors), the Reynolds Numbers must be maintained approximately the same (within the limits of Table 9.8). For Reynolds Numbers greater than those limits but within the larger ranges given in Table 9.8, PTC 10 provides for Reynolds

Number corrections to μ_p based in part on the work of Carter *et al.* (1960). Those corrections may be expressed as follows:

$$\frac{\mu_{psp}}{\mu_{pt}} = \frac{1-\left(1-\eta_{pt}\right)\left(\mathrm{R}_{et}/\mathrm{R}_{esp}\right)^x}{\eta_{pt}} \tag{9.151}$$

in which $x = 0.1$ for centrifugal compressors and $x = 0.2$ for axial compressors, and η_{pt} is the polytropic efficiency under test conditions. The adjustment to μ_p resulting from the above equation is usually small but can approach 1.5 in the extreme permissible case of $R_{et}/R_{esp} = 0.1$ (see Table 9.8) with $\eta_{pt} = 0.6$. Having thus obtained any necessary adjustments to μ_s, the pressure-ratio adjustment can be made using Equation 9.148b. In a full-blown Class II test, the test speed may require adjustment, as explained in PTC 10, in order to stay within the volume ratio (q_i/q_d) limits given in Table 9.8.

The shaft power required to drive a blower can be related to the isentropic work of compression W_s (Equation 9.145) as follows:

$$P_{sh} = \frac{g\rho W_s}{\eta_{shs}} \tag{9.152}$$

in which η_{shs} is the isentropic efficiency. In a Class I test, the isentropic efficiency remains approximately the same at test and specified conditions. Then from Equations 9.145 and 9.152, we have:

$$P_{sh} \propto q\rho T_1\left[\left(\frac{p_2}{p_1}\right)^{(k-1)/k} - 1\right] \tag{9.153}$$

With kinematic similarity (Equation 9.146a) and constant isentropic work coefficient (Equation 9.146b), the above equation gives:

$$P_{sh} \propto \rho N^3 \tag{9.154}$$

The power conversion for a Class I test is thus obtained as follows:

$$(P_{sh})_{sp} = \left(\frac{N_{sp}}{N_t}\right)^3 \frac{\rho_{sp}}{\rho_t}(P_{sh})_t \tag{9.155}$$

For Class II tests, the isentropic efficiency may be corrected by a relationship resembling Equation 9.151 as given in PTC 10. The power conversion for Class II tests is somewhat more complex, requiring separate consideration of mechanical losses and seal losses, as discussed in PTC 10.

The above relationship may be used to make certain qualitative deductions regarding the effect of operational variables on blower capacity and discharge pressure. Approximate quantitative extrapolations may also be made. A typical manufacturer's curve for a centrifugal blower will have all the pertinent variables discussed above. We are now in a position to assess the effects of various operational variables on blower performance.

For a blower piping system supplying a submerged diffuser, the system curve may be superimposed on the blower curve. The intercept of the system curve at zero flow represents the static pressure, which is proportional to the depth of water submergence. The system curve above the static pressure represents friction losses on the discharge side. The system curve may vary as a result of static pressure variations caused by changes in submergence depth and/or changes in friction losses, such as those that might be caused by valve operation in the discharge line. The blower curve is unaffected by changes in discharge pressure (provided that the inlet pressure is approximately equal to atmospheric pressure). Discharge throttling provides one method of controlling blower discharge.

We can now investigate the effect of changes in inlet conditions for a constant-speed air blower by noting that, for this case, Equation 9.144 reduces to:

$$\Phi\left(Q_1, \frac{p_2}{p_1}, T_1, \mathbf{R}_1\right) = 0 \tag{9.156}$$

For high Reynolds Number, **R** is of secondary importance and will be deleted. The above relationship may then be cast in the following form:

$$Q_i = \Phi\left(\frac{p_d}{p_i}, T_i\right) \tag{9.157}$$

or, for dynamic similarity:

$$Q_i = \text{const} \tag{9.158a}$$

$$T_i = \text{const} \tag{9.158b}$$

$$\frac{p_d}{p_i} = \text{const} \tag{9.158c}$$

Now, if p_i is changed from point 1 to point 2 while Q_i and T_i are held constant, then dynamic similarity will give the new operating point from Equations 9.158(a), 9.158(b), and 9.158(c), expressed as follows:

$$p_{d2} = \left(\frac{p_d}{p_i}\right)_1 p_{i2} \tag{9.159}$$

A decrease (increase) in inlet pressure will decrease (increase) the discharge pressure. The blower curve corresponding to a change in inlet pressure alone can be established.

Blower inlet pressure will change with changes in barometric pressure or inlet throttling. The effect of either may be visualized by superimposing the system curve. Inlet pressure changes will cause the operating point to move to the intersection of the system curve and the new blower curve established in accordance with the dynamic similarity requirements given above. Inlet valve throttling is thus seen to provide another means of controlling blower discharge Q_i. An advantage of inlet throttling is that $Q_i > Q_d$ so that there is greater turndown toward surge with inlet throttling.

Simultaneous pressure and temperature adjustments may be made by means of dynamic similarity requirements, derived from Equations 9.148 for constant speed and high Reynolds Number. In terms of a change from point 1 to point 2, these relationships may be expressed by:

$$Q_{i2} = Q_{i1} \tag{9.160a}$$

$$T_{i2}\left[\left(\frac{p_d}{p_i}\right)_2^{(k-1)/k} - 1\right] = T_{i1}\left[\left(\frac{p_d}{p_i}\right)_1^{(k-1)/k} - 1\right] \tag{9.160b}$$

These relationships are useful in adjusting manufacturers' performance curves to expected inlet pressure and temperature conditions.

For constant temperature and inlet pressure (and "high" Reynolds Number), the following equations provide the means for determining the effect on discharge pressure of changes in blower speed:

$$Q_{i2} = \frac{N_2}{N_1} Q_{i1} \tag{9.161a}$$

$$\left(\frac{p_{d2}}{p_i}\right)^{(k-1)/k} - 1 = \left(\frac{N_2}{N_1}\right)^2 \left[\left(\frac{p_{d1}}{p_i}\right)^{(k-1)/k} - 1\right] \tag{9.161b}$$

The above relationships indicate that a reduction in speed alone reduces the discharge pressure. Through appropriate successional use of the above relationships, the effects of changes in inlet pressures, inlet temperature, and speed can be estimated.

For the 23 mgd Cranston, Rhode Island Wastewater Treatment Plant, the author specified three multistage centrifugal blowers for aeration of activated sludge tanks. Each blower delivers air at a rated capacity of 5400 scfm measured at 14.7 psi absolute, 68°F, and 36% relative humidity when operating at the following conditions: inlet temperature 97°F, inlet pressure at blower flange 14.3 psia, discharge pressure at blower flange 22.2 psia, relative humidity 85%, normal barometer 14.7 psia. Each blower is driven by a 300 hp, 3600 rpm squirrel-cage induction motor, having sufficient capacity to allow full-load blower torque over the range of blower operating conditions. One blower was required to be witness tested according to the ASME Power Test Code for Compressors and Exhausters (PTC 10) discussed above. Maximum sound levels and noise tests were also specified. Inlet butterfly valves perform the functions of unloading the blower in conjunction with the discharge blow-off valves during startup and then controlling the flow through the blower by inlet throttling. At design conditions, the inlet valve is specified to control volume from 100 to 50% flow, with the blowers remaining stable throughout this range. Inlet throttling is important for this type of blower to prevent surge conditions. Inlet and discharge silencers and noise tests were specified. Accessories include ammeters calibrated in scfm, dry washable inlet filters, inlet and discharge air pressure manometers, discharge thermometers, high bearing temperature and vibration switches, and inlet and discharge silencers required to meet specified sound levels.

For the Rochester, New Hampshire, 3.9 mgd Advanced WWTP, the author specified three electric motor-driven multistage centrifugal blowers for supplying air to activated sludge aeration basins. The blowers have a rated capacity of 1930 scfm with an inlet air temperature of 100°F and an inlet pressure of 13.95 psia. Inlet butterfly valves perform the functions of unloading the blower in conjunction with the discharge blow-off valve during startup and then controlling the flow by inlet throttling from 100% inlet flow to approximately 66% inlet flow. The blowers are driven by 125 hp 3600 rpm open drip-proof squirrel-cage induction motors. Specified discharge pressure at the blower flange is 23.05 psia. Accessories include surge control devices, vibration switches, inlet and discharge air pressure manometers, discharge temperature gauges, and inlet and discharge silencers to meet specified sound levels.

For the Tallman Island, New York City, 80 mgd WWTP, the author specified four multistage centrifugal blowers for supplying air to activated-sludge aeration basins and for channel agitation. Two blowers have maximum rated capacities of 30,000 scfm, one of 20,000 scfm maximum rated capacity, and one of 10,000 scfm maximum rated capacity. One 30,000 scfm blower is driven by a gas turbine with the rest of the blowers driven by electric motors. Specifications include inlet air temperature of 95°F and inlet pressure at the compressor flange of 14.21 psia. Inlet guide vanes perform the functions of unloading the blowers in conjunction with the discharge blow-off valve during startup and then controlling the flow by inlet throttling from 100% inlet flow to approximately 50% inlet flow. The electric motors have rated horsepowers of 500, 1000, and 1500 and are 3600 or 1800 rpm drip-proof squirrel-cage induction motors. Specified discharge pressure at the blower flanges is 24.36 psia. Various accessories and maximum sound levels according to the type of drive-compressor combination are included.

9.5.2 Rotary Positive-displacement Blowers

Rotary positive-displacement blowers are used in wastewater treatment plants to supply air for mixing and channel agitation. An example is the Cranston, Rhode Island, 23 mgd Wastewater Treatment Plant, for which the author specified such blowers for sludge blend tank mixing. These rotary blowers are of the two impeller, straight-lobed, positive-displacement type. Each blower delivers air over the range of 400–905 scfm at a pressure rise of 4.5 psi with an inlet air temperature of 68°F and an inlet pressure of 14.7 psia. Each blower is driven by a three-speed motor with maximum rotative speed of 3500 rpm and with capacities of 905 and 600 scfm at the top two speeds. Accessories include dry-type inlet filters, inlet vacuum and discharge pressure gauges, inlet and discharge temperature gauges, discharge relief valves set for 5 psig, and inlet and discharge silencers of the chamber type.

Another example is for the Rochester, New Hampshire, 3.9 mgd Advanced WWTP, for which the author specified four such blowers for supplying air to equalization basins and waste-activated sludge holding tanks (the latter for aeration and mixing). These rotary blowers are of the two impeller, straight-lobed, positive-displacement type. Two of the blowers have a rated capacity of 820 scfm with turndown to 410 scfm, and two have a rated capacity of 250 scfm, both types with an inlet air temperature of 100°F and an inlet pressure of 14.3 psia. The larger blowers are driven by two-speed motors, while the smaller blowers are driven by single-speed motors, all of the open drip-proof squirrel-cage induction types with maximum rotative speeds of 1800 rpm. Specified discharge pressures at the blower flange are 21.2 and 22.7 psia for the larger and smaller blowers, respectively. Accessories include dry-type inlet filters for outdoor installation, inlet compound gauges and discharge pressure and temperature gauges, weighted discharge pressure relief valves set for 1 psi above the maximum specified discharge pressure of the blower, and inlet and discharge silencers to meet specified sound levels.

10

Concepts of Spatially Varied Flow

10.1 Introduction

Most of the material in the present chapter is taken from Graber (2004c, 2023a). By presenting relationships for spatially varied flow with combined increasing and decreasing discharge and relating those relationships to prior work, a clearer exposition of spatially varied flow concepts is provided. The applicability and usefulness of energy (First Law of Thermodynamics) and momentum (Newton's Second Law) principles to spatially varied flows are addressed. The history of the debate initiated in the nineteenth century over the role of the First Law of Thermodynamics in general and hydrodynamics specifically is discussed. That history is related to current issues regarding spatially varied flows, and hopefully resolved in the context of such flows and more generally. Also presented is a discussion of certain energy concepts regarding spillway flow. Following the presentation of concepts in the present chapter, Chapters 11 and 12 address, respectively, additional details of distribution conduits with spatially decreasing flow and collection conduits with spatially increasing flow.

10.2 Background

More than three centuries after Newton's Second Law was presented and 100 years after the First Law of Thermodynamics was stated in its modern form, surprising disagreement continues regarding their applicability in fluid mechanics. Spatially varied free-surface flow is a useful and important context within which to discuss this controversy. The history of the controversy is briefly discussed below, with particular emphasis on free-surface flows. Analysis follows, which should resolve this controversy in the context of such flows and more generally. Comprehensive reviews of spatially varied flows in environmental engineering are found in the publications of Camp and Graber (1968, 1969, 1970, 1972) and Graber (1981, 2004b). That work is built upon and generalized through consideration of combined increasing and decreasing discharge. The confusion sometimes encountered in practice and in the literature concerning spatially varied flow problems is addressed, including the application of energy and momentum principles to such flows with increasing and decreasing discharge. Additional useful concepts following from the relationships given below include correction of a published difference formulation for open-channel spatially decreasing flow, and slope invariance for spatially decreasing full-pipe flow (Graber 2004d).

10.2.1 Early History

Isaac Newton's *Principia*, the first edition of which was published in 1687, presented Newton's three Laws of Motion. Preliminary discussion of those laws may be found in Section 1.5. Gottfried Wilhelm von Leibniz (the same Leibniz who contended with Newton over supremacy for development of the calculus) expressed disagreement with Newton's momentum principle (Rouse and Ince 1957, pp. 87–88). He essentially showed that energy principles (at least those related to kinetic energy and potential energy) applied and regarded that as contrary to rather than (as is now known) consistent with and supplementary to the momentum principle. Thus, a prelude to the First Law of Thermodynamics was born in controversy. The complete First Law of Thermodynamics was largely the result of the nineteenth-century work of Carnot, Mayer, Joule, Helmholtz, Thompson (Kelvin), and Clausius, and the work of Poincaré and Carothéodory extending into the first decade of the twentieth century (Hatsopoulus and Keenan 1965; Cropper 2001).

However, appreciation of the independence and significance of the First Law of Thermodynamics was not universal in the scientific community. The notion that the First Law of Thermodynamics derives from Newton's Second Law persisted. The original source of that notion appears to be a book published in 1877 by Peter Guthrie Tait entitled *Sketch of Thermodynamics*. Tait offered as proof a highly "creative" translation from the Latin of a portion of Isaac Newton's classical *Principia*. Tait was known as a "polemicist par excellence" (Cropper 2001: Rudolf Clausius, Chapter 8; Willard Gibbs, Chapter 9; James Clerk Maxwell, Chapter 12), who was on the wrong side of a number of the big issues of his day. He disagreed with Rudolph Clausius on the now-accepted concept of entropy and contradicted both the First and Second Laws of Thermodynamics. Tait also disagreed with Willard Gibbs' concept of vectors, which is now standard mathematics. Cohen, in his "Guide to Newton's Principia" (Newton 1999), provides a side-by-side comparison of an objective translation and Tait's "creative" translation of a portion of Newton's *Principia* and, in that context, offers discussion that clearly debunks Tait's claim that the First Law of Thermodynamics derives from Newton's Second Law.

10.2.2 Initial Aspects of Misunderstanding in Free-Surface Flow

Despite thermodynamists' understanding of the relationship between Newton's Second Law of Motion and the First Law of Thermodynamics since the nineteenth century, misunderstanding has cropped up from time to time in the hydrodynamics literature, specifically relative to free-surface or open-channel flow. A 1945 paper by Eisenlohr (1945) on coefficients for velocity distribution in open-channel flow generated interesting discussion. That discussion was precipitated in part by Eisenlohr's statements that "[t]he basic equation [for flow through a single stream tube] by the momentum theorem or energy principle is that of Newton's second law of motion, $F = ma$ ...." and that "[m]any ideas frequently thought of in terms of the energy relation are really momentum concepts." Kalinske (1945), in discussing some of Eisenlohr's ideas, stated, "since basically the energy and momentum laws stem from Newton's law, these can be used to derive Bernoulli's equation." Bernoulli's equation for frictionless, incompressible flow can, in fact, be derived from either energy or momentum principles, as clearly elucidated by Shames (1962) and others. However, the notion that energy laws derive from Newton's law is restricted to specific formulations of energy principles and is, of course, not true in the general case. Taylor (1945) referred specifically to the first portion of the statement by Eisenlohr quoted above. He properly took issue with it within the context of Eisenlohr's paper. Rouse and McNown (1945) noted that "[a]ttention is again focused by this paper [Eisenlohr's] on the century-old controversy concerning the correct interpretation of the energy and momentum principles in hydraulics." Rouse and McNown stated that "it is scarcely logical to attempt to show their similarity [general momentum and energy relationships], but rather to emphasize the distinctions which make them such powerful complementary tools."

10.3 Spatially Varied Free-Surface Flows

Despite the wise counsel of Rouse and McNown, continuing confusion manifested itself in connection with spatially varied free-surface flows. Hinds (1926) correctly applied momentum principles in reporting his early work on spatially increasing flow in side-channel spillways and noted the limitations of the energy principle for that purpose. However, misunderstanding was again evident when Yen and Wenzel (1970), in their paper on spatially varied free-surface flow, recognized certain inherent difference between the energy and momentum approaches [citing Rouse and McNown (1945) in that regard] but stated erroneously that "both principles are derived from Newton's second law." A similar statement was made in a subsequent paper by Yen (1973) on open-channel flow. In that paper, Yen appears to derive energy equations for spatially varied open-channel flow from the Navier-Stokes equations, which incorporate Newton's Second Law but not the First Law of Thermodynamics (see, e.g., Schlichting 1960, Chapter XIV). However, the First Law of Thermodynamics is an unacknowledged part of the derivation, as is demonstrated below in the context of a more recent publication. In commenting on Yen (1973), Contractor (1974) indicated continued misunderstanding: "The energy and momentum equations are different from one another even though they are derived from Newton's law." In response, Yen (1975) concurred and added that "the equations of momentum and energy are two different equations derived from the same Newton's second law."

The view persists that a "mechanical-energy equation," especially applicable to spatially varied flows, can be derived solely from Newton's Second Law. Jain (2001) suggests that open-channel flow problems do not require the First Law of Thermodynamics and that "there is a class of problems (e.g., lateral outflow problems...) to which the momentum equations are not readily applicable; such problems are solved using the mechanical-energy equation, which is derived from the momentum equations." Furthermore, Jain (2001) erroneously states that the momentum equation is unsuitable for the problem of lateral outflow, i.e., spatially decreasing flow (he uses the energy equation, as does Chow 1959, Chapter 12).

Thus, misunderstandings discussed above regarding the role of the energy and momentum principles for spatially varied free-surface flows (of which spatially constant flow is a special case) take several forms. The first relates to confusion about the independence of Newton's Second Law and the First Law of Thermodynamics. A second concern the applicability of energy and momentum principles to spatially increasing vs. spatially decreasing flows. A third form of misunderstanding, which is particularly important from a practical standpoint, concerns the proper treatment of inflow and outflow and the proper equations to use for spatially increasing and spatially decreasing flows. The derivations presented next will assist in clarifying these issues.

10.3.1 Basic Relationships

The basic relationships will be derived by considering a spatially varied flow with both spatial inflow and outflow and explicitly considering each through separate variables. To make this more tangible, consider the application depicted in Figure 10.1, which is a sheet flow with lateral inflow due to rainfall and lateral outflow due to infiltration. The concepts are applicable to a wider range of applications that more commonly have only inflow or only outflow (to which the more general case can be specialized), including flow through side weirs, multi-port distribution conduits, etc.

10.3.2 Continuity

The continuity relationship is given by:

$$V\frac{dA}{dx} + A\frac{dV}{dx} = q_i - q_o \tag{10.1}$$

in which V is the cross-sectional average of main flow velocity, A is the cross-sectional area, x is the distance along the channel in the direction of flow, and q_i and q_o are lateral inflow and outflow, respectively [in considering a similar problem, Yen *et al.* (1972) explicitly considered only the lateral inflow].

10.3.3 Momentum Principle

The momentum principle is applied by equating the total change of momentum flux through a control volume (open system) to the sum of forces acting on the surface of the control volume (control surface). Consider first the control volume defined by two sections parallel to the y-axis, and the lateral boundaries of the main flow, depicted as CV1 in Figure 10.1. The momentum flux due to lateral inflow is given by $\rho q_i \Delta x U_i \cos\phi_i$, in which ρ is fluid density, Δx is the distance between the two sections parallel to the y-axis, U_i is velocity of lateral inflow, and ϕ_i is the angle between the velocity vector of lateral inflow and direction of main flow.

CV1 in Figure 10.1 intersects the pervious bed at a depth where the lateral outflow velocity vector has angle ϕ_o relative to the x direction. A component force T_x is required (exerted by the pervious media on the lateral outflow) to turn the outflow to the direction ϕ_o and accelerate it to velocity U_o. Now consider CV2 in Figure 10.1, the curved portion of which is defined by the imaginary surface separating the lateral discharge $q_o \Delta x$ from the main flow continuing across Section 2. The momentum principle applied to this control volume demonstrates that $-T_x = \rho q_o \Delta x (U_o \cos\phi_o - \beta_o V) - F_D$, in which V is the velocity of the main flow, β_o is the momentum-flux correction factor associated with the lateral outflow, and F_D is the drag force exerted by the main flow on the lateral outflow. Assuming $F_D = 0$, one obtains $-T_x = \rho q_o \Delta x (U_o \cos\phi_o - \beta_o V) - F_D$, with the result that $M_L + T_x = -\rho q_o \Delta x \beta_o V$ for the lateral outflow. By a judicious choice of control volume, such as CV3 in Figure 10.1, the force T_x may be excluded, and the momentum flux of the lateral outflow is obtained more directly. By either approach, the complete expression for the momentum flux due to the lateral flow, including both inflow and outflow, becomes:

$$M_L = \rho q_i \Delta x U_i \cos\phi_i - \rho q_o \Delta x \beta_o V \tag{10.2}$$

This momentum term is sometimes given as $M_L = \rho q \Delta x U \cos\phi$ in which $U \cos\phi$ is said to be the x–direction component of either inflow or outflow and q is said to be positive for inflow and negative for outflow (e.g., Yen and Wenzel 1970), Chaudhry (1993, p. 281). However, that is only generally valid for lateral inflow, as can be seen from the above equation and its development. In their analysis of spatially decreasing flows, Camp and Graber (1968, 1969) reasoned, as above, that in the most general case, the lateral outflow could be represented by $-\rho q_o \Delta x U_o \cos\phi_o$ only if a component force T_x acting on

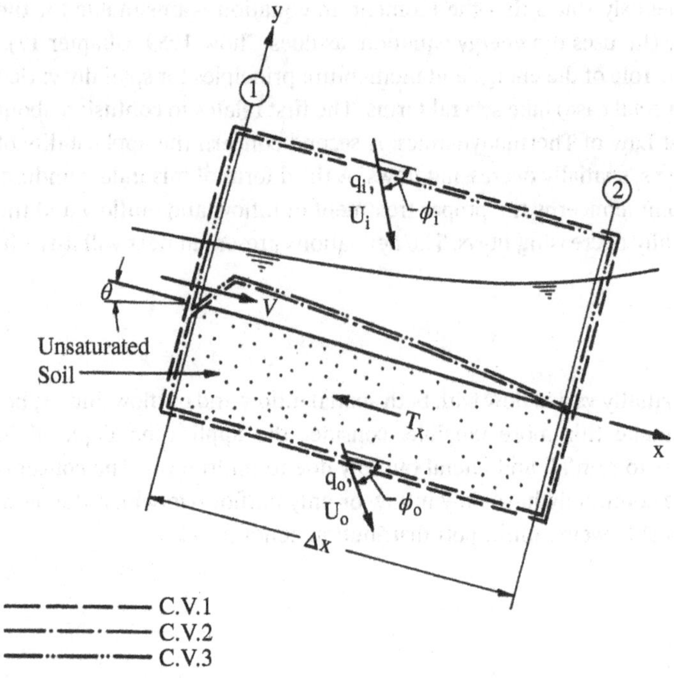

Figure 10.1 Sheet Flow with Lateral Inflow Due to Rainfall and Lateral Outflow Due to Infiltration (Graber 2004c/ American Society of Mechanical Engineers)

the control volume is recognized (and included in the momentum equation) as required to turn the outflow to the direction ϕ_o and accelerate it to velocity U_o. That component force can be circumvented, as described above.

Denote y as the depth of flow, R as the hydraulic radius, D as hydraulic depth, g as the acceleration of gravity, γ as fluid specific weight, $S_o = \tan\theta$ as the channel bottom slope, τ_o as the boundary shear stress in the x direction, β as the momentum-flux correction factor for the main flow, and K as pressure correction factor (accounting for deviations from hydrostatic pressure). Then, based on the momentum principle and employing Equations 10.1 and 10.2 as appropriate, one can follow through the derivation to obtain the differential equation for spatially varied flow analogous to that given by Yen and Wenzel (1970) [their Equation (12)]:

$$\frac{dy}{dx} = \frac{S_o - \dfrac{\tau_o}{\gamma R} + \dfrac{q_i}{gA}(U_i\cos\phi_i - 2\beta V) - \dfrac{q_o}{gA}(\beta_o V - 2\beta V) - \dfrac{V^2}{g}\dfrac{d\beta}{dx} - \dfrac{yd(K\cos\theta)}{dx}}{K\cos\theta\left(1 + \dfrac{y}{D}\right) - \dfrac{\beta V^2}{gD}} \tag{10.3}$$

If θ is constant, one can substitute $d(K\cos\theta)/dx = \cos\theta(dK/dx) = \cos\theta(dK/dy)(dy/dx)$ in Equation 10.3 and rearrange to give:

$$\frac{dy}{dx} = \frac{S_o - \dfrac{\tau_o}{\gamma R} + \dfrac{q_i}{gA}(U_i\cos\phi_i - 2\beta V) - \dfrac{q_o}{gA}(\beta_o V - 2\beta V) - \dfrac{V^2}{g}\dfrac{d\beta}{dx}}{\cos\theta\left[K\left(1 + \dfrac{y}{D}\right) + y\dfrac{dK}{dy}\right] - \dfrac{\beta V^2}{gD}} \tag{10.4}$$

which is also analogous to Yen and Wenzel (1970) [their Equation (13)]. By further assuming hydrostatic pressure distribution, $U_i\cos\phi_i = 0$, $\beta = \beta_o = 1$, and $\cos\theta = 1$, a modified form of an equation given by Yen and Wenzel (1970) [their Equation (14)] is obtained:

$$\frac{dy}{dx} = \frac{S_o - \dfrac{\tau_o}{\gamma R} - \dfrac{2q_i V}{gA} + \dfrac{q_o V}{gA}}{1 - \dfrac{V^2}{gD}} \tag{10.5}$$

By setting $q_o = 0$ in Equation 10.5, an equation identical to that given by Yen and Wenzel (1970) [their Equation (14)] is obtained, although Yen and Wenzel erroneously regarded their equation as being applicable to spatially decreasing flow (by also taking q_i as q with "q being a positive value if it is lateral inflow and negative for lateral outflow"). The resulting equation is also the same as Equation 12-4 in Chow (Chapter 12) and the equation derived (for $\partial A/\partial x = 0$) by Smith (1972), which were developed specifically for spatially increasing flow. By setting $q_i = 0$ in Equation 10.5, Equation 12-8 in Chow (1959) is obtained, which is specifically applicable to spatially decreasing flow. Jain (2001) states that "the momentum equation [is] unsuitable for lateral outflow" [he uses the energy equation, as does Chow (1959, Chapter 12)]. The author has demonstrated above the successful use of the momentum equation for lateral outflow. Use and implications of the energy equation are discussed below.

In the case of outflow over side weirs or through thin-plate square-edged orifices, one may expect T_x to be negligible which, from the above, implies $U_o \cos\phi_o = V$. From this relationship, Subramanya and Awasthy (1972) derived an expression for the side-weir coefficient, which they substantiated experimentally. The relationship $U_o \cos\phi_o = V$, or conversely $T_x = 0$, has also been shown by Subramanya and Awasthy (1970) to provide a useful approximation to outflow from multi-port diffusers. Their work establishes a theoretical basis for the use of the piezometric head rather than the energy head in the orifice equation for multi-port diffusers, in support of simplified design procedures proposed by Camp and Graber (1968, 1969, 1970, 1972) based on momentum principles.

10.3.4 Energy

Consistent with the assumption $F_D = 0$ made in treating the momentum flux due to lateral outflow is the assumption that the outflow leaves the main flow with the same energy per unit mass as the main flow. That is, there is no exchange of energy between the outflow and the continuing main flow between Sections 1 and 2 of Figure 10.1. Chow (1959, Chapter 12) has expressed this by stating that "the diverted water does not affect the energy head" (per unit mass of the continuing main flow). Under this assumption and the assumption that the convective kinetic energy flux correction factor for the outflow equals that of the main flow, the energy flux of the lateral flow is given by the following modification of the equation by Yen and Wenzel (1970) [analogous to their Equation (16)]:

$$E_L = \gamma q_i \Delta x\left(y_i \cos\theta + z_i + \frac{U_i^2}{2g}\right) - \gamma q_o \Delta x\left(\bar{y}\cos\theta + \bar{z} + \alpha_o \frac{\overline{V}^2}{2g}\right) \tag{10.6}$$

in which z is the channel bottom elevation above a horizontal datum; $\bar{y}$, $\bar{z}$, and $\overline{V}$ are the average values of y, z, and V between Sections 1 and 2; and α is the kinetic energy correction factor (with the subscript "o" denoting the outflow). The continuity relationship becomes:

$$Q_2 = Q_1 + q_i \Delta x - q_o \Delta x \tag{10.7}$$

in which Q is volumetric rate of main flow.

Denote S_e as the energy gradient, H_{pL_1} as the piezometric head of lateral inflow relative to the channel bottom, and η as the potential energy flux correction factor. Then, based on the energy principle (First Law of Thermodynamics) and employing Equations 10.1, 10.6, and 10.7, one can follow through the derivation to obtain the differential energy equation for spatially varied flow analogous to that given by Yen and Wenzel (1970) [their Equation (33)], but with explicitly separate terms for q_i and q_o:

$$\frac{dy}{dx} = \frac{S_o - S_e + \dfrac{q_i}{VA}\left(H_{pL_1} + \dfrac{U_i^2}{2g} - \eta y\cos\theta - \dfrac{3\alpha V^2}{2g}\right) + \dfrac{\alpha V q_o}{gA} - \dfrac{V^2}{g}\dfrac{d\alpha}{dx} - \dfrac{y\,d(\eta\cos\theta)}{dx}}{\eta\cos\theta - \dfrac{\alpha V^2}{gD}} \tag{10.8}$$

For the special case of $H_{pL_1} = y\cos\theta, \eta = 1$, $\cos\theta \cong 1$, and $d\alpha/dx$ negligible, Equation 10.8 reduces to:

$$\frac{dy}{dx} = \frac{S_o - S_e + \dfrac{q_i}{VA}\left(\dfrac{U_i^2}{2g} - \dfrac{3\alpha V^2}{2g}\right) + \dfrac{\alpha V q_o}{gA}}{\eta\cos\theta - \dfrac{\alpha V^2}{gD}} \tag{10.9}$$

For $q_i = 0$, Equation 10.9 becomes:

$$\frac{dy}{dx} = \frac{S_o - S_e + \dfrac{\alpha V q_o}{gA}}{1 - \dfrac{\alpha V^2}{gD}} \tag{10.10}$$

which is Equation 12-8 of Chow (1959) (noting that Chow's q_* equals $-q_o$).

10.4 Relation Between Momentum and Energy Formulations

The salient differences between the momentum and energy formulations can be seen by first denoting by S_f the energy loss per unit channel length due to wall friction (commonly referred to as "friction slope") given by $\tau_o/(\gamma R)$ in Equations 10.3 to 10.5, and combining terms from Equations 10.5 and 10.9 with the added assumptions $\alpha = 1$ and $U_i = 0$ to obtain:

$$S_e = S_f + \frac{q_i V}{2gA} \tag{10.11}$$

For $q_i = 0$, Equation 10.11 gives $S_e = S_f$, which, with $\alpha = 1$, has Equations 10.5 and 10.9 reducing to identical equations. This shows that the momentum and energy principles are equally suitable for spatially decreasing flows if the principles are properly applied.

The total head gradient S_H can now be defined in terms of the total head H (Yen and Wenzel 1970, Yen *et al.* 1972):

$$H = \alpha\frac{V^2}{2g} + y\cos\theta + z \tag{10.12a}$$

$$S_H = -\frac{dH}{dx} \tag{10.12b}$$

Using Equations 10.1, 10.9, and 10.12 together with the added assumptions $\alpha = 1$ and $U_i = 0$, the following relationship is obtained:

$$S_H = S_e + \frac{q_i V}{2gA} \tag{10.13}$$

The difference between S_f and S_H is related to the fact that non-conservative energy exchange between the inflow and main flow, accompanying their exchange of momentum, causes an energy loss in addition to that due to wall friction alone. The important point here, which is occasionally overlooked in practice, is that, even if the boundary friction is negligible (i.e., $S_f = 0$), an appreciable energy loss may still be incurred in spatially increasing flow. The difference between S_H and S_e (the latter representing energy dissipation) stems from the fact that the flow energy is associated with an increasing amount of flow. Hence, the energy per unit mass of fluid, to which S_H refers, must decrease at a greater rate than energy is dissipated. For spatially decreasing flow, $S_f = S_H = S_e$ is a useful approximation, often sufficiently accurate for engineering purposes.

Yen *et al.* (1972) defined terms analogous to the Weisbach resistance coefficient, such that $f_e = 8gRS_e/V^2$, $f_f = 8gRS_f/V^2$, and $f_H = 8gRS_H/V^2$, and experimentally determined values of these coefficients under various conditions. The differences between f_f, f_H, and f_e in spatially varied two-dimensional steady Stokes flow (Yen *et al.* 1972) are essential as expected based on the interpretations provided here. An appreciation of the theoretical significance of each of these terms should lead one to expect that f_f would most nearly represent the Weisbach f. The deviation in magnitude of f_f from the corresponding Weisbach f reflects the effect of the lateral flow on the velocity distribution of the main flow and the resulting distortion of boundary shear stresses. This is the effect that Camp (1940) measured in his studies of lateral spillway channels and which led to the recommendation by Fair and Geyer (1954, p. 691) that the roughness factor (taken to mean the Weisbach f) should be increased up to twofold in determining a value of f to be used in washwater trough design.

10.4.1 Newton's Second Law and the First Law of Thermodynamics

The above provides a perspective that enables further discussion of the relation between first principles as they apply to free-surface flows. For spatially decreasing flows, Equations 10.5 and 10.9 (for $\alpha = 1$) become identical for $q_i = 0$ [as discussed above, $S_e = \tau_o/(\gamma R)$ in this case]. For spatially increasing flow, Equations 10.5 and 10.9 (for $\alpha = 1$) become identical for $q_o = 0$ and $U_i = 0$ only if, as discussed by Babb and Ross (1968), S_o is taken as $S_f + Qq_i/(2gA^2)$ (as in Equation 10.11), in recognition of the fact that the energy approach [i.e., the First Law of Thermodynamics (Hatsopoulus and Keenan 1965)] is incapable of independently evaluating the energy exchange between the inflow and the main flow (just as the energy principle alone is incapable of relating the conditions upstream and downstream of a hydraulic jump, sudden pipe expansion, or normal shock wave, but does usefully provide the energy loss once the problem has been solved by the momentum equation). This limitation of the energy principle was discussed by Hinds (1926) in connection with side-channel spillways mentioned above, but has remained a source of confusion. Contrary to the statements and views cited above by Kalinske (1945), Yen and Wenzel (1970), Yen (1973), Contractor (1974), and Yen (1975) that momentum and energy approaches are both derived from Newton's Second Law, the momentum and energy principles are derived from Newton's Second Law and the First Law of Thermodynamics (sometimes in conjunction with Newton's Second Law), respectively, totally independent fundamental laws of nature each having its own inherent value in fluid flow problems.

Returning now to Jain (2001) in the present context, he derives one-dimensional differential equations of motion using three different approaches, which he calls the (1) elementary approach, (2) simplified approach, and (3) rigorous approach. The "elementary" approach consists of applying continuity and momentum principles to a control volume or open system bounded by the channel-wetted perimeter, the free surface, and sections perpendicular to the direction of flow an infinitesimal distance apart. The "simplified" approach integrates over infinitesimal elements (closed systems), each of which has width equal to the channel width at a particular elevation and of infinitesimal vertical and longitudinal dimensions. The "rigorous" approach integrates over infinitesimal cubical elements (closed systems). Jain claims that the "mechanical-energy equation cannot be derived with the elementary approach," which is incorrect as discussed below.

In the "simplified" and "rigorous" approaches, Jain derives the differential "mechanical-energy" equations by using differential momentum (Newton's Second Law) equations, introducing kinetic and potential energies per unit volume in terms of variables found in the differential momentum equations, and employing continuity relationships. With the "simplified" approach, a certain group of terms is said to represent energy loss without substantiation. The problem with this approach can be made most clear by focusing on the corresponding portion of the "rigorous" approach. The equations resulting from the "rigorous" approach contain terms that are said to be "customarily called the dissipation function Φ." However, the dissipation function is derived using the First Law of Thermodynamics.

Schlichting (1960, Chapter XIV) derives the general differential energy equation for fluid flow by first applying the First Law of Thermodynamics to a moving fluid element. He then utilizes the momentum equation (Newton's Second Law) to remove certain of the velocity terms related to the motion of the fluid element. Schlichting (in specifically considering the boundary layer) neglects body forces (in this case, the gravitational force field), but they are easily included and canceled from the final relationship (e.g., Lamb 1945; Eagleson and Dean 1966). Taking advantage of the fact that the energy of a system moving through a force field (e.g., gravitational) is relative to the observer (Hatsopoulus and Keenan 1965, Rohsenow and Choi 1961), Graber (1994) simplified the derivation of the differential energy equation by expressing the energy in terms of an observer moving with the fluid element. When the derivation is carried out in that manner, it is not necessary to bring Newton's Second Law into the derivation (i.e., only the First Law of Thermodynamics is utilized). The dissipation function thus derived represents the energy expended per unit volume, caused by fluid stresses in distorting elements of fluid, that is converted to internal energy, heat, and sound waves, which are exchanged with the environment. By then expressing the stress terms using Stoke's stress-strain relationships, the resulting relationship shows that energy is ultimately dissipated at the microscopic scale through the action of viscosity (e.g., Schlichting 1960, Chapter XIV; Rohsenow and Choi 1961; Graber 1994).

The interpretation of Φ as energy dissipation cannot be properly made without recourse to the First Law of Thermodynamics. Thus, Jain's "mechanical-energy" equations have been derived using Newton's Second Law and the First Law of Thermodynamics. Without the First Law of Thermodynamics, the physical significance of the terms described as the dissipation function would not be known. Furthermore, and not surprisingly, when all is said and done, the equations reduce to ones identical to the energy equations presented above. They are different from the momentum equations in the same manner as described above.

Jain used the "simplified" and "rigorous" approaches to derive the spatially varied flow equations because the "mechanical-energy equation cannot be derived with the elementary approach." However, the spatially varied flow equations can indeed be derived from a one-dimensional "elementary" approach if the First Law of Thermodynamics is used. In fact, that is precisely what Yen and Wenzel (1970) have done (as utilized and built upon above). They derived their energy equation by considering essentially the same control volume as CV1 in Figure 10.1 except with Δx changed to dx and the bottom boundary taken as the channel bottom. Yen and Wenzel derived the energy equation for that control volume by setting energy dissipation within the control volume equal to the difference in energy flux through Section 2 and the sum of the energy flux through Section 1 and that of the lateral inflow, with the energy fluxes consisting of potential and kinetic energy. In doing so, they have, without stating it, applied the First Law of Thermodynamics (Shames 1962). The author has added explicit and separate consideration of inflow and outflow terms in the above. As would be expected when the same first principles have been utilized, Jain's equation for $\partial/\partial t = 0$ (steady flow) and Equation 10.8 become identical for $H_{pL_1} = y\cos\theta$ and $\eta = 1$ in Equation 10.8, Jain's varying flow is taken as $q_i - q_o$, and the α associated with q_o in Equation 10.8 is taken as unity.

Jain's (2001) "rigorous" derivation of the general differential energy equation for fluid flow starts with the momentum equations (Newton's Second Law) and claims to not utilize the First Law of Thermodynamics. However, it should be clear from the above that the First Law of Thermodynamics is in fact utilized. The same can be said for Jain's (2001) "simplified" derivation, which also leads to an energy equation for spatially varied flow. If, in fact, Jain had used only Newton's Second Law (and Conservation of Mass) in deriving his energy equation, the derivation should not yield a spatially varied-flow energy equation differing from the spatially varied flow momentum equation, which he derived using the same first principles. It does differ; however, the difference being essentially as represented by Equation 10.13 and the discussion immediately following that equation.

10.5 Difference Formulation

Equation 10.5 may be cast into difference form by using standard relationships between hydraulic elements and changing differentials to difference terms. After a somewhat substantial algebraic endeavor, the following relationship is obtained:

$$\Delta y' = -\Delta y + S_o\Delta x = \frac{Q_1(V_1+V_2)\Delta V}{g(Q_1+Q_2)}\left[1+\frac{V_2\Delta Q_i}{Q_1\Delta V}-\frac{1}{2}\frac{\Delta Q_o}{Q_1}\right] + S_f\Delta x \tag{10.14}$$

in which subscripts "1" and "2" denote, respectively, upstream Section 1 and downstream Section 2. The Δ-delta terms represent the value of Section 2 minus the value at Section 1, such as $\Delta V = V_2 - V_1$ (this example is of particular significance as explained below); $\Delta y'$ is drop of water surface elevation between Section 1 and 2 [i.e., $\Delta y' = -(y_2' - y_1')$]; $\Delta Q_i = q_i\Delta x$; and $\Delta Q_o = q_o\Delta x$.

For $\Delta Q_o = 0$, Equation 10.14 reduces to:

$$\Delta y' = \frac{Q_1(V_1+V_2)}{g(Q_1+Q_2)}\left(\Delta V + \frac{V_2\Delta Q_i}{Q_1}\right) + S_f\Delta x \tag{10.15}$$

which is identical to Chow's (1959) well-known Equation 12–38 with $\alpha = 1$ and Chow's $\Delta Q = Q_i$. This form or slight variations have been extensively used by the author and others (see, e.g., Graber 2004b). Except for the friction term, Equation 10.15 is identical to the equation given by Hinds (1926).

For $\Delta Q_i = 0$, Equation 10.14 reduces to:

$$\Delta y' = \frac{Q_1(V_1+V_2)\Delta V}{g(Q_1+Q_2)}\left(1-\frac{1}{2}\frac{\Delta Q_o}{Q_1}\right) + S_f\Delta x \tag{10.16}$$

which is identical to Chow's (1959) Equation 12–40 with $\alpha = 1$ and Chow's $\Delta Q = Q_o$, *with one important exception.* Chow defines $\Delta V = V_1 - V_2$ (all of Chow's other terms are as defined above), whereas, as noted above, the author has $\Delta V = V_2 - V_1 = -(V_1 - V_2)$. Chow's equation is erroneous. The error may give inaccurate results without the error being easily noticed in some cases. However, in the case of constant A the error can be demonstrated both by physical reasoning and analytically, as is done next.

Equation 10.16 can be applied to full-flowing prismatic conduits with spatially increasing flow if y' is interpreted as the piezometric or hydraulic-grade-line relation to any arbitrary horizontal datum. Viewed in that light, one consequence of Chow's definition is that, for constant A, one can only have positive values of $\Delta y'$. This is so because: for constant A with decreasing flow one has $V_1 > V_2$; $\Delta Q_o/(2\Delta Q_1) < 1$ for reasonably small Δx; and all other terms in Equation 10.16 are positive. However, physical reasoning suggests that when friction is relatively small $\Delta y' = -(y_2' - y_1') < 0$ or $y_2' > y_1'$ because there will be recovery of velocity head.

A comparison can be made with an analytical solution for uniformly decreasing flow with circular cross section of constant and zero slope, given by Camp and Graber (1970), as follows:

$$\Delta h_x = h - h_e = \frac{V_e^2}{2g}\left\{1 - \left(1 - \frac{x}{L}\right)^2 - \frac{FL}{3}\left[1 - \left(1 - \frac{x}{L}\right)^3\right]\right\} \tag{10.17}$$

in which x is the distance from the upstream end of the conduit of length L over which the flow is fully dispersed, h_o and V_o are, respectively, the piezometric head and the velocity at the conduit entrance, h is the piezometric head at x, and F is a gross friction factor, which equals f/d_o with d_o as pipe diameter when pipe friction is the only form of resistance (as is assumed below for simplicity and familiarity). Equation 10.16 may be applied to this problem by taking $S_f = [f/(2gd_o)][(V_1 + V_2)/2]^2$, $V_1/V_e = Q_1/Q_e$, and $\Delta V/V_e = -\Delta Q_i/Q_e$ in which Q_e is rate of flow at the conduit entrance, giving:

$$\frac{\Delta y'}{V_e^2/(2g)} = 2\frac{(2Q_1/Q_e - \Delta Q_i/Q_e)(-\Delta Q_i/Q_e)}{[2 - (\Delta Q_i/Q_e)/(Q_1/Q_e)]}\left(1 - \frac{1}{2}\frac{\Delta Q_i/Q_e}{Q_1/Q_e}\right) + \frac{fL}{4d_o}\left(2\frac{Q_1}{Q_e} - \frac{\Delta Q_i}{Q_e}\right)\frac{\Delta x}{L} \tag{10.18a}$$

$$\Delta h_x = -\sum_{x=0}^{x} \Delta y' \tag{10.18b}$$

in which, since, $Q = Q_e(1 - x/L)$, we can set $\Delta Q_i = Q_1 - Q_2 = Q_i(+\Delta x/L)$.

The result is compared to the analytical solution (Camp and Graber 1970) on Fig. 1 in Graber (2004d), which is plotted in the same format as given by Camp and Graber (1970). [The curve there labeled x_m/L gives the location of the extreme value of Δh_x, as determined analytically by Camp and Graber (1970)]. The numerical solution, based on Equations 10.18, agrees very well with the analytical solution. Also shown on the cited Fig. 1 is the numerical solution based on Equation 10.16 with ΔV taken incorrectly as $V_1 - V_2$ for which the parenthetic term $(-\Delta Q_i/Q_e)$ at its single location in Equations 10.16 becomes $(+\Delta Q_i/Q_e)$. The result (denoted "Incorrect Numerical Solution," and with the *FL* curves not labeled, but having the same values and following the same order as for the analytical and correct numerical solutions) is seen on the cited Fig. 1 to deviate substantially from the analytical solution and to give only negative values of Δh_x (as foreseen in the discussion above). This demonstrates clearly that the equation given by Chow (1959) requires the correction given above.

An interesting comparison was provided by comparing the use of the gross friction factor to an "exact" method used by Benami (1968). Therein, a step-by-step procedure was used to analyze a lateral 3 inches in diameter and 216 meters long, which had equally spaced couplings with known constant loss coefficients. By using $F = f/D + K/\ell$ for the gross friction factor, ℓ being the spacing between couplings, and K being the loss coefficient, excellent agreement between the method of Camp and Graber (1968) and Benami's more exact method was obtained.

The Vigander *et al.* (1971) characterization of the method of Camp and Graber (1970) as an "analysis of errors" provides a good perspective. Camp and Graber did not systematically evaluate the effect of deviations from uniform discharge on the validity of their method. However, in addition to the evaluation mentioned in the previous paragraph and the evaluation presented in Section 10.5, they did compare results of their method to a detailed study of a given design by a more exact method and found that both methods predicted an identical deviation of 7%. More important is the fact that, as implied in Camp and Graber (1970), m is insensitive to reasonable approximations used in design.

In his discussion of Camp and Graber (1968), Ordon (1968) referenced his work (Ordon 1966), wherein he employed a simple procedure for handling dispersion conduits (defined and discussed thoroughly in Chapter 11) with abrupt changes in cross section. The author has used that procedure for evaluating process air headers in activated sludge aeration basins and finds it convenient and practical. It involves the use of an "effective length," as defined by Camp and Graber (1970), for each section of constant diameter. That method has also been applied to tapered conduits, as discussed in Camp and Graber (1969).

10.6 Spillway Flow

Chanson (2006) addressed minimum specific energy and critical flow conditions in open channels. The tangible problem on which he specifically focused is for flow over a spillway crest. For that problem, he derived the following third-order polynomial:

$$\left(\frac{d_c}{E_{\min}}\right)^3 - \left(\frac{d_c}{E_{\min}}\right)^2 * \frac{1}{\Lambda_{crest}} + \frac{1}{2} * \frac{\beta_{crest} * C_D^2}{\Lambda_{crest}} * \left(\frac{2}{3}\right)^3 = 0 \tag{10.19}$$

in which d_c = critical flow depth, i.e., flow depth at minimum specific energy; $E_{\min}$ = minimum specific energy; Λ_{crest} = pressure correction coefficient at the crest; β_{crest} = momentum correction coefficient at the crest; and C_D = dimensionless discharge coefficient. Equation (10.19) has one, two, or three real solutions depending on whether the discriminant, given below, is positive, zero, or negative, for which the real solutions are denoted $S1$, $S2$, and $S3$, respectively.

$$\Delta = \frac{1}{\Lambda_{crest}^6} * \frac{4}{3^6} * \beta_{crest} * C_D^2 * \Lambda_{crest}^2 * + \left(\beta_{crest} * C_D^2 * \Lambda_{crest}^2 - 1\right) \tag{10.20}$$

The $S2$ solution is negative and thus not physically meaningful. The $S1$ and $S3$ solutions are positive, but the experimental data plotted by Chanson (2006) only correspond to (and closely match) the $S3$ solution. [That data must be inspected closely. The pertinent data [Figure 3 in Chanson (2012) and Figure 2 in Chanson (2013)] are those represented by square blocks. The data points extending onto the S3 curve represent an undular flow in a venturi flume, which is not related to the present concern.] Chanson says it is "unclear why experimental data do not follow the solution $S1$, although it is conceivable that $S1$ might be an unstable solution." The reason, determined by the author, is given below.

The $S1$ solution is given by:

$$\frac{d_c}{E_{\min}} * \Lambda_{crest} = \frac{2}{3} * \left[\frac{1}{2} + \cos\left(\frac{\delta}{3}\right)\right] \tag{10.21a}$$

$$\delta = \cos^{-1}\left(1 - 2 * \beta_{crest} * C_D^2 * \Lambda_{crest}^2\right) \tag{10.21b}$$

And the $S3$ solution is given by:

$$\frac{d_c}{E_{\min}} * \Lambda_{crest} = \frac{2}{3} * \frac{1 - \cos(\delta/3) + \sqrt{3 * \{1 - [\cos(\delta/3)]^2\}}}{2} \tag{10.21c}$$

with δ the same as above.

The S1 and S3 solutions can be expressed in the following functional form, with the function differing for each solution:

$$\frac{d_c}{E_{\min}} * \Lambda_{crest} = \mathtt{fn}\left(\beta_{crest} * C_D^2 * \Lambda_{crest}^2\right) \tag{10.22}$$

The parameters at the crest are related to the minimum energy at the crest by $d_c + V_c^2/(2g) + h_L = E_{\min}$, in which new parameters are critical velocity at the crest, gravitational acceleration, and head loss. The Froude Number at the crest is unity, corresponding to $V_c^2/(gd_c) = 1$ so that $V_c^2/(2g) = [gd_c/(2g)] = d_c/2$. The following equation can then be written:

$$\frac{h_L}{E_{\min}} = 1 - \frac{3d_c}{2E_{\min}} \tag{10.23}$$

The energy loss from upstream to the crest is converted to heat, which is denoted by ΔQ. The entropy change (Hatsopoulus and Keenan 1965) is $\Delta s = \Delta Q/T$ and we have:

$$\frac{\Delta Q/T}{E_{\min}} = \frac{1}{T}\left(1 - \frac{3d_c}{2E_{\min}}\right) \tag{10.24}$$

From this, an Entropy Parameter is defined equal to the variable portion of the right-hand side of Equation (10.24):

$$\text{Entropy Parameter} = 1 - \frac{3d_c}{2E_{\min}} \tag{10.25}$$

Then, using Equation (10.25) with the $S1$ and $S3$ solutions given above, Figure 10.2 plots the Entropy Parameter vs. the abscissa for each of those solutions. Since the Entropy Parameter is negative for the $S1$ solution, that solution is inadmissible; only the $S3$ solution is physically viable.

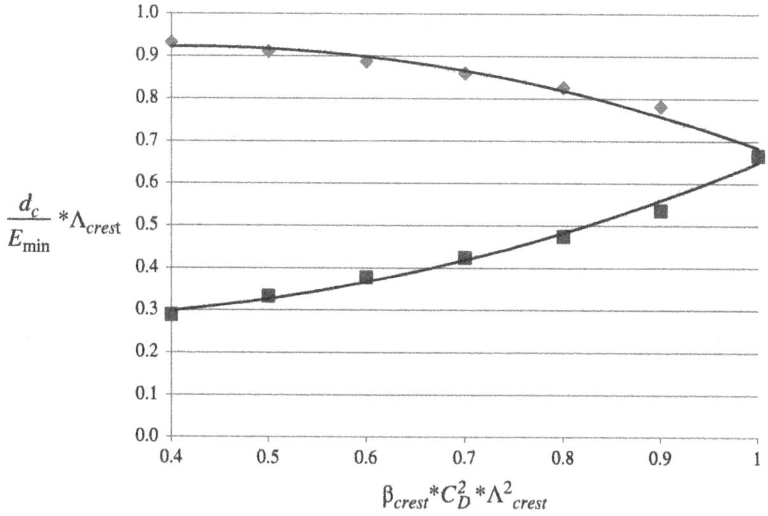

Figure 10.2 Entropy Parameter vs. Crest Parameter

10.7 Recapitulation

By presenting relationships for spatially varied flow with combined increasing and decreasing discharge and relating those relationships to prior work, a clearer exposition of spatially varied flow concepts is provided. With the relationships given in forms that separate the role of spatial inflow and spatial outflow, the proper role of each is clarified, and we avoid the shortcoming found in numerous references of using one term to represent either inflow or outflow without distinguishing their important differences. Certain incorrect statements in the literature regarding spatially varied flows are corrected with reliance on the more general methods presented. The more general relationships are simplified and related to practical equations found in the literature. The relationships between the friction, energy dissipation, and total head gradients, as well as between corresponding resistance coefficients, are elucidated.

Although both energy (First Law of Thermodynamics) and momentum (Newton's Second Law) principles are applicable to spatially varied flows, they differ in their usefulness, particularly for spatially increasing flow. The momentum equation provides a complete solution for spatially decreasing as well as spatially increasing flows and provides fundamental analytical insight. Both principles are usefully applied to spatially decreasing flow, but careful selection of control volumes and interpretation are required. The history of the debate initiated in the nineteenth century over the role of the First Law of Thermodynamics in general and hydrodynamics specifically is discussed, related to current issues regarding spatially varied flows, and hopefully resolved in the context of such flows and more generally. It is shown that those who have claimed to have avoided use of the First Law of Thermodynamics in deriving spatially varied free-surface flow equations have, in fact, used that Law in all but name. Newton's Second Law and the First Law of Thermodynamics are totally independent, fundamental laws of nature, each having its own inherent value in fluid flow problems.

Entropy considerations are employed to demonstrate admissible solutions for spillway flow.

11

Distribution Conduits

11.1 Introduction

Channels and pipes with spatially decreasing flows in environmental engineering applications are referred to as distribution or dispersion conduits, as addressed by, e.g., Camp and Graber (1968, 1969, 1970, 1972), Yao (1972), and Graber (1981). In the design of water and wastewater unit processes of many types, it becomes necessary to provide for the distribution of a fluid across the width, length, or depth of a processing unit or battery of parallel units. In some cases, equal lengths of hydraulically identical conduit can be provided, branching out from a common location, to affect a uniform distribution. In other cases, a splitting structure with multiple critical-depth controls can be employed to divide the flow into separate conduits in a prescribed manner. Sometimes feedback meter/control valve arrangements are necessary (e.g., Hross 2019). A great many situations, however, lend themselves to the use of a single pipe or channel, herein called a *distribution conduit*, to distribute the flow. Examples of such conduits include influent channels to parallel water or wastewater treatment units, chemical-solution diffusers, spray-water (foam-control) systems, air manifolds, filter bottoms (during backwashing), chlorine-generation electrolysis units, filter manifolds, sprinkler systems, and outfall diffusers. Uniformity of distribution is generally desired, in order to assure efficient use of the product being distributed or to assure that each parallel unit plays a nearly equal part in a treatment process. Failure to design for adequate distribution can not only result in process inefficiencies but can also be implicated in serious failures, such as in the case of filter collapse due to improper wastewater distribution (Becker 1963).

The ideal distribution conduit will discharge its fluid via gates, orifices, branch pipes, nozzles, gas diffusers, weirs, or other outlet devices at the same rate per unit outlet. For a real distribution conduit, the head available for discharge through the outlets varies along the length of the conduit, causing some degree of nonuniformity in the distribution. In this chapter, the two fundamental ingredients of the theory of distribution conduits, the distribution principle and the differential equation for spatially decreasing flow, are examined together with their use in design. Additional situations and methods of analysis are also addressed. Criticism of erroneous methods of design of wastewater pressure distribution systems, presented in numerous references (among them federal and state publications), is presented by Graber (2010c). Improvements to manifold design methods are suggested therein, which will result in more accurate designs. Additional, useful perspective is provided by Benefield *et al.* (1984, Chapter 2).

11.2 Distribution Principle

The *distribution principle* was first published by Camp (Camp 1942) and presented more recently by Camp and Graber (1968). A uniform distribution of discharge through the outlets of a system of dispersion conduits may be approached by making the outlet head loss large as compared to the maximum difference of available head, Δh_{max}. The head loss at the outlet where the available head is greatest is $h_o = kq^2$, in which q = the outlet discharge. The head loss at the outlet with minimum available head is $h_o - \sum \Delta h = k(mq)^2$, in which m = the ratio of the rate of discharge at the latter outlet to the rate at the former outlet. Then

$$h_o = \frac{\sum \Delta h}{1 - m^2} \tag{11.1}$$

Hydraulics and Pneumatics in Environmental Engineering, First Edition. S. David Graber.

In explaining this concept, a specialized form of the distribution principle will first be applied to a simple situation to aid in understanding the concept and to demonstrate its practical value. Then a *generalized* version of the distribution principle will be presented.

11.2.1 Introductory Case

Consider the case of a short, horizontal pressure pipe distributing water into the atmosphere through numerous orifices and/or valves equally spaced along its length (a short spray-water header is an example). If the pipe is sufficiently short, friction losses in the pipe will be small, the specific energy will remain approximately constant, and the change in pressure in the pipe will be due to recovery of velocity head. The flow q through each orifice of area a and discharge coefficient C_D is related to the piezometric head h (see Chapter 6) in the pipe adjacent to that orifice according to:

$$q = C_D a\sqrt{2gh} \tag{11.2}$$

In this particular case, the minimum flow occurs through the upstream-most orifice, where h is lowest, and the maximum flow occurs through the downstream-most orifice, where h is the highest. The ratio between these minimum and maximum flows is denoted by:

$$m = q_{\min}/q_{\max} \tag{11.3}$$

For the present case, this ratio is obtained using Equation 11.3 as:

$$m = \sqrt{\frac{h_o}{h_\ell}} \tag{11.4}$$

in which the subscripts "o" and "ℓ" denote conditions in the pipe adjacent to the upstream-most ($x = 0$) and downstream-most ($x = \ell$) orifices, respectively.

The terms h_ℓ and h_o are related approximately (in this case) by:

$$h_\ell \cong h_o + \frac{V_o^2}{2g} \tag{11.5}$$

The approximation improves as the number of orifices increases, but is conservative in any event for the short (frictionless) pipe. Substituting h_o from Equation 11.5 into Equation 11.3 results in:

$$m = \sqrt{1 - \frac{V_o^2/(2g)}{h_\ell}} \tag{11.6a}$$

or

$$h_\ell = \frac{V_o^2/(2g)}{1 - m^2} \tag{11.6b}$$

11.2.2 General Distribution Principle

A general statement of the distribution principle will now be presented. The significance of all aspects of the generalized statement may not be immediately apparent but should become so as specific types of distribution conduits and examples are considered later in this chapter.

The flow distributed through each outlet of a distribution conduit may be thought of as taking parallel flow paths between points of *equal piezometric head.* It is important to identify the upstream and downstream points of equal piezometric head at the outset. The upstream point will generally be a single point at the upstream end of the distribution conduit, upstream of the first outlet. The downstream points are not necessarily coincident but often separate points of equal piezometric head, such as free (unsubmerged) discharges of water over weirs at the same elevation (weirs of unequal height are discussed later), discharge points of water into channels or tanks of equal water surface elevation (the distribution conduit need not be parallel to the water surface if the distributed liquid is of the same density as that in the receiving channel or tank), or discharge points of gas from horizontal manifolds into channels or tanks of equal water surface elevation. Examples of varying complexity are given below.

The next step consists of determining the functional relationship between outlet flow and available piezometric head at the outlets. The available piezometric head at the outlets refers to the piezometric head at points having identical hydraulic characteristics relative to the downstream points of equal piezometric head. Letting h be that available piezometric head relative to the downstream points of equal piezometric head and letting q be the flow per outlet, then q is a function of h denoted by $q = f(h)$.

Denoting by subscripts "+" and "−" the outlets where the available head and flow are greatest and least, respectively, the following relationships pertain:

$$q_+ = f(h_+) \tag{11.7a}$$

$$q_- = f(h_-) \tag{11.7b}$$

$$h_+ = h_- + \Delta h \tag{11.7c}$$

$$m = \frac{q_-}{q_+} = \frac{f(h_-)}{f(h_+)} \tag{11.7d}$$

$$m = \frac{f(h_+ - \Delta h)}{f(h_+)} \tag{11.7e}$$

In the above, m is the ratio of minimum outlet discharge to maximum outlet discharge and Δh is the maximum difference in available piezometric head along the distribution conduit. Equation 11.7e expresses the generalized distribution principle. The value m is closest to 1 (perfectly uniform distribution) when h_+ is large compared to Δh. Various functional relations for $q = f(h)$ are of practical importance, and are discussed below.

11.2.2.1 Square-Root Outlets

Gates, orifices, branch pipes, and nozzles will generally have "square-root" relations such that $q = f(h) = C\sqrt{h}$. The distribution equation then becomes:

$$m = \frac{C_-}{C_+}\sqrt{1 - \frac{\Delta h}{h_+}} \tag{11.8a}$$

$$h_+ = \frac{\Delta h}{1 - \left(m\frac{C_+}{C_-}\right)^2} \tag{11.8b}$$

The principal reason for using the piezometric head rather than the energy head in the above relation is that it renders the coefficient C relatively constant. That is discussed in detail in Chapter 6. In many cases (and Chapter 6 provides the basis for determining when), it is acceptable to assume $C_+/C_- = 1$.

As an example for a case where $C_+/C_- \cong 1$, if a maximum deviation in outlet discharge of 3% is desired, m is 0.97, $1 - m^2$ is 0.059, and the head loss h_+ must be not less than 1/0.059 or 16.9 times Δh_{max}. Corresponding values of h_+ for m values of 0.95 and 0.90 are 10.3 Δh_{max} and 5.3 Δh_{max}, respectively.

Chapter 6 discusses means for determining C as a function of $V^2/(2gE)$. The coefficient C is more constant for some types of orifice outlets (such as bell mouthed) than for others, such as sharp-edged orifices. Camp and Graber (1970, Figure 19) provides a good demonstration of this in graphic form. Note from Equation 11.8a that C_+/C_- provides a limiting value of attainable m (when $\Delta h/h_+ \rightarrow 0$). As m approaches C_-/C_+, the value of h_+ becomes very large. The Chapter 6 results will be used in conjunction with the distribution principle in the examples given below.

The author designed air dispersion conduits for four different purposes at the Rochester, New Hampshire, Wastewater Treatment Plant (WWTP): waste-activated sludge holding tanks, sand filter-channel aeration headers, postaeration diffuser headers, and submerged turbine aerators. The latter is discussed in the following example.

Example 11.1

Submerged turbine aerators with air sparge rings are to be used in an equalization tank with a water depth varying from 5 to 15 ft (Rochester, New Hampshire WWTP). A sparge ring, to be installed under each submerged turbine aerator, is proposed by the manufacturer to consist of a 4-in. I.D. steel pipe formed into a circular ring of 60-in. diameter 2 ft above the tank bottom. The sparge ring is proposed to have a 1-psi pressure drop at 500 scfm air flow rate. The average barometric pressure at the plant site is 14.6 psia. Determine the attainable distribution uniformity and the suitability of 1-psi pressure drop.

Solution

The water depth range is important as a factor in determining the system supply pressure but has no bearing on the immediate problem. For 4-in. pipe, Schedule 40, with 0.025 mil interior protective coating, we have internal diameter $D = 4.026-2(0.025) = 3.976$ in. and cross-sectional area $A = (\pi/4)(3.976/12)^2 = 0.0860$ ft^2. With the 500 scfm flow split in two directions from the diametrical inlet pipe, the inlet velocity $V_o = (500/2)/0.0.0860 = 291$ ft/min ÷ 60 = 48.4 ft/sec and corresponding velocity head $V_o^2/(2g) = 36.4$ ft. These values seem high, but we will continue to check. With viscosity $\mu = 4.3(10)^{-7}$ lbf sec/ft^2 at 68°F, the inlet Reynolds Number is $\mathbf{R}_o = \gamma VD/(g\mu) = (0.07516)(48.4)(3.976/12)/[32.2(4.3)(10)^{-7}] = 8.71(10)^4$. The roughness $\varepsilon = 0.0004$ for coated steel, giving $\varepsilon/D = 0.0004/(3.976/12) = 0.00121$. From the Moody Diagram or regression equation (Section 5.3), the friction factor $f = 0.024$. The flow in the ring feeds from the entrance to the null point, one-half of the circumference over the length $L = (1/2)\pi(60/12) = 7.85$ ft, for which $fL/D = 0.024[7.85/(3.976/12)] = 0.569 < 2$. Therefore, from Section 11.3 (Equation 11.39a), the pressure rises along the pipe, and the maximum pressure difference along the pipe is given by:

$$\Delta h = \frac{V_o^2}{2g}\left(1 - \frac{fL}{3D}\right) = 36.4\left(1 - \frac{0.569}{3}\right) = 29.5 \text{ ft}$$

The manufacturer's proposed 1-psi pressure drop (equivalent to 2.31 ft of water) occurs predominantly through the sparger-ring orifices, and being less than Δh cannot correspond to reasonable distribution uniformity.

In fact, with a 1 psi pressure drop, no reasonable sparger-ring I.D. can give reasonable distribution uniformity. So, we will consider combinations of larger sparger-ring internal diameters and greater sparger-ring pressure drops. We will first consider an 8-in. sparger ring of the same Schedule 40 and interior coating, for which I.D. = 6.065 − 2(0.025) = 6.015 in. and cross-sectional area $A = (\pi/4)(6.015/12)^2 = 0.1973$ ft^2. The inlet velocity is then $V_o = (500/2)/0.1973 = 1267$ ft/min ÷ 60 = 21.1 ft/sec and corresponding velocity head $V_o^2/(2g) = 6.91$ ft. We now have $\mathbf{R}_o = (0.07516)(21.1)(6.015/12)/[32.2(4.3)(10)^{-7}] = 5.71(10)^4$, $\varepsilon/D = 0.0004/(6.015/12) = 0.000798$, $f = 0.023$, and $fL/D = 0.023[7.85/(6.015/12)] = 0.361 < 2$ giving:

$$\Delta h = \frac{V_o^2}{2g}\left(1 - \frac{fL}{3D}\right) = 6.91\left(1 - \frac{0.361}{3}\right) = 6.08 \text{ ft}$$

For a 95% distribution uniformity, the controlling head loss is given by $h_o = \Delta h/(1-m^2) = 6.08/[1-(0.95)^2] = 62.3$ ft ÷ 2.31 = 27.0 psi.

Now the following equations are employed:

$$Q_o = nq$$

$$q = C_D A_o \sqrt{2gh_o}$$

$$A_o = \frac{\pi}{4} d_o^2$$

$$\ell = \frac{\pi D_R}{n}$$

in which Q_o, n, q, C_D, A_o, h_o, d_o, ℓ, and D_R are, respectively, total orifice discharge rate, number of orifices, individual orifice flow rate, orifice discharge coefficient, individual orifice area, head loss across orifice, orifice diameter, orifice spacing, and circular ring diameter.

With $C_D = 0.61$, these equations are combined to give:

$$\ell = \frac{\pi D}{Q_o} C_D \frac{\pi}{4} d_o^2 \sqrt{2gh_o} = 1.51 \frac{D d_o^2 \sqrt{2gh_o}}{Q_o}$$

Substituting values given above into the above equation and including conversion factors with d_o and ℓ in inches and h_o in ft gives:

$$\ell = 1.51 \frac{60 d_o^2}{500} \sqrt{2(32.2)(62.3)} \frac{60}{144} = 4.78 d_o^2$$

Values given by the above relationship are tabulated below:

d_o (in.)	ℓ (in.)
0.25	0.299
0.38	0.672
0.50	1.195
0.63	1.867
0.75	2.689
1.00	4.780

Values of ℓ less than $2d_o$ may be accommodated by staggering the orifices along the sparger. However, for this application, the 0.50-in. diameter orifices were selected.

11.2.2.2 Rectangular Weir Outlets

Weirs along open-channel distribution conduits are usually rectangular, freely discharging, *side weirs*, of the type considered in Chapter 7. However, due to its complexity, the side weir equation (Equation 7.23) is usually approximated by a rectangular weir formula of the following form:

$$q = f(h) = C(h - S)^{3/2} \tag{11.9}$$

in which h is channel depth and S is the height of the weir above the channel bottom. The corresponding distribution equation is:

$$m = \frac{C_-}{C_+}\left(1 - \frac{\Delta h}{h_+ - S}\right)^{3/2} \tag{11.10a}$$

or

$$\frac{h_+ - S}{\Delta h} = \frac{1}{1 - \left(m\dfrac{C_+}{C_-}\right)^{2/3}} \tag{11.10b}$$

At this point, one might be tempted to approximate the ratio C_-/C_+ by using the Froude number correction from Equation 7.20, and this could be done (Graber 1981). However, that would be inconsistent with the use of Equation 11.9, in which the side weir equation reduces only for y/E close to 1 or $\mathbf{F} \ll 1$, in which case the Froude number correction for C_w becomes negligible. Thus, it is conceptually more consistent to take $C_-/C_+ = 1$ in Equations 11.10 so that:

$$m = \left(1 - \frac{\Delta h}{h_+ - S}\right)^{3/2} \tag{11.11a}$$

$$\frac{h_+ - S}{\Delta h} = \frac{1}{1 - m^{2/3}} \tag{11.11b}$$

11.2.2.3 Triangular Weir Outlets

Triangular (V-notch) weir outlets are sometimes used. From Section 7.2, the flow through each notch is approximately related to the head on the notch as follows:

$$q = f(h) = C(h - S)^{5/2} \tag{11.12}$$

in which h is channel depth and S is the height of the notch above the channel bottom. The corresponding distribution equation is:

$$m = \frac{C_-}{C_+}\left(1 - \frac{\Delta h}{h_+ - S}\right)^{5/2} \tag{11.13a}$$

or

$$\frac{h_+ - S}{\Delta h} = \frac{1}{1 - \left(m\frac{C_+}{C_-}\right)^{2/5}} \tag{11.13b}$$

The ratio C_-/C_+ is usually best taken to be unity as for rectangular weirs.

11.2.2.4 Combination Orifice Outlets and Weirs

It should be emphasized that in the above discussion of square-root orifices, an essentially constant water surface level downstream of the outlets was assumed. In other words, h in the distribution conduit is dependent on flow only as a result of the square-root outlets and Δh along the distribution conduit. Although the approximation of a constant downstream water surface is often adequate, the downstream water surface is commonly a function of flow, and not always such a weak function of flow that it can properly be assumed to be negligible.

A common configuration, for example, is that of a distribution conduit supplying parallel tanks, the effluent of which discharges over weirs. The functional relation in this case becomes:

$$h = \left(\frac{q}{C_1}\right)^{2/3} + C_2 q^2 \tag{11.14}$$

11.2.2.5 Combination Square-Root and Porous-Media Outlets

Quasi-linear head—flow relationships are encountered in flow through porous media, such as those used for fine-bubble gas diffusion. Such diffusers are commonly used in conjunction with orifices, such as in the air diffusers used for spiral-flow-activated sludge aeration. The corresponding functional relationship is:

$$h = C_1 + C_2 q + C_3 q^2 \tag{11.15}$$

For isentropic (reversible adiabatic) flow, the Mach Number is a function of p/p_o in accordance with Equation 9.108, in which p_o is the pressure in the manifold assumed to be equal to the static pressure (as defined in Section 9.4). The discharge will be subsonic ($\mathbf{M} < 1$) or sonic ($\mathbf{M} = 1$) through the orifice. In either case, the discharge pressure is equal to the absolute water pressure at the depth of the orifice plus any pressure drop across the diffuser element following the orifice. If the latter is negligible and the manifold is horizontal, then p is constant and

$$m \cong \frac{\mathbf{M}_-}{\mathbf{M}_+} = \frac{\mathbf{M}_+ - \Delta\mathbf{M}}{\mathbf{M}_+} \tag{11.16}$$

in which $\Delta\mathbf{M} = \mathbf{M}_{max} - \mathbf{M}_{min}$. It is desired to express $\mathbf{M}$ and $\Delta\mathbf{M}$ in terms of appropriate pressure terms. However, reference to Equation 9.108 suggests the infeasibility of obtaining an explicit expression for $\mathbf{M}$ in terms of p/p_o. An approximate expression may be obtained, however, by first expressing $\Delta\mathbf{M}$ in terms of a Taylor expansion as follows:

$$\Delta\mathbf{M} = \frac{d\mathbf{M}}{d(p/p_o)}\Delta(p/p_o) + \frac{d^2\mathbf{M}}{d(p/p_o)^2}\frac{\Delta(p/p_o)^2}{2!} + \cdots \tag{11.17}$$

The pressure drop along the manifold is generally small enough that higher order terms of the expansion may be neglected. Then, from Equations 11.16 and 11.17:

$$m \cong 1 - \frac{d\mathbf{M}}{d(p/p_o)}\frac{\Delta(p/p_o)}{\mathbf{M}} \tag{11.18}$$

Assuming constant p, as discussed previously, the above equation may be rearranged as follows:

$$\frac{\Delta(p_o/p)}{1-m} = \mathbf{M}\frac{d(p_o/p)}{d\mathbf{M}} \tag{11.19}$$

Differentiating Equation 9.108 with respect to $\mathbf{M}$ gives:

$$\frac{d(p_o/p)}{d\mathbf{M}} = k\left(1 + \frac{k-1}{2}\mathbf{M}^2\right)^{\frac{1}{k-1}}\mathbf{M} \tag{11.20}$$

in which k is the ratio of specific heat at constant pressure to specific heat at constant volume. The above equation may be substituted into Equation 11.19 to give:

$$\frac{\Delta(p_o/p)}{1-m} = k\mathbf{M}^2\left(1 + \frac{k-1}{2}\mathbf{M}^2\right)^{\frac{1}{k-1}} \tag{11.21}$$

Employing Equation 9.109, the above equation may be rewritten as

$$\frac{\Delta(p_o/p)}{1-m} = k\mathbf{M}^2\frac{\rho_o}{\rho} \tag{11.22}$$

in which ρ_o/ρ is the density ratio and is a commonly tabulated compressible flow function (e.g., Mooney 1959, TABLE B.2).

Equation 11.22 can be tabulated as a new compressible flow function, which is done for air ($k = 1.4$) in Table 11.1. Equation 11.22 is depicted graphically in Figure 11.1. By setting $\rho_o/\rho = 1$ in Equation 11.22, the relationship for incompressible flow is obtained and is plotted in Figure 11.1 for comparative purposes. A 1.0% error occurs at $\mathbf{M} = 0.35$, well above the range of interest in many diffused air systems if incompressible flow is assumed (at smaller $\mathbf{M}$'s).

Table 11.1 New Compressible Flow Function and Comparison to Incompressible Flow

M	p_o/p	ρ_o/ρ	$\Delta(p_o/p)/(1-m)$	$\Delta(p_o/p)/(1-m)$ with $\rho_o/\rho = 1$
0	1.00000	1.00000	0.00000	0.00000
0.1	0.99303	0.99502	0.01393	0.01400
0.14	0.98640	0.99027	0.02717	0.02744
0.2	0.97250	0.98028	0.05490	0.05600
0.29	0.94329	0.95916	0.11293	0.11774
0.3	0.93947	0.95638	0.12050	0.12600
0.349	0.91921	0.94160	0.16056	0.17052
0.3494	0.91904	0.94148	0.16091	0.17091
0.35	0.91877	0.94128	0.16143	0.17150
0.4	0.89561	0.92427	0.20704	0.22400
0.5	0.84302	0.88517	0.30981	0.35000
0.6	0.78400	0.84045	0.42359	0.50400
0.7	0.72093	0.79158	0.54302	0.68600
0.8	0.65602	0.73999	0.66303	0.89600
0.9	0.59126	0.68704	0.77911	1.13400
1	0.52828	0.63394	0.88751	1.40000
1.1	0.46835	0.58170	0.98539	1.69400
1.2	0.41238	0.53114	1.07078	2.01600
1.3	0.36091	0.48290	1.14255	2.36600
1.4	0.31424	0.43742	1.20029	2.74400
1.5	0.27240	0.39498	1.24420	3.15000
1.6	0.23527	0.35573	1.27494	3.58400
1.7	0.20259	0.31969	1.29348	4.04600
1.8	0.17404	0.28682	1.30101	4.53600
1.9	0.14924	0.25699	1.29883	5.05400
2	0.12780	0.23005	1.28827	5.60000

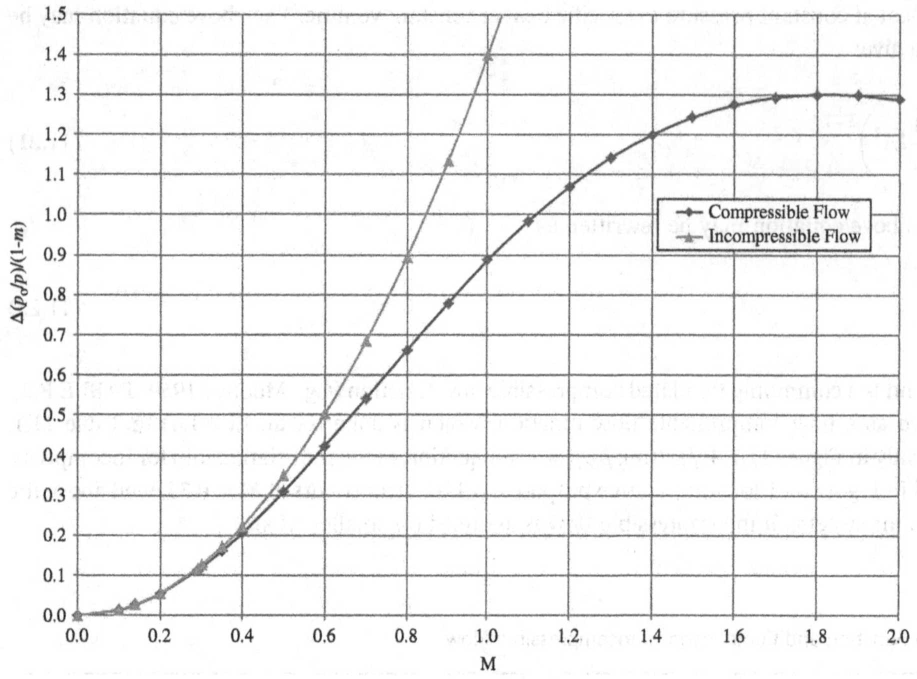

Figure 11.1 Plot of New Compressible Flow Function and Comparison to Incompressible Flow

Equation 11.22 may be reduced to the previously presented relationship for incompressible flow. Since the speed of sound waves in an isentropic flow is given by $c = \sqrt{kRT}$ (Equation 9.56), Equation 11.22 may be expressed as follows:

$$\frac{\Delta p_o}{1-m} = \rho V^2 \tag{11.23}$$

Noting that V in the above equation is a theoretical velocity, the analogous incompressible flow relation may be used to express Equation 11.23 as follows:

$$\frac{\Delta p_o}{1-m} = 2(p_o - p) \tag{11.24}$$

or, letting $\Delta h = \Delta p_o/(\rho g)$ and $h_+ = (p_o - p)/(\rho g)$,

$$\frac{\Delta h}{h_+} = 2(1-m) \tag{11.25}$$

For small values of $1 - m$, we have $2(1 - m) \cong (1 - m)(1 + m)$, and Equation 11.25 can be rearranged to give:

$$\frac{\Delta h}{h_+} = 1 - m^2 \tag{11.26}$$

which is the incompressible flow relationship.

There are numerous related examples from the author's practice, including aeration tanks with dome diffusers and aeration of waste-activated sludge storage tanks, such as at the Rochester, New Hampshire Advanced Wastewater-Treatment Plant. Pertinent variables for Rochester are $T = 693°\text{R}$, $V_o = 44.9$ ft/sec, $V_o^2/(2g) = 31.3$ ft, $c = 1290$ ft/sec, $\mathbf{M} = V_o/c = 44.9/1290 = 0.0347$. This Mach Number is an order of magnitude below the value of 0.35 mentioned above, at which compressibility becomes important.

11.2.3 Comparative Characteristics of Outlet Controls

Certain general comparisons may be made of the various outlet devices, which may be helpful in choosing among them when more than one type is feasible. First is the observation that for a given uniformity m, the required controlling head loss (h_+ or $h_+ - S$) will be least for square-root outlets. That is followed by rectangular weir outlets, then by V-notch outlets. In rare instances of weir-type outlets with partially submerged discharges (see Chapter 7), the controlling head loss may be partially recoverable (Yao 1972), modifying the relative advantages. The head–flow relationships are generally of the form $h \propto q^n$, and the higher the exponent n the lower the required controlling head loss. For example, the values of $h_+/\Delta h$ or $(h_+ - S)/\Delta h$ for $m = 0.95$ and $C_+/C_- = 1$ are 10.3, 29.7, and 49.2 for square-root, rectangular weir, and V-notch weir outlets, respectively. The corresponding values for $m = 0.90$ are 5.26, 14.7, and 24.2, respectively. Combination orifice outlets and weirs and combination square-root and linear outlets are intermediate between the square-root and weir outlets.

It will be seen from the next section that the maximum head loss Δh_{max} usually varies in proportion to the velocity head at the inlet of the distribution conduit. For pressure distribution conduits, since A_o is constant, Δh_{max} thus varies as the square of the flow entering the distribution conduit. For square-root outlets, h_+ is also approximately proportional to the square of the flow entering the distribution conduit. Therefore, by Equation 11.8a, m is essentially independent of flow for pressure distribution conduits with square-root outlets. For open-channel distribution conduits, however, A_o is not constant, and Δh_{max} also varies as the change in depth in the channel. For a rectangular channel, this relation is expressed by:

$$\Delta h_{\max} \propto \frac{V_o^2}{2g} = \frac{1}{2g}\left(\frac{Q_o}{bh_o}\right)^2 \propto \frac{Q_o^2}{h_o^2} \tag{11.27}$$

For square-root outlets, $h_+ \propto q^2 \propto Q_o^2$ which, with the above equation, gives:

$$\frac{\Delta h_{\max}}{h_+} \propto \frac{Q_o^2/h_o^2}{Q_o^2} \propto \frac{1}{h_o^2} \tilde{\propto} \frac{1}{Q_o^4} \tag{11.28}$$

Therefore, from Equation 11.8a and the above, m will decrease slightly with decreasing flow for open-channel distribution conduits with square-root outlets. Thus, the distribution uniformity of square-root outlets on open channels should be checked at reduced flows. From a practical standpoint, it will most often be found that the reduction in uniformity with flow reduction will be acceptably small when submerged square-root outlets are used.

If rectangular weir outlets with $S = 0$ (S being the height of the weir above the channel bottom) are employed, we have:

$$\Delta h_{\max} \propto \frac{Q_o^2}{h_o^2} \propto \frac{Q_o^2}{\left(q_o^{2/3}\right)^2} \tilde{\propto} \frac{Q_o^2}{Q_o^{4/3}} = Q_o^{2/3} \tag{11.29}$$

Since h_+ also varies approximately as $Q_o^{2/3}$, then $\Delta h_{\max}/h_+ \cong$ constant. Therefore, from Equation 11.10a with $S = 0$, m is essentially independent of flow for rectangular open-channel distribution conduits with rectangular outlets extending to the channel bottom ($S = 0$).

Yao (1972) has further investigated the effect of flow variations on open-channel distribution conduits, including $S > 0$. He also noted that the distribution uniformity of V-notch weir outlets improves as flow rate decreases.

Numerous other factors are important in selecting the type of outlet control for a particular design situation. When water or wastewater carries solids, the outlets are usually best located with bottoms matching the channel bottom elevation to minimize buildup of sediment deposits. In open channels, the provision of velocities sufficient to maintain solids in suspension is usually impractical or uneconomical, and air or water agitation (see Chapter 13) is commonly provided. Submerged gates (square-root outlets) are most amenable to effective baffling and introduction of solids near the bottom of settling tanks. V-notch weir outlets may be used to obtain a nearly uniform longitudinal distribution where vertical mixing is provided by other means (e.g., in multi-cell aeration basins).

11.3 Spatially Decreasing Flows in Distribution Conduits

Use of the distribution principle requires a determination of Δh in the distribution conduit. The determination of Δh can be carried out by methods which can be divided into two types. First are those that are structured around the discrete points of outflow. The distribution conduit can then be analyzed as a series of constant flow conduits, with the techniques of

Chapter 6 (for pressure conduits) and Chapter 7 (for open channels) applied to each such conduit and the adjoining "tee" junctions. Such methods are most appropriate when the outflow points are relatively few or unevenly spaced. The second type of method is structured around the conceptualization of the conduit as one of continuously decreasing flow. The latter method is adequate for the vast majority of distribution conduits encountered in environmental engineering applications. It is generally most practical and is the method emphasized below.

11.3.1 Pressure Conduits

Section 2.7 presented an analysis of a pressure distribution conduit with an assumed continuous (not necessarily uniform) decrease in incompressible flow along its length. The equation resulting from that analysis is repeated here:

$$\frac{dh_p}{dx} = -\frac{C_f}{R}\frac{V^2}{2g} - \frac{\beta Q}{gA^2}\frac{dQ}{dx} \tag{2.109}$$

in which h_p is the piezometric head, C_f is the wall shear coefficient, V is average pipe velocity, R is hydraulic radius, β is the momentum coefficient, A is cross-sectional area, and Q is volumetric rate of flow. From Chapter 6, we have $C_f = f/4$ in which f is the Darcy–Weisbach friction factor. Camp and Graber (1968) derived a more general form of the drag term, which includes, in addition to the friction drag along the walls of the conduit, friction and form drag past fixed branch pipes, nozzles, columns, or other equally spaced obstructions in the conduit. Their slightly more general form of Equation 2.109 is given by:

$$\frac{dh_p}{dx} = -F\frac{V^2}{2g} - \frac{\beta Q}{gA^2}\frac{dQ}{dx} \tag{11.31a}$$

$$F = \frac{f}{4R} + \frac{Ca}{sA} \tag{11.31b}$$

The term F is called the gross friction factor, in which C, a, and s are, respectively, the drag coefficient, projected area, and longitudinal spacing of the obstructions in the conduit. The drag coefficient is a function of the shape and geometrical position of the obstructions, the Reynolds Number, and the spacing if s is small.

11.3.1.1 Uniform Distribution from Prismatic Conduits

When uniform distribution is the objective, a useful point of departure is to assume the uniform distribution that is desired and determine the resulting variation in h. A uniform distribution is expressed by:

$$Q = Q_o\left[1 - \left(1 - \frac{Q_\ell}{Q_o}\right)\frac{x}{\ell}\right] \tag{11.32}$$

in which Q_o and Q_ℓ are the flows at the upstream ($x = 0$) and downstream ($x = \ell$) ends of the distribution conduit, respectively. For simplicity of notation, the above equation is rewritten as:

$$Q = Q_o\left(1 - \frac{x}{L}\right) \tag{11.33a}$$

$$L = \ell/(1 - Q_\ell/Q_o) \tag{11.33b}$$

in which L is the *effective* length of the distribution conduit, defined by $V = V_o(1 - x/L)$ with V_o = velocity at the upstream end of the conduit and V = velocity at distance x, and is equal to its actual length for the common case of $Q_\ell = 0$.

Assuming $\beta = 1$, substituting $V = Q/A$ and Equation 11.33a into Equation 11.30, and rearranging gives:

$$\frac{dh}{dx} = \frac{Q_o^2}{gA^2L}\left(1 - \frac{x}{L}\right) - \frac{Q_o^2}{2gA^2}\left(1 - \frac{x}{L}\right)^2 F \tag{11.34}$$

Equation 11.34 may be solved for F assumed constant, to give the change in piezometric head from its value at the diffuser inlet to a distance x from the diffuser inlet as follows:

$$\Delta h_x = \frac{V_o^2}{2g}\left\{1 - \left(1 - \frac{x}{L}\right)^2 - \frac{FL}{3}\left[1 - \left(1 - \frac{x}{L}\right)^3\right]\right\} \tag{11.35}$$

From Equation 11.35, by considering first and second derivatives of h with respect to x, it can be demonstrated (Camp and Graber 1968, 1969) that a relative minimum value of h exists within the diffuser length when FL exceeds 2, at the position x_m given by:

$$\frac{x_m}{L} = 1 - \frac{2}{FL} \tag{11.36}$$

From Equations 11.35 and 11.36, the difference in piezometric head from the inlet to $x = L$ and $x = x_m$ are given, respectively, by:

$$\Delta h_L = \frac{V_o^2}{2g}\left(1 - \frac{FL}{3}\right) \tag{11.37}$$

$$\Delta h_m = \frac{V_o^2}{2g}\left(1 - \frac{FL}{3} - \frac{4}{3F^2L^2}\right) \tag{11.38}$$

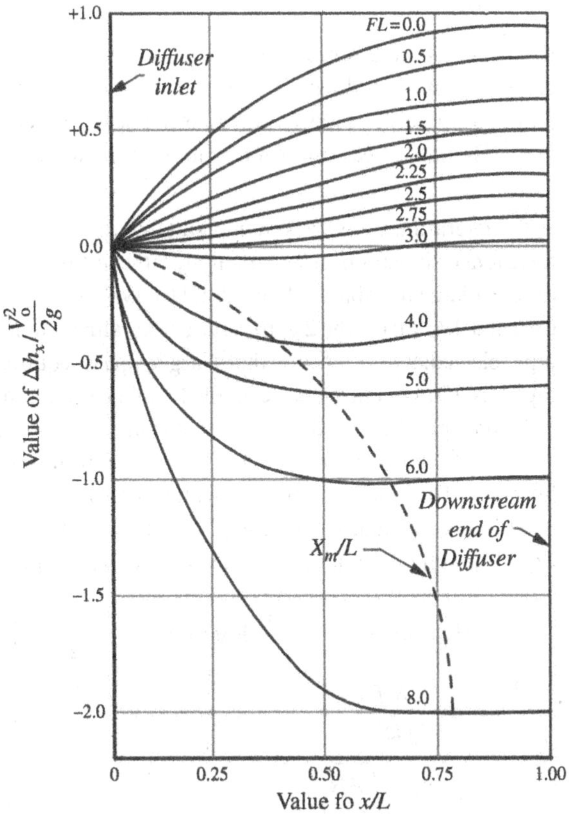

Figure 11.2 Values of $\Delta h_x / [V_o^2/(2g)]$ for Various Values of FL

Values of $\Delta h_x / [V_o^2/(2g)]$ computed for selected values of FL are presented in Figure 11.2, together with values of $\Delta h_m / [V_o^2/(2g)]$ and the corresponding values of x_m/L. It will be noted that $\Delta h_m / [V_o^2/(2g)]$ and x_m/L are both zero when FL is 2.0; and that for all values of FL greater than 2.0 there is a depression in the hydraulic grade line between the ends of the diffuser, the location for which x_m/L moves downstream with increasing values of FL. The maximum difference in Δh_x, called Δh_{max}, is $+\Delta h_L$ (between the upstream and downstream ends) for values of FL from 0 to 2.0, is $\Delta h_L - \Delta h_m$ for values of FL from 2.0 to 3.0, and is $-\Delta h_m$ (between the upstream end and an intermediate location) for values of FL of 3.0 and greater. Note that the hydraulic grade line at the downstream end of the diffuser is at the same elevation as the upstream end when FL is 3.0 and is lower for greater values of FL. The relationships for the maximum difference in Δh_x may be summarized as follows:

$$\Delta h_{max} = \begin{cases} \frac{V_o^2}{2g}\left(1 - \frac{FL}{3}\right) & 0 \le FL \le 2 \quad (11.39a) \\ \frac{V_o^2}{2g}\left(\frac{4}{3F^2L^2}\right) & 2 \le FL \le 3 \quad (11.39b) \\ -\frac{V_o^2}{2g}\left(1 - \frac{FL}{3} - \frac{4}{3F^2L^2}\right) & FL \ge 3 \quad (11.39c) \end{cases}$$

The above equations may be used, in conjunction with Equations 11.8 for orifice outlets or other appropriate expression of the distribution principle, to determine m for a given distribution conduit or to determine h_+ required for a prescribed uniformity, m. It should be observed, with reference to Equations 11.8 and 11.39, that m will change with discharge to a degree depending only on the change in f and C. That change is usually very small, so pressure distribution conduits with orifice outlets experience little change in distribution uniformity with change in flow. This is in contrast to other types of conduits discussed below.

Example 11.2

The following is quoted from WPCF/ASCE (1977, p. 391): "Perforations for both large pipe and open-channel diffusers are to be designed for chlorine solution flow and velocity of, respectively, 0.06–0.12 L/s (1–2 gpm) and 3 m/s (10 fps). The perforations normally are not smaller than 10 mm (0.38 in.). Holes are sized to permit an even hydraulic distribution of solution." Determine the suitability of these prescriptions for a representative case of 10 holes spaced equally over a 5-foot diffuser length. If unsuitable, determine characteristics giving suitable uniformity.

Solution

Calculations are given in Table 11.2 using Equation 11.39a with $F = f/D$, since $FL < 2$, and other equations that should be self-explanatory. Values of uniformity m for the stated values are incalculably low for the 1 gpm case and only 0.670 for the 2 gpm case. A value as low as 0.670 cannot be accurately determined by the method used, which performs essentially an error analysis of assumed uniform distribution. However, it does indicate very poor distribution uniformity.

To determine pipe sizes giving suitable distribution uniformity, one must use pipe inlet velocities less than the 3 m/s mentioned above and correspondingly larger diffuser pipe diameters. Trial-and-error values giving 95% distribution uniformity are calculated in Table 11.2 (only final trial values are shown). The determined diffuser pipe diameters are 2.87 in. for the 1 gpm case and 2.15 in. for the 2 gpm case. These values would, of course, be rounded up to available diameters for the actual type of pipe selected. It is interesting that a larger pipe is required for the 1 gpm case than for the 2 gpm case. That is because the smaller pipe has a lower entrance Reynolds Number **R** and, in the transitional zone, an accompanying higher friction factor f.

A conceptual contradiction results from the procedure of assuming a uniform distribution to determine Δh, and then using that Δh to determine the degree of nonuniformity. This conceptual contradiction is of little practical consequence, however. That may be seen by evaluating the results of Zsák's (1971) sophisticated, closed-form solution of the energy equation for spatially-decreasing pipe flow with the inertia term neglected. Zsák's solution, which he corroborated by experimental data and is not limited by an assumption regarding flow distribution, is given as follows:

$$3\alpha L = \ln\frac{\sqrt{\varepsilon^2+\varepsilon-1}}{1-\varepsilon} + \sqrt{3}\tan^{-1}\left(\sqrt{3}\frac{\varepsilon}{2+\varepsilon}\right) \tag{11.40a}$$

$$\alpha^3 = \frac{\tilde{a}^2 f}{4gds^2A^2} \tag{11.40b}$$

$$\varepsilon = \frac{V_o}{\sqrt{h_o}}\sqrt[3]{\frac{Afs}{4\tilde{a}gd}} \tag{11.40c}$$

in which s is the orifice or valve spacing, $\tilde{a} = C_D a\sqrt{2g}$, and other terms are as defined previously. Zsák gives the uniformity (denoted here by m and equal to y_{min}/y_L) by:

$$m = \sqrt[3]{1-\varepsilon^3} \tag{11.41}$$

Zsák's neglecting of the inertia terms is analogous to assuming large FL. When $FL \gg 3$, Equations 11.39c and 11.31b (for $C = 0$, $R = 4d$) give $\Delta h_{max} \cong \frac{fL}{3d}\frac{V_o^2}{2g}$. Using this relationship for Δh_{max} and other aspects of the development of this relationship, Camp and Graber (1972) showed the relations analogous to Equations 11.40 and 11.41 based on the uniform flow assumption to be as follows:

$$\alpha L = \varepsilon \tag{11.42a}$$

$$m = \sqrt{1-(2/3)^{1/2}\varepsilon^3} \tag{11.42b}$$

Figure 11.3(a) compares Equation 11.42a with Equation 11.40a [Zsák's Eq. (17)]; the relationships give nearly the same numerical results for αL up to 0.6. Figure 11.3(b) compares Equation 11.42b with Zsák's Eq. (20).

The comparison in Figure 11.3(b) of $m(\alpha L)$ by the two methods is the most significant comparison since the parameter αL completely characterizes the manifold system [noting from Equation 11.42a that Camp and Graber's $m(\varepsilon) = m(\alpha L)$]. The results are approximately the same by both methods up to $\alpha L = 0.6$, corresponding to $m = 0.90$. At greater values of αL conservatively low values of m are predicted by Camp and Graber's method.

This is to be expected since the redistribution of flow accounted for by the author's method results in a lower Δh_{max}. Since the velocity head recovered between the pipe inlet and pipe end is the same regardless of the magnitude of m, the assumed uniform flow case, in which the change in piezometric is due entirely to friction loss, provides the harshest test of Camp and Graber's method of assuming a uniform flow and then evaluating the departure from that assumption. This leads to the conclusion that Camp and Graber's assumption of uniform outflow provides an accurate approximation for all fL/D, provided $m > 0.90$, which covers most practical situations. For $m < 0.90$, Camp and Graber's method provides a conservative design.

Table 11.2 Tabulations for Example 11.2

$V = 10$ fps

$V^2/(2g) = 0.3106$ ft

For $d = 3/8''$, a (ft^2) = 7.670E-04

Q (gpm)	Q (cfs)	Q (l/s)	Flow per hole q for 10 holes (cfs)	Nominal Velocity through each hole v_o (fps)	Head loss through each hole h_o (ft)	Manifold Inlet Velocity Head $V^2/(2g)$ (ft)	Manifold Cross-Sectional Area A (ft^2)	Manifold Diameter D (ft)	Manifold Diameter D (in.)	Manifold Entrance **R**	Approx. f	Approx. fL/D	$h_m - h_o$ (ft)	$h_o - h_m$ (ft)	m
1	0.002228	0.06309	0.000222816	2.905	0.35218	0.31056	0.0002228	0.01684	0.2021	16,843	0.05	14.8	−1.226	1.226	#NUM!
2	0.004456	0.12619	0.000445633	5.810	1.40873	0.31056	0.0004456	0.02382	0.2858	23,820	0.05	10.5	−0.777	0.777	0.670

Cannot have reasonable distribution uniformity. Determine pipe diameters giving $m = 0.95$.

$\Sigma h/h_o = 0.0975$

Assumed V = 5 fps

$V^2/(2g) = 0.1553$ ft

For $d = 3/8''$, a (ft^2) = 7.670E-04

Q (gpm)	Q (cfs)	Q (l/s)	Flow per hole q for 10 holes (cfs)	Nominal Velocity through each hole v_o (fps)	Head loss through each hole h_o (ft)	Σh (ft)	Manifold Inlet Velocity Head $V^2/(2g)$ (ft)	Manifold Cross-Sectional Area A (ft^2)	Mani fold Diameter D (ft)	Mani fold Diameter D (in.)	Manifold Entrance **R**	Approx. f	Approx. fL/D	$h_L - h_o$ (ft)	Actual m
1	0.002228	0.06309	0.000222816	2.905	0.35218	0.0343	0.15528	0.01262	0.12676	1.5211	63,378	0.014	0.5522	0.1267	0.800
2	0.004456	0.12619	0.000445633	5.810	1.40873	0.1374	0.15528	0.02524	0.17926	2.1511	89,629	0.014	0.3905	0.1351	0.951

$\Sigma h/h_o = 0.0975$

Assumed V = 1.4 fps

$V^2/V(2g) = 0.0435$ ft

For $d = 3/8''$, a (ft^2) = 7.670E-04

Q (gpm)	Q (cfs)	Q (l/s)	Flow per hole q for 10 holes (cfs)	Nominal Velocity through each hole v_o (fps)	Head loss through each hole h_o (ft)	Σh (ft)	Manifold Inlet Velocity Head $V^2/(2g)$ (ft)	Manifold Cross-Sectional Area A (ft^2)	Manfold Diameter D (ft)	Manfold Diameter D (in.)	Manifold Entrance **R**	Approx. f	Approx. fL/D	$h_L - h_o$ (ft)	Actual m
1	0.002228	0.06309	0.000222816	2.905	0.35218	0.0343	0.0435	0.04507	0.23954	2.8745	33,536	0.023	0.4801	0.0365	0.947

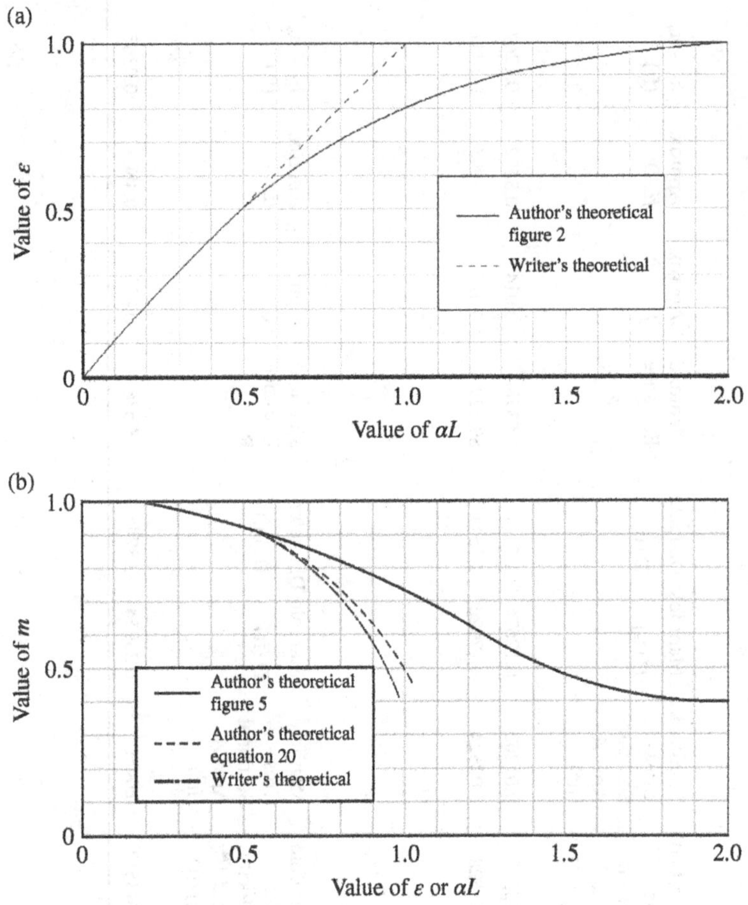

Figure 11.3 (a) Coefficient of Inlet Flow ε as Function of Similarity Number αL (Camp and Graber 1972) (b) Distribution Uniformity m as Function of ε and αL (Camp and Graber 1972/American Society of Civil Engineers)

A practical advantage of the method suggested herein over Zsák's method and other published approximations known to the author is that the problem of spacing and sizing orifices is uncoupled from the problem of determining the manifold material and diameter. Given the common problem of designing a manifold for a given Q, L, m and allowable h_o, Equations 11.3 and 11.49 may be used to establish f and d by a trial-and-error process in which only values of d need be assumed. Allowable values of $s/\tilde{\alpha}$ are thus established in accordance with:

$$s = \frac{\tilde{\alpha}\sqrt{h_o}L}{AV_o} \tag{11.43}$$

11.3.1.2 Uniform Distribution from Uniformly-Tapered Conduits

Uniformly tapered *square* conduits have been used to distribute flow to settling tanks when a velocity sufficient to maintain solids in suspension is required (such as the facility which is the subject of the example given below). Ordon (1968) presented results of Mayer's work on uniformly tapered *circular* conduits.

Noting that, in Equation 11.30,

$$\frac{Q}{g}\frac{dQ}{dx} = \frac{1}{g}\frac{Q}{A}\frac{dV}{dx} = \frac{1}{g}V\frac{dV}{dx} = \frac{d}{dx}\left(\frac{V^2}{2g}\right) \tag{11.44}$$

Equation 11.31 may be expressed in a form useful for tapered conduits as follows:

$$\frac{dh}{dx} = -\frac{d}{dx}\left(\frac{V^2}{2g}\right) - \frac{V^2}{2g}F \tag{11.45}$$

A rigorous analysis is required to demonstrate the assumptions involved here. The analysis by momentum methods is similar to that leading to Equation 2.109 (particularly with regard to the pressure terms) and results in the following equation:

$$\frac{dh_p}{dx} = -F\frac{V^2}{2g} - \frac{\beta Q}{gA^2}\left(\frac{dQ}{dx} - \frac{Q}{A}\frac{dA}{dx}\right) \tag{11.46}$$

The above equation is approximated by Equation 11.31a when $dA/A \ll dQ/Q$, or roughly stated, when the rate of taper is small compared to the rate of flow change.

For uniformly tapered circular or square conduits, substituting $F = f/4R$, $R = R_o(1 - fx/L)$; $A = A_o(1 - rx/L)$, $V = Q/A$, and Equation 11.40a into Equation 11.42 gives:

$$\frac{dh}{dx} = -\frac{V_o^2}{2g}\frac{d}{dx}\left[\frac{(1-x/L)^2}{(1-rx/L)^4}\right] - S_{Q_oR_o}\frac{(1-x/L)^2}{(1-rx/L)^5} \tag{11.47a}$$

$$S_{Q_oR_o} = \frac{f}{4R_o}\frac{V_o^2}{2g} \tag{11.47b}$$

To determine if an extreme value of the HGL (Hydraulic Grade Line) exists within the length L, dh/dx is equated to zero in Equation 11.47a, and the resulting equation is solved for x/L giving:

$$\frac{x_m}{L} = 1 - \frac{2-2r}{2r + fL/(4R_o)} \tag{11.48}$$

This relation reduces to Equation 11.46 for untapered conduits when $r = 0$, and indicates that for an extreme value of the HGL to exist within the length L, the term $4r + fL/(4R_o)$ must exceed 2.

The solution to Equation 11.47a is obtained by integrating between the limits $x = 0$ and x, giving:

$$\Delta h_x = \frac{V_o^2}{2g}\left[1 - \frac{(1-x/L)^2}{(1-rx/L)^4}\right] - S_{Q_oR_o}L\cdot$$
$$\left\{\frac{1}{4r}\left(1 - \frac{2}{r} + \frac{1}{r^2}\right)\left[\frac{1}{(1-rx/L)^4} - 1\right] + \frac{2}{3r^2}\left(1 - \frac{1}{r}\right)\left[\frac{1}{(1-rx/L)^3} - 1\right] + \frac{1}{2r^3}\left[\frac{1}{(1-rx/L)^2} - 1\right]\right\} \tag{11.49}$$

For $x = L$, this reduces to simply

$$\Delta h_L = \frac{V_o^2}{2g} - S_{Q_oR_o}L\frac{4-3r}{12(1-r)^2} \tag{11.50}$$

which reduces to Equation 11.39a for $r = 0$.

As in the case of the constant area conduit, depending on the conduit geometry and friction factor, the maximum difference in HGL may equal Δh_L, Δh_m, or $\Delta h_L - \Delta h_m$, in which Δh_m is the value of Δh_x at the location given by Equation 11.48. The latter occurs in the example given below.

Letting H_T denote the energy loss in the tapered dispersion conduit and defining $H = (1/3)S_{Q_oR_o}L$, we have:

$$\frac{H_T}{H} = \frac{4-3r}{4(1-r)^2} \tag{11.51}$$

When $r = 0$ the energy loss is the same as for a prismatic conduit (A = constant), but as the degree of taper increases, the energy loss increases.

A solution can similarly be obtained for rectangular conduits tapered in height and/or width. The case of a rectangular conduit tapered only in height will be considered here. Letting the height and width be denoted by z and b, respectively, consider the case in which b is constant and z decreases uniformly. (Note that the role of b and z could be easily reversed.) The following geometric relationships apply:

$$z = z_o\left(1 - r\frac{x}{L}\right) \tag{11.52a}$$

$$A = bz_o\left(1 - r\frac{x}{L}\right) = A_o\left(1 - r\frac{x}{L}\right) \tag{11.52b}$$

$$P = 2b + 2z_o\left(1 - r\frac{x}{L}\right) \tag{11.52c}$$

$$R = \frac{A_o(1 - rx/L)}{2b + 2z_o(1 - rx/L)} \tag{11.52d}$$

Substituting these geometric relationships, $F = f/(4R)$, $V = Q/A$, and Equation 11.40a into Equation 11.51 and proceeding as in the derivation of Equation 11.49 results in:

$$\Delta h_L = \frac{V_o^2}{2g} - \frac{S_{Q_oR_o}L}{(1 + z_o/b)(1 - r)}\left[\frac{2 - r}{2(1 - r)} + \frac{z_o}{b}\right] \tag{11.53}$$

in which $S_{Q_oR_o}$ is as defined in Equation 11.43b.

Example 11.3

A battery of primary settling tanks in an existing WWTP is fed by two tapered influent conduits (Tallman Island WWTP East Battery, New York City). These concrete (n = 0.014) conduits have a 4.5 ft square inlet, which tapers to a 2.5 ft square section in a length of 74 ft. The crowns of the distribution conduits are always below the tank weir elevation, so the conduits always flow full. Nine equally spaced 2 × 2 ft square sluice gates along each conduit distribute the flow to the tanks. The gates are normally in their full-open positions. Make a rough estimate of the distribution uniformity. Determine the modifications, if any, necessary to give a distribution uniformity of at least 90% and the corresponding total head loss with a maximum total flow to the battery of 88 mgd.

Solution

The applicable relationships in this case are Equations 11.8, and 11.48 to 11.50. The first step consists of determining whether a low point in the HGL occurs within the conduit length. This will be the case if $4r + [fL/(4R_o)] > 2$. Pertinent values are determined as follows:

$$r = \frac{z_o - z_L}{z_o} = \frac{4.5 - 2.5}{4.5} = 0.444$$

$$R_o = 4.5/4 = 1.125 ft$$

$$R_o^{1/3} = (1.125)^{1/3} = 1.04$$

$$f = 117\frac{n^2}{R^{1/3}} = 117\frac{(0.014)^2}{1.04} = 0.0221$$

$$4r + \frac{fL}{4R_o} = 4(0.444) + 0.0221(74)/[4(1.125)] = 2.14 > 2$$

Therefore, there is a low point in the HGL within the conduit, the position of which can be calculated from Equation 11.48 as follows:

$$\frac{x_m}{L} = 1 - \frac{2 - 2(0.444)}{2(0.444) + (0.0221)(74)/[4(1.125)]} = 0.111$$

Equations 11.49 and 11.50 can now be used to determine the values of Δh at $x = x_m$ and $x = L$. The flow to each distribution conduit is $Q_o = 44(1.547) = 68.1$ cfs, which, with an inlet area of $A_o = 4.5(4.5) = 20.25$ sq ft, gives an inlet velocity $V_o = 68.1/20.25 = 3.36$ ft/sec. Thus $V_o^2/(2g) = (3.36)^2/[2(32.2)] = 0.175$ ft and Equation 11.43b gives:

$$S_{Q_oR_o} = \frac{0.0221}{4(1.125)}(0.175) = 0.000861$$

Equation 11.49 evaluated at $x_m/L = 0.111$ gives:

$$(1 - rx/L) = [1 - 0.444(0.111)] = 0.9507$$

$$\frac{1}{4r}\left(1 - \frac{2}{r} + \frac{1}{r^2}\right) = \frac{1}{4(0.444)}\left[1 - \frac{2}{0.444} + \frac{1}{(0.444)^2}\right] = 0.8830$$

$$\Delta h_m = 0.175\left[1-\frac{(1-0.111)^2}{(0.9507)^4}\right]-0.000861(74)\cdot$$

$$\left\{\frac{0.8830}{4(0.444)}\left[\frac{1}{(0.9507)^4}-1\right]+\frac{2}{3(0.444)^2}\left(1-\frac{1}{0.444}\right)\left[\frac{1}{(0.9507)^3}-1\right]+\frac{1}{2(0.444)^3}\left[\frac{1}{(0.9507)^2}-1\right]\right\} = -0.00144\text{ ft}$$

Equation 11.50 gives:

$$\Delta h_L = 0.175-0.000861(74)\frac{4-3(0.444)}{12(1-0.444)^2} = 0.1292\text{ ft}$$

The maximum difference in HGL occurs between the intermediate section $x_m = 0.111L$ and the downstream end $x = L$, and is given by: $\Delta h_{max} = 0.00144 + 0.1292 = 0.131$ ft.

A rough estimate of the distribution uniformity may be made by using Equation 11.8b with C_+/C_- taken as 1. First, the approximate head loss across the sluice gate is obtained as follows: Q/gate = 68.1/9 = 7.57 cfs, $v = Q/A = 7.57/[2(2)] = 1.89$ ft/sec, $h = KV^2/(2g) = 2.69(1.89)^2/64.4 = 0.150$ ft. Then, $m \cong \sqrt{1-0.131/0.150} = 0.36$. Thus, the discharge through the gate of minimum flow is on the order of 36% of the discharge through the gate of maximum flow. Although a rough approximation (not only because of the assumption $C_+/C_- \cong 1$but also because of the uniform flow assumption in estimating Δh), the distribution uniformity is clearly unacceptable.

Equation 11.8 can be used in conjunction with the approximate relationship for $C_+/C_- = 1$ to obtain h_+. The relationship by Vigander *et al.* (1970) is most applicable and is given by:

$$C = \frac{0.63-0.58V^2/(2gE)}{\sqrt{1-V^2/(2gE)}} \tag{11.54}$$

in which E is local energy head = $h + V^2/(2g)$. A plot of the above equation reveals that, although $V^2/(2gE)$ appears in both the numerator and denominator, C decreases with increasing $V^2/(2gE)$.

Since $V^2/(2gE)$ becomes very small at the downstream-most gate through which the flow is largest, C_+ can be taken as 0.63 (USBR 1965, Table 30), so that:

$$\frac{C-}{C+} = \frac{0.63-0.58V_o^2/(2gE_o)}{0.63\sqrt{1-V_o^2/(2gE_o)}}$$

The term E_o can be taken as $E_o = h_+ - \Delta h_L + V_o^2/(2g) = h_+ - 0.129 + 0.175 = h_+ + 0.046 \cong h_+$.
Substituting the above relationship for C_-/C_+ (with $E_o \cong h_+$) into Equation 11.8a gives:

$$m = \frac{0.63-0.58V_o^2/(2gh_+)}{0.63\sqrt{1-V_o^2/(2gh_+)}}\sqrt{1-\frac{\Delta h}{h_+}}$$

For $V_o^2/(2g) = 0.175$ ft, $\Delta h = 0.131$ ft, and $m = 0.90$, this relationship becomes:

$$0.90 = \frac{0.63-0.58(0.175)/h_+}{0.63\sqrt{1-0.175/h_+}}\sqrt{1-\frac{0.131}{h_+}}$$

This relationship is solved by trial-and-error (easily done on a programmable calculator) to obtain $h_+ = 1.38$ ft.

The sluice gate opening (area a) required to obtain this 1.38 ft head loss is determined as follows: Q/gate = 68.1/9 = 7.57 cfs, $h = Kv^2/(2g) + v^2/(2g) = h/k = 1.38/2.69 = 0.513$, $v = \sqrt{2(32.2)(0.513)} = 5.75$ ft/sec, $a = (Q/gate)/v = 7.57/5.75 = 1.32$ sq ft.

The calculated area of 1.32 sq ft is less than the 4 sq ft full opening of the gate (as expected), indicating that the gates should be throttled to obtain the desired distribution uniformity when all tanks are in operation. The desired gate opening is calculated as the desired area divided by the gate width, or $1.32/2 = 0.66$ ft $\times 12 = 7.9$, say 8 in. The corresponding total head loss between the settling tank and inlet of each distribution conduit can be conservatively taken as $h_+ = 1.38$ ft from above.

In this particular case, position indicators were added to the sluice gates to enable the prescribed opening to be maintained. The head loss was considered acceptable and factored into the determination of the design TDH (Total Dynamic Head) for upstream pumping station modifications.

11.3.1.3 Rotating Manifolds

Rotating manifolds are commonly used in conjunction with circular wastewater trickling filters. Ordon (1966) has analyzed the piezometric head variations in such manifolds, the objective of which he states as follows: "The objective of this type of manifold is to spray sewage over the filter at a uniform rate over the entire area. As the sewage travels from the center of the filter, the area serviced per foot of pipe length increases linearly. This means that the rate at which sewage flows from the pipe through the ports increases along the length of the pipe and the rate of flow of that sewage remaining in the pipe decreases at the square of the length." Mathematical details are given by Ordon.

Other Methods. The relatively large spacing between sprinklers has prompted some hydraulic engineers to base analyses of sprinkler laterals on "lumped-parameter" approaches. An example of this approach, in which each length of pipe between takeoffs is treated as a discrete uniform flow section, is given by Benami (1968). This approach is distinguished from this author's "distributed-parameter" approach, in which flows and head losses are assumed to be distributed continuously along the length of the conduit. The results of the distributed-parameter approach are easier to apply and can often be as accurate as lumped-parameter methods when couplings and outlets are regularly spaced.

A computer-based numerical solution has been developed by Vigander *et al.* (1970) and applied to thermal-discharge diffusers that would be equally applicable to certain types of wastewater diffusers. A comparison of the results of the computer-based methods and the simplified methods presented above was made for one particular case (Camp and Graber 1970). The results agreed closely, and the different methods should result in a similar design. For major diffuser structures, particularly for configurations having a nonconstant value of C, the more exact numerical procedure might be preferred for final evaluation.

11.3.2 Open Channels

Camp and Graber (1968, 1969) presented a general formulation of the open-channel distribution conduit problem analogous to that presented above for pressure conduits. For a *prismatic* open conduit, they demonstrated that a relative *minimum* value of the hydraulic grade line exists within the length L when FL exceeds 2, at the position given by Equation 11.36. Closed-form solutions of the spatially decreasing flow problem, including friction, are generally not attainable for open channels. An incremental formulation can be used to facilitate numerical solutions of the problem (Camp and Graber 1968).

However, the dimensions and flow characteristics of open channels used for distribution conduits are, in most cases, such that friction loss is small and the recovery of velocity is the dominant factor influencing the piezometric head. In such a case, the outlets of minimum and maximum discharge are the first and last outlets along the conduit, respectively. The exact nature of the piezometric head variation along the conduit need not be determined because the change in piezometric head between the outlets of minimum and maximum discharge is given approximately (and somewhat conservatively) by $V_o^2/(2g)$. The exceptions are sufficiently few that it is worthwhile to start with the premises stated above and to evaluate their applicability and deviations therefrom.

If the velocity head recovery does, in fact, dominate, then a conservatively high estimate of the overall friction loss can easily be obtained by assuming a constant depth of flow equal to the upstream depth in the channel. Considering a horizontal, uniformly tapered channel, an analysis similar to that given above for the rectangular conduit tapering in one direction can be made to calculate Δh_L. Changing Equation 11.52c to $P = 2d + b_o(1 - rx/L)$ to allow for the absence of wetted perimeter at the top of the conduit, and otherwise following through the same derivation that led to Equation 11.53, gives the following equation for Δh_L (note h now equals channel depth):

$$\Delta h_L = \frac{V_o^2}{2g}\left(1 - \frac{1}{3}\frac{fL}{4\overline{R}}\right) \tag{11.55a}$$

$$\frac{1}{\overline{R}} \cong \frac{3}{r^3 y_o}\left\{2\left[(1-r) - \frac{y_o}{b_o}\right]\ln(1-r) + 2r\left(1 - \frac{y_o}{b_o}\right) - r^2\left(1 + \frac{y_o}{b_o}\right)\right\} \tag{11.55b}$$

In the above equation, Δh_L is the rise in water surface over the channel length, $\overline{R}$ is the "effective" hydraulic radius, b is channel width, r is the taper $(b_o - b_L)/b_o$, and subscripts "o" and "L" represent conditions at the upstream and downstream ends of the channel, respectively. In almost all practical cases, $fL/(4\overline{R}) \ll 1$ so that Equation 11.55a can be approximated by:

$$\Delta h_L \cong \frac{V_o^2}{2g} \tag{11.56}$$

The term $\overline{R}$ may be thought of as the "effective" hydraulic radius and approaches $R_o = 2d + b_o$ as $(b_o - b_L)/b_o$ approaches zero. In almost all practical cases of effective open-channel distribution conduits, it will be found that $[(1/3)fL/(4\overline{R})] \ll 1$ and $\Delta h_L < V_o^2/(2g)$ (see the examples below). Note that tapering then, other than as it may influence V_o (e.g., by enabling a larger b_o), does not influence Δh_L since the initial velocity head is recovered in its entirety. In theory, a channel can be tapered in such a way that a flat water surface is maintained throughout. Although an interesting mathematical exercise, such a channel is impractical because of its complex shape and the need to make $b = 0$ at the downstream end unless some of the flow is to continue beyond the distribution channel.

As for pressure conduits, tapering can maintain sediment-scouring velocities in distribution channels. Tapered distribution channels are, however, less common than untapered channels, primarily because: (1) tapering is often not sufficient to maintain scouring velocities over the range of influent flows; (2) if channel agitation for solids suspension (see Chapter 13) is necessary, tapering makes such agitation more difficult; and (3) tapering limits the possibilities for future expansion.

Other reasons for the adequacy of the simple estimate of Δh_L as suggested above are given later in this chapter. However, in rare cases, a detailed analysis of the water surface profile along a distribution channel may be warranted. This may be accomplished by treating each outlet as discrete takeoff and calculating the backwater profile between outlets by a step method (see Chapter 7). A computer-based approach is discussed by Camp and Graber (1970).

As discussed above, open-channel distribution conduits generally have a water surface rising in the direction of flow, with the first and last outlets having the minimum and maximum outflow, respectively. Thus, Equation 11.11 becomes:

$$m = \left(1 - \frac{\Delta y_L}{y_L - S}\right)^{3/2} \tag{11.57a}$$

$$\frac{y_L - S}{\Delta y_L} = \frac{1}{1 - m^{2/3}} \tag{11.57b}$$

in which Δy_L is the rise in water surface over the channel length L.

For an existing or assumed channel and weir configuration, the flow over the downstream-most weir can be approximated by $q = Q/N$, where N is the number of weirs, with which the sharp-crested weir equation (see Section 7.2) can be used to calculate y_L. Then Δy_L can be calculated from $\Delta y_L = Q_o^2/[2gb_o^2(y_L - \Delta y_L)^2]$. Equation 11.57a then gives the uniformity ratio m. The upstream y_o should be taken conservatively as y_L for subsequent hydraulic profile determinations. Conversely, if m and $E \cong y_L$ are selected (the latter corresponding to selected weir characteristics), then Equation 11.57a gives Δy_L, which, in turn, gives $y_o = y_L - \Delta y_L$. The upstream channel width can then be calculated. Other variations on the procedure are possible.

It should be noted that the above procedure essentially approximates the effective head as being h at the upstream end of the first weir and the downstream end of the last weir and neglects the variation in h along the length of each weir. The side weir relationships given in Section 7.2 can be used for a more precise determination, if needed. However, the relationships given in this section are adequate for most practical purposes.

Example 11.4

Thirty-six distribution channels distribute influent across the passes of activated sludge tanks in an existing wastewater treatment plant (Fort Worth, Texas, Village Creek). The flow is divided nearly equally among each distribution channel. Each channel has a width of 2.5 ft and has seventy 90° V-notches along its sides, the V-notch vertices being 2.5 ft above the channel bottom. Determine the distribution uniformity of each channel for a peak total flow of 120 mgd.

Solution

Equation 11.13a applies with $S = 2.5$ ft and $C_-/C_+ \cong 1$. The flow per channel is 120 mgd/36 = 3.33 mgd × 1.547 = 5.16 cfs. The flow per notch is $q = 5.16/70 = 0.0737$ cfs/notch. Using the V-notch weir equation in Section 7.2 with coefficient and exponent rounded to 2.5, the head on each weir is calculated as follows:

$$(h - S)^{5/2} = \frac{q}{C} = \frac{0.0737}{2.5} = 0.0295$$

$$h - S = (0.0295)^{2/5} = 0.244\ \text{ft}$$

The corresponding channel depth is $h = 0.244 + S = 0.244 + 2.5 = 2.74$ ft, at which the corresponding inlet velocity and velocity head are given by:

$$V_o \cong \frac{5.16}{2.5(2.74)} = 0.753\ \text{ft/sec}$$

$$\frac{V_o^2}{2g} = 0.00881\ \text{ft}$$

In this case, $\Delta h_{max} \cong V_o^2/(2g)$ and the distribution uniformity is given by:

$$m = \left(1 - \frac{0.00881}{0.244}\right)^{5/2} = 0.91$$

The distribution uniformity of 91% and better was considered acceptable for this application.

11.4 Slope Invariance for Pressure Conduits

Pressure distribution systems for distributing partially treated wastewater to on-site subsurface drain fields have come into widespread use over the past 30 years. A major impetus for this was the publication of simplified design methods by Otis (1981, 1982). The pressure distribution can be accomplished by means of a storage chamber and pumps or dosing siphons. Numerous states, as well as the US Environmental Protection Agency, have encouraged the use of pressure distribution systems, and some have prepared related guidelines or standards. These systems are the subsurface analogy of the intermittent sand filters discussed in Chapter 12. An important benefit of such systems is the return to groundwater, and subsequently to rivers during periods of low flow, of water that has commonly been withdrawn from wells in the same vicinity or otherwise. Unfortunately, those guidelines and related literature are erroneous, as discussed with citations, corrections, and correct applications to sloped beds by Graber (2010b, 2010c).

The following is taken from Graber (2010c) regarding the type of distribution systems discussed in the previous paragraph. A primary purpose of the distribution system is to dose intermittently at a flow rate higher than the actual wastewater inflow. The distribution system is designed for approximately equal flow distribution over the bed once the distribution system is full; this greatly reduces spatially-progressing clogging and substantially increases the longevity of the field. By providing a dose volume, which is approximately 5–10 times the volume of the distribution system, most of the dosing period occurs when the distribution system is full, and the duration of unequal distribution during the beginning of the dosing period, when the distribution system is filling, is made acceptably small. The period of unequal distribution, or the effective volume of the distribution system that must be filled during the period of unequal distribution, may be reduced by having the laterals containing the dosing orifices placed above the manifolds which supply them, with the manifolds connected to the laterals by tees and vertical risers. The systems thus have manifolds connected at tees to risers to which flow must be distributed to an acceptable degree of uniformity. The risers, in turn, connect to tees, which supply laterals along which flow must be distributed to orifices with acceptable uniformity. [A figure in Graber (2010c) gives an example taken from actual design plans prepared by the author.] Manifolds supplying the laterals may have either center-feed or end-feed configurations. A tee on the manifold will have its branch connected to the vertical riser, which in turn connects to the branch of a tee on the lateral. The tee on the lateral can be a reducing tee, a bushing, a reducer, or a combination of these, which can affect a reduction in the lateral diameter.

In some cases, the distribution uniformity of the manifold may not be adequate considering head losses imposed by head losses in the laterals alone. In such cases, an orifice plate can be placed in the vertical riser to increase the head loss affecting the manifold. The placement of the orifice and the length of the riser must be such as to allow full expansion of the jet within the riser. Figure 11.4 depicts the riser connection with an orifice as designed by the author for a large, sloped leaching field in Littleton, Massachusetts (9975 gpd system). The length of the riser was determined based on the jet expansion distance for the smaller orifice. A 1:8 expansion was estimated for the orifice jet (Exley and Brighton 1969). The flow per riser is 33.0 gpm × 1/448.831 cfs/gpm = 0.0735 cfs, and the riser area is $(\pi/4)(2/12)^2 = 0.0218$ ft^2, giving the velocity in the riser = 0.0735/0.0218 = 3.37 ft/sec. The orifice diameter is 0.697 in. and the orifice area is

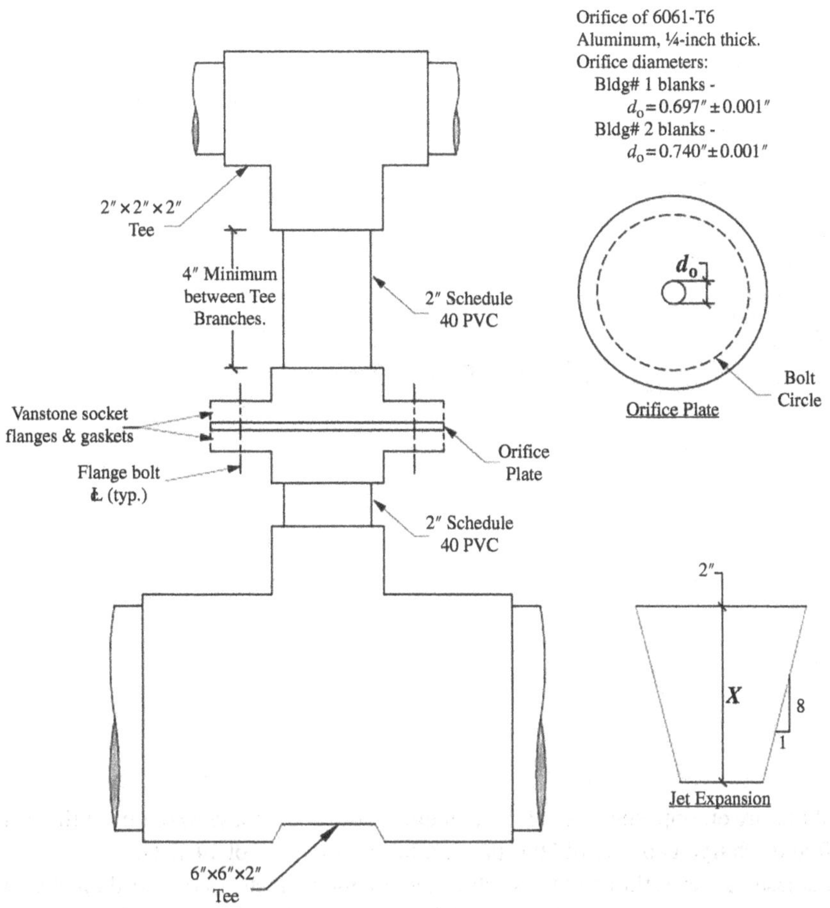

Figure 11.4 Riser Connection Between Manifold and Lateral.

$(\pi/4)(0.697)^2 = 0.382$ in.$^2 = 0.00265$ ft^2. The Reynolds Number in the riser is $\mathbf{R} = VD/\nu = [3.37(2/12)]/10^{-5} = 5.62(10)^4$ and the ratio of orifice area to riser area is 0.00265/0.0218 = 0.1216; from these two values, Streeter (1962, Figure 9–17) gives the orifice discharge coefficient $C = 0.62$. The initial diameter of the jet is then $D \cong \sqrt{C}D_o = \sqrt{0.62}(0.697) = 0.549$ in. Figure 11.4 shows the relevant dimensions from which $x = 8(2/2 - 0.549/2) = 5.80$ in. From this, we subtract the distance from inside the wall of the pipe connecting to the upper tee down to the end of the upper tee run of 1.955 in., giving the distance available for jet expansion of 5.80 − 1.955 = 3.45 in. Because the orifice is at least 0.48 in. into the socket, the clear spacing of 3 in. between tees allows for full expansion of the jet within the riser.

The author designed a system incorporating a riser connection similar to that shown in Figure 11.4 for an 8800 gpd system for Franklin Place in Acton, Massachusetts. In that case, a check for cavitation with the original design iteration with a 6-in. diameter manifold showed the low pressure near the orifice to be below vapor pressure. A 4-in. diameter manifold was selected, which indicated no cavitation with the necessary price of additional system head loss.

Similar methods of dosing are used for intermittent sand filters, which are the surface analogy of subsurface drain fields. That older technology is believed to have been first built in 1889 by S.C. Heald in Framingham, Massachusetts (WPCF/ASCE 1959, p. 25). Open channels may be used for flow distribution in such cases, with design for uniform flow distribution using the methods suggested by Graber (1981, 2004d) [the more recent of those publications corrects the difference formulation in a widely used reference]. The following example pertains:

The author designed two dosing siphons for the Bridgewater Massachusetts Correctional Facility to enable channels to distribute flow more equally to sand filter beds without the need for pumping. The dosing siphons include a cast-iron deep-seal trap with a blowoff trap, a steel bell, and an auxiliary bell to automatically and alternately discharge to their respective

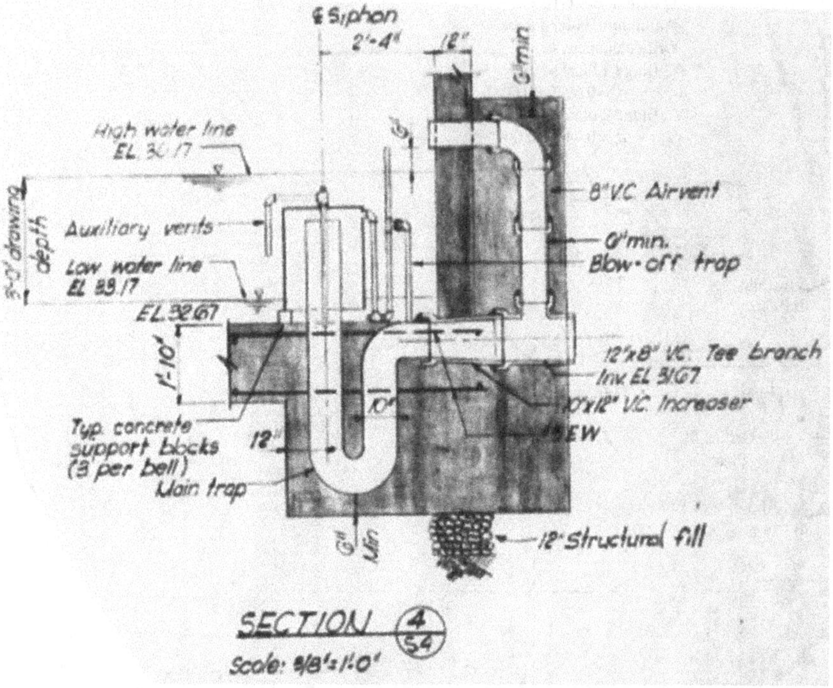

Figure 11.5 Bridgewater MCI Dosing Siphon

sand filter beds in proper sequence. The siphons are on opposite sides of the effluent basin. A section view of one of them is shown in Figure 11.5. Each siphon has a free-discharge capacity of 700 gpm at a minimum head of 14-1/2 in.

The branch entrance losses generally increase as the ratio of the takeoff (riser) diameter D_l to manifold diameter D decreases (Graber 2010b). But it is important to recognize that branch entrance losses can be significant even for $D_l/D = 1$ (Graber 2010b). For the most common case of horizontal wastewater distribution manifolds, it is common (and not disadvantageous) to have $D_l/D = 1$. A useful fit to McNown's (1954) data (see Graber 2010b) is the following fifth-order polynomial:

$$\frac{h_f'}{V^2/(2g)} = 5.300\left(\frac{Q_l}{Q}\right)^5 - 9.234\left(\frac{Q_l}{Q}\right)^4 + 3.605\left(\frac{Q_l}{Q}\right)^3 + 2.072\left(\frac{Q_l}{Q}\right)^2 - 1.605\left(\frac{Q_l}{Q}\right) + 1 \tag{11.58}$$

in which h_f = pressure head loss between the main conduit and branch (riser); V = velocity in the main conduit upstream of the flow division; Q_l = flow rate in the branch; and Q = flow rate in the main conduit upstream of the branch. This relationship has an average- and maximum- absolute difference from the data of about 0.40% and 5% (essentially within the range of scatter of the data points about a faired curve), respectively, with the maximum difference decreasing to about 2% when only Q_l/Q values less than 0.5 and equal to 1.0 are considered (the range of values of Q_l/Q when distribution is uniform). A similar polynomial can be fit to the curves for D_l/D equal to 0.25 and 0.5 (McNown's other data sets), and interpolation (parabolic or second-order polynomial) can be made between the three polynomials for a particular value of Q_l/Q. All of this can be readily implemented by computers once the polynomial fits into the curves for the three D_l/D values are obtained.

Unless an orifice is installed in the branch (see below), the branch (riser) should be at least five branch diameters in length (but sufficiently short that entrance losses dominate over wall friction). As discussed in Graber (2010b), this will help insure stability and the accuracy of the estimated branch entrance loss. For wastewater pressure distribution manifolds, which are sloped to supply laterals that are at different elevations (sloped beds), orifices may be installed in the risers to create head losses for improved flow distribution. In such cases, the author has found it preferable to install the orifice in a riser of reduced diameter (e.g., use a reducing tee to go from the manifold to the riser). The reason for this is that without the riser diameter reduction, the same pressure distribution across the orifice plate (with a given orifice opening) will result in a larger, and generally excessive, force acting on the orifice due to the larger orifice plate surface area. (The possibility of cavitation at the orifice must also be considered.) In this situation with orifices and $D_l/D < 1$, the branch entrance loss can be negligibly small compared to other losses within the branch. The assumption of negligibility should be checked in each case.

To avoid potentially excessive forces on the upper lateral tee and additional, unaccounted-for head losses, the riser length between the orifice and lateral tee should enable full expansion of the orifice jet within the riser. The contracted jet diameter can be approximated (ASME 1971, pp. 198–201), along with the diametric expansion rate (Exley and Brighton 1969).

A convenient device for specializing free-surface flow equations, such as Equation 10.4, to full-flowing conduits takes advantage of the fact that $D = (A/T) \rightarrow \infty$ as $T \rightarrow 0$ and that $y\cos\theta$ becomes the pressure head p/γ. Then Equation 10.4 reduces to:

$$\frac{d(p/\gamma + z)}{dx} = \frac{\tau_o}{\gamma R} + \frac{q_i}{gA}(U_i\cos\phi_i - 2\beta V) - \frac{q_o}{gA}(\beta_o V - 2\beta V) - \frac{V^2}{g}\frac{d\beta}{dx} \tag{11.59}$$

in which $(p/\gamma + z)$ is the piezometric head or hydraulic grade line.

Consider a specific physical situation in which the inflow and outflow terms, denoted respectively by the subscripts "i" and "o" in the above equation, are independent of conduit slope. Then all remaining terms in the above equation will also be independent of slope. The inflow q_i must enter the conduit at an angle ϕ_i, which remains constant relative to the conduit as the conduit slope varies. For spatially decreasing flow, it is only necessary that the conduit slope is constant or changes slowly enough that pressure remains approximately hydrostatic. All the terms on the right-hand side of the above equation are either fixed for the specific system (i.e., A, R, γ, g), or variable with no explicit (or implicit) dependency on slope (i.e., V, β, τ_o). This becomes apparent for τ_o by recognizing that the shear stress term in Equation 11.59 is a function only of V, R, ρ, and a roughness coefficient or friction factor (whether by Chezy, Darcy–Weisbach, Manning, or Hazen–Williams). The gradient of piezometric head will therefore also be independent of conduit slope. The equation may, in concept, be integrated over the fixed length of the specific configuration to obtain the piezometric head profile, which will also be independent of conduit slope.

In essence, the static pressure is superimposed on the flow field without affecting the dynamics of the system. The slope invariance also extends to the energy grade line because it is obtained by adding the velocity head (times any kinetic energy correction factor) to the hydraulic grade line (per Equation 10.12a). Awareness of this slope invariance, which can at times be counter-intuitive, has useful practical ramifications. For spatially decreasing flow in pressure distribution manifolds used for on-site wastewater disposal, it avoids the type of misinterpretation discussed above and enables correct design methods (Graber 2004d). As another example, the same spatially increasing flow equations applicable to vertical well screens can be applied to horizontal or sloped submerged effluent collectors, as discussed in Section 12.5. For such examples, the applicable equations can be appropriately specialized from Equation 11.59, or identically derived by applying momentum and continuity principles to control volumes encompassing conduit segments representing the specific application.

Some appreciation of the importance of velocity head recovery can be gained from Figure 10.2 based on Camp and Graber (1970) and Graber (2004d), which gives the variation in piezometric head (HGL) from the upstream end of the conduit for continuous, uniform discharge. The curves are for different values of FL, which for our purposes is fL/D. The curve labeled x_m/L gives the location of the extreme value of Δh_x. The relative significance of friction loss and velocity head recovery can be seen to depend upon the parameter FL. For smaller values of fL/D, the velocity head recovery dominates.

11.5 Kármán Vortex Street

The Kármán vortex street is a fluid phenomenon with an interesting history, and one to be aware of insofar as forces on certain hydraulic components are concerned. Such components of interest here include submerged pipeline crossings in rivers and water storage tanks. It is discussed in the present chapter because of its role in a particular distribution conduit. The phenomenon is also of interest in connection with flow measurement devices that make use of the phenomenon, as discussed below. We will first discuss the interesting history of the discovery of the Kármán vortex street and then turn attention to a practical application.

Numerous authors have written about the Kármán vortex street (e.g., Prandtl 1952, §13,14; Den Hartog 1956, §7.6; Schlichting 1960, §I.e and §II.b; Lugt 1983, §6.3; etc.). A particularly succinct history of its discovery is given by Govardhan and Ramesh (2005), from which the following is taken:

> In 1911, the celebrated German fluid dynamicist Ludwig Prandtl was investigating the static pressure distribution along the surface of a circular cylinder when it was placed in a steady stream of water. Hiemenz, a doctoral candidate, was given the task of making the measurements. Hiemenz was very frustrated to repeatedly find unsteady fluctuations in the channel. When Hiemenz reported this to Prandtl, he was told that the cylinder was probably not circular. Despite spending considerable effort in polishing and smoothing the cylinder by employing the famed German fastidiousness for precision, Hiemenz still could not get rid of the flow oscillation. When he reported this to Prandtl, he was told that his channel was probably not symmetrical! Hiemenz started perfecting the channel. Kármán, who at the time was a graduate assistant working with Prandtl, would every morning religiously ask Hiemenz, "Herr Hiemenz, is the flow steady now?", and a crestfallen Hiemenz used to sadly reply, "It always oscillates."
>
> Kármán eventually figured out that the unsteady phenomenon must be intrinsic to the flow. Thoroughly absorbed by the problem, over a weekend, he took it upon himself to calculate the stability of a system of vortices. He showed that the observed vortex system of staggered asymmetrical vortices was the only stable system. And it is only a stable configuration that one gets to observe in practice. This contribution, which was presented to the Göttingen Academy by Prandtl and a subsequent paper, is in fact Kármán's contribution to this phenomenon. Over time, this contribution of Kármán ... resulted in the pattern being called the Kármán Vortex Street.
>
> .... Clearly, the most dangerous aspect of the Kármán vortex street is its periodic nature. It leads to possibilities of resonance with natural structural modes of flexible bodies and, hence, results in failure of structures that may otherwise have been safe. The infamous Tacoma Narrows Bridge disaster in the United States in 1940 is an example where the culprit was ... the structural vibration and resonance induced due to Kármán vortex shedding phenomenon.
>
> Strouhal, in 1878, made the first measurements of the frequencies associated with the shedding of vortices ....

An important correction needs to be made to the above. Delatte (1976) notes that although the Federal Works Agency initially attributed the bridge failure to resonance, "it is now recognized that the failure of the bridge was due to aeroelastic instability and flutter, not due to either resonance or vortex shedding."

Resonance is, however, the cause of failures of trashracks that have occurred at numerous installations (Sell 1971, Crandall *et al.* 1975, Rao *et al.* 1987, Nascimento *et al.* 2006, Morabito *et al.* 2020).

The dimensionless number denoted by **S**, named after Strouhal, has $\mathbf{S} = \omega L/V$, in which ω is the frequency of vortex shedding, L is a characteristic length of the body shedding the vortices, and V is the velocity of the approaching flow. The Strouhal Number is a function of the Reynolds Number and body shape.

One of the most conspicuous and common examples of Kármán vortex streets is observation of stop signs shown twisting back and forth in high winds as the vortices are alternately shed. The posts of such signs are designed for the *endurance limit*, which is the strength that levels off after a very large number of stress cycles.

The eddy shedding on alternate sides of a cylinder causes a harmonically varying force on the cylinder in a direction perpendicular to that of the main flow, with a maximum intensity given by Equation 6.55 expressed as follows:

$$(F_k)_{\max} = C_k \frac{1}{2} \rho V^2 A \tag{11.60}$$

in which the C_k, called the Kármán force coefficient, has a value of 1 with sufficient accuracy over a large range of Reynolds Numbers from 10^2 to 10^7 (Den Hartog 1956, §7.6; Schlichting 1960, §II.b). The intensity of the alternating force per unit of sidewise projected area is thus about equal to the stagnation pressure of the flow.

Example 11.5

A fluoride distribution pipe of monel with an outer diameter $D = 4.5$ in. is to be placed diametrically across the 11.6-ft inner diameter of a water pipe and built-in at both ends (Lawrence, Massachusetts). The maximum velocity V in the water pipe is 10 ft/sec. The distribution pipe is Schedule 80 monel, which has the following additional characteristics: I.D. = 3.826 in., modulus of elasticity $E = 26(10)^6$ psi, and weight per unit length $\mu_1 g = 16.88$ lb/ft. Determine the external forces on the pipe.

Solution

The drag force on the cylindrical pipe is determined first. The Reynolds Number for the flow over the cylinder is given by $\mathbf{R} = VD/\nu = 10(4.5/12)/10^{-5} = 3.75(10)^5$ for which the drag coefficient C_D is approximately 1.0 (Streeter 1962, p. 207). The drag force D_f is given by $D_f = C_D A \gamma V^2/(2g)$ in which A is the projected area of the fluoride pipe and γ is water density. Thus, $D_f = 1.0(4.5/12)(11.5)(62.4)(10)^2/[2(32.2)]\ (\text{lbf/ft}^3)(\text{ft}^2)(\text{ft/sec})^2/(\text{ft/sec}^2) = 418\ \text{lbf}$.

Next, determine the maximum force on the water pipe associated with the Kármán vortices and compare the frequency of that force to the natural vibrational frequency to determine whether a harmful resonant condition could occur. From Equation 11.60, with $C_k \cong 1$, and since $\rho = \gamma/g$, the alternating vertical nonresonant force associated with the Kármán vortices has maximum magnitude approximately equal to 418 lbf. The natural circular frequency ω_n for a beam clamped (or free) at both ends is given by:

$$\omega_n = n\pi\sqrt{\frac{A_p E}{\mu_1 L^2}}, \quad n = 1, 2, 3, \ldots$$

in which n is the mode (first, second, etc.), A_p is the pipe wall area, E is the modulus of elasticity, μ_1 is pipe weight per unit length, and L is the pipe length. The pipe wall area is calculated from values given above as $A_p = (\pi/4)[(4.50)^2 - (3.826)^2] = 4.41\ \text{in.}^2$. Substituting that value and others given above into the above equation for ω_n with $n = 1$ gives:

$$\omega_n = (1)\pi\sqrt{\frac{4.41\ \text{in}^2\left[26(10)^6\ \text{psi}\right]}{(16.88\ \text{lb/ft}/32.2\ \text{ft/sec}^2)(10)^2\ \text{ft}^2}} = 4640\ \text{cps}$$

From Schlichting (1960, p. 31, Figure 2.9), the Strouhal Number $\mathbf{S} = \omega D/V$ equals 0.21 for circular cylinders with Reynolds Numbers $>10^3$ (as in our case per above), in which ω is the frequency of vortex shedding and D and V are as defined above. Thus, $\omega = \mathbf{S}V/D = 0.21(10)/(4.5/12) = 5.6$ cps, which is well below the first-mode resonant frequency. So, it is not necessary to consider higher modes.

The horizontal drag force and alternating maximum vertical force are approximately equal at 418 lbf. A detailed structural analysis, which is beyond the scope of this book, led to the conclusion that there was ample factor of safety relative to the yield strength of the pipe.

An interesting application of the Kármán vortex street phenomenon is in flow measurement (O'Connnor 1991). From the above, the frequency of vortex shedding is a function of the Strouhal Number, characteristic length of cross-sectional shape, and fluid velocity. For circular cylinders, the Strouhal Number varies with the Reynolds Number, and the separation point shifts as the Reynolds Number varies. By using a diamond-shaped probe or "annubar" (the latter can also be round-shaped with multiple sampling points, which functions similar to a pitot tube with averaging), the separation point remains fixed, as shown in Figure 11.6. The Strouhal Number then remains nearly constant, so the frequency of vortex shedding is closely a

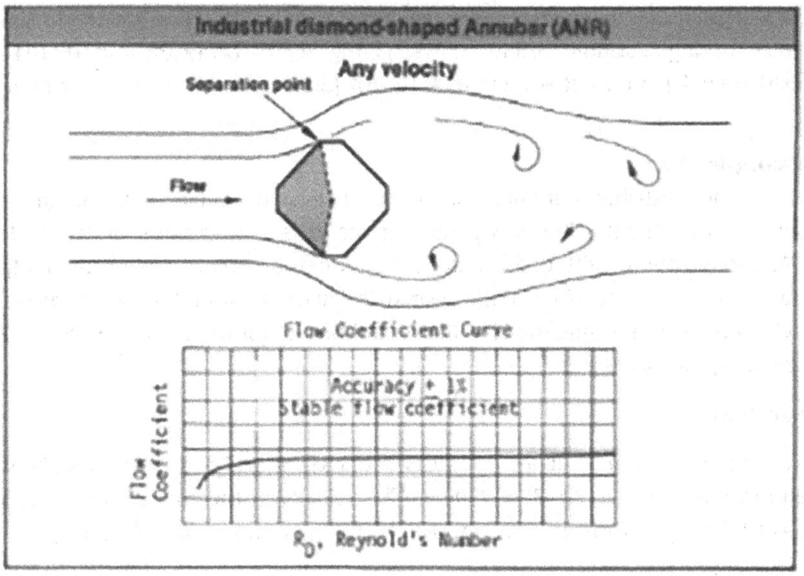

Figure 11.6 Diamond-Shaped Annubar by Dietrich Standard Corporation

function only of flow velocity. Vortex shedding flow meters are provided by Dietrich Standard, Inc., Boulder, Colorado; Georg Fischer Signet, Inc., Irvine, California; Yewflo, Yokogawa Corp., Newman, Georgia; and those once manufactured by Emco Controls can still be purchased.

Example 11.6

Chlorine solution diffusers in existing chlorine-contact tanks are to discharge a flow of 125 gpm per contact tank through four diffusers (Tallman Island WWTP, New York City). The diffusers have ½-in. holes, two holes per foot, sixteen ½-in. holes per diffuser, 3-in. diameter pipe. Determine adequacy of distribution uniformity.

Solution

Internal cross-sectional area, initial velocity, and velocity head are, respectively, $A_o = (\pi/4)(3)^2 = 7.07$ sq ft, $V_o = (31.25/449)/(7.07/144) = 1.418$ ft/sec, $V_o^2/(2g) = (1.418)^2/[2(32.2)] = 0.0312$ ft. The Reynolds Number $\mathbf{R} = VD/\nu = 1.418(3)/[12(10)^{-5}] = 3.55(10)^4$ at which the Moody Diagram (Figure 5.1) gives, for smooth pipe, $f = 0.023$. Next, we have $fL/D = 0.023(8)(12)/3 = 0.739 < 20.023(8)(12)/3 = 0.739 < 2$, for which Equation 11.39a gives $\Delta h_{\max} = 0.0312(1 - 0.739/3) = 0.0235$ ft. The flow per orifice and orifice area are, respectively, $q_o = 31.25/16 = 1.95$ gpm $= 0.00434$ cfs and $A_o = (\pi/4)(1/2)^2 = \pi/16 = 0.1964$ sq in. The contracted orifice discharge velocity is$q_o/(0.61A_o) = 0.00434(16)(144)/[0.61(0.1964)] = 5.20$ ft/ sec, corresponding to a head loss of $h_o = (5.20)^2/[2(32.2)] = 0.378$ ft, which is $>10V_o^2/(2g) = 0.312$. The actual distribution uniformity is $m = \sqrt{1 - \Delta h_{\max}/h_o} = \sqrt{1 - 0.0235/0.378} = 0.97$, which is good.

Example 11.7

Chlorine solution diffusers for new chlorine-contact tanks are to discharge a flow of 125 gpm per contact tank through two diffusers (Tallman Island WWTP). Initial design assumes ½-in. holes, two every foot over a 22-foot length, 3-in. Schedule 40 pipe with 1/16-in. rubber lining. Determine suitability for distribution uniformity.

Solution

The prescribed pipe has an internal diameter of 3.068 - 2(0.0625) = 2.943 in. Internal cross-sectional area, velocity, and velocity head are, respectively, $A = (\pi/4)(2.943)^2 = 6.80$ sq ft, $V = (62.5/449)/(6.80/144) = 2.95$ ft/sec, $V^2/(2g) = (2.95)^2/[2(32.2)] = 0.1351$ ft. The Reynolds Number $\mathbf{R} = VD/\nu = 2.95(2.943)/[12(10)^{-5}] = 7.23(10)^4$ at which the Moody Diagram (Figure 5.1) gives, for smooth pipe, $f = 0.019$. Next, we have $fL/D = 0.019(22)(12)/2.943 = 1.70 < 2$, for which Equation 11.39a gives $\Delta h_{\max} = 0.1351(1 - 1.70/3) = 0.0585$ ft. For 95% uniformity, we require controlling head loss $h_o = \Delta h_{\max}/(1 - m^2) = 10.3\Delta h_{\max} = 10.3(0.0585) = 0.603$ ft. The flow per orifice and orifice area are, respectively, $q_o = 62.5/[2(22)] = 1.42$ gpm $= 0.00316$ cfs and $A_o = (\pi/4)(1/2)^2 = \pi/16 = 0.1964$ sq in. The contracted orifice discharge velocity is $q_o/(0.61A_o) = 0.00316(16)(144)/[0.61\pi] = 3.80$ ft/sec, corresponding to a head loss of $h_o = (3.80)^2/[2(32.2)] = 0.224$ ft, which is less than 0.603 so is not acceptable. Smaller holes or a larger pipe is required for the prescribed hole spacing. Assume 3/8-in. holes, giving $A_o = (\pi/4)(3/8)^2 = 0.1104$ sq in. Then $q_o/(0.61A_o) = 0.00316(144)/[0.61(0.1104)] = 6.76$ ft/sec and $h_o = (6.76)^2/[2(32.2)] = 0.710 > 0.603$, which is acceptable.

Example 11.8

A channel distributes influent across ten activated sludge tanks in an existing 65 mgd wastewater-treatment plant (Providence, Rhode Island). A plant upgrade calls for a maximum design flow of 142 mgd. The distribution channel is 374 ft long with a width of 6 ft, and each tank has four 36-in. wide × 24-in. high gates along its bottom sides. The channel invert elevation is 123.83, and the desired design (maximum flow) water surface elevation at the upstream end of the channel is 129.28. Determine the distribution uniformity for the peak total flow. Modify the design to improve the distribution uniformity as necessary.

Solution

The channel entrance depth = (129.28 - 123.83) = 5.45 ft, and entrance flow area $A_o = 5.45(6) = 32.7$ ft². The channel entrance flow is 142(1.547) = 220 cfs. The velocity at the channel entrance $V_o = 220/32.7 = 6.72$ ft/sec, giving velocity head $V_o^2/(2g) = 0.701$ ft. The entrance hydraulic radius is $R = 32.7/[6 + 2(5.45)] = 1.935$ ft. Then $R^{1/3} = 1.246$, and the

Darcy–Weisbach friction factor at the entrance is $f = 117n^2/R^{1/3} = 117(0.015)^2/1.246 = 0.021$. Since $FL = fL/(4R) = 0.021(374)/[4(1.935)] = 1.015 < 2$, the result is:

$$\Delta h_{\max} \cong \frac{V_o^2}{2g}\left(1 - \frac{1}{3}\frac{fL}{4R}\right) = 0.701(1 - 1.015/3) = 0.464 \text{ ft}$$

For equal distribution, the flow per port/gate is $q = (1/10)(1/4)(220) = 5.50$ cfs. With $q = C_D A\sqrt{2gh}$ and $C_D \cong 0.6$, we have:

$$h \cong \left(\frac{1}{C_D}\right)^2 \left(\frac{q}{A}\right)^2 \frac{1}{2g} = \left(\frac{1}{0.6}\right)^2 \left[\frac{5.50}{3(2)}\right]^2 \frac{1}{2(32.2)} = 0.0357 \text{ ft}$$

Using Equation 11.8a with $C_-/C_+ \cong 1$:

$$m \cong \sqrt{1 - \frac{0.464}{0.0357}} = \text{imaginary number}$$

The predicted distribution uniformity is so low that there will possibly be backflow through some ports. It will be necessary to enlarge the channel and/or reduce port sizes. The value of Δh_{max} is too large to reduce ports alone. If the channel alone is changed, require at least the following for 90% distribution uniformity:

$$\Delta h = (1 - m^2)h_+ = [1 - (0.9)^2](0.0357 = 0.00678) \cong V_o^2/(2g)$$

$$V_o = \sqrt{2(32.2)(0.00678)} = 0.661 \text{ ft/sec}$$

$$A_o = Q/V_o = 220/0.661 = 333 \text{ ft}^2$$

That area is about 10 times the present port area. It was deemed necessary to enlarge the channel *and* reduce the port areas. This had to be done in such a way that any increase in upstream channel water elevation provides adequate freeboard and would be acceptable in terms of upstream (and overall) plant hydraulics.

Skipping some of the trial-and-error steps, consider 14″ × 14″ gates, four along the bottom sides of each tank. Then the open area per gate is 1.361 ft², and:

$$h_+ \cong \left(\frac{1}{0.6}\right)^2 \left(\frac{5.50}{1.361}\right)^2 \frac{1}{2(32.2)} = 0.705 \text{ ft}$$

Assume $m = 0.90$ @ $Q_{\max}$ and $\frac{\Delta h}{V_o^2/(2g)} = 1$ (corresponding to $FL = 0$). Equation 11.1 gives $\Delta h = h_o(1 - m^2) = 0.705[1 - (0.90)^2] = 0.134$ and Equation 11.39a gives $V_o = \sqrt{2(32.2)(0.134)} = 2.94$ ft/sec and $A = Q/V_o = 220/2.94 = 74.8$ ft². For the 6-ft width, this corresponds to a flow depth of 74.8/6 = 12.47 ft.

After various computational iterations, the channel invert was lowered by 9.5 ft (from El. 123.83 to El. 114.38), and we consider 12″ × 12″ gates, four along the bottom sides of each tank. That results in:

$$h_+ \cong \left(\frac{1}{0.6}\right)^2 \left[\frac{5.50}{12(12)/144}\right]^2 \frac{1}{2(32.2)} = 1.305 \text{ ft}$$

We assume the same design (maximum flow) water surface elevation at the upstream end of the channel as before of El. 129.28. Then the channel entrance depth = (129.28 – 114.38) = 14.9 ft, and entrance flow area $A_o = 14.9(6) = 89.4$ ft². The velocity at the channel entrance $V_o = 220/89.4 = 2.46$ ft/sec, giving velocity head $V_o^2/(2g) = 0.0940$ ft. The entrance hydraulic radius is $R = 89.4/[6 + 2(14.9)] = 2.50$ ft. Then $R^{1/3} = 1.357$, and the friction factor at the entrance is $f = 117(0.015)^2/1.357 = 0.0194$. Since $fL/(4R) = 0.0194(374)/[4(2.50)] = 0.726 < 2$, Equation 11.39a applies:

$$\Delta h_{\max} \cong \frac{V_o^2}{2g}\left(1 - \frac{1}{3}\frac{fL}{4R}\right) = 0.0940(1 - 0.726/3) = 0.0713 \text{ ft}$$

Using Equation 11.8a with $C_+/C_- \cong 1$ as a first approximation,

$$m = \frac{C_-}{C_+}\sqrt{1 - \frac{\Delta h}{h_+}} = (1)\sqrt{1 - \frac{0.0713}{1.305}} = 0.97$$

which is a good value.

We now adjust for the greater upstream flow depth$\cong$ 129.28 + 1.305 − 114.38 = 16.21 ft. Corresponding values are $A_o = 16.21(6) = 97.3\ \text{ft}^2$, $V_o = 220/97.3 = 2.26$ ft/sec, $V_o^2/(2g) = 0.0794$ ft, $R = 97.3/[6 + 2(16.21)] = 2.53$ ft, $R^{1/3} = 1.363$, $f = 117(0.015)^2/1.363 = 0.0193$, $fL/(4R) = 0.0193(374)/[4(2.50)] = 0.722 < 2$, then,

$$\Delta h_{\max} \cong \frac{V_o^2}{2g}\left(1 - \frac{1}{3}\frac{fL}{4R}\right) = 0.0794(1 - 0.722/3) = 0.0603 \text{ ft}$$

Reiterating using downstream depth for = 16.21 + 0.0603 = 16.29 ft,

$\overline{R} = 16.29(6)/[6 + 2(16.29)] = 2.53$ ft, the same as before.

Using Equation 11.1, the uniformity $m = \sqrt{1 - \Delta h/h_o} = \sqrt{1 - 0.0603/1.305} = 0.98$, which is good.

Example 11.9

Spray (foam) water manifolds and headers are to be added to an existing wastewater treatment plant (Providence, Rhode Island). The manifolds will run along the long length (184 ft) on each side of each aeration tank. The headers are to supply the manifolds. The design spray-water flow rate per manifold (two manifolds per tank) is 75 gpm/449 = 0.1670 cfs. A 4-in. nominal diameter manifold is assumed. The manifold is planned to rise from upstream elevation 130.85 to downstream elevation 131.00. The pressure drop per nozzle at the intended nozzle spacing and number of nozzles is 5 psi at the prescribed flow rate per nozzle of 2.1 gpm. Determine the adequacy of flow distribution. Manning's $n = 0.013$ for the pipe.

Solution

The manifold velocity and velocity head at the upstream end are, respectively, $V_o \cong 0.1670/\left[(\pi/4)(4/12)^2\right] = 1.914$ ft/sec and $V_o^2/2g = 1.914/[2(32.2)] = 0.0569$ ft. Hydraulic radius $R = (4/12)/4 = 0.0833$ ft, and friction factor $f = 117n^2/R^{1/3} = 117(0.013^2)/0.0833^{1/3} = 0.0453$. Then $FL = fL/(4R) = fL/d_o = 0.0453(184)/(4/12) = 25.0 > 3$. Thus, the piezometric head h drops continuously along the length of the manifold, and the maximum difference in head occurs between the upstream end of the manifold and close to the downstream end of the conduits. Since $C_- \cong 0.63$ is largest near the downstream end, C_-/C_+ will be >1 improving distribution uniformity (i.e., rise in C offsets drop in h along the length of the conduit, reducing the decrease in $q \propto C\sqrt{h}$). Therefore $C_-/C_+ \cong 1$ will conservatively predict m in Equation 11.8a. For $FL > 3$, Equation 11.39c gives:

$$\Delta h_{\max} = -0.0569\left[1 - \frac{24.84}{3} - \frac{4}{3(24.8)}\right] = 0.414 \text{ ft} \times 0.434 = 0.180 \text{ psi}$$

Temporarily neglecting the slope effect, the distribution uniformity $m \cong \sqrt{1 - \Delta h/h_o} = \sqrt{1 - 0.180/5} = 0.98$. Since the pipe rises by about 0.15 ft in the direction of the flow path, the slope will compensate for friction loss, improving distribution uniformity. At this high uniformity, there is no need to refine for the actual d_o.

12

Collection Conduits

12.1 Introduction

Pipes and channels with spatially increasing flows in environmental engineering are referred to as collection conduits. Many water and wastewater treatment processes and other environmental engineering systems employ conduits which collect flow over the width or circumference of a basin, battery of parallel basins, or other defined length. Examples in water and wastewater treatment plants include tank effluent channels (settling tanks, aeration tanks, thickening tanks, etc.), skimming pipes, filter wash-water troughs, filter underdrains, submerged effluent collectors, and well screens. Roof gutters provide another example. Overland flow under rainfall is a related problem of concern to environmental engineers (see, e.g., Yoon and Wenzel 1971), as are street gutters and grates. The flow may enter the conduit at numerous discrete points or continuously along its length. Section 14.6 covers water wells (other than well screens), surface drains, and related topics.

Such conduits are characterized by spatially increasing flows. An important, and sometimes overlooked, hydraulic feature of such flows is the exchange of momentum associated with the acceleration of the inflowing water to the velocity of the flow in the conduit. In the case of a short (frictionless), flat, rectangular channel discharging freely at its downstream end, the momentum exchange alone will cause the upstream depth to be greater than the downstream depth by a factor of $\sqrt{3} = 1.732$ (see Equation 12.5).

In addition to providing adequate hydraulic capacity, properly functioning collector conduits can be important to process performance. Control of outflow from process tanks is commonly provided by a weir attached to one or both sides of single or multiple collector conduits. If the weirs in different tanks are placed at the same elevation and discharge freely to the collector conduit, the load distribution in parallel basins may be more readily kept within acceptable limits. If the effluent weir is submerged by the flow in the collector conduit, the degree of submergence will vary along the conduit. Draw-off then becomes unequal, which can worsen the distribution and induce short-circuiting, adversely affecting process efficiency.

In the large majority of collector conduits, the inflow either enters continuously along the conduit length or at a sufficient number of equally spaced points that the inflow may be accurately approximated as continuous. This chapter presents methods for the design and analysis of collection conduits, with the emphasis on continuous inflow systems. Although a more concise mathematical presentation is easily developed, the author has chosen a presentation that gradually introduces increasing complexity so that the relationship with existing literature and techniques may be more readily established. Additional details may be found in Graber (2004b).

12.1.1 General Differential Equation

In Chapter 2, the general differential equation for nonprismatic spatially-increasing flow channels is derived and given as follows (Equation 2.137 with $\alpha = 1$):

$$\frac{dy}{dx} = \frac{S_o - S_f - \dfrac{2Q}{gA^2}\dfrac{dQ}{dx} + \dfrac{Q^2}{gA^3}\dfrac{\partial A}{\partial x}}{1 - \dfrac{Q^2 T}{gA^3}} \tag{12.1}$$

The terms in the above equation are as defined in Section 2.7.

Hydraulics and Pneumatics in Environmental Engineering, First Edition. S. David Graber.

12.1.2 Friction Factor

The one-dimensional momentum formulation of spatially increasing flows has been confirmed experimentally by several researchers (Hinds 1926; Beij 1934; Camp 1940; Li 1955; Brutsaert 1971). Two types of indirect measurements suggest that the friction factors in spatially-increasing flow channels should be increased above corresponding values for uniform-flow channels. First are observed cross-sectional velocity measurements, such as those of Babb and Ross (1968). Such measurements suggest that the turbulence and lateral velocities induced by the inflow distort the cross-sectional velocity profile. That results in high-velocity gradients near the walls and, by inference, higher wall shear stresses. This distortion can be accounted for by terms in Equation 12.1 containing a momentum coefficient β, which accounts for nonuniformity of the velocity distribution on the one-dimensional momentum flux. Babb and Ross reported β values up to 1.07, decreasing in the downstream direction. They did not consider the effect of friction factor but concluded that the use of a β value of unity (as assumed in Equation 12.1) is probably satisfactory in most applications.

Second is the observation by Camp (1940) and others that an increase in the friction factor produces a better match between predicted and observed water surface profiles. Yen and Wenzel (1970) suggested that the alteration of boundary shear stress is a function of the ratio of the momentum flux of the lateral flow to that of the main flow. Camp (1940) made a similar suggestion. However, no systematic studies have quantified such relationships. The related discussion in Section 10.3 suggested that the Darcy–Weisbach friction factor be increased up to twofold for wash-water trough design. Since f and Manning's n are related by $f = 117n^2/R^{1/3}$, doubling f is approximately equivalent to multiplying n by $\sqrt{2}$. This approach has been adopted by the author and, in the absence of further information, is recommended for the channels considered herein. Comments under Section 12.7 below regarding work of Khiadani *et al.* (2005, 2007) and Beecham *et al.* (2005) also apply in the present context.

12.1.3 Location of Control

In virtually all environmental engineering collection-conduit applications, the control section (starting point at which the water surface elevation is known) is located at the downstream end of the conduit, and subcritical flow exists throughout. Although those assumptions are made herein, it is prudent to check the Froude Number along channels with steep slopes. For civil engineering applications more generally, and in particular for side-channel spillways, supercritical flow may occur and should not necessarily be avoided. If the Froude Number exceeds 1 (indicating supercritical flow), the control section may be located along the channel length. Methods have been developed for locating the control section in such cases (Chow 1959, Chapter 12; Hinds 1926; Smith 1967, 1972).

12.2 Rectangular Channels

In environmental-engineering applications, tapered rectangular channels are virtually the only nonprismatic open channels encountered. Such channels are tapered primarily to maintain velocities necessary to keep solids in suspension. Common applications are activated-sludge collection channels in secondary wastewater treatment plants and collection channels on detritor-type grit-removal units.

For a uniformly tapered rectangular channel of width b, we have $T = b$, and $\partial A/\partial x = y db/dx$. Equation 12.1 then becomes:

$$\frac{dy}{dx} = \frac{S_o - S_f - \dfrac{2Q}{gA^2}\dfrac{dQ}{dx} + \dfrac{Q^2}{gA^3}\dfrac{ydb}{dx}}{1 - \dfrac{Q^2 b}{gA^3}} \tag{12.2}$$

in which $A = by$.

12.2.1 Exact Solutions

For a prismatic, rectangular channel with uniform inflow ($db/dx = 0$, $dQ/dx = q^* =$ constant), Equation 12.2 reduces to:

$$\frac{dy}{dx} = \frac{S_o - S_f - \dfrac{2Qq_*}{g(by)^2}}{1 - \dfrac{Q^2}{g(by)^2 y}} \tag{12.3}$$

An exact solution to Equation 12.3 may be obtained by assuming zero slope and negligible friction. With the additional assumption that the flow is zero at the upstream end of the channel, Camp (1940) obtained the following solution to this problem:

$$y_o = \sqrt{y_L^2 + \frac{2Q_L^2}{gb^2 y_L}} \tag{12.4a}$$

in which the subscripts "o" and "L" refer to conditions at the upstream and downstream ends of the channel, respectively. Equation 12.4a can also be written in terms of the downstream Froude Number $\mathbf{F}_L = Q_L/\sqrt{gb^2 y_L^3}$ as follows:

$$\frac{y_o}{y_L} = \sqrt{1 + 2\mathbf{F}_L^2} \tag{12.4b}$$

For the case of a free overfall at the downstream end, $\mathbf{F}_L = 1$ and Equation 12.4b reduces to (Camp 1940):

$$\frac{y_o}{y_c} = \sqrt{3} \tag{12.5a}$$

in which y_o denotes the critical depth, given by:

$$y_c = \sqrt[3]{Q^2/(gb^2)} \tag{12.5b}$$

Equation 12.5b was presented by Beij (1934) in "Flow in Roof Gutters."

It is instructive to determine the contribution of momentum exchange to the depth increase indicated by Equation 12.5a. If momentum exchange is neglected, then conservation of energy tells us that the difference in the upstream and downstream depths in this case would be the velocity head of the flow at the downstream end (there being zero velocity head at the upstream end). The downstream velocity head is $y_c/2$ [because $\mathbf{F}_L^2 = V_c^2/(gy_c) = 1$], so the upstream depth would be $(y_c + y_c/2) = 3y_c/2$ in the absence of momentum transfer. Comparing that to Equation 12.5a, one finds that momentum transfer in this case increases the change in depth by $\{[(\sqrt{3}-1)y_c - (3/2-1)y_c]/[(3/2-1)y_c]\} \cdot 100 = 46\%$. Although there are cases of spatially increasing flow in which the relative contribution of momentum transfer is less significant, its neglect is generally not advisable.

Equations 12.4 and 12.5 may also be derived by applying the momentum equation to the entire channel. This will be demonstrated after considering the more general case of a horizontal effluent channel with flow at its upstream end. Introducing the appropriate pressure and momentum terms, the governing equation becomes:

$$\gamma b\frac{y_o^2}{2} - \gamma b\frac{y_L^2}{2} = \frac{\gamma Q_L V_L}{g} - \frac{\gamma Q_o V_o}{g} \tag{12.5c}$$

Noting that $V = Q/(by)$, Equation 12.5c can be rewritten in the following form:

$$y_o^3 - [y_L^2(1 + 2\mathbf{F}_L^2)]y_o + \frac{2Q_o^2}{gb^2} = 0 \tag{12.5d}$$

With the other terms in Equation 12.5d known, this equation can be solved for y. Note that Equation 12.5d reduces to Equation 12.4b when $Q_o = 0$. Since no assumption was made regarding dQ/dx in deriving Equation 12.4b, it becomes apparent that the manner in which the flow enters along the length of the channel is not important in determining y_2 (provided that the entering flow has negligible momentum in the $x-$ direction).

Equations 12.4 and 12.5 are practical for short, horizontal channels. For more complex problems involving channel friction, bottom slopes, and tapered channels, methods of solution may be divided into three categories: (1) approximate solutions, (2) generalized numerical solutions, and (3) special-case numerical solutions. These are discussed in the following subsection and in Section 12.6.

12.2.2 Approximate Solutions

Several approximate solutions have been developed for rectangular collector channels. Camp's (1940) equation, which includes the effect of slope and friction, is as follows:

$$y_o = \sqrt{y_L^2 + \frac{2Q_L^2}{gb^2 y_L} - 2SL\bar{y} + \frac{fLQ_L^2}{12gb^2\bar{R}\bar{y}}} \tag{12.6}$$

in which $\bar{y}$ and $\bar{R}$ are average values of the flow depth and hydraulic radius in the channel.

Although the author has found that Equation 12.6 does not always provide an acceptable approximation when slope and friction effects are appreciable, it can provide a useful estimate of the significance of friction. Note that the first three terms under the radical in Equation 12.6 are related to the frictionless solution such that:

$$\frac{y_o}{y_L} = \sqrt{\left(\frac{y_o}{y_L}\right)^2_{S_f=0} + \frac{fLQ_L^2}{12gb^2\bar{R}\bar{y}y_L^2}} \tag{12.7}$$

The term $(y_o/y_L)_{S_f=0}$ represents the frictionless solution such as obtained by using Li's (1955) solution.

To determine the significance of friction in a conservative manner using Equation 12.7, it is suggested that $\bar{y}$ and $\bar{R}$ be based on the minimum depth section (which may occur at the upstream or downstream end of the channel, depending on channel slope) from the frictionless solution. Then y_o/y_L can be adjusted in accordance with Equation 12.7. If $y_o/y_L \cong (y_o/y_L)_{S_f=0}$, then friction effects are obviously negligible. If friction effects are significant, Equation 12.7 will provide an acceptable approximation if the friction effect does not increase the change in water surface elevation by more than 10%. Otherwise, numerical methods should be used as discussed below.

A convenient design device that has been used in conjunction with Equation 12.6 entails neglecting both the slope and friction terms and providing a slope which compensates for the friction effect. Although this does not usually provide the most economical design, it can be an appropriate, conservative convenience for minor collection conduits. If critical depth conditions are maintained at the downstream end, Equation 12.6 gives:

$$y_o \cong \sqrt{y_L^2 + \frac{2Q_L^2}{gb^2 y_L}} \tag{12.8a}$$

$$S_o \cong \frac{fQ_L^2}{24gb^2\bar{R}\bar{y}^2} \tag{12.8b}$$

For the case of $y_L = y_o$, assuming average values for $\bar{y}$ and $\bar{R}$, Carpenter *et al.* (1940) cast Equation 12.8a into the form of an equality which allows quick calculation of the principle dimensions of the conduit. The accuracy of that procedure is questionable. The author suggests it be used with $\bar{y}$ and $\bar{R}$ taken as minimum values to assure conservatism. For other than minor structures, the design should be refined and checked by other methods discussed herein.

The calculus of variations, used in conjunction with Equations 12.5a,b, shows that the perimeter $2y_o + b$ of a short rectangular channel with negligible friction, and hence its cost, is minimal when $y_o/b = 3/4$ (an exercise left to the reader).

Using Equation 12.6 without access to digital computers in the 1960s, the author developed simplified curves for the selection of rectangular roof gutters used as part of a Caribbean Island water supply system (San Blas Island of Ailigandi off the coast of Panamá). In that case, the curves were developed in terms of gutter length vs. roof area with commercially available cross-sectional dimensions as a parametric variable for a specified rainfall intensity and friction factor. Critical depth conditions were assumed at the downstream end, which can be realized by proper downspout design. Although a minimum slope on the order of 1/16 in. per ft (0.52 cm/m) is recommended to prevent ponding in such systems, construction control is often marginal. Therefore, zero slope was assumed. With $S_o = 0$ and $y_L = y_c$, Equation 12.6 may be rearranged in the following form:

$$L = b\frac{12(\bar{R}/b)(\bar{y}/b)}{(y_c/b)^3 f}\left[\left(\frac{y_o}{b}\right)^2 - 3\left(\frac{y_c}{b}\right)^2\right] \tag{12.9}$$

The term y_c/b is related to rainfall intensity i and tributary roof area a as follows:

$$\frac{y_c}{b} = \left[\frac{(ia)^2}{gb^5}\right]^{1/3} \tag{12.10}$$

Equation 12.9 can be used in conjunction with Equation 12.10 to calculate values of L given y_o, b, a, i, and f, without the need for a trial-and-error solution. Gutter selection curves developed by this procedure are presented in Figure 12.1. The plot is for minimum-cost cross sections mentioned above with $y_o/b = 3/4$. The significance of friction is clearly depicted by this figure. The frictionless solution corresponds to the abscissa for which $L = 0$.

The dashed parabolic lines in Figure 12.1 are plots of $a = L^2$ and $a = L^2/10$ which bracket practical roof dimensions. Beij (1934) has presented an empirical equation based on experiments with relatively short lengths of copper roof channel. Comparison of his empirical equation with the curves of Figure 12.1 within the practical range of roof dimensions showed agreement within 9–13%, with the curves of Figure 12.1 being on the conservative side. The author has developed curves similar to those of Figure 12.1 for half-round gutters. More precise curves for rectangular or half-round gutters could be developed by means of generalized numerical solutions discussed later in this chapter.

Gutter downspouts are often sized according to building codes, such as cited by Fair and Geyer (1954, §14-9), who give a table of downspout diameter vs. roof area drained. The junction between downspout and gutter should be designed to maintain a flow depth at the downstream end of the roof gutter which is less than or equal to the critical-flow depth. The downspout should not project into the gutter. A conservative design for a downspout which discharges directly to the ground will have a square downspout with sides equaling the roof channel width as closely as is practical, or a circular downspout having a perimeter of adequate length. In this latter case, an adequate length will exist if the downspout diameter is at least two-thirds of the roof channel width. The downspout perimeter P will then be $P = \pi D > 2\pi b/3 > 2b$ which will allow the required critical-flow depth to develop at the downstream end of the roof.

Before leaving this topic, it should be noted that there are potential health risks associated with consumption of untreated rainwater from roof catchments (Lye 2002; van der Sterren *et al.* 2013; O'Connor and Amin 2014). In the case referred to in

Figure 12.1 Roof Gutter Curves for Ailigandi

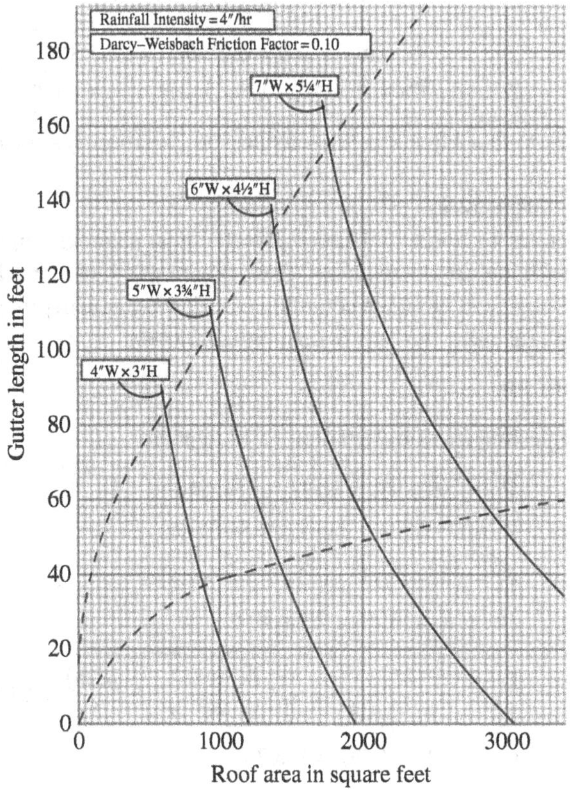

the previous paragraph, that source was used for boiling corn kernels to make a beer called "chicha"; for potable water consumption, the author designed and arranged for construction of a filtration system. If the downspout, instead of discharging at its vertical terminus, continues to, e.g., a storage tank, then a hydraulic analysis is necessary to assure that the downspout is not submerged at its connection to the gutter, so as to enable free, unrestricted discharge at the gutter.

An interesting variation of Camp's approximation was developed by Thomas (1940) and subsequently suggested by Fair and Geyer (1954, §24-10) and Fair *et al.* (1968, §27-21). Thomas' approximate solution for the case of negligible friction may be expressed in the following form:

$$\frac{y_o}{y_L} = -\frac{2}{3}\frac{SL}{y_L} + \sqrt{\left(1 - \frac{1}{3}\frac{S_oL}{y_L}\right)^2 + 2\mathbf{F}_L^2} \tag{12.11}$$

Equation 12.11 has the same functional form as the numerical solution by Li discussed below. Thomas' approximation gives values of y_o/y_L that are less than Li's for $y_o/y_L > 0.8$ and greater than Li's for $y_o/y_L < 0.8$. The discrepancy is greatest at higher values of S_oL/y_o and $\mathbf{F}_L$. For $y_o/y_L > 0.6$, the difference is less than 5%. Li's solution is more generally applicable.

Another interesting approximation can be developed for a tapered rectangular channel having zero slope and negligible friction. To obtain this approximate solution, the momentum equation is applied to a control volume encompassing the entire channel. The following equation results:

$$\gamma b_o\frac{y_o^2}{2} - \gamma b_L\frac{y_L^2}{2} + \gamma\left(\frac{y_o + y_L}{2}\right)^2 L\frac{\sin\theta}{\cos\theta} = \frac{\gamma}{g}Q_LV_L \tag{12.12}$$

The approximate nature of Equation 12.12 stems from the assumption that the third term to represent the average component of the pressure force exerted by the tapered walls. Noting that $\sin\theta/\cos\theta = \tan\theta = (b_o - b_L)/(2L)$, and substituting $\mathbf{F}_L^2 = Q_L^2/(gb_L^2y_L^3)$, Equation 12.12 can be rearranged to give the following quadratic equation:

$$\left(\frac{3b_o + b_L}{4}\right)y_o^2 + \left[\frac{(b_L - b_o)y_L}{2}\right]y_o - y_L^2\left(2b_L\mathbf{F}_L^2 + \frac{3b_o + b_L}{4}\right) = 0 \tag{12.13}$$

The solution to this equation is given by:

$$\frac{y_o}{y_L} = -\frac{b_L - b_o}{3b_o + b_L} + \sqrt{4\left(\frac{b_o + b_L}{3b_o + b_L}\right)^2 + \frac{8b_L}{3b_o + b_L}\mathbf{F}_L^2} \tag{12.14}$$

Note that Equation 12.14 reduces to Equation 12.4b when $b_L = b_o$.

Example 12.1

A concrete, horizontal, tapered, rectangular channel proposed to collect a uniform inflow of activated sludge along its 748 ft length has an upstream and downstream width of 5 and 25 ft, respectively (Montreal WWTP). The maximum total flow is 506 mgd, under which conditions the downstream depth is 13.39 ft. Determine the upstream depth, first neglecting friction, then including an approximation for friction.

Solution

Neglecting friction, Equation 12.14 applies:

$$Q_L = 506 \text{ mgd} \times 1.547 = 783 \text{ cfs}$$

$$V_L = 783/[25(13.39)] = 2.34 \text{ ft/sec}$$

$$\mathbf{F}_L^2 = (2.34)^2/[32.2(13.39)] = 0.0127$$

$$3b_o + b_L = 3(5) + 25 = 40 \text{ ft}$$

$$\frac{y_o}{y_L} = -\frac{25 - 5}{3(5) + 25} + \sqrt{4\left(\frac{5 + 25}{40}\right)^2 + \frac{8(25)}{40}(0.0127)} = 1.021$$

$$y_o = 1.021(13.39) = 13.67 \text{ ft}$$

An approximation of the additional head loss h_f due to friction may be obtained by estimating a representative average friction slope and multiplying it by the channel length. Using the Manning equation (from Section 7.3),

$$h_f = S_f L = \left(\frac{nV}{1.487R^{2/3}}\right)^2 L$$

$$R = \frac{by}{b+2y} = \frac{y}{1+2y/b}$$

Note that $V = Q/(yb)$ increases in the downstream direction (primarily since Q increases faster than b), attaining an absolute maximum at $x = L$. Simultaneously, R increases in the downstream direction, attaining an absolute maximum at $x = L$. The term $V/R^{2/3}$ has a relative maximum about 60% of the way down the channel, but decreases very slowly the rest of the distance. A reasonable approximation is obtained if S_f is based on characteristics at the downstream end of the channel:

$$R_L = \frac{25(13.39)}{25+2(13.39)} = 6.46\ \text{ft}$$

$$R_L^{2/3} = 3.47$$

$$V_L = 2.34\ \text{ft/sec (from above)}$$

Manning's n is multiplied by $\sqrt{2}$ to account for inflow turbulence; use $n = 0.013\sqrt{2} = 0.0184$. Then,

$$h_f = \left[\frac{0.0184(2.34)}{1.487(3.47)}\right]^2 (748) = 0.05\ \text{ft}$$

The upstream depth including the approximation for friction is $y_o = 13.67 + 0.05 = 13.72$ ft. The answers with and without friction are compared to a more precise numerical solution later in this chapter.

12.3 Nonrectangular Channels with Constant Top Width

Nonrectangular channels with constant top width are frequently used for wash-water troughs in water filtration practice. Semicircular, semihexagonal, and semioctagonal shapes have been employed (Fair *et al.* 1954, p. 691). The semicircular trough has become particularly common with the use of plastic materials.

For a prismatic channel, we have $dA = Tdy$, in which A, T, and y are flow cross section, top width, and depth, respectively. For a channel of constant top width, $dA = TdD$ in which D is the hydraulic depth. Therefore, for a prismatic channel of constant top width, we have $dT = dD$, and Equation 12.1 becomes:

$$\frac{dD}{dx} = \frac{S_o - S_f - \dfrac{2Q}{gA^2}\dfrac{dQ}{dx}}{1 - \dfrac{Q^2 T}{gA^3}} \tag{12.15}$$

in which $A = TD$. Equation 12.15 applies to channels having side walls that are parallel (not necessarily vertical) and a depth that is sufficient throughout the channel length to keep the water surface between the parallel walls.

12.3.1 Exact Solutions

Note that Equation 12.15 has the same form as Equation 12.2 with y replaced by D and b replaced by T. Thus, for zero slope, negligible friction, and $dQ/dx = Q_o/L$, the exact solution analogous to Equation 12.4b is as follows:

$$\frac{D_o}{D_L} = \sqrt{1 + 2\mathbf{F}_L^2} \tag{12.16}$$

in which $\mathbf{F}_L = Q_L/\sqrt{gT^2D_L^3}$. For the case of a free overfall at the downstream end, $\mathbf{F}_L = 1$ and Equation 12.16 reduces to:

$$\frac{D_o}{D_c} = \sqrt{3} \tag{12.17}$$

in which D_c denotes the critical hydraulic depth (for any cross-sectional shape) given by:

$$D_c = \sqrt[3]{\frac{Q^2}{gT^2}} \tag{12.18}$$

Equation 12.5d is also applicable, with y and b replaced by D and T.

12.3.2 Approximate Solutions

Camp (1940) demonstrated that his approximate solution was applicable to cross sections other than rectangular if y is changed to the hydraulic depth D. His proof was based on consideration of channels with V-shaped and semicircular bottoms, both surmounted by parallel walls and thus having constant top widths. The constant top width limitation has not always been recognized. One reference (WPCF/ASCE 1959, p. 25) erroneously applied Camp's equation to channels of varying top width. For Equation 11.6 to apply (with y replaced by D), it is necessary that the channel side walls be parallel and the flow depth be such that the free surface is between the parallel walls throughout the channel length.

12.4 Prismatic Channels with Varying Top Width

Circular channels are the most common channels with varying top width. Such channels are used for the underdrains of intermittent sand filters and skimming pipes in sedimentation basins. Another situation encountered is that of nonrectangular channels with parallel side walls but with the water surface in the varying top width section of the channel.

12.4.1 Exact Solutions

Exact solutions to Equation 12.1 have been obtained for prismatic channels with zero slope and negligible friction. For such channels, $\partial A/\partial x$, S_o, and S_f are all equal to zero, and Equation 12.1 reduces to:

$$\frac{dy}{dx} = \frac{-\dfrac{2Q}{gA^2}\dfrac{dQ}{dx}}{1 - \dfrac{Q^2T}{gA^3}} \tag{12.19}$$

Beij (1934) obtained a general solution to Equation 12.19, which is applicable to any cross-sectional shape. Beij assumed zero flow at the upstream end of the channel and uniform inflow along its length. Substituting $dQ/dx = q_*$ and $Q = q_*x$ into Equation 12.19 and rearranging gives:

$$\frac{2xAdx}{A^2} - \frac{x^2Tdy}{A^2} + \frac{gA}{q_*^2}dy = 0 \tag{12.20}$$

Since $dA = Tdy$, Equation 12.20 can be rewritten as follows:

$$\frac{2xAdx - x^2dA}{A^2} + \frac{gA}{q_*^2}dy = 0 \tag{12.21}$$

Noting that the first term in Equation 12.21 is an exact differential, Beij rewrote Equation 12.21 as:

$$d\left(\frac{x^2}{A}\right) + \frac{gA}{q_*^2}dy = 0 \tag{12.22}$$

in which form he integrated the equation as follows:

$$\frac{x^2}{A} = -\frac{g}{q_*^2}\int_{y_o}^{y} Ady \tag{12.23}$$

in which y_o denotes the depth at $x = 0$. Since A is a known function of y for any channel shape, Equation 12.23 can be solved in closed form. Setting $x = L$ and $A = A_L$ in Equation 12.23, and noting that q_*L equals the total flow in the channel Q_L, the equation becomes:

$$\frac{Q_L^2}{gA_L} = \int_{y_L}^{y_o} Ady \tag{12.24}$$

in which the subscript "L" denotes conditions at $x = L$. Equation 12.24 can be rewritten in terms of the downstream Froude Number as follows:

$$\mathbf{F}_L^2 = \frac{y_L}{D_L}\int_{y_L}^{y_o}\frac{A}{A_L}d\left(\frac{y}{y_L}\right) \tag{12.25}$$

For a rectangular channel, $A/A_L = y/y_L$, $y_L = D_L$, and Equation 12.25 gives Equation 12.4b. For a triangular channel, $y_L = 2D_L$, $A/A_L = (y/y_L)^2$, and Equation 12.25 gives:

$$\frac{y_o}{y_L} = \left(1 + \frac{3}{2}\mathbf{F}_L^2\right)^{1/3} \tag{12.26}$$

The solutions presented by Beij (1934) were for the specialized case of a channel with a free downstream discharge, for which y_L equals the critical depth y_c corresponding to $\mathbf{F}_L = 1$. For the rectangular channel, the resulting equation is Equation 12.5a. For the triangular channel:

$$\frac{y_o}{y_c} = \left(\frac{5}{2}\right)^{1/3} \tag{12.27}$$

Beij also presented solutions for trapezoidal and semicircular channels with $\mathbf{F}_L = 1$.

The error resulting from the application of Equation 12.16 to a channel of varying top width may be demonstrated with reference to a triangular channel. Since $D = A/T = y/2$ for a triangular channel, Equation 12.16 suggests that Equations 12.4b and 12.5a would apply. But that is clearly contrary to Equations 12.26 and 12.27. As previously stated, Equation 12.26 is limited to channels of constant top width.

An exact solution similar to Equation 12.5d may also be derived. Derived in a similar fashion, but using centroidal pressure terms, the equation analogous to Equation 12.5c is given by:

$$\gamma\bar{z}_o A_o - \gamma\bar{z}_L A_L = \frac{\gamma Q_L V_L}{g} - \frac{\gamma Q_o V_o}{g} \tag{12.28}$$

The area A and depth of the centroid $\bar{z}$ are functions of the depth of flow. Convenient tables for certain cross sections are given in Brater and King's *Handbook of Hydraulics* (1976, Tables 7-3 and 7-8). Thus, if the downstream flow depth and hence $\bar{z}_L$, A_L, and V_L are known, the upstream depth can be determined from Equation 12.38 by a trial-and-error process.

In the case where $Q_o = 0$, Equation 12.28 reduces to:

$$\bar{z}_o A_o = \bar{z}_L A_L + \mathbf{F}_L^2 A_L D_L \tag{12.29a}$$

$$\mathbf{F}_L^2 = \frac{Q_L^2}{gA_L^2 D_L} \tag{12.29b}$$

Letting σ represent a characteristic length of the cross section (diameter in the case of a circular cross sections, bottom width in the case of a trapezoidal cross section, etc.), Equation 12.29a can be rewritten as follows:

$$\frac{\bar{z}_o}{\sigma}\frac{A_o}{\sigma^2} = \frac{A_L}{\sigma^2}\left(\frac{\bar{z}_L}{\sigma} + \mathbf{F}_L^2\frac{D_L}{\sigma}\right) \tag{12.30}$$

The terms $\bar{z}_o/\sigma, A/\sigma^2$, and D_L/σ are each functions of y/σ plus any additional variables necessary to characterize the cross section (such as side slope in the case of a trapezoidal cross section). In concept, then, the above equation relates y_o/σ to y_L/σ, $\mathbf{F}_L$, and the additional fixed cross-sectional characteristics. Only in simple cases of varying top width, such as the triangular channel (for which there is no characteristic length), will an explicit relation for y_o (such as Equation 12.26) be attainable. In more complex cases, it will be necessary to obtain y_o/σ as the root of an equation by trial-and-error or numerical methods. Then, for any given cross section, curves relating y_o/σ to y_L/σ and $\mathbf{F}_L$ can be prepared.

The cross sections of practical interest here are circular and semicircular surmounted by vertical side walls. Skimming pipes are an example application of the former, and wash-water troughs are an example application of the latter. For y_o/d less than 1/2, the two cross sections are identical. For y_o/d less than about 3/4, there is no significant difference between the results, because the additional cross-sectional area between the circle and parallel side walls is small. There is, of course, no upper limit on y_L/d or y_o/d for the semicircular channel with parallel side walls, and the curves for that case have been extended above $y_o/d = 1$ only for clarity. The reader should recall that in that case when $y_L/d > 1/2$ the simple relationships for constant top width may be conveniently employed.

Most of the material given here regarding skimming pipes is taken from Graber (2023b). Skimming pipes are used to remove floating scum and grease from the surface of tanks in wastewater treatment plants. They typically consist of a horizontal circular pipe with long, wide slots cut symmetrically about the vertical axis of the pipe (as shown on Figure 1 of Graber 2023b). The slot typically comprises about 60° of arc. The edges of the slot serve as weirs, over which scum and/or surface grease flows into the pipe when the pipe is rotated about its axis. The skimming pipes are commonly connected in series (and operated individually) when parallel rectangular basins are employed, and the downstream-most skimmer discharges freely into a scum or grease sump or box. Pipes typically range from 8 to 18 in. in diameter (in 2-in. increments). The water head on the weir, when the pipe is rotated, is usually on the order of 1/2 to 3/4 in., so that the high velocity of flow over the weir will efficiently remove floating scum and grease while minimizing water volume.

The edge of the slot in the upright position must be above the maximum water level. The skimmer must be capable of tipping the slot edge the necessary amount below the minimum water level, under which conditions the pipe sector below the weir must have sufficient hydraulic capacity. The skimmer pipe design is thus related to the basin outlet design insofar as the latter affects the water surface level variations.

The Keifer–Chu method for constant flow (Keifer and Chu 1955) is not applicable to channels of zero slope (although it is for negative slope) because, as noted by Chow (1959, p. 267), for zero slope the varied-flow functions become meaningless. Hager's (1991) explicit approach for circular cross sections with constant flow also applies only for nonzero slopes.

The importance of proper hydraulic design is made clear by the report of problems associated with the Massachusetts Water Resources Authority's newly-modified Deer Island Wastewater Treatment Plant (Laquidara *et al.* 1998): "The [secondary clarifier scum] system was found to be inadequate for removing scum from the clarifier furthest away from each scum drain box. The scum tube was found to be hydraulically overloaded, because it relied on 60 feet of scum tube to move collected scum from the far clarifier to the scum box....[Control system] changes...did not fully correct the underlying hydraulic problem."

12.4.2 Free Discharge

The equation for constant flow in a horizontal channel with free discharge (critical depth at the downstream end) is of the following functional form [Graber 2004b, Equation (38b)]:

$$\frac{y_o}{\sigma} = \Phi\left(\frac{fL}{\sigma}, \frac{y_c}{\sigma}\right) \tag{12.31}$$

in which y = depth of flow, y_o = value of y at the upstream end of the channel; y_c = critical depth; σ = characteristic cross-sectional dimension, here taken to be the skimmer diameter d_o; f = Darcy–Weisbach friction factor; and L = channel length. There follows specifically:

$$\frac{y_o}{d_o} = \Phi\left(\frac{fL}{d_o}, \frac{y_c}{d_o}\right) \text{ for } \mathbf{F}_L = 1 \tag{12.32}$$

in which $\mathbf{F}_L = (Q_{crit}d_o^{2.5})/[(A/d_o^2)\sqrt{32.2(D/d_o)}]$ is the Froude Number at the downstream end of the channel.

A plot of y_o/d_o vs. fL/d_o with y_c/d_o as a parametric variable is presented as Figure 12.2. The procedure that led to the generalized solution is as follows: for values of the central angle θ of the water surface from 0 to 180°, successive values were

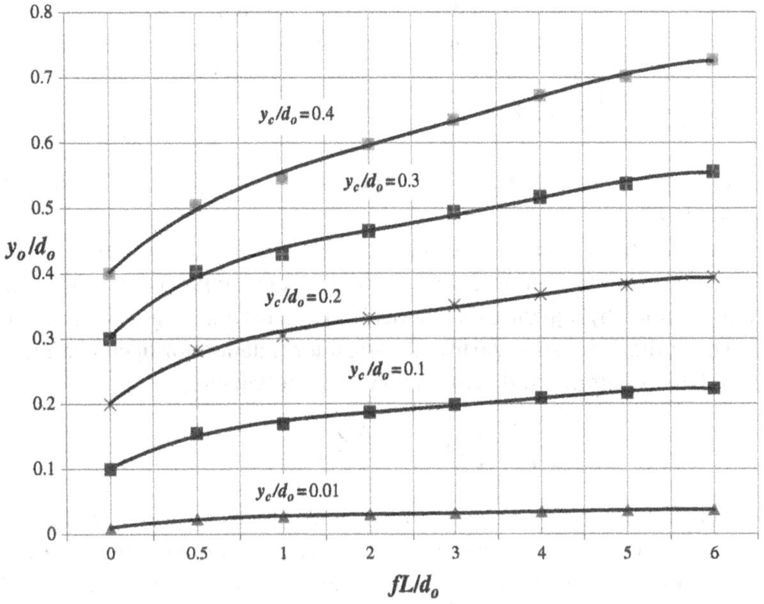

Figure 12.2 Circular Conduit Chart for Constant Flow with Free Discharge

tabulated using geometric elements formulae of $\cos(\theta/2)$; y/d_o; D/d_o where D is hydraulic depth; A/d_o^2 where A is cross-sectional area of flow; and $Z/d_o^{2.5} = Q_{crit}/\left(\sqrt{g}d_o^{2.5}\right)$ in which Z is the critical-flow section factor, Q_{crit} is critical-flow rate, and g is acceleration of gravity. Then, with no loss in generality, we assume a fixed value of d_o and calculate y_c, $Q_{crit} = \left[Q_{crit}/\left(\sqrt{g}d_o^{2.5}\right)\right]\sqrt{32.2}(d_o)^{2.5}$, and $\mathbf{F}_L$. The latter was done as a check to assure $\mathbf{F}_L = 1$ for every value of θ, which it exactly was. Then, using a formulation given by Graber (2004b) for numerical solution of open-channel conduits and the Newton–Raphson method (also described in Graber 2004b), the author computed, using suitably small intervals along the channel length (e.g., 100 ft channel length and 1 ft intervals) upstream from critical depth with given values of f and corresponding values of Manning's n, successive values of upstream depth y_o and corresponding y_o/d_o.

More recently, Shang *et al.* (2019) derived explicit dimensionless relationships for critical depth in closed conduits of various cross sections. For circular sections, they provide a relationship that is accurate to within $\leq$0.182% over the range of y_c/d_o equal to 0.005 to 1. That relationship is given by:

$$\frac{y_c}{d_o} = \left(1 + 3.83\varepsilon_c^{-2.1454} - 3.2\varepsilon_c^{-2.1}\right)^{-0.115}, \quad \varepsilon_c = \frac{Q^2}{gd_o^5} \tag{12.33}$$

The author found that the above relationship gives results virtually identical to those shown in Figure 12.2. By using Equation 12.33, some of the steps described in the preceding paragraph are eliminated (those prior to $Z/d_o^{2.5}$).

Although the author has not found a need to do so, Graber (2004b, p. 73) gives a simple device that may be used to extend the generalized solution for constant flow to horizontal channels with submerged discharges ($\mathbf{F}_L < 1$).

12.4.3 Skimming Pipe

For the skimming pipe with flow increasing along its length, the edge of the slot in the upright position must be above the maximum water level. The skimmer must be capable of tipping the slot edge the necessary amount below the minimum water level, under which conditions the pipe sector below the weir must have sufficient hydraulic capacity. The skimmer pipe design is thus related to the basin outlet design insofar as the latter affects the water surface level variations.

In some cases, the skimming pipe discharges directly and freely into the downstream structure (e.g., a scum manhole). In other cases, the skimming pipe *per se* is followed by a freely discharging pipe, which imposes a tailwater depth. Although the relatively short length of the skimming pipe itself is such that friction can generally be neglected (more about which is said below), there is thus the added complexity of having a subcritical downstream depth, i.e., Froude Number $\mathbf{F}_L < 1$. The functional form of the equation in that case is given by [Graber 2004b, Equation (38c)]:

$$\frac{y_o}{d_o} = \Phi\left\{\frac{y_L}{d_o}, \mathbf{F}_L\right\} \text{ for } S_o = 0, f = 0 \tag{12.34}$$

$\mathbf{F}_L$ in the above equation can be replaced by $Q_L/\sqrt{g}d_o^{5/2}$, giving a convenient conceptual relationship in terms of pipe capacity Q_L.

Graber (2004b) obtained a convenient, generalized uniform inflow solution corresponding to the above equation for the case of horizontal, frictionless channels of circular cross section (such as those applicable to skimming pipes). That solution was obtained by applying the momentum equation in large (i.e., to the entire channel in one step), as presented below.

The author's (Graber 2004b) numerical formulation for open-channel collector conduits can be reduced for zero slope, zero lateral outflow, and zero friction to the following:

$$\Delta y = \frac{Q_1(V_1 + V_2)}{g(Q_1 + Q_2)}\left(\Delta V + \frac{V_2 \Delta Q_i}{Q_1}\right) \tag{12.35}$$

in which subscripts "1" and "2" denote, respectively, upstream section 1 and downstream section 2 (upstream being in the direction of the ultimate discharge and downstream being toward the zero inflow end of the conduit); the "Δ" terms represent the value at section 2 minus the value at section 1, such as $\Delta V = (V_2 - V_1)$; Δy is the increase of water surface elevation between sections 2 and 1; and $\Delta Q_i = q_i \Delta x$ in which q_i = inflow per unit channel length. The flow terms Q_1 and Q_2 are prescribed based on the uniform inflow for which $Q = Q_i(x/L)$, and V_1 and V_2 are calculated by continuity based on the prescribed flows at the corresponding sections and the flow areas, which are known functions of corresponding flow depths.

Consider a single step with the usual conditions of $Q_1 = 0$ and $V_1 = 0$, for which Equation 12.35, using the notation of Equation 12.34, reduces to:

$$y_o = y_L + \frac{V_L^2 q_i \Delta x}{g Q_L} \tag{12.36}$$

in which the terms on the right-hand side are known. Using the terms of Equation 12.34, the above equation becomes:

$$\frac{y_o}{d_o} = \frac{y_L}{d_o} + \mathbf{F}_L^2 \frac{D_L}{d_o} \tag{12.37}$$

in which D_L/d_o is a function of y_L/d_o.

A plot of Equation 12.37 reveals an excellent fit by straight lines according to:

$$\frac{y_o}{d_o} = \mathrm{m}(\mathbf{F}_L)\frac{y_L}{d_o} \tag{12.38}$$

Furthermore, a parabola provides a good fit of m as a function of $\mathbf{F}_L$: $\mathrm{m} = 1.339\mathbf{F}_L^2 + 1$. Equation 12.38 can then be written as:

$$\frac{y_o}{d_o} = \left(1.339\mathbf{F}_L^2 + 1\right)\frac{y_L}{d_o} \tag{12.39}$$

Example 12.2

A skimming pipe is to be designed (Tallman Island WWTP, New York City) to skim surface scum in a deep channel 3-ft–6-in wide with the water surface elevation varying from 14.43 to 13.72 ft. The slot along the pipe will be 3 ft long. The pipe will discharge freely into a scum manhole and will have other characteristics of skimming pipes as described above. Provide for 1 in. of freeboard F and 3/4-in. head h of flow over the pipe slot, and determine the required pipe diameter and centerline elevation. The outer diameter of the pipe equals 1.072 times the inner diameter. Referring to the geometric relationships in Figure 1 in Graber (2023b), we have: $d > 1.072[14.43 - 13.72 + 1/12 + y_o + (3/4)/12] = 0.917 + 1.072\,y_o$ in units of ft.

Solution

The required flow capacity is $Q = CLh^{3/2} \cong 3.3(3)(0.75/12)^{3/2} = 0.155$ cfs. Assume d = 14 in. ÷ 12 = 1.167 ft. Then:

$$\frac{Q}{d_o^{2.5}\sqrt{g}} = \frac{Z}{d_o^{2.5}} = \frac{0.155}{(1.167)^{2.5}\sqrt{32.2}} = 0.0186$$

and, from a geometric elements table (e.g., Chow 1959, Appendix A, noting that the section factor $Z = Q/\sqrt{g}$), $y_c/d_o = 0.132$. For $y_L/d_o = y_c/d_o = 0.132$ and $\mathbf{F}_L = 1$, Equation 12.39 gives $y_o/d = 2.339(0.132) = 0.309$ and $y_o = 0.309(1.167) = 0.360$ ft = 4.32 in. From the above inequality, $d_o > 0.917 + 1.072(4.32)/12 = 1.303$ ft = 15.6 in.

The 14-in. pipe is too small. Assume $d_o = 16$ in. ÷ 12 = 1.333 ft; then:

$$\frac{Q}{d_o^{2.5}\sqrt{g}} = \frac{Z}{d_o^{2.5}} \frac{0.155}{(16/12)^{2.5}\sqrt{32.2}} = 0.0133$$

and, from geometric elements, $y_c/d_o = 0.112$. For $y_L/d_o = y_c/d_o = 0.112$ and $\mathbf{F}_L = 1$, Equation 12.39 gives $y_o/d_o = 2.339(0.112) = 0.262$ and $y_o = 0.262(1.333) = 0.349$ ft = 4.19 in. From the above inequality, $d_o > 0.917 + 1.072(4.19)/12 = 1.291$ ft = 15.5 in.

Then, from Figure 1 in Graber (2023b):

$$14.43 + \frac{1}{12} - \frac{\sqrt{3}}{4}\left(\frac{16}{12}\right) < \text{℄ El.} < 13.72 - \frac{0.75}{12} - 0.227 + \frac{16/12}{2}$$

$13.94 < \text{℄ El.} < 14.10$

A centerline elevation of 13.95 ft was chosen. The skimming pipe as designed and constructed is shown on Figure 1 of Graber (2023b).

Example 12.3

Two adjacent rectangular final settling tanks operating in parallel, each of 60-ft width, are to have a line of cast iron skimming pipes connected in series and discharging freely into a scum manhole (Tallman Island WWTP, New York City). Each tank has two intermediate flight support beams. The skimming pipes will span the flight support beams and tank walls in such a way that six 20-ft skimming pipes will make up the series. Each skimming pipe will have a total slot length approaching 20 ft and will be required to operate with a water surface elevation varying from 16.10 to 15.77 ft. The pipes are to have the typical characteristics of skimming pipes described above and provide for a 3/4-in. head h of flow over the pipe slot. The scum manhole has a decant device and can drain continuously into the plant drain system, so any small positive value of freeboard is acceptable. Determine the required pipe diameter and centerline elevation.

Solution

The maximum depth in the skimmer occurs when the upstream-most skim pipe is rotated and its flow is conducted to the scum manhole via the five downstream pipe segments totaling 5 × 20 = 100 ft in length. The flow in the 100-ft length is constant, and Figure 12.2 can be used to determine the backwater depth. The flow entering the upstream section is given by $Q = CLH^{3/2} \cong 3.3(20)(0.75/12)^{3/2} = 1.03$ cfs. Assume $d_o = 16$ in. = 1.33 ft. Then:

$$\frac{Q}{d_o^{2.5}\sqrt{g}} = \frac{1.03}{(16/12)^{2.5}\sqrt{32.2}} = 0.0885$$

From geometric elements, at the downstream end of the pipe $y_c/d_o = 0.294$, $R_c/do = 0.168$, and $A_c/d_o^2 = 0.193$. Determine f at the downstream end as follows:

For cast iron, $\dfrac{\varepsilon}{4R} = \dfrac{0.00085}{4(0.168)(16/12)} = 0.00095$

$$V_c = Q/A = 1.03/\left[0.193(16/12)^2\right] = 3.00 \text{ ft/ sec}$$

$$\mathbf{R} = \frac{4RV}{\upsilon} = \frac{4(0.168)(1.33)(3.00)}{10^{-5}} = 2.68(10)^5$$

From a Moody Diagram (Moody 1944) or, e.g., Wood's (1966) trivariate regression relationship, we have $f = 0.022$. Then for the constant flow segment $fL/d_o = 0.022(100)/1.33 = 1.65$. From Figure 12.2 for $y_o/d_o = 0.294$ and $fL/d_o = 1.65$, we obtain $y_o/d_o = 0.46$, giving $y_o = 0.46(16/12) = 0.61$ ft. (For comparison, this is within 1% of the value of 0.603 ft calculated using the Newton–Raphson method employed above.) The value of y_o at the upstream end of the 100-ft constant flow segment becomes the y_L at the downstream end of the spatially increasing flow segment. Neglecting friction in that 20-ft segment, Equation 12.39 can be employed in the following fashion: for $y_L/d_o = 0.46$, a geometric elements table gives $A/d_o^2 = 0.3527$

and $D/d_o = 0.3538$. Then, $A = 0.3527(16/12)^2 = 0.627$ sq ft, $D = 0.3538(16/12) = 0.472$ ft, $V = Q/A = 1.03/0.627 = 1.64$ ft/sec, $\mathbf{F}_L = V\sqrt{gD} = 1.64/\sqrt{32.2(0.472)} = 0.421$. From Equation 12.39 for $y_L/d_o = 0.46$ and $\mathbf{F}_L = 0.42$, the following is obtained: $y_o/d_o = [1.339(0.421)^2 + 1](0.46) = 0.57$ and $y_o = 0.57(16/12) = 0.76$ ft. Then, from the geometric relationships analogous to those shown on Figure 3 of Graber (2023b), $d_o > 1.072(16.10–15.77 + 0 + 0.76 + 0.75/12) = 1.235$ ft × 12 = 14.8 in. Therefore, the 16-in. skimmer is the best selection. The centerline elevation is then determined as follows:

$$16.10 + 0 - \frac{\sqrt{3}}{4}\left(\frac{16}{12}\right) < \text{℄ El.} < 15.77 - \frac{0.75}{12} - 0.69 + \frac{16/2}{12}$$

15.52 < ℄ El. < 15.68

A centerline elevation of 15.56 was chosen.

It is informative in this case to compare the results of the conservative frictionless analysis above to a more precise numerical analysis. This is the best example with which to do this because the flow is fully developed due to the downstream length. For the numerical analysis, the bisection method is used (Gullberg 1997; Hornbeck 1975). Intervals of 20 ft/100 = 0.2 ft were used. The resulting calculated upstream value is 0.694 ft, compared to 0.76 given by the aforementioned conservative analysis. The latter is greater than the numerical analysis by (0.76–0.694)/0.694 = 0.0868 or approximately 9%, which is reasonable for practical purposes. Momentum exchange is the dominant factor in the spatially increasing flows.

A common configuration is wash-water troughs having semicircular bottoms surmounted by parallel side walls. A simple formula can be derived for the practical (and economical) case when y_c/d (and therefore y_o/d) ≥0.5. Then the top width is constant, and Equations 12.17 and 12.18 apply, giving:

$$D_o = \sqrt{3}\left(\frac{Q^2}{gT^2}\right)^{1/3} \tag{12.40}$$

D and H are now related as follows:

$$D = \frac{A}{T} = \frac{(1/2)\pi T^2/4 + T(H - T/2)}{T} = H - \frac{T}{2}(1 - \pi/4) = H - 0.1073T \tag{12.41}$$

The above equations then give:

$$D_o = H_o - 0.1073W = \sqrt{3}\left(\frac{Q^2}{gW^2}\right)^{1/3} \tag{12.42}$$

$$H_c = D_c + 0.1073W = \left(\frac{Q^2}{gW^2}\right)^{1/3} + 0.1073W > \frac{W}{2} \tag{12.43}$$

Rearranging:

$$Q = \frac{\sqrt{g}}{(3)^{3/4}}W(H_o - 0.1073W)^{3/2} \tag{12.44}$$

$$W^5 < \frac{Q^2}{(0.3927)^3 g} \tag{12.45}$$

Assigning units of ft to W and D_o,

$$Q = \frac{\sqrt{32.2}}{(3)^{3/4}}W(H_o - 0.1073W)^{3/2} = 2.489W(H_o - 0.1073W)^{3/2} \tag{12.46a}$$

$$GPM = Q \times 448.8 \tag{12.46b}$$

$$W^5 < \frac{(GPM/448.8)^2}{(0.3927)^3(32.2)} \tag{12.46c}$$

The sizing formulae become:

$$GPM = 1117W(H_o - 0.1073W)^{3/2} \quad (12.47a)$$

$$W < 0.076(GPM)^{2/5} \quad (12.47b)$$

Equation 12.46 is more accurate and only slightly more complex than the equation evaluated in Example 12.4a.

The effect of friction is small in wash-water troughs of usual length and cross-sectional dimensions. That results in an increase in H_o, which is negligible compared to any reasonable freeboard. The effect of friction is further discussed in Section 12.6.

Example 12.4a

A manufacturer of fiberglass-reinforced plastic wash-water troughs (Warminster Fiberglass Company, Southampton, Pennsylvania) suggests the use of the following formula to determine the trough height H_o (excluding freeboard) of horizontal troughs with semicircular bottoms and vertical side walls:

$$\text{GPM} = 857WH_o^{3/2}$$

where GPM is maximum flow in gallons per minute and W and H_o are trough width and height, respectively, in ft. Such troughs are usually sufficiently short that friction may be neglected. Determine the conditions under which the manufacturer's formula provides sufficient trough height.

Solution

Since the manufacturer's formula includes no terms for downstream conditions, a free discharge is assumed to be intended. Column 1 of Table 12.1 gives critical depth ratios y_c/d, and Column 2 gives corresponding values (see Chapter 7) of the parameter $Q/(d^{2.5}\sqrt{g})$ where Q is the free discharge flow.

Noting that $d = W$ and

$$\frac{Q}{d^{2.5}\sqrt{g}} = \frac{\text{GPM}/448.8}{W^{2.5}\sqrt{32.2}}$$

we have

$$\frac{\text{GPM}}{W^{5/2}} = 2547\frac{Q}{d^{2.5}\sqrt{g}}$$

from which values of $\text{GPM}/W^{5/2}$ corresponding to the Column 2 values are tabulated in Column 3.

Table 12.1 Wash-water Trough Example

(1) y_c/d	(2) $Q/(d^{2.5}g^{1/2})$	(3) $GPM/W^{5/2}$ (from Col. 2)	(4) $y_o/d = H_o/W$	(5) $GPM/W^{5/2}$ (from mfr.)
0.1	0.0107	27.3	0.1488	49.2
0.2	0.0418	107	0.3012	142
0.3	0.0921	235	0.4566	264
0.4	0.1603	408	0.6184	417
0.43	0.1844	470	0.6700	470
0.44	0.1927	491	0.6870	488
0.45	0.2011	512	0.7040	506
0.5	0.2459	626	0.7875	599
0.6	0.3458	881	0.9607	807
1.0	0.8434	2148	1.654	1823

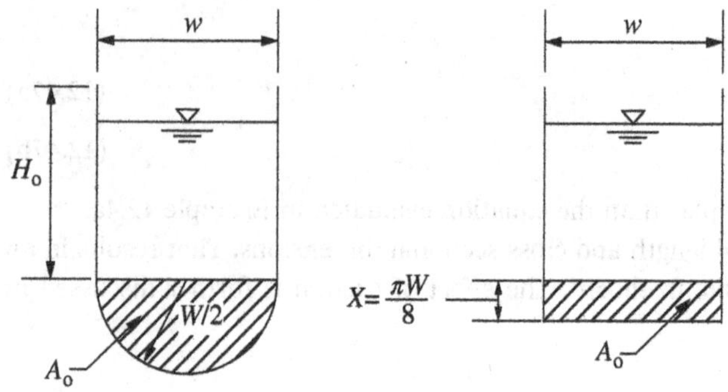

Figure 12.3 Wash-Water Trough Cross Section

From the computed values used to generate Figure 12.2 with $\mathbf{F}_L = 1$, values of $y_o/D = H_o/W$ corresponding to the values of y_c/d in Column 1 are determined and given in Column 4. From the Column 4 values of H_o/W, the manufacturer's formula is used to give corresponding values of $GPM/W^{5/2}$ in Column 5. Comparing Columns 3 and 5, the manufacturer's formula is seen to give a larger-than-actual channel capacity for $y_c/d < 0.43$ and a smaller-than-actual channel capacity for $y_c/d > 0.43$. Thus, the manufacturer's formula provides a sufficient trough height for $y_c/d > 0.43$, corresponding to $H_o/W > 0.67$.

Example 12.4b

Warminster Fiberglass Company submitted proposed shop drawings for semicircular fiberglass-reinforced plastic wash-water troughs (Andover, Massachusetts Water Treatment Plant). Each filter has a maximum flow rate of 12,000 gpm and three troughs of 22-ft length. Referring to notation in Figure 12.3, the shop drawings have $W = 24$ in. and vertical side $H_o = 12$ in. Determine the adequacy of the proposed sizes for a free discharge.

Solution

The equivalent rectangular channel is determined by equating the area of the semicircular section to that of the rectangle. Using the notation in Figure 12.3, we have $A_o = (1/2)\pi(W/2)^2 = xW \Rightarrow x = \pi W/8 = \pi(24)/8 = 3\pi\,\text{ft}/12 = \pi/4 = 0.785\,\text{ft}$. For each trough, Q = 4000/449 = 8.91 cfs, the trough-outlet critical depth $y_c = \left[Q/\left(W\sqrt{g}\right)\right]^{2/3} = \{8.91/[2(5.67)]\}^{2/3} = 0.854\,\text{ft}$, and the upstream trough flow depth $y_u = \sqrt{3}y_c = 1.48\,\text{ft}$. The latter is compared with the total trough depth of 1.785 ft, which leaves 20% for friction loss and freeboard, which is conservative for a channel of this length.

12.5 Conduits with Spatially-Increasing Flow

Full-flowing prismatic conduits with spatially-increasing flow are described by a simplified form of the momentum equation obtained by setting $\partial A/\partial x = 0$ and $T = 0$ in Equation 12.1:

$$\frac{dh}{dx} = S_o - S_f - \frac{2Q}{gA^2}\frac{dQ}{dx} \tag{12.48}$$

in which h is the piezometric head at the pipe centerline. Although less complex geometrically than their open-channel counterparts, spatially increasing flows in full conduits are often made more complex by the nonuniformity of the inflow (dQ/dx not constant). Both uniform and nonuniform inflow situations are discussed below.

12.5.1 Uniform Inflow

In some cases, uniform inflow is a desirable objective. An example is in hypochlorite-generation systems such as shown in Example 12.5, in which flow to parallel generating cells is distributed by a manifold pipe and process flow from the cells is collected by another manifold pipe (USEPA 1972). In order to ensure approximately equal flow per generating cell, uniform inflow to the collection pipe (of diameter d_o and length L) may be assumed, and the deviation therefrom is estimated by the distribution principle (see Chapter 11). Using this uniform-flow assumption for the spatially-increasing flow ($dQ/dx = Q_L/L$ and $Q = Q_L x/L = AV_L x/L$, V_L = velocity at the downstream end of the collection pipe) and the Darcy–Weisbach friction factor f (Equation 5.1 and $S_f = h_f/L$), Equation 12.48 yields the following differential equation:

$$dh = -\frac{V_L^2}{g}\left[\frac{fL}{2d_o}\left(\frac{x}{L}\right)^2 + 2\frac{x}{L}\right]d\frac{x}{L} \tag{12.49}$$

Integration of Equation 12.49 results in the following:

$$-\Delta h_x = \frac{V_L^2}{2g}\left[\frac{fL}{3d_o}\left(\frac{x}{L}\right)^3 + 2\left(\frac{x}{L}\right)^2\right] \tag{12.50}$$

in which Δh_x is the change in piezometric head from $x = 0$ (the upstream end of the collector conduit) to a distance x downstream. The maximum value of Δh_x occurs between $x = 0$ and $x = L$, and is given by:

$$-\Delta h_L = \frac{V_L^2}{2g}\left(2 + \frac{fL}{3d_o}\right) \tag{12.51}$$

The flow distribution through the overall system can be evaluated by coupling the above to the hydraulics of the influent pipe. For $fL/d_o < 2$ in the influent pipe, the piezometric head will increase in the direction of flow by an amount given by Camp and Graber (1968):

$$\Delta h' = \frac{V_o'^2}{2g}\left(1 - \frac{f'L}{3d_o'}\right) \tag{12.52}$$

Primed and unprimed variables refer to the influent and effluent pipes, respectively.

The maximum difference in available head across the generating cells depends on the relative directions of flow in the influent and effluent pipes. For countercurrent flow, the maximum difference is obtained by subtracting Equation 12.52 from Equation 12.51. If the influent and effluent pipes are of equal diameter and the flow into the influent pipe equals the flow out of the effluent pipe ($d_o = d_o'$, $V_L = V_o'$, $f = f'$), then:

$$\sum \Delta h = -\Delta h_L - \Delta h' = \frac{V_L^2}{2g}\left(1 + \frac{2}{3}\frac{fL}{d_o}\right) \tag{12.53}$$

For concurrent flow, the maximum difference, subject to the same conditions just stated, is obtained by adding Equations 12.51 and 12.52, which, for pipes of equal diameter, gives:

$$\sum \Delta h = -\Delta h_L + \Delta h' = 3\frac{V_L^2}{2g} \tag{12.54}$$

The head loss h_o across each generating cell required to give distribution uniformity m (defined as the ratio of flows through the first and last generating cells) is given by Equation 11.1:

$$h_o = \frac{\sum \Delta h}{1 - m^2} \tag{12.55}$$

Other examples where similar coupling of influent-effluent conduit hydraulics occurs and affects distribution uniformity include batteries of parallel tanks in which: (1) effluent channels submerge the weirs over which the tank effluent flows; and (2) effluent channels supplied by submerged effluent collectors, such as proposed by Lutge (1969). An example of submerged effluent collectors is given below.

Example 12.5

A system is being planned to provide on-site generation of sodium hypochlorite for effluent disinfection at a wastewater treatment plant (New Haven, Connecticut, East Shore Wastewater Treatment Plant). The system is proprietary, and the manufacturer has yet to develop hydraulic design methods for the integration of the generating cells with the inlet and outlet manifolds. A schematic of one of the multiple units that will act in parallel is shown below, including dimensions proposed by the manufacturer. The flow through the generating cells is from bottom to top of the page. Seawater drawn from offshore flows through the bottom pipe, and secondary-treatment wastewater flows into the top pipe. The chlorinated wastewater flows out from the top pipe, from which it continues to join the flow from other similar units to dose the full wastewater discharge. The pipes are PVC, and the dimensions shown are as proposed by the manufacturer. Equal distribution would supply 8 gpm to each cell. The 40 gpm diverted to the unit is to be divided into two 2-in.-diameter headers. Check the proposed design and modify it as necessary.

Solution

The uniformity of flow distribution will be evaluated first. The flow is concurrent, but the flow into the influent pipe is not equal to the flow in the effluent pipe, so Equations 12.51 and 12.52 apply, but rather than Equation 12.54 we have:

$$\sum \Delta h = -\Delta h_L - \Delta h' = \frac{V_L^2}{2g}\left(2 + \frac{fL}{3d_o}\right) - \frac{V_o'^2}{2g}\left(1 - \frac{f'L}{3d_o'}\right)$$

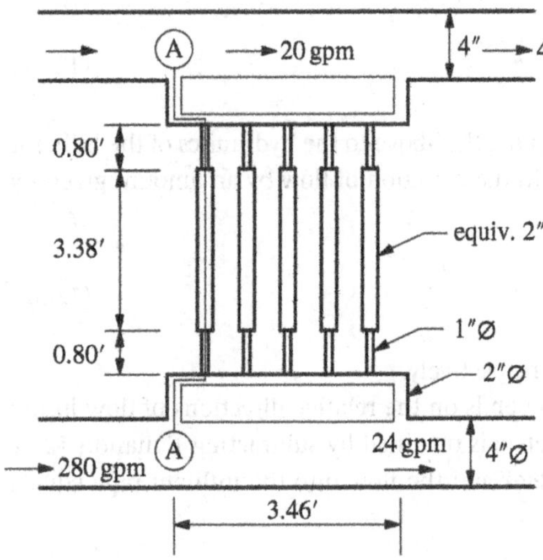

The header and collector cross-sectional areas equal $(\pi/4)(4/12)^2 = 0.0872$ ft². The flow entering the header equals 280 gpm/448.831 = 0.624 cfs, and the corresponding V_o' equals 0.624/0.0872 = 7.16 ft/sec. The flow exiting the collector equals 40 gpm/448.831 = 0.0891 cfs, and the corresponding V_L equals 0.0891/0.0872 = 1.022 ft/sec. Then,

$$\sum \Delta h \cong 2\frac{V_L^2}{2g} - \frac{V_o'^2}{2g} = 2\frac{(1.022)^2}{2(32.2)} - \frac{(7.16)^2}{2(32.2)} = |-0.764| = 0.764$$

The 40 gpm flow (= 40 gpm/448.831 = 0.0891 cfs) is assumed to be divided equally into each of the branches to the lower 2-in.-diameter header, and the flow is assumed to be divided equally in the upper 2-in.-diameter collector. The flow in the lower 2-in. header is further assumed to be divided equally into each of the five generating cells and its adjacent 1-in.-diameter pipes (40/5 = 8 gpm each). The head loss between the bottom and top 4-in.-diameter pipes through one generating cell is estimated (considering friction and form losses without giving details) to be 0.24 ft. Referring to Equation 12.55, with $\sum \Delta h > h_o$ the uniformity would be poor.

The manufacturer was contacted regarding the feasibility of increasing the sizes of the 4-in. header and collector or the alternative of installing orifices in the 1-in.-diameter pipes to increase the controlling head losses. The manufacturer wanted to keep the header and collector sizes down because of "snowflake" and sulfide gas problems but had no problem with orifices. The project team wanted to keep the orifices greater than 1/8-in. diameter to avoid possible clogging.

For $m = 0.95$ and $\sum \Delta h = 0.764$ (from above), Equation 12.55 gives:

$$h_o = \frac{\sum \Delta h}{1 - m^2} = \frac{0.764}{1 - (0.95)^2} = 7.84 \text{ ft}$$

As noted above, the relevant head loss without the orifice is 0.24 ft, so the orifice must provide 7.84–0.24 = 7.60 ft head loss.

The orifice equation given as Equation 6.20 of Section 6.3 is sufficiently accurate if taken as an equality and rearranged as follows:

$$A_o = \frac{A_1}{C_c\sqrt{2gh_o}/V_1 + 1}$$

in which the orifice area A_1 is the area of the 1-in. pipe containing the orifice, V_1 is the velocity in that pipe, and C_c is the orifice discharge coefficient, which will be taken as 0.61 (Brater and King 1976, p. 4–6). With $A_1 = (\pi/4)(1/12)^2 = 0.00545$ ft² $V_1 = (8/448.831)/0.00545 = 3.27$ ft/sec. The orifice's open area A_o and diameter d_o can then be determined as follows:

$$A_o = \frac{0.00545}{0.61\sqrt{2(32.2)(7.60)}/3.27 + 1} = 0.001063\ \text{ft}^2$$

$$d_o = \sqrt{\frac{4(0.001063)}{\pi}} = 0.0368\ \text{ft} \times 12 = 0.442,\ \text{say } 0.45\ \text{in.}$$

A final check on distribution uniformity shows that the objective is achieved:

$$h_o = \left[\frac{V_1}{C_c}\left(\frac{A_1}{A_o} - 1\right)\right]^2/(2g) = \left[\frac{3.27}{0.61}\left(\frac{0.00545}{0.001063} - 1\right)\right]^2/[2(32.2)] = 7.60\ \text{ft}$$

$$m = \sqrt{1 - \frac{\sum h}{h_o}} = \sqrt{1 - \frac{0.764}{7.60}} = 0.95$$

The high velocities and attendant low pressures at the orifice suggest the need to check for possible cavitation, i.e., the creation of water vapor bubbles, which can implode as the downstream pressure increases, causing harmful pitting of pipe walls and appurtenances. The flow rate of 8 gpm/448.831 = 0.01782 cfs through the orifice is conservatively assumed to pass through a downstream contracted area, giving a velocity and corresponding maximum possible pressure head differential of:

$$V_c = \frac{0.01782}{(\pi/4)(0.45/12)^2(0.61)} = 26.4\ \text{ft/sec}$$

$$\frac{\Delta p}{\gamma} = \frac{V_c^2}{2g} = \frac{(26.4)^2}{2(32.2)} = 10.8\ \text{ft}$$

The maximum possible absolute pressure head is then 33.9 (atmospheric pressure) − 10.8 = 23.1 ft. This is well above the vapor pressure of water (about 1 ft per Section 8.2), so no cavitation is expected.

12.5.2 Nonuniform Inflow

In some instances, uniform flow into the collection conduit is not required. Three such situations include (1) well screens, (2) submerged effluent collectors, and (3) inboard weir configurations of certain types (Graber 2007a).

Petersen *et al.* (1955) conducted theoretical and experimental studies of flow into well screens. The piezometric head outside of the well screen is constant due to the hydrostatic pressure distribution. However, the piezometric head inside the well screen decreases in the upward (downstream) direction due to energy losses associated with the exchange of momentum between the inflow and main flow. Neglecting wall friction inside the well screen, Petersen *et al.* derived the following equation for the maximum difference in piezometric head between the outside and inside of the well screen, with this maximum difference occurring at the downstream end:

$$\Delta\tilde{h}_L = 2\frac{V_L^2}{2g}\frac{\cosh\tilde{C} + 1}{\cosh\tilde{C} - 1} \tag{12.56}$$

$$\tilde{C} = 8\sqrt{2}\frac{C_D A_p L}{d_o} \tag{12.57}$$

in which V_L is the downstream velocity in the well screen; $\tilde{C}$ is referred to herein as the collector parameter; C_D is the orifice discharge coefficient of the well screen openings; A_p is the open fraction of the cylindrical surface of the well screen; and L and d_o are, respectively, the length and diameter of the well screen. Petersen *et al.* noted that $\Delta\tilde{h}_L \cong 2V_L^2/(2g)$ when $\tilde{C}$ is greater than 6.

The hydraulics of submerged effluent collectors is essentially identical to the hydraulics of flow in a well screen. The governing equation can be derived by analogy with Equation 12.56. However, its derivation is instructive and is presented here. With negligible S_f, Equation 12.48 gives the momentum equation for the flow inside the collector as follows:

$$-A^2g(dh - S_o dx) = d(Q^2) \tag{12.58}$$

in which S_o is the collector slope. The increment of flow dQ through an increment of collector length dx is given as follows:

$$dQ = C_D a\sqrt{2g\Delta\tilde{h}dx} \cdot \frac{n}{L} \tag{12.59}$$

in which C_D is the orifice discharge coefficient, a is the area of each orifice, $\Delta\tilde{h}$ is the difference in piezometric head between the outside and inside of the collector, and n is the total number of orifices spaced equally over the length L.

The difference in piezometric head between the outside and inside of the collector at a distance x along its length is given by $\Delta\tilde{h} = z - h$, in which z is the depth of the collector beneath the water surface. Differentiating this relationship results in:

$$d(\Delta\tilde{h}) = S_o dx - dh \tag{12.60}$$

Substituting Equation 12.60 into Equation 12.58 gives the general momentum equation for the submerged collector as follows:

$$d(Q^2) = A^2 g d(\Delta\tilde{h}) \tag{12.61}$$

Although submerged effluent collectors are generally parallel to the water surface so that $S_o = 0$, the disappearance of S_o in deriving Equation 12.61 shows that the hydraulics of such collectors is independent of their slope. This provides proof of the earlier statement regarding the hydraulic equivalency of well screens and submerged effluent collectors.

Integrating Equation 12.61 results in:

$$Q^2 = A^2 g(\Delta\tilde{h} - \Delta\tilde{h}_o) \tag{12.62}$$

in which $\Delta\tilde{h}_o$ is the difference in piezometric head between the inside and outside of the collector at its upstream end. Differentiating Equation 12.62 with respect to the collector length results in:

$$\frac{dQ}{dx} = \frac{A^2 g}{2Q}\frac{d(\Delta\tilde{h})}{dx} \tag{12.63}$$

Equations 12.59, 12.62, and 12.63 can be combined into the following dimensionless relationship:

$$\tilde{C}\frac{dx}{L} = \frac{d(\Delta\tilde{h})}{\sqrt{(\Delta\tilde{h})^2 - \Delta\tilde{h}\Delta\tilde{h}_o}} \tag{12.64}$$

in which:

$$\tilde{C} = 2\sqrt{2}\frac{C_D a n}{A} \tag{12.65}$$

Equation 12.64 can be integrated from $x = 0$ to L to give:

$$\tilde{C} = \cosh^{-1}\left(\frac{2\Delta\tilde{h}_L - \Delta\tilde{h}_o}{\Delta\tilde{h}_o}\right) \tag{12.66}$$

in which $\Delta\tilde{h}_L$ is the difference in piezometric head between the inside and the outside of the collector conduit at its downstream end. Substituting $\Delta\tilde{h}_o$ from Equation 12.62 into Equation 12.66 and rearranging results in Equation 12.56 with $\tilde{C}$ given by Equation 12.65. This further demonstrates the hydraulic equivalency of well screens and submerged effluent collectors mentioned earlier.

Well screens are subject to additional complications relative to the coupling of well screens and porous-medium hydraulics. Cyr (2007) is an important reference addressing this coupling.

To determine the total energy loss across the collector, the energy equation is applied between a point outside of the collector and the section inside of the collector at its downstream end, resulting in:

$$\Delta H = \Delta \tilde{h}_L - \frac{V_L^2}{2g} \tag{12.67}$$

Substituting Equation 12.56 into Equation 12.67 results in:

$$\frac{\Delta H}{V_L^2/(2g)} = \frac{\cosh \tilde{C} + 3}{\cosh \tilde{C} - 1} \tag{12.68}$$

for $\tilde{C} > 6$, $\Delta H \cong V_L^2/(2g)$. For a collector having n uniformly spaced orifices, $\tilde{C}$ is given by Equation 12.65. For a collector having a continuous slot of width w, $\tilde{C}$ is given by:

$$\tilde{C} = 2\sqrt{2}\frac{C_D w L}{A} \tag{12.69}$$

A similar configuration has been used to collect the overflow from final settling tanks. The inboard weirs discharge to a weir trough sized to produce a negligible hydraulic gradient within the trough, which in turn discharges through a series of equally spaced effluent pipes to an effluent collector pipe which is located to avoid upturning density currents.

The weir trough is sized to produce a negligible hydraulic gradient within the trough. Therefore, the piezometric head outside of the collector conduit is approximately constant with the collector conduit flowing full. The system is hydraulically analogous to the submerged effluent collector analyzed above. The relationship analogous to Equation 12.59 is as follows:

$$dQ = \frac{A_o}{\sqrt{K_c + K_e + f\ell/d_o}}\sqrt{2g\Delta\tilde{h}dx}\cdot\frac{n}{L} \tag{12.70}$$

in which K_c and K_e are, respectively, the contraction and expansion coefficients for the flow entering and leaving the effluent pipes; A_o, f, ℓ, and d_o are the cross-sectional area, friction factor, length, and diameter of each effluent pipe; $\Delta\tilde{h}$ is the head differential across the effluent pipes; and n is the total number of effluent pipes spaced equally over the length L of the effluent conduit. Equations 12.60 and 12.63 apply without modification. Continuing the analogy, Equations 12.56, 12.67, and 12.68 apply with $\tilde{C}$ given by the following:

$$\tilde{C} = 2n\frac{A_o}{A}\sqrt{\frac{2}{K_c + K_e + f\ell/d_o}} \tag{12.71}$$

in which A is the cross-sectional area of the collector pipe.

The significance of friction in the effluent collectors can be quickly assessed in an approximate way. The friction loss in an incremental length dx of the collector is given by:

$$dh_f = \frac{f}{4R}\frac{V^2}{2g}dx \tag{12.72}$$

Neglecting the effect of friction on the flow distribution, the variation of velocity head along the collector can be obtained from Equations 12.56 and 12.62 as follows:

$$\frac{V^2}{2g} = \frac{\Delta\tilde{h}_o}{4}\left(\cosh\tilde{C} - 1\right) \tag{12.73}$$

Substituting Equation 12.73 into Equation 12.72 and integrating gives the total friction loss within the collector:

$$h_f = \frac{fL}{4R}\frac{\Delta\tilde{h}_o}{4}\left(\frac{\sinh\tilde{C}}{\tilde{C}} - 1\right) \tag{12.74}$$

With the aid of Equations 12.56 and 12.62, Equation 12.74 becomes:

$$h_f = \frac{fL}{4R}\frac{V_L^2}{2g}\frac{\sinh\tilde{C} - \tilde{C}}{\tilde{C}\left(\cosh\tilde{C} - 1\right)} \tag{12.75}$$

Note that the friction loss increases the gradient of h inside the collector, thus augmenting the effect of momentum exchange. Then h_f increases the nonuniformity of inflow with still more flow coming into the openings near the collector's downstream end and less flow entering the openings near the upstream end. This in turn will increase $\Delta\tilde{h}_L$ but decrease $\Delta\tilde{h}_o$. This redistribution of flow will also reduce h_f below the value given by Equation 12.75. Therefore, a conservative approximation (the reader should verify this) to ΔH will be obtained by adding piezometric head differentials and friction loss as follows:

$$\Delta\tilde{h}_L \cong \left(\Delta\tilde{h}_o\right)_{f=0} + \frac{V_L^2}{g} + h_f \tag{12.76}$$

Noting that the first two terms on the right-hand side are equal $\left(\Delta\tilde{h}_L\right)_{f=0}$, the total head loss across the collector is conservatively given as follows:

$$\Delta H = \Delta\tilde{h}_L - \frac{V_L^2}{2g} \cong \left(\Delta\tilde{h}_L\right)_{f=0} + h_f - \frac{V_L^2}{2g} \tag{12.77}$$

In some cases, frictional effects may be sufficiently large to justify the use of numerical methods discussed in Section 12.6.

Example 12.6
This example from Graber (2007a) demonstrates the importance of checking inflow uniformity. The preliminary design of a submerged effluent collector system in a WWTP of 1012 mgd (44.4 m³/s) maximum flow capacity (Montreal, Canada) called for five 70-ft (21.3 m) cast iron collectors in each of 14 primary settling tanks. Each collector was proposed to have an inside diameter of 24 in. (0.610 m) and 220 equally-spaced orifices of 2¼-in. (5.71 cm) diameter and an orifice coefficient of $C_D = 0.6$. One of the main purposes of the collectors was to achieve a more uniform rise rate in the outlet zone of the tanks, in the belief (but see the following) that this would improve sedimentation efficiency. Demonstrate that an analysis of the collectors based on the assumption of uniformly distributed inflow would be inadequate. Then, properly determine the total head loss across each collector, first by neglecting and then by including wall friction inside the collector.

Solution

The maximum flow per tank is 1012/14 × 1.547 = 111.8 cfs, and the maximum flow per collector is 111.8/5 = 22.4 cfs. If the inflow were uniform, the flow per orifice would be 22.4/220 = 0.1017 cfs, with a head loss per orifice determined as follows:

$$\text{Orifice area, } a = \frac{\pi}{4}\left(\frac{2.25}{12}\right)^2 = 0.0276\ \text{ft}^2$$

$$\frac{Q}{a} = \frac{0.1017}{0.0276} = 3.68\ \text{ft/sec}$$

$$\text{Head loss per orifice, } \Delta h = \frac{1}{2g}\left(\frac{Q}{C_D a}\right)^2 = \frac{1}{64.4}\left(\frac{3.68}{0.6}\right)^2 = 0.584\ \text{ft}$$

The internal pressure difference along the collector can be determined by applying the momentum equation between sections 1 and 2 to obtain:

$$p_1 A - p_2 A = \frac{\gamma Q V_2}{g}$$

or

$$\frac{p_1 - p_2}{\gamma} = \frac{V_2^2}{g}$$

Then:

$$V = \frac{22.4}{\pi(2)^2/4} = 7.13\ \text{ft/sec}$$

$$\frac{p_1 - p_2}{\gamma} = \frac{(7.13)^2}{32.2} = 1.58\ \text{ft}$$

The 1.58 ft (0.482 m) piezometric head differential inside the collector is large compared to the 0.585 ft (0.178 m) head loss across the orifices, assuming equal flow through each orifice. Considering the distribution principle of Chapter 11, the inflow is highly nonuniform.

Neglecting wall friction inside the collector, Equations 12.65 and 12.56 apply and give the following:

$$A = \frac{\pi}{4}(2)^2 = \pi\ \text{ft}^2$$

$$\tilde{C} = \frac{2\sqrt{2}(0.6)(0.0276)(220)}{\pi} = 3.28$$

$$\cosh \tilde{C} = 13.31$$

$$\frac{V_L^2}{2g} = \frac{(7.13)^2}{64.4} = 0.789\ \text{ft}$$

$$\Delta\tilde{h}_L = 2(0.789)\frac{13.31+1}{13.31-1} = 1.835\ \text{ft}$$

The total energy loss across the collector is then obtained from Equation 12.77 as $\Delta H = 1.835 - 0.789 = 1.046$ ft. The total head loss, including friction, can be conservatively estimated by Equations 12.75 and 12.77. The Reynolds Number and roughness ratio are used to obtain the friction factor, as follows:

$$\mathbf{R} = \frac{VD}{\upsilon} = \frac{7.13(2)}{10^{-5}} = 1.426(10)^6$$

$$\frac{\varepsilon}{D} = \frac{0.00085}{2} = 0.000425$$

$f = 0.017$ (from Moody Diagram or Equation 5.1)

The friction factor is doubled to account for inflow momentum exchange: use $f = 0.034$. Then,

$$\frac{fL}{D} = \frac{0.034(70)}{2} = 1.19$$

$$\sinh \tilde{C} = \sinh(3.28) = 13.27$$

$$h_f = 1.19(0.789)\frac{13.27-3.28}{3.28(13.31-1)} = 0.232\ \text{ft}$$

$$\Delta H = 1.835 + 0.232 - 0.789 = 1.278\ \text{ft}$$

Based on the stated variables, calculated values are $A_p = 0.01381$, $\tilde{C} = 3.281$, $fL/(4R_f) = 1.190$, and $V_L^2/g = 0.480$ m (1.574 ft). These calculated values were used in the computational procedure described earlier in the section entitled "Difference Formulation." The numerically-computed values include $V_L^2/(g\Delta\tilde{h}_L) = 0.774$ and $\Delta\tilde{h}_o/\Delta\tilde{h}_L = 0.1155$. The resulting value of $\Delta\tilde{h}_L = (V_L^2/g)/[V_L^2/(g\Delta\tilde{h}_L)] = 0.480/0.774 = 0.620$ m(2.03 ft) was considered acceptable. However, the value of $\Delta\tilde{h}_o/\Delta\tilde{h}_L$ (which is independent of the assumed value of $\tilde{h}_o$) is 0.157, which corresponds to an inflow uniformity of $m = \sqrt{\Delta\tilde{h}_o/\Delta\tilde{h}_L} = \sqrt{0.157} \cong 0.4$. Such a low inflow uniformity cannot correspond to a uniform rise rate of the type desired. The important point is that inflow uniformity should be an explicit part of the design of collection conduits of this type and others of the types considered herein. Incidentally, but importantly, although there can be other reasons to use submerged effluent collectors [e.g., odor control (Pincince 1991) and reducing basin inlet velocities by transferring distribution-controlling head loss from the basin inlet to the outlet], primary sedimentation efficiency does not require and can be adversely affected by departures from conventional end weirs (Graber 1974a; WEF/ASCE 1998; Graber 2005).

12.6 Numerical Solutions

Collector conduits with combinations of slope, taper, varying top width, and/or significant friction are occasionally encountered. Although their preliminary design is often best accomplished by methods discussed above, final design and analysis may merit numerical methods.

Numerical solutions of collection-conduit problems can be divided into two types. First are those that are structured around discrete points of inflow; such methods are particularly appropriate where the inflow points are relatively few and/or unequally spaced so that the assumption of continuous distribution of the inflow becomes shaky. The second type of numerical solution is structured around the actual or assumed continuous nature of the inflow. The two types of numerical solutions are referred to as lumped-parameter and distributed-parameter methods, respectively. The latter type adequately covers collection conduits of the type considered in this chapter and is elaborated upon below.

12.6.1 Open Channels

A well-known formulation of the problem for numerical solution is that presented by Chow (1959) for open channels, given as follows:

$$\Delta y' = \frac{Q_1(V_1 + V_2)}{g(Q_1 + Q_2)}\left(\Delta V + \frac{V_2}{Q_1}\Delta Q\right) + S_f\Delta x \qquad (12.78a)$$

in which $\Delta y'$ is the drop in water surface between upstream section 1 and downstream section 2 a distance Δx apart. The term $\Delta y'$ is related to the change in depth Δy as follows:

$$\Delta y' = -\Delta y + S_o\Delta x \qquad (12.78b)$$

Except for the frictional term, Equation 12.78a is identical to one presented by Hinds (1926).

An important, and sometimes misunderstood, point pertaining to Equation 12.78 is that it is applicable to *all* of the categories of collection channels discussed in this chapter, including tapered channels. Equation 12.78 stems from Equation 12.1, which is the same as Equation 2.137, the derivation of which has no restriction on cross-sectional variations. The effect of such variations is taken into account if the terms V_1, V_2, and S_f in Equation 12.78 are based on the proper cross-sectional characteristics. This should be obvious insofar as the momentum terms in the derivation are concerned. The incorporation of the side pressure forces in the tapered channel is less apparent but is explained by the discussion of pressure forces in Chapter 2. Hinds (1926), in fact, applied Equation 12.78a (with $S_f = 0$) to a tapered side-channel spillway. The application of a precise numerical method to Example 12.1 gave a result that agreed within 1% with the change in depth obtained using Equation 12.14 for the frictionless case. The approximate solution in Example 12.1 for the change in depth, including friction, was only 3% higher than the corresponding numerical result.

Setting $\alpha = 1$ (the friction factor should be increased to compensate), Equation 12.78 can be rewritten as follows:

$$y_1 - y_2 = \frac{(V_1 + V_2)}{g(Q_1 + Q_2)}(V_2Q_2 - V_1Q_1) + S_f\Delta x - S_o\Delta x \qquad (12.79)$$

This equation has the advantage (over Equation 12.78a) of avoiding a singularity at the upstream end of the conduit when $Q_1 = 0$. The trial-and-error process employing the above equation is demonstrated by the example given below.

Example 12.7

Two existing precast-concrete collection channels receive effluent from an activated sludge aeration tank (Tallman Island WWTP, New York City). Each channel is 12 ft–10½ in. (12.875 ft) long, with an upstream width of 2 ft, downstream width of 3 ft, and upstream and downstream invert elevations of 11.31 and 9.16 ft, respectively. Flow enters the channel over rectangular weirs along the tapered sides of the channel. The WWTP is to be expanded, such that the two channels would receive a total flow of 39.6 mgd. Downstream modifications will reduce the downstream WSL (Water Surface Level) in the channels to El. 13.40 ft at the maximum flow. Determine if the existing channels will have adequate capacity to prevent submergence of the weirs at El. 13.68 ft. Use $n = 0.03$, including the allowance for inflow turbulence.

Solution

Table 12.2 shows the numerical solution based on Equation 12.82. Each column is explained below.

Table 12.2 Subsurface Drain Design Example

x (1)	y (2)	Q (3)	$Q_1 + Q_2$ (4)	b (5)	y (6)	A (7)	V (8)	$V_1 + V_2$ (9)	R (10)	$S_f\Delta x$ (11)	$S_o\Delta x$ (12)	$y_1 - y_2$ (13)	D (14)	F (15)	WSL (16)
12.875	....	30.6	....	3.00	4.24	12.7	2.41	....	1.108	....	0.167	....	4.24	0.206	13.40
11.875	1.0	28.2	58.8	2.92	4.09	11.9	2.36	4.77	1.076	0.00206	0.167	−0.147	4.09	0.206	13.42
10.875	1.0	25.8	54.0	2.84	3.95	11.2	2.30	4.66	1.045	0.00294	0.167	−0.146	3.95	0.204	13.44
9.875	1.0	23.5	49.3	2.77	3.80	10.5	2.23	4.53	1.015	0.00201	0.167	−0.145	3.80	0.202	13.46
8.875	1.0	21.1	44.6	2.69	3.66	9.85	2.14	4.37	0.984	0.00194	0.167	−0.143	3.66	0.197	13.49
7.875	1.0	18.7	39.8	2.61	3.52	9.19	2.04	4.18	0.952	0.00186	0.167	−0.142	3.52	0.192	13.52
6.875	1.0	16.3	35.0	2.53	3.38	8.55	1.91	3.95	0.920	0.00173	0.167	−0.141	3.38	0.183	13.54
5.875	1.0	14.0	30.3	2.46	3.24	7.97	1.76	3.67	0.892	0.00156	0.167	−0.141	3.24	0.172	13.57
4.875	1.0	11.6	25.6	2.38	3.10	7.38	1.57	3.33	0.860	0.00135	0.167	−0.140	3.10	0.157	13.60
3.875	1.0	9.21	20.8	2.30	2.95	6.79	1.36	2.93	0.827	0.00110	0.167	−0.141	2.95	0.140	13.61
2.875	1.0	6.83	16.0	2.22	2.81	6.24	1.09	2.45	0.796	0.000813	0.167	−0.142	2.81	0.115	13.64
1.875	1.0	4.46	11.3	2.15	2.67	5.74	0.777	1.867	0.766	0.000503	0.167	−0.146	2.67	0.084	13.67
0.875	1.0	2.08	6.54	2.07	2.51	5.20	0.400	1.177	0.733	0.000225	0.167	−0.152	2.51	0.044	13.67
0.000	0.875	0	2.08	2.00	2.37	4.74	0	0.400	0.703	0.0000431	0.146	−0.141	2.37	0	13.68

Col. 1. Distance of the station along the channel measured from the upstream end, in ft
Col. 2. Distance increment
Col. 3. Channel flow, equal to $Q_L - q\Sigma\Delta x$, where $Q_L = 39.6/2 \times 1.547 = 30.6$ cfs and $q = 30.6/12.875 = 2.377$ cfs
Col. 4. Sum of the flow Q_2 at the previous station and Q_1 at the station under consideration
Col. 5. Width of the station under consideration, equal to $b_L - [(b_L - b_o)/L]\Sigma\Delta x$
Col. 6. Trial value of flow depth at the station under consideration; only final trial values are tabulated
Col. 7. Flow area, equal to the width in Col. 5 times the depth in Col. 6
Col. 8. Velocity, equal to the flow in Col. 3 divided by the area in Col. 7
Col. 9. Sum of the velocity V_2 at the previous station and V_1 at the station under consideration
Col. 10. Hydraulic radius corresponding to the width in Col. 5 and depth in Col. 6: $R = by/(b + 2y)$
Col. 11. Friction loss, based on $S_f = (S_{f1} + S_{f2})/2$ where $S_f = [nV/(1.487R^{2/3})]^2$. The values of V and R are obtained from Col's. 8 and 10, respectively, from the row above for the previous section and the adjacent row for the station under consideration. Col. 2 provides the value of Δx.
Col. 12. The slope S_o is uniform, with a value of (11.31 – 9.16)/12.875 = 0.167. Col. 2 provides the value of Δx.
Col. 13. Depth change given by Equation 12.82 based on values in Col's. 3, 4, 8, 9, 11, and 12. The resulting y_1 is compared to the trial value in Col. 6, and the iterative procedure is repeated; only *final* trial values are tabulated.
Col. 14. The hydraulic depth in this case is equal to y.
Col. 15. Froude Number given by $V/\sqrt{gD}$ based on final trial values in Col's. 8 and 14. The Froude Number must be ≤ 1 at all stations for the procedure to be valid. [Chow (1959) presents an analogous procedure for the more complex case of supercritical flow.]
Col. 16. Water surface elevation equal to $y + S_o \sum \Delta x + z_L$ (invert El. @ $x = L$)

The upstream WSL is calculated to be El. 13.68 ft, the same as the weir elevation. In this case, due to the extreme nature of the maximum flow and the relative certainty of the maximum downstream WSL, the existing channels were considered acceptable.

The procedure exemplified above can be readily applied to other channel shapes. For example, a tapered trapezoidal channel is dealt with by using the more general relations for such channels when calculating the flow area, hydraulic radius, and hydraulic depth in Col's. 7, 10, and 14, respectively. A circular channel, a semicircular channel with vertical side walls, or any other mathematically describable channel cross section can be similarly accommodated. (Methods for natural cross sections not amenable to mathematical expression are discussed in Chapter 7.)

In developing and applying computer code using Equation 12.79, the author has used the graphical bisection and Newton–Raphson methods for approximating the root (y_1) of Equation 12.79. Computer speed has reached the point where the reduced efficiency of graphical bisection can be tolerated in view of its greater robustness for these problems. The lower and upper limits can be selected as the critical depth and pipe crown or overflow elevation, as a means of testing for supercritical conditions and surcharging or overflowing. When there is no root within those limits, smaller solution intervals and the downstream slope will help reveal the flow conditions.

Although Equation 12.78 is the form presented by Chow (1959, Chapter 12) based on a more limited form of Equation 12.1 for prismatic channels, and Equation 12.79 is derived from Equation 12.78, it can be shown that Equations 12.78 and 12.79 are applicable to the more general case of nonprismatic flow represented by Equation 12.1. Equation 12.79 is the form that has been used by the author in the work presented below. The depth y_1 is found by iterative calculation for each Δx step. The depth y_2 is known from the first Δx step and is the previously calculated y_1 for each subsequent Δx step. The flow terms Q_1 and Q_2 are prescribed based on a known or assumed inflow (e.g., uniform inflow, for which $Q = Q_L x/L$), and V_1 and V_2 are calculated by continuity based on the prescribed flows at the corresponding sections and the flow areas, which are known functions of corresponding flow depths.

12.6.2 Full Conduits

For full conduits with nonuniform inflow hydraulically coupled to the pressure gradient in the conduit, a numerical solution may be justified if conduit friction is significant and the conservative approximation resulting from Equations 12.75–12.77 is deemed inadequate. Then Equations 12.48 and 12.60 can be used to give the following formulation of the problem:

$$\Delta\tilde{h}_2 - \Delta\tilde{h}_1 = \frac{(Q_1 + Q_2)}{gA^2}\Delta Q + S_f\Delta x \tag{12.80a}$$

The term ΔQ is given by:

$$\Delta Q = \frac{\tilde{C}A}{2\sqrt{2}}\sqrt{2g\frac{(\Delta\tilde{h}_1 + \Delta\tilde{h}_2)}{2}\frac{\Delta x}{L}} \tag{12.80b}$$

with $\tilde{C}$ given by Equations 12.57, 12.65, 12.69, or 12.71, as appropriate. The trial-and-error process employing Equation 12.80 consists of assuming $\Delta\tilde{h}_o$, which becomes the first $\Delta\tilde{h}_1$, solving Equations 12.80a and 12.80b by trial and error for $\Delta\tilde{h}_2$, then proceeding through successive Δx's to obtain $\Delta\tilde{h}_L$ and Q_L. This value of Q_L is then compared with the desired total flow in the collector, and the entire process is repeated until the process converges on the desired Q_L and corresponding values of $\Delta\tilde{h}_o$ and $\Delta\tilde{h}_L$.

12.6.3 Automatic Computation

The computations associated with the numerical procedures discussed above are simple but too numerous for convenient hand computation, thus favoring automatic computation. The procedures may be programmed on a programmable calculator or digital computer, and several such programs have been developed. The Newton–Raphson method is an efficient method for approximating the root y_1 of Equation 12.79 and generally works well. Two such programs were developed and used to obtain the generalized solutions discussed in the next subsection.

Uniform inflow (the most common situation by far) can be expressed as follows for computational purposes:

$$Q = Q_o + qx \tag{12.81a}$$

$$q = \frac{Q_L - Q_o}{L} \tag{12.81b}$$

$$x = n\Delta x \tag{12.81c}$$

in which n is the number of Δx increments counted from the downstream end. By setting $q = 0$ (equivalent to setting $Q_1 = Q_2 = Q$ in Equation 12.79), the conventional backwater problem (see Chapter 7) may be handled by the same program. The standard-step backwater method would proceed essentially as in the example above but with constant Q.

Figure 2 of Graber (2004b) provides a method comparison for circular collector conduits. It can be used for certain design problems and demonstrates certain general characteristics of subsurface drain hydraulics. However, it is usually as convenient to use a computer program in cases for which that figure is applicable, and use of such a program is essential for accurate design in more complex cases. Use of a computer program in a more complex design is illustrated by the following example:

Example 12.8

A 1829-m (6000-ft) interceptor drain is to be designed for a total flow of 0.0680 m^3/s (2.40 cfs) uniformly distributed along its length, a constant grade of 0.20%, and Manning's $n = 0.011$ (Londonderry, New Hampshire). The length of the pipe is such that it would be more economical to design it to have multiple diameters. The computer program used employed the numerical computational procedure described in the section "Numerical Formulation" of Graber (2004b), with geometric elements for a circular conduit. That enabled determination by trial and error of the pipe size required from the upstream end of the 1829-m (6000-ft) length to the location [within 15.2-m (50-ft) increments] downstream to which a particular pipe diameter would suffice without being surcharged assuming critical depth at its downstream end. If a pipe is surcharged along its length (or if supercritical flow occurs along its length), the program so indicates and states the distance from the downstream end where the surcharge (or supercritical flow) first occurs. The diameters and lengths of pipe thus determined are: 0.356-m (14-in.) over 1829 m (6000 ft), 0.305-m (12-in.) over 1676 m (5500 ft), 0.254 m (10-in.) over 1036 m (3400 ft), 0.203 m (8-in.) over 579 m (1900 ft), 0.152 m (6-in.) over 274 m (900 ft). That gives a tentative design, as shown in the table below.

The next step entails applying the computer program to each individual length of pipe in the table below. That can be done in different ways. The one chosen here was to first apply the program to each length, on the assumption of a free discharge at the downstream end of each pipe. Actual computer input and output (with SI units added) for one such length is given below.

The output includes the values of the maximum and minimum velocities (and their corresponding locations and flow rates), which can be considered in relation to velocity-design criteria. That includes its relation to drain filter or envelope design, as given by SCS (1986, pp. 4–81 to 4–82 and 4–91 to 4–92). For each pipe length, the computed downstream critical depth and upstream depth are tabulated, and the downstream critical depth is compared to the depth, relative to the same datum, imposed by the next downstream pipe's computed upstream depth for the type of reducing coupling used (concentric, eccentric with matching inverts, eccentric with matching crowns). In this case, assuming concentric reducing couplings, the 0.305-m (12-in.) and 0.254-m (10-in.) pipes had their outlets' critical depth elevations submerged by the downstream pipe's water surface elevation, whereas the 0.203-m (8-in.) and 0.152-m (6-in.) pipes did not.

Pipe size (m) (in.)	Downstream extent (m) (ft)	Flow rate at downstream end (m^3/s) (cft)	Length of pipe required (m) (ft)	Maximum velocity (m/s) (ft/sec)	Minimum velocity (m/s) (ft/sec)
0.152 (6)	274 (900)	0.0102 (0.36)	274 (900)	0.826 (2.71)	0 (0)
0.203 (8)	579 (1900)	0.0215 (0.76)	305 (1000)	0.966 (3.17)	0.521 (1.71)
0.254 (10)	1036 (3400)	0.0385 (1.36)	457 (1500)	1.039 (3.41)	0.652 (2.14)
0.305 (12)	1676 (5500)	0.0623 (2.20)	640 (2100)	1.134 (3.72)	0.744 (2.44)
0.356 (14)	1829 (6000)	0.0680 (2.40)	152 (500)	1.161 (3.81)	0.875 (2.87)

Note: Maximum velocity occurs at downstream end and minimum velocity occurs at upstream end of each pipe length, except that minimum velocity in 0.305-m (12-in.) pipe occurs 270 m (887 ft) from downstream end of its length

Enter Downstream (Starting) Depth [m (ft)] (or 0 to start at Critical Depth):		0 (0)
Enter Conduit Length [m (ft)]:		640 (2100)
Enter Conduit Slope:		0.002
Enter Downstream Flow [m³/s (cfs)]		0.0623 (2.2)
Enter Conduit Diameter [m (ft)]:		0.305 (1)
Enter Manning's n [or 0 for Darcy f]:		0.011
Enter Spatial Inflow [m³/s-m (cfs/ft)]:		3.71EE-5 (0.0004)
Enter Number of Intervals or ⟨CR⟩ for default (L/Do):		
Do you want to suppress intermediate output (⟨Y⟩ or ⟨N⟩)?		*y*
CUMULATIVE		
DISTANCE [m (ft)]	FLOW [m³/s (cfs)]	DEPTH [m (ft)]
0.00 (0.00)	0.0623 (2.200)	0.203 (0.666) (CRITICAL DEPTH)
146.91 (482.00)	0.0568 (2 007)	0.294 (0.964) (MAXIMUM DEPTH)
640.05 (2100.00)	0.0385 (1.360)	0.200 (0.655)
0.00 (0.00)	0.0623 (2.200)	0.203 (0.666) [VMAX = 1.207 M/SEC (3.961 ft/s)]
265.16 (870.00)	0.0524 (1.852)	0.280 (0.920) [VMIN = 0.746 M/S (2.449 ft/s)]

Accordingly, the program was applied again to the 0.305-m (12-in.) pipe considering the actual downstream (starting) depth. It was found that the pipe did not surcharge, and the upstream depth was the same as determined previously. The same was found for the 0.254-m (10-in.) pipe. (Expansion and contraction losses between pipe sections can be considered where significant.) The results tabulated above can thus be accepted. The above example is actually taken from a *National Engineering Handbook* (SCS 1986) and gives a design differing from and more economical than that provided in the handbook, which based the design on Manning's equation alone. The differences are not as large as might have been the case, however, because of the relatively high value of S_oL/d_o (weighted average of 14.3, at which Manning's equation alone and spatially varied-flow theory will agree fairly closely per Figure 2 of Graber 2004b). The maximum and minimum velocities computed for each pipe length are included in the above table; such velocities are not included in the handbook example (and cannot be calculated by the handbook method).

Some additional features of the program are: (1) if a starting depth is less than critical depth in input, a critical depth fixup is conspicuously used; (2) if zero is entered for the roughness coefficient, the program prompts to enter a value of the Darcy–Weisbach friction factor (in which case zero can then be entered if desired for the frictionless case); and (3) nondefault values (other than L/d_o) can be entered for the computational distance interval (it is a good idea to test different intervals and gain familiarity with the effect of such intervals on computational accuracy).

The assumption of free discharge ($\mathbf{F}_L = 1$) is often useful in design. It allows decoupling of the collection-conduit design from downstream hydraulics, may allow use of generalized design charts, and will often give an economical design. When free discharge is assumed, downstream hydraulics should be checked to assure that the free discharge is not submerged. In some cases, submerging the discharge may give a more economical design, which can be analyzed using a computer program such as that described above. In other cases, it may be found that the submergence does not require design changes, such as for some of the pipe lengths in the above example.

12.6.4 Generalized Numerical Solutions

Section 5.4 presented a dimensional analysis which led to useful deductions regarding generalized solutions of the basic equation of spatially-increasing flow in open channels. That analysis, coupled with the feasibility of obtaining numerical solutions to virtually any type of spatially-increasing flow problem, suggests that generalized or "once-and-for-all" numerical solutions can be obtained for certain categories of collector conduits. Such solutions enable the use of a design chart or other convenient device to obtain a ready solution to a problem that would otherwise require a special-case (individual) numerical solution for comparable accuracy.

The categories of collector conduits for which generalized solutions are possible include the following (also see Section 12.7 regarding gutters).

First, prismatic channels of simple cross section [i.e., $D \propto y$ and $A^2D/(A_L^2D_L) = \mathrm{f}(y/y_L)$] and negligible friction loss, for which

$$\frac{y_o}{y_L} = \phi\left(\frac{S_oL}{y_L}, \mathbf{F}_L\right) \tag{12.82}$$

Second, prismatic channels of any cross section with negligible friction loss and free downstream discharge ($\mathbf{F}_L = 1$), for which:

$$\frac{y_o}{\sigma} = \phi\left(\frac{y_c}{\sigma}, \frac{S_oL}{\sigma}\right) \tag{12.83}$$

Third, prismatic channels of any cross section with $S_o = 0$ and free downstream discharge ($\mathbf{F}_L = 1$), for which Graber (2004b) gives:

$$\frac{y_o}{\sigma} = \phi\left(\frac{y_c}{\sigma}, \frac{fL}{\sigma}\right) \tag{12.84}$$

The generalized numerical solution for a sloped channel with negligible friction (the first category) was obtained by Li (1955) for rectangular and triangular channels. Li presented graphical representations of his solutions. Li's numerical solution for the rectangular channel can be readily extended to any channel having constant top width. To do so, depth y and width b should be replaced by hydraulic depth D and top width T in Equation 12.82. Li (1955) suggested the convenient device of transforming the channel into a rectangular one of equal area. Such a transformation was discussed in Section 12.4 for a semicircular channel with vertical side walls.

The second category is similar to the first, but for more general cross sections and uses different dimensionless variables. It appears to be of little practical value and is not considered further. The reader should be aware of the possibility of generalizing a numerical solution in this case, should the need arise.

The author has obtained generalized numerical solutions for horizontal channels ($S_o = 0$) with a free discharge ($\mathbf{F}_L = 1$, $y_L = y_c$) (category 3). Using a computer-based numerical method, the relation between the variables of Equation 12.84 was established for rectangular channels ($\sigma = b$) and plotted over a typical range of values. For each value of y_c/b, the relation between y_o/b and fL/b is approximated by a straight line, which suggests the following relationship:

$$\frac{y_o}{b} = \sqrt{3}\frac{y_c}{b} + \mathrm{M}\cdot f\frac{L}{b} \tag{12.85a}$$

$$\mathrm{M} = \mathrm{M}\left(\frac{y_c}{b}\right) \tag{12.85b}$$

The slope M is plotted against y_c/b in Figure 12.4.

Example 12.9

A final clarifier in a WWTP (Massachusetts Correctional Institute, Bridgewater) has a 20 ft diameter and a maximum flow of 0.72 mgd. Determine the dimensions of an economically proportioned peripheral, concrete, rectangular collection channel discharging freely at its downstream end.

Solution

As noted above, the perimeter of a horizontal, freely discharging, short (frictionless) rectangular channel is minimized when $y_o/b = 3/4$, corresponding to (by Equation 12.5a) $y_c/b = y_o/(\sqrt{3}b) = 3/(4\sqrt{3}) = \sqrt{3}/4 = 0.433$. This provides a good initial assumption and, by Equation 12.5b, gives a channel width of $b = 1.652(Q^2/g)^{1/5}$. For $Q = 0.72(1.547) = 1.114$ cfs per tank and a flow $Q_L = 1.114/2 = 0.557$ cfs (note the flow goes in two directions from the null point in the collection channel to the drop structure), $b = 1.652[(0.557)^2/32.2]^{1/5} = 0.653$ ft $= 7.83$ in., say 8 in. Equation 12.85 and Figure 12.4 then give:

$$\frac{y_o}{b} = \frac{3}{4} + 0.022f\frac{L}{b}$$

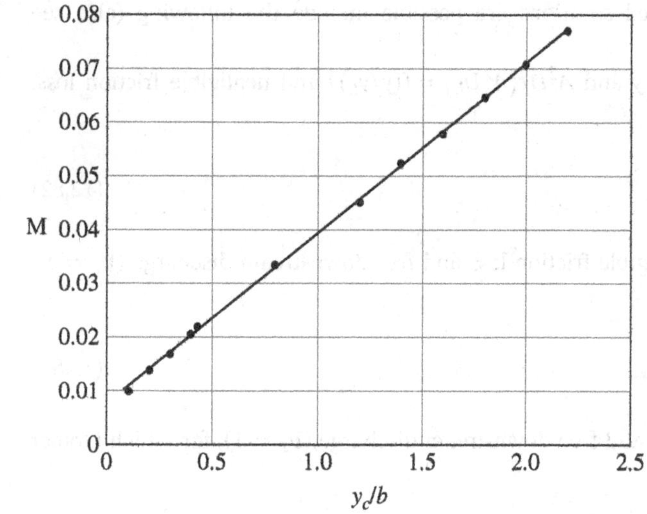

Figure 12.4 Plot of Coefficient M vs. y_c/b

$$\frac{L}{b} = \frac{20\pi/2}{8/12} = 15\pi = 47.1$$

The friction factor should be conservatively based on the minimum hydraulic radius, which occurs at the free overfall: $R = 2y_c + b \cong [2(0.433) + 1]8/12 = 1.244$ ft and $R^{1/3} = 1.075$ For $n = 0.013, f = 117n^2/R^{1/3} = 117(0.013)^2/1.075 = 0.0184$. The friction factor is doubled giving $f = 2 \times 0.0184 = 0.037$ and y_o calculated as follows:

$$\frac{y_o}{b} = \frac{3}{4} + 0.022(0.037)(47.1) = 0.75 + 0.038 = 0.788$$

$$y_o = 0.788(8) = 6.3, \text{say 7 inches}$$

The channel as designed and constructed included an allowance for free fall below the weir crest. The effluent box is 2 ft–3 in. square in plan, and the effluent pipe of 14-in. diameter extends vertically from that box. The 22-in. half perimeter of the 14-in. effluent pipe is well in excess of the 8-in. channel width, assuring free discharge conditions.

Example 12.10

An intermittent sand filter preceded by an effluent pond (lagoon) at a wastewater treatment plant (Massachusetts Correctional Institution, Bridgewater, Massachusetts) is planned to have multiple cells, each of which will be 30.5 m (100 ft) long and have ten perforated clay-pipe underdrains connected to a collector. Each underdrain is to be designed hydraulically for the minimum discharge during the dosing siphon flow cycle on the rectangular area contributing to that underdrain of 0.00442 m^3/s (70 gpm, 0.156 cfs) (the higher discharges during the dosing cycle are to be stored in the filter bed). The underdrains will discharge freely to the collector, and the upstream-most underdrain will have a slope of 0.0014. With Manning's n taken as 0.015, the size of the underdrains is to be determined.

Solution

Assuming a 0.102-m (4-in.) pipe diameter, $S_oL/d_o = 0.420$ and $gn^2L/\left(K_n^2 d_o^{4/3}\right) = 1.42$ for which Figure 2 of Graber (2004b) gives $Q_L/\left(\sqrt{g}d_o^{5/2}\right) \cong 0.28$ so that $Q_L = 0.28\sqrt{9.81}(0.102)^{5/2} = 0.0029$ m^3/s(0.10 cfs), which is less than the required design capacity. For a 0.152-m (6-in.) pipe diameter, $S_oL/d_o = 0.280$ and $gn^2L/\left(K_n^2 d_o^{4/3}\right) = 0.827$ for which Figure 2 of Graber (2004b) gives $Q_L/\left(\sqrt{g}d_o^{5/2}\right) \cong 0.30$ so that $Q_L = 0.30\sqrt{9.81}(0.152)^{5/2} = 0.0085$ m^3/s(0.30 cfs), which provides the necessary design capacity. As noted above, Figure 2 of Graber (2004b) is based on the assumption of a free discharge (i.e., unsubmerged critical depth at the outlet) and that the pipe fills to a depth equal to 0.95 of the pipe diameter. The layout (as actually designed by the author) was checked to assure that the unsubmerged-discharge assumption was satisfied by calculating the hydraulic profile in the collector and comparing it to the critical depth elevation at each adjacent underdrain.

12.6.5 Relation to Other Design Methods

Most of the following is taken from Graber (2004b). Manning's equation has been used above as one means of determining the friction slope in the differential or numerically spatially varied-flow equation. The use of Manning's equation alone for determining the capacity of subsurface drains is suggested in numerous publications cited by Graber (2004b). Manning's equation for full-pipe uniform flow can be expressed as follows, with Q_{LUF} denoting the flow:

$$\frac{Q_{LUF}}{\sqrt{g}d_o^{5/2}} = 0.312\sqrt{\frac{S_oL}{d_o}}\sqrt{\frac{K_n^2 d_o^{4/3}}{gn^2L}} \tag{12.86}$$

Some curves representing this equation (for full-pipe conditions) are superimposed on the generalized plot of Graber (2004b). For zero slope, Manning's equation gives zero flow capacity, so the curve for $S_oL/d_o = 0$ is coincident with the abscissa. For certain of the nonzero slopes plotted, the curves representing Manning's equation alone intersect the generalized curves at one point. However, no significance should be assigned to that. For example, the intersection point for $S_oL/d_o = 1.0$ corresponds to the maximum depth occurring at $x/L = 0.73$ for the spatially varied flow, at which location the rate of flow is 73/100 of the flow at the downstream end of the pipe. If Manning's equation alone had any significance, it would be because the backwater curve is such that the uniform-flow depth is reached close to the downstream end of the pipe where the flow is close to the total flow in the pipe. For $S_oL/d_o = 10.0$, the curve for spatially-varied flow fairly closely "parallels" the Manning's-alone curve, indicating that for that condition, Manning's equation alone provides a more reasonable approximation; in that case, the maximum depth occurs fairly close to the downstream end [e.g., at $x/L = 0.917$ for $gn^2L/\left(K_n^2 d_o^{4/3}\right) = 5.1$].

At the selected value of 0.95 for $y_{\max}/d_o$ the uniform-flow capacity approximately equals the maximum value of 1.07 times the full-pipe capacity for constant n, and if the uniform-flow value was taken at $y_{\max}/d_o = 0.95$ (rather than $y/d_o = 1.0$) the curves at large S_oL/d_o would be closer. Less generalized plots in other forms (e.g., Q_L vs. L/d_o and d_o for constant values of n and S_o, or Q_L vs. L/d_o and n for constant values of d_o and S_o) prepared by the author also generally show Manning's equation alone substantially underpredicting or overpredicting the subsurface drain capacity. A subsurface drain on zero slopes will have the maximum depth occur at the upstream end of the pipe. And as the slope increases, the maximum depth location moves away from the upstream end of the pipe. Manning's equation alone implies that the depth increases in the downstream direction, which is the opposite of what actually occurs overall (zero slope), much (smaller slopes), or at least some (larger slopes) of the pipe length. Furthermore, if the tailwater depth is allowed to increase to the uniform-flow depth predicted by Manning's equation at the downstream end of the pipe, then a spatially varied-flow analysis shows that the pipe will often surcharge (hydraulic grade line above the pipe) a short distance upstream from the downstream end. Manning's equation alone also implies (incorrectly) that subsurface drain capacity is independent of L/d_o. In general, Manning's equation alone provides an inadequate basis for estimating subsurface drain capacity.

Considering prismatic pipes with uniform inflow, Van der Beken (1969, 1970) suggested that the term $(dy/dx)[1 - Q^2T/(gA^3)]$ in Equation 12.1 was negligible compared to the friction term S_f. Hung (1971) suggested a similar approximation for modeling the overland flow of urban stormwater runoff. In accordance with the above, Van der Beken suggested the following approximation:

$$S_o \cong S_f + \frac{2Q}{gA^2}\frac{dQ}{dx} \tag{12.87}$$

Using Manning's equation for S_f and incorporating the uniform inflow assumption, the above equation becomes:

$$S_o \cong \left(\frac{nQ}{K_nAR^{2/3}}\right)^2 + \frac{2Q^2}{gLA^2} \tag{12.88}$$

This equation suggests that the depth of flow increases in the downstream direction and that the flow can be analyzed as a uniform flow with the slope increased to account for the energy loss due to momentum exchange associated with the uniform inflow. The above equation can be recast in terms of Q_{LUF} in Equation 12.86, with the full-pipe flow according to Van der Beken's method denoted by Q_{LVF}

$$\frac{Q_{LVF}}{\sqrt{g}d_o^{5/2}} = \frac{\left(Q_{LUF}/\sqrt{g}d_o^{5/2}\right)}{\sqrt{1 + \frac{2}{(\pi/4)^2}\left(\frac{Q_{LUF}}{\sqrt{g}d_o^{5/2}}\right)^2\left(\frac{d_o}{S_oL}\right)}} \tag{12.89}$$

Some curves representing this equation are superimposed **on** Figure 2 of Graber (2004b). For zero slope, Van der Beken gives zero flow capacity, as did Manning's equation alone. For the nonzero slopes plotted, the Van der Beken method nearly coincides with the curves representing Manning's equation alone at higher values of $gn^2L/\left(K_n^2 d_o^{4/3}\right)$, and over some of the range of values comes closer (but not necessarily adequately) to the spatially varied-flow curves than did Manning's equation alone. For $S_oL/d_o = 10.0$, where the curve for spatially varied flow fairly closely "parallels" Manning's curve, Van der Beken provides no better an approximation than does Manning's equation alone. Van der Beken's method is additionally subject to some of the same limitations as discussed above for Manning's equation alone. In general, the Van der Beken method provides an inadequate basis for estimating subsurface drain capacity.

For collector channels of constant top width, Equation 12.85 and Figure 12.4 will give a slightly conservative approximation when applied to the equivalent rectangular section. That is because the wetted perimeter is increased in the transformed section, reducing R and increasing S_f.

The author has obtained a similar generalized solution for horizontal circular channels ($\sigma = d$). The results, in this case, do not lend themselves to a relationship analogous to Equation 12.85. The results can be expressed by an equation or in terms of Figure 2 of Graber (2004b) for $S_oL/d_o = 0$. The range shown should more than cover the range of practical applications.

12.7 Street Gutters and Grates

Most of the following is taken from Graber (2013a). Although there are other street gutter shapes (FHwA 1996, 2001b), the focus here is on the most common shape, i.e., triangular with one vertical side (curb). Four issues are addressed: (1) the proper coefficient in Manning's formula for uniform flow and friction loss in gutters; (2) adjustment of frictional resistance for spatial inflow; (3) whether uniform flow occurs in gutters with spatially increasing flow; and (4) interaction with inlets. The first of these issues is addressed by considering historical and modern information to recommend the best one of two different equations that have been put forward for determining the uniform-flow capacity and friction loss in gutters. The second issue is addressed for subcritical and supercritical gutter flow, for which practical, generalized numerical solutions are derived. The range of parameters for which uniform flow provides an adequate approximation is demonstrated theoretically and by example. Among the important conclusions is that the common practice of using Manning's equation alone for such problems is not always adequate. Practical recommendations are made to address the third issue. The fourth issue is addressed by building on the material thus presented and using literature information. The concepts developed here apply not only to gutters more widely but also to the characteristics and relationships for subcritical and supercritical flow in open channels more generally.

12.7.1 Uniform Flow – Coefficient in Manning's Formula

Manning's formula, applied to a triangular gutter cross section in a conventional manner with the hydraulic radius taken to be the full cross-sectional area divided by the full wetted perimeter, gives the form given by Conner (1945), Izzard (1946), and Chow (1959, p. 151):

$$Q = \left(\frac{1}{2}\right)^{5/3} \frac{K_n}{n} \frac{z^{5/3} y_x^{8/3}}{\left(1 + \sqrt{1 + z^2}\right)^{2/3}} S_f^{1/2} = \frac{K_n'}{n} \frac{z^{5/3} y_x^{8/3}}{\left(1 + \sqrt{1 + z^2}\right)^{2/3}} S_f^{1/2} \tag{12.90}$$

in which Q = volumetric rate of flow; y_x = depth at the curb; S_f = friction slope, which, for uniform flow, equals the longitudinal channel slope S_o; n = Manning's roughness coefficient; and $K_n = 1$ for SI units (length terms in meters, time in seconds) and $K_n = 1.486$ for customary units (length terms in feet, time in seconds). For customary units $K_n' = (1/2)^{5/3} K_n = (1/2)^{5/3}(1.486) = 0.468$, and for SI units, $K_n' = (1/2)^{5/3} K_n = (1/2)^{5/3}(1) = 0.315$. The term z is the inverse of the lateral bottom slope $S_x \doteq y_x/T_x$ in which T_x = flow top width. Values of S_x are generally <1 for gutters, with z having values generally >1.

Although in the original paper Izzard (1946) used Manning's equation for gutter flow in a triangular cross section based on the use of the full cross-sectional area and wetted perimeter, in his closure he revised his use of the Manning equation in the original paper, which he called "not correct." Izzard introduced the notion, which he attributed to the National Hydraulic Laboratory Staff of the Bureau of Standards, of applying Manning's equation to the incremental discharge in an incremental cross-sectional width of the gutter flow with the hydraulic radius R taken as the depth of flow y at a distance from the curb.

Integration then results in an equation similar to Equation 12.90 but with different numerical coefficients: for SI units, $K'_n = (3/8)K_n = (3/8)(1) = 0.375$, and for customary units $K'_n = (3/8)K_n = (3/8)(1.486) = 0.557$. Izzard (1946) noted that using this form of Manning's equation gives a discharge about 19% greater (0.557/0.468 = 1.19) for an assumed value of n in a given gutter. A similar derivation is given by Burke and Burke (1994, 2008). Others using the revised Izzard form of the equation include Jens and McPherson (1964), ASCE/WEF (1992), FHwA (1996, 2001b), Virginia Department of Transportation (VDOT 2002), and Guo (2000, 2006). Selecting $R = y$ in the derivation is equivalent to taking the wetted perimeter of the fluid element as only the channel bottom. That ignores the shear stresses on the sides of the element resulting from the transverse velocity gradient, which is caused by the vertical gutter side and sloped gutter bottom. Thus a lower K'_n results, for which no experimental justification has been provided.

Conner (1945) measured rate of discharge (constant along the gutter length) and maximum depth of flow at various slopes from 0.5% to 10% in a model concrete gutter of single cross section having a transverse slope of 2/24. For the known cross-sectional dimensions, Conner calculated A, P, and R. Then average cross-sectional velocity was calculated as $V = Q/A$, and Manning's roughness coefficient n was then calculated from Manning's equation: $n = K_n R^{2/3} S^{1/2}/V$. Conner's calculated n values ranged from 0.0141 to 0.0194. Analysis of the plotted data allows outliers to be identified and a smaller range of n values to be reasonably determined. However, Conner (1945) suggested that these data be used with caution because the gutter section was built in lengths of 10 ft (3.05 m) with separating strips between each section. The measurement of flow depth and width was difficult mainly because of "the roughness of the water surface and the fact that the depth varied from time to time even though the rate of discharge remained constant." Izzard (1946) interpreted Conner's (1945) experiments as giving an average value of $n = 0.017$ for a concrete gutter with constant flow and suggested that n be increased to the "commonly used" value of 0.02 to account for lateral inflow and rainfall impact. Izzard (1946) stated that "[i]t is reasonable to assume [for gutters] that rainfall impact and lateral inflow will both operate to increase the effective resistance to flow." Urbonas and Roesner (1992) applied the factor 1.2, saying that "[w]ithout this factor the theoretical gutter capacity is underestimated."

The logical basis is not apparent for the application of reduction factors to uniform-flow conditions by the Denver Urban Drainage and Flood Control District and in the Las Vegas area (Urbonas and Roesner 1992, ASCE/WEF 1992, Guo 2000). Guo (2000) stated that Denver's reduction factors were "derived from field experience" but also stated that there are no documents that present the details of the derivation. The application of such factors to uniform-flow conditions is made even more questionable by the analyses of spatially varied gutter flow discussed below.

Numerous references suggest limitations of the concept of hydraulic radius in shallow channels, associated with the non-uniform distribution of shear stresses on the wetted perimeter. However, most offer little in the way of practical or experimentally verified suggestions for dealing with the limitations of a gutter-shaped channel. Shih and Grigg (1967) found, experimentally for rectangular channels, that when the aspect ratio (width:depth) exceeds 2, the hydraulic radius can be applied in the customary way to uniform flow. If this is extrapolated to the gutter shape considered here by taking the aspect ratio as top width:hydraulic depth (hydraulic depth being the same as average depth in this case). Then $S_x < 1$satisfies this criterion. Since $S_x < 1$ will generally be the case for practical triangular gutter cross sections, this provides reasonable support for Equation 12.90 with $K'_n = 0.468$.

Khiadani *et al.* (2005, 2007) and Beecham *et al.* (2005) observed the increase in flow resistance for spatially-increasing flow relative to uniform-flow values. If extended systematically to higher Reynolds numbers typical of gutters (fully turbulent flow), such research could be of quantitative value for estimating friction factors in gutters.

From Graber (2004c), the component of lateral inflow in the main flow direction acts opposite to the friction term in the momentum equation for spatially-increasing flow. The lateral inflow term is given by $q_i U_i \cos\phi_i/(gA)$, in which q_i = lateral inflow per unit length of channel, U_i = velocity of lateral inflow, ϕ_i = angle between velocity vector of lateral flow and direction of main flow, g = gravitational acceleration, and A is as defined previously. Since U_i and ϕ_i are not generally known for runoff and direct rainfall, all that can be said is that the lateral inflow term may offset to some extent the increase in friction due to lateral inflow and rainfall impact. Graber (2004b) discusses the basis for multiplying Manning's n by $\sqrt{2}$but notes that no systematic studies are known to have quantified such relationships.

It is recommended here that to avoid ignoring certain shear stresses as discussed above, because of the logic of Shih and Grigg (1967) as also discussed above, and for conservatism, the smaller values of K'_n be used (0.315 in SI units, 0.468 in customary units) in Equation 12.90 when calculating uniform-flow gutter capacities. However, the applicability of uniform-flow capacities requires scrutiny, as discussed below. In the absence of better information, it is further recommended that Manning's n for gutter flow be increased by the multiplier of $\sqrt{2}$ per the discussion referenced in the previous paragraph.

12.7.2 Subcritical Spatially Increasing Flow

In subcritical spatially increasing flow, the Froude Number is less than 1 throughout the channel length. Beij (1934) provided a solution for the case of uniform inflow in a frictionless, horizontal, triangular channel with equal or unequal side slopes and free discharge [$\mathbf{F}_L = 1$ and $y_L = y_c$, in which $\mathbf{F}$ = Froude Number, y_L = flow depth at the downstream end, y_c = critical-flow depth, and the subscript "L" denotes conditions at the downstream end ($x = L$)]. Its relationship to generalized, dimensionless solutions is noted by Graber (2004b). Specialized to the case of a triangular gutter (with vertical curb in which y = flow depth at curb), the solution can be expressed as:

$$\frac{y_o}{y_c} = \sqrt[3]{\frac{5}{2}} \tag{12.91a}$$

$$\frac{Q_L}{\sqrt{g}y_o^{5/2}} = \frac{1}{2^{2/3}5^{5/6}S_x} \tag{12.91b}$$

in which y_o = flow depth at the upstream end, Q_L = flow rate at the downstream end, and other terms are as previously defined.

Li (1955) presented the following analytical solution for a frictionless, horizontal, triangular channel for the more general case of $\mathbf{F}_L > 1$:

$$\frac{y_o}{y_L} = \sqrt[3]{\frac{3}{2}\mathbf{F}_L^2 + 1} \tag{12.92a}$$

That solution can be expressed in a form analogous to Equation 12.91b as:

$$\frac{Q_L}{\sqrt{g}y_o^{5/2}} = \frac{\mathbf{F}_L}{\left(3\mathbf{F}_L^2/2 + 1\right)^{5/6}}\frac{\sqrt{2}}{4S_x} \tag{12.92b}$$

Equations 12.92a and 12.92b reduce to Equations 12.91a and 12.91b, respectively, for $\mathbf{F}_L = 1$ and $y_L = y_c$.

Li (1955) also presented a generalized numerical solution in graphical form for frictionless, sloped, triangular channels with uniform inflow in terms of y_o/y_c and other variables. However, as noted by Graber (2004b), and contrary to assertions of Li (1955) and Chow (1959, Chapter 12), the value of that solution is limited, in that y_o will generally be the maximum depth in the channel only in the case of $S_o = 0$. For $S_o > 0$ the maximum depth will generally be at an intermediate location along the channel.

For the real case with slope and friction, dimensional analysis (Graber 2004b) indicates that generalized numerical solutions involving three variables at most can be obtained for triangular channels, according to the following functional form (Graber 2004b):

$$\frac{y_{max}}{y_L} = \Phi\left\{\frac{S_oL}{y_L}, \frac{fL}{y_L}\right\} \quad \text{for } \mathbf{F}_L = 1, \text{fixed } z \tag{12.93}$$

in which y_{max} is the maximum depth in the channel and f is the Darcy–Weisbach friction factor.

By following through the derivation in Graber (2004b) that leads to Equation 12.92, but without separating out the z term, one finds that the z term is associated only with the friction term in the following functional form:

$$\frac{y_{max}}{y_L} = \Phi\left\{\frac{S_oL}{y_L}, \frac{fL}{y_L}\left(\frac{1+\sqrt{1+z^2}}{z}\right)\right\} \quad \text{for } \mathbf{F}_L = 1 \tag{12.94}$$

Taking $y_L = y_c$ (since $\mathbf{F}_L = 1$), performing permissible operations on the dimensionless variables, expressing y_c in terms of Q_L and other variables, taking $z = 1/S_x$, and converting from f to Manning's n as in deriving Equation (53) of Graber (2004b), Equation 12.93 can be changed to the following form:

$$\frac{Q_LS_x}{\sqrt{g}y_{max}^{5/2}} = \psi\left\{\frac{S_oL}{y_{max}}, \frac{gn^2L}{K_n^2y_{max}^{4/3}}\left(S_x + \sqrt{1+S_x^2}\right)^{4/3}\right\} \quad \text{for } \mathbf{F}_L = 1 \tag{12.95}$$

The term on the left-hand side of Equation 12.95 is a form of modified Froude Number with which the transverse slope term S_x is associated. For $S_x \ll 1$, the transverse slope drops out of the friction term.

To study the hydraulic characteristics of gutters in the present case, the following variation of the difference formula given under "Numerical Formulation" of Graber (2004b) was used:

$$y_1 - y_2 = \frac{(V_1 + V_2)}{g(Q_1 + Q_2)}(V_2Q_2 - V_1Q_1) - \delta S_f \Delta x + \delta S_o \Delta x \tag{12.96}$$

in which $\delta = 1$ for subcritical flow with computations proceeding upstream from the control section (the downstream end of the gutter for subcritical flow), and $\delta = -1$ for supercritical flow with computations proceeding downstream from the control section. The subscripts "1" and "2" denote, respectively, current Section 1 and prior Section 2, a distance Δx apart. Other terms are as defined previously. Equation 12.96 is based on the assumption that the channel slope is small enough that its effects on the assumed hydrostatic pressure head and forces on channel sections are negligible. That requires $S_o < 0.15$ (Graber 2004b), which is generally met with gutters for which a common regulatory maximum slope is 0.12. The Newton–Raphson method was used for approximating the root (y_1) of Equation 12.96 to an accuracy of 0.0000305 m (0.000001 ft).

The subcritical-flow portion on Figure 2 of Graber (2004b) provides generalized curves prepared in terms of the functional relationship given by Equation 12.95 and using Equation 12.96. Testing verified that all results for a given S_oL/y_{max} but different values of transverse slope S_x collapse onto a single S_oL/y_{max} curve. The water surface profiles increase from critical depth at the downstream end of the gutter to the maximum depth at a section between the ends of the gutter length except for $S_oL/y_{max} = 0$ for which the maximum depth occurs at the upstream end of the gutter. For $S_oL/y_{max} = 0$, the value of $Q_LS_x/\left(\sqrt{g}y_{max}^{5/2}\right)$ at the left-hand side of Figure 2 of Graber (2004b) equals the value of 0.165 calculated for $n = 0$ from Equation 12.91b, providing additional confirmation of the numerical procedure.

The assumption $\mathbf{F}_L = 1$ requires free discharge, i.e., that the tailwater be at or below the critical depth elevation at the downstream end. This critical depth is given by:

$$y_{cL} = \left[\frac{Q_L^2}{g(1/2)^3 z^2}\right]^{1/5} \tag{12.97}$$

For more complex systems involving, for example, downstream submergence, a computer program may be used for design and analysis purposes (see Graber 2004b).

For subcritical flow, all curves for $S_oL/y_{max} > 1$ were found to terminate with $\mathbf{F}_L$ close to 1 at a value of $Q_LS_x/\left(\sqrt{g}y_{max}^{5/2}\right) = \sqrt{2}/4 = 0.354$, the significance of which will be discussed below. The curve for $S_oL/y_{max} = 1$ is asymptotic to that maximum value, the significance of which will also be discussed below.

In the development leading to Figure 2 of Graber (2004b), Manning's equation has been used as the means of determining the friction slope. The use of Manning's equation alone (uniform flow) for determining gutter capacity is suggested in numerous publications, some of which have been cited above. Manning's equation for uniform gutter flow, given by Equation 12.90, can be expressed in terms of the same variables used in Equation 12.95 and Figure 2 of Graber (2004b), with Q_{LUF} denoting the flow in that case:

$$\frac{Q_{LUF}S_x}{\sqrt{g}y_{max}^{5/2}} = \frac{(1/2)^{5/3}}{\left(S_x + \sqrt{1 + S_x^2}\right)^{2/3}}\sqrt{\frac{S_oL}{y_{max}}}\sqrt{\frac{K_n^2 y_{max}^{4/3}}{gn^2L}} \tag{12.98}$$

Some curves representing Equation 12.98 are superimposed in Figure 2 of Graber (2004b). For zero slope, Manning's equation gives zero flow capacity, so the curve for S_oL/y_{max} is coincident with the abscissa. For certain of the nonzero slopes plotted, the curves representing Manning's equation alone intersect the generalized curves at one point. However, no significance should be assigned to that. For example, the intersection point for $S_oL/y_{max} = 1.0$ corresponds to the maximum depth occurring at $x/l = 0.88$ for the spatially-varied flow, at which location the rate of flow is 88/100 of the flow at the downstream end of the gutter. If Manning's equation alone had any significance, it would be because the backwater curve is such that the uniform-flow depth is reached close to the downstream end of the gutter where the flow is close to the total flow in the gutter. For $S_oL/y_{max} = 10.0$, the curve for spatially-varied flow fairly closely "parallels" the Manning's-alone curve, indicating that for that condition Manning's equation alone provides a more reasonable approximation. In that case, the maximum depth occurs very close to the downstream end (i.e., at $x/L \cong 1$) over the full range of the curve. A gutter on

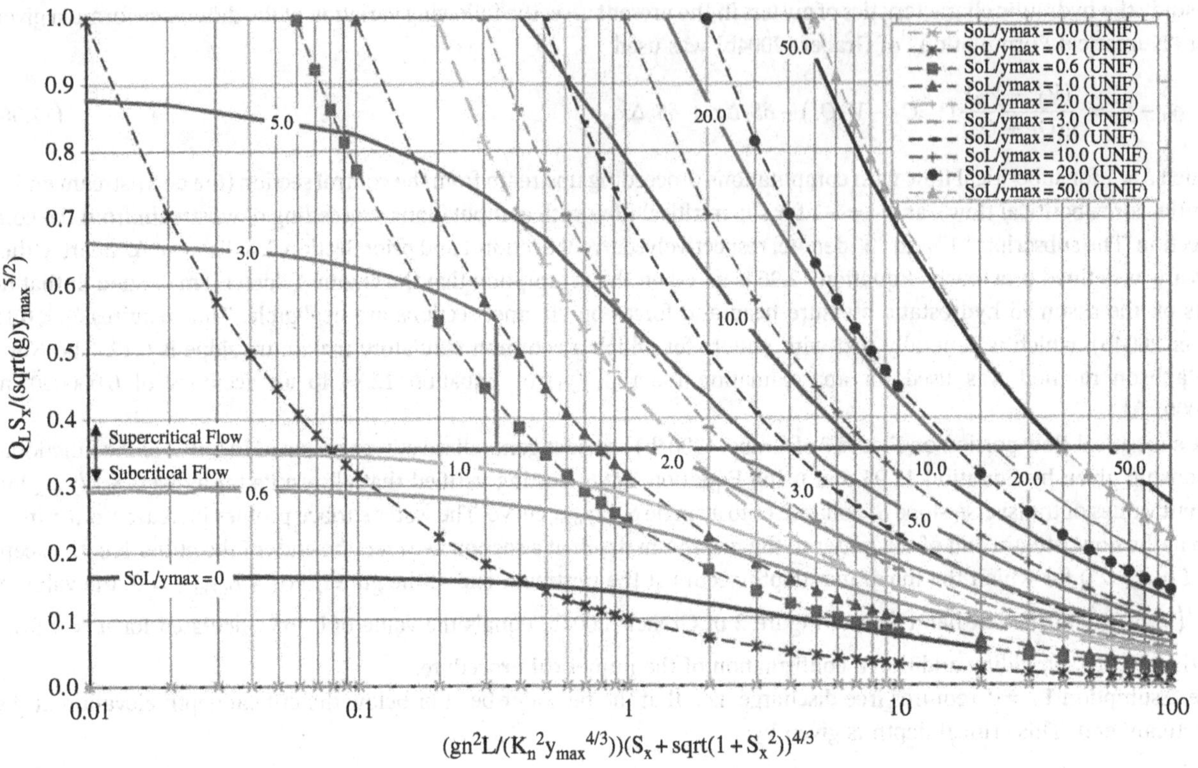

Figure 12.5 Triangular Gutter Chart for Uniform Inflow with Free Discharge (from Graber 2013a). The color rendition of this image is available on the book's landing page on wiley.com.

zero slope (of only conceptual interest) will have the maximum depth occur at the upstream end of the gutter. As the slope increases, the maximum depth location moves away from the upstream end of the gutter. Manning's equation alone implies that the depth increases in the downstream direction, which is the opposite of what actually occurs overall (zero slope), much (smaller slopes), or at least some (larger slopes) of the gutter length. Furthermore, if the tailwater depth is allowed to increase to the uniform-flow depth predicted by Manning's equation at the downstream end of the gutter, then a spatially varied-flow analysis shows that the gutter will often overflow a short distance upstream from the downstream end. The use of Manning's equation alone also implies (incorrectly) that gutter capacity is independent of L/y_{max}. Manning's equation alone does not always provide an adequate basis for estimating gutter capacity. Figure 12.5 shows that the uniform-flow approximation is always conservative for subcritical flow with $S_oL/y_{max} > 5$ and $\mathbf{F}_L = 1$.

12.7.3 Subcritical Flow Example

An example given by the Virginia Department of Transportation (VDOT 2002) can be shown to be subcritical based on $\mathbf{F}_L < 1$. That example assumes uniform flow and does not give a gutter length L. Nevertheless, certain uses can be made of that example, which has $S_o = 0.0035$, $S_x = 0.0208$, $n = 0.015$, and $T = 4.27$ m (14 ft). The uniform-flow solution is given as $Q_{LUF} = 0.112$ m^3/s (3.96 cfs), and the downstream flow depth can be calculated as $y_{max} = S_xT = 0.0208(4.27) = 0.0888$ m (0.291 ft). These values give $Q_{LUF}S_x/\left(\sqrt{g}y_{max}^{5/2}\right) = 0.317 < 0.354$ in Equation 12.99 below; the values also enable calculation of L for each of the S_oL/y_{max} values plotted in Figure 12.5, and values of $\left[gn^2L/\left(K_n^2 y_{max}^{4/3}\right)\right]\left(S_x + \sqrt{1 + S_x^2}\right)^{4/3}$ from which spatially-varied-flow values can be determined from Figure 12.5. For nonzero values of L, with $S_oL/Y_{max} = 0.6$, 1.0, 2.0, 5.0, and 10.0, the corresponding values of L are 15 m (50 ft), 25 m (83 ft), 51 m (166 ft), 127 m (416 ft), and 254 m (832 ft), respectively, and the corresponding spatially varied-flow values of $Q_LS_x/\left(\sqrt{g}y_{max}^{5/2}\right)$ are 0.246, 0.263, 0.287, 0.301, and 0.310, respectively. Comparing these latter values to the 0.317 value given above, for this example, the uniform-flow assumption results, for the same y_{max}, in gutter capacities being overpredicted by 29%, 21%, 10%, 5%, and 2%, respectively.

The last two results (5% and 2%) differ from the statement in the previous section about conservatism of the uniform-flow assumption for $S_oL/y_{max} > 5$ only because the example was done based on the fixed uniform-flow $Q_{LUF}S_x/\left(\sqrt{g}y_{max}^{5/2}\right) = 0.317$ based on VDOT (2002).

12.7.4 Supercritical Spatially-Increasing Flow

Supercritical flow often occurs in steeper gutters (e.g., McEnroe *et al.* 1999, Haestad 2010). It is readily evidenced by cross-waves where there is debris in the gutter. The examples given by Burke and Burke (2008, Chapter 5) and FHwA (2001b, pp. 4–11, 4–75–81) have supercritical flow, based on $\mathbf{F}_L > 1$. When supercritical flow occurs, a hydraulic control point will be located at an intermediate position along the channel, in addition to or instead of the critical-flow control point at the downstream end. Several researchers have proposed methods for locating the control point in channels with spatially-increasing flow (Hinds 1926; Chow 1959, Chapter 12; Smith 1967, 1972; Lopes and Shirley 1993).

The onset of supercritical-flow conditions was determined by error traps in the subcritical-flow program used for generating the subcritical spatially-varied-flow curves in Figure 12.5. That onset is represented by the upper limit of subcritical flow for those curves when $S_oL/y_{max} > 1$. That upper limit is plotted in Figure 12.5.

More generally, at a control section along the channel length:

$$\frac{Q_cS_x}{\sqrt{g}y_c^{5/2}} = \frac{\sqrt{2}}{4} = 0.354 \tag{12.99}$$

in which y_c = depth and Q_c = flow rate at the critical-control section.

Lopes and Shirley (1993) derived equations for calculating the depth at the critical-control section for a triangular channel. Specializing to the case of a triangular gutter cross section, their equations compress to (in terms of S_x and with the term K_n added for metrification):

$$\left(\frac{gn^2}{K_n^2}\right)\left(2^{1/3}\right)\left(S_x + \sqrt{1+S_x^2}\right)^{4/3} y_c^{-1/3} + \frac{(Q_L/L)S_x}{\sqrt{g}}\left(2^{3/2}\right)y_c^{-3/2} - S_o = 0 \tag{12.100}$$

Equation 12.100 must be solved for y_c numerically in the general case (Lopes and Shirley 1993).

Rearranging Equation 12.100 enables it to be expressed in a form similar to the functional form of Equation 12.95:

$$\left(\frac{gn^2L}{K_n^2y_c^{4/3}}\right)\left(2^{1/3}\right)\left(S_x + \sqrt{1+S_x^2}\right)^{4/3} + \frac{Q_LS_x}{\sqrt{g}y_c^{5/2}}\left(2^{3/2}\right) - \frac{S_oL}{y_c} = 0 \tag{12.101}$$

Considering the critical depth to be the maximum depth at the downstream end of the channel, substituting $y_c = y_{max}$ into Equation 12.101, and solving for the flow term gives:

$$\frac{Q_LS_x}{\sqrt{g}y_{max}^{5/2}} = \left(2^{-3/2}\right)\frac{S_oL}{y_{max}} - \left(2^{-7/6}\right)\left(S_x + \sqrt{1+S_x^2}\right)^{4/3}\left(\frac{gn^2L}{K_n^2y_{max}^{4/3}}\right) \tag{12.102}$$

Setting the flow term in Equation 12.102 equal to $\sqrt{2}/4$ from Equation 12.99 and taking $n = 0$ in Equation 12.102 gives $S_oL/y_{max} = 1$. This is the maximum value of the slope term for which supercritical flow does not occur at lower n values, in precise agreement with Figure 12.5.

It is instructive in the case of supercritical gutter flow to consider an example first. For the abovementioned example from Burke and Burke (2008, Chapter 5), given values are $S_o = 0.01$, $S_x = 0.03$, $n = 0.013$, $K_o = 1.486$, $Q_L = 0.5\,\text{cfs}(0.142\,\text{m}^3/\text{s})$, and $L = 235\,\text{ft}$ (71.6 m). Equation 12.100 gives $y_c = 0.065\,\text{ft}$ (0.01966 m). Equation 12.99 then gives $Q_c = 0.0707\,\text{cfs}$ (0.00200 m³/s), from which the distance from the upstream end of the gutter to the control section $x_c = (Q_c/Q_L)L = (0.0707/0.5)235 = 33.2\,\text{ft}(10.1\,\text{m})$. Starting at that control section and using Equation 12.96 upstream and downstream of the control section to determine the flow depth and computing the corresponding Froude numbers gives results shown in Figure 12.6. As is typical for such channels, conditions are subcritical ($\mathbf{F} < 1$) upstream of the control station and supercritical ($\mathbf{F} > 1$) downstream of the control station. The depth of flow increases to a maximum at the downstream end where, in this case, it is nearly equal to the uniform-flow depth. Similar results were found for the abovementioned example from FHwA (2001b, pp. 4–75–81).

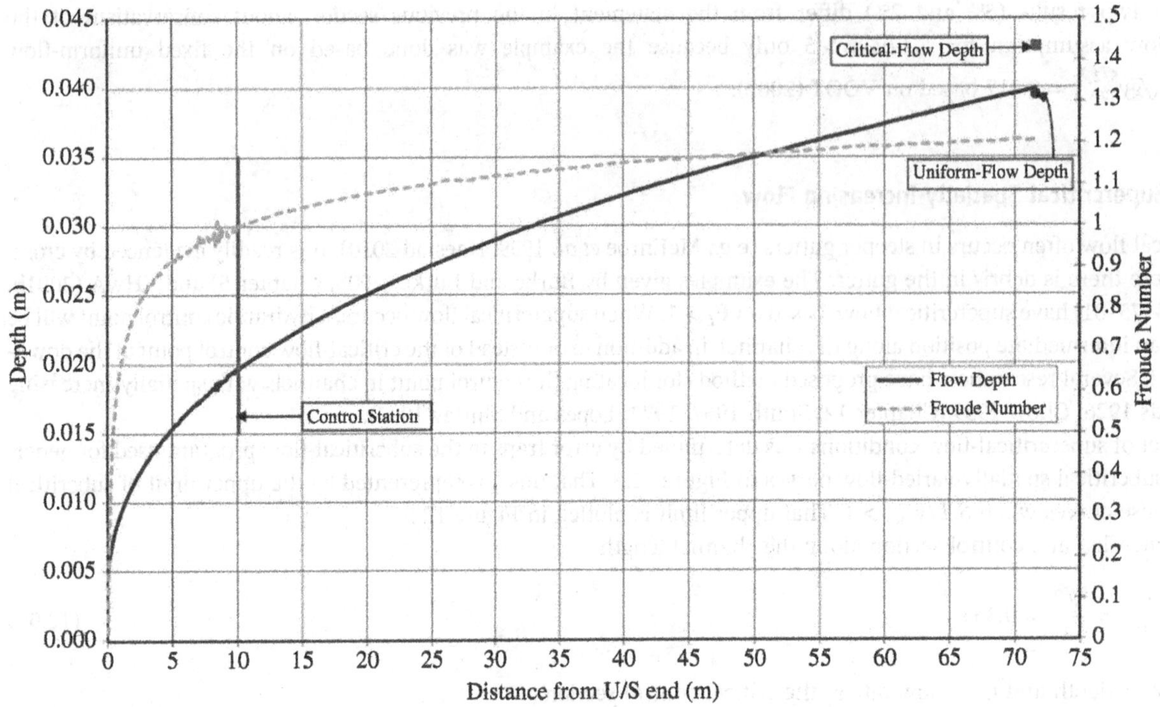

Figure 12.6 Example for Supercritical Gutter Flow (from Graber 2013a)

Generalized results for supercritical gutter flow can be obtained by, first, using Equation 12.99 to express Equation 12.101 in the following form:

$$\left(\frac{gn^2L}{K_n^2 y_{\max}^{4/3}}\right)\left(S_x+\sqrt{1+S_x^2}\right)^{4/3}\left(\frac{\sqrt{g}y_{\max}^{5/2}}{Q_L S_x}\right)^{8/15}\left(\frac{1}{2}\right)^{7/15}\left(\frac{Q_L}{Q_c}\right)^{8/15}+\frac{Q_L}{Q_c}-\left(\frac{S_o L}{y_{\max}}\right)\left(\frac{\sqrt{g}y_{\max}^{5/2}}{Q_L S_x}\right)^{2/5}\left(\frac{1}{8}\right)^{1/5}\left(\frac{Q_L}{Q_c}\right)^{2/5}=0 \tag{12.103}$$

The relationship analogous to Equation 12.93 is given by:

$$\frac{Q_L S_x}{\sqrt{g}y_{\max}^{5/2}}=\psi\left\{\frac{S_o L}{y_{\max}},\ \frac{gn^2L}{K_n^2 y_{\max}^{4/3}}\left(S_x+\sqrt{1+S_x^2}\right)^{4/3},\ \frac{Q_c}{Q_L}\right\}\quad \text{for } \mathbf{F}_c=1 \tag{12.104}$$

The same variables as in Equation 12.104 are in Equation 12.103, giving two relationships in four dimensionless variables. Conceptually, Q_c/Q_L can be eliminated between Equations 12.103 and 12.104, resulting in a single relationship for the three remaining variables.

The supercritical-flow portion of Figure 12.5 provides generalized curves prepared with the relationships and methods described in the previous paragraph and using Equation 12.96. A discontinuity occurs when proceeding from subcritical to supercritical flow at the subcritical–supercritical dividing line ($Q_L S_x/\left(\sqrt{g}y_{\max}^{5/2}\right)=\sqrt{2}/4$), with the supercritical-flow points "jumping" to the uniform-flow curve, then proceeding to the left of the uniform-flow curve as $Q_L S_x/\left(\sqrt{g}y_{\max}^{5/2}\right)$ increases.

An additional feature can be seen by substituting $Q_L S_x/\left(\sqrt{g}y_{\max}^{5/2}\right)=\sqrt{2}/4$ into Equation 12.103; substituting, as an approximation, $Q_{LUF} S_x/\left(\sqrt{g}y_{\max}^{5/2}\right)=\sqrt{2}/4$ into Equation 12.98; and combining the results to obtain:

$$\frac{S_o L}{y_{\max}}=\frac{(Q_c/Q_L)^{-7/15}}{\left[1-(Q_c/Q_L)^{2/5}\right]} \tag{12.105}$$

A minimum value of $S_o L/y_{\max}$ is obtained by setting $d(S_o L/y_{\max})/d(Q_c/Q_L)=0$, which gives $Q_c/Q_L=(7/13)^{5/2}$. Substituting that value into Equation 12.105 gives minimum $S_o L/y_{\max}=4.46$. For $S_o L/y_{\max}$ approximately < 4.46, there is no single control point in the channel at the maximum subcritical-flow line. However, $S_oL/y_{max} < 4.46$ with $Q_LS_x/\left(\sqrt{g}y_{max}^{5/2}\right) = \sqrt{2}/4$ is clearly a physically realizable situation. The logical conclusion is that for S_oL/y_{max} < approximately 4.46 and $Q_LS_x/\left(\sqrt{g}y_{max}^{5/2}\right) = \sqrt{2}/4$ all points along the channel are control points, meaning that critical flow occurs along the entire length of the channel. [Essentially, the same notion is stated by Lopes and Shirley (1993).] An example of this is given by computations for $S_oL/y_{max} = 3$, for which, with increasing $\left[gn^2L/\left(K_n^2y_{max}^{4/3}\right)\right]\left(S_x + \sqrt{1 + S_x^2}\right)^{4/3}$, a value is reached beyond which no root to Equation 12.103 is found. The result, depicted in Figure 12.5 for $S_oL/y_{max} = 3$, is an abrupt drop to the subcritical–supercritical dividing line. For $S_oL/y_{max} = 2$, no root is found for Q_c/Q_L for any $\left[gn^2L/\left(K_n^2y_{max}^{4/3}\right)\right]\left(S_x + \sqrt{1 + S_x^2}\right)^{4/3} \geq 0$.

For supercritical-flow curves, which do proceed smoothly to the subcritical–supercritical dividing line (for $S_oL/y_{max} \geq 5$ in Figure 12.5), the location of the control point moves toward the downstream end of the channel, reaching the downstream end of the channel at the dividing line. Substituting $Q_LS_x/\left(\sqrt{g}y_{max}^{5/2}\right) = \sqrt{2}/4$ and $Q_c/Q_L = 1$ (recall $Q_c/Q_L = x_c/x_L$) into Equation 12.103, gives $\left[gn^2L/\left(K_n^2y_{max}^{4/3}\right)\right]\left(S_x + \sqrt{1 + S_x^2}\right)^{4/3} = 2^{-1/3}(S_oL/y_{max} - 1)$. The resulting values of the friction parameter for $S_oL/y_{max} = 5,\ 10,\ 20,$ and 50 correspond, as they should, to the intersections of the supercritical-flow curves and the dividing line.

Figure 12.5 shows that at the higher values of S_oL/y_{max} the curve for spatially varied flow fairly closely parallels the Manning's-alone curve, indicating when Manning's equation alone provides a more reasonable approximation. In those cases, the curves get closer as $Q_LS_x/\left(\sqrt{g}y_{max}^{5/2}\right)$ decrease, becoming nearly convergent for $S_oL/y_{max} = 50$. In no case is the Manning's-alone curve conservative for supercritical flow. Figure 12.5 shows the conditions for which the chart values should be used.

The Burke and Burke (2008, Chapter 5) example mentioned above, with $S_oL/y_{max} = 17.6$, $\left[gn^2L/\left(K_n^2y_{max}^{4/3}\right)\right]\left(S_x + \sqrt{1 + S_x^2}\right)^{4/3} = 8.71$, and $Q_LS_x/\left(\sqrt{g}y_{max}^{5/2}\right) = 0.433$, comports closely with Figure 12.5. The same is true for the FHwA (2001b, pp. 4–75–81) example mentioned above, for which $S_oL/y_{max} = 44.5$, $\left[gn^2L/\left(K_n^2y_{max}^{4/3}\right)\right]\left(S_x + \sqrt{1 + S_x^2}\right)^{4/3} = 8.88$, and $Q_LS_x/\left(\sqrt{g}y_{max}^{5/2}\right) = 0.685$.

12.7.5 Interaction with Inlets

FHwA (2001b) provides a procedure for merging theory and experimental data to estimate the gutter flow intercepted by drainage inlets. The procedure assumes uniform flow using Manning's equation. Figure 12.7 provides curves (solid lines) of intercepted flow Q_i vs. longitudinal slope and approaching flow Q, for a particular type and size (width W and length L) of grate and cross slope, calculated by the author according to the FHwA procedure, which matches the curves given in a specific example (FHwA 2001b, pp. B-14–16). A more logical approach for subcritical flow is to assume critical-flow approach conditions, which are the conditions expected at the entrance to an inlet provided that the inlet and conditions downstream of the inlet create a backwater depth not greater than the critical depth of the gutter flow just upstream of the inlet (likely conditions with a reasonable design). As expected, the downstream critical depth of flow in the gutter is seen from Figure 12.7 to be less than the maximum depth in the gutter for subcritical flow by comparing $Q_LS_x/\left(\sqrt{g}y_{max}^{5/2}\right)$ for conditions below the maximum subcritical-flow line to conditions at that line for the same Q_L and S_x giving $y_{max} \geq y_c$. By similar reasoning for the uniform-flow depth y_{unif}, Figure 12.7 shows for subcritical uniform-flow conditions $y_{unif} \geq y_c$, with implications discussed below.

For supercritical flow, at the downstream end of the gutter $y_{unif} < y_{max} < y_c$, as can be seen from Figure 12.7 and reasoning given above. It is therefore not adequate to use the uniform-flow depth for inlet design with supercritical approach flow, impractical to use each y_{max} such as determined from Figure 12.7, and conservative to use y_c. The latter is thus recommended, and the same procedure as described below for subcritical flow can be used for supercritical flow.

The 9-step FHwA (2001b) procedure may be modified for critical-flow approach conditions by changing Steps 1 and 3 of the procedure. For Step 1, the top width for critical-flow conditions is calculated for a given flow and cross slope by

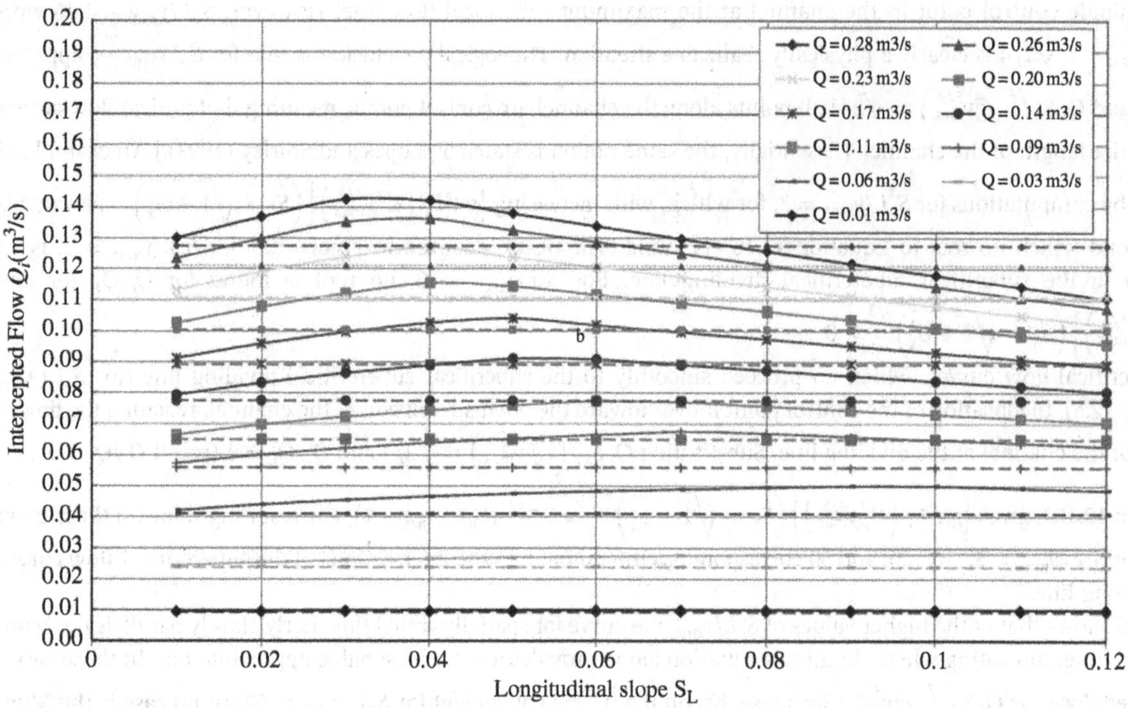

Figure 12.7 P-30 GRATE, S_x = 0.04, W = 0.6 m, L = 0.6 m – Solid Lines Based on Manning's Equation (n = 0.016), Dashed Lines Based on Critical-Flow Conditions (from Graber 2013a). The color rendition of this image is available on the book's landing page on wiley.com.

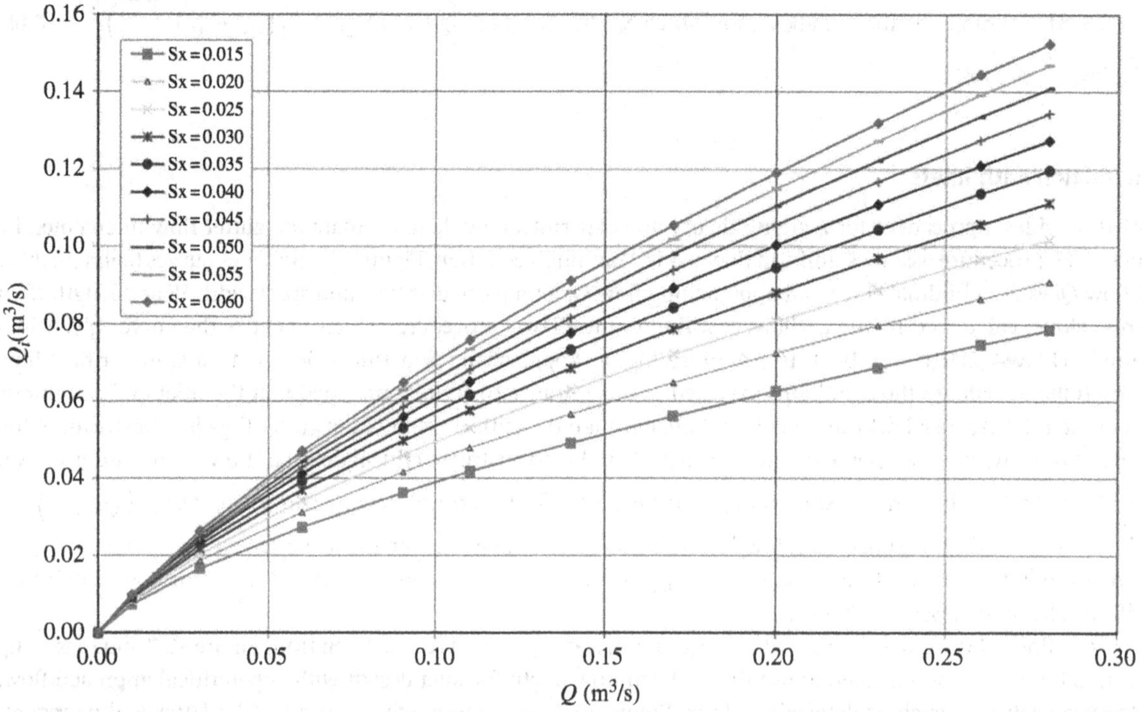

Figure 12.8 P-30 GRATE, W = 0.6 m, L = 0.6 m (from Graber 2013a). The color rendition of this image is available on the book's landing page on wiley.com.

$T = \left[8Q^2/\left(gS_x^3\right)\right]^{1/5}$. For Step 3, the velocity approaching the inlet is calculated by $V = 2Q/(T^2S_x)$. Other steps then remain unchanged. The dashed lines in Figure 12.7 are for critical-flow conditions with values of S_x and Q corresponding to the uniform-flow curves. The dashed lines are independent of longitudinal slope because the critical-flow depth is independent of longitudinal slope. This independence allows a more general figure, such as Figure 12.7, to be prepared for a particular inlet. The range of S_x values on Figure 12.8 are those considered acceptable by FHwA (2001b) which cites AASHTO policy.

12.7.6 Equivalent Length

For flow continuing on beyond an inlet to the next inlet, Figure 12.5 can be used again as it was for the flow to the inlet just considered, provided an equivalent gutter length is used. That equivalent length L_e can be readily derived and is given by:

$$L_e = L_n + \frac{Q_n}{q_n} \tag{12.106}$$

in which L_n = distance to the next inlet from the inlet just considered, Q_n = flow continuing to the next inlet (equaling $Q - Q_i$ for the inlet just considered), and q_n = uniform inflow per unit length over the distance to the next inlet from the one just considered. The downstream critical depth can then be used for the inlet design. The procedure can be repeated for successive gutter lengths and inlets.

13

Sediment Transport: Flow of Sludges and Slurries

13.1 Introduction

Many of the liquids encountered in environmental engineering practice are sediment-laden. Examples are raw sewage, chemical slurries used in water and wastewater treatment, activated sludge, and other water and wastewater sludges. Processes that separate solids from liquids by sedimentation and other methods are vital elements of water and wastewater treatment. However, it is very important to the proper operation of water and wastewater facilities that sediments be transported to and between unit processes without inadvertent sedimentation that could clog pipes and open channels and have other deleterious effects. The transport of sediments in water and wastewater engineering applications is the principal concern of this chapter. The topics addressed include means of maintaining solids in suspension, determination of pressure losses in sludge and slurry flows, and the use of additives to reduce the frictional drag of liquids in conduits.

13.2 Limiting Velocity Concepts

Requirements for sediment transport in pipes and open channels are often empirical in nature. For example, the suggestion (WPCF 1977, p. 36) that "a minimum velocity of 0.6 m/s (2.0 fps) at design average flow is usually considered adequate [to maintain solids in suspension] for channel flow of unsettled wastewater" is soundly based on practical experience rather than theoretical considerations. Another example for transport of sludge in pipelines is "velocities of flow between 5 and 8 fps are, in general, found to be satisfactory" for transport of sludge (WPCF/ASCE 1959, p. 200). More recently, (WPCF 1977, p. 104) says that velocity range was suggested for incinerated sludge and gives lesser velocities for other types of sludges. The upper range is still 8 fps for primary sludge in the 0.2–1% solids range, but the recommended values are put into question because a smaller velocity range is suggested for primary sludge in the 4–10% solids range. The use of a velocity criterion is common for a wide variety of sediment-laden flows and does have a theoretical basis, as discussed below.

Camp (1946a, 1952), noting the then-common practice of designing sewers for a full velocity of not less than 2 ft/sec for self-cleansing, employed some observations in conjunction with the Shields equation for large Reynolds Numbers to investigate the self-cleansing ability of sewers flowing partially full. He gave the Shields equation as:

$$V = \sqrt{\frac{8\beta}{f} g(s-1)d} \tag{13.1}$$

in which V is the mean velocity in the conduit, d is the diameter of the particle (or stone, as discussed in Section 7.7) to be transported, s is the specific gravity of the particle, g is the acceleration of gravity, f is the Darcy–Weisbach friction factor for the conduit, and β is a constant whose value is about 0.04 to start to scour and 0.8+ for more adequate cleaning (Camp 1946a).

Terms that are used in the present context are first the "scouring velocity" which is the velocity required to initiate particle motion. Second is the "limit deposit velocity," which is defined as the velocity below which some of the suspended particles begin to settle out and move along the bottom of the conduit as bed load. The latter term is essentially synonymous with the term "cleansing velocity," as used below.

Hydraulics and Pneumatics in Environmental Engineering, First Edition. S. David Graber.

For sediment particles of a given size and specific gravity, Equation 13.1 (see discussion of limitation below) suggests that the cleansing velocity is proportional to the inverse square root of f (Committee to Study Limiting Velocities of Flow in Sewers 1942, App. 4). For cleansing equal to that under full pipe conditions, the velocity V_{ec} under partially full conditions must then be as follows:

$$\frac{V_{ec}}{V_{ecf}} = \left(\frac{f_f}{f}\right)^{1/2} \tag{13.2}$$

in which the subscript "f" denotes full flow. Using the above equation together with the friction factor curve of Figure 5.2, Camp developed the curve labeled "Velocity for Equal Cleansing" on that figure. The curve indicates that less velocity is required for self-cleansing of a partially full pipe than for a full pipe. A sewer adequately designed for self-cleansing when full will, according to the curve, be self-cleansing when the sewer flows more than half full. Experience indicates that a 2 ft/sec full velocity is adequate for self-cleansing with typical municipal sewage if the sewer flows more than half full (Camp 1946a). Pomeroy (1967) collected data that supported 2 ft/sec as necessary to avoid excessive accumulation of debris.

The accuracy of the above approach is limited by the fact that the Reynolds Number $\mathbf{R}_s$ for sewers is often too small for Equation 13.1 (with constant β) to be valid. The $\mathbf{R}_s$ for sewers typically puts one on the variable portion of the Shields curve of Figure 13.1 {on which the "Critical Shields Parameter" is $\tau_c/[(\rho_s - \rho_w)gd]$}. It is further limited by the fact that shear stresses are not uniformly distributed over the wetted perimeter. However, we will momentarily disregard these limitations and carry the above concepts through to their practical conclusions. After that, we shall return to consideration of approaches which address the above-mentioned limitations.

For sewer design based on a minimum flowing-full velocity, Equation 7.37b allows the minimum slope to be expressed as follows:

$$S_{\min} = \frac{n_f^2 V_f^2}{K_n^2 R_f^{4/3}} \tag{13.3}$$

in which $K_n = 1.486$ (customary units) or 1.0 (metric units).

For the common example of a circular sewer with $n_f = 0.013$, $V_f = 2$ ft/sec, and $R_f = d_o/4$, the above equation in customary units results in:

$$S_{\min} = 0.00194/[d_o(\text{ft})]^{4/3} \tag{13.4a}$$

$$S_{\min} = 0.0534/[d_o(\text{in.})]^{4/3} \tag{13.4b}$$

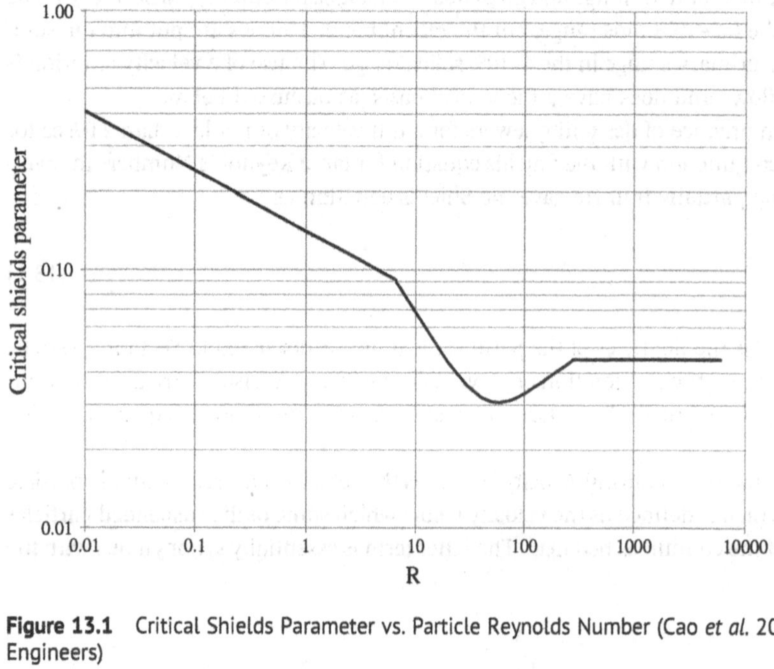

Figure 13.1 Critical Shields Parameter vs. Particle Reynolds Number (Cao *et al.* 2006/with permission of American Society of Civil Engineers)

Table 13.1 Minimum Slopes for Sanitary Sewers

Diameter (in.)	Slope (ft/ft)	V_f (ft/s)	Q_f (mgd)
8	0.004	2.19	0.494
10	0.003	2.20	0.775
12	0.0022	2.13	1.08
15	0.0015	2.04	1.62
18	0.0012	2.06	2.35
21	0.0010	2.08	3.24
24	0.0009	2.16	4.39
≳27	0.0008	≳2.20	≳5.66

Thus, for any size sewer, a minimum slope can be calculated. A well-known relationship (from Metcalf and Eddy, Inc. 1972, Table 4-1) between diameter and minimum slope for sanitary sewers, which gives full velocities slightly in excess of 2 ft/sec for pipe diameters up to 27 in. (0.686 m), is given in Table 13.1; velocities and full flows for $n = 0.013$ have been added by the author.

A minimum diameter of 8 in. is also recommended to help avoid clogging (Metcalf and Eddy, Inc. 1972, p. 110). Sewers laid on the minimum slope discussed in the preceding paragraph would, according to the equal-cleansing velocity notion, be self-cleansing when flowing more than half full. At lower flows, sediment accumulation would be expected. One approach is to accept such accumulations and provide for their removal by periodic cleaning or flushing in the early years of operation when the amount of wastewater is insufficient for self-cleansing. This may be a cost-effective approach if the costs of such cleaning or flushing in the early years are less than the equivalent of the increased construction necessary to maintain cleansing velocities at the beginning of the design period. Such increased construction could include steeper slopes with the attendant deeper excavations and possibly additional pumping stations. Another possible approach entails the construction of a smaller sewer, which could be self-cleansing from the start to serve for a shorter period of time and be replaced by a larger sewer when the service area has developed further (Metcalf and Eddy, Inc. 1972, p. 110). Still, another meritorious approach is to use flushing methods discussed in Section 7.9. In any event, serious consideration should be given to the alternatives in each situation rather than using all-encompassing design expediencies.

To address the limitations of the above, we shall first address incipient motion of sediments more accurately. C. M. White (1940) presented an analysis, the essence of which can be presented as follows: The forces on a bed particle are shown in Figure 13.2. The horizontal force on the particle is given by $C_1\tau_o D^2$, in which the C values here and below are proportionality constants, τ_o is shear stress on the particle due to water flow, and D is particle equivalent diameter. With a moment arm of C_2D from the point of action on the particle to the point of rotation, the moment acting on the particle is $M = C_1C_2\tau_o D^3$. The submerged weight of the particle equals $C_3(\gamma_s - \gamma_f)D^3$, in which γ_s and γ_f are the specific weight of sediment and fluid, respectively, and has a moment arm of C_4D. The moment-resisting particle movement is thus equal to $C_3C_4(\gamma_s - \gamma_f)D^4$. Motion begins when the critical shear force τ_c satisfies $\tau_c C_1 C_2 D^3 = C_3C_4(\gamma_s - \gamma_f)D^4$, which reduces to $\tau_c = C_5(\gamma_s - \gamma_f)D$.

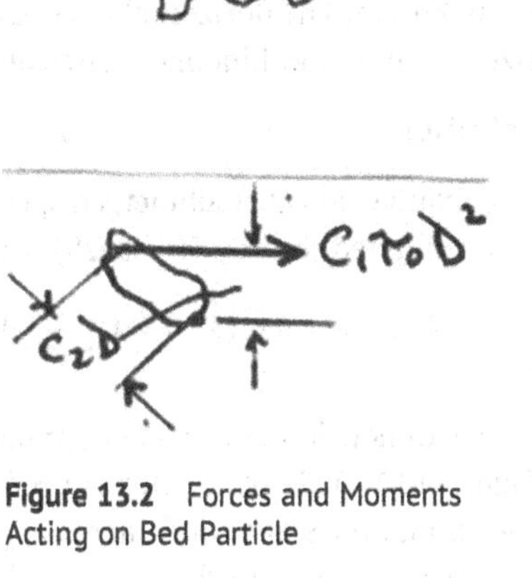

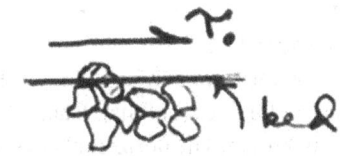

Figure 13.2 Forces and Moments Acting on Bed Particle

Shields presented a more complicated analysis, which can be presented in a more simplified form by using dimensional analysis. The pertinent variables are D, $(\gamma_s - \gamma_f)$, γ_f, τ_c, and kinematic viscosity of water ν. With the basic dimensions being length, mass, and time (γ being equal to ρg), the Buckingham π-Theorem gives 5 - 3 = two dimensionless groups, which Shields gave as $\tau_c/[(\gamma_s - \gamma_f)D]$ and u_*D/ν with $u_* = \sqrt{\tau_c/\rho_f}$. Shields was the first to express the relationship between the variables using nondimensional parameters. At large values of u_*D/ν, $\tau_c/[(\gamma_s - \gamma_f)D]$ is constant and matches White's relationship.

Cao *et al.* (2006) used a logarithmic matching method and more accurate recent data to derive the following empirical relationships for the Shields parameter.

$$\frac{\tau_c}{(\rho_s-\rho_w)gd} = 0.1414\ \mathbf{R}^{-0.2306}\quad \mathbf{R}\le 6.61 \tag{13.5a}$$

$$\frac{\tau_c}{(\rho_s-\rho_w)gd} = \frac{\left[1+(0.0223\mathbf{R})^{2.8358}\right]^{0.3542}}{3.0946\mathbf{R}^{0.6769}}\quad 6.61<\mathbf{R}\le 282.84 \tag{13.5b}$$

$$\frac{\tau_c}{(\rho_s-\rho_w)gd} = 0.045\quad \mathbf{R}\ge 282.84 \tag{13.5c}$$

in which τ_c is the threshold value of the bed shear stress, ρ_s and ρ_w are the density of the sediment and the density of water, respectively, d is the mean particle diameter, and the "particle Reynolds Number" is given by $\mathbf{R} = d\sqrt{(s-1)gd}/\nu$ in which $s=\rho_s/\rho_w$ is submerged specific gravity of sediment. $\mathbf{R}$ is determined solely by the fluid and sediment characteristics. *Flow depth should not be important because the particles are in the inner layer of a turbulent boundary layer in which only the local structure of the flow governs the forces felt by the bed particles.*

With the Shields formulation (Committee on Hydromechanics 1963), the ordinate is the same as in the formulation by Cao *et al.*, but the abscissa is $\mathbf{R}_* = U_*D/\nu = \sqrt{\tau_c/\rho}D/\nu$ known as the Boundary Reynolds Number. That abscissa has the disadvantage of having both τ_c and D appearing in both of the axis variables. The preference for the Cao *et al.* formulation is apparent.

An understanding of the mechanism of scour is important not only for the design of scour-protection measures but is also basic to understanding aspects of sediment transport (present chapter) and sedimentation (Section 14.3).

In Figure 13.1, the magnitude of the Boundary Reynolds Number may be estimated by noting that $U_c^* = \sqrt{gRS}$ and $S = fV^2/(8gR)$, from which $U_c^*d_s/\nu = (Vd_s/\nu)\sqrt{f/8}$, in which V is the average velocity. For velocities on the order of 1 ft/sec, $\nu = 10^{-5}$ ft²/sec, and $f = 0.01$, we have $U_c^*d_s/\nu = 3500d_s$ in which d_s is the sediment diameter in feet. For grit and sediment particles, the value of d_s is typically on the order of 1 mm for which the Boundary Reynolds Number is on the order of 10, putting us on the curved portion of the Shields curve for which the value of $f(U_c^*d_s/\nu)$ is as low as 0.03. For riprap (cobbles or stones from 3 in. diameter or more), the Boundary Reynolds Number is on the order of 1000 or more, putting us out on the flat portion of the Shields curve (Figure 13.1) for which $f(U_c^*d_s/\nu)$ is constant.

The relationships presented above have an interesting background of theoretical and experimental development. The reader is referred to *Sedimentation Engineering* (Vanoni 1975, §II-E) for a thorough discussion. Here, the author will just state that experimental studies have established the adequacy of the Shields relationships for practical applications.

Although the author prefers the method discussed above, others should be mentioned. Yao (1974) presented a method for sewer line design based on critical shear stress. Hager (1999, Chapter 3) has expanded on that method and made it more practical. Hager also discusses the ATV (Sewage Technical Union – Abwassertechnische Vereinigung) method for avoiding the deposition of material in sewers.

Example 13.1

A pipeline is to convey primary sludge with a maximum 1% solids concentration to a cyclone degritter via vertical and horizontal lines (Tallman Island WWTP, New York City). The pipe is to be ANSI Schedule 40 steel with absolute roughness $\varepsilon = 0.0018$ in. The horizontal lines both precede and follow the vertical line. Design flow rate is 1180 gpm, maximum particle size is 2 mm, and kinematic viscosity $\nu = 1(10)^{-5}$ ft²/sec. Determine line sizes sufficient to maintain cleansing velocity.

Solution

The relative density (submerged specific gravity) of grit is taken as $s = 2.65$ (Metcalf and Eddy, Inc., 1972). The particle size d is converted to ft by 2(0.00328) = 0.00656 ft. The particle Reynolds Number as expressed above is:

$$\mathbf{R}_{particle} = d\sqrt{(s-1)gd}/\nu = 0.00656\sqrt{(2.65-1)(32.2)(0.00256)}/10^{-5} = 242$$

The dimensionless Shields parameter $\tau_c/[(\rho_s-\rho_w)gd]$ equals 0.0429 using Equation 13.5b, which agrees well with Figure 13.1. This parameter is used in conjunction with the equation for the Darcy–Weisbach friction factor f, using, e.g., Equation 9.66b, and increasing f by 1% per Section 13.4: $\tau_w = 1.01(f/4)\gamma V^2/(2g)$. Equating τ_c and τ_w, and taking $\gamma = \rho_w g$ yields: $0.0429[(\rho_s-\rho_w)gd] = 1.01(f/4)\gamma V^2/(2g) \Rightarrow 0.0429[(\rho_s/\rho_w-1)gd] = 1.01(f/4)(\gamma/\rho_w)V^2/(2g) \Rightarrow (f/8)V^2 = (1/1.01)0.0429(2.65-1)(32.2)(0.00656) = 0.01480$ ft²/sec².

An iterative process is now used in which it is assumed that pipe internal diameter D using standard specifications for the type of pipe, which gives pipe area A and with the known Q gives $V = Q/A$. We then calculate ε/D and the pipe Reynolds Number, which gives f from the Moody Diagram, and iteratively use the above equation to calculate f until convergence (the two f values are approximately equal). Only the final iteration is given here: $D = 13.13$ in. (14-in. nominal), $A = 0.778$ sq ft, $V = 2.80$ ft/sec, $\varepsilon/D = 0.0001371$, pipe Reynolds Number $\mathbf{R} = 2.80(13.13/12)/[1(10)^{-5}] = 30{,}6000$, Moody $f = 0.0160$ from Figure 5.1, calculated $f = 0.0153$ from Equation 5.2.

For vertical pipes, it is only necessary that the vertical velocity exceed the settling velocity. The vertical pipes are initially assumed to be the same size as the horizontal pipes (the reader should explain why). For those pipes, we initially assume Stokes flow for which Equation 4.93b gives the settling velocity as $v = (g/18\nu)(s-1)d_s^2 = \{32.2/[(18)(10^{-5})]\}(1.65)(0.00656)^2 = 12.7$ ft/sec. The corresponding particle Reynolds Number is $\mathbf{R}_{particle} = vd/\nu = 12.7(0.00656)/10^{-5} = 8{,}330$, which is in the turbulentregion, exceeding the range for Stokes flow. Figure 13.3 applies for turbulent flow and gives for the 2 mm particle $v \cong 30$ cm/sec $\div$ 30.48 cm/ft $= 0.984$ ft/sec $\cong 1$ ft/sec. This fall velocity is less than the 2.80 ft/sec vertical pipe velocity determined above, so sufficient velocity will be maintained.

An interesting application of critical water velocity is given by Jobin *et al.* (1984) and Jobin (1999), who studied the dislodgement or immobilization of harmful aquatic snails. These particular snails are vectors in the transmission of the dangerous tropical parasitic disease bilharzia, also known as schistosomiasis (Courrier and Eckholm 1977). The mean of water velocities measured in a trapezoidal canal with a current meter at 20% and 80% of the depth in the center of the canal were found to closely approximate the mean velocity of flow. The number of snails was found to be inversely related to the mean velocity, with a plot indicating there would be no snails if that velocity exceeded 1.8 ft/sec (0.55 m/s). This field study agreed well with results of a laboratory study.

Continuing our development, Vanoni (1975; Chapter II, §9) presents a convenient, single-step, graphical means of determining the fall velocity of a spherical particle. That consists of a plot of velocity in cm/sec vs. diameter in mm with

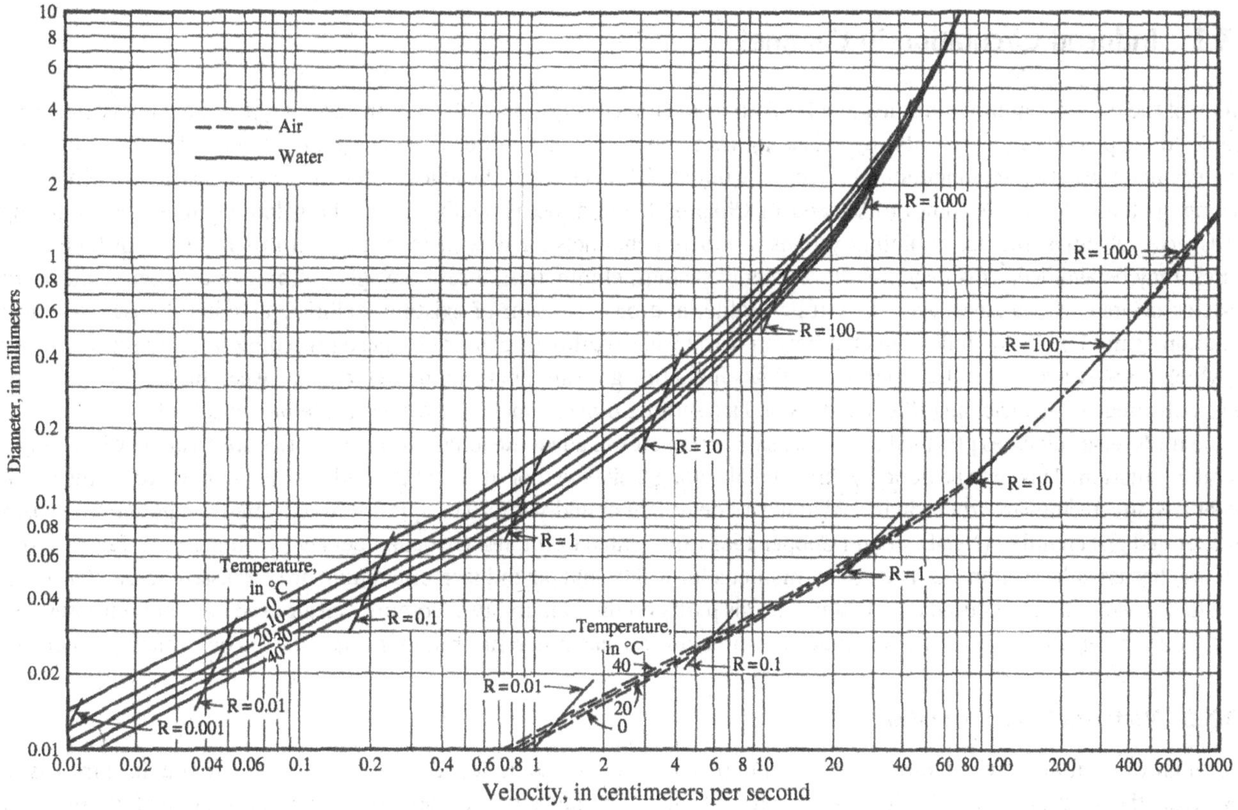

Figure 13.3 Fall Velocity of Quartz Spheres in Air and Water

temperature as a parameter in increments from 0 to 40°C in 10°C. That plot is given in Figure 13.3 and is used in the following example.

Example 13.2
A 48-in. diameter intercepting sewer along the Connecticut River in Springfield, Massachusetts, conveys combined sewage along a relatively flat grade for a distance of about 1000 ft. Deposits of sand and other dense material have accumulated in this length to an average of approximately half full. Self-cleansing velocities are not obtained under any conditions. Considerable overflows of sewage into the Connecticut River occur during high sewage flows and even minor storm flows. The velocities that result due to the half-full conditions, although not sufficient to scour deposited material, do convey sand downstream to the York Street Pumping Station, causing extensive wear on the pumps. The recommended solution for these problems is to intercept the sand in a sand removal facility, which will be installed in the upstream portion of the flat sewer section. Grit will be removed from the sand trap by a jet-eductor truck. Determine the size of the facility based on the 16.0 mgd flowing-full capacity of the sewer and the removal of relatively coarse grit (35 mesh).

Solution
The grit size of 35 mesh corresponds to 0.42 mm = 0.001378 ft. For sand (specific gravity of 2.65) of this size and temperature of 10°C (50°F), ASCE Manual of Practice 54 Figure 2.2 gives a fall velocity of approximately 5 cm/sec = 0.1645 ft/sec. That equates to 5(2835) (ft/day)/(cm/sec) = 14,175 ft/day = 14,175 cu ft/(sq ft-day) × 7.48 gal/(cu ft) = 106,000 gpd/(sq ft). The corresponding surface area is $A = [16(10)^6 \text{ gpd}] / (106{,}000 \text{ gpd/sq ft})\} = 151$ sq ft. The sand trap was actually designed to be 25 ft long by 8 ft wide giving 200 sq ft, which is conservative by these calculations.

The kinematic viscosity of water at 50°F is approximately $1.5(10)^{-5}$ ft^2/sec (Shames 1962, Fig. B-2). The Reynolds Number of the falling grit particles is $\mathbf{R} = dV/\nu = 0.001378(0.1645)/1.5(10)^{-5} = 15.1$, well in excess of the Reynolds Number of creeping flows discussed in Section 4.5.

13.3 Induced Circulation in Channels

Much of the discussion under this heading is taken from Graber (1994, pp. 539–543). In certain types of channels, it is preferable to keep solids in suspension by providing transverse velocities rather than longitudinal velocities. This situation occurs when the necessary longitudinal velocities would result in excessive head loss, such as in many distribution channels having spatially decreasing flows (discussed in Chapter 11) and, occasionally, in constant-flow channels. By inducing transverse velocities, the use of piping systems or tapered channels for flow distribution, with their attendant high head losses, may be avoided. This may also allow energy savings, elimination of raw-sewage pumping, reduction of site-work costs associated with a steep plant hydraulic profile, and/or expansion of existing facilities at reduced cost. The most common examples are the following channels in wastewater treatment plants: (1) those preceding primary settling tanks in which raw-sewage suspended solids (sometimes including grit) are present; and (2) those between biological treatment (e.g., activated sludge, trickling filter) and secondary settling tanks, which convey settleable biological solids.

The transverse velocities provided to compensate for the lack of adequate longitudinal velocities are created by inducing a spiral circulation. This may be done by diffusing air along a side wall or at the mid-width of the channel, or by means of water jets directed inward from a manifold along a side wall or at mid-width. The spiral-flow activated sludge aeration basin and aerated grit chamber (Neighbor and Cooper 1965) are related examples. Grit chambers with longitudinal velocity control are discussed in Section 7.2. The design methods for water- and air-induced circulation are discussed below. The decision to use water- or air-induced circulation depends partly on the availability of compressed air used for other purposes and partly on the velocities required. As will be shown below, air has inherent limitations as a means of inducing velocities.

13.3.1 Water-Induced Circulation

Although water-induced circulation is less commonly employed, it can be economically advantageous and, in some cases, even essential to proper channel agitation. A good study of water-induced circulation in grit chambers is that of Brenner and Diskin (1991). Particularly in shallower channels, air-induced circulation may be inadequate. Water-induced circulation is more fully understood and provides a basis for at least a qualitative understanding of air-induced circulation. Water-induced circulation is, therefore, discussed first.

The studies by Iamandi and Rouse (1969) provide a useful theoretical formulation for water-induced circulation. They experimentally studied the circulation induced in air by an air jet in a rectangular chamber under conditions of negligible compressibility. The stated purpose of their study was to obtain results that could be extrapolated to the air-induced circulation of water, as in aerated basins. The immediate practical value of Iamandi and Rouse's work, however, stems from its applicability to the use of water jets to induce circulation in water. Their work is discussed below in this context, followed later by a discussion of the difficulties associated with the application and extrapolation of these studies to aerated channels.

Iamandi and Rouse's experiments on the air-induced circulation of air showed no influence of Reynolds Number $\sqrt{Dm\rho}/\mu$ for values greater than 30,000 (note change in meaning of D). In this Reynolds Number, m is the jet momentum flux per unit length of channel given by $\rho q_o v_o$ in which v_o is the jet discharge velocity and q_o is the jet volumetric flow rate per unit length of channel, D is the channel dimension in the initial direction of the jet, ρ is water density, and μ is absolute water viscosity. (This form of the Reynolds Number can be derived by taking the characteristic length and velocity, respectively, as D and v_o, and considering geometric similarity.) Similitude principles discussed in Chapter 5 thus suggest that the results are equally applicable to the water-induced circulation of water under geometrically similar conditions exceeding the same Reynolds Number.

In the studies by Iamandi and Rouse (1969), a continuous slot jet was directed parallel to one wall of the channel, as represented in Figure 13.4. Included in that figure are the streamlines (see Chapter 4) as experimentally measured. Using dimensional analysis and experimental studies of air-induced circulation of air, which is *similar* to water-induced circulation of water, Iamandi and Rouse were able to characterize the system by means of the dimensionless variables in the following functional relation:

$$\Phi\left(\psi/\sqrt{mD/\rho}, y/D, x/D, H/D\right) = 0 \tag{13.6}$$

In the above relationship, m, the jet momentum flux per unit length of channel, is normal to the paper and dimensions are as shown in Figure 13.4.

The term ψ in the above relationship is the stream function as defined for two-dimensional flows in Section 4.2. The fluid velocity v in the direction of the streamlines (Ψ = constant) is related to Ψ by $v = d\psi/dn$ in which n is a measure of distance normal to the streamlines. Noting that:

$$v \cong \Delta\psi/\Delta n = \sqrt{mD/\rho}\frac{1}{D}\frac{\Delta\left(\psi/\sqrt{mD/\rho}\right)}{\Delta(n/D)} \tag{13.7}$$

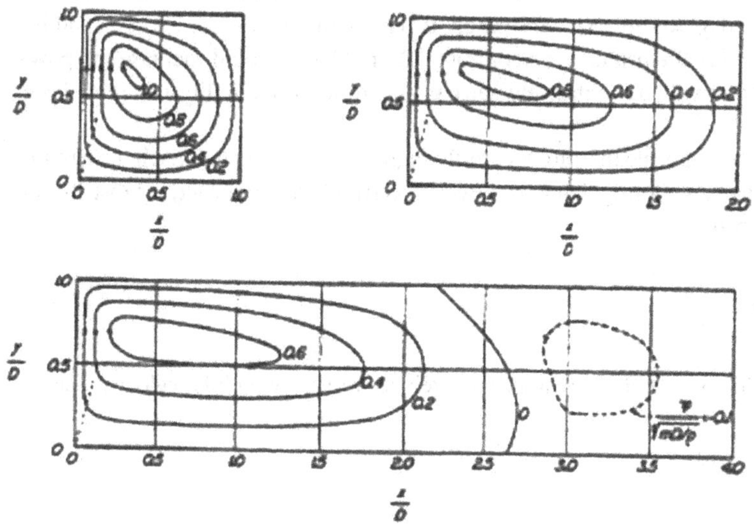

Figure 13.4 Circulation Induced by a Continuous Slot Jet (Iamandi and Rouse 1969/with permission of American Society of Civil Engineers)

the velocity at any point in the cross section, such as near the bottom of the channel center, may be determined by the H/D ratios, which Iamandi and Rouse studied as indicated in Figure 13.4. Thus, the transverse bottom velocity v_b at the midpoint of the channel ($y/D = 0.5$) is given by:

$$v_b = K_I\sqrt{\frac{m/\rho}{D}} = K_I\sqrt{q_o v_o/D} \tag{13.8}$$

in which K_I is a function of H/D and is constant for a given channel cross section. The values of K_I range from about 4 for a 2:1 channel height: width ratio to 5 for a 1:1 channel height: width ratio.

In practical applications, a series of closely spaced nozzles along one or more manifolds will generally be used rather than a continuous slot, as used by Iamandi and Rouse. If the nozzles are closely spaced, the effect on the flow pattern should be essentially the same. Furthermore, since the test section used by Iamandi and Rouse employed no copings or fillets, the above should be somewhat conservative for channels in which corner copings or fillets are used.

As an additional practical point, water-induced circulation systems once installed should be operated continuously. In one case of intermittent operation observed by the author (Pittsfield, Massachusetts WWTP), a primary sedimentation basin influent channel experienced severe sedimentation, with 3-1/2 ft of sediment buildup in a 5-foot-deep channel. One apparent cause was clogging of the manifold ports during periods when the circulation system was off. Another more basic reason was design based on G-value, as critiqued in Chapter 3 and below.

Example 13.3

A primary-settling-tank influent channel of 4.5-ft width, 7-ft–9-in. depth, and 152-ft length (Tallman Island WWTP East Battery, New York City) requires a channel-agitation system to keep organics and small grit particles in suspension. Determine the basic design characteristics of a water-induced circulation system that will maintain a minimum $v_b = 2$ ft/sec. Thickener overflow is the source of the agitation water, reliably available in the amount of 2450 gpm = 5.46 cfs. The piping configuration allows backup water to be available from plant final effluent or directly from the primary settling tanks.

Solution

The channel has a height:width ratio of 7.75:4.5 = 1.7:1, which is intermediate between the 2:1 and 1:1 ratios corresponding to the data reported above (Figure 13.4). For conservatism, a K_I value of 4 is selected. For $v_b = 2$ ft/sec, Equation 13.8 then gives $q_o v_o = v_b^2 D/K_I^2 = (2)^2(4.5)/(4)^2 = 1.125\ \text{ft}^3/\text{sec}^2$.

The applicability of Equation 13.8 is contingent on the Reynolds Number **R** being greater than 30,000. Noting that $\mathbf{R} = \sqrt{Dm\rho}/\mu = \sqrt{D\rho q_o v_o \rho}/\mu = \sqrt{Dq_o v_o}/\upsilon$, we have $\mathbf{R} = \sqrt{4.5(1.125)}/10^{-5} = 225{,}000$, well in excess of 30,000. The available agitation flow and channel length correspond to $q_o = 5.46/152 = 0.0359$ cfs/ft. The required velocity v_o corresponding to that q_o is $v_o = q_o v_o/q_o = 1.125/0.0359 = 31.3$ ft/sec at the orifice vena contracta. A smaller v_o would increase the power required. Thus, the values of q_o and v_o given above will be tentatively selected pending acceptability of other design variables.

In order to facilitate manifold removal for maintenance, nine manifolds of 16-ft length each are selected. Each distribution pipe has a length of 8 ft and a flow of 5.46 cfs/8 = 0.683 cfs. Noting that v_o is the contracted velocity of the jet, the distribution pipe controlling head loss (see Chapter 11) is given by:

$$h_o = \frac{1}{2g}\left(\frac{q_o}{C_D A_o}\right)^2 \cong \frac{v_o^2}{2g} = \frac{(31.3)^2}{2(32.2)} = 15.2\ \text{ft}$$

Four-in. nominal cast-iron pipe is considered (with an actual I.D. of approximately four in.), giving inlet velocity, velocity head, and Reynolds Number as follows:

$$V_o = 0.683/\left[(\pi/4)(4/12)^2\right] = 7.83\ \text{ft/sec}$$

$$V_o^2/(2g) = 0.951\ \text{ft}$$

$$\mathbf{R} = V_o D/\upsilon = 7.83(4/12)/10^{-5} = 2.61(10)^5$$

For cast-iron pipe ($\varepsilon = 0.00085$), the Moody Diagram (Figure 5.1) or Equation 5.2 gives $f = 0.020$. Then the parameter FL (see Equation 10.16) has the value $fL/D = 0.020(8)/(4/12) = 0.480$, which is less than 2, so that Equation 11.39a applies for the head loss along the distribution conduit:

$$\Delta h_{\text{max}} = \frac{V_o^2}{2g}\left(1 - \frac{FL}{3}\right) = 0.951(1 - 0.480/3) = 0.799 \text{ ft}$$

The resulting distribution uniformity (from Equation 11.1) is good:

$$m = \sqrt{1 - 0.799/15.2} = 0.973$$

By trial and error, the number of orifices per each of the nine manifolds was established as $n = 26$ (accommodating actual manifold lengths and fittings) and orifice spacing $\ell = 16(12)/26 = 7.39$ in., corresponding to flow per orifice $Q_o = C_D A_o v_o = 0.61 A_o v_o = 5.46 \text{ cfs}/[9(26)] = 0.02\overline{33}$ cfs, orifice diameter $d_o = 13/32 = 0.40625$ in., orifice area $A_o = 0.00090015$ sq ft. Corner fillets are included to reduce the possibility of sedimentation and to add conservatism to the spiral motion. The final design is depicted in Figure 13.5 below.

G-*Value*. The author is aware of the use of G-value (defined in Section 3.4) for the design of water-induced channel-circulation systems. The conceptual fallacy of this approach can be seen by noting that power per unit volume (equal to μG^2) is proportional to $q_o v_o^2/(WD)$, which by Equation 13.8 equals $(v_b/K_I)^2 v_o/W$. For a given channel (fixed D and K_I), the choice of q_o and v_o must be made to give the required value of v_b (see Equation 13.8). A given value of G thus corresponds to a wide range of cleansing v_b's, depending on the values of the additional variables v_o, W and channel shape (on which K_I is dependent). In other words, since the power of the jet is equal to $mv_o/2 = \rho q_o v_o^2/2$, it is apparent that a wide range of momentum flux values and resulting values of v_b can correspond to given power or G-values. This adds to the reasons that G-value is inappropriate for channel agitation. If the water is substantially free of solids, there is essentially no upper limit on the velocity v_o. The availability of water must also be considered, and the flow should not be sufficient to upset the design of those process units receiving the additional flow.

13.3.2 Air-Induced Circulation

The valid information available for air-induced circulation is entirely of an empirical nature. Methods that have been proposed for the theoretical design of air-induced circulation do not give reasonable results in all cases and must be used with caution. This section discusses empirical and theoretical studies, leading to conclusions regarding the applicability of various techniques. It is important to recognize that air has inherent limitations as a means of inducing circulation.

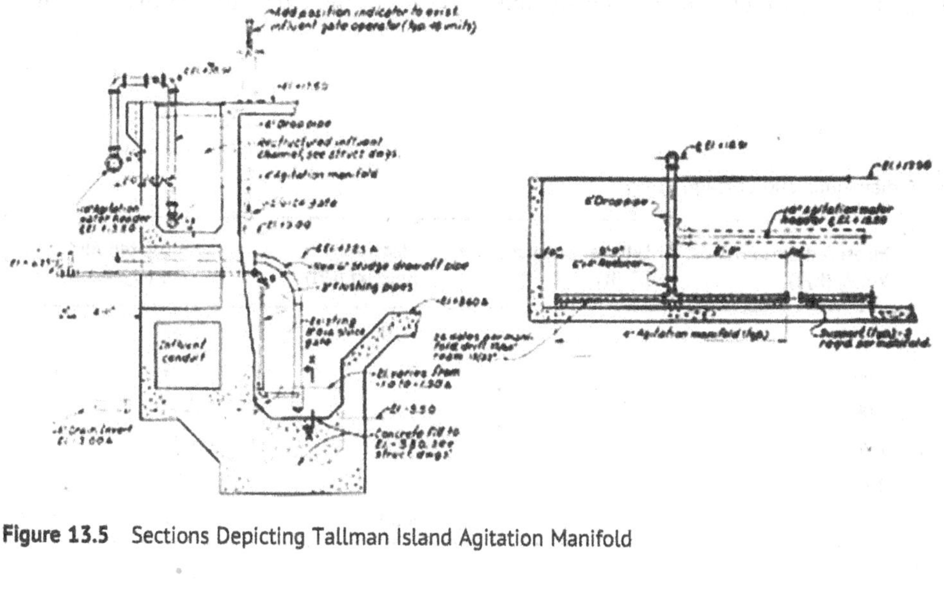

Figure 13.5 Sections Depicting Tallman Island Agitation Manifold

13.3.2.1 Early Empirical Studies and Guidelines

Early empirical studies of air-induced circulation were oriented toward the spiral-flow diffused air tanks employed in activated sludge systems. These include the studies of air-induced velocities in rectangular basins made in 1922 at Indianapolis (Hurd 1923) and in 1924 and 1930 at Chicago (WPCF 1952, p. 32 – Table II). More recent measurements have been obtained in tests conducted at the University of Iowa (Lemke 1969) for the Chicago Pump Company (now the Grundfos Co.). The primary purpose of these studies was to assess the adequacy of the induced currents to maintain mixed liquor suspended solids in suspension. The Indianapolis and University of Iowa tests were conducted in short model sections of 5-ft length, with widths ranging from 15 to 30 ft and depths of 8 to 15 ft, while the tests at Chicago were conducted in actual aeration tanks. The Iowa tests reported by Bewtra and Nicholas (1964) and discussed below were carried out in tanks of full-sized depth and width, but short model sections of 4-ft (1.2-m) length.

Data from the Indianapolis and Chicago tests are summarized in Table 13.2. One observation that may be made from that data and the Iowa tests is that the Indianapolis and Iowa tests indicate significantly higher air rates for the same induced velocities in basins of comparable dimensions, or, conversely, lower induced velocities at the same air rates, than the Chicago tests. This is probably due to the proximity of the end walls of the test sections in the Indianapolis and Iowa tests. The Indianapolis and Iowa test data are thus probably somewhat conservative when applied to actual channels of similar cross-sectional dimensions, and those data are useful in establishing the relative effects of air rates, geometry, and corner baffles.

The type of diffuser is seen to have a significant influence on the velocity induced at a given air rate. The Chicago tests demonstrate that the wider the diffuser band and/or the greater the distance from the wall to diffuser center, the lower the induced velocity. The Iowa tests suggest that saran tubes, which are fine-bubble diffusers, are more effective in inducing velocities than spargers, which are coarse-bubble diffusers. The effect of tank width is variable; the induced velocity sometimes decreases and sometimes increases with decreasing width, all other factors (including total air rate) being the same. A particularly important observation, about which more will be said below, is that more air is required for the same bottom velocity in shallower tanks of similar geometry (including the same depth:width ratio). The quantitative use of the above data should be strictly limited to tanks having depth, width, filleting, and air rates within the ranges of the test data. Extrapolation outside this range should be made with great caution. Reasons for this are given below.

As a practical matter, bottom velocities sufficient to prevent sediment deposition in spiral-flow aeration tanks correspond to spiral-flow characteristics which result in a fairly homogenous mixture of suspended solids. Fair et al. (1968, §35-12) recommended 1.0 fps bottom velocities to avoid the settling of floc (i.e., keeping the activated sludge in suspension) resulting in septic conditions, and suggested that this will also perform the two additional functions of supplying the needed oxygen and fostering contact between wastewater and sludge. Greeley et al. (1969a, p. 41–38) suggested for spiral-flow tanks a transverse velocity of 1 to 1½ fps to prevent solids deposition.

It was once speculated that a "dead zone" would occur in the center of spiral-flow aeration tanks, but available evidence refutes that notion (WPCF/ASCE 1959, pp. 30–31). As another practical matter, in wide tanks an air header at bottom mid-width with diffusers on both sides may be an economical choice. The absence of copings at the mid-width may reasonably be assumed to be compensated for by the absence of a solid boundary and the fact that the circulation of the two cells is

Table 13.2 Summary of Data from Indianapolis and Chicago Tests

Tank Depth D (ft)	Tank Width (ft)	K_l	Measured Bottom Vel v_b (ft/s)	Diffuser Submergence D^* (ft)	Measured Air Rate (cfm/ft)	Calculated Air Rate (cfm/ft)
15.5	33.3	2	1.45	15.0	3.10	2.65
"	"	"	1.84	"	6.19	5.04
"	"	"	2.09	"	9.29	7.14
"	16.1	3.33	1.70	"	1.50	1.47
"	"	"	2.15	"	3.00	2.78
"	"	"	2.34	"	4.50	3.51

mutually reinforcing at the mid-width so that the tank halves can be analyzed as separate tank sections. An example incorporating a mid-width header and making appropriate use of the above data is as follows:

Example 13.4

Related to Example 8.5, we have two equalization basins (Rochester, New Hampshire Advanced Wastewater Treatment Facility), each having a width of 60 feet with three fillet walls (i.e., moderate bottom copings) per tank, 325-ft length, and maximum water depth of 10 feet. An aeration header is to run the long direction of the tank at the side of each 15-foot pass, with air diffusers on one side. Process oxygen-transfer requirements established the need to supply a maximum of 20.8 scfm per diffuser at 2-ft spacing on each header. The particular diffusers to be used are considered to be intermediate between spargers and saran tubes in terms of their ability to induce velocities. The mixed liquor suspended solids concentration is not expected to exceed 1%. Determine whether this air rate will be sufficient to keep solids in suspension with a required bottom velocity of 1.2 fps.

Solution

The most proximate available data for a 1.2 fps bottom velocity is from Chicago Pump Company for fine-bubble diffusers (Saran tubes) with 4.1 scfm per linear foot at 8-ft depth and 16-ft width. The airflow rate of 20.8 scfm/2 ft = 10.4 scfm per linear ft of tank length is thus more than adequate to maintain solids in suspension.

Aside from aeration tanks, sludge holding tanks, and similarly sized *tanks*, many of the actual or contemplated applications of air-induced circulation are for *channels* which are considerably shallower and/or narrower than those for which test data are available. A standard reference (WPCF 1977, p. 113) suggests that "more air will be required if deep and wide tanks are used than for tanks that are comparatively shallow." That is not always the case, as indicated by the data cited above. It is further suggested (WPCF 1977, p. 113) that "If air is supplied at a minimal rate, usually between 1.5 and 6.2 L/s per meter [of channel] (1.0–4.0 cfm/lin ft), the required agitation to maintain solids in suspension will be accomplished." The latter guideline must be heavily qualified, for reasons discussed below in the context of an actual design situation studied by the author for the City of New York in 1969.

In the late 1960s, New York City decided to eliminate grit chambers used for preliminary treatment at some of the city's wastewater treatment plants. The grit chambers had become overloaded or were expected to become overloaded with projected population increases, and room for expansion of the chambers was limited. In the Tallman Island Facility, the existing grit chambers had become inefficient to the point that grit had to be regularly removed from the digesters. It was decided to abandon the existing grit chambers and transport the grit into the primary settling basins, where it would settle with the primary sludge and be subsequently separated from the organic fraction in cyclone (vortex) degritters (discussed by Chasick and Burger 1964 and Neighbor and Cooper 1965). It was necessary to assure that the grit would not deposit in the screening channels or the primary sedimentation basin influent channels. The author was given the task of designing the system for New York City's Tallman Island Water Pollution Control Facility. The design of the sedimentation basin influent channel agitation system is discussed first.

The use of air to prevent deposition of de-gritted sludge and activated sludge in feed channels to primary and secondary settling basins, respectively, had been used in the Tallman Island plant and other New York City facilities for many years. Early systems for this purpose made use of the ridge-and-furrow configuration (WPCF/ASCE 1959, Chapter 10; Leary *et al.* 1969) common to activated sludge plants of this time and later relied on spiral-flow aeration. In the Tallman Island Facility, another alternative was employed of tapering some of the influent channels to maintain cleansing velocities, but these were to be changed to open channels in order to facilitate plant expansion. One of the primary influent channels had existing air diffusers, which effectively prevented partially degritted primary sludge from depositing in it. This channel was 5-ft–9-in. deep and 9-ft wide, with 12-in. porous tube (fine-bubble ceramic) air diffusers along one side, which had originally been designed to induce velocities sufficient to provide uniform distribution of grit-free solids to the primary settling basins.

The feasibility of using air-induced circulation to keep the grit in suspension was investigated. A criterion was established of 2.0 ft/sec minimum transverse bottom velocity. At first glance, this would not seem unreasonable, since in the Chicago studies of a full-scale spiral-flow aeration basin approximately 15-ft deep by 16-ft wide (see Table 13.2), velocities measured at the tank center, 0.5 ft above the bottom, varied from 1.7 to 2.3 ft/sec as the air rate varied from 1.5 to 4.5 cubic feet per second per linear foot of tank. Fortunately, a respect for the complexity of two-phase (air–water) flows led the author to conduct a test program rather than extrapolate from the data presented above for larger tanks. The existing fine-bubble diffusers were used, and the header supplying the air diffusers was fitted with orifices, pressure taps, and a thermometer to allow accurate measurement of the air rate (see Chapter 9). A series of three carefully machined orifices was used to cover the flow range of interest. The orifice dimensions, instrumentation, and calculation of air rates from measured variables were in accordance with ASME "Fluid Meters" (ASME 1971). A Price current meter (Gurley Precision Instruments, Troy, New York) was employed to measure the velocity at the channel center, 0.5 ft from the bottom.

Measurements gave the following results for bottom velocity vs. air rate: 0.6 ft/sec at 5.5 scfm per ft, 0.7 ft per sec at 18 scfm per ft, and 0.9 ft per sec at 30 scfm per ft. The maximum measured air rate (scfm per ft) was more than 6 times the highest supplied to the larger Chicago tank just mentioned, yet the corresponding velocity was less than half that measured in the Chicago tank. Further increases in air rate lead to no further increase in bottom velocity. At the higher air rates, the air appeared to be "piping" through the water to the free surface. The air quantities were impractically high, and the induced velocities were considerably less than required, proving air diffusion unfeasible for the intended purpose. Water agitation was designed and implemented as discussed above.

A similar piping phenomenon, also referred to as "flooding" or "chimney effect," was found to restrict the effectiveness of mechanical aerators at the Cranston, Rhode Island Wastewater Treatment Plant. Stanbury *et al.* (2016) discuss similar flooding in an aeration sparger installed in a fermentation lagoon.

The inadequacy of existing predictive techniques and the limitations of diffused air for inducing velocities in shallow channels are evident from the above discussion. Air-induced velocities are important, particularly with respect to their relationship to oxygenation efficiency, as discussed below. The author has incorporated the effects of compressibility into an analysis, which agrees essentially with results derived independently by Kobus (1968, 1970) and has developed a semi-empirical energy approach, which indicates a constant energy transfer efficiency from air to water over the range of the Chicago tests (WPCF 1952).

Example 13.5

Agitation of a WWTP primary-effluent channel (Deer Island WWTP, Boston Harbor, Massachusetts) was proposed by the designer to prevent deposition of solids. It was proposed to use waste nitrogen from a pure oxygen generation system (used for activated sludge oxygenation) for this purpose.

Solution

The author reviewed this proposal and commented that the scouring requirements of such channels are generally met by maintaining small longitudinal velocities. Further, by gas mixing such channels the volumetric capacity of the planned odor control facilities would be significantly increased due to the greater vent-gas quantity. The author also questioned the effect that agitation of the primary-effluent channels with waste nitrogen would have on the oxygen saturation concentration, and hence driving force for oxygenation in the secondary reactor. Based on the author's comments, the plan to agitate the channels with nitrogen was dropped. However, problems did not end there, as discussed in the next example.

Example 13.6

Having dropped plans for agitation of primary-effluent channels with nitrogen as discussed in the previous example, the Massachusetts Water Resources Authority (MWRA) proposed to agitate those channels with air. Again, the author commented, as above, that such agitation was unnecessary. MWRA proceeded anyway and immediately observed that the entrained air diminished secondary clarifier efficiencies and increased effluent suspended solids. It also resulted in the buildup of excess scum that was difficult to remove. Mixers were then placed in the channels so that air agitation could be discontinued. No tests were run to determine if, as the author proposed, mixing in these channels could have been avoided altogether.

Air-induced circulation is capable of providing bottom velocities of 0.5–1.0 ft/sec, sufficient to maintain in suspension the solids in degritted domestic wastewaters, waste-activated sludges, and comparable sediment-laden flows. The data presented above together with information presented below may be used to guide the design of such systems.

Employing earlier work by Rouse, Yih, and Humphreys (1952), Iamandi and Rouse (1969) equated the momentum flux of the water jet inducing the circulation to the "buoyancy flux" of air inducing circulation, in an attempt to extrapolate their results to the air/water system. They conducted no experiments on the air/water system. The author (Graber 1982) found their theory to be inadequate for the air/water system. This was in part because it neglected compressibility of the rising air bubbles, variations in the air–water flow regime, and the influence of the induced water velocity on the absolute bubble rise velocity. These factors are considered below.

For air systems, questionable empirical guidelines have been developed in terms of air rate per unit length of channel for channels conveying degritted sewage to primary settling tanks and activated sludge mixed liquor. The method of Iamandi and Rouse underestimates by an order of magnitude the air requirements for particular bottom velocities determined experimentally in Indianapolis and Chicago, as discussed above. However, under certain circumstances, the author has found that the following *functional* relation between Iamandi and Rouse applies:

$$\frac{v_s^3}{q_a g} = \phi\left(\frac{W}{D}\right) \tag{13.9}$$

in which v_s is the induced velocity near the tank bottom, q_a is the volumetric airflow per unit length of channel, g is the acceleration of gravity, W is tank width, and D is tank depth. The above relationship indicates that for a given tank geometry (W/D) and desired bottom velocity, the volumetric airflow per unit length is the basic design parameter.

The buoyancy flux of air bubbles in water, analogous to the momentum flux discussed previously, is equal to the weight of water displaced by the bubbles. Mathematically,

$$m = \rho_w g \forall_{disp} \tag{13.10}$$

$$\forall_{disp} = \int_0^{D*} d\forall_{disp} \tag{13.11}$$

in which $d\forall_{disp}$ is the displaced volume of water per unit length of channel in an incremental vertical element dx over the diffusers, and $\forall_{disp}$ is the total displaced volume in the total depth D^* of water over the diffusers.

The incremental displaced volume is related to the actual volumetric air flow rate at x in an increment of time dt by:

$$d\forall_{disp} = q_a dt \tag{13.12}$$

Neglecting temperature changes in the rising air bubbles, the actual air rate q_a is related to the air rate at the water surface q_o by:

$$q_a = q_o \frac{h_{atm}}{D^* + h_{atm} - x} \tag{13.13}$$

in which h_{atm} is the atmospheric pressure head. The incremental time dt is related to the incremental height dx and absolute bubble rise velocity u_b as follows:

$$dt = dx/u_b \tag{13.14}$$

From Equations, 13.12, 13.13, and 13.14, the following is obtained:

$$\forall_{disp} = \int_0^{D*} q_a dt = \int_0^{D*} q_o \frac{h_{atm}}{D^* + h_{atm} - x} \frac{dx}{u_b} = q_o h_{atm} \int_0^{D*} \frac{1}{u_b} \frac{dx}{D^* + h_{atm} - x} \tag{13.15}$$

If u_b is assumed to be constant, the integration may be carried out to obtain:

$$\forall_{disp} = \frac{h_{atm} q_o}{u_b} \ln \frac{D^* + h_{atm}}{h_{atm}} \tag{13.16}$$

Substituting the above into Equation 13.10 gives:

$$m = \frac{\rho_w g h_{atm} q_o}{u_b} \ln \frac{D^* + h_{atm}}{h_{atm}} \tag{13.17}$$

The above equation is essentially the same as that given by Kobus (1968, 1970). [Kobus used the air rate at the diffuser as the reference value rather than the air rate at the water surface (at atmospheric pressure)].

As before (see Equation 13.8), the bottom velocity is given by:

$$v_b = K_I \sqrt{m/(\rho_w D)} \tag{13.18}$$

Substituting Equation 13.10 into the above results in:

$$v_b = K_I \sqrt{g V_{disp}/D} \tag{13.19}$$

Then substituting Equation 13.16 for V_{disp} into the above equation gives:

$$v_b = K_I \sqrt{\frac{g h_{atm} q_o}{u_b D} \ln \frac{D^* + h_{atm}}{h_{atm}}} \tag{13.20}$$

The term K_I is assumed to have the same values as given previously. The other terms in the above equation are readily established, with the exception of u_b.

The bubble rise velocity u_b determines the magnitude of the driving force by influencing the amount of air and water displaced in the water column above the diffuser at any instant. The slower the rise velocity, the greater the time spent (this is referred to as "hold-up time" in the chemical engineering literature) by each bubble rising through the water column and the greater the number of bubbles remaining in the water column. The rate of rise of individual air bubbles (minimally influenced by each other) is primarily a function of bubble size, as shown in Figure 13.6 based on measurements by Haberman and Morton (1956, as presented by Camp 1963) in tap water at 21°C. The data in Figure 13.6 are equally applicable to the rise of bubbles for gases other than air because the density of the gas is generally very small compared to the density of water. Haberman and Morton found that the bubbles were essentially spherical in shape when bubble diameters were less than 0.6 mm. As the size of the bubbles increased, a change from spherical to ellipsoidal to cap-shaped was observed. The larger ellipsoidal and cap-shaped bubbles did not rise in a straight vertical path but rather in a helical or rectilinear path with a rocking motion of the bubble. The flat shape of the larger bubbles probably accounts for the relatively constant velocity for all sizes between 2 and 10 mm, as shown in Figure 13.6. The velocity in this range is about 22 cm/sec or 0.72 ft/sec.

One important point that can be made here is that the bubble rise velocities in Figure 13.6 are those observed in a quiescent water column with little or no induced velocity. The observed velocities are of the same order of magnitude as those to be induced in our applications. The ***absolute*** bubble rise velocity is our concern for use in Equation 13.20 and is the sum of the bubble rise velocity of the bubbles relative to the water around the bubbles and the velocity of that water. It should also be noted that the velocities in Figure 13.6 are those for individual single bubbles. When many bubbles are rising concurrently, in close proximity to each other, the velocities may be larger or smaller. The tendency for the velocities to increase is caused by the assistance provided by the wake of preceding bubbles (Marks 1972), while the tendency to decrease the velocity may be caused by the increase in effective velocity due to adjacent bubbles (an effect analogous to hindered settling – see Section 13.2).

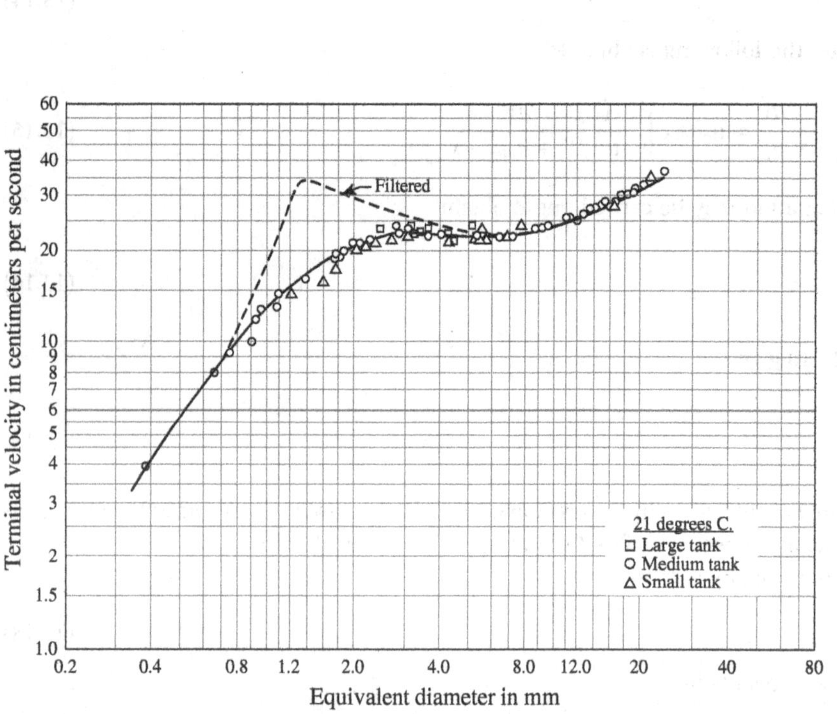

Figure 13.6 Terminal Velocity of Air Bubbles in Tap Water as a Function of Bubble Size (Camp et al. 1963/with permission of National Association of Biology Teachers)

A rough estimate of the velocity induced under certain conditions may be made by setting the absolute bubble rise velocity u_b equal to the sum of the average relative bubble rise velocity $\bar{v}_b$ plus the average upward velocity of water over the diffusers (denoted by $\bar{v}_\ell$):

$$u_b = \bar{v}_b + \bar{v}_\ell \tag{13.21}$$

Figure 13.4 suggests that $\bar{v}_\ell$ is the same order of magnitude as the bottom velocity at mid-width v_b; for simplicity, assume $\bar{v}_\ell = v_b$. Thus:

$$u_b \cong \bar{v}_b + v_b \tag{13.22}$$

Substituting the above equation into Equation 13.20 and solving for q_o gives:

$$q_o = \frac{Dv_b^2(v_b + \bar{v}_b)}{K_I^2 g h_{atm} \ln[(D^* + h_{atm})/h_{atm}]} \tag{13.23}$$

The Chicago data for the narrower diffuser widths and diffusers closest to the wall, as given by the first two sets of Chicago data in the 15.5-ft-deep tank of Table 13.2, provide a good basis for comparison with Equation 13.23. We will assume h_{atm} = 33.9 feet and μ_b = 0.72 ft/sec (from above). The remaining parameters are given in Table 13.2. The calculated and actual (measured) air rates are reasonably close, particularly at the lower air rates, suggesting at least qualitatively that the mechanism of air-induced circulation is fairly well understood for the larger tanks at low to moderate air rates.

The question then arises as to what causes the major disparity for smaller tanks or channels, as observed in the author's studies mentioned above. The channel studied by the author had D = 5.83 ft, D_* = 5.33 ft, and a depth:width ratio corresponding roughly to $K_I = 2$, for which Equation 13.23 gives a value for q_o of about 6 cfs/ft for a bottom velocity v_b = 2 ft/sec. As noted previously, air rates well in excess of 30 cfs/ft failed to produce a bottom velocity of 1 ft/sec. The author's explanation of this large disparity is given below.

Equation 13.16 indicates that if all other variables are held constant, the volume of water displaced by air per unit length of channel is proportional to $\ln(D^*/h_{atm} + 1)$. For the usual case of $D^*/h_{atm} < 1$ (note $h_{atm} \approx$ 33.9 ft), series expansion of the ln function indicates to a first approximation that this displaced volume is proportional to the channel depth. The volume of the air–water column over the diffusers retains approximate geometrical similarity as the tank depth is varied, so the volume of the air–water column per unit length of channel is approximately proportional to the square of the channel depth. Therefore, the "void fraction," defined as the volume displaced divided by the volume of the air–water column, is inversely proportional to depth. The "void fraction" thus increases with decreasing depth. A high void fraction is associated with transition to the "slug" and "semi-annular" flow regimes observed in two-phase flows, in which the air is, to a large degree, "piped" through the water (Wallis 1961). The result is a greatly reduced airlift effect.

13.3.2.2 Full-Width Air Diffusion

One final aspect that merits consideration with respect to air agitation of channels concerns the use of plate diffusers across the full width of the channel bottom. In this case, circulation is not involved, but rather, the settled particles are scoured by the shear created by the rising air bubbles. The shearing ability of the air bubbles can be estimated by comparing the shear created by the bubbles to that required to scour different-sized sediment particles.

From Section 4.5, the maximum shear stress created by an air bubble in the Stokes flow regime is given by:

$$\tau_{\max} = \frac{3}{2}\mu\frac{U_\infty}{R} \tag{13.24}$$

in which μ = water viscosity, U_∞ = bubble rise velocity, and R = bubble radius. Equating the above to the shear required to scour a particle of diameter d, as given by Equation 13.1 (with $\beta = 0.06$ for noncohesive particles), we obtain:

$$\frac{3}{2}\mu\frac{U_\infty}{R} = 0.06\gamma(s-1)d \tag{13.25}$$

Solving for d gives the maximum size particles that would be scoured:

$$d = 25\frac{\upsilon U_\infty}{(s-1)gR} \tag{13.26}$$

Assuming a 2 mm air bubble diameter [a common size from commercially available porous plates (Camp 1955)] and a terminal bubble velocity of 20 cm/sec (Haberman and Morton 1956), Equation 13.26 gives $d \approx 25(10)^{-5}(20)/[1.65(32.2)(0.1)] = 0.00094$ ft × 305 = 0.29 mm. Particles larger than that size would not be scoured. Grit-sized particles in domestic wastewater are typically in the range of 0.2–1.0 mm (Camp 1942), so full-width porous plates would not be expected to be effective in scouring grit. This was borne out by independent tests conducted by the City of New York in 1969.

G-Value. G-value was critiqued above in connection with its suggested applicability to water-induced circulation. Walker (1969) has suggested that "the amount of air ... to stir solids-bearing liquids under sluggish flow through channels is best designed on the criterion G-value..." (see Section 3.4). For deep channels with moderate air rates, Equation 13.23 has been shown above to provide a fair approximation. For a given tank geometry and degree of "stirring" (as measured by bottom velocity), Equation 13.23 gives the following approximation:

$$q_o \cong C_{prop} \frac{D}{\ln[(D^* + h_{atm})/h_{atm}]} \tag{13.27}$$

in which C_{prop} is a proportionality constant. Since the energy per unit length of the air influx is equal to $q_o p$ in which p is the air pressure and is proportional to $q_o p D^*$, it is apparent that a wide range of values of C_{prop} and corresponding values of v_b can correspond to a given energy or G-value. Furthermore, the author is aware of no experimental evidence, for shallow or deep channels, to support the use of G-value as a basis for design of air-induced circulation.

13.4 Pressure Drop in Sludge and Slurry Flows

Fair and Geyer (1954, §26-9) suggested that "[w]hen the flow is turbulent, the loss of head of fairly homogeneous sludges (digested sludges and activated sludge) is increased by not more than 1% for each percent of solids in the sludge (this is accomplished by increasing f accordingly when using the Darcy-Weisbach friction factor). Fresh, plain sedimentation sludge is transported at losses that are 1.5 to 4 times those of water." These recommendations have been successfully employed by the author. An example of their use is given in Section 6.7 (Example 6.8).

It is interesting to compare the above to suggestions gleaned from other references. Hedström's (1952) method requires a viscometer measurement of the yield stress and plastic viscosity for the particular slurry, which limits its use here. The method of Vocadlo and Sagoo (1972) requires the use of parameters for which their determination is not explained, and no example is given.

Aude *et al.* (1971) give the following equation, which they attribute to Durand:

$$f = f_w \left[1 + 82\left(\frac{gD}{V^2} \cdot \frac{\rho - \rho_w}{\rho_w}\right)^{3/2} \frac{C_v}{C_d^{3/4}}\right] \tag{13.28}$$

in which f is the adjusted Fanning friction factor equal to $\tau/(\rho V^2/2)$ (dimensionless) with τ being shear stress, V is the bulk flow velocity, and ρ is the mixture density; f_w *is the* Fanning friction factor for clean water; g is gravitational acceleration; D is pipe inside diameter; ρ_w is water density; C_v is the slurry volume concentration; and C_d is drag coefficient for which a "quick, conservative estimate can be made by assuming the minimum value of the drag coefficient (0.44) for all sizes," giving $C_d^{3/4} = (0.44)^{3/4} = 0.540$.

Assuming a specific gravity of sand, the term $(\rho - \rho_w)/\rho_w = (2.65 - 1)/2.65 = 0.623$. The above equation can then be written as:

$$\frac{f}{f_w} = \left[1 + 82\left(0.623\frac{gD}{V^2}\right)^{3/2} \frac{C_v}{0.540}\right] \tag{13.29}$$

Thomas (1977) critiques Durand but then proposes an unwieldy variation involving a multitude of relationships and lack of clarity.

Hughmark (1961) compiled data from 14 data sources with 27 mean particle diameters, which he presented in terms of a "Froude Number" $V/\sqrt{gD}$ vs. $C[(\rho_s - \rho_f)/\rho_f]F_d$ in which V is average fluid velocity in the pipe, g is gravitational acceleration, D is pipe inside diameter, C is concentration of solids in percent, ρ_s is density of solid, ρ_f is density of medium (water in our case), and F_d is a "correction factor" for the particle diameter. The latter increases linearly from 0 to 1.0 for particle diameters from 0 to about 0.0015 ft and is 1.0 for particle diameters larger than 0.0015 ft. Hughmark fitted a curve to the data plotted on logarithmic scales. Since sludge particle sizes fall within the range of variables F_d, these results are of limited value here.

Mulbarger et al. (1981) provide extensive test results in the inconvenient form of many charts and only for one type of pipe – cast iron, without any attempt to make the results more convenient or generalized.

Extensive test results are presented by Murakami *et al.* (2001) for laminar and turbulent flow of sludges with solids concentrations between 2% and 5%. The critical velocity for turbulent flow is given by:

$$V_c(\mathrm{m/s}) = 1.20\frac{C_H}{100}k^{0.52} \tag{13.30}$$

in which C_H and k are the Hazen–Williams coefficient and viscosity coefficient, respectively. The pipe friction head loss for turbulent flow is:

$$H_f(\mathrm{m}) = 9.06\left(\frac{1}{C_H}\right)^{1.93(1-C/100)}\frac{L}{D^{1.18}}V^{1.82} \tag{13.31}$$

in which L, D, and V are pipe length in meters, pipe I.D. in meters, and pipe velocity in m/s, respectively.

Babbitt and Caldwell (1942) presented relationships for critical velocity for turbulent flow that are limited to black steel pipe.

Babbitt and Baumann (1958, §24-12) gave relationships for the critical velocity for turbulent flow and cited ranges of Hazen and Williams friction factors, but without sufficient specificity for practical use or any attempt at curve fitting.

13.5 Drag Reduction in Conduits

Polymers and other additives have been used to decrease drag and increase flow conveyance in pipes and open channels (e.g., Mignot *et al.* 2019, Overfield *et al.* 1969, Pazwash 1984, and Proeh *et al.* 1970). Mignot *et al.* provide a comprehensive review, and indicate that the exact mechanism is not understood. However, the overlooked work of Hansen (1972) merits consideration in this regard.

One interesting practical application in environmental engineering was the use of such an additive to temporarily increase the capacity of the sewer network of Whistler, Canada, during the high population increase at the 2010 Winter Olympic Games (Hart *et al.* 2011). Peters and McNeill (2003) reported on the use of a polymer to temporarily increase lift-station capacity in Denver, Colorado. Although polymers are the most common additives for drag reduction, Proeh *et al.* found a complex soap with the tongue-twisting name of cetyl trimethylammonium bromide (CTAB-1 Naphthol) to be more effective than polymers. Details may be found in the cited references. A more complete discussion of the work by Overfield *et al.* may be found in their full report (FWPCA 1969). A good discussion of sludge-pumping/ polymer-metering systems is given by Follest (2000).

Mignot *et al.* noted the importance of considering the impact on receiving waters of the injection of drag reduction additives into sewers. Tests by Overfield *et al.* (1969, FWPCA 1969) showed no adverse effects of their polymer additions on fish and algae or on wastewater bacteria, sludge drying, and solids settling. Bioassays should be required as necessary to supplement available information.

14

Process Applications

14.1 Introduction

The applications of hydraulics to various processes are discussed in this chapter. Such knowledge is a very important part of understanding and designing such processes. Those discussed here include treatment plant tanks such as chemical contact tanks and sedimentation basins, sedimentation and thickening, mixing applications, chemical contacts such as chlorination and dechlorination, flocculation and floc breakup, aeration and postaeration, and filtration and fluidized beds.

14.2 Basic Reactor Concepts

Treatment plant tanks of various types are designed to have *plug flow* at one end of the spectrum and *complete mixing* at the other. Complete mixing and plug flow can be stepped in successive compartments or channels. In plug flow, it is desired to minimize *short-circuiting*, the term applied to deviations from the ideal of plug flow. Discussion of these concepts is presented immediately below, in Section 14.5 in the context of chemical contact, and in Section 14.7 in conjunction with aeration.

Some of the discussion under this heading is taken from Graber (1994, pp. 544–545). It has been suggested (Kleinschmidt 1961) that G-value be used as a criterion for controlling hydraulic short-circuiting in detention tank design. Kleinschmidt distinguished appropriately between detention tanks (referring to tanks in which reactions take place, such as contact tanks) and sedimentation basins. In fact, contrary to views occasionally seen in the literature and as discussed in Section 14.3, hydraulic short-circuiting is not necessarily detrimental to sedimentation basin performance (Graber 1974a, 1975a).

Kleinschmidt reasoned that "mixing" must be provided to prevent stratification, and G-value is a measure of mixing [with specific reference to flocculation and the work of Camp and Stein (1943)]. He suggested specifically that G-value be used as a criterion for determining whether there is sufficient mixing to reduce short-circuiting due to temperature stratification and associated density currents. Using the first trace on flow-through tracer curves as a measure of short-circuiting, he established a fairly good correlation between tank performance and G-value for a number of model and prototype basins. He found that short-circuiting decreased (i.e., the first trace occurred closer to the theoretical detention time) as G-value increased and proposed that detention tanks be designed with a G-value greater than 0.3.

Kleinschmidt gave the equation for G-value from the formulation by Camp and Stein (1943) for one-dimensional energy dissipation due to wall friction as follows:

$$G = \sqrt{\frac{fv^3}{\upsilon 8R}} \tag{14.1}$$

in which f is the Darcy–Weisbach friction factor, v is mean velocity in ft per sec, υ is kinematic viscosity in ft^2 per sec, and R is hydraulic radius in ft (the equation is actually dimensionless). Kleinschmidt then used the Blasius formula to express f in terms of Reynolds Number, which leads to an expression for G-value in terms of v, υ, and R. Since the Blasius formula is applicable to hydraulically smooth conduits over a relatively low range of Reynolds Numbers, a more useful formulation for

Hydraulics and Pneumatics in Environmental Engineering, First Edition. S. David Graber.

the conduits encountered in practice would be based on the relationship between f and Manning's friction coefficient n, given as $f = 117n^2/R^{4/3}$. Equation 14.1 can then be written as follows:

$$G = \sqrt{\frac{117n^2v^3}{8\upsilon R^{7/3}}} \tag{14.2}$$

For basins in which stratification is minimal, the dispersion number provides a suitable basis for design to reduce short-circuiting (Graber 1972). The dispersion number has the advantages of representing the full tracer curve, relating to the underlying hydrodynamic phenomenon, and being directly incorporated into quantitative reactor theory. The author has found that chlorine contact tanks, which he has successfully designed based on dispersion number (e.g., Graber 1972), have G-values, which are at least four times the Kleinschmidt criterion.

Kleinschmidt also proposed that G-value may be used as a similitude parameter, relating model and prototype results. The evidence presented for this is weak. Kleinschmidt observed that the flow-through curves for a prototype and 1:32 scale model were similar when operated at the same G-value. However, this could also be the case if dispersion was the controlling variable and only geometric similarity was maintained. The dispersion number for an open channel can be given as $d = C(\mathbf{R}, W/H)\sqrt{fR/(8L)}$ in which $C(\mathbf{R}, W/H)$ denotes a proportionality factor, which is a function of the Reynolds Number and channel width:depth ratio, L is a channel length, and f and R are as defined previously (Graber 1972; this equation is derived in Section 14.5). With geometric similarity between model and prototype, the dispersion number will vary only with **R** and f. At high Reynolds Numbers and the hydraulically smooth conditions assumed by Kleinschmidt, one would expect **R** and f to become relatively unimportant. In that case, only geometric similarity would be required for similar flow-through curves.

14.3 Sedimentation and Thickening

14.3.1 Basic Concepts of Sedimentation

We will begin our discussion by considering the simplest case of a low concentration of discrete, non-cohesive particles (such as sand-sized particles in raw sewage) being conveyed in a pipe or channel. Consider further a constant, prismatic, turbulent flow for which the liquid velocity remains relatively constant over much of the cross section in the direction of flow. A particle as described above will move in the direction of flow at essentially the same velocity as the fluid. If the particle density exceeds that of the fluid, it will have an additional velocity component in the direction of the gravity vector, referred to as the settling velocity. If the flow were laminar, the settling velocity would be the same as if the fluid were quiescent.

Denoting the "quiescent" settling velocity (about which more will be said below) by v_s, and idealizing the situation by assuming: (1) that the particle actually settles at this rate and (2) that the velocity in the direction of flow may be assumed to be constant U, then the trajectory of a particle in the direction of flow may be determined. For a horizontal conduit of rectangular cross section, a particle of settling velocity v_s starting at a vertical distance h above the conduit bottom will settle to the conduit bottom at a distance L as determined by:

$$\frac{v_s}{U} = \frac{h}{L}$$

Particles of the same or greater v_s entering the conduit length L at vertical heights less than h will settle within the length L, whereas those of lesser v_s entering at vertical heights greater than h will not settle within the length L. Setting $h = H$ gives the settling velocity for which all particles of that or larger v_s will settle within the length L:

$$v_s = U\frac{H}{L} \tag{14.3}$$

Noting that $U = Q/A = Q/(HW)$ in which W is channel width, we have:

$$v_s = \frac{Q}{HW}\frac{H}{L} = \frac{Q}{WL} \tag{14.4}$$

Since WL is the surface area of an open rectangular conduit (such as a rectangular grit chamber or settling basin), the term $Q/(WL) = Q/A_{surface}$ is referred to as the overflow rate. A so-called *idealized sedimentation theory* (which can be extended to other basin geometries) says that particles having a settling velocity less than or equal to the overflow rate will be fully removed.

One cause of departure from idealized sedimentation theory is *turbulence*. Camp has discussed turbulence in that context in some detail (Camp 1944, 1946b, 1947, 1953). However, ultimately, design is based on resilient criteria for overflow rate based on experience, as discussed below.

Some states set limits on both maximum overflow rate and minimum detention time. A figure of 3 hours of detention for primary tanks is common, although shorter times are permitted for primary tanks ahead of activated sludge. For secondary tanks, 2 hours of detention is a common specification. The influential Great Lakes – Upper Mississippi River Board of State and Provincial Public Health and Environmental Managers (2014) does not specify detention times but sensibly does set minimum depth and overflow rate standards for primary and secondary settling tanks. This indirectly gives minimum detention times, since detention time is proportional to tank depth divided by overflow rate.

Solids loading owes its currency as a design variable for final settling (secondary) tanks to a combination of empiricism and theoretical efforts, exemplified by Dick (1970) and Dick and Young (1972). Detention time is also important for final settling, even though it is not important for primary settling, as discussed above. This is because activated sludge suspensions continue to coagulate or flocculate. Minimum tank depths are specified to insure desired detention times in conjunction with overflow rates. Through empiricism and studies of the internal hydraulic characteristics of secondary clarifiers, such as those by Anderson (1945) and Gould (1945), relative dimensions (depth:radius) have also been suggested for design.

14.3.1.1 Settling Tank Configuration Controversy

There has been controversy, reflected in the literature, regarding the sedimentation efficiency of different configurations of settling tanks: rectangular vs. circular peripheral feed vs. circular center feed. The author joined that controversy in Graber (1975a) and in animated discussions with a representative of Lakeside Equipment Company's promotion of peripheral-feed clarifiers. In Graber (1975a), the author (also Silberman 1975) disagreed with an assertion that sedimentation is adversely affected in such a tank by inherent instability that gives rise to boundary layer separation (discussed in Section 4.7). The disagreement was specifically with (1) the manner in which the offending conclusion was reached and (2) the notion that separation adversely affects sedimentation.

The incorrect assertion (by Chiu 1974) was due to reasoning based on the fact that the flow in central-feed circular tanks encounters adverse pressure gradients that results in a positive value of $\partial^2 u_x/\partial y^2$, in which u_x is velocity in the direction of flow and y is upward, perpendicular to the direction of flow. This implies that a point of inflexion occurs in the velocity profile. From this, the conclusion was drawn that separation occurs. However, an adverse pressure gradient and positive values of $\partial^2 u_x/\partial y^2$constitute necessary, but not sufficient, conditions for the occurrence of separation (Prandtl 1952, pp. 137–138; Schlichting 1960, pp. 27–33, 112–114). To prove the existence of separation, it would be necessary to show that $\partial u_x/\partial y = 0$, which has not been done. Further, the author suggested that, in a truly two-dimensional radially outward flow, the adverse pressure gradient would not be great enough to cause separation from the tank bottom.

However, separation and eddying do occur in settling tanks and are not restricted to central-feed circular tanks. That may be inferred from the results of tracer studies in circular (Villemonte and Rohlich 1962, Murphy 1963, Tekippe and Cleasby 1968) and rectangular tanks (Burgess *et al.* 1960, Willis and Davis 1962). Such separation is mostly attributable to inlet conditions, temperature differences, and the adverse pressure gradient (theoretically infinite) created by the flow pattern in the vicinity of the effluent weirs. As regards the latter, note that the flow has to negotiate a 90° corner at the junction of the tank bottom and the plane of the effluent weir, tank wall, or tank center. Dye tests do suggest (Dague and Baumann 1961) that the resulting eddy zone is more extensive in central-feed than in peripheral-feed circular tanks of comparable dimensions, presumably because of the additional adverse pressure gradient associated with the radially outward flow. Rectangular tanks would be expected to have less of their tank volume occupied by the eddy zone because their length:depth ratio is generally much larger than the radius:depth ratio of circular tanks.

The author disagrees with the contention that "in settling tanks an eddy formation is absolutely unfavorable" (Chiu 1974). The author (Graber 1974a) used sedimentation theory in conjunction with flow patterns with and without eddy zones to demonstrate that an eddy zone can actually increase sedimentation efficiency. Some of the lighter particles, which would flow over the tank weirs in an eddy-free tank impinge on the eddy zone in which they are trapped and settle to the tank bottom. This effect is probably of great significance in central-feed circular tanks, particularly for smaller tanks having low radius:depth ratios. The sedimentation efficiency of tanks with eddy zones may actually exceed that of the so-called "ideal" sedimentation theory (Graber 1974a). Fitch (1957) also pointed out that absence of short-circuiting is not necessarily important to good tank performance.

The author's contentions are supported by experience with central-feed circular settling tanks, which are found to perform no less efficiently than rectangular tanks having comparable overflow rates (Subcommittee on Sewage Treatment 1946; WPCF/ASCE 1959, pp. 93, 103). The author has similarly found that circular tanks are no less efficient than rectangular tanks and has preferred the former due to their lower cost for smaller facilities. Recommended or required design overflow rates are typically similar to the following "Ten State Standards" (Great Lakes – Upper Mississippi River Board of State and Provincial Public Health and Environmental Managers 2014, Section 72.21):

Type of Primary Settling Tank	Surface Overflow Rates at:*	
	Design Average Flow gpd/ft^2 [m^3/(m^2·d)]	Design Peal Hourly Flow gpd/ft^2 [m^3/(m^2·d)]
Tanks not receiving waste activated sludge**	1000 (41)	1500–2000 (61–81)
Tanks receiving waste activated sludge	700 (29)	1200 (49)

*Surface overflow rates shall be calculated with all flows received at the settling tanks. Primary settling of normal domestic wastewater can be expected to remove approximately one-third of the influent BOD when operating at an overflow rate of 1000 gallons per day/square foot [41 m^3/(m^2·d)].

**Anticipated BOD removal should be determined by laboratory tests and should consider the character of the wastes. Significant reduction in BOD removal efficiency will result when the peak hourly overflow rate exceeds 1500 gallons per day/square foot [61 m^3/(m^2·d)].

The above also suggests that other researchers (Thirumurthi 1973) have related the hydraulic efficiency and sedimentation efficiency in an all-too-simplistic way. It is necessary to recognize the hydraulic complexity of settling tanks to adequately understand the sedimentation process.

An interesting but ultimately inconclusive debate about the merits of rectangular vs. circular clarifiers was published in the same journal by Wilson (1991) and Parker (1991).

As a postscript to the above, the author recently observed that Lakeside Equipment Corporation (2018) still misleadingly states that their "peripheral-feed design provides the best hydraulic flow pattern and performs two to four times better hydraulically than centerfeed clarifiers..." (Water Environment and Technology, June 2018).

High-rate settling or tube-settler systems, also known as Lamella clarifiers, are also worthy of mention here. Yao (1970) has done fundamental work on such settlers.

14.3.1.2 Weir Loadings

In 1974, the author published a paper (Graber 1974a) critiquing requirements for limiting weir loadings in settling tanks. That resulted in some changes in those requirements, but not enough based on more current references. So the following is an updated reiteration of the author's efforts.

Various outlet weir configurations have been provided for settling tanks. Those configurations are often more complex and costly than the simple full-width end weirs or peripheral weirs, which are naturally suggested by the shape of rectangular or circular settling tanks, respectively. The purpose of this practice has generally been to limit weir "loadings," defined as the volume rate of flow per unit length of weir.

Designs for limiting weir loadings have often been based on the requirements of regulatory agencies. An example of such requirements was found in the 1971 edition of the influential "Ten States Standards" (Great Lakes – Upper Mississippi River Board of State Sanitary Engineers 1971), which stipulated "weir loadings should not exceed 10,000 gallons per day per linear foot (124 cu m/day/m) for plants designed for average flows of 1.0 mgd (3785 cu m/day) or less. Special consideration will be given to weir loadings for flows in excess of 1.0 mgd (3785 cu m/day), but such loadings should preferably not exceed 15,000 gallons per day per linear foot (186 cu m/day/m)." In larger plants, compliance with these requirements becomes impractical. A compromise frequently resulted in which, through the use of costly launders, segmented weirs, or inboard weirs, the weir loadings were reduced, but the resulting loadings were still 10 times those suggested by the regulatory agency.

The results of an investigation of the rationale for weir loading limitations are presented here. The investigation included a thorough literature review, critical evaluation of existing theories, and the development of an improved theory.

According to Giles (1943), concern for weir loadings existed in the years just before World War II, when a few wastewater treatment plants were built with outlet weirs located away from the outlet end of the tank. Weirs could then be provided on both sides of the effluent channel as well as along the sides of the tank; this design, according to Giles, proved helpful in avoiding the escape of light solids in secondary settling tanks. Anderson (1945) definitively established the importance of careful weir design for secondary settling tanks by presenting the results of parallel tests at the Chicago Southwest Plant. Anderson showed that improper weir location and higher weir loading rates resulted in higher effluent suspended solids. He related this to the existence of density currents, whose importance in secondary settling tanks was established at the same time by Gould (1945) in New York City. Anderson recommended weir loading rates for secondary settling as follows: for effluent weirs located away from the upturn of the density current, the weir loading rates should not exceed 20,000 gpd/ft (248 cu m/day/m) of weir. For weirs located within the upturn zone, the rate should not exceed 15,000 gpd/ft (186 cu m/day/m). A 1946 report by the National Research Council Subcommittee on Sewage Treatment (Subcommittee on Sewage Treatment 1946) presented a theory implying that weir loading rates were of significance in primary settling tanks as well as in secondary settling tanks. This theory is discussed in detail below.

Camp (1953) reported on hydraulic laboratory experiments from which he concluded that, with long settling tanks (length 20 or more times the depth), the outlet design becomes unimportant because of the relatively small extent of the outlet zone. Ingersoll *et al.* (1956), in their comprehensive study of rectangular settling tanks, criticized the National Research Council theory, pointing out that the assumed radial velocity pattern does not enable a comparison with Camp's (1946b) ideal sedimentation theory. The authors proposed the use of potential-flow theory to establish the effect of the outlet weir on the flow pattern throughout the tank and related the flow pattern to settling efficiency. Flow nets were obtained by sketching orthogonal curves representing the velocity potential and stream function (defined in Section 4.2) within the tank. The authors reported that the distortion of the flow was negligible at distances from the weir equal to a few multiples of the depth and that, for long tanks, differences in outlet types are of small importance compared to differences in inlet types.

Definitive testing of the effect of the weir loading rate on the performance of primary settling tanks was conducted in Los Angeles by Theroux and Betz (1959). Parallel tests were run in which about 40% of the weir length was blocked off in one settling tank, increasing the weir loading from 40,000 gpd/ft (494 cu m/day/m) to 70,000 gpd/ft (868 cu m/day/m) and the tank performance compared with that of an identical, unmodified tank. The experiments showed a negligible difference in the removal of suspended solids. The investigators cited performance data at other plants that, together with their experiments, led them to conclude that "there appears to be no obvious advantage, as far as suspended solids is concerned, in using the longer weirs."

The Water Pollution Control Federation's Manual of Practice on Sewage Treatment Plant Design (WPCF 1952) acknowledged the "Ten State Standards" requirement for weir loading rates and commented on them. Besides noting that the significance of weir location, as established in the Chicago studies by Anderson (1945), is not mentioned, the Manual of Practice (1952) presented the following criticisms:

1) "In primary sedimentation tanks, there is no positive evidence that weir rate per se has any significant effect on removals."
2) "Although some authorities recommend maximum weir loadings, except on final tanks for activated sludge, there is little evidence of any limitation."

The Manual of Practice cites circular and rectangular tanks that have weir rates over 10 times those recommended in the "Ten State Standards" and exhibit excellent SS removal in actual operation.

In 1959, Rankin published a paper entitled "Weirs on Sedimentation Tanks" from which the following excerpts are taken:

1) "The length of weir required for sedimentation tanks in sewage treatment plants is a matter of fairly exact calculation according to the prescribed rules of some regulatory agencies. As standards or guide rules of such agencies are usually based on experience one naturally expects to find abundant operating records confirming the prescription. But any search will not be rewarding, because plant recordings contain little if any evidence to establish requirements for the length of weir for a given flow. One may logically question the reason for such rules and their justification."
2) "In rectangular tanks, compliance with these rules is also a serious and expensive problem because a single outlet weir across the end of the tank which one would expect to normally provide, is seldom more than 25–30% of the required length...to comply, multiple weirs in total length of between three and four times the tank width are required also at added cost of construction, not to mention the increased operating burden of cleaning the weirs."

3) "There is no evidence in published operating records of numerous plants to indicates that weir rates have any influence whatsoever on primary tank efficiencies. Reference to tables of performance in the Manual of Practice of Sewage Treatment Plant Design shows circular tanks with weir rates of 58,000 gpd per lin ft in 170 ft diam tanks removing 61% of suspended solids and rectangular tanks with weir rates of 215,000 gpd per lin ft in 50 ft by 124 ft units removing 55% of suspended solids. Several other examples in the tables show units with weir rates far above the stipulated maximums, and any careful study of these as well as numerous other published records will show the major influence on performance to be overflow rate with weir rate of practically no significance."
4) "How this regulation originated is, of course, unknown but the theoretical dissertation on weir rates in the NRC Report on Sewage Treatment at Military Installations was one influence which aroused interest in this hitherto disregarded factor. An equation developed in this report shows the theoretical zone of influence of the weir and the limiting rates for best performance. However, one interesting observation is the disparity between the actual performance of primary plants and the theoretical calculated influence of the weir rates. In plants listed in the NRC report, no apparent effect of weir rates on removals of suspended solids is evident."
5) "Anyone who is familiar with the long rectangular Imhoff tanks quite common 30–40 years ago will recall that most of them had flow-through settling compartments from 10 to 20 ft in width with a single outlet weir across the end. Weir rates were obviously many times those now required to comply with the foregoing rules yet removals of suspended solids in those earlier units were usually better than 60%."

The previously mentioned 1946 theoretical study by the National Research Council (NRC – Subcommittee on Sewage Treatment 1946) is apparently the only evidence that attempts to support the practice of limiting weir loadings in primary settling tanks, it will be referred to as the "NRC theory" herein. The NRC theory is examined below.

In the derivation leading to the NRC theory, the assumption was made that the liquid velocity in the quadrant bounded by the weir and free surface is directed radially toward the weir. The behavior of a particle settling in this radial flow field was studied. In the notation shown for the rectangular tank in Figure 14.1(a), the motion of a particle at point (x, y) settling unhindered at velocity v_s relative to the fluid is determined by $-dx/dt = u$ and $-dy/dt = v - v_s$, in which u and v are, respectively, the x and y components of the fluid velocity. Eliminating dt gives the slope of the particle trajectory at a point, as follows:

$$\frac{dy}{dx} = \frac{v - v_s}{u} \tag{14.5}$$

For a uniform horizontal velocity ($u = U$ and $v = 0$), Equation 14.5 leads to Camp's (1946b) ideal sedimentation theory, which established the basis for using overflow rate as the primary parameter for settling tank design.

The velocity field assumed in the NRC theory has components $u = v_r \cos\theta$ and $v = v_r \sin\theta$ that, with Equation 14.5, yields:

$$\frac{dy}{dx} = \frac{v_r \sin\theta - v_s}{v_r \cos\theta} \tag{14.6}$$

By expressing the left-hand side of Equation 14.6 in polar coordinates r and θ and setting $v_r = 2q/(\pi r)$, in which q is the weir loading, a differential equation is obtained that may be solved (Subcommittee on Sewage Treatment 1946) to give:

$$r = \left(C - \frac{2q}{v_s \pi}\theta\right)\sec\theta \tag{14.7}$$

in which C is a constant of integration equal to the x intercept of the particle trajectory. It can be shown that, with $C = q/v_s$, Equation 14.7 gives the limiting trajectory of particles that will just overflow the weir. The x and y intercepts of this trajectory are $M = q/v_s$ and $N = 2q/(\pi v_s)$, respectively. The volume within the limiting trajectory is given by the integral $\frac{1}{2}\int_0^{\pi/2} r^2 d\theta$ and is termed the ineffective volume by the Subcommittee on Sewage Treatment (1946).

An example was presented in the Subcommittee report in which a weir loading rate of 35,000 gpd/ft (434 cu m/day/m) and settling velocity of 5 ft/hr (1.5 m/hr) were shown to correspond to an ineffective volume of 670 cu ft/ft (62,000 l/m) of weir. This was seemingly a substantial value, particularly at a time (1946) when detention time was still accepted by some as a basic parameter in primary settling tank design. [Jin *et al.* 2000 specifically investigated the role of overflow rate and

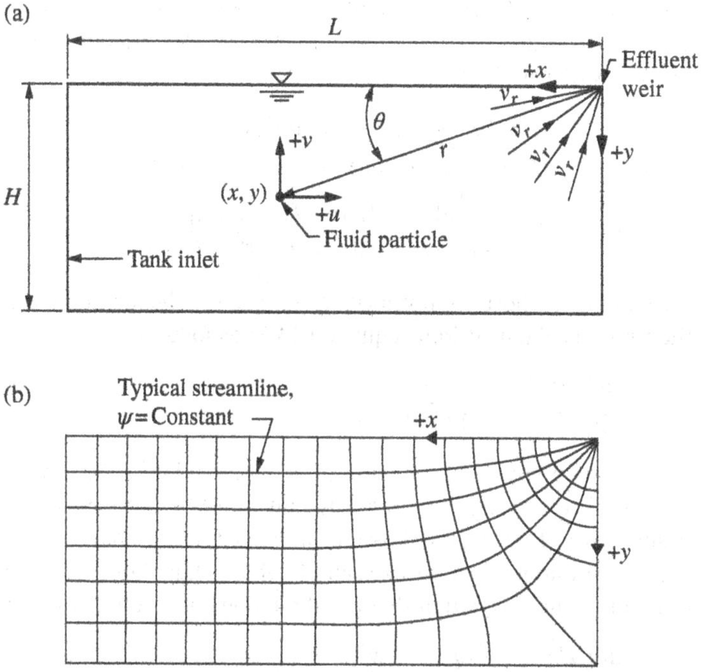

Figure 14.1 (a) Particle Settling in Tank with Radial Flow Pattern Assumed in Subcommittee on Sewage Treatment (1946); (b) Potential-flow Pattern Approximating Flow in Rectangular Tank (Graber 1974a)

detention time for settling tanks, concluding that "tanks with the same overflow rate...have almost the same removal efficiency no matter how large they are" and "no clear relationship exists between the detention time and removal efficiency."]

However, by calculating the intercepts of the limiting trajectory at the water surface and the plane of the weir (M and N, respectively), it was found by the author that the critical zone extends along the surface 39 ft (12 m) from the weir and extends vertically 25 ft (7.6 m) below the weir. For a tank 10 ft (3 m) deep, which is the average depth of tanks considered in the Subcommittee report, approximately one-half of the ineffective volume would lie below the floor of the tank.

The Subcommittee theory is obviously unable to account for the influence of the tank bottom on the flow pattern. Furthermore, were the tank bottom effect properly accounted for, the ineffective zone approach would still not be meaningful since much of this zone lies in a portion of the tank that is not significant in the sedimentation process (as is made apparent below). Therefore, the Subcommittee theory, which purports to establish the importance of weir loadings as a basic design parameter, is erroneous.

Potential-flow theory (Rouse 1959, Shames 1962) provides a sound theoretical basis for demonstrating the effect of weir loading on settling efficiency. The theory is applicable to two-dimensional flow in regions where large velocity gradients are not encountered (Shames 1962). For the turbulent flow of water, such gradients exist primarily in the immediate vicinity of solid boundaries. Potential-flow theory provides a reasonable approximation to the flow pattern in a rectangular settling tank under assumed conditions of uniform inflow. Although, for various reasons, uniform inflow is rarely achieved in practice (Camp 1936, p. 749; Burgess *et al.* 1960), the deviations therefrom are of little significance to the analysis presented herein.

In the case of a rectangular tank with a single effluent end weir, generalized results may be obtained by using a potential-flow solution obtained analytically as presented by Streeter (1948). Figure 14.1(b) depicts the corresponding potential-flow lines that result from a series of equal and equidistant "sinks" along the y axis. The stream function defined in Section 4.2 applied to the flow pattern of Figure 14.1(b) is given by (the signs are consistently and acceptably reversed from those used in Section 4.2):

$$\psi = C \tan^{-1} \frac{\tan[\pi y/(2H)]}{\tanh[\pi x/(2H)]} \tag{14.8}$$

where H = tank depth and C = constant of integration.

By setting $C = 2q/\pi$, the potential-flow solution for the rectangular tank is obtained. Taking partial derivatives of ψ in accordance with Equations 4.5a and b gives the velocity components u and v in terms of the x, y coordinates of Figure 14.1(b), as follows:

$$u = -\frac{q}{H}\frac{\{\sec[\pi y/(2H)]\}^2\{\tanh[\pi x/(2H)]\}}{\{\tanh[\pi x/(2H)]\}^2 + \{\tan[\pi y/(2H)]\}^2} \tag{14.9a}$$

$$v = -\frac{q}{H}\frac{\tan[\pi y/(2H)]\sec\{[\pi x/(2H)]\}^2}{\tanh\{[\pi x/(2H)]\}^2 + \tan\{[\pi y/(2H)]\}^2} \tag{14.9b}$$

Denoting the horizontal velocity component at the top and bottom of the tank inlet by $u(L, 0)$ and $u(L, H)$, respectively, their ratio is obtained from Equation 14.9a as follows:

$$\frac{u(L,0)}{u(L,H)} = \frac{1}{[\tanh(\pi L/2H)]^2} \tag{14.10}$$

This ratio is equal to 0.99, that is, the horizontal velocities are equal to within 1.0% at $L/H = 1.69$. Therefore, for $L/H \geq 1.69$, the flow pattern represented by Equation 14.9 corresponds closely to the uniform inlet flow assumed in Camp's ideal basin theory. Tesařík (1967) used a similar solution in his studies of flow in sludge-blanket clarifiers.

By substituting Equation 14.9a and 14.9b into Equation 14.5 and nondimensionalizing, one obtains a fairly complex first-order differential equation that has the following functional form:

$$\frac{d(y/H)}{d(x/H)} = F\left(\frac{q}{Hv_s}, \frac{x}{H}, \frac{y}{H}\right) \tag{14.11}$$

Numerical solutions to Equation 14.11 were obtained using the Runge–Kutta method (Hildebrand 1956). By starting at $(x, y) = (L, 0)$, the trajectory of particles entering at the top of the tank may be traced. The results for one particular tank are depicted in Figure 14.2. The trajectory of the particle that comes closest to the bottom rear corner of the tank corresponds to $q/Hv_s = 1/0.038$ for this particular tank having $L/H = 26.3$. As explained below, this value of q/Hv_s is within 0.1% of the value corresponding to ideal tank performance.

The author computed values of q/Hv_s corresponding to the particle that enters at the top of the tank and reaches the bottom rear corner for a range of tank length:depth ratios. The results are plotted in Figure 14.3, together with the line representing ideal settling tank theory for which $q/Hv_s = L/H$. The effect of weir loading q is seen to be negligible for

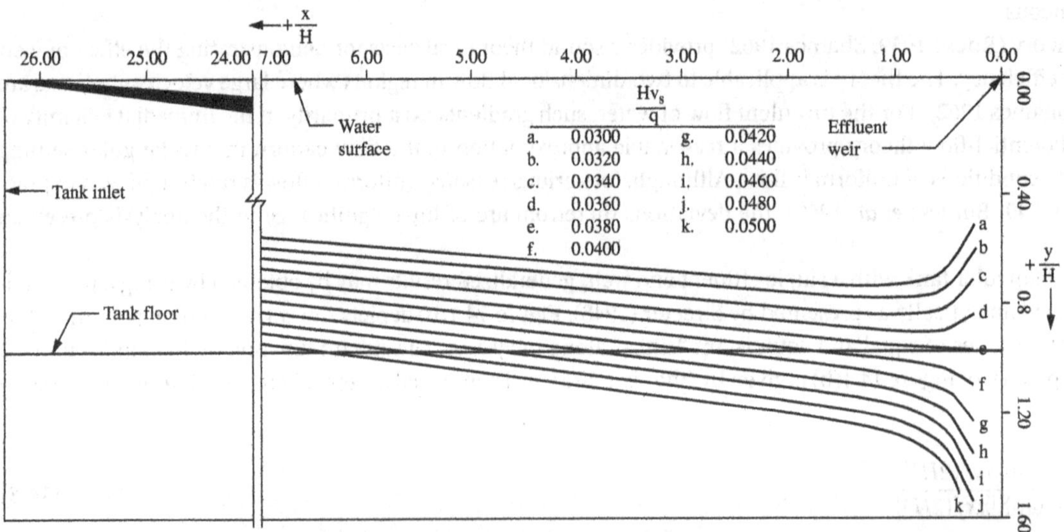

Figure 14.2 Trajectories of Particles Settling in Tank Having Flow Pattern of Figure 14.1b, L/H = 26.3 (Graber 1974a)

all practical values of L/H. Thus, the potential-flow analysis supports the experimental evidence that weir loadings are of no direct consequence in primary settling tank design.

An additional interesting result may be obtained by considering the approximate flow pattern derived by Yih (1959), presented in Section 4.3 and plotted in Figure 4.10. That flow pattern accounts for the eddy or dead zone that exists in the bottom rear corner of a rectangular settling tank. For that flow pattern, the results of an analysis similar to that presented above are also plotted in Figure 14.3, which shows that the settling efficiency is actually better than that predicted by "ideal" sedimentation theory.

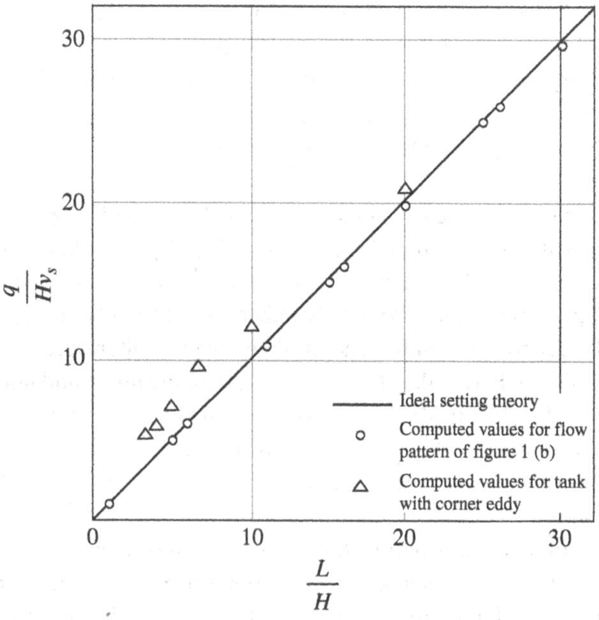

Figure 14.3 Effect of Weir Loading and Corner Eddy on Settling Efficiency (Graber 1974a)

14.3.1.3 Subsequent Follow-Up

So what has transpired since the author's initial challenge (Graber 1974a) of the notion of limiting weir loadings that had been enshrined in various "standards"? *Design of Municipal Wastewater Treatment Plants* (WEF/ASCE 1998) cited Graber (1998) and concurred that "Weir rates have little effect on the performance of primary sedimentation tanks, especially with side wall depths in excess of 3.7 m (12 ft)." But that publication still found it necessary to follow that statement with a table summarizing typical weir rates for primary sedimentation tanks.

In commenting (Graber 2005) on a reasonable application of computational fluid dynamics (CFD) to sedimentation tanks, the author noted that "CFD can be used in questionable ways similar to the manner in which experimental studies have been misused in investigating sedimentation (e.g., the reliance on short-circuiting investigations as mentioned above)." The author cited an example of a paper in which CFD was used to estimate the detention time (time to first appearance of fluid "marker particles") to assess alternative clarifier enhancements, thus focusing solely on "hydraulic efficiency." The shortcomings of such an approach should be apparent from the above.

More recently, the successor to the "Ten State Standards" (Great Lakes – Upper Mississippi River Board of State and Provincial Public Health and Environmental Managers 2014, Section 72.43) has only slightly revised the weir loading "standards," still erroneously, to the following: for an average plant capacity equal to or less than 1 mgd (3785 m^3/day), a loading rate at design peak hourly flow of 20,000 gpd/lin ft (250 m^3/m•d) and a loading rate for an average plant capacity greater than 1 mgd (3785 m^3/day), a loading rate at design peak hourly flow of 30,000 gpd/lin ft (375 m^3/m•d). Once again, there is no justification for such a requirement.

Section 14.3 discusses a related problem pertaining to a misguided plan for controlling algae in a pond by installing a weir on a pond outlet.

14.4 Mixing

Some of the discussion under this heading is taken from Graber (1994, pp. 545–547). G-value was proposed by Camp (1969) for the design of mixing vessels, and it is still used occasionally for that purpose. Such vessels include rapid mixers (preceding flocculators in water treatment facilities), anaerobic sludge digesters, and mechanical aerators in the activated sludge process. The concepts of dimensional analysis are helpful in assessing the applicability of G-value to mixing.

As discussed in Chapter 5, dimensional analysis is a well-proven technique whereby the significant variables in a process are grouped into a set of dimensionless parameters which completely describe the process. If all but one of the dimensionless parameters are equal in a model and full-scale process and complete geometric similarity is maintained, then similitude is said to exist and the process results (appropriately defined) will be equal.

Chemical engineers have used dimensional analysis to determine power requirements in batch mixing (i.e., fill-and-draw mixers in which there is no throughput during mixing). The resulting variables (Johnstone and Thring 1957, Chapter 14; Bates *et al.* 1966) are given in the following functional relationship:

$$\frac{Pg_c}{\rho N^3 D^5} = \Phi\left(\frac{D^2 N\rho}{\mu}, \frac{DN^2}{g}, \text{impeller geometry, tank geometry}\right) \tag{14.12}$$

in which P is the power consumption of the impeller, g_c is Newton's law conversion factor (e.g., 1.0 lbm ft/sec^2-lbf), ρ is fluid density, N is rotational speed of impeller, D is impeller diameter (or any other characteristic dimension), μ is absolute fluid viscosity, and g is gravitational acceleration. The first and second dimensionless terms on the right-hand side of the above equation are the Reynolds Number and Froude Number, respectively. The term left of the equal sign is referred to as the Power Number in the chemical engineering literature.

When there is flow throughput, as in many mixer and flocculation applications, dimensional analysis yields the additional variable TN, in which T is theoretical detention time (tank volume divided by volumetric rate of flow). The interpretation is that, if the impeller geometry, tank geometry, and parameters $D^2N\rho/\mu$, DN^2/g, and TN are the same in model and prototype, then the power parameter will be the same. This allows the power required for a prototype mixer to be predicted from a laboratory-scale model.

Under certain circumstances, some of the dimensionless parameters may be deleted, simplifying the situation. For example, for closed mixers or well-baffled open mixers, in which swirl is avoided so that any free surface remains essentially horizontal, the pressure parameter DN^2/g may be deleted (Johnstone and Thring 1957, p. 175). If, in addition, fully developed turbulence exists, the parameter $D^2N\rho/\mu$ may be deleted (Johnstone and Thring 1957, p. 175). The validity of this latter deletion was noted by Camp (1969) in his laboratory experiments with mixing tanks. A further simplification may be made if the pump circulation rate is greater than about five times the throughput rate (Olson and Stout 1966, p. 120). Then the power requirement is not affected by the throughput, and the parameter TN may be deleted. Therefore, in a closed or well-baffled open mixer with high rate of pumpage and fully developed turbulence, the dimensionless parameters reduce to:

$$\frac{Pg_c}{\rho N^3 D^5} = \Phi(\text{impeller geometry, tank geometry}) \tag{14.13}$$

For geometrically similar mixers, this reduces to:

$$\frac{Pg_c}{\rho N^3 D^5} = \text{constant} \tag{14.14}$$

The relationship between the above and G-value can now be established. Considering the case of a well-baffled or closed vessel, the pertinent parameters can be expressed in the following functional form:

$$\psi\left(\frac{Pg_c}{\rho N^3 D^5}, \frac{D^2 N\rho}{\mu}, TN, \text{impeller geometry, tank geometry}\right) = 0 \tag{14.15}$$

Multiplying the square root of the first two parameters by the third parameter results in a new parameter $[Pg_c/(\mu D^3)]^{1/2}T$, which (since tank volume is proportional to D^3 when geometric similarity is maintained) is equivalent to GT. In order to retain the density and rotational terms, the second and third parameters must still be retained. The new functional relationship becomes:

$$\psi\left(GT, \frac{D^2 N\rho}{\mu}, TN, \text{impeller geometry, tank geometry}\right) = 0 \tag{14.16}$$

In this case, since the density and impeller-speed terms are excluded from the parameter GT, and since both of these terms are important in turbulent mixing, the parameters $D^2N\rho/\mu$ and TN cannot be eliminated as before. In fact, even though the viscosity term is not significant in determining the impeller power for fully developed turbulent flow, it cannot be deleted from Equation 14.16 because the remaining terms D, N, ρ, and T cannot form a dimensionless grouping (due to the presence of mass in the density term). Therefore, Equation 14.16 cannot be reduced further and indicates that for GT to be used as a parameter for scaling the hydraulics of a mixer, two other parameters, $D^2N\rho/\mu$ and TN, must also be held constant. This is in contrast to Equation 14.14, which is a much more convenient result. Therefore, the G-value, as represented in the parameter GT, has little merit for scaling the hydraulic performance of a mixer or flocculator.

Up to this point, only hydraulic performance has been considered. In the process of rapid mixing (and flocculation, as discussed below), the addition of chemicals and the physico-chemical reactions have essentially no effect on the hydraulics. Therefore, the functional relationships presented above remain valid. However, variables related to the physico-chemical reactions must also be considered. In particular, the parameter *TN* affects the process not only insofar as it affects the power used but also as it affects the residence time distribution of the chemicals, floc, etc.

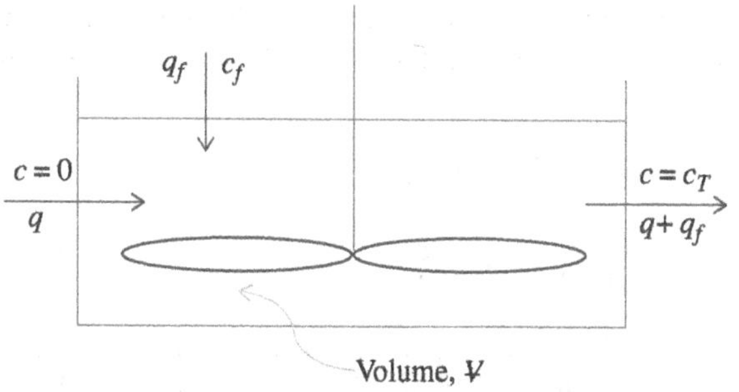

Figure 14.4 Single-Stage Mixer

Let us consider the ramifications of all of this for the process of rapid mixing. Even though the reaction is known to be very fast, it has been shown (Vrale and Jorden 1971, Stenquist and Kaufman 1972) that backmixing can have a detrimental effect on the coagulation reaction, and, therefore, the residence time distribution is significant. However, Stenquist and Kaufman (1972, pp. 91–92) imply that, given the inherent limitations of the impeller-type rapid mixer, any difference in residence times may not have much influence, so the parameter *TN* would not be of great significance. Therefore, subject to the assumptions noted above (all of which should be capable of being satisfied), it should be possible to scale rapid mixer performance by means of Equation 14.14. That is, provided that complete geometric similarity is maintained between model and prototype (or between different full-scale units), the importance of which cannot be overemphasized. In scaling rapid mixing, purely chemical variables such as water temperature and pH should also be held constant or appropriate adjustments made.

We will derive the "perfect" mixing relationships for single- and two-stage mixers with abrupt, continuous addition of tracer or dilute liquid chemical. The perfect mixing assumption implies that the outflow concentration equals the uniform concentration inside the mixer. Notation is shown in Figure 14.4, in which c, q, and V are concentration, flow rate, and volume, respectively; and subscripts f and T denote the added tracer/chemical and outflow, respectively. A mass balance gives:

$$q_f c_f = \left(q + q_f\right)c + V\frac{dc}{dt} \tag{14.17a}$$

$$\frac{dc}{dt} + \frac{\left(q + q_f\right)}{V}c = \frac{q_f c_f}{V} \tag{14.17b}$$

in which the second equation is a rearrangement of the first. Denoting by c_∞ the steady-state concentration at $t \to \infty$, the above equation gives:

$$c_\infty = \frac{q_f c_f}{\left(q + q_f\right)} \tag{14.18}$$

Using the above equations, we can then write:

$$\frac{d(c/c_\infty)}{dt} + \frac{\left(q + q_f\right)}{V}\frac{c}{c_\infty} = \frac{q_f}{V}\frac{c_f}{c_\infty} \tag{14.19}$$

$$\frac{d(c/c_\infty)}{dt} + \frac{\left(q + q_f\right)}{V}\frac{c}{c_\infty} = \frac{\left(q + q_f\right)}{V} \tag{14.20}$$

The above equation can be solved by the method of integrating factor (Hildebrand 1962, §1.4) together with the initial condition $c/c_\infty = 0$ at $t = 0$ to give:

$$\frac{c}{c_\infty} = 1 - \exp\left(-\frac{q + q_f}{V}t\right) \tag{14.21}$$

The above equation is the same as that given by Danckwerts (1953) for $q_f \ll q$.

Continuing with this latter inequality for two tanks in series (which can be readily extended if q_f is significant), and denoting the effluent from the first tank by c_1 and that from the second tank by c_2, we have:

$$qc_1 = qc_2 + \mathcal{V}\frac{dc_2}{dt} \tag{14.22}$$

$$\frac{dc_2}{dt} + \frac{q}{\mathcal{V}}c_2 = \frac{qc_1}{\mathcal{V}} \tag{14.23}$$

$$\frac{d(c_2/c_\infty)}{dt} + \frac{q}{\mathcal{V}}\frac{c_2}{c_\infty} = \frac{q}{\mathcal{V}}\frac{c_1}{c_\infty} = \frac{q}{\mathcal{V}}\left[1 - \exp\left(-\frac{q}{\mathcal{V}}t\right)\right] \tag{14.24}$$

Solving again by the method of integrating factor,

$$\frac{c_2}{c_\infty} = 1 - \left(1 + \frac{q}{\mathcal{V}}t\right)\exp\left(-\frac{q}{\mathcal{V}}t\right) \tag{14.25}$$

Calculations using Equations 14.21 and 14.25 give results matching exactly the numerical values tabulated by an equipment manufacturer (Lightnin®) whose methods are proprietary.

Example 14.1

Twenty-eight single-state rapid mixers are to be provided at the Springfield, Massachusetts Water Treatment Plant, each for a tank of depth 10.48 ft at maximum water surface elevation and plan dimensions of 9 × 8.5 ft. Total plant design flow rate is 78,680 gpm, and for this system $q_f \ll q$. Determine the time required for 99% mixture uniformity. Also, determine the time required for 99% mixture uniformity with two-stage rapid mixers having the same total volume as in the single-stage case.

Solution

For the single stage, design-flow rate per mixer is 78,680/28 = 2,810 gpm. Each mixer has volume $\mathcal{V} = 10.48(9)(8.5) = 802$ cu ft. Then $q/\mathcal{V} = 2810/[802(7.48)] = 0.469\ \text{min}^{-1}$. For this case, Equation 14.21 gives $c/c_\infty = 1 - \exp(-0.469t)$. For $c/c_\infty = 0.99$, we have $t = [\ln(1-0.99)]/(-0.469) = 9.82$ min. For the two-stage case, Equation 14.25 gives $c_2/c_\infty = 1 - [1 + 0.469t)] \exp[-0.469t)]$. Then, for $c_2/c_\infty = 0.99$, we have by trial-and-error $t = 8.29$ min as the total time for the two stages.

Air mixing has been suggested to reduce or prevent the deposition of solids in chlorine contact tanks. However, Graber (1972) has suggested that this does not seem justified. The provision of duplicate contact facilities, such as those required by the U.S. Environmental Protection Agency, makes tank draining, inspection, and sludge removal a simple operation, even while maintaining one unit in service. Plug-drain valves may be set in slight depressions at the end of shallow longitudinal gutters to which the tank floor is pitched to allow complete drainage. In unbaffled aeration basins, Murphy and Boyko (1970) found that the dispersion index increased with increased air-flow rate. The beneficial effects of air mixing in a baffled model tank critiqued by Graber (1972) stem from the resulting reduction of dead volume. Air mixing in a chlorine contact tank might more justifiably be done in conjunction with postaeration to increase effluent dissolved oxygen (DO), as discussed in Section 14.7.

Static mixers are useful in certain applications. With no moving parts, under suitable circumstances, static mixers offer reduced power requirements, simple scale-up, and, relatively little maintenance. Manufacturers include Kenics/Chemineer Inc., Dayton, Ohio; Koch Engineering Company Inc., Wichita, Kansas; and Sulzer Engineering Company, Winterthur, Switzerland. The author specified static mixers constructed of PVC for the Cranston, Rhode Island Wastewater Treatment Plant in conjunction with feed systems of various chemicals and capacities.

Mixing in pipes requires estimation of distance to which sufficient mixing has occurred for accurate sampling. French (2015) derived criteria for effective mixing based on the 3-sigma statistical rule (Pukelsheim 1994). For the case of a single longitudinal dosing point centrally located in a circular conduit, he gives the following equation for the most-conservative mixing distance (A, R, D, and f are, respectively, pipe area, hydraulic radius, diameter, and Darcy–Weisbach friction factor):

$$x_m = 7.5A/\left(Rf^{0.5}\right) = 7.5\pi D/f^{0.5} \tag{14.26}$$

Antifoulant chemicals at seawater intakes are most effectively applied just around the inside wall of the pipe; French (2017) addresses the application of ring diffusers for that purpose.

Example 14.2

MetroWest Water Supply Tunnel of the Massachusetts Water Resources Authority (MWRA) supplies water to the greater Boston area. The author was tasked with designing fluoride dosing in the tunnel. Following a stilling basin, a 12 foot horizontal tunnel of 340 ft length leads to a 14 ft diameter, 262 ft-long vertical shaft, and thence to a 17 mile-long, 14 ft diameter horizontal tunnel. Tunnels are bored and shotcrete-lined (roughness $e \cong 0.0004$ ft). A single dosing pipe was to be provided at the 12 foot diameter horizontal tunnel. Design is to be based on the initial minimum flow of 300 mgd. Because the author was asked to estimate the required distance before the dosing arrangement had been designed, a single-point discharge was assumed. The nearest sampling location is at the top of the Hosmer riser that supplies the Town of Southborough, approximately 4 miles beyond the abovementioned vertical shaft. Determine if that is a suitable location for the sampling point.

Solution

For conservatism, a 12 ft diameter pipe is assumed throughout. It is then reasonable to neglect the effect of bends in determining mixing distance. Pertinent terms are $A = (\pi/4)(12)^2 = 113.1$ sq ft, $Q = 300(1.547) = 464$ cfs, $V = 464/113.1 = 4.10$ ft/sec, $\mathbf{R} = VD/\upsilon = 4.10(12)/10^{-5} = 4.93(10)^6$. For that **R** and $\varepsilon/D \cong 0.0004/12 = 0.000, 033.3$, the Moody Diagram gives $f \cong 0.0105$. Equation 14.26 gives $x_m = 7.5\pi(12)/(0.0105^{0.5}) = 2760$ ft (0.506 miles). This is within the 4 mile distance to the sampling point.

14.5 Chemical Contact

Numerous processes require a period of chemical contact following mixing in order for rate reactions to take place. Chlorination for disinfection of wastewater effluents and subsequent dechlorination are the predominant applications in which contact reactors must be specifically designed for this purpose, and those applications will be emphasized here. The importance of good mixing prior to chlorination and dechlorination has been discussed in the previous section. Application of the principles presented to other applications should be apparent.

It has long been acknowledged that plug flow provides the most efficient chemical contact. The significance of short-circuiting in contact facilities can be appreciated by considering the combined effects of short-circuiting and the dynamics of the chlorine–microorganism reaction (Graber 1972). In a study on the "Investigation of the Laws of Disinfection," Harriet Chick (1908; as cited by Fair and Geyer 1968, p. 798) proposed a first-rate reaction (with negative K) for the number of organisms N remaining of an initial number N_o after time t, i.e., $N = N_o e^{-Kt}$. Although this relationship is still being improved upon (see below), it provides a useful basis for gaining a qualitative understanding of short-circuiting effects on chlorination effectiveness. Consider a contact tank having a flow-through curve in which c is the concentration of a conservative tracer. The area under the flow-through curve is given by $\int_0^\infty c\,dt$ and is proportional to the mass of the conservative tracer instantaneously introduced at the tank inlet or to the number of specific microorganisms recovered at the tank outlet following instantaneous introduction at the tank inlet in the absence of chlorination. When chlorinating, the number of specific microorganisms recovered under otherwise similar test conditions is proportional to $\int_0^\infty e^{-Kt} c\,dt$. If plug flow existed, this latter quantity would be equal to $e^{-KT} \int_0^\infty c\,dt$ in which T is the theoretical detention time. Following Warriner (1967), if T_c is the contact time that would produce the same amount of microorganism recovery under plug-flow conditions as in the actual tank, then

$$e^{-KT_c} \int_0^\infty c\,dt = \int_0^\infty e^{-KT_c} c\,dt \tag{14.27}$$

Using a numerical form of Equation 14.27 to evaluate his data, Warriner concluded that, for virus disinfection, contact basins should be twice as large as would be considered necessary if plug-flow conditions existed. A similar method was used by Mangelson and Watters (1972) for evaluating stabilization pond performance. An analogous numerical or graphical procedure may be used to determine the additional volume required for adequate contact based on any relationship for the chorine–microorganism reaction.

14.5.1 Dispersion

Graber (1972) mentions various methods that have been used to evaluate tank flow-through curves. The author strongly favors the variance of the flow-through curve and the dispersion index to which the variance is related, as described by Thirumurthi (1969). The term "dispersion," as used by the hydraulic engineer, is the combined effect of relative displacement associated with velocity gradients and molecular or turbulent diffusion (Holley 1969). For tanks, the dispersion model allows the evaluation of flow-through curves and tank design to be related to the hydraulic engineer's understanding of the basic flow phenomena involved. Marske and Boyle (1973) support the dispersion model based on their extensive field evaluations and provide additional useful design information.

Using one-dimensional dispersion theory and assuming a first-order chemical reaction, Wehner and Wilhelm (1956) derived the following relationship for a continuous flow-through reactor:

$$\frac{C_e}{C_i} = \frac{4a\exp[1/(2d)]}{(1+a)^2\exp[a/(2d)] - (1-a)^2\exp[-a/(2d)]} \tag{14.28}$$

in which C_e = effluent concentration, C_i = influent concentration, dispersion number $a = \sqrt{1+4Ktd}$, $d = D_L/(UL)$, K = first-order reaction constant, D_L = longitudinal dispersion coefficient, U = average velocity, and L = length of reactor. The above equation was used by Murphy and Timpany (1967) in their study of mixing in aeration tanks and by Thirumurthi (1969) and Polprasett *et al.* (1983) in their studies of waste stabilization ponds. Murphy and Timpany provide a particularly good discussion of short-circuiting models and the influence of dispersion (plug flow, complete mixing, etc.) on efficiency of the activated sludge process. Useful, more recent work on stabilization pond design is given by Uhlmann *et al.* (1983), limited to ponds in series.

L'Hospitale's Rule can be used as a check to demonstrate that for $d \to 0$ and $x \to \infty$ Equation 14.28 correctly gives $C_e/C_i \to 1$ (an exercise left to the reader). For plug-flow conditions corresponding to $d = 0$ and with T denoting the plug-flow travel time-through the reactor, Equation 14.28 gives:

$$\frac{C_e}{C_i} = \exp(-KT) \tag{14.29}$$

Setting $T = T_c$ in Equation 14.29, C_e/C_i may be eliminated from Equations 14.28 and 14.29, and the resulting equation may be rearranged as follows:

$$\frac{T_c}{T} = \frac{1}{KT}\ln\left\{\frac{(1+a)^2\exp[a/(2d)] - (1-a)^2\exp[-a/(2d)]}{4a\exp[1/(2d)]}\right\} \tag{14.30}$$

Equation 14.30 is plotted in Figure 14.5 and demonstrates the effect of dispersion.

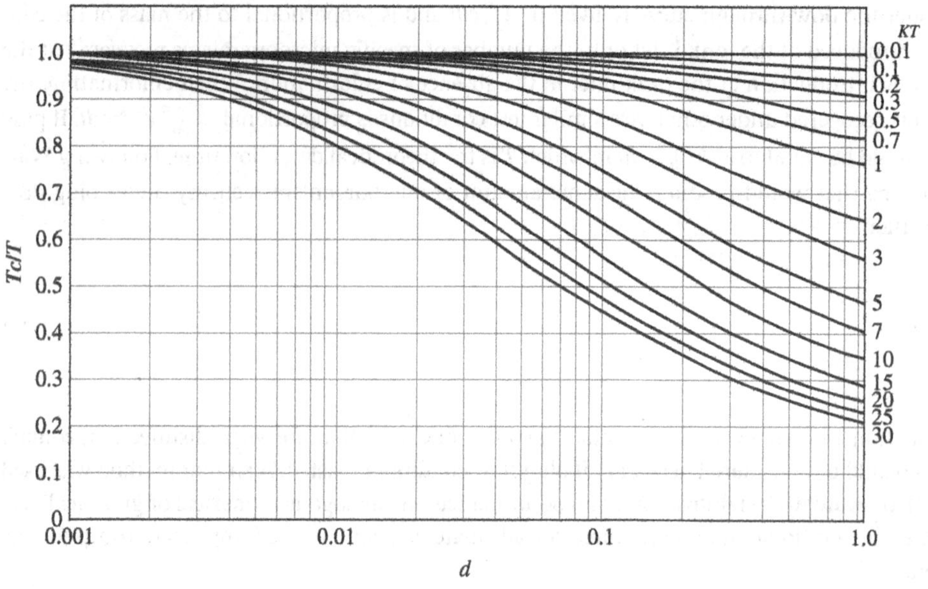

Figure 14.5 Ratio of Effective:Theoretical Contact Times vs. Dispersion Number and Dimensionless Reaction Rate Constant (Graber 1972)

14.5.2 Dispersion in Pipes

Taylor (1953, 1954) analytically derived and experimentally verified the dispersion coefficients for laminar and turbulent pipes. Taylor's equation for turbulent flow is:

$$D_L = 5.05 D u_* \tag{14.31}$$

in which D is the pipe diameter and u_* is the shear velocity defined by $\sqrt{\tau_o/\rho}$ where τ_o is the average shear stress on the pipe wall. Because $\tau_o = (f/4)(\rho U^2/2)$ in which f is the Darcy–Weisbach friction factor, Equation 14.31 may be used to express the dispersion number for turbulent pipe flow as follows:

$$d = 5.05 \frac{D}{L} \sqrt{\frac{f}{8}} \tag{14.32}$$

Example 14.3

The author proposed twin outfall pipes for chlorine contact of tertiary effluent as part of an upgrade of a water pollution control facility for the Massachusetts Correctional Institution at Bridgewater, Massachusetts. A distance of 375 ft (114 m) was available from the last treatment process to the point of discharge into Saw Mill Brook. Determine pipe sizes that will provide 30 minutes of detention time at the average design flow of 380,000 gpd (1440 cu m/day).

Solution

Thirty minutes detention will nominally be provided by pipes of diameter:$D = \sqrt{4QT/(\pi L)} = \sqrt{4\left(\frac{380,000}{2}\frac{\text{gal}}{\text{day}}\right)(30\text{min})\left[\frac{1}{24(60)}\frac{\text{day}}{\text{min}}\right]\left(\frac{1}{7.48}\frac{\text{ft}^3}{\text{gal}}\right)/[\pi(375\ \text{ft})]} = 1.340\ \text{ft}\ 12 = \ 16.08$ in. (the reader should verify the above relationship). Vitrified-clay pipe of 18 in. (1.5 ft, 45.7 cm) diameter is selected (actual inside diameter equal to nominal diameter per ASTM Specification C-13), giving the average velocity as: $U = Q/A = \frac{380,000\ \text{gal/day}}{2} \frac{4}{\pi(18)^2\text{in}^2}\frac{1\ \text{ft}^3}{7.48\ \text{gal}}\frac{1\ \text{min}}{60\ \text{sec}}\frac{144\ \text{in}^2}{\text{ft}^2}\frac{1\ \text{day}}{24(60)\ \text{min}} = 0.167$ fps (5.09 cm/ sec), which results in a Reynolds Number $\mathbf{R} = UD/\upsilon = \ 0.167(1.5)/10^{-5} = 25,000$. At this Reynolds Number, the flow will be predominately turbulent, and Equation 14.32 is applicable. With relative roughness $\varepsilon/D = \ 0.01/1.5 \ = 0.00667$, the Moody Diagram (Figure 5.1) gives a friction factor $f = 0.032$. The calculated dispersion number from Equation 14.32 is:

$$d = 5.05 \frac{1.5}{375}\sqrt{\frac{0.032}{8}} = 0.001278$$

From Figure 14.5 at this value of d, T_c is within 99% of T for KT values up to 10. A value of $KT = 10$ corresponds to a kill of $(1 - e^{-10}) * 100 = 99.99\%$. Therefore, negligible short-circuiting is expected to occur with this design. Additional advantages of this arrangement include a 25% savings in cost as compared to a contact tank and single 12 in. (30.5 cm) outfall, as well as improved initial mixing of chlorine and wastewater, which is of considerable importance. It is of interest to note that with 10 ft (3.05 m) contact pipes of comparable size to the contact tank considered, Equation 14.32 gives $d = 0.259$, which from Figure 14.5 gives $T_c/T = 0.52$ for $KT = 10$. The 10 ft (3.05 m) contact pipes would have approximately 100 times more microorganisms in the effluent than the 375 ft (114 m) pipes, based on $KT = 10$.

For the aforementioned Bridgewater facility, the author provided the reference (Graber 1972) to convince the Massachusetts Department of Environmental Protection (MDEP) to accept the first application of the use of outfall pipes for chlorine contact. That system has functioned well. MDEP subsequently accepted chlorine contact pipes for projects per the author's recommendations for the Haverhill, Massachusetts Wastewater Treatment Plant (Ringer 1974) and the Plymouth Manomet Wastewater Treatment Plant. Also, again citing Graber (1972), the author convinced the MWRA to include vanes in the design of the chlorine contact tanks for the Deer Island Wastewater Treatment Plant, Boston Harbor, Massachusetts. The author similarly convinced MWRA to dechlorinate within the 24 ft-diameter 9.5 mile outfall pipe following dechlorination (and using sodium bisulfite for that purpose as a safer alternative to gaseous sulfur dioxide). It is noteworthy that with pipe contactors, chlorination-dechlorination can become more economically favorable than ozonation.

14.5.3 Dispersion in Open Channels

Sooky (1969) has elucidated the difficulties associated with predicting the dispersion coefficient for open channels of finite width. He suggested a relationship of the form

$$D_L = CRu_* \tag{14.33}$$

in which C is a function of channel geometry and the Reynolds Number, as yet undetermined for an open rectangular channel. Following a similar procedure as that employed in going from Equation 14.31 to Equation 14.32, Equation 14.33 may be used to obtain an expression for the dispersion number

$$d = C\left(\mathbf{R}, \frac{W}{H}\right)\sqrt{\frac{f}{8}\frac{R}{L}} \tag{14.34a}$$

in which $C(\mathbf{R}, W/H)$ denotes a proportionality factor which is a function of the Reynolds Number and channel width:depth ratio. For "fully rough" turbulent flow, occurring at a high Reynolds Number, f in Equations 14.32 and 14.34a becomes a constant (Committee on Hydromechanics 1963).

The effect of Reynolds Number on the proportionality factor C of Equation 14.34a is caused mainly by the high-velocity gradients existing in the viscous sublayer at the walls of channels in turbulent flow. Observed discrepancies between Equation 14.32 and experimental studies are also attributed to the viscous sublayer. This effect persists under highly turbulent conditions, decreasing as the Reynolds Number increases (Yen 1969). Because the Reynolds Number of a prototype exceeds that of the model when Froude Number is used as the scaling parameter, one would expect short-circuiting in the prototype to be less than that in the model.

It is suggested (Graber 1972) that C in Equation 14.34a be tentatively taken equal to 20, based on substitution of $D = 4R$ in Equation 14.32, giving:

$$d = 20UL\sqrt{\frac{f}{8}\frac{R}{L}} \tag{14.34b}$$

Such an extrapolation from pipes to open channels has been justly criticized (Sooky 1969), and fundamental research directed toward quantifying dispersion phenomena in rectangular man-made channels should be performed. Additional perspective is provided by Holley and Harleman (1965, Chapter 3).

Example 14.4

The author reviewed a proposal to construct a concrete chlorine contact tank consisting of six 40 ft passes each of 4.5 ft width and 10 ft depth (Hanover, New Hampshire Water Pollution Control Plant). Average design wastewater flow is 2.3 mgd (×1.547 = 3.56 cfs). Determine the effective contact time and its efficacy.

Solution

The hydraulic radius and velocity are given, respectively, by R = 4.5(10)/[2(10) + 4.5] = 1.84 ft and V = 3.56/[4.5 (10)] = 0.079 ft/sec. The Reynolds Number is $\mathbf{R} = 4RV/\nu = 4(1.84)(0.079)/10^{-5}$ = 58, 144, so the flow is predominately turbulent. The Darcy–Weisbach friction factor $f = 117n^2/R^{4/3} = 117(0.013)^2/(1.84)^{4/3} = 0.0088$. The dispersion index is given by Equation 14.34b: $d = 20\sqrt{0.0088/8}[1.84/6(40)] = 0.0051$.

From Figure 14.5, at this value of d, T_c is within 95% of T for KT values up to 10. A value of $KT = 10$ corresponds to a kill of $(1 - e^{-10}) * 100 = 99.99\%$. Therefore, negligible short-circuiting is expected to occur with this design. It is noteworthy that Marske and Boyle (1973) suggest that for chlorine contact tanks of rectangular cross section, the length:width ratio should be greater than 40 if feasible for good chlorination efficiency; that ratio is our case if 240/4.5 = 53.3.

It is also noteworthy from Trussell and Chao (1977) that little additional disinfection is achieved by reducing the dispersion index below 0.01.

14.5.4 Relation to Earlier Work

Camp (1936) reported a correlation between short-circuiting and a modified form of the Froude Number, which he defined as $V^2/(gR)$. He observed that as $V^2/(gR)$ increased, the time of initial appearance of a tracer and the time to the centroid of the flow-through curve approached the theoretical detention time. His observations led him to suggest (Camp 1946b) that long, narrow basins are superior with respect to short-circuiting. It may be seen from Camp's data that an increase in $V^2/(gR)$ from $7.5(10)^{-7}$ to $4.53(10)^{-5}$ corresponds to any increase in L/R for first appearance from approximately 20 to 10,000. Of further interest is the fact that the parameter V/R was used as the basis for reducing short-circuiting in early sedimentation basin design (Camp 1936). The parameter V/R is equal to $(1/T)(L/R)$. Because sedimentation basins were designed for a specified detention time T prior to general adoption, as a result largely of Camp's efforts, of the overflow rate for this purpose, the design parameter V/R was directly related to L/R in Equation 14.34a.

14.5.5 Baffling and Dead Zones

The value of baffling is apparent from Equation 14.34a: baffling decreases the hydraulic radius R and increases the flow path L, decreasing the dispersion number. However, baffling accomplishes this at the expense of effective tank volume because of the dead zones which result. These dead zones may occupy as much as 40% of the tank volume (Louie and Fohrman 1968). Referring to Figure 14.5, the decrease in d caused by baffling is accompanied by a decrease in T, the combined effect of which may or may not decrease rather than increase T, but will certainly not permit the value of decreasing d to be fully realized. Thackston and Schnelle (1970) have studied the effects of dead zones on the flow-through curve and related these effects to the dispersion model. Using hydraulic measurements or predicted hydraulic behavior in the absence of tracer tests, their procedure should permit prediction of the flow-through curve and hence T_c by Equation 14.27.

The presence and extent of dead zones in existing contact tanks can often be ascertained when the tank is dewatered. Fine sediments will generally cover the tank bottom except where dead zones occur. This may seem to be the opposite of what one would expect. Note, however, that the overflow rate is significantly below that in final settling tanks, and the velocity will not scour settled solids. Thus, solids will settle out in the active portion of the tank and will not enter dead zones where they could settle out.

Claims that mixing the contents of a chlorine contact tank improves performance by providing more intimate contact between the microorganisms and chlorine or the beneficial effect of turbulence created by baffling have been refuted by Camp (1962) and Graber (1972). Camp's studies indicate little increase in coliform kill as a result of increased violence from mixing with raw wastewater; the effect on treated wastewater should still be smaller.

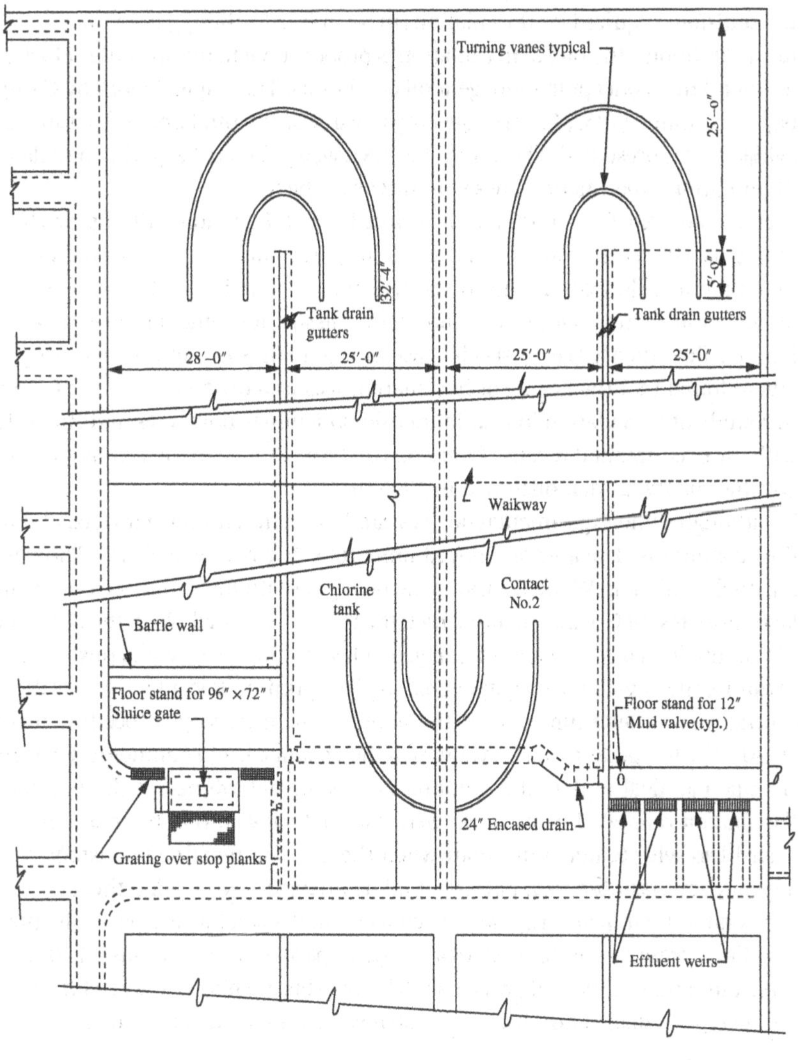

Figure 14.6 Multipass Chlorine Contact Tank (Graber 1972)

14.5.6 Multipass Contact Tanks

Unbaffled basins of sufficiently large L/R to minimize dispersion would be ideal for processes such as chlorine contact, but not always practical. A practical compromise is the multipass configuration. The chlorine contact tank shown in Figure 14.6, designed by the author for New York City's Tallman Island Water Pollution Control Facility, is an example of the multipass arrangement. Turning vanes are provided at the bends to minimize flow separation and the attendant dead volume and to establish a nearly plug-flow pattern at the head of each pass. The flow is distributed uniformly across the entrance to the first pass by baffling in accordance with the principles of dispersion conduits (Camp and Graber 1968) and the experimental studies of Kleinschmidt (1961). An interesting retrofit with baffle modifications that greatly reduced short-circuiting is discussed by Dismukes (1966). More recently, hydraulic modeling using CFD has proven useful for improving existing contact tanks and designing new ones (Hannoun *et al.* 1998).

14.6 Flocculation and Floc Breakup

14.6.1 Flocculation

Flocculation is the process by which particulate matter in water or wastewater is caused to collide and form a smaller number of "floc" particles conditioned for subsequent processing (usually sedimentation and/or filtration). As used herein, the term "flocculation" refers to the agglomeration of particles by hydraulic gradients as distinct from "coagulation" by which process the particles are destabilized by physico-chemical reactions. Normally, coagulation occurs in rapid-mixing basins immediately upstream of the flocculation units. Much of the discussion under this heading and under **Floc Breakup** below is taken from Graber (1994, pp. 547–551). Camp and Stein (1943) noted the significance of velocity gradients in promoting the collisions required for flocculation (more precisely, the agglomeration portion of the flocculation process). Under certain ideal conditions, the rate of flocculation is proportional to the absolute velocity gradient at a point. Camp and Stein proposed to extend that concept to more general conditions. The original work by Camp and Stein (1943) was followed by additional papers by Camp (1955, 1968) suggesting G-value as the fundamental parameter for flocculation design, but no experimental evidence was presented. The author has reviewed published experimental data and was responsible for instigating one set of illuminating experiments. These are discussed next.

Under laminar-flow conditions in a closed vessel, Equation 3.17 suggests that power dissipation is associated with velocity gradients. However, it does not suggest a simple relation. The author has evaluated two experimental studies of flocculation under laminar flow conditions. It is interesting to note that in the case of one-dimensional laminar flow ("Couette" flow as discussed in Section 4.4), Camp and Stein's dissipation function does reduce to the same formula as that of conventional laminar flow theory (i.e., a special case of Equation 3.16). In a study using a Couette flow device, Swift and Friedlander (1964) found that the degree of flocculation was proportional to GT, as Camp and Stein's theory would predict. However, in a study of flocculation in multidimensional laminar flow (Ives and Bhole 1973, 1977), G-value theory did not agree well with the experimental results. The latter can be attributed to the fact that the acceleration terms are improperly dismissed in G-value theory, as demonstrated in Chapter 3.

Although some experimental devices can have laminar flow, jar tests, and full-scale flocculators operate under turbulent flow conditions. The author enjoyed numerous discussions with Dr. Thomas R. Camp on the subject of G-value and turbulent flocculation. While Dr. Camp did not agree with the author's views relative to the validity of G-value, he did agree to a laboratory test of G-value theory which the author proposed. With the assistance of Dr. Camp's laboratory assistant, a series of flocculation jar tests were run (all geometrically identical) with constant concentrations of kaolin and alum. At one temperature, the G-value, which appeared to give optimum flocculation, was determined. The corresponding floc volume concentration was determined, using the stereoscopic microscope procedure reported by Camp (1968) and Camp and Conklin (1970). The temperature of the water was then changed, and another test was run with the power input changed to result in a G-value identical to that of the optimum G-value determined at the first temperature. Under these conditions, *no visible flocculation resulted.* Further tests were run that showed that by maintaining a constant power input, floc characteristics varied less with temperature than when the G-value remained constant. With constant power input, some temperature dependency was observed, but not at all as predicted by G-value theory.

Insofar as their *hydrodynamics* are concerned, flocculators are basically mixers with throughput, to which discussion and equations presented in Section 14.4 apply. In particular, Equations 14.10–14.12 would apply, the latter two subject to the same limitations discussed previously. The variable TN may or may not have an effect on hydraulic performance, depending on the type of flocculator. It would probably have more effect on the flow pattern in a paddle flocculator than in a vertical turbine flocculator due to the high pumping rate of the latter. Geometric similarity is again important. The sensitivity of flocculator performance to geometry has been pointed out by Bean (1953) and Drobny (1963) (whose work applies equally, with minor exceptions to rapid mixing). However, the floc size is also a variable, which must be considered in any attempt to model flocculation. For flocculation, the question of what additional process variables must be scaled is considerably more complex than in the case of rapid mixing. Is it necessary to scale the turbidity particles to maintain geometric similarity? Can the complex surface chemistry be properly scaled? If flocculation were fully scaled, how would one interpret the results for direct filtration (filtration following immediately after flocculation without sedimentation basins), recognizing that the filter is not geometrically scaled but is a full-sized element of the prototype filter? Is it not possible or even likely that proper scaling of the flocculation process is not achievable?

Early work on turbulent flocculation [e.g., Harris *et al.* (1966), Argaman and Kaufman (1970)] found no substantial agreement with G-value theory. Subsequent researchers [e.g., Cleasby (1984) and Clark (1985)] have voiced explicit disagreement with G-value theory. However, other researchers, including more recent ones, have seemed compelled to attempt to work G-value into their methods, probably on the assumption that its success in practice demands it. That assumption is not warranted.

One manufacturer's testing of their walking-beam flocculators led to the following conclusions (Albano and Grieshaber 1970): (1) the "equipment was performing well at G and GT levels significantly below the minima recommended in the literature"; (2) "the commonly used criteria (G and GT parameters) for application of flocculating mechanisms should not be applied to...[their walking beam] equipment"; (3) "increasing G beyond that which results from...[their] application criteria does not improve treatment performance, but it does increase first cost at least linearly with increase in G"; and (4) "the design criteria is sensitive to tank configuration." Georgia (1949) provides further useful information on the walking-beam flocculator.

During a Lightnin® (equipment manufacturer) seminar session on flocculation attended by the author on April 11, 1973, the presenter, Jerry Locke, roundly criticized the "conventional" approach, stating that G-value does not adequately describe flocculation (and that tip speed limitations are not meaningful). During a private discussion on April 12, 1973, James Y. Oldshue, Ph. D., of Lightnin® suggested that the HP per unit volume criterion generally provides a "safe over-design." Treybal (1959) suggested that GT can give good results but not economical ones. And further (Treybal 1961), "it is insufficient to know that by mere application of large amounts of power we can get a job done ... Rather, we need to know how little power is required and how best to apply it."

That some flocculators have been designed using G-value and found to function adequately may be attributed to the following: (1) some flocculators are merely of empirically established design for which G, GT, or any other parameter, which incorporates the duplicated variables will work; (2) many flocculators are probably designed with a great deal of unintentional conservatism; and (3) flocculation is a fairly insensitive process in many field applications.

Cleasby's (1984) critique of the application of G-value to turbulent flocculation is particularly comprehensive. The diminished role of temperature and the practical ramifications of that are clearly elucidated by Cleasby. The author would take Cleasby's criticism a step further. Cleasby suggested that G-value is only a valid parameter for flocculation of particles smaller than the Kolmogoroff microscale, a size characterization of turbulence which he pointed out is not typical of common flocculation practice. The analysis presented above indicates that G-value *per se* is valid only for *one-dimensional laminar* flow.

How then should pilot plant or jar test results be interpreted as far as the flocculation process is concerned? Such test results are useful in demonstrating the physico-chemical feasibility of flocculation of a particular raw water with a particular chemical flocculant. If we can obtain the required flocculation in the jar test or pilot plant, then it can also be done in the full-scale plant. However, the jar or pilot-plant flocculator will not necessarily give us G-values or any other parameters that should be duplicated at full scale. The full-scale flocculators should be designed on the basis of the designer's and the manufacturer's experience with experimental or operating full-scale units, with sufficient adjustability built in (e.g., variable flocculator speed) to achieve the desired result.

Graber (1997) discussed a more recent misuse of G-value, which was applied theoretically to a flocculating baffled channel without any experimental support. The author reiterated that the application of G-value to flocculation was "a misapplication except under very limited conditions that are not encountered in practice." The author subsequently offered similar comments (Graber 2004a) on another proposed application of G-value to flocculators.

Graber (1997) also commented on the "noteworthy critique by Clark...[that] also challenges the theoretical basis of G-value." Graber (1998) elaborated by recognizing a follow-up paper by Kramer and Clark but took issue with their contention that G-value was applicable to 2D flow. Whereas the author's theoretical critique (see Chapter 3) was presented in terms of the historical derivation of the G-value, Kramer and Clark used a tensor method, which appears to have gone somewhat astray.

14.6.2 Floc Breakup

Once formed, floc particles are somewhat fragile and can be broken up if subjected to excessive shear stresses. This appears to imply that it is necessary not only to create suitable conditions in the flocculator but also to provide acceptable conditions downstream of the flocculation process. This also pertains to biomass flocculation occurring in the activated sludge process of wastewater treatment as well as flocculation in water treatment plants. Related theoretical and laboratory studies are discussed below. The last paragraph in this section provides practical suggestions related to actual, full-scale facilities.

Some of the more noteworthy studies of floc breakup include those of Bradley and Krone (1971), Parker *et al.* (1971, 1972), Delichatsios and Probstein (1975), and Das *et al.* (1993). Some of these authors have also used G-value inappropriately to quantify floc breakup. Das *et al.* discuss some of the limitations of G-value for this purpose, but then use G-value "based on its ease of determination, simplicity, and widespread acceptance." Das *et al.* found a qualitative correlation between dispersed concentration (supernatant suspended solids in a standard settling column) and G-value calculated *individually* for aeration basins mixed by mechanical and diffused aeration and for conveyance pipes and elbows. This has potentially important practical ramifications, but it is not surprising that increasing floc agitation as measured by a parameter that includes power per unit volume would reveal such an effect.

It seems reasonable to accept shear stress as the parameter that most affects floc breakup. Relatively few studies have directly considered the shear stresses acting on floc particles, although some studies allow the calculation of those stresses. The study by Bradley and Krone (1971) is one such. Bradley and Krone constructed a concentric-cylinder device in which they tested an activated sludge obtained from a laboratory batch culture that had been seeded with sludge from a wastewater treatment plant. They subjected the activated sludge floc to controlled velocity gradients and determined the effect on aggregation and settling and aggregate breakup. The test device created a simple, laminar shear flow for which the shear stress on the floc is readily related to the velocity gradient, or G-value. Note that this is a case of one-dimensional laminar flow to which G-value is applicable. However, the G-value is of little direct use in extrapolating the results to the complex turbulent flows occurring in actual applications.

Bradley and Krone oversimplified the relations for the flow between their rotating cylinders. The velocity distribution between such cylinders is given more precisely in Section 4.4:

$$u = \frac{\omega r_2^2}{r_2^2 - r_1^2}\left(r - \frac{r_1^2}{r}\right) \tag{14.35}$$

in which u is fluid velocity, ω is the rotational speed, and r_1 and r_2 are, respectively, the radii of the inner and outer cylinders. Parameters incorporating the G-value, differential G-value, and mean G-value are then given as follows:

$$G = \frac{du}{dr} = \frac{\omega r_2^2}{r_2^2 - r_1^2}\left(1 + \frac{r_1^2}{r}\right) \tag{14.36a}$$

$$\Delta G = G_{\max} - G_{\min} = \omega \tag{14.36b}$$

$$G_m = \frac{\omega r_2}{r_2 - r_1} \tag{14.36c}$$

$$\frac{\Delta G}{G_m} = \frac{r_2 - r_1}{r_2} \tag{14.36d}$$

For the dimensions of Bradley and Krone's apparatus, the above equations give $\Delta G/G_m = 0.13$, less than half the value given by Bradley and Krone's corresponding equation, and values for G_m 8% higher than calculated by Bradley and Krone.

The practical value of the above is that the shear stress in one-dimensional laminar flow can be calculated from $\tau = \mu G$. The shear stress thus calculated can then, in concept, be related to shear stresses acting on the floc particles. However, this is not simply a matter of calculating or measuring full-scale shear stresses. Delichatsios and Probstein (1975) addressed two sources of shear stress in a turbulent flow field. First are the shear stresses associated with the relative motion of the fluid particles with respect to the continuous flow field. That is physically analogous to the shear stress causing breakup in laminar flow. Second are the shear forces due to the unsteady, turbulent stress fluctuations across the particle diameter. Delichatsios and Probstein (1975) asserted that, for particles dispersed in water, the second stress is the larger of the two. Other modes of breakup are discussed by Argaman and Kaufman (1970) and Parker *et al.* (1971).

The mean value of the turbulent stresses across the floc diameter (d) depends on whether the particle diameter is greater than or less than the Kolmogorov microscale (λ) of the turbulence (Parker *et al.* 1971, 1972; Delichatsios and Probstein 1975).

The critical shear stress can be related to the local energy dissipation in a turbulent flow system as follows (Delichatsios and Probstein 1975):

$$\tau \cong \mu G \text{ for } d < \lambda \tag{14.37a}$$

$$\tau \cong \rho(\varepsilon d)^{2/3} \text{ for } d > \lambda \tag{14.37b}$$

in which ε is the energy dissipation per unit mass.

Although the shear stress in a turbulent pipe is maximum at the pipe wall, Delichatsios and Probstein (1975) note that 90% of the particles at any given cross section are found at the *core* of the pipe. Their experiments indicate that core energy dissipation is the effective variable for flocculation. That energy dissipation can be calculated for a pipe from the equations given by Delichatsios and Probstein and substituted into Equation 14.37b to give the following relationship for critical shear stress:

$$\tau = \frac{1}{2}\rho f U^2 \left(\frac{d}{D}\right)^{2/3} \tag{14.38}$$

in which f is the Darcy–Weisbach friction factor, U is mean velocity, D is pipe diameter, and other terms are as defined above. By equating the critical shear stress, as determined from a laminar-flow device as discussed above, with the shear stress given by Equation 14.37 and knowing the floc diameter, it should be possible, in theory, to determine the parameters of a pipe that will safely convey that floc. (However, see below.)

In the above-cited study of Bradley and Krone, the test device created a simple laminar shear flow and gave a "G" value of 16.2/sec, above which the maximum particle size of the mixed liquor decreased and settling was adversely affected. The "G" value is of little direct use in extrapolating the results to the turbulent flows common to actual applications. However, the results may be used to estimate the effect of shearing on activated sludge floc as follows: For the flow existing in the test device, shear stress τ may be related to G by $\tau = \mu G$. The temperature in the Bradley-Krone experiments was held constant at 20°C, for which $\mu = 0.01$ gm mass/cm-sec. The limiting shear stress is given by $\tau = 0.01(16.2)$ times the 1.08 correction discussed above = 0.162 dyne/sq cm = 0.175 kg/sq cm = $2.54(10)^{-6}$ psi.

The shear stress can be estimated for many turbulent flow situations. The shear stress varies from a maximum at the walls to zero at the pipe centerline. For flow in a pipe, the wall shear stress is given by (Equation 9.66b with change of notation):

$$\tau_o = \frac{f}{4}\gamma\frac{U^2}{2g} \tag{14.39}$$

in which γ = weight density of fluid, 62.4 lbf/cu ft. The turbulent shear stress varies linearly, according to Schlichting (1960, p. 510):

$$\tau = \tau_o\left(1 - \frac{y}{R}\right) \tag{14.40}$$

from which the average shear stress $\bar{\tau}$ is obtained as follows:

$$\bar{\tau} = \frac{\int_0^R \tau dy}{\int_0^R dy} = \frac{1}{R}\int_0^R \tau_o\left(1-\frac{y}{R}\right)dy = \int_0^1 \tau_o\left(1-\frac{y}{R}\right)d\left(\frac{y}{R}\right) = \tau_o\left[\frac{y}{R} - \frac{1}{2}\left(\frac{y}{R}\right)^2\right]_0^1 = \frac{1}{2}\tau_o \tag{14.41}$$

Example 14.5

Two 84-inch concrete-lined pipes are planned to convey activated sludge to circular final clarifiers at a maximum velocity of 4.8 ft/sec (Bondi Island Wastewater Treatment Plant, Springfield, Massachusetts). Determine the suitability of those pipes regarding floc breakup.

Solution

For the proposed pipe, $f \cong 0.012$ (Davis 1942, Figure 11). The velocity head $U^2/(2g) = (4.8)^2/[2(32.2)] = 0.358$ ft The wall shear stress from Equation 14.39 is $\tau_o = (0.012/4)(62.4)(0.358)(\text{lbf/ft}^3)(\text{ft}) = 0.0670\,\text{lbf/ft}^2 \div 144 = 465(10)^{-6}\text{psi}$. From

Equation 14.41, the average shear stress is $\bar{\tau} = \tau_o/2 = 233(10)^{-6}$psi. This value is substantially greater than the limiting shear stress of $2.54(10)^{-6}$ psi given above. Pipes giving average shear stresses within the limiting value would have to be approximately three times the 84 inch diameters. However, the system performed well with the 84 inch pipes – see below.

A dynamic process occurs in conveyance of activated sludge, in which floc breakup and reflocculation occur simultaneously. And, if some undoing of flocculation occurs when mixed liquor is conveyed to clarifiers following flocculation in the activated sludge process, experience suggests that reflocculation occurs quickly in final clarification so that breakup during conveyance is not a concern. Dick (1971) addressed this process in the context of his discussion of the above-cited paper by Bradley and Krone: "If the dissipation of velocity gradients in the final settling tank occurs in proper fashion and in the presence of a high concentration of solids, then the suspension may 'heal' and nullify the adverse effects of prior shearing." Following extensive theoretical development, Parker *et al.* (1972) concluded that it is "the balance of the opposing processes of aggregation and breakup that determine flocculator performance."

14.7 Aeration and Postaeration

14.7.1 Aeration

14.7.1.1 Atmospheric Aeration

In a seminal paper, Tuček and Chudoba (1969) effectively resolved the debate regarding plug flow (aka piston flow) vs. compete-mixing aeration. Plug flow advocates had noted the more favorable flow-through characteristics such as demonstrated by dye testing, and the prevention of some of the influent from going directly to the outlet. Complete-mixing advocates noted the uniform concentration of organic matter and rate of oxygen uptake among other advantages. From experimental and theoretical studies, Tuček and Chudoba found that the reaction rate in the complete-mixing system was about seven times higher than in the plug-flow system and that high rates of BOD removal in the complete-mixing system fully eliminate the unfavorable effects of short-circuiting. Reasons for that are: (1) in the complete-mixing system, the concentration of organic matter is the same throughout the tank and the rate of oxygen uptake is thus the same; (2) the microorganisms are under more favorable conditions because they are not exposed to spatial variations and less exposed to temporal variations; and (3) the entire contents of the tank are utilized for dilution of the tank influent.

Since dispersion in open-channel flow results from the combined effect of relative displacement associated with velocity gradients and turbulent diffusion (Holley 1969), the diffusion of air into the flow may be expected to influence the dispersion coefficient. Most of the experimental work on the influence of aeration on dispersion has been done in spiral-flow tanks (Murphy and Boyko 1970). For such tanks, they experimentally established the following relationship:

$$D_L = 3.118W^2 q_a^{0.346} \tag{14.42}$$

in which D_L is the dispersion coefficient in ft^2/hr, W is the tank width in ft, and q_a is the airflow per unit tank volume in scfm/1,000 cu ft.

It is informative to compare the above equation with Equation 14.34b for the full-scale tank used in Murphy and Boyko's studies. Their tank had a width of 27 ft and a depth of 15 ft, which corresponds to a hydraulic radius of 7.11 ft. The tank length was 270 ft and detention times ranged from 4.8 to 13.3 hours, which corresponds to a velocity U ranging from 56.3 to 20.3 ft/hr. The greatest dispersion occurs at the maximum velocity at which the Reynolds Number $\mathbf{R} = 4RU/\nu = 4(7.11)(56.3)/[10 - 5(3600)] = 44,500$. At this Reynolds Number and relative roughness $\varepsilon/(4R) = 0.01/[4(7.11)] = 0.00035$, the Moody Diagram (Figure 5.1) gives a friction factor $f = 0.023$.

Substituting the above values of U, L, f and R into Equation 14.34b gives a value for d of 2660 ft^2/hr. That is less than the values of 5000–8000 ft^2/hr determined by Murphy and Boyko in their spiral-flow aerated tank, substantiating the above hypothesis that spiral-flow aeration will increase dispersion. Mattingly (per Whipple *et al.* 1970a) also found that air diffusion significantly increased dispersion in the Delaware River.

If full-width aeration were actually achieved, it would be adequate to use Equation 14.34b to predict the dispersion coefficient. However, wide-band configurations such as dome-diffuser systems and ridge-and-furrow diffuser systems only approach full-width aeration and may create a number of individual transverse spirals over the tank cross section. Therefore, it would be prudent to use Equation 14.34a for all air-diffuser systems, except when Equation 14.34b gives a larger value for the dispersion coefficient. Equation 14.34b could give a larger value at low values of q_a, since Equation 14.34a implies that D_L approaches zero as q_a approaches zero, which is not actually the case. Section 13.3 discusses the related topics of full width and spiral-flow aeration in connection with induced circulation in constructed channels.

The author specified dome-diffuser systems for the activated sludge aeration tanks at New York City's 80 mgd Tallman Island Water Pollution Control Facility; the 65 mgd Providence, Rhode Island Wastewater Treatment Plant; and the Rochester, New Hampshire, 3.9 mgd Advanced Wastewater Treatment Plant.

It is useful to mention river and lagoon aeration in the context of the present section; pertinent references are McWhirter (1966), Amberg *et al.* (1969, which discusses both air and molecular oxygenation), Hogan *et al.* (1970), and Murphy and Wilson (1974, which uses a variation on Equation 14.28). Improvement of hydropower release of DO by turbine venting can also be significant (Wilhelms *et al.* 1987).

14.7.1.2 Molecular Oxygenation

i.e., the use of pure oxygen in the activated sludge process has the primary advantage of requiring less space. The higher DO concentration can also be advantageous in the plant effluent, although it does not necessarily obviate the need for postaeration. An example of where molecular oxygenation is followed by atmospheric postaeration is at the MWRA's Deer Island Wastewater Treatment Plant. Shelef and Green (1980) found no difference in pertinent kinetic parameters in experiments comparing two activated sludge systems, one with aeration by air and the other by pure oxygen. The results proved that the advantages of pure oxygen, other than those just mentioned, lie solely in the higher partial pressure and the corresponding driving force of its dissolution. The above discussion of aeration hydraulics applies equally to molecular oxygenation.

14.7.1.3 Ozonation

has certain advantages as an alternative to chlorination in the disinfection of drinking water because chlorination byproducts can be carcinogenic. Ozone concentration is much more important than contact time in determining the efficiency of the disinfection process (Zhou and Smith 1994): "In other words, the disinfectant concentration is the predominant factor and contact time is less important. Similar phenomena have been noted in many previous investigations on the ozone inactivation over a wide range of organisms.... [I]n the disinfection of wastewaters ... an excellent correlation exists between the log transferred ozone dosage and log coliform survival ratio. Once the transferred ozone dosage is established, the level of disinfection performance can be predicted for a certain ozone contact basin, regardless of the contact time.... [O]zone concentration is a much more important factor than the contact time in determining the efficiency of the disinfection processes." Zhou and Smith cite numerous references in support of these and similar statements. In the context of the present chapter, the disinfection–time relationship will be discussed in connection with chlorination.

14.7.2 Postaeration

The primary purpose of postaeration is to add DO to the effluent of wastewater treatment plants. The most common reason for doing so is to prevent contravention of the receiving water (e.g., river) DO standard at the point where the effluent and receiving water intermix. As a side benefit, postaeration may strip out some of the carbon dioxide produced in biological oxidation (Kalinske 1968), thus raising the pH and reducing an algae nutrient. It may also strip out some of the ammonia in the plant effluent, thus reducing ammonia toxicity and nitrogenous oxygen demand. Postaeration (particularly with pure oxygen) can also be used to raise the river DO above the standard so as to prevent contravention at some point downstream. That is akin to instream aeration, which has been found, in some cases, to achieve overall economy by reducing point-source treatment requirements and also to offset deoxygenation due to nonpoint sources such as urban runoff.

There is little information available on which to base the design of postaeration systems. Literature references on the subject are often contradictory. The importance of hydraulic dispersion is often overlooked. Here a critical review of the literature is presented, design procedures are proposed, and suggestions are made for further research in certain areas. Three basic methods of postaeration are considered: (1) diffused air aeration; (2) mechanical aeration; and (3) hydraulic aeration, such as results from weirs and cascades.

14.7.2.1 Diffused Air Aeration

Postaeration in basins requires the addition of oxygen to a flowing stream. The situation is somewhat more complex than aeration in the activated sludge process, in which a nearly constant DO level is maintained and the "driving force" remains constant. (The "driving force" is basically the difference between the DO saturation concentration for existing conditions and the actual DO concentration of the water.) In postaeration, the DO level increases as the water flows through the basin, and it becomes necessary to consider the interaction between the aeration devices and flow pattern.

The need for postaeration by any method can be determined from the DO of the effluent-river water mixture given as follows:

$$C_m = \frac{C_r Q_r + C_e Q_e}{Q_r + Q_e} \tag{14.43}$$

in which C_m = DO concentration of mixed effluent and river water, C_r = minimum river DO concentration upstream of the plant outfall, C_e = minimum effluent DO concentration without postaeration (typically 0–1 mg/L), Q_e = wastewater effluent flow rate, Q_r = critical river flow rate (e.g., 7-day, 10-year low flow).

If C_m is less than the DO standard, then contravention is indicated. In such a case, the effluent DO concentration required to achieve the river standard is given by:

$$C_{er} = \frac{(Q_e + Q_r)S - Q_r C_r}{Q_e} \tag{14.44}$$

in which C_{er} = required effluent DO concentration, S = river DO concentration standard (typically 4.0–6.0 mg/L), and other terms are as defined above. Denoting the DO concentration of the flow entering the postaeration unit by C_i and that leaving the postaeration unit as C_o, these values should be set equal to C_e and C_{er} (or a value somewhat in excess of C_{er}), respectively.

In designing a diffused air unit for postaeration, it is usually desirable to develop plug-flow conditions. Plug flow is the most efficient flow pattern for postaeration, and is approached in spiral-flow or wide-band aeration systems. The only feasible exception is the multiple-cell configuration, achieved by a spiral-flow diffused air system in which a series of air manifolds runs across the tank in a direction perpendicular to the flow. The City of Fort Worth, Texas, has such a configuration in their activated sludge aeration basins; it is amenable to wide passes, which minimizes dividing walls and air headers. A postaeration system having the Fort Worth configuration can be designed as a series of completely mixed "cells." Since the plug-flow pattern would be employed in most postaeration systems, it is the diffused air system considered herein. (The multiple-cell configuration for postaeration would be designed in a manner similar to that for postaeration by mechanical aeration, as discussed in the next section.)

Equations for plug-flow postaeration have been derived by Kormanik (1970) and the author (Graber 1972). The derivations are similar, but Kormanik's equations rely on an exponent n and constant C_a to characterize the air diffusion equipment. Graber's equations make use of the standard oxygen transfer efficiency, representative values of which are more well-established for air diffusion equipment.

Under certain circumstances, the deviation from plug-flow conditions could be sufficient to affect the performance of the postaeration system. Such deviations can be accounted for in a dispersion model of postaeration, developed as follows. The general equation for the steady-state transport of DO is given by:

$$U\frac{dC}{dx} - D\frac{d^2C}{dx^2} = r \tag{14.45}$$

in which C = DO concentration, U = mean velocity of flow, D = dispersion coefficient, r = time rate of DO mass addition per unit volume. If the exertion of BOD is negligible, r is the rate at which DO is dissolved by oxygen addition. For a diffused air (or oxygen) system, r is given by:

$$r = \frac{dm/dt}{Adx} \tag{14.46}$$

in which dm/dt is the mass rate of oxygen dissolution per unit volume and A is the basin cross-sectional area. For a diffused air system, the rate of oxygen dissolution is given as follows (Graber 1972):

$$\frac{dm}{dt} = K_1(C_s - C)d\dot{M} \tag{14.47a}$$

$$K_1 = \frac{E(1.024)^{T-20}}{(C_s)_{20}} \tag{14.47b}$$

The temperature adjustment in the above equation will appear in numerous contexts in addition to the present one in this chapter and Chapter 15 and is given by Elmore and West (1961).

Substituting Equations 14.46 and 14.47 into Equation 14.45 and rearranging:

$$\frac{d^2C}{dx^2} - \frac{U}{D}\frac{dC}{dx} + \frac{K_1(C_s - C)d\dot{M}}{DAdx} = 0 \tag{14.48}$$

If the air is introduced uniformly along the length of the basin, then $d\dot{M}/dx$ is a constant,$\dot{M}/L$. Transforming to the dependent variable $C_s - C$, Equation 14.48 can be rewritten as follows:

$$\frac{d^2(C_s - C)}{dx^2} - \frac{U}{D}\frac{d(C_s - C)}{dx} - \frac{K_2}{D}(C_s - C) = 0 \tag{14.49a}$$

$$K_2 = K_1\frac{d\dot{M}}{dx} \tag{14.49b}$$

Equation 14.49 is a linear, second-order differential equation with constant coefficients, the solution of which is as follows:

$$(C_s - C) = K_3 \exp\left[\left(\frac{1}{2}\frac{U}{D} + \sqrt{\left(\frac{U}{2D}\right)^2 + \frac{K_2}{D}}\right)x\right] + K_4 \exp\left[\left(\frac{1}{2}\frac{U}{D} - \sqrt{\left(\frac{U}{2D}\right)^2 + \frac{K_2}{D}}\right)x\right] \tag{14.50}$$

in which K_3 and K_4 are constants of integration. These constants can be evaluated using the following boundary conditions: $C = C_i$ @ $x = 0$, $C = C_s$ @ $x = \infty$. Substituting these boundary conditions into Equation 14.50 gives: $K_3 = 0$, $K_4 = (C_s - C_i)$. Then Equation 14.50 becomes:

$$\frac{C_s - C}{C_s - C_u} = \exp\left\{\left[\frac{U}{2D} - \sqrt{\left(\frac{U}{2D}\right)^2 + \frac{K_2}{D}}\right]x\right\} \tag{14.51}$$

The value of the DO at the outlet of the basin of length L is obtained by substituting $x = L$ and $C = C_d$ into the above equation, which may then be rearranged to give the following:

$$\frac{C_s - C_d}{C_s - C_u} = \exp\left[\frac{UL}{2D}\left(1 - \sqrt{1 + \frac{4K_2D}{U^2}}\right)\right] \tag{14.52}$$

The constant K_2 is given by:

$$K_2 = \frac{E(1.024)^{T-20}}{(C_s)_{20}}\frac{\dot{M}}{AL} \tag{14.53}$$

If D, L, and U are in units of ft²/sec, ft, and ft/sec, respectively, and the correction factor α is taken into account, Equation 14.53 may be expressed in the following dimensional form:

$$K_2 = \frac{E\alpha(1.024)^{T-20}}{(C_s)_{20}}\frac{Q_a}{0.214AL} \tag{14.54}$$

in which Q_a is the air requirement in scfm.

By appropriate manipulation of basic relationships, the equation for determining the postaeration diffused air requirement may be determined (Graber 1972). The oxygen solution rate is given by:

$$\frac{dm}{dt} = K_{La}(C_s - C) \tag{14.55}$$

in which C_s and C are the saturation and prevailing DO concentrations, respectively, and K_{La} is a solution constant related to temperature T in degrees Centigrade by:

$$(K_{La})_{T°C} = (K_{La})_{20°C} \times (1.024)^{T-20} \tag{14.56}$$

The oxygen transfer efficiency is defined by:

$$E = \frac{(dm/dt)_{20°C,C=0}}{\dot{M}} \tag{14.57}$$

in which $(dm/dt)_{20°C,C=0}$ is the oxygen solution rate at 20°C and $\dot{M}$ is the rate at which oxygen is introduced.

Equations 14.55–14.57 may be combined to give:

$$\dot{M} = \frac{1}{E}\left(\frac{dm}{dt}\right)_T \times \frac{(C_s)_{20}}{(C_s - C)_T(1.024)^{T-20}} \tag{14.58}$$

Equation 14.58 may be applied to an infinitesimal transverse segment of the tank, noting that $dm/dt = d\mathcal{V}(dc/dt) = Qdc$, as follows:

$$d\dot{M} = \frac{Q}{E}\frac{(C_s)_{20}}{(1.024)^{T-20}}\left(\frac{dc}{C_s - C}\right)_T \tag{14.59}$$

Integrating Equation 14.59 from the inlet to the outlet of the tank, where $C = C_i$ and C_o, respectively, the total oxygen input requirement may be obtained:

$$\dot{M} = \frac{Q}{E}\frac{(C_s)_{20}}{(1.024)^{T-20}}\left(\ln\frac{C_s - C_i}{C_s - C_o}\right)_T \tag{14.60}$$

Considering that each cubic foot of air contains 0.0175 lb (0.0075 kg) of oxygen, and taking the correction α into account, the air requirement may be determined from Equation 14.60 and expressed in the following dimensional form:

$$Q_a = \frac{0.214}{E\alpha}\frac{(C_s)_{20}}{(1.024)^{T-20}}\ln\frac{C_s - C_i}{C_s - C_o} \tag{14.61}$$

in which Q_a is the air requirement in scfm and Q is the water flow rate in cfs. The factor α is assumed to be equal to 0.9 and $(C_s)_{20} = 9.02$ mg/L.

Some discussion of the feasibility of providing postaeration in chlorine contact tanks is useful. Presenting an example from Graber (1972), at 29°C water temperature, assuming a 6% oxygen transfer efficiency, the air requirement is approximately 20 scfm/mgd (0.000148 cu m/min/cu m/day) to increase DO from zero to 5 mg/L. A chlorine detention time of 15 minutes requires a tank volume of 1400 cu ft/mgd (0.0104 cu m/cu m/day) or 140 sq ft (13.0 sq m) plan area for a 10 ft (3.05 m) depth. This requires an air loading of 20/140 = 1/7 scfm/sq ft (0.0431 cum/min/sq m) of tank area. With an air rate of 3 scfm/sq ft (0.905 cu m/min/sq m) of diffuser area to tank area of 1:20, which is within the range of good design practice. Therefore, the two purposes are compatible with respect to tank dimensions. Kothandaraman and Evans (1972) demonstrated that such agitation has no effect on loss of chlorine at pH values of 6–9.

14.7.2.2 Mechanical Aeration

With mechanical surface-type aerators, each aerator has associated with it a "cell" of water, which is mixed and recirculated to varying degrees. The actual oxygen dissolution rate in the cell (completely mixed or otherwise) is:

$$\dot{M} = \rho Q_w(C_d - C_u)/10^6 \tag{14.62a}$$

in which $\dot{M}$ is in units of lbm/day, fluid density ρ is in units of lbm/ft^3, Q_w is flow rate in cu ft/day, and C_d C_u are the downstream and upstream DO concentrations, respectively, in mg/L. Using conversions of 8.35 lb/gal and 24 hours/day, one obtains:

$$\dot{M} = \frac{8.35}{24}Q_w(C_d - C_u)/10^6 = 0.348Q_w(C_d - C_u)/10^6 \tag{14.62b}$$

in which $\dot{M}$ is in lbm/day and Q_w is in mgd.

The relationship between $(dm/dt)_{20°C,C=0}$ and the oxygen input at other temperatures and concentrations is approximated by:

$$\frac{(dm/dt)_{T,C}}{(dm/dt)_{20°C,C=0}} = (1.024)^{T-20}\frac{(C_s - C)}{(C_s)_{20°C}} \tag{14.62c}$$

in which T denotes the prevailing water temperature in degrees Centigrade and C_s and C are the saturation and prevailing DO concentrations in mg/L, respectively. Since $(dm/dt)_{T,C} = \dot{M}$, the above two equations may be combined to give:

$$P = \frac{(dm/dt)_{20°C,C=0}}{\eta} = \frac{\dot{M}}{\eta}\frac{(C_s)_{20°C}}{(1.024)^{T-20}(C_s - C)} = 0.348\frac{Q_w(C_d - C_u)}{\eta}\frac{(C_s)_{20°C}}{(1.024)^{T-20}(C_s - C)} \tag{14.62d}$$

in which η is aerator oxygen transfer efficiency in lb O_2 per hp-hr (a function of basin geometry), $(C_s - C)$ is the driving force "seen" by the aerator, and P is the power required by each aerator in hp-hr/day.

For the completely mixed aerator, $C = C_d$ and the above equation for the power required by each aerator becomes:

$$P = \frac{0.348Q_w(C_d - C_u)}{\eta\left[(C_s - C_d)/(C_s)_{20}\right](1.024)^{T-20}} \tag{14.63}$$

The aerator "sees" the DO content of the flow entering the aerator. Since the DO is increasing from an upstream value of C_u to a downstream value of C_D, it is necessary to determine the DO concentration C against which the aerator does its work. Susag *et al.* (1966) studied three possibilities: (1) the upstream DO, (2) the arithmetic average of upstream and downstream DO, and (3) the DO based on plug flow through the aeration zone (although the assumption of plug flow is not apparent from Susag's "derivation," one must have recourse to this assumption to properly derive his equation). Kalinske (1965, 1968) suggested that the second of these be used. Kormanik (1969) suggested a fourth possibility: the downstream DO based on the assumption of complete mixing in the aeration cell. Under the complete-mixing assumption, the flow entering each cell has its DO concentration raised abruptly to the uniform concentration in the cell, at which concentration the flow is assumed to leave the cell.

Susag *et al.* (1966) obtained experimental data using a single aerator in a flowing stream. They computed oxygen dissolution rates based on the first three of the abovementioned assumptions and compared them to actual dissolution rates based on the average of DO samples taken within the aeration zone. Interestingly, the oxygen transfer rates calculated by the three methods are all lower than those based on the estimate of DO at the aeration unit, with closer approximation provided by the method giving a higher value of C. Thus, the C value based on the plug-flow assumption was better than that based on the average of upstream and downstream DO concentrations, and the latter was better than that based on the upstream DO. If the C value based on the complete-mixing assumption had been used, the agreement would presumably have been better still (or the calculated transfer rate would have been higher than the actual).

Inspection of the basic relationships shows that, for the same increase in DO, i.e., $(C_d - C_u)$, the calculated power requirement is higher if the complete-mixing assumption is used. Complete mixing is, in fact, considered desirable to assure distribution of the oxygen throughout the cell (Kalinske 1965, Kormanik 1969). Unless actual measurements are available to show that the DO concentration of the liquid entering the aerator is less than C_d, it is recommended that Equation 14.63 be used for design.

The relative merits of a series of aerators vs. a single aerator or aerators operating in parallel may now be addressed. For a given wastewater flow rate, temperature, and aerator efficiency, the above equation may be expressed by the following proportionality:

$$P \propto \frac{C_d - C_u}{C_s - C_d} \tag{14.64}$$

For a series of complete-mixing aerators in series, the power input of the ith aerator and the total power input are given, respectively, by:

$$P_i \propto \frac{C_{d,i} - C_{d,i-1}}{C_s - C_{d,i}} \tag{14.65}$$

$$P_{tot} = \sum_{i=1}^{n} P_i \propto \sum_{i=1}^{n} \frac{C_{d,i} - C_{d,i-1}}{C_s - C_{d,i}} \tag{14.66}$$

From Equation 14.65,

$$(C_{d,i} - C_{d,i-1}) \propto P_i(C_s - C_{d,i}) \tag{14.67}$$

and

$$\sum_{i=1}^{n}(C_{d,i} - C_{d,i-1}) \propto \sum_{i=1}^{n} P_i(C_s - C_{d,i}) \tag{14.68}$$

If each aerator is equally powered ($P_i^{'}s$ all equal), then Equation 14.68 becomes:

$$\sum_{i=1}^{n}(C_{d,i} - C_{d,i-1}) \propto P_i \sum_{i=1}^{n}(C_s - C_{d,i}) \tag{14.69}$$

Since

$$\sum_{i=1}^{n}(C_{d,i} - C_{d,i-1}) = C_d - C_u \tag{14.70}$$

Equation 14.69 can be expressed as:

$$P_i \propto \frac{C_d - C_u}{\sum_{i=1}^{n}(C_s - C_{d,i})} \tag{14.71}$$

and the total power is given by:

$$P_{tot} = nP_i \propto n \frac{C_d - C_u}{\sum_{i=1}^{n}(C_s - C_{d,i})} \tag{14.72}$$

which is the specialized form of Equation 14.66 for an equally powered series of aerators.

Noting that

$$\frac{1}{n}\sum_{i=1}^{n}(C_s - C_{d,i}) = C_s - \frac{\sum_{i=1}^{n} C_{d,i}}{n} \tag{14.73}$$

Equation 14.71 can be usefully modified to:

$$P_{tot} \propto \frac{C_d - C_u}{C_s - (1/n)\sum_{i=1}^{n} C_{d,i}} \tag{14.74}$$

in which form it can be conveniently compared to Equation 14.64 for the single aerator. Since the average of the DO levels in each of the aerator cells of the series system is less than the final effluent DO, then

$$\left[C_s - (1/n)\sum_{i=1}^{n} C_{d,i}\right] > (C_s - C_d) \tag{14.75}$$

which implies that $P_{tot} < P_{single}$. Therefore, the total power required for a series of two or more aerators is less than that required for a single aerator for the same degree of postaeration.

It is useful to ascertain the circumstances in which the effective driving force for the series of aerators, given by $C_s - (1/n)\sum_{i=1}^{n} C_{d,i}$ can be approximated by $C_s - (C_u + C_d)/2$. It can be shown mathematically (or graphically) that these two relationships for the driving force are equal only when: (1) the power input by each aerator in series is the same;

(2) the number of aerators in series is large; and (3) the DO level is increased by an equal amount in each aeration cell (this would be approximated only when $C_u - C_d$ is small). This is commensurate with the conclusion by Ottengraf and Rietema (1969) that plug flow can be approximated by a large number of aerators in series (14 in their example). Since these conditions cannot be assured, it would be prudent to design the postaeration system by applying Equation 14.63 to each aerator and taking the effluent DO concentration from one cell C_d to be the influent concentration C_u for the downstream cell. The best number of aerators should be established individually for each application, based on total operating and annual costs. The economic analysis should consider the number of days per year that postaeration would be required. Infrequent use would favor minimizing capital costs.

14.7.2.3 Hydraulic Aeration

In diffused air and mechanical aeration, the energy for aeration is input to water in basins by aeration devices. In hydraulic aeration, the energy for aeration is provided by the kinetic and/or potential energy of the flowing water itself.

Early studies of hydraulic aeration were concerned primarily with aeration at dams on polluted rivers. Gameson (1957) was one of the first to study this phenomenon in detail and attribute it to the entrainment of air in the splash zone at the base of the fall (rather than entrainment during the fall). Aside from this, little is known of the mechanism of hydraulic aeration, and subsequent researchers (Gameson *et al.* 1958, Barrett *et al.* 1960, Gannon 1967, Mastropietro 1968, Preul and Holler 1969, NCSI 1969, Quirk and Eder 1970, Huckabay and Keller 1970) have relied on various combinations of hypothesis, theory, and empiricism in developing predictive equations.

The variations in the results obtained with the different equations, the lack of a theoretical understanding, the scatter of data supporting the various equations proposed, the differences in the drop structures considered by different researchers, and other factors make it difficult to favor one particular equation. This led the author to undertake an independent analysis as part of a study of aeration at dams on the Nashua River in Massachusetts (Graber 1974b, 1975b).

In the absence of a good theoretical model, a predominantly empirical approach was taken. Dimensional analysis (see Chapter 5) provides a sound basis for such an approach. The only other analysis known to the author of aeration at dams based explicitly on dimensional analysis is that of Preul and Holler (1969), who studied aeration at navigation dams on the Ohio River. However, at those dams, the flow was discharged through underflow gates, and the aeration resulted from a downstream hydraulic jump. Their analysis is limited to a configuration that is not amenable to postaeration.

The first step in dimensional analysis is to define the variables associated with a particular phenomenon. The variables associated with dam aeration can be expressed by the following functional relationship:

$$\phi(C_s, C_u, C_d, \text{dam geometry, hydraulic variables, water quality}) = 0 \tag{14.76}$$

in which C_u and C_d are the DO concentrations upstream and downstream of the dam, respectively, and C_s is the saturation concentration. Since the saturation concentration is involved in establishing the "driving force" to which the mass diffusion of air is proportional, and since the saturation concentration is a function of temperature only [at sea level and for a given concentration of chloride (Fair and Geyer 1954, Table A-6)], it is reasonable to express the above relationship in terms of DO deficits as follows:

$$\phi(C_s - C_u, C_s - C_d, T, \text{dam geometry, hydraulic variables, water quality}) = 0 \tag{14.77}$$

in which $(C_s - C_u)$ and $(C_s - C_d)$ are the DO deficits upstream and downstream of the dam, respectively.

The formulation by Gameson (1957) is compatible with the above. Gameson defined the deficit ratio, r, as:

$$r = \frac{C_s - C_u}{C_s - C_d} \tag{14.78}$$

in terms of r the above functional relationship becomes:

$$r = \phi(T, \text{dam geometry, hydraulic variables, water quality}) \tag{14.79}$$

The different formulations of various researchers can be rearranged and expressed in terms of the deficit ratio expressed as a function of temperature, dam geometry, water quality, and hydraulic variables.

In conjunction with the author's Nashua River study, the Massachusetts Division of Water Pollution Control measured DO levels upstream and downstream of 12 dams on the river in August 1973. Various reaeration relationships were screened to determine their applicability to these dams. Although it has been applied to overflow dams, the relationship developed by Quirk and Eder was primarily for vented turbine and submerged gate aeration (Quirk and Eder 1970). This latter relationship considers the rate of flow to be the primary variable affecting the aeration at dams, whereas Gameson (1957) and Mastropietro (1968) found, for the overflow weirs and dams which they tested, that there was no apparent relation between aeration and flow rate. Accordingly, only the relationships of Gameson (1957) [as also used by Gameson *et al.* (1958), Barrett *et al.* (1960), NCSI (1969), and Huckabay and Keller (1970)] and Mastropietro (1968) were considered to be potentially applicable to the dams on the Nashua River.

Two methods have been developed to account for temperature effects in dam aeration. One was developed empirically by Gameson *et al.* (1958). The other is based on the Van't Hoff–Arrhenius equation, which has been widely applied to biochemical reactions and stream aeration and has been applied to dam aeration by Preul and Holler (1969) and Quirk and Eder (1970). It has also been incorporated into the Massachusetts Model used in the study of the Nashua River and was therefore selected. The equation is as follows:

$$r_T = r_{20}\exp[a(T-20)] \tag{14.80a}$$

or

$$r_T = r_{20}\theta^{(T-20)} \tag{14.80b}$$

in which r_T and r_{20} are the deficit ratios at the actual temperature and $\exp(a) = \theta$ is the temperature correction factor.

The deficit ratio was calculated from the Nashua River dam aeration data. That data was corrected to 20°C by means of Equation 14.80b and a θ-value of 1.024. The corresponding fall of water over each dam was calculated as the difference in elevation between the dam crest and the tailwater surface. The results were plotted in terms of r_{20} vs. fall of water and compared to the curves of Mastropietro and Gameson, the latter being a linear relationship. The data showed quite a bit of scatter, which is believed to result primarily from sampling difficulties associated with inaccessibility and the entrainment of small bubbles (which causes some of the aeration). Nevertheless, the curve corresponding to the Mastropietro relationship provided a much better fit of the data.

The same phenomenon that results in aeration at weirs can also provide valuable aeration at wastewater treatment plant outfall structures [in fact, two of the abovementioned studies, those of Gannon (1967) and Huckabay and Keller (1970), were concerned primarily with that application.]

From 1993 to 1995, the U.S. Army Corps of Engineers sponsored numerous studies of aeration by weirs, spillways, and other hydraulic structures. One of the most recent ones is that of Hibbs *et al.* (1995), which provides references to some of the other sponsored studies. More recent references on aeration performance of weirs are by Baylar and Bagatur (2000, 2001). The first of those references found by experiment was that "differing weir discharges were far less significant than ... drop height." The latter reference suggests the important relation of tailwater depth to weir aeration, although the scalability of their laboratory data is uncertain. The lack of scalability for another air–water application, namely air-induced circulation of water in channels, is discussed in Section 13.3. Aras and Berkun (2010) give interesting information for aeration on steps, although again, the scalability of their laboratory data is uncertain. At present, one must use judgment in selecting the most applicable information in a particular application.

For the example given below, the results of Gannon (1967) [also given by Hogan *et al.* (1970, Chapter V)] were deemed most applicable. Those results were for aeration at a wastewater treatment plant provided by the overflow weir at the outlet of a chlorine contact tank. Their data are expressed in terms of the following equation:

$$\frac{C_S - C_A}{C_S - C_B} = 1 + 0.11ab(1 + 0.046T)h \tag{14.81a}$$

where C_S is DO saturation concentration, C_A is DO concentration above the weir, C_B is DO concentration below the weir, T is water temperature (°C); h is height through which the water falls (ft); a is 1.25 for clean river water, 1.0 for polluted river water, 0.8 for sewage effluents; and b is 1.0 for a free weir and 1.3 for step weirs. Solving the above equation for C_B:

$$C_B = C_S - \frac{C_S - C_A}{[1 + 0.11ab(1 + 0.046T)h]} \tag{14.81b}$$

There are limitations on the height of drop for effective aeration. Baylar and Bagatur (2000) noted that "[w]ith increasing drop height, the nappe eventually breaks up into discrete droplets" and that "breakup of the nappe was observed as the drop height increased to more than 90 cm [3 ft]." That is quantitatively consistent with observations at waterfalls, as discussed below.

Example 14.6

Steps are to be provided to aerate the secondary treatment discharge from a wastewater treatment plant before discharge at 20°C and 2 mg/L to a stream (French Stream in Rockland, Massachusetts). Assuming 1 ft step rises, determine the number of steps required to achieve a DO level of 5 mg/L. Also, determine the step run.

Solution

At 20°C, the DO saturation concentration is 9.2 mg/L. For $h = 1$ (1 ft steps), the DO concentration after the first step using the above equation and pertinent values is:

$$C_B = 9.2 - \frac{9.2 - 2.0}{1 + 0.11(0.8)(1.3)[1 + 0.046(20)](1)} = 9.2 - \frac{7.2}{1.220} = 3.30 \text{ mg/L}$$

After a second step:

$$C_B = 9.2 - \frac{9.2 - 3.30}{1.220} = 9.2 - \frac{5.90}{1.220} = 4.36 \text{ mg/L}$$

After a third step:

$$C_B = 9.2 - \frac{9.2 - 4.36}{1.220} = 9.2 - \frac{4.84}{1.220} = 5.23 \text{ mg/L}$$

Thus, the desired DO level is achieved with three 1 ft steps.

For the peak flow of 6 mgd (9.28 cfs) and crest length of 6 ft, the broad-crested weir formulas from Example 7.4 with C_{bc} taken conservatively as 2.67 give: $q = 9.28/6 = 1.547$ cfs/ft, $H \doteq y = (q/C_{bc})^{2/3} = (1.547/2.67)^{2/3} = 0.695$ ft for the crest flow depth. A conservative estimate of the upper nappe coordinates is obtained by calculating the trajectory using Equation 7.24 and adding the crest depth. Thus, $v_o = 9.28/[0.695(6)] = 2.23$ ft/sec and $x = \sqrt{2(v_o^2/g)y} = \sqrt{2[(2.23)^2/32.2]0.695} = 0.556 + 0.695 = 1.251$ ft. A step length (run) of 1.5 ft = 18 in. is provided.

This is a good place to discuss aeration of waterfalls and other interesting features of waterfalls. James Prescott Joule, a pioneer in the field of thermodynamics (Reynolds 1892), is also known for a famous story that during his honeymoon he measured the differences in water temperature at the top and bottom of a waterfall (Cropper 2001, Chapter 5). Joule attempted to demonstrate that water at the bottom of a waterfall should be warmer than water at the top. That is a reasonable expectation since kinetic energy would be converted to heat. It would be expected to be partially converted to noise as well, and such conversion was the subject of a study by Al-Musawi (2012). Al-Musawi measured sound spectra for both laboratory and field waterfalls. The flow rates were only measured for the laboratory waterfalls, but the sound levels were found to be similar for field waterfalls and some of the laboratory ones.

Quantitative results of waterfall noise measurements were provided by Lilly (2007) for cascading and free-falling (plunge) waterfalls and indoor and outdoor man-made waterfalls. He found that waterfall noise is greatest in the free fall configuration, that the sound level is independent of height if it exceeds 3 feet (due to breakup of the plunging water into discontinuous flows), and that the overall noise level is dependent only on the volumetric flow rate. Lilly gives the most abundant noise measurements for the free-falling Snoqualmie Falls in the State of Washington, with a drop of 267 feet. A USGS gaging station provided daily mean discharges. The photograph below (Figure 14.7) was taken when the flow rate was 862 cfs on July 25, 2007. Another photograph by Lilly (2007) shows water cascading over the full width of the falls with a flow rate of 23,700 cfs on January 13, 2003.

Lilly reports sound power levels, which correlate well with flow rate over a wide range of flows for Snoqualmie Falls and others, covering a flow range of 0.1 cfs to about 500,000 cfs. Making any reasonable assumption about the terminal fall rate of waterfalls based on Gunn and Kinzer (1949), it can be shown that the water power associated with the waterfalls is orders of magnitude greater than the sound power levels. Most of that water power is converted to heat. Similar results are found in Schalko and Boes (2021). The reader can make this comparison using the cited references.

14.8 Filtration and Fluidized Beds

Figure 14.7 Snoqualmie Falls. Lilly, J. G. (2007). "Waterfall Noise", 154th Meeting, Acoustical Society of America, New Orleans, Louisiana, November 28, 2007.

14.8.1 Filtration

Seminal papers on the theory of water filtration are those by Camp (1964, 1965). The 1964 paper also discusses backwashing, which is addressed below under **Fluidized Beds** section. For filtration, Camp extended the Kozeny equation of laminar flow through granular materials (see Section 4.6) to the transition region by means of a chart of the friction factor vs. the Reynolds Number. Small floc particles deposit in a filter, as a sheath, around the sand grains. Camp presented a chart for the easy computation of the deposit ratio from the sand size and hydraulic gradient. The rate of removal for a given volumetric concentration of floc in the water at each depth is found to decrease greatly with depth. This decrease in effectiveness is related to the adhesiveness of the floc to the sand. Young (1985) provides a thorough review of wastewater filtration with granular-media filters.

Filtration of wastewater sludge is important in the sludge dewatering process. Quantitative measurements of the filterability of sludge by determination of its average specific resistance to filtration are made by the modified Buchner funnel and pressure methods attributed to Carman (1938) by Cassel and Johnson (1979). The specific resistance is defined as the pressure difference required to produce a unit rate of filtrate flow of unit viscosity through a unit weight of cake. The derivation of the formula for calculating the specific resistance is presented below following Pratsinis (2016) using notation similar to that of Cassel and Johnson.

Darcy's law is given by Equation 4.95 as $v = ki$ with the terms as defined in Section 4.6. The discharge velocity (aka face velocity) is equal to the suspension volume flow rate V_f^* divided by the filter area A_F, and the hydraulic gradient i is equal to the pressure drop over the filter and cake Δp (filtration pressure) divided by the time-dependent cake thickness h_K so that:

$$\frac{V_f^*}{A_F} = \frac{k\Delta p}{h_K} = \frac{k\Delta p}{\mu[h_K/(k\mu)]} \tag{14.82}$$

The term $\mu[h_K/(k\mu)]$ gives the total resistance to flow through the filter and describes the buildup of filter cake. The term μ is the viscosity of the liquid filtrate, as used below.

The filter medium resistance is considered negligible compared to the filter-cake resistance, and the filter-cake resistance is expressed in terms of the unit height cake resistance α_c as follows:

$$[h_K/(k\mu)] = R_c h_k \tag{14.83}$$

Setting $V_f^* = dV_f/dt$ in which V_f is the suspension volume and combining Equations 14.82 and 14.83 gives:

$$\frac{1}{A_F}\frac{dV_f}{dt} = \frac{\Delta p}{\mu(R_c h_K)} \tag{14.84}$$

A mass balance over the solid material can be expressed as follows:

$$A_F h_K(1-\phi)\rho_{solid} = V_f c \tag{14.85}$$

in which ϕ is the porosity or void ratio as employed in Section 4.6, and c is the cake solids concentration or mass per unit volume of filtrate.

Substituting h_K from Equation 14.85 into Equation 14.84 gives:

$$\frac{1}{A_F}\frac{dV_f}{dt} = \frac{\Delta p}{\mu\left[\frac{R_c}{(1-\phi)\rho_{solid}}\frac{cV_f}{A_F}\right]} \tag{14.86}$$

We replace the term $R_c/[(1-\phi)\rho_{solid}]$ by "area-specific cake resistance" R to obtain:

$$\frac{1}{A_F}\frac{dV_f}{dt} = \frac{\Delta p}{\mu RcV_f/A_F} \tag{14.87}$$

Letting b equal the slope of the filtrate discharge curve time/volume vs. volume, i.e., $b = (dt/dV_f)/V_f$, the above equation becomes:

$$R = \frac{\Delta p A_F^2 b}{\mu c} \tag{14.88}$$

The above equation differs from that given (without derivation) by Cassel and Johnson, in that they have a two in the numerator of the above equation, which the above derivation shows to be incorrect.

14.8.2 Fluidized Beds

Much of the discussion under this heading is taken from Camp (1964), Camp *et al.* (1971, 1973), and Graber (1994, pp. 551–553). We will first discuss the misapplication of G-value to fluidized beds, followed by discussion of sound concepts of backwashing.

14.8.2.1 G-Value

Cleasby (1972), building on a suggested use of G-value, which he attributed to Camp (1965), proposed a theory of optimum water backwashing of water filters. Cleasby's results, properly interpreted, are useful, but their derivation and original interpretation are incorrect. Most particularly, they do not represent an application of G-value in the sense reported by Cleasby.

Cleasby derived the following equation for the power dissipated by shear forces in a fluidized bed:

$$P_1 = \mu\left(\frac{dV'}{dz}\right)^2 \Delta x \Delta y \Delta z \tag{14.89}$$

in which V' is the average fluid velocity within the pores of the backwashed bed, x and y are Cartesian coordinates perpendicular to the direct of V', and z is the coordinate in the direction of V'. Cleasby's two-dimensional derivation is incorrect, primarily because of erroneous use of coordinate directions. The error can be demonstrated with reference to the original derivation or by using the power dissipation relationships given above. Using the latter approach, Equation 3.16b for the case of one-dimensional flow ($v_x = v_y = \partial/\partial x = \partial/\partial y = 0$) gives the power dissipation per unit volume as $2\mu(dv_z/dz)^2$. That is twice the power dissipation per unit volume $[P_1/(\Delta x\Delta y\Delta z)]$ given by Equation 14.89. From the summary derivation leading up to Equation 3.16b, it can be seen that $2\mu(dv_z/dz)^2$ represents normal stresses (pressure differences) rather than shear stresses, as suggested by Cleasby. Furthermore, $\mu(dV'/dz)^2$ cannot be equated to Camp and Stein's G-value, as claimed by Cleasby, because, as discussed earlier, the normal stress terms were deleted in Camp and Stein's derivation of G-value. Proper interpretation of what $\mu(dV'/dz)^2$ represents is not only of academic interest; it has important practical ramifications, as discussed below.

By essentially applying the one-dimensional energy equation to the fluidized bed, Cleasby derived a second relationship for power dissipation, given as follows:

$$P_2 = \rho g \cdot V' \Delta x \Delta y \cdot \left(\frac{dh}{dz}\right)\Delta z \tag{14.90}$$

Cleasby equated P_1 and P_2 (Equations 14.89 and 14.90) to derive an incorrect equation for "shear intensity" [incorrect because $\mu(dV'/dz)$ is not shear stress], which he then maximized to obtain what he thought was the porosity at which shear stress was maximized in a fluidized bed. The analysis yielded a potentially important result, but one that requires revised interpretation. Proper interpretation is revealed by modifying Cleasby's derivation to obtain essentially the same result as described next.

The head loss in a fluidized bed is given by the following form of the equation by Fair and Hatch (1933, Camp *et al.* 1971):

$$g\rho dh = gdz\left(\rho_g - \rho\right)(1-\varepsilon) \tag{14.91}$$

in which ρ_g is the mass density of the fluidized grains and ε is the expanded porosity at the selected level. Richardson and Zaki's equation, as modified by Amirtharajah and Cleasby (1972), relates the velocity and porosity at a level in a fluidized bed as follows:

$$V = K\varepsilon^n \tag{14.92}$$

in which the coefficient K and exponent n are functions of the size and other characteristics of the fluidized grains as given by Cleasby (1972).

Substituting dh from Equation 14.91 and V from Equation 14.92 into Equation 14.90 (and noting that $V' = V/\varepsilon$) results in the following equation, giving power dissipation per unit volume:

$$\frac{P_2}{\Delta x \Delta y \Delta z} = gK\left(\rho_g - \rho\right)\left(\varepsilon^{n-1} - \varepsilon^n\right) \tag{14.93}$$

For a particular system (K and ρ_g constant), the differential of Equation 14.93 with respect to ε can be set equal to zero to obtain a stationary point for ε (which can be shown from the second derivative to be a maximum) with the following result:

$$\varepsilon = \frac{n-1}{n} \tag{14.94}$$

The above equation is the same as that derived by Cleasby (1972) with an important difference in derivation and associated interpretation. It represents the expanded porosity at which power per unit volume is maximized, not that at which shear stress is maximized. At greater backwash rates, velocity V (and bed expansion) will be greater but dh/dz will be less, and the power per unit volume (proportional to Vdh/dz) will be less. The practical importance of the distinction between maximum power per unit volume and maximum shear stress is discussed below.

Cleasby (1972) offered experimental support for his theory that cleansing was maximized when power per unit volume was maximized in the fluidized bed. For the fluidized grains with which he experimented, the porosity at which power per unit volume was maximized was calculated to be about 0.70. Cleasby found that this was close to the porosity in the top layer of the expanded bed (i.e., the layer containing the maximum amount of filtered solids), at which he observed maximum cleansing in his experiments. [It is not clear how Cleasby measured porosity within the fluidized beds, although he appears to have had instrumentation with which to use the techniques developed by Camp (1965) and subsequently employed by Camp *et al.* (1971).] Cleasby's findings tied in nicely with earlier work by Amirtharajah (1970), who suggested that turbulence is the principal mode of cleansing. Amirtharajah presented an excellent literature review in which he cited studies using four different experimental techniques, each indicating that the scale and intensity of turbulence, particle motion, and fluid motion are all maximum at a porosity of approximately 0.70. He also suggested that filter cleansing can be optimized by backwashing so that the section of the bed where most of the suspended matter is found is expanded to a porosity approximately 0.70.

Although Camp (1965) discussed G-value as a parameter in backwashing, his eventual notions of backwashing clearly did not incorporate G-value (Camp *et al.* 1971). Camp *et al.* (1971) suggested that shear forces acting on the grains in a fluidized bed were responsible for cleansing during backwashing and that: "The maximum wash rate to be made available must be adequate to expand the bed sufficiently at the highest water temperature to fully fluidize the bed in the area of the bed where the wash rate is least.... When fluidized, the friction force past the grains is equal to the weight of the grains in water and is not increased with further expansion at higher wash rates." In other words, cleansing is a function of the shear drag. Argaman and Kaufman (1970), in their studies of the erosion of floc particles, proposed that the rate at which primary particles are stripped from the surface of a floc particle should be proportional to the floc surface area and surface shearing stresses. That is analogous to the notion of Camp *et al.* (1971), as expressed above.

The question of whether cleansing during backwashing is determined by power per unit volume or shear stress has important practical ramifications. The two parameters have different implications for the role that particle size plays in determining cleansing efficiency. Thus, the selection of media is partly dependent on the answer to this question. Furthermore, there are important practical ramifications to the suggestion by Cleasby (1972) on the one hand that maximum power per unit volume be applied to optimize cleansing, and the suggestion by Camp *et al.* (1971) on the other hand that the bed only be fully fluidized. Cleasby's proposal not only requires that the bed be expanded to a greater degree in some cases, but also that the input power (power per unit volume times bed area times *expanded bed depth*) be maximized. Not only are there operating cost implications (some of which were addressed by Cleasby) but also capital cost implications (e.g., a deeper vessel is required for a larger expanded bed depth). It seems likely that the economically optimal backwash would differ from that corresponding to maximum power per unit volume.

14.8.2.2 Correct Backwashing Concepts

With some supplementation, the following is taken primarily from Camp *et al.* (1971).

14.8.3 Introduction

The grains of a rapid water filter are best cleansed by backwashing upward with water consistently at a rate sufficient to fluidize the entire bed and for a period sufficient to wash the dislodged floc particles out of the bed and into the wash water gutters (typically 5–10 minutes). As noted above, when fluidized the friction force past the grains is equal to the weight of the grains in water and is not increased with further expansion at higher wash rates. The friction loss past the grains and the effectiveness of cleansing may be increased by increasing the grain size or density. Surface wash is of no help in fluidizing the bed. When air is applied to a bed, which is fluidized with water, the bed expansion is reduced. Air is of value in dislodging floc particles if applied to a fluidized bed, but should be applied at infrequent intervals because it compacts the bed and destroys stratification. The rubbing of the grains together during backwashing is a negligible factor in cleansing because nearly all of the energy of water wash is required to suspend the grains. Pulsation and vibration are also negligible factors in cleansing, and they are to be avoided because they tend to compact the bed and shorten filter runs by reducing the pore space available for floc deposits.

Effective washing cannot be accomplished in operation unless the means are provided in the design and construction of the filter plant. The grain size must be large enough and the density great enough to provide cleansing shear forces for all grains when suspended. As removal by filtration is increased by increased bed depth or by decreased grain size, a choice must be made in design between added depth or smaller grain size within the limitations that the grains must not be too small for cleansing, that small grains result in shorter filter runs, and that grains that are too large require excessive wash rates for suspension. Multibed filters with coarse-light grains supported by fine-heavy grains comprise an ingenious device for obtaining both long filter runs and good removal. For any grain size and depth of bed, the designer may determine by laboratory experiments the wash rates required at various water temperatures to fully fluidize the bed and obtain various bed expansions.

The filter bottom should be designed as dispersion conduits (Camp and Graber 1968, 1969; Ordon 1968) to minimize this variation in wash rates over the filter area and to effect vertical flow upward to the bed within the least vertical height from the filter bottom. The apertures that distribute the wash water to the underside of the bed should be smaller than any of the grains of the filter medium and should be fixed in position. Because they clog slowly with floc, the filter bottom and filter should be designed to facilitate cleansing without removing the bed or causing damage to walls or parts with strong solutions of sulfuric acid (or alkali for alum floc) at intervals of several months to two years.

The purpose herein is to review the facts developed from previous studies of filter backwashing and to present experimental procedures and data used to develop or substantiate these relationships. All of this is directed toward facilitating the design, evaluation, and operation of granular water filters.

14.8.4 Facts and Fancy

The technical literature on water filtration abounds with concepts which fail to meet the test of scientific examination. These fancies must be replaced with facts if progress is to be made in filtration practice.

14.8.4.1 Floc Penetration

It is widely believed that floc is deposited within the bed only near the top and that no removal takes place below, thus obviating the need for cleansing the bottom grains. This belief is a legacy of slow-sand-filtration practice, where only the top sand layer was removed for cleansing. It was furthered by studies of rapid filtration in glass tubes, in which one of the observations was depth of floc penetration, i.e., the depth down to the level where the floc had made no appreciable change in the color of the sand. A conclusion to be drawn from this concept is that, at constant filter rate, the head loss does not increase during the filter run in the layers below the depth of floc penetration. The facts are that floc penetrates completely through the bed and filter bottom, as proved by the presence of some filterable impurities in the filter effluent, and that deposits occur throughout the depth and in the filter bottom, as proved by measured increases in head loss.

14.8.4.2 Breakthrough

It is true (Camp 1964, p. 28) that the greatest rate of removal per unit depth occurs at the top of the bed at the beginning of a run and that the level of greatest removal moves downward into the bed as the deposits increase in the upper layers. If the bed is deep enough and effectively washed, the quality of the effluent is best at the beginning of a run and slowly deteriorates throughout the run. If the bed is not deep enough, the level of greatest removal may approach the bottom too closely before the end of a run and cause a more rapid deterioration in the quality of the effluent. The term breakthrough implies dislodgement of floc from the grains and its discharge in high concentrations in the effluent. No such breakthrough occurs (Camp 1964, p. 28) because the shearing force on the floc sheaths covering the grains is inadequate to tear the floc gel apart. Such tearing does not occur even at the top of the bed at the end of a run, when the shear force is greatest. That is evidenced by the fact that the head loss continues to increase because of continued deposits, albeit at sharply decreased rates. In the early use of rapid filters, it was common practice to filter to waste for a few minutes at the start of a run to purge sediment remaining after the wash. It was also common practice to use wash rates of 12–18 in. rise per minute, now known to be inadequate to completely fluidize the bed in the summer months.

The concept of breakthrough was originated by Hudson (1959) and was publicized by Fair *et al.* (1968, §27-15). Hudson suggested the existence of a phenomenon that he called "terminal breakthrough" in which floc deposited in the filter is sheared and suddenly breaks through causing a large increase in the filtrate turbidity. Camp (1964) presented convincing evidence against the existence of breakthrough. He noted that if breakthrough is significant, it should be reflected in the shape of the head loss curve, i.e., the hydraulic gradient should decrease at the filter level where breakdown occurs and increase at the levels where the floc mass is intercepted. Camp's studies of head loss curves for granular filters found no such effect (see, e.g., Camp 1964, Figure 9). Camp (1964, p. 13) also cites studies with radioactive floc, which support his contention.

Camp *et al.* (1971, p. 904) explained Hudson's interpretation by pointing out that "... the greatest rate of removal per unit depth occurs at the top of the beginning of a run and ... the level of greatest removal moves downward into the bed as the deposits increase in the upper layers .... If the bed is not deep enough, the level of greatest removal may approach the bottom too closely before the end of a run and cause a more rapid deterioration in the quality of the effluent." The variation in water quality and floc strength that have been attributed to "breakthrough" can be explained by the initial reception of floc in the filter.

14.8.4.3 Mud Balls Within Beds

Mud balls remaining in the bed after washing are the result of failure to fluidize the entire bed. The belief is widespread that mud balls in the bed are formed by the breakup of the top layer as expansion begins and settling of the fragments and mud within the bed. In many sand filters, the surface sand grains are too small to be washed effectively by suspension alone, but it is unlikely that fragments of the top layer can settle into a fully fluidized bed without disintegration of the fragment so that the mud will be washed away. A much more probable cause is failure to fluidize the bed. If a bed is not fully fluidized, the wash water will short-circuit at high velocity through the path of least resistance, drawing bottom grains into this path and creating rollers which will carry the top mud laterally, then down into and under the bed. In order to observe firsthand what actually happens with mud balls in a fluidized bed, about 20 mud balls which had been collected a year previously from a filter which had not been properly fluidized in washing were placed in an experimental filter (then provided with a nozzle for both air and water wash) at intervals of depth throughout the sand and coal. The filter was then washed with water at a rate sufficient to fluidize the bed, whereupon all of the mud balls rose to the surface of the coal, being broken up to 1/4–1/2 in. on passage through the bed. Some of the smaller fragments were carried away with the waste wash water. The water was then shut off and air was applied, which resulted in much smaller fragmentation of the mud balls on the surface. The air was shut off, and the bed was again fluidized with water. Many of the smaller fragments were washed out of the tube, but the larger and denser particles remained, floating on the top of the coal. Some of these particles were not removed with a wash rate of 65 in./min, well beyond practical rates.

Mud balls will not settle into a fluidized bed unless their density is greater than the density of the mixture of grains and fluid. The density of the mixture at any level is

$$\rho_m = (1 - p_e)\rho_g + p_e\rho_f \tag{14.95}$$

in which ρ_m, ρ_g, and ρ_f = the mass densities of the mixture, grains, and fluid, respectively, and p_e is the expanded porosity at that level. For example, with coal in water at 48 in./min. rise, ρ_g at 1.54, ρ_f at 1.0, and p_e at 0.63, ρ_m is 1.20 g/cc. Mud balls must have densities greater than 1.20 to settle into this bed. Conversely, the 20 mud balls used in the previously described experiment must have been lighter, containing few, if any, grains.

14.8.4.4 Mud Balls on Top of Bed

Mud balls on top of filters usually consist of sediment permitted to settle back on top of the bed after a wash because either the wash rate or time of wash was inadequate to carry it into the gutters. These floating mud ball fragments have no influence on either filtration or backwashing, but they are unsightly and are the only mud balls visible to the operator and the public. They must, therefore, be removed. If a sufficient wash rate is provided in design to fluidize the bed at all times, a small increase in wash rate may wash out these mud balls, provided however they have not been allowed to remain too long and increase too much in density. If air wash is available, it may be used to break up these mud balls so that they may be washed out when the bed is fluidized by water. It is not feasible to provide a high enough water wash rate to wash away floating mud balls accumulated under the worst conditions. In the experiments described above, with the filter at rest, a 1/4 inch glass tube attached to a 1/4 inch rubber tube was held 1/2 inch above the coal, and water was sucked up into the glass tube at selected velocities. The minimum velocity which would remove all of the mud ball fragments was found to be 30 fpm, about 10 times the maximum wash rate usually provided. A simple and inexpensive means for removing mud balls from the top of filters is a portable vacuum cleaner, such as used for swimming pools, but with the nozzle equipped with runners at both ends to keep the opening about 1/2 inch off the top of the bed.

14.8.5 Beds Fluidized by Water

When the grains of a filter bed are fully suspended in water, the force of friction supporting the grains is equal to the weight of the grains in water. Following the method of Fair and Hatch (1933):

$$g\rho dh = gdl\left(\rho_g - \rho\right)(1 - p_e) \tag{14.96}$$

in which dh = the head loss through the expanded layer of thickness dl, and ρ = the mass density of the water. For the bed as a whole, Equation 14.96 becomes:

$$g\rho h = gl\left(\rho_g - \rho\right)(1 - p_{av}) \tag{14.97}$$

in which h = the head loss through the depth l of fluidized bed, and p_{av} = the average porosity throughout the depth l. As ρ_g and p_e may vary throughout l, the right side of Equation 14.96 is not very useful. The left side is very useful, however, for weighing a fluidized bed in place by means of a piezometer column. The weight in the fluid per unit bed area is essential to estimates of bed porosity, and no such information is available from construction since the depth and not the weight are called for by the designer. The method has been used by Camp (1965) and is quite accurate, as shown hereinafter. If the piezometers are closely spaced, much information may be derived about the other parameters in washing. Equation 14.96 may be written as:

$$i = \frac{\rho_g - \rho}{\rho}(1 - p_e) \tag{14.98}$$

in which i = the hydraulic slope dh/dl at any level in the fluidized bed, and p_e is the porosity at that level. This value of i may be introduced into the Kozeny equation to produce a relation between the wash rate and the expanded porosity at any level in a fluidized bed, as follows

$$q_w = \frac{g\left(\rho_g - \rho\right)}{\beta \upsilon \rho} \frac{p_e^3}{1 - p_e} d^2 \tag{14.99}$$

in which q_w = the water wash rate per unit area of bed; β = a dimensionless factor depending upon the porosity, pore shape, and flow rate; ν = the kinematic viscosity of the water; and d = the count-and-weigh average grain size at that level. The grains tend to be stratified in a fluidized bed, from coarse at the bottom to fine at the top, but stratification is poor even with highly nonuniform grain sizes, and at each level the grains vary in size. The less the variation in grain size in a bed, the smaller will be the gradation in average gain size with depth.

14.8.6 Beds Fluidized by Air–Water Mixture

A comparison may be made with Equation 14.99 of the relative effectiveness of air (alone in a dry bed) and water in fluidizing a bed. Assuming β is the same, it is found that the air wash rate must be about 84 times the water wash rate for the same expanded porosity. Although the kinematic viscosity of air is approximately 15 times that of water, the density of water is about 800 times that of the air in the bed.

More general relationships, analogous to Equations 14.96 and 14.97 for beds fluidized by water, may be developed for beds fluidized by an air–water mixture. The effective density of an air–water mixture depends upon the rise velocity of air with respect to the water, which in turn depends on the size of the air bubbles. The effective density of an air–water mixture is determined by the volume of each in the pores of the bed and the porosity at a selected level. The wash rates in the bed are

$$q_w = V_w p_w \text{ for water} \tag{14.100}$$

and

$$q_a = (V_w + v)(p - p_w) \text{ for air} \tag{14.101}$$

in which V_w = the water velocity in the water pore fraction p_w; v = the rise velocity of the air bubbles with respect to the water; $p - p_w$ = the pore fraction occupied by the air; and p = the porosity in the bed.

The effective density, ρ_{aw}, of the air–water mixture is

$$\rho_{aw} = \frac{p_w}{p}\rho + \frac{p - p_w}{p}\rho_a \tag{14.102}$$

As $\rho_a \ll \rho$, the second term of Equation 14.102 may usually be deleted (unless $p_w \ll p$).

Neglecting the change in momentum caused by expansion of the air, a force balance for an expanded layer of thickness dl may be expressed by

$$-dP = gdl\left[\rho_g(1 - p_e) + p_e\rho_{aw}\right] \tag{14.103}$$

in which dP = the pressure differential across the layer.

Neglecting the change in kinetic energy in the rising air–water mixture caused by expansion of the air, the energy principle requires that the energy head loss dh be equal to minus the differential in piezometric head, dh_p, the piezometric head being defined by

$$h_p = \frac{P}{g\rho_{aw}} + z \tag{14.104}$$

in which z = the elevation above an arbitrary datum at a level in the bed. Therefore, Equation 14.103 may be written in terms of the energy head loss as

$$g\rho_{aw}dh = gdl\left(\rho_g - \rho_{aw}\right)(1 - p_e) \tag{14.105}$$

which reduces to Equation 14.96 for the special case of a water wash, $\rho_{aw} = \rho$. For the bed as a whole, assuming ρ_{aw} constant:

$$g\rho_{aw}h = gl\left(\rho_g - \rho_{aw}\right)(1 - p_{av}) \tag{14.106}$$

The right-hand side of Equation 14.106 represents the weight per unit area of the filter medium in the air–water mixture.

Fluid statics must be properly applied to water manometers used to measure differentials in a bed fluidized by an air–water mixture. The pressure differential is related to the water-manometer differential, dh_M, by:

$$-dP = \rho g(dh_M + dl) \tag{14.107}$$

Eliminating dP from Equations 14.104 and 14.107 and taking $dh = -dh_p$, as explained above, results in the following relationship between dh and dh_M:

$$\rho g(dh_M + dl) = \rho_{aw}g(dh + dl) \tag{14.108}$$

Using Equations 14.108 and 14.102 to express h in Equation 14.106 in terms of the manometer reading h_M, the following expression is obtained:

$$h_M = l\left(\rho_g/\rho - 1\right)(1 - p_{av}) - l(p_{av} - p_w) \tag{14.109}$$

The first term on the right-hand side of Equation 14.109 is a constant property of the bed, equal to the head loss across the bed fluidized in water alone. The manometer h_M is therefore equal to the head loss across the bed fluidized in water alone minus the term $l(p_{av} - p_w)$, which represents the volume of air per unit area of bed. By means of Equation 14.109, the pore fraction occupied by air and the water pore fraction may be calculated from backwash test data. Then, through the use of Equations 14.100 and 14.101, the water velocity and bubble rise velocity within the bed may be determined.

14.8.7 Multimedia Beds

In a fluidized bed, the grains of a light filter medium will not settle into and mix with the grains of heavier material unless the density of the light grains exceeds the density of the mixture of heavy grains and water. A precise criterion for the mixing of grains of unequal density may be developed by considering the forces acting on the grains.

The buoyant force on a grain in a fluidized bed is equal to the weight of the mixture displaced, given by

$$F_b = V_d g \rho_m \tag{14.110}$$

in which V_d = the volume of the mixture displaced, and ρ_m = the mass density of the mixture.

A floating particle within a mixture, having the volume V_d, must have a net impelling force downward equal to its weight less the drag force of the washwater past the particle. The net impelling force is

$$F_i = V_d g \rho_m - C_D \frac{3}{2d} V_d \rho \frac{q_w^2}{2p_e^2} \tag{14.111}$$

in which $V_d g \rho_p$ = the weight of the particle of density ρ_p suspended within the mixture; C_D = the drag coefficient of the particle of diameter d and cross-sectional area $[3/(2d)]V_d$; q_w = the wash rate; and $\rho\left[q_w^2/(2p_e^2)\right]$ = the dynamic pressure.

For equilibrium, F_i equals F_b and the following results:

$$\rho_m = \rho_p - C_D \frac{3}{2d} \rho \frac{q_w^2}{2gp_e^2} \tag{14.112}$$

The Reynolds Number for the friction drag past the particle is

$$\mathbf{R} = \frac{\rho \frac{q_w}{p_e} d}{\mu} \tag{14.113}$$

from which

$$\frac{q_w^2}{p_e^2} = \frac{\mu^2}{\rho^2 d^2} \mathbf{R}^2 \tag{14.114}$$

Substituting this value into Equation 14.111:

$$\rho_m = \rho_p - \frac{3\mu^2}{4g\rho d^3} C_D \mathbf{R}^2 \tag{14.115}$$

Experimental values of the drag coefficient C_D for spheres as a function of the Reynolds Number **R** are given by Camp (1946b; 1963, Figure 9). It is shown later that the friction drag term on the right side of Equation 14.115 is small compared with the density ρ_p and may be neglected in some cases without appreciable error. This, of course, is not true when mixing grains of equal density, in which case the friction drag term is important.

Consider coal with a density of 1.5 and a fluidized bed of sand, ρ_g of 2.65. Because from Equation 14.95 ρ_m is $(1 - p_e)2.65 + p_e \times 1.0$, Equation 14.115 gives $p_e > 0.70$, and the expansion $e = (p_e - p_o)/(1 - p_e) > 0.83$, in which the porosity p_o before expansion is assumed to be 0.45 for the sand. Coal will float on top of fluidized sand or heavier media without mixing because the expansion required for mixing is too great for practical application. Similarly, an activated carbon bed will float on top of a fluidized sand bed because the carbon density is less than that of anthracite coal.

Consider sand with a density of 2.65 and a fluidized bed of garnet grains of density 4.00. From Equations 14.95 and 14.115, $\rho_m = (1 - p_e)4.00 + p_e$ and $p_e > 0.45$. As the porosity of a garnet bed with a moderate wash rate exceeds this value, sand will fall into and mix with a fluidized garnet bed.

The density of a mixed-media fluidized bed at a level with the porosity p_e is:

$$\rho_m = p_1\rho_1 + p_2\rho_2 + p_e\rho_f \tag{14.116}$$

in which p_1 = the volume fraction of heavy grains with density ρ_1; p_2 = the volume fraction of lighter grains with density ρ_2; and ρ_f = fluid density. Now

$$\frac{p_2}{p_1} = \frac{w_2}{w_1}\frac{\rho_1}{\rho_2} \tag{14.117}$$

in which w_2 and w_1 = the fractional weights (per unit volume of bed) of the light and heavier grains, respectively. But

$$p_1 + p_2 = 1 - p_e \tag{14.118a}$$

from which

$$\frac{p_2}{p_1} = \frac{1 - p_e}{p_1} - 1 \tag{14.118b}$$

Equating the values of p_2/p_1 in Equations 14.117 and 14.118b and solving for p_1,

$$p_1 = \frac{1 - pe}{1 + \dfrac{w_2}{w_1}\dfrac{\rho_1}{\rho_2}} \tag{14.119}$$

Substituting the value of $p_2\rho_2$ from Equation 14.117 into Equation 14.116, and the value of p_1 from Equation 14.119 into the resulting equation, ρ_m becomes

$$\rho_m = (1 - p_e)\rho_1 \frac{1 + \dfrac{w_2}{w_1}}{1 + \dfrac{w_2}{w_1}\dfrac{\rho_1}{\rho_2}} + p_e\rho_f \tag{14.120}$$

The effective density of the solid grain is

$$\rho_s = \frac{p_1\rho_1 + p_2\rho_2}{1 - p_e} = \rho_1 \frac{1 + w_2/w_1}{1 + (w_2/w_1)(\rho_1/\rho_2)} \tag{14.121}$$

Thus, Equation 14.120 becomes

$$\rho_m = (1 - p_e)\rho_s + p_e\rho_f \tag{14.122}$$

Equating the value of ρ_g in Equation 14.98 to ρ_s in Equation 14.121, and solving for $1 - p_e$:

$$1 - p_e = \frac{i}{\dfrac{1 + (w_2/w_1)}{1 + w_2\rho_1/(w_1\rho_2)} - 1} \tag{14.123}$$

In order to solve for β at any level by means of Equation 14.99, it is necessary to have an average value of d_2 for both heavy and light grains. From the definition of count-and-weigh diameter:

$$d_{av}^3 = \frac{6}{\pi}\frac{w}{\rho_s g n} \tag{14.124}$$

in which w is the fractional weight of the grain and n = number of grains per unit volume of bed. Because $w/(\rho_s g) = p = p_1 + p_2$ and $n = n_1 + n_2$, Equation 14.124 can be rewritten as

$$d_{av}^3 = \frac{6}{\pi}\frac{p_1 + p_2}{n_1 + n_2} \tag{14.125}$$

in which n_1 and n_2 are the numbers of grains per cubic centimeter of fluidized bed having count-and-weigh sizes d_1 and d_2 and densities ρ_1 and ρ_2, respectively. As p_1 and p_2 are the volumes per cubic centimeter of bed occupied by the heavy and light grains, respectively, the number of grains per cubic centimeter of bed are

$$n_1 = \frac{p_1}{(\pi/6)d_1^3} \tag{14.126}$$

and

$$n_2 = \frac{p_2}{(\pi/6)d_2^3} \tag{14.127}$$

Combining Equations 14.125–14.127 gives the average value of d as follows:

$$d_{av}^2 = \left(\frac{n_1 d_1^3 + n_2 d_2^3}{n_1 + n_2}\right)^{2/3} \tag{14.128}$$

14.8.8 Laboratory Experiments

Experimental procedures, air–water backwash experiments, and water backwash experiments are described in Camp *et al.* (1971).

14.8.9 Follow-Up

The paper discussed above (Camp *et al.* 1971) solicited substantial discussion, which enabled the author, following Camp's death, to provide important substantiation and refutation accompanied by presentation of numerous additional references (Camp *et al.* 1973).

14.9 Washwater Trough External Flow

Spatially-increasing flow within filter washwater troughs during backwashing is addressed in Chapter 12. The external flow in the vicinity of round-bottom filter washwater troughs is discussed here. The author has encountered situations in which the designer, wishing to minimize the depth of the filter structure, plans to fluidize the bed to a level between the washwater troughs [contrary to recommendations of Fair and Geyer (1954, §24-10)]. The designer assumes that the rise velocity between the troughs is equal to that below the troughs increased by the ratio of the flow area between the troughs divided by the flow area below the troughs (based on simple continuity). However, the actual velocity near the edge of the troughs will be substantially higher than given by that assumption and is likely to result in loss of media flowing into the troughs. If one approximates the velocity on the side of the troughs as equal to that on the side of a cylinder as given in Section 4.2, we have a velocity on the side of the trough of two times the average velocity that would occur between the troughs based on simple bulk continuity.

We can use the Section 4.2 analysis more generally by casting the stream function in Cartesian coordinates (Equation 4.42 with $\mu = 2\pi R^2 V_\infty$):

$$\psi = V_\infty y\left(1 - \frac{R^2}{x^2 + y^2}\right) \tag{14.129}$$

then obtaining the x and y components of velocity as follows:'

$$u = \frac{\partial\psi}{\partial y} = V_\infty\left[1 - R^2\frac{(x^2 - y^2)}{(x^2 + y^2)^2}\right] \tag{14.130a}$$

$$v = -\frac{\partial\psi}{\partial x} = -V_\infty R^2\frac{2yx}{(x^2 + y^2)^2} \tag{14.130b}$$

In nondimensional form, these equations become:

$$\frac{u}{V_\infty} = 1 - \frac{\left[\left(\frac{x}{R}\right)^2 - \left(\frac{y}{R}\right)^2\right]}{\left[\left(\frac{x}{R}\right)^2 + \left(\frac{y}{R}\right)^2\right]^2} \tag{14.131a}$$

$$\frac{v}{V_\infty} = -\frac{2(y/R)(x/R)}{\left[\left(\frac{x}{R}\right)^2 + \left(\frac{y}{R}\right)^2\right]^2} \tag{14.131b}$$

Equation 14.131a can be used to represent the upflow to washwater troughs. With R being the radius of the trough bottom, x the distance measured downward from the bottom of the trough, and L the clear distance between adjacent troughs, and m the uniformity of upflow given by:

$$m = \frac{u(y=0)}{u(y=L)} \tag{14.132}$$

Using Equation 14.131a, the terms in Equation 14.132 are given by:

$$\frac{u(y=0)}{V_\infty} = 1 - \frac{1}{(x/R)^2} \tag{14.133a}$$

$$\frac{u(y=L)}{V_\infty} = 1 - \frac{\left[(x/R)^2 - (L/R)^2\right]}{\left[(x/R)^2 + (L/R)^2\right]^2} \tag{14.133b}$$

Using the above equations, the uniformity m as a function of x/R and L/R can be plotted over pertinent ranges, as shown in the following Figure 14.8.

The uniformity is seen to be relatively independent of trough spacing and to decrease as the filter bed gets closer to the bottom of the trough. If the bed expands to a level above the bottom of the trough, then we approach the condition mentioned above, in which the velocity near the side of the trough approaches a condition corresponding to a uniformity m of 0.5, which is clearly unacceptable. A good criterion for reasonable bed uniformity is the bed not being allowed to expand to less than $x/R = 3$.

Rectangular filter troughs will be briefly mentioned. Semicircular troughs are "much superior to the flat-bottomed ... (troughs) commonly used. Flat-bottomed troughs tend to set up eddy currents in the rising wash water. Such currents are not in evidence with the circular section and a much better flow of the wash water to the troughs is obtained" (Georgia 1949, 1950). Vortices form under the bottom and along the sides near the bottom of rectangular troughs, adding to the nonuniformity. A more recent use in Chicago of rectangular filter troughs and add-ons to reduce media loss is extensively reported (Kawamura *et al.* 1997), but the major takeaway in the present author's opinion is the relative deficiencies of rectangular troughs.

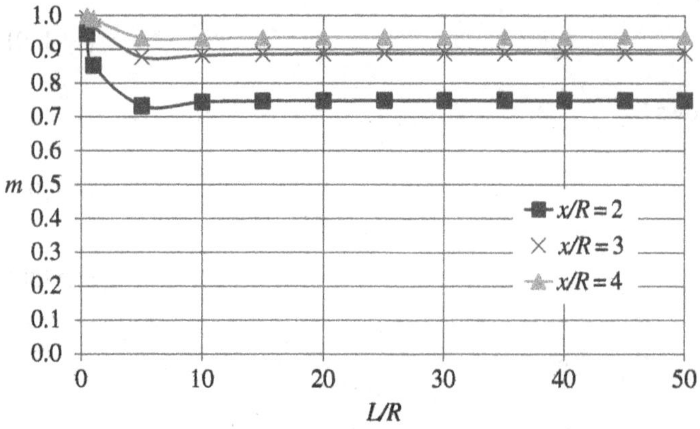

Figure 14.8 Chart of m vs. L/R and x/R

Example 14.7

The author was asked to investigate the loss of filter media into the washwater troughs during backwashing at a water-filtration plant (Andover, Massachusetts). The troughs have semicircular bottoms and vertical sides, 30 inch trough top widths, and with centerlines spaced 70 in. apart. The plant operator reported that he could not backwash the filters above a rate of 8 mgd (5600 gpm) without losing filter coal. The clear distance between adjacent troughs is $L = 70 - 30 = 40$ in. giving $L/R = 40/15 = 2.67$. From Figure 14.8, the uniformity is very good. The filter bed consists of 24 in. of anthracite coal over 16 in. of sand resting on the filter bottom. The top of the coal bed during filtration is 7 in. below the bottom of the troughs. Full fluidization of the bed necessary for adequate backwash cleansing could not be accomplished without losing coal into the troughs. In retrospect, an initial recommendation of reducing the sand bed depth to 9 in. would have helped by increasing the fluidized bed expansion from 17.5 to 22%. Column testing of the type presented in Camp *et al.* (1971) would have been desirable, and that reference does suggest that a 22% expansion would have sufficed.

Example 14.8

For New Bedford, Massachusetts, a water filtration plant is to be designed using troughs with semicircular bottoms and vertical sides spanning the 24 ft width of the 24 × 36 ft filter. The backwash rate is to expand the anthracite over sand bed to an elevation under the bottom of the troughs. During filtration, the depths of the sand and anthracite are 6 in. and 36 in., respectively, totaling 3.5 ft. Over the 36 ft length, there will be four troughs of 2′–6″ width with clear distance between the troughs and side walls of $L = 6$ ft-6 in. Other design parameters are expansion of 25%, equaling $0.25(6 + 36) = 10.5$ in. $= 0.875$ ft. Determine the appropriate distance of the trough bottoms above the expanded bed to achieve 90% uniformity of upflow during backwashing.

Solution

With $R = 2.5/2 = 1.25$ ft, we have $L / R = 6.5/1.25 = 5.2$. We will have 90% uniformity using the criterion of $x / R = 3.5$ mentioned above. With $L/R = 5.2$, we have $x = 3(1.25) = 3.75$ ft as the distance from the top of the troughs to the top of the expanded bed. The distance of the trough bottoms above the expanded bed is then $3.75–2.5/2 = 2.50$ ft. The reader is encouraged to sketch the dimensions to get the overall picture.

15

Introduction to Hydraulic Aspects of Natural Systems

15.1 Introduction

The term "natural" is used somewhat loosely here. Although the systems addressed do occur naturally, in some cases, they can also be modified and even created by man. Natural systems of concern to the environmental engineer can be categorized as rivers (and streams), lakes (and ponds and freshwater wetlands), estuaries, ocean waters, and groundwater. The areas of concern include the assessment of pollution impacts and their use as receiving waters; their use as water sources (rivers, lakes, and groundwater); the protection of engineered structures against their destructive effects; water wells, subsurface drainage, and related groundwater topics. Waterfalls are discussed in Section 14.7 on Aeration and Postaeration.

Natural systems are usually more complex hydraulically than the man-made systems emphasized in previous chapters. This complexity is a source of pleasure to the naturalist-engineer, and a challenge to the engineer faced with a practical problem. The latter is a key reason for the greater involvement of hydraulic specialists in such systems. There is often less good technical literature for natural systems than there is for man-made hydraulic systems most often encountered by environmental engineers.

This chapter is, as the title suggests, introductory in nature. The intention is largely to provide a sense of perspective and an appreciation of the complex character of natural systems. The reader should gain the ability to solve certain problems and to understand the type of assistance and need for assistance that can be provided by appropriate specialists.

15.2 Rivers

The hydraulic aspects of rivers addressed here are dealt with under the headings of *backwater analyses, asymmetric flow in symmetric expansions*, *meanders*, *water-quality modeling*, and *river aeration*. An interesting and enjoyable article about the names and epithets of rivers, including numerous of those mentioned below, is given by Ettema (2005).

15.2.1 Backwater Analyses

Backwater calculations are used to establish hydraulic characteristics along rivers for several purposes. The most common purpose in the U.S. is for flood plain studies used to establish flood zones for the National Flood Insurance Program under the Federal Emergency Management Agency (FEMA). Such calculations are also used for the related purpose of determining whether a project proposed in the flood zone (e.g., interceptors or levees) will significantly increase flood levels for which the U.S. Army Corps of Engineers has approval authority. Similar studies are undertaken on smaller waterways not subject to federal jurisdiction but under local jurisdiction (e.g., Conservation Commissions) to determine if proposed encroachments on the floodplain will worsen flooding. Although in most cases such studies are not directly related to environmental engineering, there are direct relationships in some cases. One such is in determining discharge conditions for wastewater treatment plants as part of determining elevations for plant components. An example for that purpose is given below for the Trinity River in Texas.

Another purpose of backwater analyses is to determine the tailwater conditions for detention basins whose outlets are submerged by a downstream waterway. That tailwater must be considered as an integral part of the flow-elevation relationship for the basin. Yet another purpose is to most accurately determine time of travel and atmospheric reaeration characteristics in conjunction with water-quality modeling under low-flow conditions. [Such as for the Nashua River in Massachusetts (Graber 1974b, 1975b) and also for the Trinity River.]

The hydrologic aspects of floods are well addressed and readily available elsewhere (Linsley *et al.* 1982, Bedient and Huber 1988, Maidment 1993), and are outside of the present scope. Also see Graber (2009b) and Gonwa (2012), which include perspectives on rainfall-intensity-frequency relationships and common misconceptions.

An important point to realize at the outset is that uniform-flow conditions rarely occur along natural streams. Rather, nonuniform flow is much more common. Accordingly, nonuniform flow methods, analogous to those discussed in Chapter 7, are necessary for satisfactory accuracy.

One wishing to assess the departure of river hydraulics from uniform-flow conditions may do so readily by analyzing the cross sections and "rating curves" available for stream gauging stations. The U.S. Geological Survey has established such stations throughout the United States. Rating curves are a fit of measured points of flow vs. water-surface elevation. Having established such a curve, flow can be inferred from depth measured above a reference datum. The U.S. Geological Survey commonly measures depth by means of a nitrogen bubbler and automatic chart recorder. A good discussion of such devices and streamflow measurement more generally is given by Boyer (1964, Section 15).

Manning's equation for uniform flow (Equation 7.36) gives the following relation between Manning's n and other hydraulic variables:

$$n = \frac{CAR^{2/3}S_o^{1/2}}{Q} \tag{15.1}$$

where C is 1.486 in customary units or 1.0 in SI units. Knowing S_o and Q, and determining A and R for a known cross section at water-surface elevations corresponding to specific flows Q, n-values calculated by the above equation can be plotted against flow or water-surface elevation. An example is shown in Figure 15.1. If uniform-flow conditions were approximated, Manning's n would be essentially constant at low water-surface elevations (corresponding to fairly uniform roughness over the wetted perimeter) and vary at higher water-surface elevations only to the channel. The Figure 15.1 variation in n is from 0.0455 to 0.00397, the largest being 11.5 times the smallest. At lower depths, not only is the extent

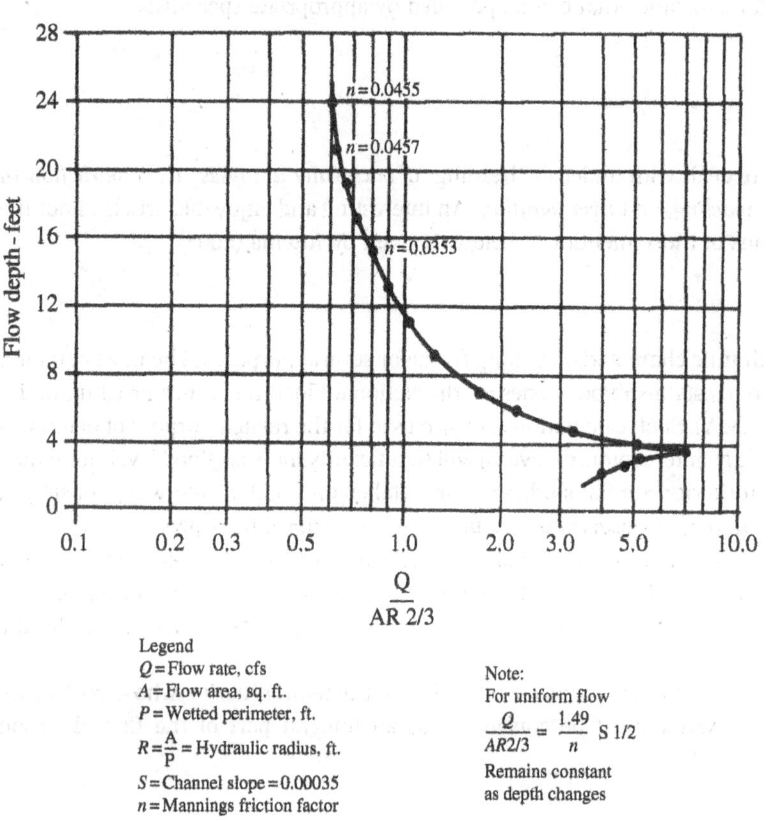

Figure 15.1 Hydraulic Analysis of USGS Grand Prairie Station, Trinity River, Texas

to which the overbank roughness affects the composite n-value for the variation much larger than could result from roughness variations alone, but the calculated value of n is also much smaller than could reasonably be associated with the channel roughness. Equation 15.1 (with S taken as the channel slope S_o as opposed to the energy gradient S_f), does not provide a reasonable approximation, because uniform-flow conditions do not occur at this cross section (except possibly at the highest depths). The reader may wish to go through a similar exercise for other rating curves.

The departure from uniform-flow conditions in natural channels is due primarily to variations in channel slope or cross section. As discussed in Section 7.3, a fairly long reach of constant slope or cross section is necessary for uniform-flow conditions to be attained. Thus, gradually-varied flow computations are required for suitable accuracy, by methods analogous to those discussed for man-made channels in Chapter 7.

The one-dimensional energy equation (Equation 2.64) can be written in terms of water-surface elevations or levels (WSL) as follows (for α assumed equal 1):

$$(WSL)_1 + V_1^2/(2g) = (WSL)_2 + V_2^2/(2g) + h_e \tag{15.2a}$$

$$h_e = \bar{S}_f L + C\left|\frac{V_2^2}{2g} - \frac{V_1^2}{2g}\right| \tag{15.2b}$$

in which subscripts "1" and "2" denote conditions at known upstream and downstream cross sections, respectively. The term h_o is the energy loss, consisting of friction loss and form loss, the latter based on the absolute value of the velocity head difference. Values of the expansion or contraction loss coefficient C equal to 0.1 for contractions $[V_2^2/(2g) > V_1^2/(2g)]$ and 0.3 for expansions $[V_2^2/(2g) < V_1^2/(2g)]$ are commonly used for gradual transitions (USACE 1990).

The friction loss is given by the product of a suitable, representative friction slope S_f and the flow length between sections "1" and "2". Several relationships have been employed for the representative friction slope in a reach, including the following (USACE 1990):

Average Friction Slope Equation

$$\bar{S}_f = \frac{S_{f1} + S_{f2}}{2} \tag{15.3a}$$

Geometric Mean Friction Slope Equation

$$\bar{S}_f = \sqrt{S_{f1} \cdot S_{f2}} \tag{15.3b}$$

Harmonic Mean Friction Slope Equation

$$\bar{S}_f = \frac{2S_{f1} \cdot S_{f2}}{S_{f1} + S_{f2}} \tag{15.3c}$$

Average Conveyance Equation

$$\bar{S}_f = \left(\frac{Q_1 + Q_2}{K_1 + K_2}\right)^2 \tag{15.3d}$$

The conveyance is defined by $Q/S_f^{1/2}$ and is given according to the Manning equation by:

$$K = \frac{C}{n}AR^{2/3} \tag{15.4}$$

The friction slope at a particular section (S_{f1} or S_{f2}) is given by Manning's equation:

$$S_f = \left(\frac{nQ}{CAR^{2/3}}\right)^2 \tag{15.5}$$

Appropriate n values are commonly determined by comparing field observations to photographs of channels having known n values (Chow 1959, §5.9, §5.10; Barnes 1977), although determination or adjustment by comparison of calculated and observed profiles is preferable where data permits. However, in the selection of n values, steep channels with natural or man-made rock linings require special consideration; our main interest is in such channels on the side slopes of some landfills, and they are discussed in Section 7.5.

Based partially on the studies of Reed and Wolfkill (1976), the Hydrologic Engineering Center has suggested the following criteria for selection of the friction equation giving the most accurate determination of the water-surface profile with the least number of cross sections. The criteria are said to not select the best friction equation when there is significant lateral expansion, such as downstream of a contracted bridge opening. The direction of change of friction slope can be readily ascertained by noting $S_f^{1/2} \propto 1/(AR^{2/3})$ for constant Q and n, and that depth increases with increasing $AR^{2/3}$. Thus if depth increases in the direction of computations (upstream for subcritical flow and downstream for supercritical flow) S_f decreases and vice versa for decreasing depth.

Criteria Utilized to Select Friction Equation (USACE 1990)

Profile Type	Is Friction Slope at Current Cross Section Greater than Friction Slope at Preceding Cross Section?	Equation Used
Subcritical (M1, S1)	Yes	15.3a
Subcritical (M2)	No	15.3c
Supercritical (S2)	Yes	15.3a
Supercritical (M3, S3)	No	15.3b

Of the three friction equations recommended by the above criteria, the average friction slope equation [Equation 15.3a] can be shown to always give the more conservative (larger) value of $\overline{S}_f$. This may be proven by the following mathematical progression, beginning with an inequality stemming from the theorem that the square of any real number is a positive real number:

$$(S_{f1} - S_{f2})^2 > 0 \tag{15.6a}$$

$$S_{f1}^2 - 2S_{f1}S_{f2} + S_{f2}^2 > 0 \tag{15.6b}$$

$$S_{f1}^2 + 2S_{f1}S_{f2} + S_{f2}^2 > 4S_{f1}S_{f2} \tag{15.6c}$$

$$(S_{f1} + S_{f2})^2 > 4S_{f1}S_{f2} \tag{15.6d}$$

From Equation 15.6d, we obtain:

$$\frac{S_{f1} + S_{f2}}{2} > \sqrt{S_{f1}S_{f2}} \tag{15.7a}$$

which proves that Equation 15.3a gives larger results than Equation 15.3b. Also, by Equation 15.6d we obtain:

$$\frac{S_{f1} + S_{f2}}{2} > \frac{2S_{f1}S_{f2}}{S_{f1} + S_{f2}} \tag{15.7b}$$

which proves that Equation 15.3a gives larger results than Equation 15.3c. This conclusion comports with the computational findings of Davis and Burnham (1988).

Knowing *WSL*, and hence *V*, at one of the end sections, *WSL* at the opposite end section may be solved by trial-and-error. Even for the simplest cases, the geometrical complexity makes a computer-based solution economically preferable if an accurate solution is desired (the *WSL* accuracy limit of the HEC-2 Program discussed below is 0.01 ft). When the full range of common complexities (overbanks of roughness and flow length different than the main channel, bridges, etc.) is added, a computerized approach becomes virtually mandatory.

A number of computer programs have been developed for backwater analysis. The most widely used such program in the United States has been the HEC-2 Water-Surface Profiles Program and its successors, developed at the U.S Army Corps of Engineers Hydrologic Engineering Center (USACE 1959, 1990). It has received widespread use in the preparation of reports for the national flood insurance program under the auspices of the Department of Housing and Urban Development (HUD) and more recently the Federal Emergency Management Agency (FEMA). The documentation of the HEC-2 Program with ancillary material has progressed to the point where it provides a good introduction to computer-based backwater analysis as well as user instructions for the particular program. In addition to its use for flood insurance studies, the author has used the program to determine flood levels adjacent to a wastewater treatment plant, the effect of encroachment by the construction of large interceptors on flood levels, and to develop hydraulic parameters for water-quality modeling (discussed later in this section). The capabilities and uses of programs such as HEC-2 include evaluation of the effects of bridges, culverts, weirs, and floodway encroachment; the determination of flood hazard zones and encroached floodways; etc.

As an example, the author used the HEC-2 Program to calculate backwater on the Trinity River from the gauging station shown in Figure 15.1 upstream approximately 11 miles (18 km) through 17 cross sections provided by the Corps of Engineers to the Village Creek Wastewater Treatment Plant serving Fort Worth, Texas. Eleven flows between 100 and 17,000 cfs were run. That enabled hydraulic design of the facility relative to flood levels. Times of travel, mean velocities, and depths from the backwater program output were used in conjunction with a water-quality model to determine the relationship between effluent biochemical oxygen demand (BOD) and river-dissolved oxygen under low-flow and augmented-flow conditions. That enabled establishing of treatment criteria for carbonaceous and nitrogenous BOD removal.

Occasionally flood levels must be approximated on minor streams for which the cost of a detailed backwater analysis seems difficult to justify (or exceeds available budget), and for which conservative, approximate results would be acceptable. A convenient semi-manual procedure for this purpose is described below. The procedure employs the standard-step method with a suitably large accuracy limit and is aided by a computer or programmable calculator. The procedure is not only useful in itself for appropriate problems but provides a good introduction to riverine backwater analyses.

Referring to Equations 15.2 and 15.3, the variables that must be known functions of *WSL* at each cross section are the cross-sectional area A (since $V = Q/A$), hydraulic radius R, and $AR^{2/3}$. To these are added the hydraulic depth D (area/top width) and $A\sqrt{D}$ for purposes of the critical depth check discussed below (note $\mathbf{F} = Q/\left(A\sqrt{D}\right)$). The above geometric elements can be readily computed for a channel cross section entered as a succession of Cartesian coordinates (x_i, y_i).

A couple of comments are necessary here regarding assumptions commonly made in flood-profile determinations. The first concerns the concept of a "flow line" which refers to a longitudinal profile along a stream showing, at all points, the maximum height attained by a flood (Lawler 1964, pp. 25–58). On larger drainage areas, in particular, the maximum n-year flood elevation at a particular location may be caused by an unsteady flood wave. The n-year flood elevation at another location along the river does not necessarily correspond to the same actual or simulated storm event. Nevertheless, a steady flow, increasing in increments as drainage area increases, is commonly assumed and analyzed by steady-state backwater methods. The actual n-year flood profile may thus be a locus of flood levels, corresponding to separate n-year events. The steady-state calculations clearly represent an *approximation* to that.

It should be understood that some "100-year" storm (as opposed to the 100-year storm for a particular duration) can occur many times during a 100-year period. A hydrologic persistence tendency notwithstanding, this can explain the occasional reports that there have been rare storms (e.g., "100-year") in each of several consecutive years. Skepticism should be directed toward unsupported claims in the popular press or by public officials that such events as roof collapses, sanitary sewer overflows (SSOs), failure of construction sediment controls, or flood-water levels are due to some n-year extreme storm. Supporting information should be in the form of direct metrics for the exceedance (such as flood levels directly rather than some rainfall metric), or such indirect metrics as rainfall characteristics that can be directly related to the planned capacity of the roof drains, sewer, sediment control, or flood level for the design storm characteristics (such as intensity, duration, frequency, and time of peak) (Graber 2009b, Gonwa 2012).

The second point concerns the occurrence of supercritical flow conditions and their treatment. Except on especially-steep channels for which supercritical flow is expected throughout, it is common to begin river backwater computations at a known downstream *WSL* and proceed upstream assuming subcritical flow conditions. At each cross section, the Froude Number is checked to determine if the resulting flow depth is subcritical or supercritical. If supercritical, a critical-depth "fix-up" is commonly taken, which means that the calculated depth is increased to critical depth. That approach is generally conservative insofar as the *WSL* is concerned (but not necessarily for erosive velocities), and avoids the complication of determining the actual transition from supercritical to subcritical flow. The latter is difficult if not impossible to do accurately by analytical methods, because the transition does not usually occur as a clear-cut hydraulic jump as in prismatic channels (see Section 7.5), but is affected by oblique waves at channel bends, obstructions, etc. Furthermore, the calculation of supercritical-flow profiles in natural channels is made tenuous by the existence of such waves, which will raise the local water surface above the calculated supercritical depth. A conservative, and usually reasonable, approach is to perform the subcritical drawdown computations with any critical-depth fix-ups for the determination of the flood profile. Where erosive velocities are important, these may be conservatively determined by supercritical drawdown computations in the reaches where supercritical flow has been suggested by the backwater analysis. In such cases, starting conditions can often be taken at critical depth at the downstream end of the next upstream reach where subcritical flow conditions were indicated.

15.2.2 Asymmetric Flow in Symmetric Expansions

We discuss here asymmetric flows occurring in symmetric river expansions and also the similar phenomenon occurring in man-made open channels. The earliest mention of this phenomenon known to the author appeared in the February 1895 issue of *Scientific American* (Bates 1895), which occurred on the Mianus River near Bedford, New York at a place known locally as the "ten foot hole," depicted below. The circular ice cake slowly revolved due to the asymmetry of flow downstream of the bridge despite the symmetry of the downstream channel.

A REVOLVING ICE CAKE.

Bates (1895) / Scientific American

Although not as good an example in the present context, another revolving ice cake was observed in 2019, is shown below, and described as follows (Dockrill 2019): "It's been compared to crop circles, or the strange handiwork of aliens. Others say it looks like the Moon, if it somehow materialised in the chilly waters of the Presumpscot River .... But this massive churning formation in the city of Westbrook, Maine isn't evidence of something from outer space, but rather a bizarre and brilliant example of a natural terrestrial phenomenon called an ice disc .... These strange, spinning spectacles are incredibly rare sightings, but even so, this one – estimated to measure about 90 m (300 ft) across – is a giant amongst its little-seen kind."

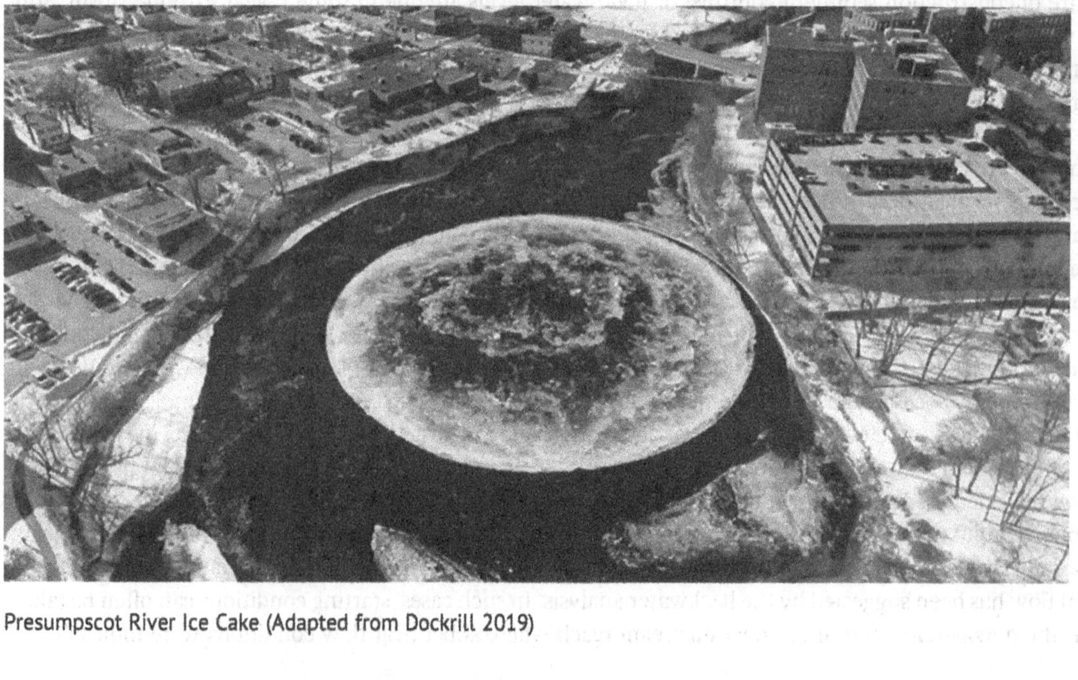

Presumpscot River Ice Cake (Adapted from Dockrill 2019)

Another example of asymmetric flow is given by Nimmo (1928) for which Graber (1982) reiterates the relevant dimensions and quotes Nimmo follows: "One would expect that the effect of the curve in the canal would be to cause the high velocity stream to diverge to the left (looking downstream), but ... it ... [was] seen diverging to the right ... on the first occasion on which the canal was filled .... After the trial run the flow was shut off, but the next time the headgates were opened the high velocity line diverged to the left. Since then, the author has observed the action of the transition on numerous occasions and has found that when the water is turned into the empty canal it is a matter of chance to which side the high velocity stream is deflected, but that having been deflected to one particular side, it remains at that side until the water is again shut off."

It is helpful to first understand the Coanda effect before explaining the type of instability mentioned above. The Coanda effect is named after Henri Coanda, who discovered it the hard way. A Romanian jet aircraft pioneer, he was taxiing for takeoff in an airplane of his design with a wood fuselage and a jet engine on each wing. He had tested the engines off the airplane to determine the spread of the issuing jet and had placed them on the wings based on that testing in an effort to insure the jets would not expand to reach the fuselage. However, the jets curved toward the fuselage setting it on fire. Coanda was able to abort the takeoff and escape with minor burns while the airplane was totally consumed. What occurred was that the entrainment of air into the jet together with the resulting pressure drop between that expanding jet and the fuselage caused the jet to curve toward the fuselage. The Coanda effect can be observed readily when poring liquid from a container without tipping it far enough such that the liquid attaches to the container (Reba 1966).

Similar to the river/canal examples given above, when a jet issues into a constructed symmetric expansion of certain dimensions, the jet can become unstable and latch onto one side or other of the expansion. An example is given in Figure 15.2 below.

Classical experiments by Abbott and Kline (1962) showed that this would occur in an abrupt rectangular two-dimensional (constant depth) expansion when the ratio of the expanded channel width to the flow width upstream of the expansion exceeded 1.5. Graber (1982) provided a theoretical analysis which predicted the identical result, attributing it to a static instability of the system that can be investigated by analyzing the forces acting on the system when it undergoes a small deflection from its symmetric position. Graber (2001, 2006) extended the analysis to open channels (without a confining top boundary) with higher subcritical Froude Numbers and to supercritical Froude Numbers (in which a hydraulic jump is one of the features). The unstable expansion ratio was shown to decrease as the Froude Number increases, from the value of 1.5 for two-dimensional flow (small Froude Number) to 1.2 for a Froude Number of unity and to unity for Supercritical Froude Numbers. The theoretical analyses were shown to agree with available observations. Graber (2006) also discusses oscillatory behavior (dynamic instability) that occurs under certain conditions.

The type of instabilities discussed above can be undesirable. The author has observed asymmetric flow behavior at wastewater treatment plant screen channels fed by pipes or symmetric channels from which the flow abruptly expands. In extreme cases, most of the flow passes through one side of the screen, with as much as two-thirds of the screen ineffective. The use of other than a straight approach channel has been suggested (WPCF/ASCE 1977, p. 129) as a possible cause, but such asymmetries may occur with straight symmetrical approach conditions. Corrective measures detailed by Graber (1981, 2001) include vanes and communication between the eddy regions by conduit or submergence.

The instability discussed above has been used beneficially in miniature devices referred to a fluid jet amplifiers or fluidic devices (Richards and Graber 1965). The Moore fluidic flowmeter (Drzewiecki and Houck 1985) is now available from other sellers; it has liquid flow capacities from about 1 to 1000 gpm and pressure drops from 0.01 to 20 ft of flowing fluid. Use at a larger scale has been proposed by Freeman (1971) for combined sewer regulation. Lager and Smith (1974, Chapter 7) provide additional discussion of fluidic regulators for combined sewer control. A fluidic-control system for stacked rapid-sand filters is described by Adelman *et al.* (2013).

Eisner and Aneshansley (1982) discuss the use of the Coanda effect by bombardier beetles when aiming their defensive spray. *Life in Moving Fluids* by Vogel (1994) provides many additional and fascinating examples of fluid mechanics in the natural world. The role of vortices in the flight of moths is also fascinating (Ashley 1997). P. S. (2001) discusses the use of the Coanda effect in improving the aerodynamics and reducing the drag of highway trucks.

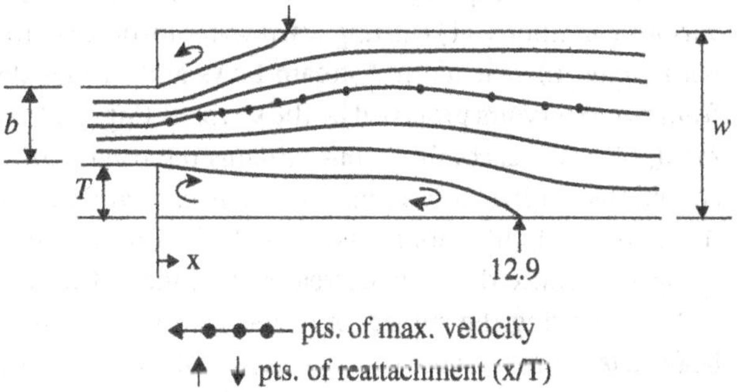

Figure 15.2 Asymmetric Flow in Rectangular Expansion: Expansion Ratio $w/b = 2.5$, $\mathbf{R} = 10^5$ (after Graber 1982 based on Mehta 1979)

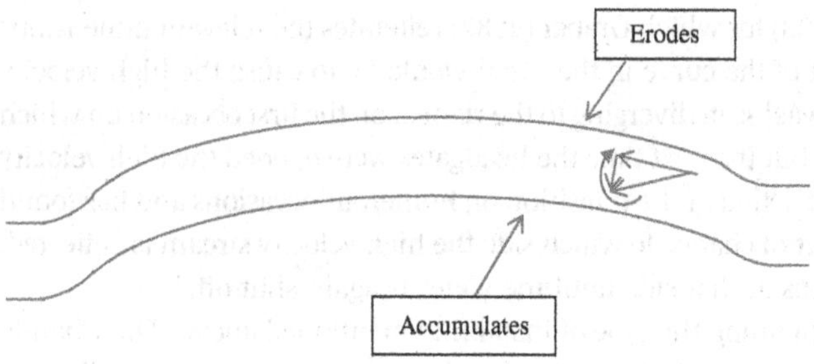

Figure 15.3 Depiction of River Meander

15.2.3 Meanders

River meanders are interesting features of sedimentary rivers that are allowed to assume their natural form. Quick (1974) provided a qualitative explanation of meander formation that remains the best known to this author. Helmholtz's second vortex theorem can be stated, with reference to Section 4.7, as fluid elements lying on a vortex line at some instant continue to lie on that vortex line and move with the fluid. Consider a small bend in a river arising from a naturally occurring asymmetry. Such a perturbation, depicted in Figure 15.3, results in superelevation on the outside of the bend due to centrifugal forces. The vorticity vector then rotates as depicted in the figure, resulting in a component in the stream-wise direction. Associated with that component is a circulation that erodes sediment from the outside of the bend and transports it to the inside of the bend. That action increases the size of the bend progressively, resulting in the meander. A wavy pattern of successive meanders results. Attainment of the final, equilibrium form of the meanders is one in which "net erosion and deposition reduce to zero and the ... [meanders] cease to develop further" (da Silva and Ebrahimi 2017). In some cases, the meander forms a complete loop and flow breaks through the loop leaving an "oxbow lake" alongside the main channel. da Silva and Ebrahimi (2017) note that "many questions still need to be addressed in order to develop a full understanding of the behavior of meandering streams". Additional aspects of river sedimentation are discussed by Einstein (1964) and Vanoni (1975).

15.2.4 Water-Quality Modeling

Water-quality modeling usually focuses on dissolved oxygen. Pertinent factors include carbonaceous and nitrogenous BOD in the water column, settling out of BOD, addition of BOD from bottom deposits, atmospheric reaeration, benthic oxygen demand, and photosynthetic production or reduction (Camp 1963, Chapter 11). A current model is the Water Quality Analysis Simulation Program (WASP) (USEPA 2020). Time of travel in connection with water quality modeling is discussed in Section 15.2.

In conducting water-quality modeling in rivers, it should be realized that more than one D.O. sag can occur. For example, in modeling the Nashua River in Massachusetts (Graber 1974b, 1975b) a minimum D.O. level occurred on the North Nashua River but a second and lower D.O. level occurred downstream in the more sluggish Nashua River into which the North Nashua flowed. That lower D.O. level was determinative of upstream wastewater treatment requirements in the North Nashua.

For the Nashua River Basin water-quality management plan, the author used the HEC-2 Program to obtain low-flow hydraulic characteristics including times of travel, velocities, water-surface widths, and depths. The hydraulic characteristics were determined for various conditions of total streamflow. The effects of dam removal on the hydraulic variables were ascertained using the HEC-2 Program, as was the hydraulic effect of turbidity-induced stratification in a large pond along the river. An existing water-quality model was modified and extended and then calibrated using data obtained from an intensive river sampling conducted for this purpose. HEC-2 and the calibrated water-quality model were then applied to various conditions: (1) existing wastewater loadings with 7-day, 10-year natural stream flows; (2) existing wastewater flows with 7-day, 10-year natural stream flows and various degrees of advanced wastewater treatment; (3) flow augmentation from two reservoirs proposed by the Corps of Engineers and resulting reductions in the need for advanced wastewater treatment; (4) various combinations of dams removed with and without flow augmentation; and (5) year 1980 and year 2020 conditions, which were found to be less critical than year 1975 conditions in terms of dissolved-oxygen criteria because the increased BOD loadings associated with future years were more than offset by the accompanying increase in flow (particularly because this flow decreases the time of travel in critical reaches of the river).

Before leaving the subject of water-quality modeling, mention will be made of the role of beavers. The importance of beavers is being an increasingly recognized part of river restoration. "The beaver, in building dams and developing ponds, increasing the supply of pasturage for livestock, creates ponds and streams where there is good fishing and recreation in mountains." Such ponds attenuate floods, increase groundwater levels, and contribute significantly to the maintenance of dry-season streamflow and mitigation of drought (and reduce costs of irrigation). Direct habitat and adjacent vegetation that result increase biodiversity, including aquatic insects and other invertebrates, bird species including ducks and songbirds, turtles and lizards, frogs, more diverse and abundant fish communities, and attract moose. Beavers are considered keystone species that support an entire biological community. Their ponds trap suspended nutrients and other pollutants and treat farm runoff. Denitrification in the pond sediments purges large amounts of nitrates, preventing dead zones from forming in downstream estuaries (Goldfarb 2018).

15.2.5 River Aeration

River aeration by means of air or oxygen diffusers or mechanical aeration has been the subject of numerous studies. Such aeration causes an abrupt increase in the dissolved oxygen concentration, similar to that resulting from the hydraulic aeration that occurs at dams on a river (discussed in Section 14.7). Wastewater treatment plants are designed to protect the river for periods which may exist for only several weeks of the year. Their design is commonly based on the most critical conditions which can be expected once in ten or twenty years and then affect only a limited portion of the river. Extremely costly and sophisticated treatment facilities may be required if reliance is placed entirely on wastewater treatment. From the standpoint of economics, one would expect that if river oxygenation is less costly under dry-weather flow conditions, it should be even more so for urban runoff due to the infrequent need for such oxygenation and the large cost of stormwater interception and storage facilities. Furthermore, the significant photosynthetic production of some rivers could be utilized to the maximum extent, river oxygenation being used only when necessary to supplement the natural aeration capacity of the river. The river oxygenation units can be automatically started when the dissolved-oxygen level falls to a predetermined value at automatic sampling stations. A feedback system, using dissolved-oxygen sensors, can be used to control the oxygen input.

River aeration has been found in numerous studies to have appreciable economic advantages as an alternative to advanced wastewater treatment. For example, Susag *et al.* (1966) performed an economic evaluation of the alternatives of additional treatment at the Minneapolis-Saint Paul Sanitary District Plant and mechanical surface aeration of the Mississippi River. It was estimated that the additional treatment or mechanical aeration would be required only 7% of the time. Total annual costs of mechanical surface aeration were found to range from 25 to 50% of those for additional wastewater treatment. A system analysis of the Potomac Estuary (Davis 1968) showed that, after 90% treatment of effluent had been achieved, instream aeration would provide a much more economical means of further improvement than would any practicable alternative.

A study (including field tests) of the Passaic River in New Jersey (Whipple *et al.* 1969, 1970a, 1970b, Yu 1970) led to similar conclusions. It was found that the annual cost to achieve a dissolved-oxygen level of 4 mg/L in the Passaic River by instream aeration (using mechanical surface aerators) would be about one-fourth the cost of advanced wastewater treatment for this purpose. Hunter and Whipple (1970) noted that this corresponds generally to actual experience on the Ruhr River in Germany. It was also noted that the Passaic River experiences higher BOD loadings during periods of high flow and that less than half of these loadings could be attributed to known wastewater discharges. This suggested that stormwater runoff was a major factor there, and led Whipple *et al.* (1969) to conclude that "induced river aeration may be not only an economical alternative, but the only practicable way of achieving [high levels of dissolved oxygen]." Studies in various parts of the country have shown that the pollution entering the river from point sources may be only one-third of the total pollution load on the river system. Thus expensive advanced wastewater treatment plants may not achieve the desired pollution control.

In a study of the Delaware River, Whipple *et al.* (1970c) found river aeration there to be a feasible alternative to advanced wastewater treatment, costing approximately half as much to achieve a given dissolved oxygen level. In that study, submerged air diffusers and oxygen diffusers were considered to have possible advantages over mechanical aerators in terms of efficiency and navigational use of the river.

The above-mentioned studies of river aeration were based on dissolved-oxygen standards which are generally lower than those which might be desired in freshwater rivers and streams capable of supporting sensitive species of fish and other aquatic life. Mechanical and diffused-air aeration are suitable in such cases because the driving force for oxygenation (which is the difference between the dissolved-oxygen saturation concentration and prevailing concentration) is sufficient. Where higher dissolved standards apply, the reduced driving force will lower the efficiency of such aeration. Oxygenation with pure (molecular) oxygen then becomes particularly suitable.

Major studies of river oxygenation with molecular oxygen have been carried out on the Pearl River in Louisiana (Amberg 1969, Amberg *et al.* 1969). There the Crown Zellerbach Corporation used an onshore, side-stream oxygenation system (Linde Division, Union Carbide Corporation) to increase the dissolved-oxygen content of the Pearl River into which their mill effluent flows. In this system, a high-pressure pump draws water out of the river. This "side-stream" water, which constitutes up to 1.7% of the river's flow, is pumped through a pipe into which oxygen is diffused by a sparger (a tube perforated with small holes). Oxygen is pumped into the water through the sparger outlets at a high pressure, which causes the water in the pipe to become supersaturated with oxygen. This oxygen-rich water is then returned to the river through a tapered diffuser header extending about 50 m along the river bottom. This header diffuses the oxygen-rich water into the river through 27 nozzles.

When the oxygen-rich water is diffused into the river, its pressure is reduced and some of the oxygen comes out of solution. The ratio by weight of oxygen absorbed in the river to that introduced to the system is termed the oxygen-transfer efficiency. In the Pearl River system, at low flows, this efficiency is about 55%. (The comparable efficiency of a diffused-air system is 5–15%.) The dissolved-oxygen level of the entire river flow was reportedly increased by 2 mg/L just downstream of the diffuser, and this resulted in an increase of 1.6 mg/L at the critical point some 9 miles downstream of the oxygenation installation.

The oxygen is introduced to the sparger in the Pearl River system as a gas but is vaporized from liquid oxygen in which form it is stored. In another approach using oxygen, liquid oxygen is injected directly into the side stream. Preliminary studies indicate that this may be an economical and effective method of river oxygenation. Amberg has pointed out that one of the major advantages of river aeration using molecular oxygen is that it lends itself to short-term use during critical river conditions. Low dissolved-oxygen levels in streams are generally associated with drought flows, high water temperatures, and urban runoff.

Another method of river oxygenation involves the direct diffusion of oxygen gas into the river. The results of tests of this method on the Pearl River have been reported (Amberg 1969, Amberg *et al.* 1969). The oxygen absorption efficiencies were quite low, compared to the "side-stream" method, ranging from 4.1 to 22.0% for single diffuser headers and 14.6–21.4% for a double diffuser header. Extrapolation of the performance of the most efficient diffusion system to critical low-flow summer months indicated a maximum oxygen absorption efficiency of about 14.5%. Molecular oxygen has also been added to the Willamette River in Oregon by venting it into water passing through a hydroelectric turbine (Amberg 1969). Additional discussion of methods of river oxygenation is provided by Graber (1978, pp. 6-25–6-40).

Fischer (1967); Thackston and Krenkel (1967); Thackston *et al.* (1967); and Bansal (1971) are important references which address longitudinal dispersion (as defined in Chapter 14) of pollutants in streams.

15.3 Lakes, Ponds, and Freshwater Wetlands

The distinction between lakes and ponds is commonly based on size and is legally formalized in, e.g., some of the New England states. There, "Ponds," "Great Ponds," and "Lakes" are commonly distinguished based on acreage, with the result that the large ones are called "lakes," such as in Maine and Massachusetts where those greater than 10 acres in area are "Great Ponds" or "Lakes," with "Lakes" being the largest water bodies which are arbitrarily called such. Aspects of the discussion of lakes below apply equally to reservoirs. The basic relationships given in Section 5.4 apply similarly to reservoirs and lakes, such as those used by the author for routing flood flows through Mare Meadow Reservoir and Bickford Pond Reservoir in Westminster, Massachusetts. Freshwater wetlands include lakes and ponds, along with additional categories discussed below.

15.3.1 Tidal Ponds

15.3.1.1 Introduction

Weir Creek, located in the Massachusetts Town of Dennis on Cape Cod, is a tidal waterway connecting to Nantucket Sound via the Bass River. At its inland end, Weir Creek exchanges tidal flows with four tidal water bodies via an interconnected arrangement of bridge waterways and culverts. As shown schematically in Figure 15.4, those four water bodies are Kelley's Pond, the Upper Basin, Uncle Stephan's Pond, and the Lower Basin. The total of five water bodies comprise an estuarine network, in which each water body responds to oceanic tides in its own way, but interactively with the others. A 266 ha (657 acre) watershed contributes freshwater runoff to the four water bodies upstream of Weir Creek. At mean tide in each of the four water bodies, their surface areas total about 24 ha (59 acres), increasing substantially as tide elevation increases.

In the late 1970s, an old timber highway bridge over the waterway connecting Weir Creek to the Lower Basin was replaced by a corrugated metal culvert. The installation of the culvert severely restricted flow, even during normal tide cycles. As a result, increased velocities and vortices scoured the channel bottom on the ocean side while creating areas of shoaling on the inland side. During Hurricane Bob in 1991, the storm surge overtopped the road, and portions of the pavement and embankment were washed away. In addition, the Town of Dennis became concerned about the safety of the rapidly flowing current through the culvert and the dangerous conditions created at the jagged ends of the culvert on each side of the roadway.

The Town of Dennis wanted to replace the culvert with a prefabricated timber bridge. However, the larger opening would increase tidal fluctuations in the four tidal ponds. After the culvert was installed, development continued along the shores of the tidal ponds. The residential properties lining the ponds have small docks or piers and have individual septic systems for wastewater treatment and disposal. The groundwater responds to tidal fluctuations in a way that could potentially be inimical to the proper functioning of the septic systems. Tidal fluctuations also affect water quality, flora, and fauna within the tidal ponds. Thus, it was considered important to determine the expected changes in tidal fluctuations associated with alternative replacement bridge designs. The following explains how that was done via hydraulic modeling. Tidal fluctuations under existing conditions were measured and a model was developed and tested for its ability to model existing conditions, then used to project future conditions with the replacement bridge. Freshwater runoff was also considered in the model.

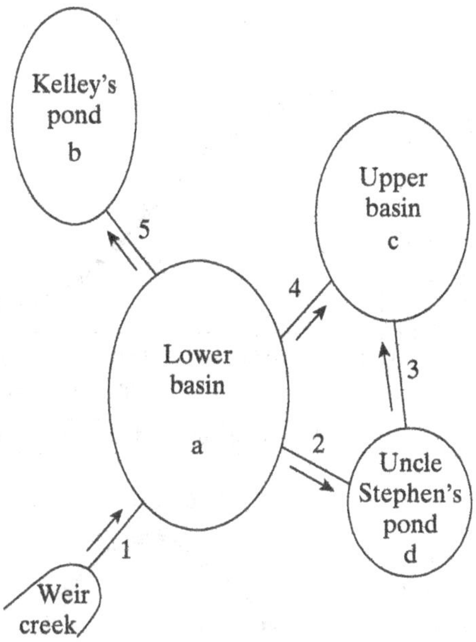

Figure 15.4 Tidal Pond Network Representation

15.3.1.2 Tidal, Rainfall, and Topographic Data

Tide elevations were measured by visual observation over most of a normal tide cycle and most of a spring tide cycle at 12 staked locations. The observations included high tides and low or near low tides at each stake and are plotted in Graber and Elkerton (1999) for the four tidal ponds shown in Figure 15.4. Figure 15.5 shows measurements in Weir Creek, along with calculated values that are discussed below.

In the Lower Basin, the tidal elevations at four widely separated points are nearly the same at any given time. Kelley's Pond behaves similarly, although the tidal range is very small. In Uncle Stephan's Pond, the two stakes within the pond proper and most representative of the pond as a whole show nearly the same tidal variation. A third stake agrees with the other two over roughly the high-tide half of the tidal cycle, then departs during the low-tide half-cycle. That stake is along a narrow channel close to the culvert connecting Uncle Stephan's Pond to the Upper Basin (Conduit 3 in Figure 15.4). The effect of the difference between that stake and the other Uncle Stephan's Pond stakes comes into play in the Conduit 3 hydraulic relationship. The channel at that stake drains to elevation 0.396 m (1.30 ft) under both tidal conditions and then drains no further, suggesting a blockage or an abrupt drop in the channel between that stake and the pond proper. Either has the same effect – when the channel has drained to elevation of 0.396 m (1.30 ft) it drains no further. The manner in which that was incorporated in the hydraulic model is discussed below. The Upper Basin's two stakes match over most of the high-tide half-cycle, and then depart during the low-tide half-cycle. One of those stakes is the more representative stake for the Upper Basin, and for the other stake similar considerations as those discussed above for Uncle Stephan's Pond apply.

National Climatic Data Center rainfall data for a nearby cooperative weather station (Hyannis) indicated that there was little or no rainfall during and at least 3 days prior to the periods of the normal and spring tide measurements. Detailed topographic surveys were conducted along transects in the Lower Basin, Uncle Stephan's Pond, and Kelley's Pond. The Upper Basin was only surveyed along transects for a certain distance upland, and so the entire basin was not surveyed. There also are dikes within the Upper Basin. Area-elevation data were derived from the topographic surveys and are shown in Figure 15.6. The area-elevation data provided for the Upper Basin incorporated assumptions due to the less complete survey data for that basin.

15.3.1.3 Network Formulation

The network model was formulated on the assumption that Weir Creek provides the tidal boundary condition, and that Kelley's Pond, the Upper Basin, Uncle Stephan's Pond, and the Lower Basin act as tidal ponds in which there is no backwater effect within each pond. In other words, the pond water surface raises and lowers but does not tilt. This assumption

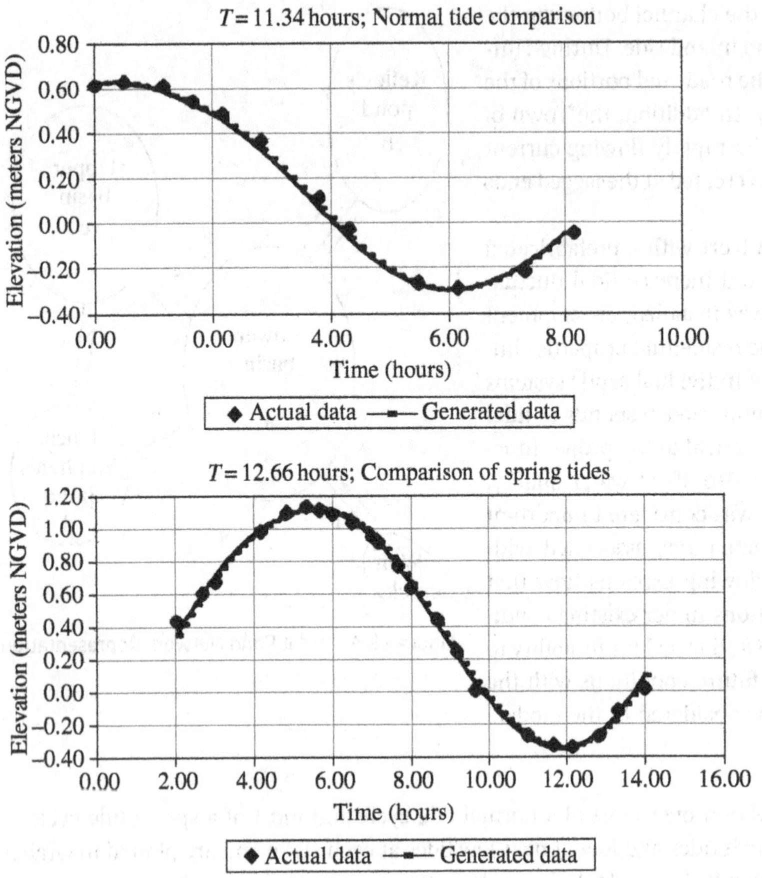

Figure 15.5 Weir Creek – Actual and Generated Tide Data

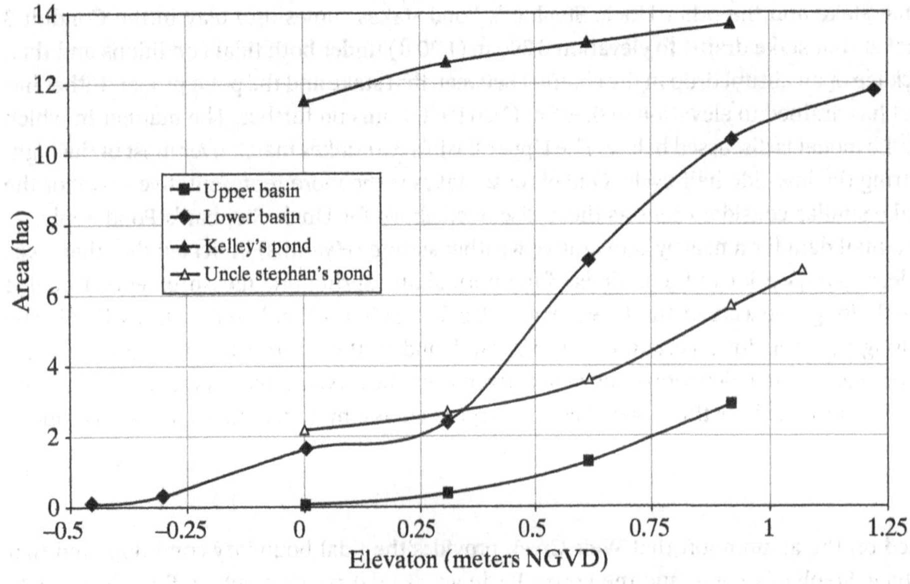

Figure 15.6 Tidal Pond Area-Elevation Relations

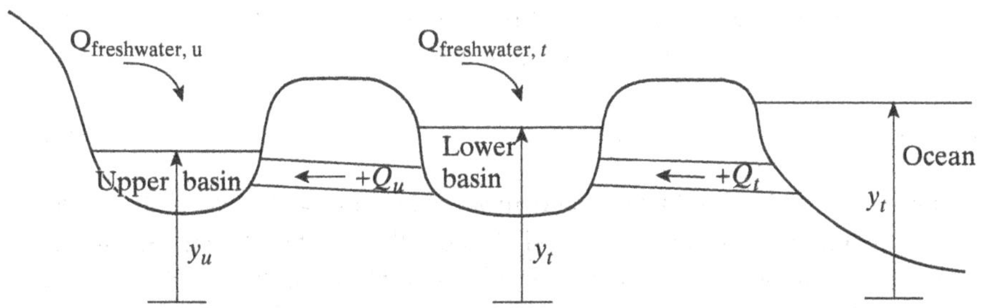

Figure 15.7 Profile of Two Tidal Ponds

requires two things. First, the tidal wavelength must be large compared with the physical dimensions of the network. This is easily satisfied: The wavelength of tides is typically measured in hundreds of miles, which is much greater than the lengths involved in the present problem. The second requirement is that flow velocities within each pond are too small and unimpeded to result in significant spatial variations in elevation within each pond. The empirical basis for that assumption is given above. The computational basis for that assumption is presented below.

The essence of the problem can be explained by first considering the system of two interconnected tidal ponds, one of which is connected to the ocean by a conduit as shown in Figure 15.7. The tidal pond nearest the ocean and its watershed will be referred to as the Lower Basin and the one furthest from the ocean and its watershed will be referred to as the Upper Basin. The conduits in Figure 15.7 are shown for simplicity as submerged culverts, but the discussion that follows is equally applicable to culverts with free-surface flow or bridge waterways.

Consider first the Lower Basin, neglecting for the moment the exchange of water with the Upper Basin. The Lower Basin then exchanges tidal water only with the ocean through one flow connection. Conservation of mass for the Lower Basin then indicates that the change in water volume during a specified time interval equals the volume of water entering its pond minus the volume of water leaving that pond during the time interval. Over a small time interval in which the inflow and outflow may be assumed constant, this may be expressed mathematically as

$$\Delta V = Q_{conduit}\Delta t + Q_{freshwater}\Delta t \tag{15.8}$$

in which ΔV = change in pond volume; Δt = time interval; and $Q_{conduit}$ and $Q_{freshwater}$ = average conduit inflow and freshwater (stormwater) inflow, respectively, over the time interval.

Denoting the pond water elevation relative to a prescribed datum by y and denoting the pond water-surface area by A, we have $\Delta V = A\Delta y$. Because the water-surface area is a function only of the pond water elevation, this can be written as $\Delta V = A(y)\Delta y$, in which the notation $A(y)$ denotes A as a function of y. Therefore, Equation 15.8 can be rewritten as:

$$A(y)\frac{\Delta y}{\Delta t} = Q \tag{15.9}$$

in which Δt = suitably small time interval; and Q = net inflow equal to $Q_{conduit} + Q_{freshwater}$.

The flow through the conduit between the Lower Basin and the ocean is a determinable function of the water level in the pond, the tide water level, and the hydraulic characteristics of the conduit connecting the pond to the ocean. The tidewater level is assumed to be a known function of time, so that functionally:

$$Q_{conduit} = \psi[y_t(t) - y] = \theta(y, t) \tag{15.10}$$

As an example, for the relatively simple case of a fully submerged culvert flow (Type I flow discussed in Section 7.7), the functional relationship becomes:

$$Q_{conduit} = A_f\sqrt{[y_t(t) - y]/\left\{\left[\left(n/\left(1.486R_f^{2/3}\right)\right)^2 L\right] + (K_e + 1)/(2g)\right\}} \tag{15.11}$$

in which A_f = full-culvert flow area; n = Manning's roughness coefficient; R_f = full-culvert hydraulic radius; L = culvert length; K_e = entrance-loss coefficient; and g = gravitational acceleration.

Any freshwater inflow to the basin is also assumed to be a determinable function of time. Therefore, Q is ultimately a function only of y and t, and Equation 15.9 can be expressed in the following functional form:

$$A(y)\frac{\Delta y}{\Delta t} = Q_{conduit} + Q_{freshwater} = \Phi(y, t) \tag{15.12}$$

Equation 15.12 consists essentially of a single equation with a single dependent variable y and single independent variable t and is, therefore, a complete equation for which a solution can be obtained. Because the equation is generally too complicated for a direct analytical solution, a numerical solution would normally be required. The most common techniques for obtaining such a numerical solution are the bisection method and the Newton–Raphson method (Hildebrand 1962, Hornbeck 1975).

The author has applied these methods to detention basins, designed to attenuate freshwater inflow, i.e., reduce the peak outflow to prescribed amounts lower than would occur in the absence of the detention basin (see Section 5.4). The author's subroutines for accomplishing that encompass virtually every culvert flow type that can occur (see further discussion below and in Section 7.7). Detention basins include a prescribed freshwater inflow, an area-elevation relationship, and an outflow relationship similar to those described above. In more complex detention basins, there is a variable tailwater, for which the tailwater elevation (and hence outflow) is a function of basin outflow. In that case, Equation 15.9 applies and the equation analogous to Equation 15.10 becomes:

$$Q_{conduit} = \psi[y - y_t(Q_{conduit})] \tag{15.13}$$

The freshwater inflow is again a determinable function of t, so that Q is again ultimately a function of y and t. Thus, Equation 15.12 is applicable in that case as well and has been used by the author in conjunction with numerical methods.

Now consider the more complex case involving the network of tidal ponds shown in the schematic plan view in Figure 15.4. Figure 15.7 can be regarded as the cross section through, e.g., the Lower Basin and Kelley's Pond, with the understanding that there are also other conduits connected to the Lower Basin. The conduits and tidal ponds are denoted with subscripts as follows: a = Lower Basin, b = Kelley's Pond, c = Upper Basin, d = Uncle Stephan's Pond, t = tidal level in Weir Creek, 1 = Loring Avenue culvert or bridge between Weir Creek and Lower Basin, 2 = Lighthouse Road bridge between Lower Basin and Uncle Stephan's Pond, 3 = Lower County Road culvert between Uncle Stephan's Pond and Upper Basin, 4 = Lower County Road culvert between Lower Basin and Upper Basin, and 5 = Fisk Street culvert between Lower Basin and Kelley's Pond.

Equation 15.9 can then be written for each pond to give the following set of equations:

$$A(y_a)\frac{\Delta y_a}{\Delta t} = Q_1 - Q_2 - Q_4 - Q_5 + Q_{Fa} \tag{15.14a}$$

$$A(y_b)\frac{\Delta y_b}{\Delta t} = Q_5 + Q_{Fb} \tag{15.14b}$$

$$A(y_c)\frac{\Delta y_c}{\Delta t} = Q_3 + Q_4 + Q_{Fc} \tag{15.14c}$$

$$A(y_d)\frac{\Delta y_d}{\Delta t} = Q_2 - Q_3 + Q_{Fd} \tag{15.14d}$$

in which Q_{Fa} = freshwater inflow to pond "a," Q_{Fb} = freshwater inflow to pond "b," etc. Similar to Equation 15.10, the conduit flow terms in Equation 15.14a–15.14d can be written as follows (with positive flow assumed to be in the directions shown in Figure 15.4):

$$Q_1 = \psi_1(y_t - y_a) \tag{15.15a}$$

$$Q_2 = \psi_2(y_a - y_d) \tag{15.15b}$$

$$Q_3 = \psi_3(y_d - y_c) \tag{15.15c}$$

$$Q_4 = \psi_4(y_a - y_c) \tag{15.15d}$$

$$Q_5 = \psi_5(y_a - y_b) \tag{15.15e}$$

Equations 15.14a–d and Equation 15.15a–e constitute a system of nine independent equations in nine unknowns (four water elevations and five flows). It is thus a complete system of equations that may theoretically be solved. In concept, Equations 15.15a–15.15e may be substituted into Equation 15.14a–15.14d to give four equations in the four unknowns y_a, y_b, y_c, and y_d. With all of the terms moved to the left side of the equation, the first such equation can be denoted by F_a and would be as follows:

$$F_a = A(y_a) - \psi_1(y_1 - y_a) + \psi_2(y_a - y_d) + \psi_4(y_a - y_c) + \psi_5(y_a - y_b) - Q_{Fa} = 0 \tag{15.16}$$

The remaining three equations are similar, so there are four equations, written in functional form as follows:

$$F_a(y_a, y_b, y_c, y_d) = 0 \tag{15.17a}$$

$$F_b(y_a, y_b, \cdot\cdot, \cdot\cdot) = 0 \tag{15.17b}$$

$$F_c(y_a, \cdot\cdot, y_c, y_d) = 0 \tag{15.17c}$$

$$F_d(y_a, \cdot\cdot, y_c, y_d) = 0 \tag{15.17d}$$

in which the double dots indicate absent terms that are not variables in the particular equation.

The secant modification of the Newton–Raphson method can be used to obtain numerical solutions to a complex, nonlinear system of equations such as this. With this method, initial approximations are provided and the method calculates successive corrections to those approximations until the equations are satisfied to within a specified degree of accuracy. In this case, the system of equations must be solved for each time increment over the tidal cycle of interest, for which the initial approximations at each time step can be the calculated water levels at the previous time step.

To outline the solution technique, additional nomenclature is necessary. Let F_{ao}, F_{bo}, F_{co}, and F_{do} denote the values of the Equation 15.17a–15.17d functions at approximation values of the y variables. Each of the four functions has up to four partial derivatives, an example of the notation for which is given by:

$$F_{aya0} = \left(\frac{\partial F_a}{\partial y_a}\right)_0 \tag{15.18}$$

The above is the partial derivative of F_a with respect to y_a, evaluated at the values of y_{a0}, y_{b0}, y_{c0}, and y_{d0}. That is, it gives the variation of F_a as the water-surface elevation of the Lower Basin y_a changes, evaluated at the values of the current approximations of the other related water-surface elevations. Some of the partial derivatives are zero. An example is F_{cyb0}, which is zero because the Upper Basin is not directly affected by variations in Kelley's Pond.

The Jacobian is denoted and defined for this problem by the following matrix determinant:

$$\frac{\partial(F_a, F_b, F_c, F_d)_0}{\partial(y_a, y_b, y_c, y_d)} = \begin{vmatrix} F_{aya0} & F_{ayb0} & F_{ayc0} & F_{ayd0} \\ F_{bya0} & F_{byb0} & F_{byc0} & F_{byd0} \\ F_{cya0} & F_{cyb0} & F_{cyc0} & F_{cyd0} \\ F_{dya0} & F_{dyb0} & F_{dyc0} & F_{dyd0} \end{vmatrix} \tag{15.19}$$

Setting appropriate terms equal to zero gives the following simplified version of Equation 15.19:

$$\frac{\partial(F_a, F_b, F_c, F_d)_0}{\partial(y_a, y_b, y_c, y_d)} = \begin{vmatrix} F_{aya0} & F_{ayb0} & F_{ayc0} & F_{ayd0} \\ F_{bya0} & F_{byb0} & 0 & 0 \\ F_{cya0} & 0 & F_{cyc0} & F_{cyd0} \\ F_{dya0} & 0 & F_{dyc0} & F_{dyd0} \end{vmatrix} \tag{15.20}$$

The adjustments or corrections for each trial value of y_{a0}, y_{b0}, y_{c0}, and y_{d0} are calculated by adding to them, respectively, values of h_{a0}, h_{b0}, h_{c0}, and h_{d0}, which are adjustments as given below. Those sums then become the new values of y_{a0}, y_{b0}, y_{c0}, and y_{d0}. The trials continue until the computed changes (i.e., the values of h_{a0}, h_{b0}, h_{c0}, and h_{d0}) in the values of the *roots* (y_{a0}, y_{b0}, y_{c0}, and y_{d0}) are all less than a specified quantity ε. Hornbeck (1975) suggests that a useful and conservative value of ε is one-tenth of the permissible error in the root. Except as otherwise noted, an ε of 0.000305 m (0.001 ft) was used.

The value of adjustment h_{a0} is given by:

$$h_{a0} = \frac{\begin{vmatrix} F_{a0} & F_{ayb0} & F_{ayc0} & F_{ayd0} \\ F_{b0} & F_{byb0} & 0 & 0 \\ F_{c0} & 0 & F_{cyc0} & F_{cyd0} \\ F_{d0} & 0 & F_{dyc0} & F_{dyd0} \end{vmatrix}}{\frac{\partial(F_a, F_b, F_c, F_d)_0}{\partial(y_a, y_b, y_c, y_d)}} \tag{15.21}$$

in which the denominator is given by Equation 15.20. The numerator determinant in Equation 15.21 has the same form as the Jacobian (Equation 15.20), except that the first column is replaced by F_{a0}, F_{b0}, F_{c0}, and F_{d0} as defined above. The adjustments h_{b0}, h_{c0}, and h_{d0} have a similar form, with the second, third, and fourth columns of the numerator determinant respectively replaced with F_{ao}, F_{bo}, F_{co}, and F_{do}.

As mentioned above, the secant modification of the Newton–Raphson method has been used. That is because the values of the partial derivatives cannot be calculated by using a direct equation (the functions are too complex to differentiate), but rather they are calculated numerically using a difference expression.

15.3.1.4 Tidal Boundary Conditions

Tides in Weir Creek, on the ocean side of Loring Avenue, are the prime force driving the problem at hand. Those tides, as measured for a normal and spring tide, constitute a boundary condition. Mathematically, the Weir Creek tide level is the value of y_t in Equation 15.15a. The model needs to represent those tides over multiple cycles to achieve equilibrium and to model the effects of freshwater inflows. Tides are periodic functions that can be represented by Fourier series (Hildebrand 1962, Eagleson and Dean 1966). The stormwater management model (SWMM) (USEPA 1994) contains a subroutine TIDCF that can perform a Fourier analysis of tide data, in which the coefficients of the following equation are determined:

$$H = A_1 + A_2 \sin(Wt) + A_3 \sin(2Wt) + A_4 \sin(3Wt) + A_5 \cos(Wt) + A_6 \cos(2Wt) + A_7 \cos(3Wt) \tag{15.22}$$

In this equation, H = elevation above a datum in m or ft; W = frequency (1/hour); t = elapsed time in hours; and A_1, A_2, etc. are the coefficients produced by the Fourier analysis. A_1 is the mean tide (for the particular tidal cycle being analyzed). The frequency is equal to $2\pi/T$, in which T is tidal period in hours. For semi-diurnal tides such as those that occur in Nantucket Sound, the wavelength is about 12.4 hours but can vary significantly for individual tidal cycles.

The SWMM TIDCF subroutine was applied to the normal and spring tidal data for Weir Creek as shown in Figure 15.5 (12 data points of tide elevation vs. time for normal tide and 22 such points for spring tide). Using the generated coefficients, the actual tide data and generated tide curves were plotted and compared. Different values of tide period T were tried (guided by the time difference between maximum and minimum measured values, which is approximately $T/2$) until the comparisons showed good fits to the data. The selected values of T and plots based on those values of T and the computed values of the Fourier coefficients are given in Figure 15.5.

15.3.1.5 Freshwater Hydrology

Based on Natural Resource Conservation Service (NRCS) soil data, other pertinent characteristics of the drainage basins, and different storm frequencies, an NRCS method (TR-55) was used to estimate the peak stormwater runoff for each basin. The NRCS method is based on a 24-hour storm duration and the results correspond to a 24-hour storm hydrograph. A hydrograph of shorter duration was of interest for our purposes. The peak runoff values can be considered to correspond to the peak runoff values of the rational method, with which shorter-duration hydrographs were derived from the perspective of the rational method (Graber 1992).

15.3.1.6 Connecting Conduits

The existing conduits between the water bodies of interest include four culverts and one bridge. Referring to Figure 15.4 and identification numbers provided earlier, Conduits 1, 3, 4, and 5 are culverts and Conduit 2 is the Lighthouse Road Bridge. Conduit 1 (Loring Avenue) is a culvert to be replaced by a bridge. Culvert survey data, Lighthouse Road Bridge design drawings, photographs, and field inspections allowed detailed characterization of the conduits. The hydraulics of culverts is fairly complex. Chow (1959) defines six types of culvert flow (see Section 7.7). Other authors define additional flow types, but they can be regarded as subtypes of those defined by Chow (1959) if the latter types are defined more generally. The author has developed subroutines which, over the course of many analyses, have evolved to the point where they can determine and analyze all of the flow types. The methods are generally based on those presented in the thorough review by the Portland Cement Association (PCA 1964, Chapters 3–5). Based on collected data, each culvert was characterized as to cross-sectional dimensions (including sediment depth if any), length, material, roughness (Manning's n), invert elevations at the two ends, and hydraulic coefficients associated with the culvert entrances. Depending on the direction of the flow, either end of each culvert can be an entrance. So both ends of the culverts were assigned separate coefficients, which differed at the opposite ends in some cases. For each entrance, those coefficients include one entrance-loss coefficient, two nonsubmerged inlet coefficients, and two submerged inlet coefficients.

Culverts ranged from 2.63 m (8.62 ft) to 0.41 m (16 in.) in effective diameter; from 15 m (50 ft) to 20 m (65 ft) in length; and were variously constructed of corrugated metal, concrete, and concrete-coated corrugated metal. Surveyed sediment depths in the culverts ranged from zero to one-half of the culvert diameter. The culverts were accurately represented as circular segments with a bottom chord to account for sediment on the culvert bottoms where applicable. That hydraulic geometry is readily handled by the author's subroutine.

The culvert flow types are part of the computer output at two stages of the analytical procedure. The first is during the generation of the culvert files as described below. The resulting files give the flow types at each condition of tailwater, flow, and headwater for each culvert with flow in both directions. The second is when those files are used during the analytical simulation of tidal activity, in which the output for each culvert at each time step includes the culvert flow type. Every culvert flow type discussed in Section 7.7 except Type II, was found to occur. (Type II differs from Type III primarily in that the culvert must be *hydraulically long* for Type II to occur.) The analyses of bridge hydraulics (existing Lighthouse Road Bridge and proposed Loring Avenue Bridge) entailed determining the backwater effect due to the constriction imposed by the bridge geometry. Historically, two approaches have been applied to this type of problem. One is based on generalized laboratory model investigations, which can in some cases be applied to full-scale bridge hydraulics. That approach is developed in publications of the U.S. Geological Survey and Federal Highway Administration and also by Chow (1959). However, the geometric configurations for which such data are available are limited, are most likely to be applicable to large bridges (with wide, flat-bottomed waterways), and are not accurately applied to the bridge configurations of interest here.

The second approach, and the one taken here, is to obtain a solution using continuity, energy, and momentum principles, applied to the specific bridge geometry. While similar analyses have been published (Ippen 1950, Graber 1982) for riverine reaches upstream and downstream of the bridge, the solution here took into account the presence of ponds immediately upstream and downstream. There are two separate flow regimes, one for subcritical flow and one for critical flow. In terms of well-known Corps of Engineers HEC-2 methods, the adopted method is most closely related to the "normal" bridge routine for subcritical flow and to the "special" bridge routine for critical flow (although the latter is not directly applicable to cases where there is a downstream pond, i.e., a zero downstream velocity). Program runs predicted critical-flow conditions and the existence of a *hydraulic drop* near low tide, which is precisely what was observed at the Lighthouse Road Bridge. [Although Chow (1959) categorizes the hydraulic drop as rapidly varied along with the hydraulic jump and tidal bore (discussed below), it actually should be regarded as gradually varied by his categorizations].

Conduit 3, the culvert and channels between the Lower Basin and Upper Stephan's Pond, was further modified to account for the observations noted previously. That was done by causing Conduit 3 flow to be zero when the tailwater falls below elevation 0.396 m (1.30 ft) during flow from the Upper Basin to Uncle Stephan's Pond and when the tailwater falls below elevation 0.472 m (1.55 ft) during flow from Uncle Stephan's Pond to the Upper Basin.

Special conduit file generation and interpolation techniques were developed to render the problem computationally feasible. Those techniques are the subject of Graber (1999).

15.3.1.7 Initial Conditions and Equilibrium

It is useful to summarize the conceptual approach at this point. For each time step, the tide elevation in Weir Creek is determined, freshwater inflow (if any) to each basin is calculated, and trial values of the pond water elevations are assumed. The corresponding flows through each conduit are then calculated using relationships represented functionally by Equations 15.15, and the pond water areas are calculated by parabolic interpolation on the area-elevation data. The corresponding values of the functions represented by Equation 15.17 are then determined (which will not yet equal or approximate zero), and adjustments to the trial pond elevations are determined using relationships typified by Equation 15.21. The process is repeated with the new trial pond elevations until the adjustments are all within the prescribed limit.

As noted previously, the tidal variations in Weir Creek on the ocean side of Loring Avenue constitute a *boundary condition* for our analyses. The boundary tidal variations are assumed to persist over multiple tidal cycles, a necessary assumption in the absence of long-term tidal data. This assumption has certain implications for interpretation of the results, as discussed below. *Initial conditions* need to be specified, which are the values of tide levels in each of the four tidal ponds at the start of each analysis (zero time). If those initial conditions are selected judiciously (aided by test runs), then a *quasi-steady* or equilibrium condition will be reached in a few tidal cycles. Equilibrium will exist when subsequent tidal variations are identical in each tidal pond. The results then meaningfully correspond to the boundary conditions. The test adopted for attainment of equilibrium is that two successive high-tide elevations are within 0.0305 m (0.1 ft) of each other in all four tidal basins. The computer program was devised to automatically determine when that is the case.

If freshwater inflow is desired, it is initiated automatically on the next tide cycle after equilibrium has been achieved. That has the advantage of allowing a single computer run to include the case with and without rainfall. The analysis continues for one tide cycle after the cycle with freshwater inflow, then automatically stops.

Initial runs illustrated the importance of using test runs for guidance in selecting initial conditions. In the first runs, mean tides (A1 values as discussed under the *Tidal Boundary Conditions* section) were assumed as initial conditions. Equilibrium was not achieved within six tide cycles because of Kelley's Pond. A better initial condition for the Kelley's Pond tide elevation leads to a faster attainment of equilibrium. To improve program speed, advantage was taken of the fact that all of the culverts exhibit fully submerged flow [essentially Chow's (1959) Type I] at higher tailwater elevations. For Type I flow, the hydraulic relationship is simple enough that one can solve explicitly for flow given tailwater and headwater values. [This is also true of Chow's (1959) Type VI flow, although that is not useful to us because if Type VI occurs it does so at lower tailwaters in the midst of other flow types.] Having determined the tailwater at which Type I flow begins for each culvert, the program was modified to calculate the flow directly for Type I flow rather than use the interpolation procedure. That turned out to be particularly convenient for Conduit 3, because only Type I flow occurs above the zero-flow tailwater elevations mentioned previously.

A computational time step of 0.1 hour (6 min) was initially selected. However, occasionally that did not result in convergence on the solution. Accordingly, the program was modified to start with a 0.1 h time step, and if convergence was not obtained within the prescribed number of iterations (100) then the time step was successively quartered until convergence was obtained.

15.3.1.8 Model Results – Existing Conditions

The program output files can be readily imported into a spreadsheet program from which charts can be prepared. Results are presented in Graber and Elkerton (1999) for existing conditions with normal tides and spring tides, respectively. They favorably compared with the observed high and low-tide elevations. For Uncle Stephan's Pond and the Upper Basin, those points were based on the most representative stakes, as discussed above. Freshwater inflow starts at time equal to 31.9 hours for the normal tide run and at time equal to 38.4 hours for the spring tide run. The effects of freshwater runoff were found most significant for Kelley's Pond and the Upper Basin.

Key features observed during the tide measurements were time lag and attenuation. Time lag refers to the observation that the tidal pond high and low tides occurred after the time of high and low tides in Weir Creek. Attenuation refers to the observation that the magnitudes of high tides and tidal ranges are smaller in the tidal ponds than in Weir Creek. Those features are properly simulated by the model, as shown in figures by Graber and Elkerton (1999). Also illustrated in those figures are the attainment of equilibrium and the subsequent increase in the high-tide level due to freshwater inflow commencing automatically at mean flood tide in the next tidal cycle. Because equilibrium is reached, the program terminates the run before the specified maximum number of cycles.

The modeled tide ranges were compared with the observed tide ranges when equilibrium was achieved and before freshwater runoff occurred. That is the tide cycle immediately preceding the start of freshwater inflow at the times given above. The Lower Basin and Uncle Stephan's Pond are highly responsive, as revealed by the quick attainment of equilibrium. That indicates that those basins reflect the contemporaneous tidal boundary conditions (at Weir Creek) with relatively little influence from prior astronomical tide cycles. The Lower Basin and Uncle Stephan's Pond show relatively little impact of freshwater inflow. The calculated high tides under both normal tide and spring tide conditions agree well with observations. The calculated low tides under spring tide agree well with observations, while the calculated low tides under normal tide conditions are lower than observed values. That is attributed to shoaling on the upland side of the Loring Avenue culvert and Lighthouse Road bridge, which prevents the basins from draining as much as they would otherwise (note that the normal low tides are lower than the spring low tides in those basins).

A correlation graph (Graber and Elkerton 1999) compared observed and computed high and low tides from the data used to generate figures mentioned above just prior to the beginning of freshwater runoff. The comparisons are for the Lower Basin, Uncle Stephan's Pond, and the Upper Basin; Kelley's Pond is excluded because of the fortuitous ease of modeling it as discussed below. The correlations are generally good.

The important thing for our analysis is the prediction of high tides, which can be seen to be unaffected by the inability to precisely predict the low normal tides. It would be fairly straightforward to survey the shoals and incorporate them in the overall hydraulic relationships for Conduits 1 and 2 (as represented functionally by Equation 15.15a and 15.15b), considering the shoals in series with the culvert or bridge. However, in this case, that would be only of academic interest, not only because it is not necessary for the ultimate purpose of the analysis, but also because the shoals will change with future coastal storms.

Kelley's Pond has a small tidal range and responds very slowly to tidal boundary conditions. Although equilibrium is achieved sufficiently to trigger freshwater inflow, that is actually because of the small tidal range in Kelley's Pond. The tide curves can be seen to be sloping to the right. The tidal range is modeled accurately, but the high and low tides are less so. That is because Kelley's Pond integrates the effects of boundary tidal fluctuations (i.e., prior astronomical and storm tides) and freshwater runoff over long periods of time. After many tidal cycles, the modeled pond would reach the theoretical equilibrium condition, but in actuality, the boundary conditions (downstream astronomical tides) would have been changed by then. The fairly close agreement between the modeled and observed normal high and low tides [figure in Graber and Elkerton (1999)] is a somewhat fortuitous result of the assumed starting conditions. The modeled and observed tides could be made to match by assuming initial conditions close to the observed tide elevation, but the match would then be artificial and the slow response of the basin would not be revealed. Kelley's Pond can be seen to respond fairly quickly to the freshwater inflow.

For the Upper Basin, the agreement between computed and observed values is not as good as for the other three basins. As noted previously, the area-elevation data for the Upper Basin incorporated assumptions owing to less complete survey data for that basin. Those limitations in the Upper Basin do not seem to affect the accuracy with which the other three basins can be modeled. The Upper Basin is particularly responsive to freshwater runoff. This is particularly so under normal tide conditions, because lower on the area-elevation relationship the water elevation increases more in response to a particular volume of runoff.

The ability to accurately predict the observed tidal variations in the Lower Basin, Uncle Stephan's Pond, and Kelley's Pond adds a computational basis to the empirical basis for the assumption, stated earlier, of small variations in elevation within each basin. The comparison of model results with observations reported above also suggests that the effect of conduit changes at Loring Avenue should be predictable with a high level of confidence for the Lower Basin, Uncle Stephan's Pond, and Kelley's Pond. Predictions of conduit changes on the Upper Basin must be interpreted more cautiously.

15.3.1.9 Model Results – Future Conditions

The network of tidal basins was modeled with the Loring Avenue culvert replaced by different bridge designs. The bridges were modeled by the same method described previously for the Lighthouse Road Bridge. The predicted high tides were observed from the output files for normal tides and spring tides under existing conditions and different future scenarios. Table 15.1 gives an example of predicted spring high tides (Weir Creek high tides are the measured boundary conditions).

Table 15.1 Predicted Spring High-Tide Elevations

			Changes Without Freshwater Inflow		Changes with Freshwater Inflow	
Tidal Basin (1)	**Existing Conditions Without Freshwater Inflow (2)**	**Existing Conditions With Freshwater Inflow (3)**	**Loring Avenue Bridge, 7.9 m (26 ft) span (4)**	**Loring Avenue Bridge, 6.1 m (20 ft) span (5)**	**Loring Avenue Bridge, 7.9 m (26 ft) span (6)**	**Loring Avenue Bridge, 6.1 m (20 ft) span (7)**
Weir Creek	1.107 m (3.633 ft)	1.107 m (3.633 ft)	0	0	0	0
Lower Basin	0.895 m (2.935 ft)	0.923 m (3.029 ft)	0.190 m (0.624 ft)	0.180 m (0.591 ft)	0.167 m (0.547 ft)	0.158 m (0.520 ft)
Kelley's Pond	0.616 m (2.020 ft)	0.668 m (2.192 ft)	0.057 m (0.188 ft)	0.057 m (0.186 ft)	0.051 m (0.166 ft)	0.503 m (0.165 ft)
Upper Basin	0.819 m (2.686 ft)	0.881 m (2.889 ft)	0.073 m (0.238 ft)	0.072 m (0.235 ft)	0.109 m (0.358 ft)	0.104 m (0.340 ft)
Uncle Stephan's Pond	0.894 m (2.934 ft)	0.923 m (3.028 ft)	0.190 m (0.623 ft)	0.180 m (0.590 ft)	0.166 m (0.546 ft)	0.158 m (0.520 ft)

Table 15.2 Changes in Predicted Spring High-Tide Elevations

Tidal Basin (1)	Existing Conditions Without Freshwater Inflow (2)	Existing Conditions With Freshwater Inflow (3)	Future Conditions Without Freshwater Inflow		Future Conditions With Freshwater Inflow	
			Loring Avenue Bridge, 7.9 m (26 ft) span (4)	Loring Avenue Bridge, 6.1 m (20 ft) span (5)	Loring Avenue Bridge, 7.9 m (26 ft) span (6)	Loring Avenue Bridge, 6.1 m (20 ft) span (7)
Weir Creek	1.107 m (3.633 ft)	1.107 m (3.633 ft)	1.107 m (3.633 ft)	1.107 m (3.633 ft)	1.107 m (3.633 ft)	1.107 m (3.633 ft)
Lower Basin	0.895 m (2.935 ft)	0.923 m (3.029 ft)	1.085 m (3.559 ft)	1.075 m (3.526 ft)	1.090 m (3.576 ft)	1.082 m (3.549 ft)
Kelley's Pond	0.616 m (2.020 ft)	0.668 m (2.192 ft)	0.673 m (2.208 ft)	0.672 m (2.206 ft)	0.719 m (2.358 ft)	0.718 m (2.357 ft)
Upper Basin	0.819 m (2.686 ft)	0.881 m (2.889 ft)	0.891 m (2.924 ft)	0.890 m (2.921 ft)	0.990 m (3.247 ft)	0.984 m (3.229 ft)
Uncle Stephan's Pond	0.894 m (2.934 ft)	0.923 m (3.028 ft)	1.084 m (3.557 ft)	1.074 m (3.524 ft)	1.089 m (3.574 ft)	1.081 m (3.548 ft)

Table 15.2 gives example results in terms of changes in spring high-tide elevations relative to the corresponding existing conditions.

As expected, all other factors being equal the high-tide elevations in each of the four basins increase with enlargement of the Loring Avenue waterway. A 6.1 m (20 ft) bridge span gave higher high tides than the existing culvert, and a 7.9 m (26 ft) bridge span gave higher high tides than the 6.1 m (20 ft) span. Also, all other factors being equal, freshwater runoff increases the high-tide elevations in each basin. Enlarging the flow conduit at the Loring Avenue Bridge will increase water levels in the upland basins due to tide effects alone, but possibly reduce water levels in the upland basins associated with extreme freshwater runoff. The modeled freshwater runoff corresponds to a one-year return period, the volume of which in each basin is small compared with the corresponding tidal prism. It is possible that with more extreme freshwater flooding (e.g., a 100-year rainstorm) the opening of the Loring Avenue waterway could reduce upland freshwater flooding (under the same Weir Creek tidal conditions). That would occur if the freshwater volume overwhelmed the tidal prism in each basin, causing the water level to exceed that in Weir Creek and thus turn the tide in the four basins.

The results in Table 15.2 show, as expected, that the change in spring high-tide elevation in any basin for any particular comparison between future and existing conditions decreases in the upland direction (the same was true for normal high tides). Thus, Uncle Stephan's Pond and Kelley's Pond have smaller changes than the Lower Basin, and the Upper Basin has smaller changes than the Lower Basin or Uncle Stephan's Pond. This allows further comment concerning the predictions for the Upper Basin, for which the model accuracy may not be as high for reasons stated earlier: the Upper Basin changes will be less than those for Uncle Stephan's Pond, and may be close to the modeled values.

15.3.1.10 Conclusions

A model has been developed and successfully applied to analyze a network of tidal ponds connected by culverts and bridges. Measured tidal fluctuations under normal and spring tide conditions at the seaward terminus of the tidal ponds were applied as boundary conditions. The model uses the secant modification of the Newton–Raphson method for solving a set of non-linear differential equations. It was necessary to develop special conduit file generation and interpolation techniques to render the model feasible in terms of required computational time. The model adequately represented tidal fluctuations measured in the tidal ponds under existing conditions. The model was then applied to predict changes in tidal fluctuations resulting from replacement of one of the culverts with alternative bridge designs with and without freshwater runoff. The predictions enabled separate assessments of impacts on shore-side properties with individual septic systems and on water quality, flora, and fauna within the tidal ponds.

15.3.2 Lake Stratification

In lakes of sufficient depth in their climatic zone, a thermocline develops as the surface waters are heated in the spring and summer, resulting in stratification with a warmer upper layer called the *epilimnion* and a lower level called the *hypolimnion*, divided by the *thermocline* or *metalimnion*. As the epilimnion is cooled in the fall and winter, a "turnover" occurs in which the layers intermix (Hutchinson 1975, Chapter 7). The author has personally experienced the stratification while scuba diving in Three Mile Pond, Windsor, Maine is shown in Figure 15.8 below, experiencing the abrupt temperature decrease upon entering the hypolimnion. Stratification has practical significance for the selective withdrawal and the vertical location of downstream reservoir releases for water quality (particularly dissolved oxygen levels), the sustaining of cold water fisheries, water supply temperature levels, prevention of lake anoxia, and phosphorus control (Brooks and Koh 1969, Gaillard 1984, Nürnberg *et al.* 1987, Rheinheimer *et al.* 2015). A comprehensive discussion of artificial aeration, oxygenation, and destratification of reservoirs is provided by Pastorok *et al.* (1982). Destratification induced by bubble plumes is addressed by Cederwall and Ditmars (1970) and Zic and Stefan (1994). Additional methods of in-lake treatment are discussed by Graber (1978, pp. 6-23–6-24).

A seiche is a tilting of the thermocline that can occur due to wind stress on the lake surface. Lake Champlain between Vermont and New York State has a large north-south length (125 miles, 201 km) and relatively narrow east-west width (averaging 14.3 miles, 23.0 km). Wind along its long direction often produces a seiche (Cohn *et al.* 2007, Winslow 2008). The seiche can tilt the thermocline sufficiently that it intersects the bottom, bringing up submerged debris such as waterlogged tree trunks. That is the likely cause of sightings of "Champ," the Lake Champlain monster (Nickell 2003).

Section 13.3 discusses the pattern of flow approaching sedimentation-tank outlet weirs. A related issue arose in connection with a proposal for algae control in Three Mile Pond located in Windsor, Augusta, and China, Maine. Members of the Three Mile Pond Association proposed that a weir be installed on the only pond outlet to allow nuisance algae at and near the surface to be skimmed off. The author cited the flow pattern discussed in Section 13.3 and shown in Graber (1974a) and on Figure 13.3 to point out that such a weir would not result in preferential surface skimming but would discharge water from the full depth above the thermocline. It was also noted that such a weir would impede the migration of anadromous alewives. The proposal was not implemented.

15.3.3 Langmuir Circulation

Langmuir circulation is an interesting phenomenon, occurring in lakes, ponds, estuaries, and oceans. In the ocean, Langmuir circulation is evidenced by multiple windrows of seaweed, as reported initially by Langmuir (Nobel laureate in chemistry) during a transatlantic voyage; he subsequently observed similar lines of foam caused by decayed algae on Lake George

Figure 15.8 Langmuir Circulation on Three Mile Pond, Windsor, Maine. Photograph by author.

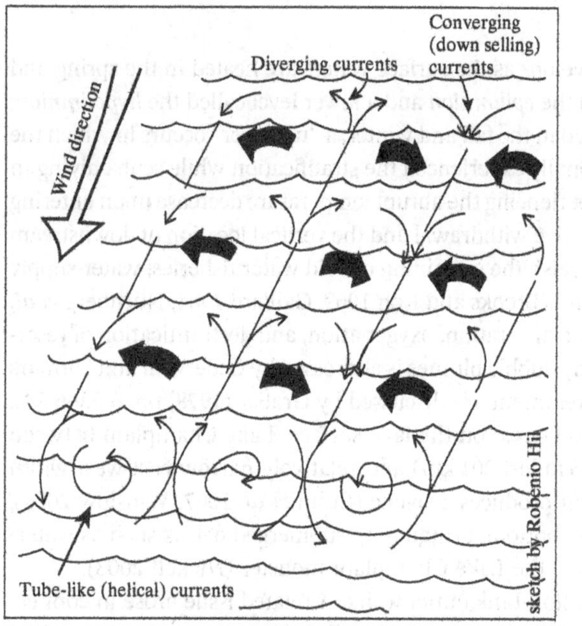

Figure 15.9 Schematic Sketch of Langmuir Currents (Williams 2007 / with permission of Lake Stewards of Maine)

in the State of New York (Langmuir 1938). Such foam lines can be desirably associated with algae die-off, with good attendant water clarity. At times, however, the foam can build up on downwind shores to a nuisance degree. A photograph by the author of Langmuir circulation on Three Mile Pond, Windsor, Maine is shown above.

Langmuir reported his observations but offered no explanation of the phenomenon. There are now two types of explanations, one being based on a complex instability theory (Nepf and Monismith 1991, Phillips 2003, 2005) said to be "a useful metaphor ... [for which] we do as yet know whether this theory is realizable physically or whether in practice it is primary responsible for LCs ..." (Phillips 2003). Lugt (1983) says "[t]he explanations for the occurrence of Langmuir vortices are controversial." Williams (2007) says "[t]he exact way in which they form is not fully understood." He depicts two currents alternating in their direction of circulation between each streak of algae, with the converging currents forming the streaks. His depiction is shown in Figure 15.9 below. Mild wind-generated waves are necessary for the Langmuir currents to form.

Langmuir circulation can be explained in terms of secondary currents of the second kind, which are discussed in Section 4.7. A longitudinal velocity profile forms with the lowest velocities at the edges of the wind fetch (particularly in the ocean) or the boundaries of the lake, pond or estuary. Rotation of the vorticity vector at the edges causes a spiral current to form at the edges, and friction with the adjacent water causes a spiral to form in the opposite direction. Each spiral causes an adjacent spiral to form until the complete set of spiral currents is formed.

Elzerman and Armstrong (1978) discuss the role lake foam plays as a natural concentration process for various chemicals which is of ecological significance.

15.3.4 Freshwater Wetlands

Categories of freshwater wetlands other than lakes and ponds are discussed here. Detention basins, including examples, are discussed in Section 5.4. Chapters 7, 8, and 15 discuss other elements of such basins, particularly outlet hydraulics. Numerous others of the author's projects have addressed so-called "Determinations of Applicability" in which hydrologic and hydraulic calculations are made to ascertain whether wetlands meet certain requirements for regulatory protection. One such category is "isolated land subject to flooding" for which the author made determinations for the Massachusetts towns of Lynnfield, Topsfield, and Randolph. Other wetland determinations on which the author has worked include perennial vs. intermittent stream classifications, e.g., for Nagog Brook in Acton, Massachusetts; effects on wetlands and alternative for school siting in Quincy, Massachusetts; wetland impacts of alternative routes for a gas pipeline route in Franklin, Massachusetts; wetland impacts for a bridge and associated housing development in Lincoln, Massachusetts; and in Stoughton, Massachusetts the certification of several vernal pools with protected status [a good reference is *Vernal Pools* by Colburn (2004)]. Good general references on important characteristics and values of wetlands are Larson (1973), U.S. Congress (1974), Ogawa and Male (1983, 1986), and Mitsch and Gosselink (2000), and other publications referenced therein. Graber (2017b) provides additional perspective regarding wetland values in the context of stormwater impoundments, particularly in Sections 4.4 and 6.1.10.3.

Of greatest relevance to the present chapter, the author has worked on a number of projects in which natural freshwater wetlands have played an important role in the design of stormwater systems for a wide range of return periods. For Westford, Massachusetts, an existing culvert and upstream wetland played a useful role in attenuating runoff from a portion of a housing development; the culvert flow regime and *effective* storage capacity of the wetland were determined.

The largest wetland evaluation was one in which the author participated in 1966 as his first professional assignment. That was for a drainage/flood control study for the South River in Salem, Massachusetts. One of the key recommendations was that it was "absolutely essential" that existing swamp storage along South River be preserved. We recommended the rezoning of 74 acres of swamp land into a new classification, "Conservation Flood Control," and that these swamp areas be purchased and removed from private ownership. One feature of the analysis concerned the fact that the drainage system terminated in the tidal waters of Massachusetts Bay. To address the conjunctive effects of storm frequency and tidal conditions, the design storm frequency was assumed to coincident with mean high-water (MHW) tidal conditions. A more recent analysis considered such coincidence in a more sophisticated way (Feng *et al.* 2017) and merits consideration.

15.4 Estuaries

An estuary is defined as the location where a river flowing into a bay or ocean is influenced by tidal conditions. The upstream limit of the estuary may be defined as the limit of tidally influenced water-surface variations or the limit of saltwater intrusion. Saltwater cannot intrude beyond the location where the elevation of the river bed equals the sea water level at high tide. The limit of intrusion can be held significantly downstream of this location by the freshwater discharge. In one case, the author conservatively chose a location upstream of the upper of those locations to site an intake for a water supply to run from that intake to the offshore island of Ailigandi in the San Blas Comarca (Region) off the Caribbean coast of Panamá.

Langmuir circulation, occurring in estuaries as well as in lakes and oceans, is discussed in Section 15.3. Pollution in estuaries is addressed here only by reference to Harleman (1966) and Metcalf & Eddy, Inc. (1972, pp. 687–691). The tidal bore is discussed below.

In most of the world's estuaries, the tide rises gradually. However, in some places, a remarkable phenomenon known as a tidal bore occurs. Then the tide comes into the estuary abruptly as a fast-moving wall of water, in some places up to 10 or more feet high. Estuarine tidal bores have been the subject or cause of legend, poetry, recreation, tourism, shipping disasters, and scientific interest for centuries. Graber (2012a, 2012b, 2018) discusses and tabulates legends and related information.

Graber presents a simplified theory based on monoclinal rising waves for the initiation of the tidal bore in a rectangular channel. The theory elucidates the mechanisms of the tidal bore more clearly and completely than more complex theories. Early studies of the tidal bore are cited and discussed in historical context. The papers should provide greater understanding of the tidal bore for a broad set of readers. The theory was favorably compared to rectangular flume data and to all field data available to the author. The roles of channel taper and friction; building, maximum height, variation, and dissipation of the bore; and standing waves are discussed. Early studies of these mechanisms are cited and discussed in historical context. Relationships developed to investigate the effects of horizontal channel taper, vertical channel taper, and channel friction agree qualitatively with and explain field observations. Building of the bore is explained in terms of the accumulation of wavelets at the bore front, and favorable comparison of methodology with flume observations is reported. The maximum height that the bore is capable of achieving is one of its most interesting features; that is explained in terms of the formation of a roller (and the mechanism of such formation) at a certain height or relative Froude number, in agreement with experimental observation. The mechanism of the formation of the roller is explained. The two forms of the jump, undular and wall-like, are discussed. The variation of the bore height as it moves into regions of greater or lesser depth is described from field observations and flume observation, and relationships are derived which explain such variation. Frictional dissipation of the bore is discussed, citing pertinent references. Standing waves that occur after the tidal bore has passed are explained as a separate phenomenon.

Present-day or historic (extinct) tidal bores occur on China's Qiantang River (see the author's photograph in Figure 15.10 below), the Shubenacadie and other rivers flowing into Canada's Bay of Fundy, England's Humber Estuary, Alaska's Knik Arm and Turnagain Arm, Brazil's Amazon River, the Colorado River in the Southwestern United States and northern Mexico, France's Seine River and Mont-Saint-Michel Bay, England's Severn and Trent Rivers, Scotland's Solway Firth, the Bamu River in New Guinea, the Indus River in Pakistan, and India's Hugli River. The bore has the fascinating appearance of currents from upstream and downstream conspicuously coming into each other along a line approximately perpendicular to the shore. On the upriver side of the bore, the tidal current goes out and on the downriver side the tidal current comes in (or goes out at a lower speed, depending on the freshwater flow), with an abrupt change in the tidal velocity across the bore.

None of the records of tidal bores and related cultural interactions approach those for the Qiantang River, which has been famous for its tidal bore, and related legends, poetry, science, engineering, and history for more than a thousand years. In the 1700s (Qian Long period of Qing Dynasty), the so-called "fish-scale" stone embankment was constructed, although

Figure 15.10 Tidal Bore of About 1 m (3 ft) on the Qiantang River, Haining, China, October 20, 2010 [S. David Graber (Book Author)]

construction on a Qiantang River embankment is thought to date back to around 100 CE (construction started as far back as the Han Dynasty and lasted until the Qing Dynasty). The embankment of the Qiantang River is considered one of the three great ancient engineering feats in China, alongside the Great Wall and the Grand Canal that connects Beijing and Hangzhou.

Eight iron oxen were also placed on the Qiantang River bank near the village of Haining to provide mythical protection against the assaults of the bore. However, they were not able to protect themselves. The power of the Qiantang tidal bore is evidenced by the fact that the eight iron oxen, 1.5 tons each, were washed away in 1953. Figure 15.11 shows a photograph of one of the two replica replacements. After the bore has passed, local shipping traveling upstream benefits from the strong current. But the ships must avoid being damaged by the bore. One means of dealing with the oncoming bore is for ships to ground themselves on platforms provided for that purpose. Figure 15.12 shows a portion of the platform and a portion of the embankment as the bore is passing.

Figure 15.11 Replica Replacement of Protective Iron Oxen along the Qiantang River, Haining, China, October 20, 2010 [S. David Graber (Book Author)]

Figure 15.12 Ship Grounding Platform and Tidal Bore on the Qiantang River, Haining, China, October 20, 2010 [S. David Graber (Book Author)]

An interesting feature of the tidal bore is its sound. The bore is not always of dramatic height. On the night of June 7, 2002, although too dark to see, the author could hear the loud roar of the tidal bore well upstream on the Shubenacadie River in Nova Scotia from a nearby cabin. However, the following morning the author observed the bore from a boat on the River at Anthony's Nose, nearer the River's mouth. At that time and location, it was little more than a few inches high and only made a sizzling sound. Chanson (2016) describes the tidal bore on the Qiantang River when crashing into the side platform with loud and powerful noises on October 12, 2014, with a bore height of about 3 m (10 ft). Section 14.7 discusses noise associated with waterfalls.

The Shubenacadie River has a particularly interesting feature. The only remaining self-sustaining population of striped bass (*Morone saxatilis*), in the Bay of Fundy spawns in the tidal bore-dominated Shubenacadie–Stewiacke watershed of Nova Scotia (Rulifson and Tull 1999). Being anadromous, the striped bass (depicted below) live in saltwater but spawn in freshwater rivers. A particularly interesting adaptation allows their eggs to remain in the Shubenacadie River during the two to three days it takes for them to hatch. The eggs have a greater density than those in other rivers, i.e., those without a tidal bore. That apparently enables the eggs to be retained within lowest salinity waters within or near the tidal bore on flood tides until hatching. After the eggs hatch, the larvae are able to swim well enough to remain in the river, where most spend their first two years of life.

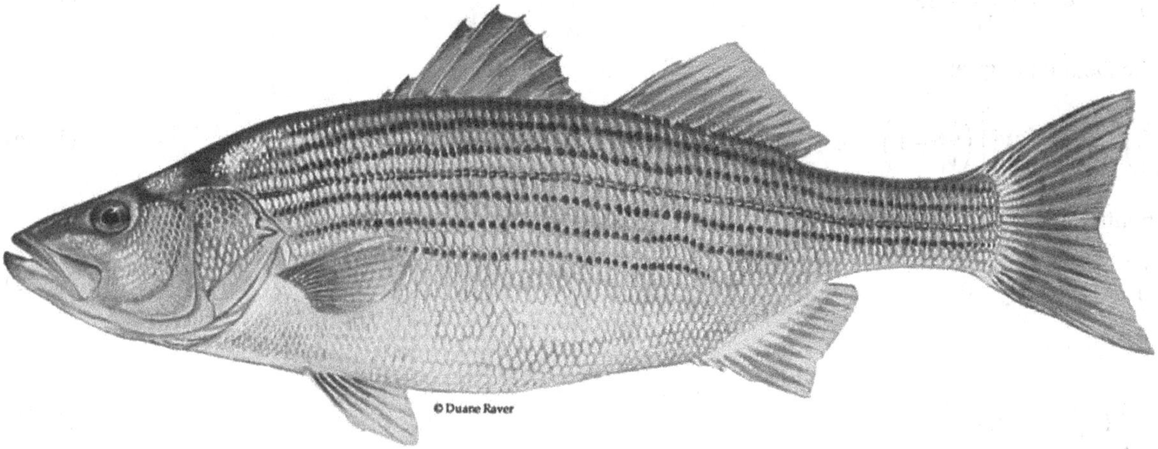

A particularly interesting feature of the Qiantang River bore described by Chanson (2016) is the lateral vortices and streaks at the base of the bore, reminiscent of the Langmuir circulations discussed in Section 15.3 and shown in Figure 15.9.

Another interesting feature of the tidal bore is that it does not reverse and head back out in a seaward direction. The author has applied the second law of thermodynamics to demonstrate why the tidal bore can only proceed upstream in an estuary (Graber 2023a). A tidal bore retreating to the sea (with a change of tide) will become unstable and flatten out in accordance with observations. The rapidly varied flow analogous to the hydraulic drop discussed above would be a reverse hydraulic jump, which leads to a discussion of its impossibility in the context of the tidal bore. Chow (1959, §'s 1-2, 3-4, 3-7, 15-2, 15-3, 15-4, 18-3, 19-1, 19-2, 19-6.) notes the dissipating role of friction, plus points out that a bore retreating to the sea (with a change of tide) will become unstable and flatten out. Schönfeld (1955) and Stoker (1957) also note the latter. Graber (2023a) addresses the reasons for this [with reference to Hornung *et al.* (1995) and Swan (2020)] by considering entropy more generally and a more definitive development.

Hornung *et al.* (1995) addressed the entropy increase across a hydraulic jump, drawing an analogy between "the manner in which dissipative effects behave in the analogous situation of a shock wave in a compressible fluid. The square of the Mach number $\mathbf{M}^2$ corresponds to the Froude number **F** in the analogy and dissipative effects manifest themselves in the form of an entropy increase in the shock wave. This entropy change increases with the cube of $\mathbf{M}^2 - 1$, just as the manifestation of dissipative effects in the hydraulic jump (namely vorticity) increases as the cube of $\mathbf{F} - 1$ here."

Swan (2020) stated: "The hydraulic jump is in many respects analogous to a shock wave arising within a compressible flow. For example, whereas the hydraulic jump provides a transition from supercritical to subcritical flow, the shock wave involves a transition from supersonic to subsonic flow. In both cases there is a critical velocity below which these transitions cannot occur, and both processes involve an increase in entropy. Indeed, in the case of a hydraulic jump the increase in entropy per unit mass is proportional to the cube of the depth change, whereas in a shock wave this increase is proportional to the cube of the pressure difference (provided this is small)."

The reason for the instability of a tidal bore retreating to the sea will now be addressed. It will be useful in the present context to derive the equation for the hydraulic jump. Consider the basic case of a hydraulic jump in a channel of rectangular cross section with weight and drag terms neglected. The steady-flow momentum equation between upstream section "1" and downstream section "2" gives:

$$\frac{y_1}{2}by_1 + \frac{Q^2}{gby_1} = \frac{y_2}{2}by_2 + \frac{Q^2}{gby_2} \tag{15.23}$$

in which y is depth of flow, b is channel width, Q is volumetric rate of flow, and g is gravitational acceleration. Incorporating the upstream Froude Number $\mathbf{F}_1 = V_1/\sqrt{gy_1}$, and continuing with the development yields:

$$\frac{y_1^2}{2} + y_1^2\mathbf{F}_1^2 = \frac{y_2^2}{2} + \frac{y_1^3}{y_2}\mathbf{F}_1^2 \tag{15.24}$$

hence:

$$\left(\frac{y_2}{y_1}\right)^3 - (2\mathbf{F}_1^2 + 1)\frac{y_2}{y_1} + 2\mathbf{F}_1^2 = 0 \tag{15.25}$$

which can be factored to give:

$$\left[\left(\frac{y_2}{y_1}\right)^2 + \frac{y_2}{y_1} - 2\mathbf{F}_1^2\right]\left(\frac{y_2}{y_1} - 1\right) = 0 \tag{15.26}$$

For the nontrivial solution, we solve the following quadratic equation:

$$\left(\frac{y_2}{y_1}\right)^2 + \frac{y_2}{y_1} - 2\mathbf{F}_1^2 = 0 \tag{15.27}$$

to obtain:

$$\frac{y_2}{y_1} = \frac{1}{2}\left(\sqrt{1 + 8\mathbf{F}_1^2} - 1\right) \tag{15.28}$$

Investigation of the reverse jump by continuing to take y_1 upstream and y_2 downstream, but now with $y_2 > y_1$ and $\mathbf{F}_2 = V_2/\sqrt{gy_2} > 1$, yields:

$$\frac{y_1}{2} b y_1 + \frac{Q^2}{g b y_1} = \frac{y_2}{2} b y_2 + \frac{Q^2}{g b y_2} \tag{15.29}$$

Following through as above gives:

$$\frac{y_1^2}{2} + y_2^2 \mathbf{F}_2^2 = \frac{y_2^2}{2} + y_2^2 \mathbf{F}_2^2 \tag{15.30}$$

and

$$\left(\frac{y_1}{y_2}\right)^2 + 2\mathbf{F}_1^2 = 1 + 2\mathbf{F}_1^2 \tag{15.31}$$

leading to:

$$\left(\frac{y_1}{y_2}\right)^2 = 1 \tag{15.32}$$

and $y_1/y_2 = 1$, a trivial solution.

Additionally, the second law of thermodynamics states that entropy can be created but it cannot be destroyed. This is called the entropy balance. Therefore, the entropy change of a system is zero if the state of the system does not change during the process. The upshot of all this is that the tidal bore cannot return to the sea.

An interesting analogy between the aurora borealis (northern lights), as depicted on the jacket of this book, and the tidal bore has been established (Robinson 2020). Numerous variations of the aurora have been observed. Dune auroras are a variation which are believed to result from a phenomenon called a "mesopheric bore," named by analogy to the tidal bore. With a mesopheric bore, oxygen atoms are pushed higher in the Earth's atmosphere, causing a distinctive frozen-wave glow.

Holley and Harleman (1965) and Holley *et al.* (1970) are important references which address dispersion (as defined in Chapter 14) of pollutants in estuaries.

15.5 Ocean Waters

Early explanations of the tides were quite creative. In Norse mythology, the tide was said to be caused by the breathing of the Earth-serpent Iormungander, an enormous monster that encircled the globe and held its tail in its mouth (Klingel 1999). There are beautiful Native American (Caduto and Bruchac 1991, pp. 73–74) and Japanese (Piggott 1982, pp. 6, 7, 14, 15, 22) myths regarding the cause of the tides. One early notion had the moonlight, i.e., the sunlight reflected off the Moon, heating the ocean causing it to expand; which failed, of course, to explain the high tide on the side of the Earth opposite the Moon, or the high tide during the new Moon, or during eclipses of the Moon. Early Chinese writers suggested that the water is the blood of the earth and that the tides are the beating of its pulse, and secondly that the tides are caused by the Earth breathing. Seafaring people in many parts of the world had made the connection between tides and the Moon and its phases. But the reasons for that connection had to await the scientific advances that started earnestly in the sixteenth century. The role of gravity in establishing the orbits of the planets around the Sun, the orbit of the Moon around the Earth, and the gravitational attraction of the Sun and Moon on the Earth's oceans was not discovered until the seventeenth century.

Scientific knowledge of the tides is based on a long, multicultural quest for understanding. One of the earliest attempts to explain the tides scientifically appeared in a ninth century work by Abu Ma'shar (or Albumasar), an Arab scholar who was born in Afghanistan and lived in Baghdad (Cartwright 1999, p. 15). Through the centuries, scientists of various countries added to the understanding, leading up to seventeenth century England where we come to the huge contributions to science of Isaac Newton [Principia (Newton 1999), which Newton may have never published if not for being convinced to do so by his friend Edmond Halley of Halley's comet fame.]. Newton provided the first complete proof of the revolution of the Earth and planets around the Sun and correct explanation of the causes of the tides.

In the nineteenth and early twentieth centuries, George Darwin, who was the British Royal Astronomer and second son of Charles Darwin of theory-of-evolution fame, added substantially to understanding of the tides. Darwin's explanation is the basis for what immediately follows. (Darwin 1962, Chapter III, Eagleson and Dean 1966).

The basic causes of the tides can be explained most clearly by focusing on the interrelationship between the Earth and the Moon. The Moon is responsible for approximately 60% of the tide-producing forces, depending on the distances of the Moon and Sun from Earth and other factors. And, of course, the role of the Sun is literally reflected in the Moon, that is to say in the phases of the Moon, as we shall discuss shortly.

It is known that the Moon "revolves" around the Earth,; although, it will be seen momentarily that it is really somewhat more complicated than that. In the Earth–Moon system, centrifugal force holds the Moon outward, and the force of gravitational attraction holds the Moon in toward the Earth. Considering a control volume encompassing the Moon alone, we see that the centrifugal force balances the gravitational force and the Moon is in equilibrium. Because of the balance of forces, the Moon does not move closer to or further away from the Earth, but rather maintains a fixed distance. (More precisely, because of tidal friction, the Earth's rotation is slowing by about two thousands of a second per century and the Moon is moving away from the Earth at about 4 m per century.)

Now consider a control volume around the Earth alone. It appears at first as though only one force is acting on the Earth, the equal and opposite gravitational attraction from the Earth to the Moon. However, the Earth would then tend to move toward the Moon and the Earth–Moon system as we know it could not be maintained. So, what happens? The answer is that the Moon does not simply revolve around the Earth, but rather the Earth and Moon both revolve around a common axis between the two bodies, kind of like a Do-Si-Do. By doing so, a centrifugal force on Earth is created that balances the gravitational attraction toward the Moon so that the Earth is now in equilibrium. For every action, there is an equal and opposite reaction. Because the Earth is so much more massive than the Moon, the common axis of revolution between them is actually through the edge of the Earth, and the Earth's motion relative to the Moon is more of a wobble than a revolution. But it is a very important wobble, creating the centrifugal force that not only keeps the Earth–Moon system in equilibrium, but that centrifugal force, on the side of the Earth opposite the Moon, is what causes the high tide on the other side of the Earth from that on which the Moon's direct gravitational attraction is causing a high tide.

So the Moon does not simply revolve around the Earth, but rather the Moon and Earth do a celestial dance which is the cause of the tides. In fact, whenever there are two bodies in space in a stable orbital configuration, if they are of approximately equal in size they will conspicuously revolve around each other. It is estimated that approximately half of the stars in our Milky Way galaxy are binary stars that revolve around each other in this manner. The nearest star system to Earth is the three-star Centauri system and two of its stars are binary stars that revolve around each other. However, when one of the bodies is much more massive than the other, the larger one just wobbles. A distant observer could detect the wobble of our Sun as its planets revolve around it. One way by which astronomers are now identifying distant stars thought to have orbiting planets is by detecting wobbles of those stars.

Along the coast of Massachusetts, as in much of the world, there are two high tides and two low tides every day, referred to as a semi-diurnal type of tide. For example, on January 11, 2007, in Boston Harbor, the high tides occurred at 4:27 a.m. and 4:51 p.m. [From the National Oceanic and Atmospheric Administration (NOAA) website for Deer Island (south end), http://www.co-ops.nos.noaa.gov.]. (In some parts of the world, particularly near the north and south poles but not only there, the tides are diurnal with only one high and low tide each day.) One of the daily high tides occurs when, in the course of the Earth's daily rotation, the Moon is over our side of the Earth and the other about 12½ hours later when the Moon is over the opposite side of the Earth. The direct gravitational pull of the Moon is the main cause of the high tide on one side of the Earth. The discussion above explains why, even though the Moon is approximately overhead only once a day, there is a high tide at, say, Boston Harbor, when the Moon is on the opposite side of the Earth.

[At least one well-known encyclopedia ("Tide", in Vol. 19 of The World Book Encyclopedia, World Book, Inc., Chicago, Illinois, 1985), a college astronomy textbook (Seeds 1986, p. 24), and a respected magazine ("Ask the Experts," *Scientific American*, November 2006, p. 108.) have given incorrect explanations of the tides, specifically what causes the high tide on the side of the Earth facing away from the Moon.]

The tidal range, i.e., the difference between high and low-tide elevations, varies from less than one foot on the Caribbean side of the Panamá Canal to over 50 ft at the heads of the Bay of Fundy in Nova Scotia and New Brunswick, Canada. Further discussion of the latter is given below. The tidal range can vary significantly over short distances, such as the 3-foot range in Nantucket Sound off Cape Cod and about 9.5 ft off Quincy, Massachusetts and in Boston Harbor.

To discuss the tides more fully, one would have to add in consideration of the role of the Sun, which together with the Moon causes the highest high tides and lowest low tides twice a month around the time of the full and new Moons, the

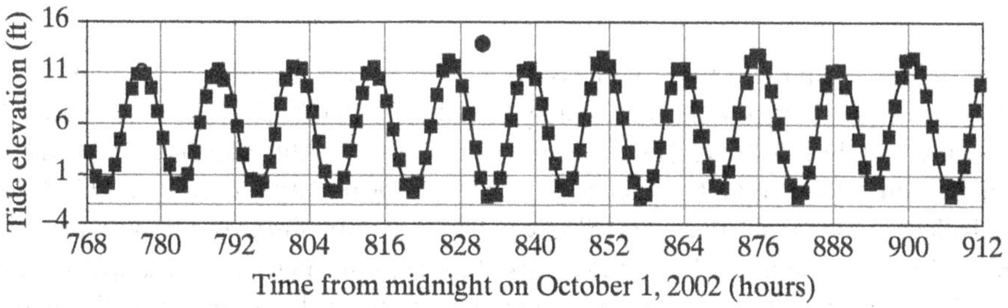

Figure 15.13 Predicted Boston Tides, November 2, 2002 to November 7, 2002

so-called spring and neap tides. (The word "spring" in "spring tides" has nothing to do with the spring season. Rather it is derived from an Old English word "springan" meaning "a welling." The word "neap" as in "neap tides" comes from the Old English word "nep," meaning "scant" or "lacking.") However, if one looks at the time of the full or new Moon in relation to the tides, one finds that the highest and lowest tides of the month do not occur at the time of the full or new Moon but rather one tidal cycle after the full or new Moon, i.e., about one-half day later. An example of this is given below in Figure 15.13.

The above figure has been prepared using a NOAA computer program together with harmonic constants for Boston Harbor and covers the period from November 2 to November 7, 2002. The diurnal inequality can be seen by looking at two successive tides. The lowest tide of the year occurs on November 5, 2002, at 5:24 p.m. The new Moon is also shown as a dot; it occurs at 3:34 p.m. on Monday, November 4, 2002. It occurs one-half tidal cycle after the high tide. This is due to the manner in which vibrating systems respond to vibrating forces, in this case, the vibrating system being the Earth's oceans and the vibrating force being the gravitational force of the Moon. The frequency of the tide-producing force of the Moon which is about one-half day is higher than the natural frequency of the Earth's oceans, and the Earth's oceans are responding 180° out of phase. So basically, a particular high tide will be in response to the gravitational pull of the Moon from one-half day ago.

At some locations, the tidal fluctuations occur at a near-resonant condition, in which the forcing frequency is close to the natural frequency of the body of water. This occurs most dramatically in Canada's Bay of Fundy, causing the world's largest tidal range.

The Moon and Sun combined result in a 19-year "metonic" cycle. It is the repetition of that cycle that allows tide predictions to be made based on prior observations. Since the author received his engineering degrees in the 1960s, advances have been made that now allow accurate tide predictions to be made entirely from theoretical modeling that considers the astronomical positions of the Earth, Moon, and Sun together with the geographic variations of the Earth's oceans and shorelines (Cartwright 1999, p. 187, etc.).

Similar principles apply to ocean outfalls as to freshwater outfalls discussed in Section 7.8 – Outfall Structures and Scour Protection. The Chapter 11 discussions of outfall diffusers pertain equally to ocean disposal of wastewater. The author played a determinative role in the siting of the outfall from Massachusetts' Deer Island Wastewater Treatment Plant with a peak capacity of 1.2 billion gal/day. The outfall begins with a deep rock tunnel of 24-foot diameter extending under Massachusetts Bay to a point about 9.5 miles east of Deer Island. In the last mile of the tunnel, 55 riser pipes carry treated effluent up from the tunnel to the ocean floor where it is dispersed by mushroom-shaped diffuser caps (McMonagle and Otoski 1985).

Density stratification can be a significant consideration in ocean disposal. Fan and Brooks (1969) is an important relative reference, and addresses inclined slot and round buoyant jets in uniform ambient fluid and ambient fluid with linear density stratification. Further information on ocean disposal is given by Metcalf & Eddy, Inc. (1972, pp. 691–700).

Langmuir circulation, occurring in oceans as well as in lakes and estuaries, is discussed in Section 15.3.

15.6 Groundwater

The topic of flow through porous media such as soil is introduced in Section 4.6. Spatially increasing flows for various man-made subsurface drainage systems are discussed in Chapter 12. In the present section, topics addressed are water wells including well hydraulics (well screens are discussed in Chapter 12), infiltration basins, various types of subsurface drains, and berm seepage.

15.6.1 Water Wells

During the drilling and initial testing of a well, it is important that data be collected concerning the lithographic nature of the geologic formations which the well penetrates, the hydraulic characteristics of the aquifer, and the physicochemical characteristics of the water. Among the purposes of these tests are: (1) to ascertain whether the quality and quantity of water available is adequate to meet design requirements, (2) to establish the yield of the well for planning purposes, and (3) to facilitate the proper design of the well. The latter entails selecting the well screen and designing the gravel pack if needed, setting the pump bowl and selecting the pump, and designing other components of the system (such as water tanks) which may be hydraulically dependent on the well characteristics. A general discussion of hydraulic well testing is provided here.

The parameters measured in a well-pumping test include the discharge from the pumped well and water levels in the pumped well and observation wells, if any. Mathematical theories have been developed to relate these measured parameters and parameters characterizing the water-bearing formation. It is essential to understand the derivation, applicability, and limitations of each of the major relationships.

The most well-known equations include the equilibrium equations for artesian and water table wells and the nonequilibrium equations of Theis and Jacob (pertinent references are cited below). The use of the equilibrium equations in analyzing the results of well-pumping tests in some cases requires that the drawdown be measured in observation wells within the radius of influence of the pumped well. Through the appropriate equilibrium equation, the data may be used to predict the permeability of the aquifer and thereby facilitate the evaluation and design of the well. However, the equilibrium equations provide meaningful results only when the characteristics of the aquifer satisfy assumptions made in deriving the equations, among them that the water-bearing formation is homogeneous and isotropic. This is often not the case, and the above-mentioned equations are then of little use and observation wells will then be of limited value.

The nonequilibrium well equations can be used to obtain useful results, under appropriate circumstances, by obtaining data from a single pumped well without the need for observation wells (e.g., Esmaili and Scott 1968). For these equations to be valid, the assumptions previously mentioned in connection with the equilibrium equations must be satisfied. In addition, the aquifer must allow water removed from storage to be discharged instantaneously as the head is lowered and the formation must receive no discharge from any source. Even when these requirements are not met the information obtained in this type of test can be used to reveal the nature of the aquifer. The nonequilibrium equations will provide a useful comparison and merit looking into in some detail. The Theis equation is given by:

$$s(r,T) = \frac{Q}{4\pi T} W(u) \tag{15.33a}$$

$$u = \frac{r^2 S}{4Tt} \tag{15.33b}$$

in which s = drawdown, t = period of pumping, r = radial distance in the aquifer, T = transmissibility, S = storage coefficient, Q = discharge, and $W(u)$ is the Theis well function. This equation is derived specifically for artesian conditions but can be extended to water table aquifers under certain circumstances. The use of the Theis equation involves the matching of "type" curves and is described in standard textbooks on groundwater hydraulics (such as those cited below). The well function $W(u)$, is an exponential integral which may be written in the form of an infinite series as follows:

$$W(u) = \int_u^\infty \frac{e^{-x}}{x} dx = -\gamma - \ln u + u - \frac{u^2}{2 \cdot 2!} + \frac{u^3}{3 \cdot 3!} \mp \cdots \tag{15.34a}$$

$$\gamma = 0.5772, \textit{Euler's constant} \tag{15.34b}$$

For values of u less than 0.05, all but the first two terms in the series can be neglected, resulting in the following:

$$s = 264 \frac{Q}{T} \log_{10} \frac{0.3Tt}{r^2 S} \tag{15.35}$$

in which s, t, r, T, and Q have units of ft, days, ft, gpd/ft, and gpm, respectively.

Therefore, for an aquifer satisfying the assumptions made in deriving the Theis equation, the Jacob equation will fit that portion of the drawdown-time curve which plots as a straight line on semi-logarithmic graph paper. The slope of the straight line will give the transmissibility of the well from:

$$T = 264 \frac{Q}{\Delta S} \tag{15.36a}$$

$$T = 264\frac{Q}{\Delta S} \tag{15.36b}$$

$$\Delta S = \frac{S_2 - S_1}{\log(t_2/t_1)} \tag{15.36c}$$

During the early period of pumping, the small value of t results in u being greater than 0.05 and the time-drawdown curve is not straight on the semi-log graph but has increasing negative slope. If the time-drawdown data is from the pumped well, r is small and u less than 0.05 will generally be satisfied within the first few minutes of pumping. The straight line fit to the data from the pumped well will pass within 1 minute of the origin of the graph. The recovery of the well level after termination of pumping is defined as the difference between the dynamic water level in the well during the recovery period and the projected dynamic level in the well at the same time that would exist if pumping had continued. If the Theis equation were valid during the test period, the recovery-time curve would be identical and the transmissibility could also be calculated by:

$$T = 264\frac{Q}{\Delta(S - S')} \tag{15.37a}$$

$$\Delta(S - S') = \frac{(S - S')_2 - (S - S')_1}{\log(t_2/t_1)} \tag{15.37b}$$

The time required for the recovery to equal the drawdown at the end of the pumping period would be equal to the duration of pumping.

Derivations of the relationships presented above may be found in several references: Theis (1935, 1952; Cooper and Jacob 1946; Jacob 1950; Linsley and Franzini 1964, Chapter 4; Aron and Scott 1965; Davis and De Wiest 1966; Kriz *et al.* 1966). The last reference uses dimensional analysis to obtain a relationship between the parameters of a homogeneous, isotropic confined aquifer.

Bailer testing is used in some places. Although this method can be used to eliminate very low productivity wells from further consideration, its value is limited for other than hand-pump wells. Pump tests are best run at a constant discharge rate, such as by fitting the pump discharge line with a gate valve to regulate flow and an orifice and manometer to provide a continuous, instantaneous flow reading. The water level should be measured while pumping and during recovery of the well after pumping; such measurements can be made by means of an air line connection at the surface to a hand pump and pressure gauge. Typical well-testing appurtenances, of the type described above, are discussed in standard references (Rohwer 1941; Wenzel 1942; Gordon 1958; Johnson 1966, Chapter 5).

The Theis equation discussed above is a useful conceptual tool in planning a data collection program and in the interpretation of well-test data as well as its use in design. From this point of departure, techniques should be developed which are appropriate to the conditions encountered in a particular region. That should be a continuously-evolving process, based on a thorough understanding and diligent evaluation of methods in use.

Having obtained the drawdown-flow relationship for equilibrium conditions, the design of the well can be carried out. A factor of safety should be incorporated to account for the fact that, for the same total production, drawdowns under cyclic pumping conditions are greater than drawdowns under continuous pumping conditions. The aquifer will be used more efficiently, from the standpoint of total productivity, if the well is pumped continuously at a low rate rather than intermittently at a high rate.

The use of side-scanning sonar from airplanes to detect fissures in consolidated aquifer formations has been promoted in various areas. The author's experience with companies performing such work in areas as diverse as Panamá and Massachusetts has been decidedly unfavorable. After considerable expense, not a single promising well site was located. In one instance where a company touted success in locating a particularly productive well location in a consolidated formation, it turned out that the well existed at a foreclosed homeowner's property that a town tested and found highly productive. Sonar had nothing to do with its discovery. This borders on the pseudoscience of water dowsing aka water witching (Ainsworth 1999, DeNileon 1999, Enright 1999, Woods 1999). Proceed with caution!

An important principle for wells near coastal areas is that of Ghyben–Herzberg (Camp and Lawler 1969a), which pertains to the saltwater-freshwater interface in an unconfined coastal aquifer under hydrostatic conditions and conditions of steady-state seaward flow. Then the weight of a unit column of fresh water extending from the water table to the interface is balanced by a unit column of saltwater extending from sea level to the same depth as the point on the interface. We have:

$$\rho_s g z_s = \rho_f g (z_s + z_f) \tag{15.38}$$

or

$$z_s = \frac{\rho_f}{\rho_s - \rho_f} z_f \tag{15.39}$$

in which z_s is the depth below sea level to the saltwater interface, z_f is the head of fresh water above sea level, ρ_s is the density of saltwater, and ρ_f is the freshwater density. For $\rho_f = 1.0$ and $\rho_s = 1.025$, $z_s = 40z_w$ is obtained. If the water table in an unconfined coastal aquifer is lowered 1 m, the saltwater interface will rise 40 m. Additional details regarding saltwater intrusion of coastal aquifers and the saltwater lens on islands (which was the reason for the author's choice of a mainland river as described in Section 15.4) is given by Todd (1964) and Johnson (1966, pp. 61, 415–416). The former also derives relationships analogous to the above for circular islands. The latter explains that the position and fluctuation of the freshwater-saltwater contact zone depends on the hydrodynamic balance of the two waters which adds complexity beyond the static balance given by the Ghyben–Herzberg principle.

In 1968, while working on the Caribbean coast of Panamá, the author encountered a World Health Organization (WHO) representative who proposed to construct a series of hand-pumped drinking water wells along a coastal beach. The author explained to him the Ghyben–Herzberg principle and why the wells would pull in saltwater. The representative disagreed and proceeded to arrange the construction of the wells. As expected, the salty wells proved to be an unusable waste of money. It is noteworthy in this context that the Cuna Indians who live off this same coast on the San Blas Comarca mentioned in Section 15.4 could not obtain drinkable groundwater from dug wells on their islands.

15.6.2 Water Mounding/Groundwater Disposal

Water mounding calculations can be important for some on-site wastewater disposal systems, particularly for larger communal systems (see Figure 15.14). The Hantush method, as given by Hantush (1967), Bouwer (1978, pp. 283–287), and Canter and Knox (1985, pp. 145–149), is useful for this purpose. The Hantush equation is given as follows:

$$\begin{aligned} h_{x,y,t} - H = \frac{v_a t}{4f} \{ & F[(W/2 + x)n, (L/2 + y)n] + F[(W/2 + x)n, (L/2 - y)n] \\ & + F[(W/2 - x)n, (L/2 + y)n] + F[(W/2 - x)n, (L/2 - y)n] \} \end{aligned} \tag{15.40}$$

where $h_{x,y,t}$ = height of water table above impermeable layer at x, y, and time t (see figure below)
H = original height of water table above impermeable layer
v_a = arrival rate at water table of water from infiltration basin
t = time since start of recharge
f = fillable porosity ($1 > f > 0$)
L = length of recharge basin (in y direction)
W = width of recharge basin (in x direction)
$n = (4T/f)^{-1/2}$
T = aquifer transmissivity (ft²/day)

$$F(\alpha, \beta) = \int_0^1 erf\left(\alpha t^{-1/2}\right) \cdot erf\left(\beta t^{-1/2}\right) dt$$

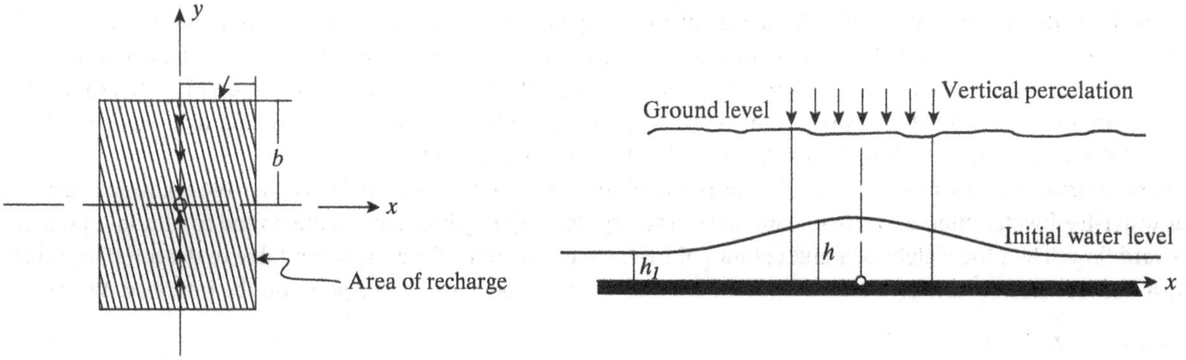

Figure 15.14 Diagrammatic Representation of Groundwater Mound Beneath Rectangular Recharging Area

For the greatest accuracy, the procedure using the above function can be programmed for automatic computation. For an approximation, a tabulation of the above function can be used as given by Hantush (1967) from which one page is presented below. For the central point ($x = y = 0$), we have $\alpha = (W/2)n$ and $\beta = (L/2)n$ and Equation 15.40 simplifies to:

$$h_{0,0,t} - H = \frac{v_a t}{f} F[\alpha_T, \beta_T] \tag{15.41}$$

in which

$$\alpha_T = (W/2)n \tag{15.42a}$$

$$\beta_T = (L/2)n \tag{15.42b}$$

The approximate method is demonstrated above in Figure 15.14.

Example 15.1

Calculate the mounding effect beneath the center of a 30 by 40 ft leach bed over silty sand with an application rate of 800 gpd (107 ft^3/day) (Topsfield, Massachusetts). Assume $t = 150$ days since start of recharge and fillable porosity $f = 0.10$ (conservative estimate). Average saturated thickness for water flowing away from the center of the bed is 10 ft.

Solution

The arrival rate $v_a = 107$ ft^3/day/(30 × 40) ft^2 = 0.0891 ft/day. A conservative estimate of hydraulic conductivity of silty sand is 2 ft/day. The estimated saturated thickness beneath the center of the bed is 11 ft.

The aquifer transmissivity T equals the product of hydraulic conductivity and estimated saturated thickness = 2 × 11 = 22 ft^2/day. The value of $n = [4(150)(22)/0.10]^{-1/2} = 2.75(10)^{-3}$. For the central point, with $L = 40$ ft and $W = 30$ ft, $h_{0,0,t} - H$ $= [(0.0891)(150)/0.10]F[(30/2)2.75(10)^{-3},\ (40/2)2.75(10)^{-3}] = 1.605(10)^5[0.04125,\ 0.05500]$. For $\alpha_T = 0.04125$ and $\beta_T = 0.05500$, the closest values in Table 15.3 are $\alpha = 0.04$ and $\beta = 0.06$, at which $F = 0.0188$. Thus $h_{0,0,t} - H = [(0.0891)(150)/0.10]0.0188$. With $H = 10$ ft, this gives $h_{0,0,t} = 12.513$ ft. Iterating again with this new estimated saturated thickness gives $F = 0.0188$ again and $h_{0,0,t} - H = 2.513$ ft and $h_{0,0,t} = 12.513$ ft again.

15.6.3 Infiltration Basins

Discussion of infiltration basins (in some cases, combined with detention), with particular emphasis on their limitations and adverse effects, may be found in Graber (2013b, Section 4.4.5) and Graber (2017b, Chapter 4). Fischer *et al.* (2003) "compared groundwater beneath storm water management basins to background groundwater.... (T)he samples collected in the basins exhibited ... greater detection frequency of petroleum hydrocarbons such as benzene and toluene. Pesticides used to control weeds along roads were also found in greater frequency beneath detention basins, while a selective herbicide was found with greater frequency in background groundwater, as were chlorinated hydrocarbons such as TCA and PCE.... Basin groundwater contained much lower levels of dissolved oxygen, which caused increased concentrations of major ions." Similar concerns are discussed later in this section under LAND APPLICATION OF WASTEWATERS.

On one project reviewed by the author (Dover, Massachusetts), a developer proposed to exclude building roof areas from the area contributing to a detention basin/recharge structure under storm conditions up to a 100-year return period. The intention was that each house have its roof downspouts connected to dry wells. Some appreciation of the difficulty in doing this can be gained by realizing that if there is minimal percolation during the rainfall event, such that the primary benefit of percolation is draining of the dry well after the rain storm [within 72 hours per the Massachusetts Stormwater Management Policy (MDEEP 1997)], then a five-foot deep dry well would have to have a diameter of over 20 ft for a 30 × 40 ft roof (or four dry wells, each of more than 10-foot diameter). The proponent was asked to provide details of the roof-recharge arrangement and other convincing assurance that such an arrangement would recharge the necessary flows (including percolation tests of which there were none), or modify the analysis of the detention basin/recharge structure to take roof drainage into account.

The author also asked the proponent to address the following design elements of the Massachusetts Stormwater Management Policy regarding dry wells:

- "Not intended for general stormwater quantity ... control, and therefore has limited applicability."
- "May experience high failure due to clogging. [N.B. Only sidewall percolation should be considered...."

Table 15.3 Selected Table from Hantush (1967)

Values of the function $S^*(\alpha,\beta) = \int_0^1 \mathrm{erf}\left(\frac{\alpha}{\sqrt{\tau}}\right) \mathrm{erf}\left(\frac{\beta}{\sqrt{\tau}}\right) d\tau$

α \ β	0.02	0.04	0.06	0.08	0.10	0.14	0.18	0.22	0.26	0.30	0.34	0.38	0.42	0.46	0.50	0.54	0.58	0.62
0.02	0.0041	0.0073	0.0101	0.0125	0.0146	0.0184	0.0216	0.0243	0.0267	0.0288	0.0306	0.0322	0.0337	0.0349	0.0361	0.0371	0.0380	0.0387
0.04	0.0073	0.0135	0.0183	0.0236	0.0278	0.0353	0.0416	0.0470	0.0518	0.0559	0.0596	0.0628	0.0657	0.0683	0.0705	0.0725	0.0743	0.0759
0.06	0.0101	0.0188	0.0266	0.0335	0.0393	0.0509	0.0602	0.0684	0.0754	0.0817	0.0871	0.0920	0.0963	0.1001	0.1035	0.1065	0.1091	0.1115
0.08	0.0125	0.0236	0.0335	0.0425	0.0508	0.0652	0.0776	0.0884	0.0978	0.1060	0.1133	0.1197	0.1254	0.1305	0.1350	0.1389	0.1425	0.1456
0.10	0.0146	0.0278	0.0393	0.0508	0.0608	0.0786	0.0939	0.1072	0.1188	0.1290	0.1381	0.1461	0.1532	0.1595	0.1650	0.1700	0.1744	0.1783
0.14	0.0184	0.0353	0.0509	0.0652	0.0786	0.1025	0.1232	0.1414	0.1573	0.1714	0.1839	0.1949	0.2048	0.2135	0.2212	0.2281	0.2343	0.2397
0.18	0.0216	0.0416	0.0602	0.0776	0.0939	0.1232	0.1490	0.1716	0.1916	0.2094	0.2251	0.2391	0.2515	0.2626	0.2724	0.2812	0.2890	0.2959
0.22	0.0243	0.0470	0.0634	0.0884	0.1072	0.1414	0.1716	0.1984	0.2222	0.2433	0.2621	0.2789	0.2938	0.3071	0.3189	0.3295	0.3389	0.3472
0.26	0.0267	0.0518	0.0754	0.0978	0.1188	0.1573	0.1916	0.2222	0.2494	0.2737	0.2954	0.3147	0.3320	0.3474	0.3612	0.3735	0.3844	0.3941
0.30	0.0288	0.0559	0.0817	0.1060	0.1290	0.1714	0.2094	0.2433	0.2737	0.3009	0.3252	0.3470	0.3665	0.3839	0.3995	0.4134	0.4257	0.4368
0.34	0.0306	0.0596	0.0871	0.1133	0.1381	0.1839	0.2251	0.2621	0.2954	0.3252	0.3520	0.3761	0.3976	0.4169	0.4341	0.4495	0.4633	0.4756
0.38	0.0322	0.0628	0.0920	0.1197	0.1461	0.1949	0.2391	0.2789	0.3147	0.3470	0.3761	0.4022	0.4256	0.4466	0.4654	0.4823	0.4973	0.5108
0.42	0.0337	0.0657	0.0963	0.1254	0.1532	0.2048	0.2515	0.2938	0.3320	0.3665	0.3976	0.4256	0.4508	0.4734	0.4937	0.5119	0.5281	0.5427
0.46	0.0349	0.0683	0.1001	0.1305	0.1595	0.2135	0.2626	0.3071	0.3474	0.3839	0.4169	0.4466	0.4734	0.4975	0.5191	0.5385	0.5559	0.5715
0.50	0.0361	0.0705	0.1035	0.1350	0.1650	0.2212	0.2724	0.3189	0.3612	0.3995	0.4341	0.4654	0.4937	0.5191	0.5420	0.5626	0.5810	0.5975
0.54	0.0371	0.0725	0.1065	0.1389	0.1700	0.2281	0.2812	0.3295	0.3735	0.4134	0.4495	0.4823	0.5119	0.5385	0.5626	0.5842	0.6036	0.6209
0.58	0.0380	0.0743	0.1091	0.1426	0.1744	0.2343	0.2890	0.3389	0.3844	0.4257	0.4633	0.4973	0.5281	0.5559	0.5810	0.6036	0.6238	0.6420
0.62	0.0387	0.0759	0.1115	0.1456	0.1783	0.2397	0.2959	0.3472	0.3941	0.4368	0.4756	0.5108	0.5427	0.5715	0.5975	0.6209	0.6420	0.6609
0.66	0.0394	0.0773	0.1136	0.1484	0.1818	0.2445	0.3020	0.3547	0.4027	0.4466	0.4865	0.5227	0.5556	0.5854	0.6122	0.6364	0.6582	0.6778
0.70	0.0401	0.0785	0.1154	0.1509	0.1849	0.2488	0.3075	0.3612	0.4104	0.4553	0.4962	0.5334	0.5672	0.5977	0.6254	0.6503	0.6728	0.6929
0.74	0.0406	0.0796	0.1171	0.1531	0.1876	0.2526	0.3123	0.3671	0.4172	0.4630	0.5048	0.5429	0.5774	0.6087	0.6371	0.6627	0.6857	0.7064
0.78	0.0411	0.0806	0.1185	0.1550	0.1900	0.2559	0.3166	0.3722	0.4232	0.4699	0.5125	0.5513	0.5865	0.6185	0.6475	0.6736	0.6972	0.7184
0.82	0.0415	0.0814	0.1198	0.1567	0.1921	0.2589	0.3203	0.3768	0.4286	0.4760	0.5192	0.5587	0.5946	0.6272	0.6567	0.6834	0.7074	0.7291
0.86	0.0419	0.0822	0.1209	0.1582	0.1940	0.2615	0.3237	0.3808	0.4333	0.4813	0.5252	0.5653	0.6017	0.6348	0.6648	0.6920	0.7165	0.7386
0.90	0.0422	0.0828	0.1219	0.1595	0.1957	0.2638	0.3266	0.3844	0.4374	0.4860	0.5305	0.5711	0.6080	0.6416	0.6721	0.6996	0.7245	0.7469
0.94	0.0425	0.0834	0.1228	0.1607	0.1971	0.2658	0.3292	0.3875	0.4411	0.4902	0.5351	0.5762	0.6136	0.6476	0.6784	0.7063	0.7316	0.7543
0.98	0.0428	0.0839	0.1236	0.1617	0.1984	0.2676	0.3314	0.3902	0.4442	0.4938	0.5392	0.5807	0.6184	0.6528	0.6840	0.7123	0.7378	0.7608
1.00	0.0429	0.0842	0.1239	0.1622	0.1990	0.2684	0.3324	0.3914	0.4457	0.4955	0.5410	0.5827	0.6206	0.6552	0.6865	0.7150	0.7406	0.7638
1.20	0.0437	0.0858	0.1263	0.1654	0.2030	0.2740	0.3396	0.4001	0.4558	0.5070	0.5540	0.5969	0.6362	0.6719	0.7044	0.7339	0.7605	0.7846
1.40	0.0441	0.0866	0.1275	0.1669	0.2049	0.2767	0.3431	0.4043	0.4608	0.5127	0.5603	0.6039	0.6433	0.6801	0.7132	0.7432	0.7704	0.7949
1.80	0.0444	0.0871	0.1283	0.1680	0.2062	0.2785	0.3454	0.4071	0.4641	0.5165	0.5645	0.6086	0.6489	0.6856	0.7190	0.7494	0.7769	0.8018
2.00	0.0444	0.0871	0.1284	0.1681	0.2064	0.2787	0.3457	0.4075	0.4645	0.5169	0.5651	0.6092	0.6495	0.6863	0.7198	0.7502	0.7778	0.8027
2.20	0.0444	0.0872	0.1284	0.1682	0.2065	0.2788	0.3458	0.4076	0.4646	0.5171	0.5653	0.6094	0.6497	0.6865	0.7200	0.7505	0.7781	0.8030
2.50	0.0444	0.0872	0.1284	0.1682	0.2065	0.2788	0.3458	0.4077	0.4647	0.5172	0.5653	0.6095	0.6498	0.6867	0.7202	0.7506	0.7782	0.8032
3.00	0.0444	0.0872	0.1284	0.1682	0.2065	0.2789	0.3458	0.4077	0.4647	0.5172	0.5654	0.6095	0.6499	0.6867	0.7202	0.7506	0.7782	0.8032

- "At least one soil sample for each dry well is recommended ... taken at the actual location of the proposed dry well where any localized soil conditions are detected."
- "The bottom of the dry well must be at least two feet above the seasonal high water table (defined as annual mean high water table) and bedrock ...."

The proponent did not address the above-mentioned items, and the project did not go forward because of this and other hydrologic and hydraulic deficiencies discussed below.

A detention basin/recharge structure was proposed as part of the above-mentioned Dover, Massachusetts project. The author addressed two issues in his review: (1) the areas over which percolation through the soil should be assumed to occur; and (2) appropriate rates of percolation. On the first issue, the proponent assumed that percolation would reliably occur not only at the sides of the structure but also over the full 18 × 90 ft bottom area of the structure. The concern with this is that the sediments that invariably enter such structures will plug the bottom area thus substantially diminishing the recharge capability. In his comments, the author cited the following in this regard:

- Stahre and Urbonas (1990) at §2.3.5 (Percolation Basins) and §17.4.4 (Effective Percolation Area). For example, the latter states: "... the bottom of a percolation pit or trench is considered impervious. The reason for this is that the bottom seals quickly by the accumulation of sediments. It is a good idea to anticipate this by assuming that all water will percolate into the ground only through the vertical sides of the basin...."
- The Massachusetts Stormwater Management Policy (MDEEP 1997) at 3.F states "...infiltration basins are prone to failure due to clogging of porous soils...," and lists the following disadvantages of infiltration basins: "High failure rates due to improper siting, design, and lack of maintenance" and "Requires frequent maintenance." The Policy (3.F) further emphasizes the need for specific and elaborate maintenance in sections on "Maintenance" and "Maintenance Criteria"; those sections include but are not limited to similar maintenance activities to those listed above for dry wells.

The proponent's resubmittal revised the percolation-storage relationship to accord with the above.

The next issue concerned the percolation-elevation relationship. Another reviewer, based on test borings, recommended a design permeability rate of 2.285×10^{-4} ft/sec. The proponent considered only three sides of the structure, based on the reviewer's recommendation that the upgradient side be excluded. The three exposed side areas then equal the product of the length of the exposed side areas (90 + 18 + 18 = 126 ft) times the height of the water above the bottom of the structure (Elevation 286 ft), i.e., the water elevation minus 286 ft for water elevations up to the top of the recharge structure (at Elevation 288 ft). Above that, the percolation rate cannot increase because there is no further effective sidewall since the structure above that elevation is sloped and would be sealed by sediment and not effective. On this basis, the author calculated the percolation-elevation relationship and compared it with that assumed by the proponent, resulting in the following tabulation:

	Percolation (cfs)	
Elevation	**Proponent**	**Reviewer**
286.00	0.000	0.000
287.00	0.075	0.029
288.00	0.150	0.058
290.00	0.150	0.058
29.200	0.150	0.058

The proponent assumed, without explanation, a much higher percolation rate at each elevation. If the lower percolation rates are used, the reduction in stormwater peak flow and volume will be significantly less than the values presented by the proponent.

Further, the proponent's submittal showed that test holes revealed groundwater within 1–1½ ft below the bottom of the recharge structure, and implied that was acceptable because he was being conservative in considering leaching through three sidewalls only. In fact, the purposes of requiring that groundwater and bedrock be a greater distance below the bottom of the recharge structure are twofold: (1) prevent the groundwater mounding that occurs below the recharge structure from extending up to the recharge structure thus diminishing the recharge capacity, and (2) reduce the likelihood of causing groundwater pollution. The author further noted that the observed groundwater levels may not be the seasonal high-water table by virtue of the time of year and possibly less-than-typical groundwater levels. Suitable groundwater levels can be estimated by methods discussed in Graber 2017b, Chapter 4.

The Massachusetts Stormwater Management Policy (MDEEP 1997) at 3.F Infiltration Basins says infiltration basins must have a minimum separation from seasonal high groundwater or bedrock of 2 ft. That requirement seems marginal; others recommend higher separation as given in the following:

- The Handbook of Hydrology (Maidment 1993) states:
  - "Infiltration basins must be located on very porous soils and the basin bottom must be at least 4 ft above the seasonal high groundwater or bedrock." [§28.6.4]
  - "Field observations of older installations indicate a very high rate of failure for infiltration basins and percolation trenches." [§28.6.4]
  - Infiltration basins may "tend to develop groundwater mounds that will not drain off laterally rapidly enough to keep the groundwater from surfacing." [§28.6.5]
- The ASCE Manual of Practice (ASCE/WEF 1992, p. 500) states that the "seasonal high-water table should be at least three feet beneath the bottom of the infiltration area to assure that stormwater pollutants are removed by the vegetation, soil, and microbes before coming into contact with the groundwater."

An additional requirement of the Massachusetts Stormwater Management Policy at 3.F stipulates a "minimum of three borings per infiltration basin." The infiltration should be calculated by applying the lowest measured permeability rate to the effective side wall areas of the recharge structure; the formula for percolation (or recharge) would be the product of that permeability rate in length units per second and the exposed side area, which results in increasing percolation as water depth increases.

Another issue at Dover, Massachusetts concerned the likelihood of percolate breaking out from the slope down gradient from the infiltration basin. This issue was addressed by the author in several ways, one of which is mentioned here. The Massachusetts Stormwater Management Policy (MDEEP 1997) at 3.F gives a requirement for infiltration basins of a minimum of 50 ft (15 m) from any slope greater than 15%. The downgradient slope in this case as proposed was approximately 33% at a distance less than 30 ft away. The location of the infiltration basin was for this and other reasons called into question.

An additional issue concerned the plan for the drainage system, which was intended to convey all stormwater runoff, up to the 100-year storm, to the detention basin/recharge structure. It was implicit that the catch basin inlets (24-inch square grates) would admit all of the larger storm flows. That is an unreasonable expectation. Runoff not captured by these inlets would flow off the site onto adjacent properties and otherwise modify the performance of the drainage system in ways inimical to achieving its objectives. A more complete discussion of street gutters and grates is provided in Section 12.7.

The deficiencies identified above and others noted by the author led to disapproval and withdrawal of the project.

Although infiltration basins are being incorporated increasingly in designs, the maintenance of the associated pretreatment facilities and the basins themselves add a burden to the municipality. In the case of private developments, maintenance can be required of the developers, but municipal oversight of that can also be burdensome. Infiltration basins can also be problematic if infiltration is relied upon as part of the control of stormwater runoff (such as in conjunction with attenuation) due to clogging. Rather than routinely requiring or permitting infiltration basins, the hydrologic need for them in relation to maintaining groundwater levels, streamflow, or wetlands should be evaluated on a case-by-case basis.

Some jurisdictions have actually banned infiltration for impoundment systems, e.g., Portland, Oregon, which requires that detention basins be lined in groundwater protection areas, and in the combined-sewer-overflow area of Chicago, Illinois, where detention basins must also be lined. Other situations include when the infiltration would be close to a coastal area where the infiltration will do little if any good. In some locations, added infiltration can have other negative effects, such as increasing invasive *Phragmites* (common reed).

Certain jurisdictions have actually begun adding requirements for greater uses of infiltration. The benefits of incorporating infiltration need to be balanced against its long-term reliability and maintenance. On sites where soils drain well and the water table is low enough, a stormwater infiltration or recharge system may be the most economical means for managing runoff. Infiltration systems can also minimize water table reductions. An interesting requirement for stormwater recharge is in a district of the City of Boston where declining water tables have contributed to the deterioration of wood piles that support old building foundations (Thomas and Vogel 2012). The stormwater recharge stabilizes the groundwater to help protect the old wooden pile-supported buildings. "Under appropriate circumstances, infiltration practices are an excellent technique for meeting the recharge requirement and may also provide stormwater retention. Extraordinary care should be taken to ensure that long-term infiltration rates are achieved through the use of performance bonds, post construction inspection and long-term maintenance." This quote and additional related advice may be found in Graber (2017b).

The recent *Standard Guidelines for Managed Aquifer Recharge* (ASCE 2020) provide additional comprehensive information relative to the present subject.

15.6.4 Subsurface Drains

Much of the discussion under this heading is taken from Graber (2013b, Chapter 6) and others of the author's publications cited below. Topics discussed in detail here include drainpipes, aggregate drains, geotextile drains, geonets, and geocomposite drains. The reader is referred to Graber (2013b, Chapter 6) for discussion of relief drains, wick drains and chimney drains, interceptor drains, perimeter drains and edge drains, and consolidation drains.

15.6.4.1 Outflow of Collected Water

The design of a subsurface drainage system includes the hydraulic design of all conduits that collect the excess water from the soil and deliver the water to the outlet for safe discharge. The hydraulic capacity of the drains should be checked at all points of size modification and at lateral connections. Subsurface drains are normally designed as open channels with a free water surface (at atmospheric pressure) inside the pipes. That incorporates the implicit assumption of decoupling of the groundwater hydrology from the drain conduit hydraulics and enables a rational, quantitative design. The purposeful or inadvertent design of the drain conduit for pressurized flow can result in nonuniform inflow at one end of the spectrum and uneconomical design at the other (Graber 2007a). The amount of water collected depends on the drainage coefficient for the area and the size of the area contributing water to the drain. If lateral seepage is occurring from outside the area drained, the conduits must be large enough to carry the additional water that will enter the system.

15.6.4.2 Drainpipes

The specification of a free water surface inside the drainpipe as a design factor means that all water moves in the system in response to gravity forces and frictional forces. The water in the saturated soil above the drain moves by gravity toward the water level inside the drain, developing a classic drawdown curve and a water table that slopes toward the drain. Once inside the drain, the water moves down gradient in the pipe according to open-channel hydraulic principles. Under free-discharge conditions, there is a progressive drop in the energy grade line in the direction of the outlets. For steady uniform flow, the water surface and the energy lines are parallel to the bottom slope of the pipe. However, in a subsurface drain, the flow rate increases in a downstream direction so the flow rate to be carried by the pipe at any point along its length is not a constant and the slope of the energy line also varies along the length of the pipe. Drainpipes are usually installed on a uniform grade. The maximum carrying capacity for the drain is calculated at the end of each section, that point at which maximum flow occurs in that pipe section having a uniform pipe size and a constant bottom slope. Consider energy losses due to friction, changes in flow direction, changes in the size or shape of the pipe, control or measurement devices, inlet and outlet head losses, and momentum transfer. Momentum transfer required to accelerate water entering the pipe at nonzero angles to the flow direction in the pipe must also be considered at junctions.

An adequate pipe bottom slope and adequate minimum flow velocity (but see below) are important in subsurface drains to minimize sediment deposition. Slopes of at least 0.001 are usually adequate. Excessively steep slopes may cause erosion of the soil outside the pipe perforations and result in failure of the drain. Some drains, such as foundation drains, may have to be laid with a zero slope. However, a hydraulic grade line will develop to remove the water as long as an adequate outlet is available and the pipe is properly designed. Information regarding drain grades and velocities and related aspects of drain filters or envelopes (which can eliminate the need for minimum velocities) can be found in Section 16 of USDA 1971 on pages 4-81–4-83, including Table 4-6, and pages 4-90–4-96.

The potential inadequacies of basing the design of conduits with spatially-increasing flow on the assumption of uniform flow (i.e., Manning's equation alone) are addressed by Graber (2004b). A spatially-varied flow calculation using numerical methods may be made using the ordinary differential equations for spatially varied flow (Chow 1959) to ensure that the hydraulic grade line will remain inside the conduit and that the design is economical. In the latter case, the Manning equation is used to calculate a friction slope at each computational section along the conduit. Figure 15.15, based on Graber (2004b), provides a chart that can be used for design of circular conduits of single pipe diameter with uniform inflow

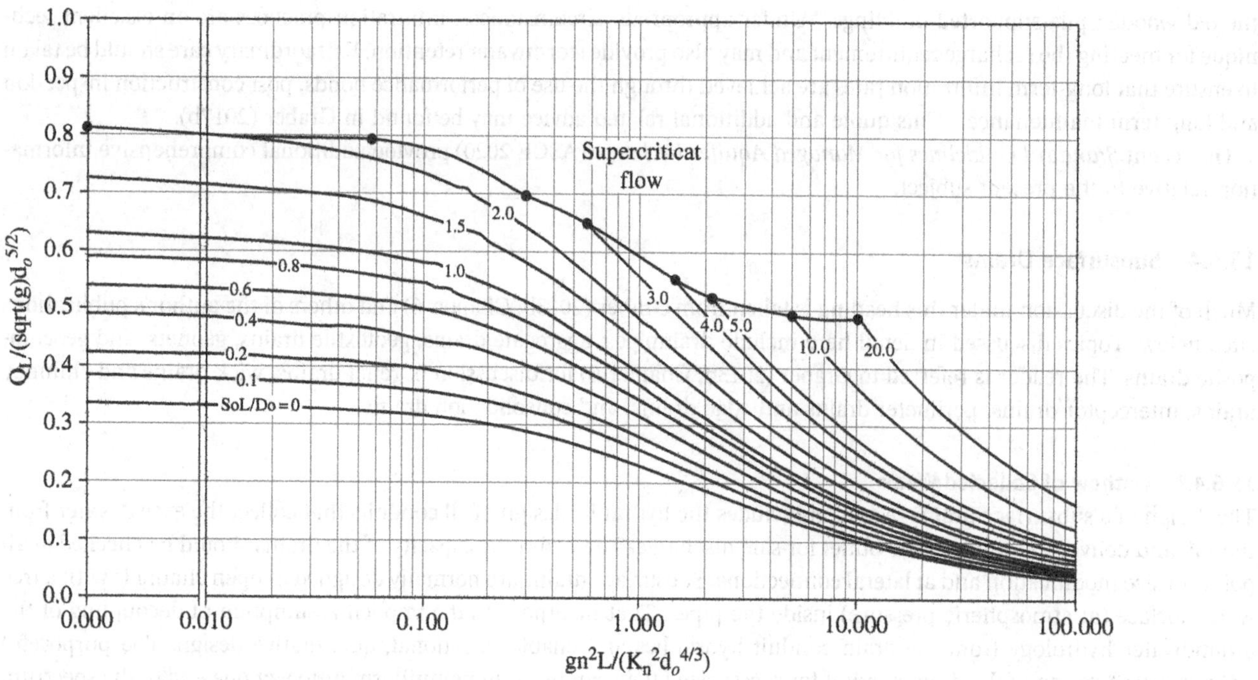

Figure 15.15 Circular Collector Conduit Chart for $y_{max}/d_o = 0.95$ with Uniform Inflow and $\mathbf{F}_L = 1$

and free-discharge conditions (tailwater elevation less than downstream critical-depth elevation). The variables in Figure 15.15 are as follows: n = Manning roughness coefficient; d_o or D_o = pipe diameter (ft, m); S_o = pipe slope; L = pipe length (ft, m); Q_L = rate of flow (cfs, m³/sec); g = gravitational acceleration (32.2 ft/sec², 9.81 m/sec²); $K_n = 1.486$ for customary units, 1 for SI units. Figure 15.15 is based on a ratio of maximum flow depth y_{max} to pipe diameter of $y_{max}/d_o = 0.95$. This figure provides another example of a generalized solution aided by dimensional analysis to enable a once-and-for-all plot.

For uniform spatial inflow, constant pipe diameter, turbulent flow, and free discharge, Graber (2004b, 2007b) shows that the uniform-flow approximation is conservative for $S_oL/d_o > 5$, $y_{max}/d_o = 0.95$, $\mathbf{F}_L = 1$ in which $\mathbf{F}_L$ = Froude number at downstream end. $\mathbf{F}_L = 1$ requires that the tailwater be at or below the critical-depth elevation at the downstream end (free discharge). An additional perspective regarding this approximation and its relation to other references is given in Graber (2007b).

For more complex systems, involving, for example, changes in pipe diameter or downstream submergence, a computer program may be used for design and analysis purposes (Graber 2004b).

15.6.4.3 Aggregate Drains

In all but extremely coarse aggregates, the flow velocity is low and laminar, thus permitting use of the basic Darcy equation:

$$Q = KiA \tag{15.43}$$

where Q = flow rate, cubic feet per second (cubic meters per second), K = hydraulic conductivity, feet per second (meters per second), i = hydraulic gradient (approximately equal to the slope ($\Delta h/l$), feet per foot (meter per meter), and A = cross-sectional area of the aggregate-filled trench, square feet (square meters).

The most significant variable is conductivity, which is principally dependent on the voids established by the aggregate's particle size and gradation. Aggregate conductivity can vary widely. Laboratory testing to confirm conductivity of selected aggregate vs. the design requirements is highly recommended.

The amount of fines in the aggregate mix affects conductivity. Generally, higher fine content results in lower conductivity, indicating the desirability of clean, washed stone. Safeguards are required to minimize passage of excessive fines from adjacent soil into the aggregate drain.

For aggregate drains with Darcian, homogeneous in-plane flow, a rational design method (Graber 2007b) can be developed by slight modification of a derivation by McEnroe (1993). For Darcian flow in a sloped medium of rectangular cross section and constant width, the following dimensionless variables are defined in terms of S_o = channel slope; $y_{\max}$ = maximum depth of flow; y_L = depth of flow at downstream end; L = horizontal projection of channel length; r = uniform rate of inflow per unit horizontal area; K = hydraulic conductivity of drain medium; $Y'_{\max} = y_{\max}/L$; $Y'_L = y_L/L$; $R' = (1 + S_o^2)r/K$; $A' = (S_o^2 - 4R')^{1/2}$; and $B' = (4R' - S_o^2)^{1/2}$. The nondimensional primed variables are so denoted to distinguish them from McEnroe's (1993) similar variables. A free-drainage Numerov boundary condition (Harr 1962, pp. 210–226; McEnroe 1993) is assumed, for which $Y'_L = R'$; this requires that the tailwater elevation not exceed an elevation corresponding to $y_L = rL(1 + S_o^2)/K$. The solution utilizing these variables, presented below, admits the case of zero slope.

$$Y'_{\max} = \left(R' - S_o R' + R'^2\right)^{1/2} \cdot \exp\left[\frac{S_o}{B'} \tan^{-1}\left(\frac{2R' - S_o}{B'}\right) - \frac{S_o}{B'} \tan^{-1}\left(\frac{2R'/S_o - S_o}{B'}\right)\right],\ S_o^2 < 4R' \qquad (15.44a)$$

$$Y'_{\max} = \frac{R'}{S_o} \frac{S_o^2 - 2R' S_o}{S_o^2 - 2R'} \exp\left[\frac{2R' S_o^2 (S_o - 1)}{\left(S_o^2 - 2R' S_o\right)\left(S_o^2 - 2R'\right)}\right],\ S_o^2 = 4R' \qquad (15.44b)$$

$$Y'_{\max} = \left(R' - S_o R' + R'^2\right)^{1/2} \left[\frac{(S_o - A' - 2R'/S_o)(S_o + A' - 2R')}{(S_o + A' - 2R'/S_o)(S_o - A' - 2R')}\right]^{S_o/(2A')},\ S_o^2 > 4R' \qquad (15.44c)$$

A plot of the above relationships is presented in Figure 15.16. The maximum depth occurs at the upstream end for $S_o = 0$, moving downstream as S_o increases until at $S_o = 1$ the maximum depth is at the downstream end and $Y'_{\max} = Y_L = R'$ in conformance with Figure 15.16. Values of $S_o > 1$ will still have $Y'_{\max} = Y_L = R'$. As suggested by Figure 15.16, values of $R' > 10$ (well within the range of practicality for aggregate drains) will have $Y'_{\max} \cong Y_L = R'$. This figure provides yet another example of a generalized solution aided by dimensional analysis to enable a once-and-for-all plot.

Denoting the measured full-flow transmissivity by T_f, the transmissivity is related to conductivity by $T_f = Kb$. Thus the test values of T_f and known b give $K = T_f/b$ which can be used in the calculation of R' for use with Equation 15.44 and Figure 15.16. The factor of safety should be selected based on the accuracy to which the hydrology (flow rate) is known, the practical importance of inaccuracies in assumed hydrology, and the accuracy to which drain-medium conductivity is known. If a permeameter or other full-flow test device of cross-sectional dimensions smaller than the full-scale cross section is used to estimate conductivity, the conductivity will be higher in the permeameter; that effect can be minimized by using a permeameter with a sufficiently high ratio of permeameter diameter to medium particle diameter or a sufficiently high Darcy number = square of permeameter hydraulic diameter divided by b (Somerton and Wood 1988). Based on Somerton and Wood, a ratio of 100:1 permeameter diameter to the largest particle diameter should give a conductivity within 1% of that for an infinite porous medium. A Darcy number > 10^5 provides an alternative criterion. For a permeameter with a circular cross section, it would be reasonable to use the least stringent of these criteria. Otherwise, a suitable factor of safety

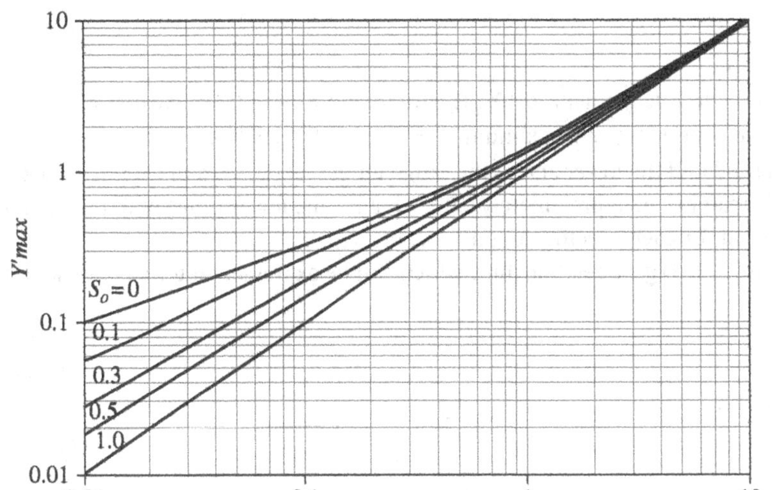

Figure 15.16 Rectangular Aggregate Drain Chart with Uniform Inflow and Free Discharge

should be used for conductivity determined by permeameter or otherwise. For design, factors of safety are discussed by Koerner (2005) in the context of flow-rate reduction factors.

Example 15.2

The author evaluated the stability of a slope proposed for the Quarry Hills project (discussed in Section 7.5) in Milton, Massachusetts. The slope was 1:2.5, estimated sand permeability 0.02 cm/sec = 56.7 ft/day, inflow per unit area $r = 2$ in./day = 0.1667 ft/day, and horizontal projection $L = 62.5$ ft. Determine the maximum depth of subsurface flow (seepage depth).

Solution

In terms of the parameters given above, we have $S_o = 1/2.5 = 0.4$, $R' = (1 + 0.4^2)0.1667/56.7 = 0.00341$, $A' = \left(S_o^2 - 4R'\right)^{1/2} = 0.383$, $4R' = 0.01364$, $S_o^2 = 0.16 > 4R'$. Equation 15.44c applies:

$$Y'_{\max} = \left[0.00341 - 0.4(0.00341) + (0.00341)^2\right]^{1/2} \cdot \left\{\frac{[0.4 - 0.383 - 2(0.00341)/0.4][0.4 + 0.383 - 2(0.00341)]}{[0.4 + 0.383 - 2(0.00341)/0.4][0.4 - 0.383 - 2(0.00341)]}\right\}^{0.523} = 0.222$$

giving $y_{\max} = Y'_{\max} L = 0.222(62.5) = 13.87$ ft

Results of this analysis were used as part of a geotechnical evaluation of slope stability.

Departure from Darcian flow occurs as an appropriately-defined Reynolds number increases, first because inertial forces become significant while the flow remains laminar and, with still higher Reynolds number, as the flow becomes turbulent in an increasing portion of the porous medium (Ward 1964, Davis and De Wiest 1966). Ward (1964) expressed the initial departure from Darcian flow in terms of a Reynolds number $\mathbf{R}_k = Vk^{1/2}/\nu$ in which V = macroscopic velocity (also referred to as Darcy velocity or specific discharge), k = permeability, and ν = kinematic viscosity. Permeability is related to conductivity by $k = \nu K/g$, in which g = gravitational acceleration. Since $V = rL/y$ with y = flow depth, and using other terms previously defined, the Reynolds number can be expressed as:

$$\mathbf{R}_k = \frac{R' K^{3/2}}{(y/L)\left(1 + S_o^2\right)(\nu g)^{1/2}} \tag{15.45}$$

The Reynolds number will be a maximum at the minimum flow depth, for which $y/L = y_L/L = Y_L = R'$, so the Reynolds number can be conservatively expressed as:

$$\mathbf{R}_k = \frac{K^{3/2}}{\left(1 + S_o^2\right)(\nu g)^{1/2}} \tag{15.46}$$

Ward (1964) suggested that to limit the error in hydraulic gradient to 10% $\mathbf{R}_k$ should not exceed 0.182. Ward's criterion, with $\nu \cong 10^{-2}$cm^2/s and g = 9.81 m/s^2 in the above equation gives maximum $K = 0.7\left(1 + S_o^2\right)$ cm/sec = O(1) cm/s, corresponding to the upper end of the range for clean sand and the lower end of the range for gravel (Davis and De Wiest 1966). For typical values of r, that will correspond to R' greater than approximately 0.01, which is the lower range of R' plotted on their Figure 6-6. The value of $\mathbf{R}_k$ should be checked for each specific design.

For the transitional and turbulent regimes, Ward (1964) gave relationships which can be expressed in terms of variables used here as follows:

$$i = f_k \frac{V^2}{\sqrt{g\nu K}} \tag{15.47}$$

$$f_k = \frac{1}{\mathbf{R}_k} + C_1 \tag{15.48}$$

in which f_k is the Fanning friction factor for porous media and C_1 is the fully-turbulent friction factor given as $C_1 = 0.550$ by Ward. Rumer and Drinker (1966) suggested on theoretical grounds that C_1 is not a universal constant as Ward indicated, but rather that it will vary slightly for different media. Venkataraman and Rao (1998) gave values of C_1 varying by a factor of more than 10 for different media. They suggested that relating C_1 to media properties needs further study, although estimates can be made in some cases from their tabulation of data together with original sources which they reference and others. Equations 15.47 and 15.48 can be used in conjunction with appropriate test data. The friction factor decreases with increasing Reynolds number per Equations 15.47 and 15.48. Therefore, a conservative and practical approach for aggregate drains is to use a test Reynolds number that is less than the Reynolds number at the maximum depth location under design conditions.

Solutions for geotextile drains or laminar-flow aggregate drains with tailwater higher than that corresponding to the Numerov condition, or for aggregate drains with transitional and turbulent flow, can be obtained for known flow distribution and downstream flow depth using the numerical computational procedures described under *Numerical Formulation* of Graber (2004b). In doing so, the friction slope S_f should be taken as i from Equation 15.47, and the velocities should be taken as the mean interstitial velocity given by $Q/(Ap)$, with Q = flow rate, A = wetted cross-sectional area, and p = medium porosity.

An additional perspective regarding these methods and their relation to other references is given in Graber (2007b).

15.6.4.4 Geotextile Drains

Fluid flows through geotextiles serving as a filter, with the primary function of the fabric being to restrain solids. The measure of a fabric's ability to permit water flow perpendicular to its plane is termed "permittivity."

Nonwoven geotextiles are capable of transmitting fluids within the plane of the fabric. Care should be taken to consider the potential for fabric clogging due to small soil or other particles in the system, or due to biological growth. Nonwoven resin-bonded and nonwoven needle-punched geotextiles are especially suitable for drainage applications (Koerner 2005). Planar flow is dependent on the nature of the fabric, the gradient (slope), and the compressive forces on the material. Planar flow properties for any geosynthetic may be determined from ASTM D4716 with suitable modifications as discussed below. Most planar flow is laminar and is expressed in gallons/minute-foot width or cubic meters/second-meter width.

Geotextile drains with Darcian flow are characterized hydraulically by manufacturers in terms of transmissivity based on ASTM D4716 or variations thereof (Koerner 2005) carried out under full-flow conditions. Transmissivity (m^3/sec-m or m^3/min-m) is measured at unit hydraulic gradient for a particular geotextile vs. normal stress (kPa) which takes into account the geotextile's compressibility under load and attendant variation in thickness. The standard is based on laminar (actually Darcian) flow tests and, although it recognizes the possibility of transitional and turbulent flow in such conduits, it stipulates that the test data for hydraulic transmissivity be obtained from the laminar test range. However, the laminar range is not generally reported; it is recommended here that such be required so that the laminar range of the test data can be known. Darcy's formula is then used to calculate the transmissivity required for the actual hydraulic gradient on the assumption that the hydraulic gradient equals the geotextile slope. Factors of safety and reduction factors (Koerner 2005) relate the required transmissivity to the ultimate transmissivity.

It is also important to note that the D4716 tests are for pressurized, full-conduit, constant flow. The applicability of these methods to spatially varied, free-surface flow applications must be considered. A useful criterion is derived by Graber (2007b) and leads to the following when the inflow is uniform (constant inflow per unit length). The first of the following inequalities applies subject to the second and third of the following inequalities:

$$\frac{d(y/b)}{d(x/L)} \geq 0, \quad \frac{S_oL}{b} \geq \frac{S_fL}{b}, \quad S_oL/b \gg 2q_L^2/\left(gpb^3\right) \tag{15.49}$$

in which x = distance along the drain in the direction of flow; y = flow depth; L = total conduit length; S_o = conduit slope; S_f = friction slope; q_L = flow per unit width at the downstream end of the drain; g = gravitational acceleration; and p = medium porosity.

The above relationship indicates that if the drain is designed for a flow equal to or less than the flowing-full capacity of the drain (i.e., $S_o \geq S_f$), the tailwater does not submerge the crown of the drain, and $S_oL/b \gg 2q_L^2/\left(gpb^3\right)$, then the water surface will remain within the drain. Thus, for Darcian flow in geotextiles, with qualifications just discussed, design can rationally be based on transmissivity measurements for full-conduit Darcian flow conditions and, where cross-sectional distortion is significant, matching normal stress. Darcy's formula is then used to extrapolate to other hydraulic gradients with the assumption that the full-conduit geotextile slope is equal to the extrapolated hydraulic gradient. This method is valid for open-channel, spatially-increasing flow design provided that the Equation 15.49 inequalities are satisfied. For this method, some conservatism results from the lesser wetted perimeter compared to the full-conduit test measurements.

Tailwater must be at or below the downstream crown, and conservatism is gained to the extent that tailwater is below the downstream crown. Conservatism is also gained if extrapolating to smaller normal stress conditions for which there is less cross-sectional distortion.

For geotextile drains with Darcian, homogeneous in-plane flow under more general conditions when the Equation 15.49 inequalities are not satisfied (or even if they are), but subject to the same restrictions as discussed above for aggregate drains, a rational design can be based on the same method as given above for aggregate drains (Graber 2007b). The tailwater must be below that corresponding to the Numerov boundary condition. Equation 15.44 and Figure 6-6 then apply. These relationships are not limited by the slope and other restrictions mentioned in the previous paragraph and can enable greater design economy. The relationship between the measured full-flow transmissivity and other variables, including R', are the same as discussed above for aggregate drains. The factor of safety should be selected based on the accuracy to which the hydrology (flow rate) is known, the practical importance of inaccuracies in assumed hydrology, and the accuracy to which drain-medium conductivity is known. Other considerations also pertain as discussed by Koerner (2005) in the context of flow-rate reduction factors.

Solutions for geotextile drains with tailwater higher than that corresponding to the Numerov condition can be obtained for known flow distribution and downstream flow depth using the numerical, spatially varied flow computational procedures such as described under *Numerical Formulation* of Graber (2004b). In doing so, the friction slope S_f should be taken as i, and the velocities should be taken as the mean interstitial velocity given by $Q/(Ap)$, with Q = flow rate, A = wetted cross-sectional area, and p = medium porosity. Should the flow be in transitional or turbulent regimes, considerations similar to those described above in connection with aggregate drains apply.

Additional perspective regarding these methods and their relation to other references is given in Graber (2007b).

15.6.4.5 Geonets and Geocomposite Drains

The following should be considered when geonets or geocomposite drains are utilized:

1) Geonets and geocomposite drains generally have more complex flow patterns than pipes, thus contributing to higher flow energy losses.
2) Certain designs are more hydraulically efficient, thus providing higher flow rates per unit width or thickness.
3) The outer geotextile wrapping on geocomposite drains may affect hydraulic properties.
4) Most geonets and geocomposite drains react noticeably to applied Earth loads.

Flow in geonets and geocomposite drains is commonly turbulent. Graber (2007b) discusses means for determining the existence of turbulent flow in geocomposite drains, which should be used as necessary. Koerner (2005) suggests that design be based on full-conduit, constant-flow test measurements of the flow rate per unit width (m^3/min-m) for a particular material vs. normal stress (kPa) and at different hydraulic gradients. The ultimate flow is then taken from the data at the design normal stress and geosynthetic slope equal to the test hydraulic gradient (interpolated as necessary), and then reduction factors and factors of safety are applied. For both the laminar- and turbulent-flow cases, the assumption that the hydraulic gradient equals the geosynthetic slope is equivalent to the uniform-flow assumption in a just full-conduit based on the downstream-most flow.

The adequacy of these assumptions for spatially-varied, free-surface flow applications can be tested using the inequalities in Equation 15.49 above and the related discussion, which applies equally to laminar, transitional, or turbulent flow (the flow regime directly affects S_f). With $p = 1$, the above equation applies to polymer-core geocomposite drains of rectangular cross section, while geonets will generally have $p < 1$. Graber (2007b) gives an example of the application of Equation 15.49 to a landfill geonet. For geonets, Equations 15.47 and 15.48 with their associated discussion may also be useful.

For the methods discussed in the previous two paragraphs, some conservatism results from the lesser wetted perimeter compared to the full-conduit test measurements. Tailwater must be at or below the downstream crown, and conservatism is gained to the extent that tailwater is below the downstream crown. Conservatism is also gained if extrapolating to smaller normal stress conditions for which there is less cross-sectional distortion

An additional consideration concerns the possibility that the distortion of the longitudinal velocity profile due to spatial inflow can result in an increase in effective wall friction (Graber 2004b). That can make the roughness coefficient determined under spatially increasing flow conditions greater than that resulting from constant-flow tests. Aided by the seminal work of McNown (1954), Graber (2007b) defines the following criterion for minimal increases in effective wall friction compared to constant flow:

$$N_o > 11 \tag{15.50}$$

in which N_o = total number of columns of holes. For fabric-covered geocomposite drains without slots or holes, N_o can be approximated by:

$$N_o = sf_o \frac{yL}{(\pi/4)(AOS)^2} \tag{15.51}$$

in which s = number of sides effectively receiving drainage, f_o = fraction of side area open for drainage, AOS = apparent opening size, and y = an appropriate flow depth.

For the design method for turbulent flow in geonets and geocomposite drains described above with qualifications, to the extent that the tailwater elevation is below the downstream crown a factor of safety is introduced. For cases in which Equation 15.49 cannot be satisfied, the design should be accomplished by numerical, spatially varied flow methods, such as discussed under *Numerical Formulation* in Graber (2004b). It should be recognized that for such flows, the maximum depth can occur at an intermediate location along the channel (Graber 2004b).

For highway edge drains, it can be important to design for a maximum free-surface depth and freeboard within the edge drain (Dempsey 1988, NCHRP 2004). Rather than the method discussed at the beginning of this subsection, that may be accomplished more economically by an alternative method. The alternative method discussed here is Graber's (2007b) modification/correction of methods presented by FHwA (1990b, 2002) and NCHRP (2004). Graber (2007b) uses Dempsey's (1988) experimental data to properly estimate a constant-flow friction factor in the form of Manning's roughness coefficient; Table 6-2 presents the n values thus calculated. Manning's equation can be used to calculate the friction slope in the numerical formulation, using properly-determined values of Manning's n such as developed for Table 6-2. Graber (2007b) presents an NCHRP (2004) example for a high-column edge drain using the numerical computational procedures for spatially varied, open-channel flow described under *Numerical Formulation* of Graber (2004b) and also incorporating the use of Equation 15.51. That example is compared to the corresponding NCHRP (2004) result. An additional perspective regarding these methods and their relation to other references is given in Graber (2007b).

Table 6-3 presents Manning's n values extrapolated from Table 6-2 for current products. Where n values in Table 6-3 are the same as those in Table 6-2, it is because the core specifications and hence surface roughness characteristics (roughness height and pattern) are the same and the geocomposite thicknesses are equal or the thicknesses in Table 6-3 are slightly greater (Marlow 2011). In the latter case, the Table 6-3 values should be conservative. Caution should be exercised in extrapolating to other surface roughness characteristics or thicknesses.

15.6.4.6 Berm Seepage

Seepage through berms is addressed by Harr (1962, §2-7), who provides Dupuit's solution beginning with the differential equation for discharge per unit width q through a vertical section of the berm as follows:

$$q = -ky\frac{dy}{dx} \tag{15.52}$$

in which k is hydraulic conductivity, y is water surface elevation in the berm above the horizontal bottom of the berm, and x is distance downgradient from the intersection of the upstream water surface and the berm (see Figure 15.17 below). Integrating with the boundary conditions $x = 0$, $y = h_1$, and $x = L$, the distance to the intersection of the downstream water surface and the berm and the downstream water surface where $y = h_2$ yields:

$$q = \frac{k\left(h_1^2 - h_2^2\right)}{2L} \tag{15.53}$$

Figure 15.17 Dupuit's Solution for Flow through an Earth Dam

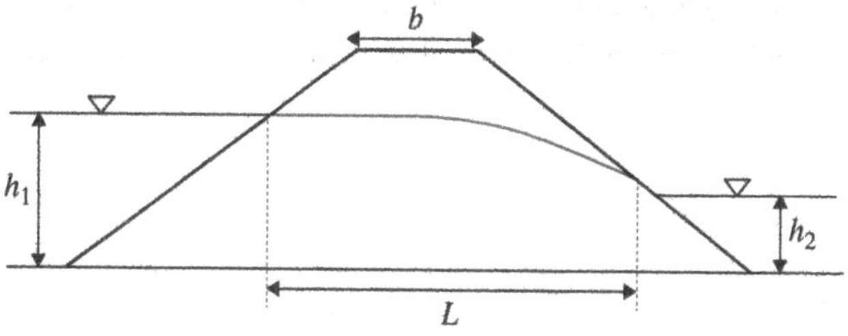

Example 15.3
A symmetrical berm of silty sand was constructed along piping through the berm to convey water accumulating behind the berm to a pumping station (Graber 2010a). The purpose of that construction was to protect homes from flooding resulting from the filling of a large basin as part of park construction (Winthrop, Massachusetts). The author was asked to review a proposal to modify the berm by installing a polyethylene film held in place by additional fill to reduce seepage from the park to the protected areas. Determine whether that construction would be worthwhile. The berm height is 2.56 ft, the total width of the berm is 15 ft, the berm slopes are 2:5, and the total berm length is 123 ft.

Solution

For conservatism, assume $h_2 = 0$ so $q = kh_1^2/(2L)$ and for h_1 use the maximum berm height of 2.56 ft. For further conservatism, with $k = 0.002$ cm/sec for silty sand (Harr 1962, p. 8), we take the upper range for fine sand (Bouwer 1978, pp. 18, 38) $k = 5$ m/day × 100 cm/m × 1/24 day/hr × 1/60 h r/min × 1/60 min/sec = 0.00578 cm/sec × 1/2.54 in/cm × 1/12 ft/in = 0.000190 ft/sec. The distance $L = 15 - (5/2)(2.56) = 8.6$ ft, and $q = 0.000190(2.56)^2/[2(8.6)] = 0.0000722$ ft²/sec. The total estimated flow rate is 0.0000722 ft²/sec × 123 = 0.00888 cfs × 448.831 = 4.0 gpm. This flow is negligible compared to the flowing-full capacity through the dual 8-inch pipes at the base of the berm of 720 gpm, and smaller still compared to the two-year runoff rate from the area behind the berm or the capacity of the pipes with greater upstream heads. The installation of the polyethylene film is not justified.

15.6.4.7 Land Application of Wastewaters

The U.S. Army Corps of Engineers, its dam construction program diminishing, sought another source of work by assuming and promoting a role in design and construction of the land application of wastewater (USACE 1983). They became active in a regional water-quality study that the author was preparing for the Nashua River Basin (Graber 1974b, 1975b). They proposed the importation of 65 mgd (101 cfs) of wastewater from outside the basin be applied to land (by the infiltration-percolation method) at Fort Devens in Ayer, Massachusetts adjacent to the Nashua River. Because such a scheme could have a significant effect on alternatives within the Nashua River Basin, it was evaluated by the author using hydraulic and water-quality models.

The 101 cfs flow would be applied to land adjacent to a 3.86 mile reach of the river and was assumed to enter the river uniformly along this length. This land-applied flow is much greater (by a factor of approximately 85) than the natural groundwater inflow entering the river along this distance. The wastewater discharge was added to the 7-day, 10-year flow conditions considered for water-quality modeling per State requirements. A backwater model was used to determine hydraulic parameters for input to the calibrated water-quality model. The water in the Nashua River upstream of the land application site was assumed to be saturated with oxygen (8.31 mg/L at 75°F) and contain no BOD. Existing wastewater discharges along the modeled length were neglected. Groundwater inflows, including the land-applied wastewater, were assumed to enter the river devoid of oxygen. The resulting minimum dissolved oxygen concentration was estimated to be 4.05 mg/L; this is less than 50% of saturation and is in violation of the Massachusetts standard for the Nashua River. Depression of the dissolved oxygen level would occur along the full distance of inflow of land-applied wastewater and would extend downstream for several miles and into a large pond. The dissolved oxygen sag would be greater still if the then-existing dissolved oxygen deficit and BOD concentration upstream of the land application site were taken into account. Other variations of the analysis are discussed in the above-mentioned report, with similar conclusions. The land application proposal was abandoned.

Additional related history, both ancient and recent, and important criticisms of land application are given by Whipple (1972) and Egeland (1973). Among the criticisms are that it is "an extremely costly method of disposal and ... that implementation on a large scale could result in devastation of vast areas of land, serious contamination of the air, accelerated depletion of energy resources, and, ultimately, more severe water pollution than that which exists today [including phosphorus pollution of adjacent waterways]. The danger of emphasis on land disposal is that time, money, and resources be wasted on an obsolete technology while far more promising programs are aborted."

References

Abbott, D. E., and Kline, S. J. (1962). "Experimental Investigation of Subsonic Turbulent Flow Over Single and Double Backward Facing Steps", *J. Basic Eng.*, Transactions, American Society of Mechanical Engineers, Series D, Vol. 84, Paper No. 61-Hyd. 15, September 1962, pp. 317-325.

Aczel, A. D. (2003). *Pendulum: Leon Foucault and the Triumph of Science*, Atria Books, New York.

Adelman, M., Weber-Shirk, M., Will, J., Cordero, A., Maher, W., and Lion, L. (2013). "Novel Fluidic Control System for Stacked Rapid Sand Filters", *J. Environ. Eng.*, 139(7), 939–946.

Afzalimehr, H., and Anctil, F. (1998). "Estimation of Gravel-Bed River Flow Resistance", *J. Hyd. Div.*, 124(10), 1054–1058. [With *devastating* discussions December 1999.]

Aguirre-Pe, J., and Fuentes, R. (1990). "Resistance to Flow in Steep Rough Streams", *J. Hyd. Div.*, 116(11), 1374–1387.

Ainsworth, L. (1999). "Water Witching Should Be Treated as Fiction", *Opflow*, 25(9), 3–6.

Albano, J., and Grieshaber, O. (1970). *Stuart-Carter Walking Beam Flocculators – Our Approach to Velocity Gradient (G, Gt Factor) and Other Considerations for Flocculators*", Ralph B Carter Company, Hackensack, New Jersey.

Albertson, M. L., Dai, Y. B., and Rouse, H. (1950). "Diffusion of Submerged Jets", *Trans. Am. Soc. Civil Eng.*, 115(1), 639–664.

Al-Musawi, T. T. A. (2012). Acoustical Design of Water Features and Their Use for Road Traffic Noise Masking, Submitted for the Degree of Doctor of Philosophy, Heriot-Watt University School of the Built Environment, Edinburgh, Scotland, October 2012.

Amberg, H. R. (1969). "Stream Reaeration Using Molecular Oxygen", *Proc. Fourth International Conf. on Water Pollution Research*, Prague.

Amberg, H. R., Wise, D. W., and Aspitarte, T. R. (1969). "Aeration of Streams with Air and Molecular Oxygen", *TAPPI J.*, 52(10), 1866–1871.

American National Standards Institute/Hydraulic Institute (ANSI/HI) (1998). *American National Standard for Pump Intake Design*, Hydraulic Institute, Parsippany, New Jersey.

American National Standards Institute/Hydraulic Institute (ANSI/HI) (2012). *Intake Design for Rotodynamic Pumps*, Parsippany, New Jersey.

American Society of Civil Engineers (ASCE). (2005). *Standard guidelines for the design of urban subsurface drainage*, American Society of Civil Engineers/ Environmental and Water Resources Institute 12-05, Reston, Virginia.

American Society of Civil Engineers (ASCE). (2006). *Minimum Design Loads for Buildings and Other Structures*, ASCE/SEI 7-05, Reston, Virginia.

American Society of Civil Engineers (ASCE). (2020). *Standard Guidelines for Managed Aquifer Recharge*, American Society of Civil Engineers/ Environmental and Water Resources Institute ASCE/EWRI 69-19, Reston, Virginia.

American Society of Civil Engineers (ASCE) and Water Environment Federation (WEF) (1992). *Design and Construction of Urban Stormwater Management Systems*, ASCE Manual of Practice No. 77, ASCE, Reston, Virginia.

American Society of Civil Engineers (ASCE) and Water Pollution Control Federation (WPCF) (1960). *Design and Construction of Sanitary and Storm Sewers*, ASCE Manual of Practice No. 37, ASCE and WPCF, New York and Washington, D.C.

American Society of Civil Engineers (ASCE) and Water Pollution Control Federation (WPCF) (1969). *Design and Construction of Sanitary and Storm Sewers*, ASCE Manual of Practice No. 37 (WPCF Manual of Practice No. 9), ASCE and WPCF, New York and Washington, D.C.

American Society of Mechanical Engineers (ASME) (1971). *Fluid Meters – Their Theory and Application*, 6th Ed., New York.

Hydraulics and Pneumatics in Environmental Engineering, First Edition. S. David Graber.

American Water Works Association (AWWA) (1969). *Water Treatment Plant Design*, New York.

Amirtharajah, A. (1970). Expansion of Graded Sand Filters During Backwashing, MS Thesis, Iowa State University, Ames, Iowa.

Amirtharajah, A., and Cleasby, J. L. (1972). "Predicting Expansion of Filters during Backwashing", *J. Am. Water Works Assoc.*, 64(1), 52–59.

Anderson, N. E. (1945). "Design of Final Settling Tanks for Activated Sludge", *Sew. Works Jour.*, 17, 50–65.

Appel, D. W., and Yu, Y-S. (1966). "Pressure Pulsations in Flow Through Branched Pipes", *J. Hyd. Div.*, 92(6), 179–197.

Aras, E., and Berkun, M., and (2010). "Comparison of Stepped and Smooth Spillway Effects on Stream Reaeration", *Water SA*, 36(3), 309–314.

Argaman, Y., and Kaufman, W. J. (1970). "Turbulence and Flocculation", *J. San. Eng. Div.*, 96(2), 223–241.

Aron, G., and Kibler, D. F. (1990). "Pond Sizing for Rational Formula Hydrographs", *Water Resour. Bull.*, 26(2), 255–258.

Aron, G., and Scott, V. H. (1965). "Simplified Solution for Decreasing Flow in Wells", *J. Hyd. Div.*, 91(5), 1–12.

Ashley, S. (1997). "Against All Odds: How Bugs Take Wing", *Mech. Eng.*, 119(3), 114.

ASME Performance Test Codes (ASME) (1973). *Centrifugal Pumps*, PTC 8.2, American Society of Mechanical Engineers, New York, 1965 (with 1973 addendum).

ASME Performance Test Codes (ASME) (1979a). *Displacement Compressors, Vacuum Pumps and Blowers*, PTC 9, American Society of Mechanical Engineers, New York, 1970 (reaffirmed 1979).

ASME Performance Test Codes (ASME) (1979b). *Compressors and Exhausters*, PTC 10, American Society of Mechanical Engineers, New York, 1965 (reaffirmed 1979).

Aude, T. C., Cowper, N. T., Thompson, T. L., and Wasp, E. J. (1971). "Slurry Piping Systems: Trends, Design Methods, Guidelines", *Chem. Eng.*, 78(14), 74–90.

Avery, S. T., and Novak, P. (1978). "Oxygen Transfer at Hydraulic Structures", *J. Hydrol. Div.*, 104(11), 1615–1630.

Babb, A. F., and Ross, J. A. (1968). "Discussion of 'Control Point in a Lateral Spillway Channel'", *J. Hyd. Div.*, 94(1), 317–321.

Babbitt, H. E., and Baumann, E. R. (1958). *Sewerage and Sewage Treatment*, John Wiley & Sons, Inc., New York.

Babbitt, H. E., and Caldwell, D. H. (1942). "The Flow of Solid-Liquid Mixtures", *Proc. Second Hydraulics Conference*, University of Iowa, Bulletin 27, No. 400, pp. 117-129.

Babcock, R. H. (1966). "Flow Measurement/1/2/3", *Water and Wastes Engineering*, New York.

Ball, J. W., and Tullis, J. P. (1973). "Cavitation in Butterfly Valves", *J. Hyd. Div.*, 99(9), 1303–1318 [with errata 100(4), 620].

Banister, F. J., Moeller, W. P., Natarius, E. M., and Sampson, K. M. (1999). "Spin Away Odor and Decay in Sewage Drop Structures", *Water Eng. Manag.*, 146(2), 13–18.

Bansal, M. K. (1971). "Dispersion in Natural Streams", *J. Hyd. Div.*, 97(11), 1867–1886.

Barnes, H. H., Jr. (1977). *Roughness Characteristics of Natural Channels*, U.S. Geological Survey Water-Supply Paper 1849, USGS, Denver, Colorado.

Barrett, M. J., Gameson, A. L. H., and Ogden, C. G. (1960). "Aeration Studies at Four Weir Systems", *Water Water Eng.*, 64(September 1961), 407–413.

Bartlett, H. A. (1999). "How to Maximize Performance of Diaphragm Metering Pumps in Water Treatment Applications", *J. N. Engl. Water Works Assoc.*, CXIII(3), 189–197.

Basha, H. A. (1994). "Nonlinear Reservoir Routing: Particular Analytical Solution", *J. Hydraulic Eng.*, 120(5), 624–632.

Bates, J. M. (1895). "A Revolving Ice Cake", *Sci. Am.*, LXXII(6), 85.

Bates, R. L., Fondy, R. L., and Fenic, J. G. (1966). "Impeller Characteristics and Power", Section 3 in *Mixing Theory and Practice*, Vol. I, V. W. Uhl and J. B. Gray (eds.), Academic Press, New York.

Bathurst, J. C. (1985). "Flow Resistance Estimation in Mountain Rivers", *J. Hyd. Div.*, 111(4), 625–643. [With discussion June 1987]

Bathurst, J. C., Li, R. M., and Simons, D. B. (1981). "Resistance Equation for Large-Scale Roughness", *J. Hyd. Div.*, 107(12), 1593–1613.

Bauer, W. J., Louie, D. S., and Voorduin, W. L. (1969). "Basic Hydraulics", Section 2 in *Handbook of Applied Hydraulics*, C. V. Davis and K. E. Sorensen (eds.), McGraw-Hill, New York, pp. 2–30.

Bauer, D. I., Nakato, T., and Ansar, M. (1996). *Vortex Suppression in Multiple-Pump Sumps*, Iowa Institute of Hydraulic Research, University of Iowa, Iowa City, Iowa.

Baumgardner, R. H. (1983). "Hydraulic Design of Stormwater Pumping Stations: The Effect of Storage", *Transp. Res. Rec.*, 948, 63–70.

Baylar, A., and Bagatur, T. (2000). "Aeration Performance of Weirs", *Water SA*, 26(4), 521–526.

Baylar, A., and Bagatur, T. (2001). "Aeration Performance of Weirs Part 2", *Water Eng. Manage.*, 2001, 14–16.

Beck, C. F. (1979). "Ethical Considerations in Computer Use", *J. Technical Council ASCE*, 105(TC2), 415–427.

Bean, E. L. (1953). "Study of Physical Factors Affecting Flocculation", *Water Works Eng.*, 106, 33-34, 65-68.

Becker, R. J. (1963). "Filter Collapse Due to Improper Wash Water Distribution", *J. Am. Water Works Assoc.*, 55(9), 1157–1164.

Bedient, P. B., and Huber, W. C. (1988). *Hydrology and Floodplain Analysis*, Addison-Wesley Publishing Company, Reading, Massachusetts.

Beecham, S., Khiadani, M. H., and Kandasamy, J. (2005). "Friction Factors for Spatially Varied Flow with Increasing Discharge", *J. Hydraul. Eng.*, 131(9), 792–799.

Beij, K. H. (1934). "Flow in Roof Gutters", *Bur. Stand. J. Res.*, 12, 193–213.

Benami, A. (1968). "New Head Loss Tables for Sprinkler Laterals", *J. Irrig. Drain. Div.*, 94(2), 185–197.

Benefield, L. D., Judkins, J. F., Jr., and Parr, A. D. (1984). *Treatment Plant Hydraulics for Environmental Engineers*, Prentice Hall, Englewood Cliffs, New Jersey.

Bengtson, H. H. (2019). "Sharp Crested Weirs for Open Channel Flow Measurement", Course No: C02-022, Continuing Education and Development, Inc., New York, https://www.cedengineering.com/userfiles/Sharp-Crested%20Weirs.pdf, accessed September 12, 2019.

Bengtson, J. (1998). "Water District Solves Instability in Dual Pressure-Reducing Station Design", *Water Eng. Manage.*, 1998, 16–17.

Benjes, H. H., Sr., and Foster, W. E. (1976). "Sewage", Section 10.2 in *Pump Handbook*, I. J. Karassik, W. C. Krutzsch, W. H. Fraser, and J. P. Messina (eds.), McGraw-Hill, New York.

Bettess, R. (1999). "Flow Resistance Equations for Gravel Bed Rivers", *Proc. 28th International Association for Hydraulic Research*, Graz, Austria.

Bewtra, J. K., and Nicholas, W. R. (1964). "Oxygenation from Diffused Air in Aeration Tanks", *J. WPCF*, 36(10), 1195–1224.

Blaisdell, F. W., and Manson, P. W. (1963). *Loss of Energy at Sharp-Edged Pipe Junctions in Water Conveyance Systems*, U.S. Department of Agriculture Technical Bulletin No. 1283, Washington, D.C.

Blaisdell, F. W., and Manson, P. W. (1967a). "Energy Loss at Pipe Junctions", *J. Irrig. Drain. Div.*, 93(3), 59–78.

Blaisdell, F. W., and Manson, P. W. (1967b). "Closure to 'Energy Loss at Pipe Junctions'", *J. Irrig. Drain. Div.*, 96(2), 218–222.

Bluman, G. W., and Kumei, S. (1996). *Symmetries and Differential Equations*, 2nd Ed., Springer Verlag, New York.

Boksiner, G., Jackson, D., Terrill, D., Young, M., Tatum, B., and Anderson, D. (2014). "Managing the Deluge", *Water Environ. Technol.*, 26(4), 28–33.

Bolton, M. M. (2011). "Lien Mailing Anger Residents." *Boston Globe*, January 9.

Bouwer, H. (1978). *Groundwater Hydrology*, McGraw-Hill Book Company, New York.

Boyce, M. P., Bayley, R. D., Sudhakar, V., and Elchuri, V. (1976). *Field Testing of Compressors, Proc. 5th Turbomachinery Symposium*, Texas A&M University, College Station, Texas.

Boyer, M. C. (1964). "Streamflow Measurement", Section 15 in *Handbook of Applied Hydrology*, V. T. Chow (ed.), McGraw-Hill Book Company, New York.

Bradley, R. A., and Krone, R. B. (1971). "Shearing Effects on Settling of Activated Sludge", *J. San. Eng. Div.*, 97(1), 59–79.

Brater, E. F., and King, H. W. (1976). *Handbook of Hydraulics*, 6th Ed., McGraw-Hill Book Company, New York.

Brenner, A., and Diskin, M. H. (1991). "Model Study of Jet-Circulated Grit Chamber", *J. Environ. Eng.*, 117(6), 782–787.

Briggs, B., and Zimmerman, C. H. (1999). "The Versatile Venturi", *J. N. Engl. Water Works Assoc.*, 113(2), 151–175.

Brock, R. R. (1970) "Periodic Permanent Roll Waves", *J. Hyd. Div.*, 96(12), 2565–2580.

Brooks, N. H., and Koh, R. C. Y. (1969). "Selective Withdrawal from Density-Stratified Reservoirs", *J. Hyd. Div.*, 95(4), 1369–1400.

Brutsaert, W. (1971). "De Saint-Venant Equations Experimentally Verified", *J. Hyd. Div.*, 97(9), 1387–1401.

Brzozowski, C. (2016). *"Water Sustainability in Prisons", Water Efficiency*, Forester Communications Inc., Santa Barbara, CA.

Buchan, A. (2006). "Addressing Ergonomic and Workplace Safety Issues in Manhole Cover Design", *Water Environ. Technol.*, 18(4), 52–53.

Buckingham, E. (1914). "On Physically Similar Systems: Illustrations of the Use of Dimensional Equations", *Phys. Rev.*, 4, 345–376.

Building Design and Construction (1979). "Roof of Auto Factory Collapses during Storm", August.

Burgess, S. G., Green, A. F., and Easterby, P. A. (1960). "More Detailed Examination of Flow in Sewage Tanks using Radioactive Tracers", *J. Inst. Sewage Purif.* 2, 184–192.

Burgreen, D. (1960). "Development of Flow in Tank Drainage", *J. Hyd. Div.*, 86(3), 13–28.

Burke, C., and Burke, T. T. (1994). *HERPICC Stormwater Drainage Manual*, Purdue University, Lafayette, Indiana.

Burke, C., and Burke, T. T. (2008). *Stormwater Drainage Manual*, Purdue Research Foundation, West Lafayette, Indiana.

Burton, K. R. (1980). "Stormwater Detention Basin Sizing", *J. Hyd. Div.*, 106(3), 437–439.

Buse, F. (1976). "Power Pumps", Section 3.1 in *Pump Handbook*, I. J. Karassik, W. C. Krutzsch, W. H. Fraser, and J. P. Messina (eds.), McGraw-Hill, New York.

Caduto, M. J., and Bruchac J. (1991). "How Raven Made the Tides," in *Native American Stories Told*, J. Bruchac (Ed.), Fulcrum Publishing, Golden, Colorado.

Callier, V. (2022). "Hard Knocks", *Sci. Am.*, 327(4), 12–14.

Camp, T. R. (1935). "Flow Control Apparatus", U.S. Patent 2,025,722, December 31, 1935.

Camp, T. R. (1936). "A Study of the Rational Design of Settling Tanks", *Sewage Works J.*, 8(5), 742–758.

Camp, T. R. (1940). "Lateral Spillway Channels", *Trans. Am. Soc. Civil Eng.*, 105(1), 606-617, 633-647.

Camp, T. R. (1942). "Grit Chamber Design", *Sewage Works J.*, 14(2), 368–381.

Camp, T. R. (1944). "Discussion of 'Effect of Turbulence on Sedimentation'", *Trans. Am. Soc. Civil Eng.*, 109(1), 660–667.

Camp, T. R. (1946a). "Design of Sewers to Facilitate Flow", *Sewage Works J.*, 18(1), 3–16.

Camp, T. R. (1946b). "Sedimentation and the Design of Settling Tanks", *Trans. Am. Soc. Civil Eng.*, 111(4), 895–957.

Camp, T. R. (1947). "The Effect of Turbulence in Retarding Settling", *Proc. Third Hydraulics Conference*, Iowa Institute of Hydraulic Research, State University of Iowa, June 10–12, 1946, Iowa City, Iowa.

Camp, T. R. (1951). "Hydraulics of Treatment Plants (Part II)", Boston Society of Civil Engineers John R. Freeman Lectures in Applied Hydraulic Design, February 5, 1951.

Camp, T. R. (1952). "Hydraulics of Sewers", *Public Works*, 83(6), 59–65.

Camp, T. R. (1953). "Studies of Sedimentation Basin Design", *Sewage Ind. Wastes*, 25(1), 1–14.

Camp, T. R. (1955). "Flocculation and Flocculation Basins", *Trans. Am. Soc. Civil Eng.*, 120(1), 1–16.

Camp, T. R. (1962). "Chlorination of Mixed Sewage and Storm Water", *Trans. Am. Soc. Civil Eng.*, 127(3), 452–473.

Camp, T. R (1963). *Water and its Impurities*, Reinhold Publishing Corporation, New York.

Camp, T. R. (1964). "Theory of Water Filtration", *J. San. Eng. Div.*, 90(4), 1–30.

Camp, T. R. (1965). "Closure to 'Theory of Water Filtration'", *J. San. Eng. Div.*, 91(5), 55–69.

Camp, T. R. (1968). "Floc Volume Concentration", *J. Am. Water Works Assoc.*, 60(6), 656–673.

Camp, T. R. (1969). "Hydraulics of Mixing Tanks", *J. Boston Soc. Civil Eng.*, 56(1), 1–28.

Camp, T. R., and Conklin, G. F. (1970). "Toward a Rational Jar Test for Coagulation", *J. New Engl. Water Works Assoc.*, 80, 325–338.

Camp, T. R., and Graber, S. D. (1968). "Dispersion Conduits", *J. San. Eng. Div.*, 94(1), 31–39.

Camp, T. R., and Graber, S. D. (1969). "Closure to 'Dispersion Conduits'", *J. San. Eng. Div.*, 95(5), 943–947.

Camp, T. R., and Graber, S. D. (1970). "Discussion of 'Internal Hydraulics of Thermal Discharge Diffusers', by S. Vigander, R. A. Elder, and N. H. Brooks", *J. Hyd. Div.*, 96(12), 2631–2635.

Camp, T. R., and Graber, S. D. (1972). "Discussion of 'Manifold Flow in Subirrigation Pipes'", *J. Hyd. Div.*, 98(4), 728–732.

Camp, T. R., Graber, S. D., and Conklin, G. F. (1971). "Backwashing of Granular Water Filters", *J. San. Eng. Div.*, 97(6), 903–926.

Camp, T. R., Graber, S. D., and Conklin, G. F. (1973). "Closure to 'Backwashing of Granular Water Filters'", *J. San. Eng. Div.*, 99(4), 547–553.

Camp, T. R., and Lawler, J. C. (1969a). "Groundwater", Section 35 in *Handbook of Applied Hydraulics*, C. V. Davis, and K. E. Sorensen (Eds.), 3rd Ed., McGraw-Hill Book Company, New York.

Camp, T. R., and Lawler, J. C. (1969b). "Water Distribution", Section 37 in *Handbook of Applied Hydraulics*, C. V. Davis and K. E. Sorensen (Eds.), 3rd Ed., McGraw-Hill Book Company, New York.

Camp, T. R., and Stein, P. C. (1943). "Velocity Gradients and Internal Work in Fluid Friction", *J. Boston Soc. Civil Eng.*, XXX(4), 219–237.

Campbell, F. B. (1966). *Hydraulic Design of Rock Riprap*, Miscellaneous Paper 2-777, Waterways Experiment Station, Vicksburg, Mississippi.

Canter, L. W., and Knox, R. C. (1985). *Septic Tank System Effects on Ground Water Quality*, Lewis Publishers, Inc., Chelsea, Michigan.

Cao, Z., Pender, G., and Meng, J. (2006). "Explicit Formulation of the Shields Diagram for Incipient Motion of Sediment", *J. Hyd. Eng.*, 132(10), 1097–1099.

Carman, P.C. (1938). "Fundamental Principles of Industrial Filtration (A Critical Review of Present Knowledge)", *Trans. Inst. Chem. Eng.*, 16, 168–188.

Carpenter, L. V., Vennard, J. K., and Whitley, F. H. (1940). "Discussion of 'Lateral Spillway Channels', by T. R. Camp", *Trans. Am. Soc. Civil Eng.*, 105(1), 621–624.

Carter, A. D. S., Moss, C. E., Green, G. R., and Annear, G. (1960). *The Effect of Reynolds Number on the Performance of a Single Stage Compressor*, Reports and Memoranda No. 3184 (19,504), Aeronautical Research Council Technical Report, Ministry of Aviation, Aeronautical Research Council, Her Majesty's Stationery Office.

Carter, T. L., and Rasmussen, T. C. (2006). "Hydrologic Behavior of Vegetated Roofs", *J. AWRA*, 2006, 1261–1274.

Cartwright, D. E. (1999). *Tides A Scientific History*, Cambridge University Press, Cambridge, United Kingdom.

Cassel, A. F., and Johnson, B. P. (1979). *Evaluation of Dewatering Devices for Producing High-Solids Sludge Cake*", EPA-600/2-79-123, Water Resources Management Administration, Washington, D.C.

Cederwall, K., and Ditmars, J. D. (1970). *Analysis of Air-Bubble Plumes*, Report No. KH-R-24, W. M. Keck Laboratory of Hydraulics and Water Resources, California Institute of Technology, Pasadena, California, September.

Cesario, L. (1995). *Modeling, Analysis, and Design of Water Distribution Systems*, American Water Works Association, Denver, Colorado.

Chandrasekaran, D., and Rao, N. S. L. (1976). "Characteristics of Proportional Weirs", *J. Hyd. Div.*, 102(11), 1677–1692.

Changnon, S. A., Jr. (1978). "Heavy Falls of Hail and Rain Leading to Roof Collapse", *J. Struct. Div.*, 104(1), 198–200.

Chanson, H. (1995). "Air-Water Gas Transfer at Hydraulic Jump with Partially Developed Inflow", *Water Res.*, 29(10), 2247–2254.

Chanson, H. (2006). "Minimum Specific Energy and Critical Flow Conditions in Open Channels." *J. Irrig. Drain. Div.*, 132(5), 498–502.

Chanson, H. (2012). "Free-Surface Profiles, Velocity and Pressure Distributions on a Broad-Crested Weir: A Physical Study." *J. Irrig. Drain. Div.*, 138(12), 1068–1074.

Chanson, H. (2013). "Explicit Equations for Critical Depth in Open Channels with Complex Compound Cross Sections. A Discussion." *Flow Measur. Instru.*, 29, 65–66.

Chanson, H. (2016). "Atmospheric Noise of a Breaking Tidal Bore", *J. Acoust. Soc. America*, 139(1), 12–20.

Chao, J.-L., and Trussell, R. R. (1977). "Hydraulic Design of Flow Distribution Channels", *J. Environ. Eng. Div.*, 106(2), 321–334.

Chasick, A. H., and Burger, T. B. (1964). "Using Graded Sand to Test Grit Removal Apparatus", *J. WPCF*, 36(7), 884–894.

Chaudhry, M. H. (1993). *Open-Channel Flow*. Prentice Hall, Englewood Cliffs, New Jersey.

Chick, H. (1908). "An Investigation of the Laws of Disinfection", *J. Hygiene*, 8, 92–158.

Chiu, Y. (1974). "Boundary Layer Separation Concept of Settling Tanks", *J. Environ. Eng. Div.*, 100(6), 1213–1218.

Chow, V. T. (1959). *Open-Channel Hydraulics*, McGraw-Hill Book Company, New York.

Christman, W. P. (1962). "Rotary-Gravity Type Screening Machines", *Ind. Water Wastes*, 7(6), 162–167.

City of Attleboro, Massachusetts (2012). "Revised Ordinances of the City of Attleboro." Section 16-18 Sewer Use Rates, July 17, 2012.

City of Manassas. (2007). "Article 8 – Storm Drainage System", Virginia, http://www.manassascity.org/documents/Public%20Works/DCSM%20Article8.pdf, accessed August 13, 2007, Section 8-890.

Clark, M. M. (1985). "Critique of Camp and Stein's RMS Velocity Gradient", *J. Environ. Eng. Div.*, 111(6), 741–754.

Cleasby, J. L. (1972). *Backwash of Granular Filters Used in Waste Water Filtration*, Submitted to Water Quality Office, Environmental Protection Agency, EPA Project 17030 DKG, Engineering Research Institute, Iowa State University, Ames, Iowa, August.

Cleasby, J. L. (1984). "Is Velocity Gradient a Valid Turbulent Flocculation Parameter", *J. Environ. Eng. Div.*, 110(5), 875–897.

Click, C. N., and Mixon, F. O. (1974). "Flow Smoothing in Sanitary Sewers", *J. WPCF*, 46(3), 522–531.

Cohn, A. B., Manley, T. O., Manley, P. L., Smeltzer, E., and Watzin, M. C. (2007). "Lake Champlain", *Lakeline*, 2007, 46–56.

Colburn, E. A. (2004). *Vernal Pools: Natural History and Conservation*, The McDonald Publishing Company, Blacksburg, Virginia.

Committee on Hydromechanics (1963). "Friction Factors in Open Channels", Prog. Rept. Task Force on Friction Factors in Open Channels, *J. Hyd. Div.*, 89(2), 97–143.

Committee on Hydraulic Structures (1965). "Factors Influencing Flow In Large Conduits", Report of the Task Force on Flow in Large Conduits, *J. Hyd. Div.*, 91(6), 123–152.

Committee to Study Limiting Velocities of Flow in Sewers (1942). "Minimum Velocities for Sewers", *J. Boston Soc. Civil Eng.*, XXIX 4, 286–363.

Conner, N. W. (1945). *Design and Capacity of Gutter Inlets*, Bulletin No. 30, Engineering Experiment Station, North Carolina State College, Raleigh, North Carolina, July.

Contractor, D. N. (1974). "Discussion of 'Open-Channel Flow Equations Revisited' by B. C. Yen", *J. Eng. Mech. Div.*, 100(5), 1059–1060.

Cooper, H. H., and Jacob, C. E. (1946). "A Generalized Graphical Method For Evaluating Formation Constants and Summarizing Well Field History", *Trans. Am. Geophys. Union*, 27(4), 526–534.

Courrier, K., and Eckholm, E. (1977). "Schistosomiasis: The Curious Bond Between Snails, People, and Disease", *Sierra* 1977, 11–14.

Crandall, S. H., Vigander, S., and March, P. A. (1975). "Destructive Vibration of Trashracks due to Fluid-Structure Interaction", *J. Manuf. Sci. Eng.*, 97(4), 1359–1365.

Crawford, M. (2019). "Vortex Shedding Flowmeters", https://www.asme.org/topics-resources/content/an-example-of-a-vortex-shedding-flowmeter, accessed March 29, 2019.

Cropper, W. H. (2001). *Great Physicists*, Oxford University Press, New York.

Cyr, M. D. (2007). An Aquifer-Well Coupled Model: A Refined Implementation of Well Bore Boundary Conditions in Three-Dimensional, Heterogeneous Formations, MS Thesis, Queen's University, Kingston, Ontario, Canada.

Czarnecki, G. J., and Lippincott, J. K. (1976). "Screw Pumps", Section 3.3 in *Pump Handbook*, I. J. Karassik, W. C. Krutzsch, W. H. Fraser, and J. P. Messina (eds.), McGraw-Hill, New York.

da Silva, A. M., and Ebrahimi, M. (2017). "Meandering Morphodynamics: Insights from Laboratory and Numerical Experiments and Beyond", *J. Hyd. Eng.*, 143(9), 1–16.

Dague, R. R., and Baumann, E. R. (1961). "Hydraulics of Circular Settling Tanks Determined by Models", *Presented at the June 1961 Annual Meeting of the Iowa Water Pollution Control Association*, Lake Okoboji, Iowa.

Danckwerts, P. V. (1953). "Continuous Flow Systems: Distribution of Residence Times", *Chem. Eng. Sci.*, 2(1), 1–13.

Darwin, G. H. (1962). *Tides and Kindred Phenomena in the Solar System*, W. H. Freeman and Company, San Francisco, California.

Das, D., Keinath, T. M., Parker, D. S., and Wahlberg, E. J. (1993). "Floc Breakup in Activated Sludge Plants", *Water Environ. Res.*, 65(2), 138–145.

Davis, C. V. (1942). *Handbook of Applied Hydraulics*, McGraw-Hill Book Company, New York.

Davis, C. V., and Sorensen, K. E. (eds) (1969), *Handbook of Applied Hydraulics*, 3rd Ed., McGraw-Hill Book Company, New York.

Davis, D. W., and Burnham, M. W. (1988). "Discussion of 'Friction Slope Averaging in Backwater Calculations'", *J. Hyd. Eng.*, 114(4), 453–455.

Davis, R. K. (1968). *The Range of Choice in Water Management*, Johns Hopkins Press, Baltimore, Maryland.

Davis, S. (1961). "Unification of Parshall Flume Data", *J. Irrig. Drain. Div.*, 87(4), 13–26.

Davis, S. N., and De Wiest, R. J. M. (1966). *Hydrogeology*, John Wiley & Sons, Inc., New York.

DeLapp, W. (1943). "Discussion of 'Entrainment of Air in Flowing Water: Closed Conduit Flow' by A. A. Kalinske and J. M. Robertson", *Trans. Am. Soc. Civil Eng.*, 108(1), 1448.

Delatte, N. (1976). "Learning from Failures", *Journal of the Boston Society of Civil Engineers – Civil Engineering Practice*, 21(2), 21–38.

Delatte, N. (2006). "Learning from Failures, Civil Engineering Practice", *J. Boston Soc. Civil Eng.*, 21(2), 21–38.

Delichatsios, M. A., and Probstein, R. F. (1975). "Scaling Laws for Coagulation and Sedimentation", *J. WPCF*, 47(5), 941–949.

Dempsey, B. J. (1988). "Core Flow-Capacity Requirements of Geocomposite Fin-Drain Materials Used in Pavement Subdrainage", *Transportation Research Record*, Transportation Research Board, Washington, D.C., 1159, 21-29.

Den Hartog, J. P. (1956). *Mechanical Vibrations*, 4th Ed., McGraw-Hill Book Company, New York.

DeNileon, G. P. (1999). "Witching Works, Say Believers", *Opflow*, 25(9), 4–5.

Dick, R. I. (1970). "Role of Activated Sludge Final Settling Tanks", *J. San. Eng. Div.*, 96(2), 423–436.

Dick, R. I. (1971). "Discussion of 'Shearing Effects on Settling of Activated Sludge'", *J. San. Eng. Div.*, 97(6), 942–943.

Dick, R. I., and Young, K. W. (1972). "Analysis of Thickening Performance of Final Settling Tanks", *Presented at the Purdue Industrial Waste Conference, 27th Annual Meeting*, Purdue University, Lafayette, Indiana, May 2-4, 1972.

Dismukes, D. E., Jr. (1966). "Process Control Applied to Disinfection", Eighth Sanitary Engineering Conference, *Illinois State Dept. of Public Health and the University of Illinois*, Urbana, Illinois, February 8-9, 1966.

Dockrill, P. (2019). "This Freaky Spinning Ice Circle in Maine Looks like a Giant Alien Landmark", *Sci. Alert*, https://www.sciencealert.com/bizarre-spinning-phenomenon-in-maine-looks-like-a-giant-alien-calling-card, accessed January 16, 2019.

Domeshek, D. 2009. "Medfield Sewer Board denies Camden Place Residents' Bid to Exempt Sprinklers from Fees." Wicked Local Medfield, November 12, 2009. https://www.wickedlocal.com/story/thepress/2009/11/12/medfield-sewer-board-denies-camden/39220649007/

Donsky, B. (1961). "Complete Pump Characteristics and the Effects of Specific Speeds on Hydraulic Transients", *J. Basic Eng.*, 83(4), 685–696.

Drobny, N. L. (1963). "Effect of Paddle Design on Flocculation", *J. San. Eng. Div.*, 89(2), 17–30.

Drzewiecki, T. M., and Houck, D. W. (1985). "Fluidics – A Control Technology for the Microprocessor Age", *Mech. Eng.*, 107(3), 28–36.

Dunn, C. J., Allaben, C. C., and Reinhold, R. A. (2002). "Model Testing Optimizes and Verifies Design of Large Trench Wet Well", *Proc. Water Environment Federation*, WEFTEC 2002.

Dwight, H. B. (1961). *Tables of Integrals and Other Mathematical Data*, 4th Ed., The MacMillan Company, New York.

Eagleson, P. S., and Dean, R. G. (1966). "Small Amplitude Wave Theory", Chapter 1 in *Estuary and Coastline Hydrodynamics*, A. T. Ippen (ed.), McGraw-Hill, New York.

Egeland, D. R. (1973). "Land Disposal I: A Giant Step Backward", *J. WPCF*, 45(7), 1465–1475.

Einstein, H. A. (1964). "Sedimentation: Part II. River Sedimentation", Section 17-II in *Handbook of Applied Hydrology*, V. T. Chow (ed.), McGraw-Hill Book Company, New York.

Einstein, H. A., and Li, H. (1958). "Secondary Currents in Straight Channels", *Trans. Am. Geophys. Union*, 39(6), 1085–1088.

Eisenlohr, W. S., Jr. (1945). "Coefficients for Velocity Distribution in Open-Channel Flow", *Trans. Am. Soc. Civ. Engr.*, 110, 633-644 & 657-668.

Eisner, T., and Aneshansley, D. J. (1982). "Spray Aiming in Bombadier Beetles: Jet Deflection by the Coanda Effect", *Science*, 215, 83–85.

Elberti, A. G. (2021). "Focus on Safety, Resiliency, and Effectiveness", *Water and Wastes Engineering*, January 2021.

Elmore, H. L., and West, W. F. (1961). "Effects of Water Temperature on Stream Reaeration", *J. San. Eng. Div.*, 85(4), 59–71.

Elzerman, A. W., and Armstrong, D. E. (1978). "Enrichment of Micronutrients, Heavy Metals, and Chlorinated Hydrocarbons in Wind-Generated Lake Foam", *Environ. Sci. Technol.*, 12(4), 413–417.

Enright, J. T. (1999). "Testing Dowsing: The Failure of the Munich Experiments," *Skep. Inq.*, 23(1), 39–46. [And additional articles cited therein.]

Epp, R., and Fowler, A. G. (1973). "Efficient Code for Steady-State Flows in Networks", *J. Hyd. Div.*, 96(1), 43–56.

Eriksen, H. A. (1972). "Surge Analysis of a Water Transmission System", *Presented at the International Conference on Pressure Surges*, Canterbury, England, September 6-8.

Eriksen, H. A. (1974). Personal Communication, Project Manager, Camp Dresser & McKee Inc., Boston, Massachusetts, February 4, 1974.

Escoffier, F. F. (1955). "Discussion of 'Backwater Functions by Numerical Integration' by C. J. Keifer and H. H. Chu", *Trans. Am. Soc. Civ. Eng.*, 120(1), 443–444.

Eshbach, O. (1936). *Handbook of Engineering Fundamentals*, John Wiley and Sons, Inc., New York.

Esmaili, H., and Scott, V. H. (1968). "Unconfined Aquifer Characteristics and Well Flow", *J. Irrig. Drain. Div.*, 94(1), 115–136.

Ettema, R. (2005). "Rivers Viewed Through Names and Epithets", *J. Hyd. Eng.*, 131(7), 535–541.

Evans, W. E., and Crawford, C. C. (1954). "Design Charts for Air Chambers on Pumps Lines", *Trans. Am. Soc. Civ. Eng.*, 119(1), 1025–1036.

Everett J. Prescott Company (EJP) (2022). "Thrust & Anchorage at Fittings & Valves", EJP Reference Manual, p. R-16, accessed December 13, 2022.

Exley, J. T., and Brighton, J. A. (1969). "Flow Separation and Reattachment in Confined Jet Mixing", *1969 ASME Applied Mechanics and Fluids Engineering Conference*, Evanston, Illinois, 69-FE-29.

Factory Mutual Insurance Company (FMIC). (2006). "Loss Prevention Data 1-54, Roof Loads for New Construction", Norwood, Massachusetts.

Fair, G. M., and Geyer, J. C. (1954). *Water Supply and Wastewater Disposal*. John Wiley & Sons, Inc., New York.

Fair, G. M., and Geyer, J. C. (1968). *Water Supply and Waste-Water Disposal*, John Wiley and Sons, Inc., New York.

Fair, G. M., Geyer, J. C., and Okun, D. A. (1968). *Water and Wastewater Engineering*, Vol. 2, John Wiley and Sons, Inc., New York.

Fair, G. M., and Hatch, L. P. (1933). "Fundamental Factors Governing the Streamline Flow of Water Through Sand", *J. Am. Water Works Assoc.*, 25(11), 1551–1565.

Fairfax County. (2007). "6-1300 Retention and Detention Facilities," Fairfax, Virginia, www.fairfaxcounty.gov/dpwes/publications/pfm/6-1300.htm, accessed August 1, 2007.

Fan, L.-N., and Brooks, N. H. (1969). *Numerical Solution of Turbulent Buoyant Jet Problems*", Report No. KH-R-18, W. M. Keck Laboratory of Hydraulics and Water Resources, California Institute of Technology, Pasadena, California, January 1969.

Federal Highway Administration (FHwA) (1965). *Hydraulic Charts for the Selection of Highway Culverts*, Hydraulic Engineering Circular No. 5, Washington, D.C.

Federal Highway Administration (FHwA) (1982). *Manual for Highway Stormwater Pumping Stations*, Vols. I and II, Publication No. FHWA-IP-82-17, FHWA, Washington, D.C.

Federal Highway Administration (FHwA) (1990a). *Highways in the River Environment*, Publication No. FHWA-Hl-90-016, FHWA, Washington, D.C.

Federal Highway Administration (FHwA) (1990b). *Highway Subdrainage Design by Microcomputer: (DAMP) Drainage Analysis & Modeling Programs*, Publication No. FHWA-IP-90-012, FHWA, Washington, D.C., 84-88, 101-108.

Federal Highway Administration (FHwA) (1996). *Urban Drainage Design Manual*, Hydraulic Engineering Circular No. 22, Publication No. FHWA-SA-96-078, FHWA, Washington, D.C., Section 4.3, Appendix B.

Federal Highway Administration (FHwA) (2001a). *Highway Stormwater Pump Station Design*, Hydraulic Engineering Circular No. 24, Publication No. FHWA-NHI-01-007, FHWA, Washington, D.C.

Federal Highway Administration (FHwA) (2001b). *Urban Drainage Design Manual*, Hydraulic Engineering Circular No. 22, Second Edition, Publication No. FHWA-NHI-01-021, FHWA, Washington, D.C., 4.2.3, 4.3, 4.4, Appendix B.

Federal Highway Administration (FHwA) (2002). *User's Guide for Drainage Requirements in Pavements – DRIP 2.0 Microcomputer Program*, FHWA, Washington, D.C., 4-21 to 29, 5-8 to 14.

Federal Highway Administration (FHwA) (2005). *Design of Roadside Channels with Flexible Linings*, Hydraulic Engineering Circular No. 15, Third Edition, FHWA-NHI-05-114, U.S. Department of Transportation, Washington, D.C.

Federal Highway Administration (FHwA) (2010). *Culvert Design for Aquatic Organism Passage*, Hydraulic Engineering Circular No. 26, Publication No. FHWA-HIF-11-008, U.S. Department of Transportation, Washington, D.C.

Federal Highway Administration (FHwA) (2012). *Hydraulic Design of Highway Culverts*, Publication No. FHWA-HIF-12-026, Hydraulic Design Series Number 5, 3rd Ed., U.S. Department of Transportation, Washington, D.C.

Federal Water Pollution Control Administration (FWPCA) (1969). *Polymers for Sewer Flow Control*, Prepared for the U.S. Department of the Interior by The Western Company, Richardson, Texas, August 1969.

Feng, Y., Brubaker, K. L., McCuen, R. H. (2017). "New View of Flood Frequency Incorporating Duration", *J. Hydrol. Eng.*, 22(11), 1–11.

Ferro, V. (1999). "Evaluating Friction Factor for Gravel-Bed Channel with High Boulder Concentration", *J. Hyd. Div.*, 125(7), 771–778.

Field, R., and Masters, H. (1977). *Swirl Device for Regulating and Treating Combined Sewer Overflows*, EPA-625/2-77-012, Technology Transfer Capsule Report, National Technical Information Service, Springfield, Virginia.

Field, R., O'Connor, T. P., and Cowan, H. (1995). "Swirl Technology: Proper Design, Application and Evaluation," *Proc. Annual Risk Reduction Engineering Laboratory Research Symposium*, Cincinnati, Ohio, April 4-6, 1995, pp. 66-84.

Fischer, D., Charles, E. G., and Baehr, A. L. (2003). "Effects of Stormwater Infiltration on Quality of Groundwater Beneath Retention and Detention Basins", *J. Environ. Eng.*, 129(5), 464–471.

Fischer, H. B. (1967). "The Mechanics of Dispersion in Natural Streams", *J. Hyd. Div.*, 93(6), 187–216

Fischetti, M. (2023). "The New Tornado Alley", *Sci. Am.*, 329(1), 70–75.

Fitch, E. B. (1957). "The Significance of Detention in Sedimentation", *Sew. Ind. Wastes*, 29(10), 1123–1133.

Follest, T. (2000). "Sludge Pumping/Polymer Metering Systems", *Water Engineering & Management*, May 2000.

Forrester, K. (2001). *Subsurface Drainage for Slope Stabilization*, ASCE, Reston, Virginia, pp. 130–133.

Forstner, G. (2007). "Screen Selection Simplified: Understanding Your Choices", *Water Environ. Technol.*, 19(10), 60–64.

Forthofer, J. M. (2019). "Fire Tornadoes", *Sci. Am.*, 321(6), 60–67.

Freeman, P. A. (1971). "Combined Sewer Regulation with Fluidic Regulators", *J. WPCF*, 43(5), 862–871.

French, J. A. (1980. "Discussion of 'Dividing-Flow Manifolds with Square-Edged Laterals', by H. E. Hudson, R. W. Bailey, and R. B. Uhler, *J. Environ. Eng. Div.*, 106(3), 675.

French, J. A. (2015). "Thorough Dispersal of Dosants in Conduits", *J. Hyd. Eng.*, 141(5), 1–7.

French, J. A. (2017). "Ring Diffuser to Dispense Antifoulants at Seawater Intakes", *J. Hyd. Eng.*, 143(9), 1–4.

Frenkel, J. (1955). *Kinetic Theory of Liquids*, Dover Publications, Inc., New York, pp. 191–208.

Froehlich, D. C. (1993). "Short-Duration-Rainfall Intensity Equations for Drainage Design", *J. Irrig. Drain. Eng.*, 119(5), 814–828.

Froehlich, D. C. (1994). "Sizing Small Stormwater Pumping Stations", *Water Resour. Bull.*, 30(6), 1055–1062.

Froehlich, D. C. (2012). "Resistance to Shallow Uniform Flow in Small, Riprap-Lined Drainage Channels", *J. Irrig. Drain. Eng.*, 138(2), 203–210.

Gaillard, J. (1984). "Multilevel Withdrawal and Water Quality", *J. Environ. Eng.*, 110(1), 123–130.

Gameson, A. L. H. (1957). "Weirs and the Aeration of Rivers", *J. Inst. Water Eng.*, 11(6), 477–490.

Gameson, A. L. H., Vandyke, K. G., and Ogden, C. G. (1958). "The Effect of Temperature on Aeration at Weirs", *Water Water Eng.*, 62, 489–491.

Gannon, J. J. (1967). "Aeration at Waste Treatment Plant Outfall Structures", *Water and Wastes Engineering*, April 1967.

Gardner, P. E. J., and Gummer, J. H. (1973a). "The Use of air Chambers to Suppress Hydraulic Resonance Part One", *Water Power* 1973, 102–105.

Gardner, P. E. J., and Gummer, J. H. (1973b). "The Use of Air Chambers to Suppress Hydraulic Resonance Part Two", *Water Power* 1973, 135–139.

Gardner-Thorpe, C., Pearn, J., and Gardner-Thorpe, D. (2007). "Alterations without Attendance", *J. Hyd. Eng.*, 133(5), 463–467.

Georgia, F. R. (1949). "Remodeled Cornell University Filter Plant", *J. Am. Water Works Assoc.*, 41(11), 1047–1052.

Georgia, F. R. (1950). "Modernization of the Cornell University Filter Plant", *Water Sew. Works*, 96(6).

Gerard, R. (1978). "Secondary Flow in Noncircular Conduits", *J. Hyd. Div.*, 104(5), 755–773.

Ghobadian, R., and Meratifashi, E. (2012). "Modified Theoretical Stage-Discharge Relation for Circular Sharp-Crested Weirs", *Water Sci. Eng.*, 5(1), 26–33.

Gilcrest, G. H. (1950). "Flood Routing", Chapter X in *Engineering Hydraulics*, H. Rouse (ed.), John Wiley & Sons, Inc., New York.

Giles, J. H. (1943). "Inlet and Outlet Designs for Sedimentation Tanks", *Sew. Works J.*, 15, 609.

Gilman, B. R. (1964). "Flood Routing," Section 9 in *Handbook of Applied Hydrology*, V.T. Chow (ed.), McGraw-Hill Book Company, New York, pp. 9-52 to 53.

Gladwell, J. S., and Tinney, E. R. (1965). "Hydraulic Studies of Large Penstock Trifurcations", *J. Power Div.*, 91(1), 59–80.

Goldfarb, B. (2018). *Eager: The Surprising, Secret Life of Beavers and Why They Matter*, Chelsea Green Publishing, White River Junction, Vermont.

Goldstein, S. (ed.) (1938). *Modern Developments in Fluid Dynamics*, Oxford University Press, New York.

Gonwa, W. S. (2012). "Has it Been 100 Years Already", *Water Environ. Technol.*, 24(8), 51–54.

Gordon, R. W. (1958). *Water Well Drilling with Cable Tools*, Bucyrus-Erie Company, South Milwaukee, Wisconsin, p. 144.

Gould, R. H. (1945). "Discussion of 'Design of Final Settling Tanks for Activated Sludge' by N. E. Anderson", *Sew. Works J.*, 17, 63.

Govardhan, R. N., and Ramesh, O. N. (2005). "A Stroll Down Karman Street", *Reson. J. Sci. Educ.*, 10(8), 25–37.

Graber, S. D. (1964). "Stall Cell Sensor", *Research and Development of Fluid Amplifiers for Turbopropulsion System Control*, Report No. APL TDR 64-82, Department of Mechanical Engineering, Massachusetts Institute of Technology, Cambridge, Massachusetts, pp. 142–166.

Graber, S. D. (1972). "Discussion/Communication on Hydraulic Model Studies of Chlorine Contact Tanks", by V. Kothandaraman and R. L. Evans (eds.). *J. WPCF*, 44(10), 2029–2035.

Graber, S. D. (1974a). "Outlet Weir Loadings for Settling Tanks'", *J. WPCF*, 46(10), 2355–2362.

Graber, S. D. (1974b). "Water Quality Management Plan for the Nashua River Basin", *Presented at the ASCE National and Environmental Engineering Meeting*, Kansas City, Missouri, October 22, 1974.

Graber, S. D. (1975a). "Discussion of 'Boundary Layer Separation Concept of Settling Tanks', by Y. Chiu", *J. Environ. Eng. Div.*, 101(5), 866–868.

Graber, S. D. (1975b). *Water Quality Management Plan Nashua River Basin*, Camp Dresser & McKee Inc., Boston, Massachusetts.

Graber, S. D. (1978). *Final Urban Stormwater Evaluation Report*, Areawide Water Quality Management Plan, Prepared for the Rhode Island Statewide Planning Program, Providence, Rhode Island.

Graber, S. D. (1981). "Discussion of 'Hydraulic Design of Flow Distribution Channels' by J.-L. Chao and R. R. Trussell", *J. Environ. Eng. Div.*, 107(1), 299–301.

Graber, S. D. (1982). "Asymmetric Flow in Symmetric Expansions", *J. Hyd. Div.*, 108(10), 1082–1101.

Graber, S. D. (1992). "Relations between Rational and SCS Runoff Coefficients and Methods", *Catchment Runoff and Rational Formula*, B. C. Yen (ed.), Water Resources Publications, Littleton, Colorado, pp. 60–73.

Graber, S. D. (1994). "A Critical Review of the Use of G-Value (RMS Velocity Gradient) in Environmental Engineering", *Developments in Theoretical and Applied Mechanics*, Vol. XVII, I. C. Jong and F. A. Akl (eds.), Louisiana Tech University and University of Arkansas, pp. 533–556.

Graber, S. D. (1997). "Discussion of 'Design of Flocculating Baffled Channel', by P. K. Swamee", *J. Environ. Eng. Div.*, 123(12), 1269.

Graber, S. D. (1998). "Discussion of 'Influence of Strain-Rate on Coagulation Kinetics', by T. A. Kramer and M. M. Clark", *J. Environ. Eng. Div.*, 124(10), 1028.

Graber, S. D. (1999). "File Generation and Interpretation for Tidal Pond Network Analysis", *J. Waterw. Port Coastal Ocean Eng.*, 125(5), 267–270.

Graber, S. D. (2001). "Discussion of 'Submerged Hydraulic Jumps below Abrupt Expansions', by I. Ohtsu, Y. Yasuda, and M. Ishikawa", *J. Hyd. Eng.*, 127(1), 84.

Graber, S. D. (2004a). "Discussion of 'Factors Affecting Ballasted Flocculation Reactions', by J. C. Young and F. G. Edwards", *Water Environ. Res.*, 76(3), 284.

Graber, S. D. (2004b). "Collection Conduits Including Subsurface Drains", *J. Environ. Eng.*, 130(1), 67–80.

Graber, S. D. (2004c). "Concepts of Spatially Varied Flow", *Proc. HT-FED04, 2004 ASME Heat Transfer/Fluids Engineering Summer Conference*, July 11-15, 2004, Charlotte, North Carolina.

Graber, S. D. (2004d). "Difference Formulation and Slope Invariance in Spatially Varied Free Surface Flow", *Proc. HT-FED04, 2004 ASME Heat Transfer/Fluids Engineering Summer Conference*, July 11-15, 2004, Charlotte, North Carolina.

Graber, S. D. (2005). "Discussion of 'Computational Study of Particle-Eddy Interaction in Sedimentation Tanks', by S. Jayanti and S. Narayanan", *J. Environ. Eng.*, 131(6), 994.

Graber, S. D. (2006). "Asymmetric Flow in Symmetric Supercritical Expansions", *J. Hyd. Eng.*, 132(2), 207–213.

Graber, S. D. (2007a). "Full-Flowing Collection Conduits with Nonuniform Inflow", *J. Environ. Eng.*, 133(6), 575–580.

Graber, S. D. (2007b). "Hydraulic Considerations in Geosynthetic and Aggregate Subsurface Drains", *J. Environ. Eng.*, 133(9), 869–878.

Graber, S. D. (2009a). "Generalized Numerical Solution for Detention Basin Design", *J. Irrig. Drain. Eng.*, 135(4), 487–492.

Graber, S. D. (2009b). "Rain Loads and Flow Attenuation on Roofs", *J. Arch. Eng.*, 15(3), 91–101.

Graber, S. D. (2010a). "Generalized Method for Stormwater Pumping Station Design", *J. Hydrol. Eng.*, 15(11), 901–908.

Graber, S. D. (2010b). "Manifold Flow in Pressure-Distribution Systems", *J. Pipeline Syst. Eng. Pract.*, 1(3), 120–126.

Graber, S. D. (2010c). "Manifold Flow in Wastewater Pressure-Distribution Systems", *Proc. World Environmental and Water Resources Congress 2010*, ASCE Environmental and Water Resources Institute, May 16-20, 2010, Providence, Rhode Island.

Graber, S. D. (2011a). "Closure to 'Generalized Method for Storm-Water Pumping Station Design'", *J. Hydrol. Eng.*, 16(9), 761–762.

Graber, S. D. (2011b). "Discussion of 'Graphical Sizing of Small Single-Outlet Detention Basins in the Semiarid Southwest' by D. C. Froehlich", *J. Irrig. Drain. Eng.*, 137(5), 340–342.

Graber, S. D. (2012a). "Perspective on the Tidal Bore: Background and Initiation", *Proc. FEDSM2012: 2012 ASME Fluids Engineering Summer Meeting*, July 8-12, 2012, Rio Grande, Puerto Rico.

Graber, S. D. (2012b). "Perspective on the Tidal Bore: Channel Taper, Friction, Building, Maximum Height, Variation, Dissipation", *Proc. FEDSM2012: 2012 ASME Fluids Engineering Summer Meeting*, July 8-12, 2012, Rio Grande, Puerto Rico.

Graber, S. D. (2013a). "Numerical Investigation of Flow in Triangular Gutters", *J. Irrig. Drain. Eng.*, 139(2), 165–172.

Graber, S. D. (Corresponding Editor and Co-author) (2013b). *Standard Guidelines for the Design, Installation, Operation and Maintenance of Urban Subsurface Drainage*, American Society of Civil Engineers/ Environmental and Water Resources Institute 12/13/14-13, ANSI/ASCE/EWRI, Reston, Virginia.

Graber, S. D. (Corresponding Editor and Co-author) (2017a). *Standard Guidelines for the Design, Installation, Operation and Maintenance of Urban Stormwater Systems*, American Society of Civil Engineers/ Environmental and Water Resources Institute 45/46/47-16, ANSI/ASCE/EWRI, Reston, Virginia.

Graber, S. D. (Corresponding Editor and Co-author) (2017b). *Standard Guidelines for the Design, Installation, and Operation & Maintenance of Stormwater Impoundments*, American Society of Civil Engineers/ Environmental and Water Resources Institute 62/63/64-16, ANSI/ASCE/EWRI, Reston, Virginia.

Graber, S. D. (2018). "Discussion of 'Application of the SPH Method to Breaking and Undular Tidal Bores on a Movable Bed' by P. Nikeghbali and P. Omidvar", *J. Waterw. Port Coastal Ocean Eng.*, 145(3), 1.

Graber, S. D. (2023a). "Thermodynamic Concepts in Civil Engineering", *IOSR J. Mech. Civil Eng.*, 20(5, Ser. 2), 8–23.

Graber, S. D. (2023b). "Backwater in Circular Channels with Zero Slope", *IOSR J. Mech. Civil Eng.*, 20(6, Ser. 3), 1–7.

Graber, S. D. (2024a). "Commonsense Water Billing", *Int. J. Eng. Sci. Invent.*, 13(3), 35–40.

Graber, S. D. (2024b). "Outdoor Noise and its Mitigation", *Int. J. Eng. Sci. Invent.*, 13(5), 21–50.

Graber, S. D. (2024c). "Mitigation Engineering for Water Pollution Control Facilities", *Int. J. Eng. Sci. Invent.*, 13(7), 14–22.

Graber, S. D., and Elkerton, S. D. (1999). "Tidal Pond Network Analysis on Cape Cod", *J. Waterw. Port Coastal Ocean Eng.*, 125(5), 256–265.

Great Lakes – Upper Mississippi River Board of State and Provincial Public Health and Environmental Managers (2014). *Recommended Standards for Wastewater Facilities*, Health Research, Inc., Health Education Services Division, Albany, New York.

Great Lakes – Upper Mississippi River Board of State Sanitary Engineers (1971). *Recommended Standards for Sewage Works*, Health Education Service, Albany, New York.

Greeley, S. A., Stanley, W. E., and Hill, K.V. (1969a). "Sewage and Waste-Waters Treatment", Section 41 in *Handbook of Applied Hydraulics*, C. V. Davis and K. E. Sorensen (eds.), McGraw-Hill, New York.

Greeley, S. A., Stanley, W. E., and Newton, D. (1969b). "Sewage Quantities, Sewers, and Pumping Stations", Section 40 in *Handbook of Applied Hydraulics*, C. V. Davis and K. E. Sorensen (eds.), McGraw-Hill, New York.

Greve, F. W. (1921). "Parabolic Weirs", *Trans. Am. Soc. Civ. Eng.*, 84(1), 486–515.

Guins, V. G. (1968). "Flow Characteristics of Butterfly and Spherical Valves", *J. Hyd. Div.*, 94(3), 675–690.

Gullberg, J. (1997). "Graphic and Iterative Methods", Section 32.3 in *Mathematics from the Birth of Numbers*, W.W Norton & Company, New York.

Gulliver, J. S. (1988). "Discussion of 'Free-Surface Air Core Vortex', by A. Odgaard", *J. Hyd. Eng.*, 114(4), 447–449.

Gunn, R., and Kinzer, G. D. (1949). "The Terminal Velocity of Fall for Water Droplets in Stagnant Air", *J. Meteorol.*, 6, 243–248.

Guo, J. C. Y. (2000). "Street Storm Water Conveyance Capacity", *J. Irrig. Drain. Eng.*, 126(2), 119–123.

Guo, J. C. Y. (2006). "Design of Street Opening Inlets Using a Decay-Based Clogging Factor", *J. Hydraul. Eng.*, 132(11), 1237–1241.

Haberman, W. L., and Morton, R. K. (1956). "An Experimental Study of Bubbles Moving in Liquids", *Trans. Am. Soc. Civil Engr.*, 121(1), 227–252.

Haestad (2010). "Subcritical and Supercritical Flow", http://www.haestad.com/library/books/FMRAS/FloodplainOnlineBook/javascript/Floodplain%20with%20HEC-RAS-11-04.html, accessed February 18, 2010.

Hager, W. H. (1991). "Backwater Curves in Circular Channels", *J. Irrig. Drain. Eng.*, 117(2), 173–183.

Hager, W. H. (1999). *Wastewater Hydraulics*, Springer-Verlag, New York.

Hall, L. S. (1943). "Open Channel Flow at High Velocities", *Trans. Am. Soc. Civ. Eng.*, 108(1), 1394–1434.

Hannoun, I. A., Boulos, P. F., and List, E. J. (1998). "Using hydraulic modeling to optimize contact time", *J. Am. Water Works Assoc.*, 90(8), 77–87.

Hansen, R. J. (1972). "The Reduced Drag of Polymer Solutions in Turbulent and Transient Laminar Shear Flows", *Winter Annual Meeting, ASME*, Paper No. 72-WA/FE-7.

Hantush, M. S. (1967). "Growth and Decay of Groundwater-Mounds in Response to Uniform Percolation", *Water Resour. Res.*, 3(1), 227–234.

Harish, A. (2023). "When NASA Lost a Spacecraft Due to a Metric Math Mistake", https://www.simscale.com/blog/author/harish, Updated December 8, 2023.

Harleman, D. R. F. (1966). "Pollution in Estuaries", Chapter 14 in *Estuary and Coastline Hydrodynamics*, A. T. Ippen (ed.), McGraw-Hill, New York.

Harr, M. E. (1962). *Groundwater and Seepage*. McGraw-Hill Book Company, New York.

Harris, G. S. (1970). "Real Time Routing of Flood Hydrographs in Storm Sewers", *J. Hyd. Div*, 96(6), 1247–1260.

Harris, H. S., Kaufman, W. J., and Krone, R. B. (1966). "Orthokinetic Flocculation in Water Purification", *J. San. Eng. Div.*, 92(6), 95–114.

Hart, J., Cotter, L., and Vine, J. (2011). "Polymer Addition to Increase Trunk Sewer Flow Capacity at the Resort Municipality of Whistler during the 2010 Winter Olympic Games", *Proc. Water Environment Federation*, No. 10, pp. 5763-5774.

Hatsopoulus, G. N., and Keenan, J. H. (1965). *Principles of General Thermodynamics*, John Wiley and Sons, Inc., New York.

Hecht, J. M. (2003). *Doubt: A History*, HarperSanFrancisco, San Francisco, pp. 318–319.

Hedström, B. O. A. (1952). "Flow of Plastics Materials in Pipes", *Ind. and Eng. Chem.*, 44(3), 651–656.

Helm, R. B., Hewitt, G., and Good, D. R. (1994). "Improving the Illinois", *Water Environ. Technol.* 6(6), 44–53.

Hibbs, D. E., Gulliver, J. S., and McDonald, J. P. (1995). "Methane Sampling Technique and the Measurement of Plunge Pool Impact on Gas Transfer Rates at Low-Head Hydraulic Structures", U.S. Department of the Army, Corps of Engineers, Miscellaneous Paper W-95-2, Washington, D.C.

Hildebrand, F. B. (1952). *Methods of Applied Mathematics*, Prentice-Hall, Inc., Englewood Cliffs, New Jersey.

Hildebrand, F. B. (1956). *Introduction to Numerical Analysis*, McGraw-Hill Book Company, New York.

Hildebrand, F. B. (1962). *Advanced Calculus for Applications*, Prentice-Hall, Inc., Englewood Cliffs, New Jersey.

Hinds, J. (1926). "Side Channel Spillways: Hydraulic Theory, Economic Factors, and Experimental Determination of Losses", *Trans. Am. Soc. Civil Eng.*, 89(1), 881–939.

Hinds, J. (1928). "The Hydraulic Design of Flume and Siphon Transitions", *Trans. Am. Soc. Civil Eng.*, 92(1), 1433–1459.

Hoff, J. W., and Libke, A. W. (1995a). "Performance Factors and Installation Procedures for AWWA Butterfly Valves", *Water Eng. Manage.*, 142(11), 17–19.

Hoff, J. W., and Libke, A. W. (1995b). "Performance Factors and Installation Procedures for AWWA Butterfly Valves: Part II", *Water Eng. Manage.*, 142(12), 29–32.

Hogan, W. T., Reed, F. E., and Starbird, A. W. (1970). *Optimum Mechanical Aeration Systems For Rivers And Ponds*, Submitted to Water Quality Office, Environmental Protection Agency, EPA Program #16080 DOO, Contract #14-12-576, Littleton Research and Engineering Corp., Littleton, Massachusetts, November 1970.

Holley, E. R. (1969). "Unified View of Diffusion and Dispersion", *J. Hyd. Div.*, 95(2), 621–631.

Holley, E. R., and Harleman, D. R. F. (1965). *Dispersion of Pollutants in Estuary Type Flows*, Report No. 74, Hydrodynamics Laboratory, Massachusetts Institute of Technology, Cambridge, Massachusetts, January 1965.

Holley, E. R., Harleman, D. R. F., and Fischer, H. B. (1970). "Dispersion in Homogeneous Estuary Flow", *J. Hyd. Div.*, 96(8), 1691–1709.

Hornbeck, R. W. (1975). Roots of Equations, Chapter 5 in *Numerical Methods*, Prentice-Hall/Quantum, Englewood Cliffs, New Jersey.

Hornung, H. G., Willert, C., and Turner, S. (1995) "The Flow Field Downstream of a Hydraulic Jump", *J. Fluid Mech.*, 287, 299–316.

Howe, J. W. (1950). "Flow Measurement", Chapter III in *Engineering Hydraulics*, H. Rouse (ed.), John Wiley & Sons, Inc., New York.

Hross, M. (2019). "Flow Distribution Improvements at the Stamford Water Pollution Control Facility", *NEWEA J.*, 53(2), 26–31.

Huckabay, H. K., and Keller, A. G. (1970). "Aeration on Inclined Transversely Corrugated Solid Surfaces", *J. WPCF*, 42(5, part 2), R202–R208.

Hudson, H. E. (1959). "Declining-Rate Filtration", *J. Am. Water Works Assoc.*, 51(11), 1455–1469.

Hudson, H. E., Bailey, R. W., and Uhler, R. B. (1979). "Dividing-Flow Manifolds with Square-Edged Laterals", *J. Environ. Eng. Div.*, 105(4), 745–755.

Hughmark, G. A. (1961). "Aqueous Transport of Settling Slurries", *Ind. Eng. Chem.*, 53(1), 389–390.

Hung, C.-Y. (1971). "Discussion of 'Computer Simulation of Urban Storm Water Runoff', by C. W. Shen and R. P. Shubinski", *J. Hyd. Div.*, 97(11), 1918–1920.

Hunt, B. E. (1956). "Gas Service Design – Final Report of Task Group", (DMC-56-114), *Proc. American Gas Association*, Washington, D.C., pp. 272-287.

Hunter, J. V., and Whipple, W., Jr. (1970). "Evaluating Instream Aeration of Polluted Rivers", *J. WPCF*, 42(8, Part 2), R249–R262.

Hurd, C. E. (1923). "Design Features of the Indianapolis Activated-Sludge Plant", *Eng. News-Record*, 91(7), 258–262.

Hutchinson, G. E. (1975). A Treatise on Limnology, Vol. I, Part 1, John Wiley & Sons, Inc., New York.

Hynes, H. B. N. (1972). *The Ecology of Running Waters*, University of Toronto Press, Toronto, Canada.

Iamandi, C., and Rouse, H. (1969). "Jet Induced Circulation and Diffusion", *J. Hyd. Div.*, 95(2), 589–601.

Ingersoll, A. C., McKee, J. E., and Brooks, N. H. (1956). "Fundamental Concepts of Rectangular Settling Tanks", *Trans. Am. Soc. Civil Eng.*, 121(1), pp. 1179–1218.

Ippen, A. T. (1950). "Channel Transitions and Controls", Chapter VIII in *Engineering Hydraulics*, H. Rouse (ed.), John Wiley & Sons, Inc., New York.

ITT Grinnell Corporation (2008). *Piping and Pipe Hanger Design and Engineering*, ITT Grinell Corporation, Providence, Rhode Island.

Ives, K. J., and Bhole, A. G. (1973). "Theory of Flocculation for Continuous Flow Systems", *J. Environ. Eng. Div.*, 99(1), 17–34.

Ives, K. J., and Bhole, A. G. (1977). "Study of Flowthrough Couette Flocculators-II. Laboratory Studies of Flocculation Kinetics", *Water Res.*, 11(2), 209–215.

Izzard, C. F. (1946). "Hydraulics of Runoff from Developed Surfaces", *Proc. Hwy. Res. Board*, 26, 129–150.

Jacob, C. E. (1950). "Flow of Groundwater", Chapter V in *Engineering Hydraulics*, H. Rouse (ed.), John Wiley & Sons, Inc., New York.

Jain, S. C. (2001). *Open-Channel Flow*. John Wiley & Sons, Inc., New York, pp. 2-3, 16-19, 24-26, 50-57, 187-188.

Jarrett, A. R., Hunt, W. F., and Berghage, R. D. (2006). *Annual and Individual-Storm Green Roof Stormwater Response Models*, Paper number 062310, Sponsored by ASABE, St. Joseph, MI, 2006 ASABE Annual International Meeting, Portland, Oregon.

Jarrett, R. D. (1984). "Hydraulics of High-Gradient Streams", *J. Hydraul. Eng.*, 110(11), 1519–1539.

Jarrett, R. D. (1992). "Hydraulics of Mountain Rivers", in *Channel Flow Resistance: Centennial of Manning's Formula*, B. C. Yen (ed.), Water Resources Publications, Littleton, Colorado, pp. 287–298.

Jens, S. W., and McPherson, M. B. (1964). "Hydrology of Urban Areas", Section 20 in *Handbook of Applied Hydrology*, V.T. Chow (ed.), McGraw-Hill Book Company, New York.

Jeppson, R. W. (1976). *Analysis of Flow in Pipe Networks*, Ann Arbor Science (Butterworth), Woburn, Massachusetts.

Jin, Y-C., Guo, Q.-C., and Viraraghavan, T. (2000). "Modeling of Class I Settling Tanks", *J. Environ. Eng. Div.*, 126(8), 754–760.

Jobin, W. (1999). *Dams and Disease: Ecological Design and Health Implications of Large Dams, Canals and Irrigation Systems*, E & FN Spon, New York.

Jobin, W. R., Laracuente, A., Mercado, R., and Negron-Aponte, H. (1984). "Critical Water Velocity for Snail Habitats in Canals", *J. Environ. Eng.*, 110(1), 279–282.

Johnson, E. E. (1966). *Ground Water and Wells*, Edward E. Johnson, Inc., St. Paul, Minnesota, Chapter 5.

Johnstone, R. E., and Thring, M. W. (1957). *Pilot Plants, Models, and Scale-Up Methods in Chemical Engineering*, McGraw-Hill Book Company, New York.

Jones, D. E. (1971). "Where is Urban Hydrology Practice Today", *J. Hyd. Div.*, ASCE, Proc. Paper No. 7917, February 1971.

Jones, G. M. (2002). "Recent Improvements to Trench-Type Self-Cleaning Wet Wells", *Proc. Water Environment Federation*, WEFTEC 2002.

Jones, G. M., and Sanks, R. L. (2008). *Pumping Station Design*, 3rd Ed., Butterworth-Heinemann, Waltham, Massachusetts.

Jorinsen, A. L. (1956). "A New Development in Flow Measurement: The Dall Flow Tube", *J. Hyd. Div.*, 82(4), 1–12.

Joseph, I., and Hamill, F. A. (1972). "Start-Up Pressures in Short Pump Discharge Lines", *J. Hyd. Div.*, 98(7), 1117–1125.

Joshua, 10:12-13; Isaiah 38:7-8, Ecclesiastes 1:5; as cited by Aczel, *op cit.*, pp, 10-11, 251.

Jumpeter, A. M. (1976). "Jet Pumps", Section 4 in *Pump Handbook*, I. J. Karassik, W. C. Krutzsch, W. H. Fraser, J. P. Messina (eds.), McGraw-Hill, New York.

Kalinske, A. A. (1945). "Discussion of 'Coefficients for Velocity Distribution in Open-Channel Flow,' by W. S. Eisenlohr, Jr.", *Trans. Am. Soc. Civil Eng.*, 110, 645–646.

Kalinske, A. A. (1965). "Evaluation of Oxygenation Capacity of Localized Aerators", *J. WPCF*, 37(11), 1521–1529.

Kalinske, A. A. (1968). "Surface Aerators for Absorption and Desorption of Gases into Water and Liquid Wastes", *Water Sew. Works*, 115(1), 33–37.

Kalinske, A. A., and Robertson, J. M. (1943). "Entrainment of Air in Flowing Water: Closed Conduit Flow", *Trans. Am. Soc. Civil Eng.*, 108(1), 1435–1447.

Kalkwijk, J. P. Th., and Kranenburg, C. (1971). "Cavitation in Horizontal Pipelines Due to Water Hammer", *J. Hyd. Div.*, 97(10), 1585–1605.

Kaltenbach, A. B. (1963). "Storm Sewer Design by the Inlet Method", *Pub. Works*, 94(1), 86–89.

Karassik, I. J., Krutzsch, W. C., Fraser, W. H., and Messina, J. P. (eds.) (1976). *Pump Handbook*, McGraw-Hill, New York.

Kawamura, S., Najm, I. N., and Gramith, K. (1997). "Modifying a Backwash Trough to Reduce Media Loss", *J. Am. Water Works Assoc.*, 89(12), 47–60.

Keenan, J. H., and Kaye, J. (1948). *Gas Tables*, John Wiley & Sons, Inc., New York.

Keenan, J. H., and Keyes, F. G. (1936). *Thermodynamic Properties of Steam*, John Wiley & Sons, Inc., New York.

Keifer, C. J., and Chu, H. H. (1955). "Backwater Functions by Numerical Integration", *Trans. Am. Soc. Civ. Engr.*, 120(1), 429–448.

Kennison, H. F. (1956). "Surge-Wave Velocity – Concrete Pressure Pipe", *Am. Soc. Mech. Eng.*, 78, 1323–1327.

Kennison, K. R. (1943). "Discussion of 'Entrainment of Air in Flowing Water': A Symposium", *Proc. ASCE*, 69(2), 306.

Keulegan, G. H. (1950). "Wave Motion", Chapter XI in *Engineering Hydraulics*, H. Rouse (ed.), John Wiley & Sons, Inc., New York. [See especially Section F.18 – Intermittent Surges on Steep Slopes.]

Khiadani, M. H., Beecham, S., Kandasamy, J., and Sivakumar, S. (2005). "Boundary Shear Stress in Spatially Varied Flow with Increasing Discharge", *J. Hydraul. Eng.*, 131(8), 705–714.

Khiadani, M. H., Kandasamy, J., and Beecham, S. (2007). "Velocity Distributions in Spatially Varied Flow with Increasing Discharge", *J. Hydraul. Eng.*, 133(7), 721–735.

King, H. R. (1944). "Design of Air Distribution Systems for Activated Sludge Plants", *Water Works & Sewage*, 91(R-205).

Kinno, H. (1968). "Water-Hammer Control in Centrifugal Pump Systems", *J. Hyd. Div.*, 94(3), 619–640.

Kinno, H., and Kennedy, J. F. (1965). "Water Hammer Charts for Centrifugal Pump Systems", *J. Hyd. Div.*, 91(3), 247–270.

Kinshella, P., Sanks, R. L., Richardson, A. W., Conway, K. D., and Cowee, J. D. (2005). "A Model Operation", *Water Environ. Technol.*, 17(6), 61–66.

Kirshner, J. M. (undated). "*Kinetic Theory of Gases*", Catholic University of America Seminar Lecture Notes.

Kittredge, C. P. (1976). "Centrifugal Pump Performance", Section 2.3 in *Pump Handbook*, I. J. Karassik, W. C. Krutzsch, W. H. Fraser, and J. P. Messina (eds.), McGraw-Hill, New York.

Kleinschmidt, R. S. (1961). "Hydraulic Design of Detention Tanks", *J. Boston Soc. Civil Eng.*, 48(4), 247–294.

Klingel, G. (1999). "The Marvel of a Tide, from Inagua", in *The Seacoast Reader*, J. A. Murray (ed.), Nature Conservancy Book, The Lyons Press, Guilford, Connecticut.

Kobus, H. E. (1968) "Analysis of the Flow Induced by Air-Bubble Systems", Chapter 65, *Proc. 11th Conference on Coastal Engineering*, Sponsored by the Coastal Engineering Research Council of ASCE, London, United Kingdom.

Kobus, H. E. (1970). "Discussion of 'Jet Induced Circulation and Diffusion', by C. Iamandi and H. Rouse", *J. Hyd. Div. ASCE*, 96(1), 280–285.

Koerner, R. M. (2005). *Designing with Geosynthetics*, 5th Ed., Pearson Prentice-Hall, Upper Saddle River, New Jersey.

Kohler, W. H., and Ball, J. W. (1969). "High-Pressure Outlets, Gates, and Valves", Section 22 in *Handbook of Applied Hydraulics*, C. V. Davis and K. E. Sorensen (eds.), 3rd Ed., McGraw-Hill Book Company, New York.

Kolf, R. C., and Zielinski, P. B. (1959). "The Vortex Chamber as an Automatic Flow Control Device", *J. Hyd. Div.*, 85(12), 1–8.

Kormanik, R. A. (1969). "Simplified Mathematical Procedure for Designing Post-Aeration Systems", *J. WPCF*, 41(11, Part 1), 1956–1958.

Kormanik, R. A. (1970). "Design of Plug-Flow Post-Aeration Basins", *J. WPCF*, 42(11), 1922–1931.

Kothandaraman, V., and Evans, R. L. (1972). "Hydraulic Model Studies of Chlorine Contact Tanks", *J. WPCF*, 44(4), 625–633.

Kozeny, J. (1927). "Ueber kapillare Leitung des Wasser sim Boden", *Sitzungsber Akad Wiss, Wien*, 136(2a), 271–306.

Kratch, K. (1998). "Bends in the Channel Remove Grit", *Water Environ. Technol.*, 10(8), 43–44.

Kriz, G. J., Scott, V. H., and Burgy, R. H. (1966). "Graphical Determination of Aquifer Parameters", *J. Hyd. Div.*, 92(5), 39–48.

Kuethe, A. M., and Schetzer, J. D. (1959). *Foundations of Aerodynamics*, 2nd Ed., John Wiley & Sons, Inc., New York.

Kurkjian, G. A. (1978). "Proper Butterfly Application Avoids Costly Repairs", *Water Sew. Works*, 125(6), 84–89.

Lager, J. A., and Smith, W. G. (1974). *Urban Stormwater Management and Technology: An Assessment*, EPA-670/2-74-040, National Environmental Research Center, Cincinnati, Ohio.

Lakeside Equipment Corporation (2018). Advertisement in *Water Environment and Technology*, June 2018.

Lamb, H. (1945). *Hydrodynamics*, 6th Ed., Dover Publications, New York.

Lane, E. W. (1955). "Design of Stable Channels", *Trans. Am. Soc. Civil Eng.*, 120(1), 1234–1260.

Langmuir, I. (1938). "Surface Motion of Water Induced by Wind", *Science*, 87, 119–123.

Laquidara, M., Garrity, G., Tyler, C., Waitt, W., and White, N. (1998). "Water Testing the Secondary Treatment Facilities at MWRA's Deer Island Wastewater Treatment Plant", *NEWEA J.*, 32(2), 126–137.

Larson, J. S. (1973). *A Guide to Important Characteristics and Values of Freshwater Wetlands in the Northeast*, Publication No. 31, Completion Report FY 74-1A, Water Resources Research Center, University of Massachusetts at Amherst.

Law, S. W., and Reynolds, A. J. (1966). "Dividing Flow in An Open Channel", *J. Hyd. Div.*, 92(2), 207–231.

Lawler, E. A. (1964). "Hydrology of Flow Control: Part II. Flood Routing", Section 25-II in *Handbook of Applied Hydrology*, V. T. Chow (ed.), McGraw-Hill Book Company, New York.

Leary, R. D., Ernest L. A., and Katz, W. J. (1969). "Full-Scale Oxygen Transfer Studies of Seven Diffuser Systems", *J. WPCF*, 41(3), 459–473.

Lee, M., and Babbitt, H. E. (1946). "Constant Velocity Grit Chamber with Parshall Flume Control", *Sew. Works J.*, 18(4), 646–650.

Lefkowitz, I. (1958). "Instruments", Section 16 in *Mark's Mechanical Engineers' Handbook – Sixth Edition*, T. Baumeister (ed.), McGraw-Hill Book Company, New York.

Leitch, J. K. (1988). "The Pressure to Change: Recent Technical Innovations in the Structural Design of Archimedes Screw Pumps", *J. New Engl. Water Poll. Contr. Assoc.*, 22(2), 129–139.

Lemke, A. A. (1952). "Flow of Air in Pipes", *Sew. Ind. Wastes*, 24(1), 24–37.

Lemke, A. A. (1969). Personal Communication, Supervisor Process Engineering Department, FMC Corporation, Hydrodynamics Division, Chicago Pump, Chicago, Illinois, December 31, 1969.

Lenhart, C. F., and McGarry, J. A. (1984). "Screening System Makes Downstream Life Simpler", *Water Eng. Manage.*, 1984, 38–41.

Levi, E. (1972). "Experiments on Unstable Vortices", *J. Eng. Mech. Div.*, 98(3), 539–559.

Levi, E. (1995). *The Science of Water: The Foundation of Modern Hydraulics*, ASCE Press, New York.

Li, W.-H. (1955). "Open Channels with Nonuniform Discharge", *Trans. Am. Soc. Civil Eng.*, 120(1), 255–274.

Lilly, J. G. (2007). "Waterfall Noise", *154th Meeting*, Acoustical Society of America, New Orleans, Louisiana, November 28, 2007.

Linsley, R. K., and Franzini, J. B. (1964). *Water Resources Engineering*, McGraw-Hill Book Company, New York.

Linsley, R. K., Kohler, M. A., and Paulhus, J. L. H. (1982). *Hydrology for Engineers*, McGraw-Hill Book Company, New York.

Liou, C. P. (1998). "Limitations and Proper Use of the Hazen-Williams Equation", *J. Hydraul. Eng.*, 124(9), 951–954.

Little, C. W. (1976). "Rotary Pumps", Section 3.4 in *Pump Handbook*, I. J. Karassik, W. C. Krutzsch, W. H. Fraser, and J. P. Messina (eds.), McGraw-Hill, New York.

Littlefield, A. (2010). "Man Says Water Bill Doesn't Wash." *Boston Globe*, November 2.

Lopes, V. L., and Shirley, E. D. (1993). "Computation of Flow Transitions in Open Channels with Steady Uniform Lateral Inflow", *J. Irrig. Drain. Eng.*, 119(1), 187–200.

Louie, D. S., and Fohrman, M. S. (1968). "Hydraulic Model Studies of Chlorine Mixing and Contact Chambers", *J. WPCF*, 40(2), 174–184.

Lugt, H. J. (1983). *Vortex Flow in Nature and Technology*, John Wiley & Sons, Inc., Somerset, New Jersey.

Lutge, T. V. (1969). "Hydraulic Control Utilizing Submerged Effluent Collectors", *J. WPCF*, 41(8, Part I), 1451–1455.

Lye, D. J. (2002). "Health Risks Associated with Consumption of Untreated Water from Household Roof Catchment Systems", *J. Am. Water Resour. Assoc.*, 38(5), 1301–1306.

Macias, R. (2006). "San Antonio Goes With the Flow", *Water Environ. Technol.*, 18(4), 54–56.

Marlow, S. (2011). Personal Communication, Technical Sales Manager, American Wick Drain Corporation, Monroe, North Carolina, January 5, 2011.

Maidment, D. R. (ed.) (1993). *Handbook of Hydrology*, McGraw-Hill Book Company, New York.

Mangelson, K. A., and Watters, G. Z. (1972). "Treatment Efficiency of Waste Stabilization Ponds", *J. San. Eng. Div.*, 98(2), 407–425.

Marchal, M. G., Flesh, G., and Suter, P. (1965). "The Calculation of Waterhammer Problems by Means of the Digital Computer", *Proc. International Symposium on Waterhammer Pumped Storage Projects*, ASME, Chicago, Illinois, November 1965.

Marks, C. H. (1972). "Measurements of the Terminal Velocity of Bubbles Rising in a Chain", *Winter Annual Meeting*, Am. Soc. Mech. Eng., Paper No. 72-WA/FE-24.

Marske, D. M., and Boyle, J. D. (1973). "Chlorine Contact Chamber Design – A Field Evaluation", *Water Sew. Works*, 120(1), 70–77.

Massachusetts Department of Energy and Environmental Protection (MDEEP) (1997). *Massachusetts Stormwater Management Policy*, November 18, 1996, minor revisions, March 1997.

Massachusetts Water Resources Authority (MWRA) (2009). "Sewer Use", 10.008(1) Monitoring Devices, August 7, 2009.

Mastropietro, M. A. (1968). "Effects of Dam Reaeration on Waste Assimilation Capacities of the Mohawk River", *Proc. 23rd Industrial Waste Conference*, Purdue University, Part I, p. 754.

Mavis, F. P., Neill, C. R., and Hallmark, D. E. (1966). "Discussion of 'Flow in Culverts and Related Design Philosophies'", *J. Hyd. Div.*, 92(5), 261–268.

Mays, L. W. (2011). *Water Resources Engineering*, 2nd Ed., John Wiley & Sons, Inc., Hoboken, NJ.

McCuen, R. H. (1982). *A Guide to Hydrologic Analysis Using SCS Methods*, Prentice-Hall, Inc., Englewood Cliffs, New Jersey.

McDonald, J. E. (1958). "Centrifugal and Axial Fans", Section 14 in *Mark's Mechanical Engineers' Handbook – Sixth Edition*, T. Baumeister (ed.), McGraw-Hill Book Company, New York.

McEnroe, B. M. (1992). "Preliminary Sizing of Detention Reservoirs to Reduce Peak Discharges", *J. Hydraul. Eng.*, 118(11), 1540–1549.

McEnroe, B. M. (1993). "Maximum Saturated Depth Over Landfill Liner", *J. Environ. Eng.*, 119(2), 262–270.

McEnroe, B. M., Wade, R. P., and Smith, A. K. (1999). *Hydraulic Performance of Curb and Gutter Inlets*, Report No. K-TRAN: KU-99-1, Kansas Department of Transportation, Topeka, Kansas, p. 8.

McMonagle, T. C., and Otoski, R. M. (1985). "Toward a Healthy Harbor", *Civil Eng., ASCE*, 67(9), 46–49.

McNamara, B., Sherony, M., and Herrick, P. (2013). "Relative Performance of Grit Removal Systems", *NEWEA J.*, 47(3), 42–48.

McNown, J. S. (1950). "Surges and Water Hammer", Chapter VII in *Engineering Hydraulics*, H. Rouse (ed.), John Wiley & Sons, Inc., New York.

McNown, J. S. (1954). "Mechanics of Manifold Flow", *Trans. Am. Soc. Civil Eng.*, 119(1), 1103–1142.

McNown, J. S., and Hsu, E.-Y. (1951). "Application of Conformal Mapping to Divided Flow", *Proc. Midwestern Conference on Fluid Dynamics*, J. W. Edwards, Ann Arbor, Michigan, pp. 143-155.

McWhirter, J. R. (1966). "Application of Aeration Concepts to Lagoons", *Proc. 3rd Annual San. Eng. Conference*, University of Missouri, Columbia, Missouri, November 9, 1966.

Mehta, P. R. (1979). "Flow Characteristics in Two-Dimensional Expansions", *J. Hyd. Div.*, 105(5), 501–516.

Menon, E. S. (2005). *Piping Calculations Manual*, McGraw-Hill Book Company, New York.

Metcalf & Eddy, Inc. (1972). *Wastewater Engineering: Collection, Treatment, Disposal*, McGraw-Hill Book Company, New York.

Michalos, J. (1972). "Hardy Cross and the Rise of Numerical Methods", *J. Struct. Div.*, 98(3), 691–694.

Mignot, E., Riviere, N., Lefevre, A., and Quillien, B. (2019). "Smoother Than Smooth: Increasing the Flow Conveyance of an Open-Channel Flow by Using Drag Reduction Methods", *J. Hyd. Eng.*, 145(4), 1–8.

Mitsch, W. J., and Gosselink, J. G. (2000). *Wetlands*, 3rd Ed., John Wiley and Sons, Inc., New York, Chapter 16 – Values and Valuation of Wetlands.

Moody, L. F. (1944). "Friction Factors for Pipe Flow", *Transactions*, 66(8), 671–684.

Mooney, D. A. (1953). *Mechanical Engineering Thermodynamics*, Prentice-Hall, Inc., Englewood Cliffs, New Jersey.

Mooney, D. A. (1959). *Introduction to Thermodynamics and Heat Transfer*, Seventh Printing, Prentice-Hall, Inc., Englewood Cliffs, New Jersey.

Morabito, A., Spriet, J., Vagnoni, E., and Hendrick, P. (2020). "Underground Pumped Storage Hydropower Case Studies in Belgium: Perspectives and Challenger", *Energies*, 13, 1–24.

Moskowitz, C. (2021). "Dynamic Planet", *Sci. Am.*, 324(1), 68–75.

Mulbarger, M. C., Copas, S. R., Kordic, J. R., and Cash, F. M. (1981). "Pipeline Friction Losses for Wastewater Sludge", *J. Water Poll. Control Fed.*, 53(8), 1303–1313.

Murakami, H., Katayama, H., and Matsuura, H. (2001). "Pipe Friction Head Loss in Transportation of High-Concentration Sludge for Centralized Solids Treatment", *Water Environ. Res.*, 73(5): 558–566.

Murphy, K. L. (1963). "Tracer Studies in Circular Sedimentation Basins", *Proc. 18th Industrial Waste Conference, Purdue University, Lafayette, Indiana, pp. 374–390.*

Murphy, K. L., and Boyko, B. I. (1970). "Longitudinal Mixing in Spiral Flow Aeration Tanks", *J. San. Eng. Div.*, 96(2), 211–221.

Murphy, K. L., and Timpany, P. L. (1967). "Design and Analysis of Mixing for an Aeration Tank", *J. San. Eng. Div.*, 93(5), 1–15.

Murphy, K. L., and Wilson, A. W. (1974). "Characterization of Mixing in Aerated Lagoons", *J. Environ. Eng.*, 100(5), 1105–1117.

Nalluri, C., and Tomlinson, J. H. (1978). "Varied Flow Functions for Circular Channels", *J. Hyd. Div.*, 104(7), 983–1000.

Naot, D., and Rodi, W. (1982). "Calculation of Secondary Currents in Channel Flow", *J. Hyd. Div.*, 108(8), 948–968.

Nascimento, L. P., Silva, J. B. C., and Di Giunta, V. (2006). "Damage of Hydroelectric Power Plant Trash-Racks Due to Fluid-Dynamic Exciting Frequencies", *Latin Am. J. Solids Struct.*, 3, 223–243.

Natarius, E. M. (2001). "Odor and Corrosion Control: Insert Assembly Improves Sewer Drops", *Water Eng. Manage.*, 25, 14–15.

National Committee for Fluids Mechanics Films (1972). *Illustrated Experiments in Fluid Mechanics*, The MIT Press, Cambridge, Massachusetts.

National Cooperative Highway Research Program (NCHRP) (2004). *Guide for Mechanistic-Empirical Design of New and Rehabilitated Pavement Structures – Appendix TT: Drainage Requirement in Pavements (DRIP) Microcomputer Program User's Guide*. National Research Council, Washington, D.C.

National Council for Stream Improvement (NCSI) (1969). *Artificial Aeration of Receiving Waters*, Technical Bulletin No. 229, May 1969.

Natural Resources Conservation Service (NRCS). (2007). "Conservation Practice Standard – Runoff Management System – Code 570", Washington, D.C., ftp://ftp.fc.sc.egov.usda.gov/NHQ/practice-standards/standards/570.doc, accessed August 1, 2007.

Neighbor, J. B., and Cooper, T. W. (1965). "Design and Operation Criteria for Aerated Grit Chambers", *Water Sew. Works*, 112(12), 448–454.

Nepf, H. M., and Monismith, S. G. (1991). "Experimental Study of Wave-induced Longitudinal Vortices", *J. Hyd. Eng.*, 117(12), 1639–1649.

Newton, I. (1999). *The Principia: Mathematical Principles of Natural Philosophy, A New Translation*, by I. B. Cohen and A. Whitman, University of California Press, Berkeley, California.

Nickell, J. (2003). "Legend of the Lake Champlain Monster", *Skeptical Inquirer*, 27(4), 18–23.

Nimmo, W. H. R. (1928). "Discussion of 'The Hydraulic Design of Flume and Siphon Transitions', by J. Hinds", *Trans. Am. Soc. Civil Eng.*, 92, 1472–1476.

Nürnberg, G. K. (1987). "Hypolimnetic Withdrawal as Lake Restoration Technique", *J. Environ. Eng.*, 113(5), 1006–1017.

O'Brien, J. E. (1995). "Reducing Corrosivity and Radon by the Venturi-Aeration Process", *J. New Engl. Water Works Assoc.*, CIX(2), 105–114.

O'Connor, L. (1991). "Vortex Meters: High-Accuracy Flow Measurement", *Mechanical Engineering*, 113(10), 46–49.

O'Connor, T. P., and Amin, M. (2014). "Rainwater Collection and Management from Roof at the Edison Environmental Center", *J. Sustain. Water Built Environ.*, 1(1), 1–9.

Odgaard, A. J. (1986). "Free-Surface Air Core Vortex", *J. Hyd. Eng.*, 112(7), 610–620.

Odgaard, A. J. (1988). "Closure to 'Free-Surface Air Core Vortex'", *J. Hyd. Eng.*, 114(4), 451–452.

Ogawa, H., and Male, J. W. (1983). *The Flood Mitigation Potential of Wetlands*, Publication No. A-136, Completion Report Project No. A-136, Water Resources Research Center, University of Massachusetts at Amherst.

Ogawa, H., and Male, J. W. (1986). "Simulating the Flood Mitigation Role of Wetlands", *J. Water Resour. Plan. Manage.*, 112(1), 114–128.

Olson, J. H., and Stout, L. E., Jr. (1966). "Mixing and Chemical Reactions", Section 7 in *Mixing Theory and Practice*, Vols. II, V. W. Uhl and J. B. Gray (eds.), Academic Press, New York.

Ongerth, J. E. (1979). *Evaluation of Flow Equalization in Municipal Wastewater Treatment, Office of Research and Development*, U.S. Environmental Protection Agency, Cincinnati, Ohio.

Ordon, C. J. (1966). "Manifolds, Rotating and Stationary", *J. San. Eng. Div.*, 92(1), 269–280.

Ordon, C. J. (1968). "Discussion of 'Dispersion Conduits', by T. R. Camp and S. D. Graber", *J. San. Eng. Div.*, 94(4), 762–764.

Orenstein, G. S. (1984). "Instant Expertise: A Danger of Small Computers", *Civil Eng., ASCE*, 66(6), 50–51.

Otis, R. J. (1981). *Design of Pressure Distribution Networks for Septic Tank-Soil Absorption Systems*, Small Scale Waste Management Project, University of Wisconsin, Madison, Wisconsin.

Otis, R. J. (1982). "Pressure Distribution Design for Septic Tank Systems", *J. Environ. Eng. Div.*, 108(1), 123–140.

Ottengraf, S. P. P., and Rietema, K. (1969). "The Influence of Mixing on the Activated Sludge Process in Industrial Aeration Basins", *J. Water Control Fed.*, 41(8, Part 2), R282–R293.

Overfield, J. L., Crawford, H. R., Baxter, J. K., Harrington, L. J., and Santry, I. W. Jr. (1969). "Increasing Wastewater Flow Velocity by Using Chemical Additives", *J. Water Control Fed.*, 41(9), 1570–1585.

Palmer, A. C. (2008). *Dimensional Analysis and Intelligent Experimentation*, World Scientific Publishing Company Pte. Ltd., Hackensack, New Jersey.

Parker, D. S. (1991). "The Case for Circular Clarifiers' and 'Rebuttal to Rectangular Clarifiers Should be Considered'", *Water Eng. Manage.*, 1991, 22–25.

Parker, D. S., Kaufman, W. J., and Jenkins, D. (1971). "Physical Conditioning of Activated Sludge Floc", *J. Water Control Fed.*, 43(9), 1817–1833.

Parker, D. S., Kaufman, W. J., and Jenkins, D. (1972). "Floc Breakup in Turbulent Flocculation Processes", *J. San. Eng. Div.*, 98(1), 79–99.

Parmakian, J. (1950). "Air Inlet Valves for Steel Pipe Lines", *Trans. Am. Soc. Civil Eng.*, 115(1), 438–443.

Parmakian, J. (1963). *Waterhammer Analysis*, Dover Publications, Inc., New York.

Parmakian, J. (1976). "Water Hammer", Section 9.4 in *Pump Handbook*, I. J. Karassik, W. C. Krutzsch, W. H. Fraser, J. P. Messina (eds.), McGraw-Hill, New York.

Parr, A. D., Judkins, J. F., and Jones, T. E. (1981). "Point-Velocity Discharge Measurement Method for Sewers", *J. WPCF*, 53(1), 113–118.

Parshall, R. L. (1926). "The Improved Venturi Flume", *Trans. Am. Soc. Civil Eng.*, 89, 841–851.

Pastorok, R. A., Lorenzen, M. W., and Finn, T. C. (1982). *Environmental Aspects of Artificial Aeration and Oxygenation of Reservoirs: A Review of Theory, Techniques, and Experiences*, Technical Report E-82-3, Waterways Experiment Station, Vicksburg, Mississippi.

Patel, T., Gill, L., and Faram, G. (2011). "Grit Removal from Wastewater Using Secondary Currents in Open-Channel Flow around Bends", *J. Environ. Eng.*, 137(11), 1026–1039.

Paynter, H. M., and Free, J. G. (1966). "Discussion of 'Water-Hammer Charts for Centrifugal Pump System', by H. Kinno and J. F. Kennedy", *J. Hyd. Div.*, 92(3), 379–382.

Pazwash, H. (1984). "Reduction of Friction Drag in Pipes by Polymers: A Review", *J. Transp. Eng.*, 110(2), 223–234.

Pedersen, F. B., and Mark, O. (1990). "Head Losses in Storm Sewer Manholes: Submerged Jet Theory", *J. Hyd. Eng.*, 116(11), 1317–1328.

Pérez, R., Martinez, F., and Vela, A. (1993). "Improved Design of Branched Networks Using Pressure-Reducing Valves", *J. Hyd. Eng.*, 119(2), 164–180.

Peterka, A. J. (1964). *Hydraulic Design of Stilling Basins and Energy Dissipators*, Engineering Monograph No. 25, U.S. Department of the Interior, Bureau of Reclamation, Denver, Colorado.

Peters, E., and McNeill, R. (2003). "Priming the Pump: Colorado Facility Uses Polymer to Increase Lift Station Capacity", *Water Environ. Technol.*, 15(5), 70–72.

Petersen, J. S., Rohwer, C., and Albertson, M. L. (1955). "Effect of Well Screens on Flow into Wells", *Trans. Am. Soc. Civil Eng.*, 120(1), 563–585.

Phillips, W. R. C. (2003). "Langmuir Circulations", *Wind Over Waves II: Forecasting and Fundamentals of Applications*, S. G. Saijadi and J. C. R. Hunt (eds.), Horwood, Chester, pp. 157–167.

Phillips, W. R. C. (2005). "On the Spacing of Langmuir Circulation in Strong Shear", *J. Fluid Mech.*, 525, 215–236.

Piggott, J. (1982). *Japanese Mythology: Library of the World's Myths and Legends*, Peter Bedrick Books, New York.

Pincince, A. B. (1970). "Wet-well Volume for Fixed-Speed Pumps", *J. Water Poll. Control Fed.*, 42(1), 126–130.

Pincince, A. B. (1991). "Transfer of Oxygen and Emissions of Volatile Organic Compounds at Clarifier Weirs", *Res. J. WPCF*, 63(2), 114–119.

Pinkayan, S. (1972). "Routing Storm Water through a Drainage System", *J. Hyd. Div.*, 98(1), 123–135.

Pisano, W. C., and Brombach, H. (1995). "Turbo Vortex Regulator Relies on Natural Flows", *Water Environ. Technol.*, 7(9), 15.

Pisano, W. C., Thibault, N., and Forbes, G. (1990). "The Vortex Solids Separator", *Water Environ. Technol.*, 2(5), 64–71.

Pletta, D. H. (1951). *Engineering Statics and Dynamics*, The Ronald Press Company, New York, pp. 150–151.

Polprasett, C., Dissanayake, M. G., and Thanh, N. C. (1983). "Bacterial Die-off Kinetics in Waste Stabilization Ponds", *J. WPCF*, 55(3), 285–296.

Pomeroy, R. D. (1967). "Flow Velocities in Small Sewers", *J. WPCF*, 39(9), 1525–1548.

Portland Cement Association (PCA) (1962). *Design Aids: An Application of U.S. Bureau of Pipe Roads Culvert Capacity Charts*, PCA, Chicago, Illinois.

Portland Cement Association (PCA) 1964. *Handbook of Concrete Culvert Pipe Hydraulics*, PCA, Skokie, Illinois.

Posey, C. J. (1950). "Gradually Varied Channel Flow", Chapter IX in H. Rouse (ed.), *Engineering Hydraulics*, John Wiley & Sons, Inc., New York.

Prandtl, L. (1952). *The Essentials of Fluid Dynamics*, Hafner Publishing Company, New York.

Prasad, R. (1970). "Numerical Method of Computing Flow Profiles", *J. Hyd. Div.*, 96(1), 75–86.

Pratsinis, S. E. (2016). *Filtration – Praktikum 'Mechanical Engineering'*, Swiss Federal Institute of Technology, Zurich, Switzerland.

Pratte, B. D., and Baines, W. D. (1967). "Profiles of the Round Turbulent Jet in a Cross Flow", *J. Hyd. Div.*, 92(6), 53–64.

Preul, H. C., and Holler, A. G. (1969). "Reaeration Through Low Dams in the Ohio River", *Proc. 24th Industrial Waste Conference*, Purdue University, Part II, p. 1249.

Prescott, S. L., and Ulanicki, B. (2003). "Dynamic Modeling of Pressure Reducing Valves", *J. Hyd. Eng.*, 129(10), 804–812.

Prescott, S. L., and Ulanicki, B. (2008). "Improved Control of Pressure Reducing Valves in Water Distribution Networks", *J. Hyd. Eng.*, 134(1), 56–65.

Proeh, M., Zakin, J. L., Brosh, A., and Warshavsky, M., Sr. (1970). "Drag Reduction in Hydraulic Transport of Solids", *J. Hyd. Div.*, 96(4), 903–909.

P. S. (2001). "Blown on Down the Highway", *Mech. Eng.*, 123(10), 18–19.

Pukelsheim, F. (1994). "The Three Sigma Rule", *Am. Stat.*, 48(2), 88–91.

Quick, M. C. (1974). "Mechanism for Streamflow Meandering", *J. Hyd. Div.*, 100(6), 741–753.

Quirk, T. P., and Eder, L. J. (1970). "Evaluation of Alternative Solutions for Achievement of River Standards", *J. WPCF*, 42(2, Part 1), 272–290.

Rajaratnam, N. (1980). "Erosion by Circular Wall Jets in a Cross Flow", *J. Hyd. Div.*, 106(11), 1867–1883.

Rajaratnam, N., and Pani, B. S. (1974). "Three-Dimensional Turbulent Wall Jets", *J. Hyd. Div.*, 100(1), 69–83.

Rajaratnam, N., and Subramanya, K. (1967). "Flow Immediately Below Submerged Sluice Gate," *J. Hyd. Div.*, 93(4), 57–77.

Ramachandran, V. S. (2006). "Creativity versus Skepticism within Science", *Skep. Inq.*, 30(6), 48–51.

Ramamurthy, A. S., Carballada, L. B., and Tran, D. M. (1988). "Combining Open Channel Flow at Right Angled Junctions", *J. Hydraul. Eng.*, 114(12), 1449–1460.

Ramamurthy, A. S., and Satish, M. G. (1988). "Division of Flow in Short Open Channel Branches", *J. Hydraul. Eng.*, 114(4), 428–438.

Rankin, R. S. (1959). "Weirs on Sedimentation Tanks", *Pub. Works*, 12, 66.

Rao, B. C., Rao, N. S. L., and Shivaswamy, M. S. (1968). "Distribution of Energy Losses at Conduit Trifurcations", *J. Hyd. Div.*, 94(6), 1363–1374.

Rao, B. C. S., Thang, N. D., and Naudascher, E. (1987). "Vibration of Trashracks With Different Incidence Angles", *Proc. International Conference on Flow Induced Vibrations*, Bowness-on-Windermere, England, 12-14 May, pp. 329-338.

Rao, N. S. L., Rao, B. C. S., and Ramaswamy, R. I. (1967). "Pressure Losses at Trifurcations in Closed Conduits", *J. Hyd. Div.*, 93(3), 51–64.

Rao, S. S., and Shukla, M. K. (1971). "Characteristics of Flow Over Weirs of Finite Crest Width", *J. Hyd. Div.*, 97(11), 1807–1816.

Reardon, N. (2008). "Water Bill is All Wet, Homeowner Says", *Patriot Ledger*, October 20, 2008.

Reba, I. (1966). "Applications of the Coanda Effect", *Sci. Am.*, 214(6), 84–92.

Reed, J. R., and Wolfkill, A. J. (1976). "Evaluation of Friction Slope Models", *Rivers '76, Proc. 3rd Ann. Symp. Waterways, Harbors and Coastal Engrg. Div.*, ASCE II, pp. 1159-1178.

Reethof, G. (1960). "Characteristics of Positive-Displacement Pumps and Motors", Chapter 4 in *Fluid Power Control* J. F. Blackburn, G. Reethof, and J. L. Shearer (eds.), The MIT Press, Cambridge, Massachusetts.

Reid, J. (2000). "The Search for Water", *AWWA Mainstream, American Water Works Association*, 44(10), 10.

Reynolds, O. (1892). *Biography of James Prescott Joule*, Manchester Literary and Philosophical Society, Manchester, England.

Rheinheimer, D. E., Null, S. E., and Lund, J. R. (2015). "Optimizing Selective Withdrawal from Reservoirs to Manage Downstream Temperatures with Climate Warming", *J. Water Resour. Plan. Manage.*, 141(4), 1–9.

Rice, L. (1971). "Reduction of Urban Runoff Peak Flows by Ponding", *J. Irrig. Drain. Div.*, 97(3), 469–482.

Rich, G. R. (1963). *Hydraulic Transients*, 2nd Ed., Dover Publications, Inc., New York.

Richards, E. F., and Graber, S. D. (1965). "Transition to Turbulence and Wall Attachment of Miniature Jets", *Proc. Fluid Amplification Symposium*, Washington, D.C., October 1965.

Ringer, W. H. (1974). "Haverhill, Mass. Plant to Treat Municipal and Industrial Wastewaters", *J. New Engl. Water Pollut. Contr. Assoc.*, 8(1), 22–45.

Robinson, R. (2020). "Researchers Discover Different Types of Aurora Borealis", *Now*, Powered by Northrop Grumman, May 1, 2020. https://now.northropgrumman.com/researchers-discover-different-types-of-aurora-borealis, accessed July 30, 2024.

Rohsenow, W. M., and Choi, H. Y. (1961). *Heat, Mass & Momentum Transfer*, Prentice-Hall, Inc., Englewood Cliffs, New Jersey.

Rohwer, C. (1941). "Putting Down and Developing Wells for Irrigation", U.S. Department of Agriculture Circular No. 546, February 1940, slightly revised March 1941, Washington, D.C.

Rollins, J. P. (ed.). (1973). *Compressed Air and Gas Handbook*, 4th Ed., Compressed Air and Gas Institute, New York.

Rossini, F. D., Pitzer, K. S., Taylor, W. J., Ebert, J. P., Kilpatrick, J. E., Beckett, C. W., Williams, M. D., and Werner, H. C. (1947). *Selected Values of Properties of Hydrocarbons*, National Bureau of Standards Circular C461, Washington, D.C.

Rouse, H. (1950). "Fundamental Principles of Flow", Chapter I in H. Rouse (ed.), *Engineering Hydraulics*, John Wiley & Sons, Inc., New York.

Rouse, H. (ed.) (1959). *Advanced Mechanics of Fluids*, John Wiley & Sons, Inc., New York.

Rouse, H., and Ince, S. (1957). *History of Hydraulics*, Dover Publications, Inc., New York.

Rouse, H., and McNown, J. S., 1945, "Discussion of 'Coefficients for Velocity Distribution in Open-Channel Flow', by W. S. Eisenlohr, Jr., *Trans. Am. Soc. Civil Eng.*, 110, 651–657.

Rouse, H., Yih, C. S., and Humphreys, H. W. (1952). "Gravitational Convection from a Boundary Source", *Tellus*, 4(3), 201–210.

Rulifson, R. A., and Tull, K. A. (1999). "Striped Bass Spawning in a Tidal Bore River: The Shubenacadie Estuary, Atlantic Canada", *Trans. Am. Fish. Soc.*, 128(4), 613–624.

Rumer, R. R., Jr., and Drinker, P. A. (1966). "Resistance to Laminar Flow through Porous Media", *J. Hyd. Div.*, 92(5), 155–163.

Ruus, E. (1970). "Head Losses in Wyes and Manifolds", *J. Hyd. Div.*, 96(3), 593–608.

Safwat, H. H. (1972). "Photographic Study of Water Column Separation", *J. Hyd. Div.*, 98(4), 739–746.

Safwat, H. H., and Van Den Polder, J. (1972). "Experimental and Analytic Data Correlation Study of Water Column Separation", *J. Fluids Eng.*, 95(1), 91–97.

Samani, Z. (2017). "Three Simple Flumes for Flow Measurement in Open Channels", *J. Irrig. Drain. Eng.*, 143(6), 1–4.

Sandvik, A. (1985). "Proportional Weirs for Stormwater Pond Outlets", *Civ. Eng.*, 55(3), 54–56.

Sanger, N. L. (1969). "An Experimental Investigation of Several Low-Area-Ratio Water Jet Pumps", *J. Basic Eng.* 69-FE-5, 1–9.

Sangster, W. M., Wood, H. W., Smerdon, E. T., and Bossy, H. G. (1961). "Pressure Changes at Open Junctions in Conduits", *Trans. Am. Soc. Civil Eng.*, 126(1), 364–393.

Sanks, R. L., Jones, G. M., and Sweeney, C. E. (1993a). "Designing Self-Cleaning Wet Wells for Wastewater Pumping", *Proc. Hydraul. Eng.'93*, 1, 180–185.

Sanks, R. L., Jones, G. M., and Sweeney, C. E. (1993b). "Self-Cleaning Wet Wells: Definition and Design", *Proc. International Conference, Pipeline Infrastructure II*, San Francisco, California, July 25-30, 1993, pp. 102–114.

Sanks, R. L., Jones, G. M., and Sweeney, C. E. (1995). *Improvements in Pump Intake Design*, Montana State University, Bozeman, Montana.

Schalko, I., and Boes, R.M. (2021). "Effect of Water Withdrawal on the Appearance and Sound Level of Waterfalls", *Water Resour. Res.*, 57(10), 1–35.

Schiff, A. J. (1991). "Operations and Maintenance of Automatic Valves Used in the Water Works Industry", *Presented at the Operator's Form, New England Water Works Association and New Hampshire Water Works Association*, Laconia, New Hampshire.

Schlichting, H. (1960). *Boundary Layer Theory*, 4th Ed., McGraw-Hill Book Company, New York.

Schönfeld, J. C. (1955). "Theoretical Considerations on an Experimental Bore", *Proc. Sixth General Meeting*, IAHR, Vol. 1, The Hague, A-15-1 to A-15-12.

Scott, A. G. (1972). "Grit Removal System Modifications", *Deed and Data*, Water Pollution Control Federation, May 1972.

Seeds, M. A. (1986). *Foundations of Astronomy*, Wadsworth Publishing Company, Belmont, California.

Segeler, C. G. (1967). "Gas System Piping", Section 13 in *Piping Handbook*, R. C. King and S. Crocker (eds.), 5th Ed., McGraw-Hill Book Company, New York.

Sell, L. E. (1971). "Hydroelectric Power Plant Trashrack Design", *J. Power Div.*, 97(1), 115–121.

Shabayek, S., Steffler, P., and Hicks, F. (2002). "Dynamic Model for Subcritical Combining Flows in Channel Junctions", *J. Hyd. Eng.*, 128(9), 821–828.

Shames, I. H. (1962). *Mechanics of Fluids*, McGraw-Hill Book Company, New York.

Shang, H., Xu, S., Zhang, K., and Zhao, L. (2019). "Explicit Solution for Critical Depth in Closed Conduits Flowing Partly Full." *Water*, 11(10), 1–17.

Shapiro, A. H. (1953). *The Dynamics and Thermodynamics of Compressible Fluid Flow*, Vol. I, The Ronald Press Company, New York.

Shapiro, A. H. (1954). *The Dynamics and Thermodynamics of Compressible Fluid Flow*, Vol. II, The Ronald Press Company, New York.

Shapiro, A. H. (1963). "M.I.T. Class Notes from Course 2.25T – Advanced Fluid Mechanics".

Shelef, G., and Green, M. (1980). "Kinetics of Pure Oxygen Activated Sludge", *J. Environ. Eng. Div.*, 106(6), 1099–1110.

Shelley, G. J., Stone, P. B., and Cullen, A. J. (1981). *Field Evaluation of a Swirl Degritter at Tamworth, New South Wales, Australia*, EPA-600/S2-81-063, Municipal Environmental Research Center, Cincinnati, Ohio.

Shih, C. C., and Grigg, N. S. (1967). "A Reconsideration of the Hydraulic Radius as a Geometrical Quantity in Open Channel Hydraulics", *I.A.H.R. Proc. 12th Congress*, Paper A-36.

Shumaker, S., Cashon, G., Cox, A., Conner, R., and Rajah, S. (2017). "Update to the Improved Approach for the Design of Thrust Blocks in Buried Pipelines", *Proc. Sessions of the Pipelines 2017 Conference*, August 6-9, 2017, Phoenix, Arizona, pp. 586-596.

Shumaker, S., Rajah, S., Bardakjian, H., Bushdiecker, K., Barnhurst, E., Luska, J., Nasr, S., Rowell, R., Shumaker, S., and Turkopp, R. (2011). "An Improved Approach for the Design of Thrust Blocks in Buried Pipelines", *Proc. ASCE Pipelines Conference*, July 23-27, 2011, Seattle, Washington, pp. 821-839.

Sietsema, J., Foster, R. E., and LP Gas (2015). "Using Coriolis Mass Flow Meters with Positive Displacement Vane Pumps", *JLP Gas*, Cleveland, Ohio, January 15, 2015.

Silberman, E. (1975). "Discussion of 'Boundary Layer Separation Concept of Settling Tanks', by Y. Chiu", *J. Environ. Eng. Div.*, 101(4), 685.

Simpson, A. R, and Marchi, A. (2013). "Evaluating the Approximation of the Affinity Laws and Improving the Efficiency Estimate for Variable Speed Pumps", *J. Hyd. Eng.*, 139(12), 1314–1317.

Singler, A., Graber, B., and Banks, C. (eds.) (2012). *Massachusetts Streams Crossing Handbook*, 2nd Ed., Division of Ecological Restoration, Executive Office of Energy and Environmental Affairs, Department of Fish And Game, Boston, Massachusetts.

Skogerboe, G. V., Hyatt, M L., England, J. D., and Johnson, J. R. (1966). *Measuring Water with Parshall Flumes*, Utah Water Research Laboratory, Logan, Utah.

Skogerboe, G. V., Hyatt, M. L., England, J. D., and Johnson, J. R. (1967a). *Design and Calibration of Submerged Open Channel Flow Measurement Structures Part 2 Parshall Flumes*, Utah Water Research Laboratory, Logan, Utah.

Skogerboe, G. V., and Hyatt, M. L. (1967b). *Analysis of Submergence in Flow Measuring Flumes, J. Hyd. Div.*, 93(4), 183–200.

Skogerboe, G. V., Hyatt, M. L., Anderson, R. K., and Eggleston, K. O. (1967). *Design and Calibration of Submerged Open Channel Flow Measurement Structures: Part 3 – Cutthroat Flumes*, Reports WG 31-5, Utah Water Res. Lab., Utah State University, Logan, Utah.

Skogerboe, G. V., Bennett, R. S., and Walker, W. R. (1972). "Generalized Discharge Relations for Cutthroat Flumes", *J. Irrig. Drain. Div.*, 98(4), 569–583.

Smith, C. D., and Korolischuk, E. M. (1973). "Modified USBR Impact Basin", *J. Hyd. Div.*, 99(1), 283–287.

Smith, C. D., and Yu, J. N. G. (1966). "Use of Baffles in Open Channel Expansions", *J. Hyd. Div.*, 92(2), 1–17.

Smith, K. V. H. (1967). "Control Point in a Lateral Spillway Channel", *J. Hyd. Div.*, 93(3), 27–34.

Smith, K. V. H. (1972). "Computer Determination of Critical Depth Control Points in Open-Channel Flow", *Proc. Instn. Civ. Eng.*, United Kingdom, pp. 461–470.

Sobelman, T. B., Sullivan, J., Chatelain, C., and Alderete, D. (2005). "Caltrans Takes Out the Trash", *Civil Eng.*, 75(10), 53–61.

Soil Conservation Service (SCS) (1986). "U.S. Department of Agriculture, "Urban Hydrology for Small Watersheds", *Technical Release No. 55*, Washington, D.C.

Somerton, C. W., and Wood, P. (1988). "Effect of Walls in Modeling Flow Through Porous Media", *J. Hydraul. Eng.*, 114(12), 1431–1448.

Sooky, A. A. (1969). "Longitudinal Dispersion in Open Channels", *J. Hyd. Div.*, 95(4), 1327–1346.

Sowby, R. (2015). "Rethinking Stream Crossings for Humans and Animals", *NEWEA J.*, 49(2), 38–42.

Sparrow, E. M., (1955). "Analysis of Laminar Forced-Convection Heat Transfer in Entrance Region of Flat Rectangular Ducts", National Advisory Committee for Aeronautics (NACA), Technical Note 3331, Lewis Flight Propulsion Laboratory, Cleveland, Ohio. https://digital.library.unt.edu/ark:/67531/metadc57449, accessed November 14, 2021.

Sreenivasulu, P., and Raghavendran, R. (1970). "Linear Proportional Weirs", *J. Hyd. Div.*, 96(2), 379–389.

Stagg Chevrolet, Inc. vs. Board of Assessors of the Town of Harwich (2006). "Commonwealth of Massachusetts Appellate Tax Board", Docket No. F266854, Promulgated: February 1, 2006. https://www.mass.gov/files/documents/2016/07/wa/06p019.doc

Stagg Chevrolet Inc. vs. Board of Water Commissioners of Harwich (2007). "Appeals Court of Massachusetts, Suffolk", No. 06-P-522, Decided: January 30, 2007. https://caselaw.findlaw.com/ma-court-of-appeals/1112101.html

Stahl, H., and Hager, W. H. (1999). "Hydraulic Jump in Circular Pipes", *Can. J. Civ. Eng.*, 26, 368–373.

Stahre, P., and Urbonas, B. (1990). *Stormwater Detention for Drainage*, Water Quality and CSO Management, Prentice Hall, Englewood Cliffs, New Jersey.

Stanbury, P. F., Whitaker, A., and Hall, S. J. (2016). "Design of a Fermenter", Chapter 7 in Principles of Fermentation Technology, 3rd Ed., Butterworth-Heinemann, Oxford, England.

Stenning, A. H., and Shearer, J. L. (1960). "Fundamentals of Fluid Flow", Chapter 3 in *Fluid Power Control*, J. F. Blackburn, G. Reethof, and J. L. Shearer (eds.), The MIT Press, Cambridge, Massachusetts.

Stenquist, R. J., and Kaufman, W. J. (1972). "*Initial Mixing in Coagulation Processes*", SERL Report No. 72-2, University of California, Berkeley, California, February 1972.

Stephenson, D. (1966). "Water-Hammer Charts Including Fluid Friction", *J. Hyd. Div.*, 92(5), 71–94.

Stephenson, D. (2002). "Simple Guide for Design of Air Vessels for Water Hammer Protectionof Pumping Lines", *J. Hydraul. Eng.*, 128(8), 792–797.

Stoker, J. J. (1957). *Water Waves: The Mathematical Theory with Applications*, Interscience Publishers, New York, §'s 10.4, 10.6, 10.7, 10.10.

Stokes, G. G. (1845). "On the Theories of the Internal Friction of Fluids in Motion, &c", Camb. *Trans.*, viii, 287.

Straub, L. G., and Lamb, O. P. (1956). "Studies of Air Entrainment in Open-Channel Flows", *Trans. Am. Soc. Civ. Eng.*, 121(1), 30–44.

Streeter, V. L. (1948). *Fluid Dynamics*, McGraw-Hill Book Company, Inc., New York.

Streeter, V. L. (1962). *Fluid Mechanics*, 3rd Ed., McGraw-Hill Book Company, Inc., New York.

Streeter, V. L. (1971). "Unsteady Flow Calculations by Numerical Methods", *J. Basic Eng.*, 94(2), 457–465.

Subcommittee on Sewage Treatment (1946). "Sewage Treatment at Military Installations", *Sew. Works J.*, 18(5), 789–1028.

Subcommittee on Slope Protection (1948). "Review of Slope Protection Methods", *Trans. Am. Soc. Civil Eng.*, 74(6), 845–866.

Subramanya, K., and Awasthy, S. C. (1970). "Discussion of 'Internal Hydraulics of Thermal Discharge Diffusers' by S. Vigander, R. A. Elder, and N. H. Brooks, *J. Hyd. Div.*, 96(12), 2635–2637.

Subramanya, K., and Awasthy, S. C. (1972). "Spatially Varied Flow Over Side Weirs", *J. Hyd. Div.*, 98(1), 1–10.

Susag, R. H., Polta, R. C., and Schroepfer, G. J. (1966). "Mechanical Surface Aeration of Receiving Waters", *J. WPCF*, 38(1), 53–68.

Sutro, H. H. (1915). "Apparatus for Measuring Liquids", U.S. Patent 1,138,700, May 11, 1915.

Swamee, P. K., and Khanna, P. (1974). "Equivalent Pipe Methods for Optimizing Water Networks – Facts and Fallacies", *J. Environ. Eng. Div.*, 100(1), 93–99.

Swan, C. (2020). "Open-Channel Flow", THERMOPEDIA A-Z Guide to Thermodynamics, Heat and Mass Transfer, and Fluids Engineering, www.thermopedia.com, accessed March 9, 2020.

Swift, D. L., and Friedlander, S. K. (1964). "The Coagulation of Hydrosols by Brownian Motion and Laminar Shear Flow", *J. Coll. Sci.*, 19, 621–647.

Switalski, R., Sharon, J. D., and Sadowski, D. J. (1992). "Big Wastewater and Collection Project Continues on Schedule and Under Budget", *Water Eng. Manage.*, 1992, 34, 38-42.

Symons, G. E. (1966). "Pumps and Pumping", *Manual of Practice Number One*, Water and Wastes Engineering, New York.

Symons, G. E. (1968). "Valves, Hydrants, and Fittings", *Manual of Practice Number Four*, Water and Wastes Engineering, New York.

Taylor, E. H. (1945). "Discussion of 'Coefficients for Velocity Distribution in Open-Channel Flow,' by W. S. Eisenlohr, Jr.", *Trans. Am. Soc. Civil Eng.*, 110, 646–648.

Tanverakul, S. A., and Juneseok, L. (2013). "Residential Water Demand Analysis Due to Water Meter Installation in California." In *ASCE 2013 World Environmental and Water Resources Congress*, ASCE, Cincinnati, Ohio.

Taylor, G. I. (1953). "Dispersion of Soluble Matter in Solvent Flowing Slowly Through a Tube", *Proc. Royal Soc. London*, A219, 186.

Taylor, G. I. (1954). "The Dispersion of Matter in Turbulent Flow through a Pipe", *Proc. Royal Soc. London*, A223, 446.

TeKippe, R. J., and Cleasby, J. L. (1968). "Model Studies of a Peripheral Feed Settling Tank", *J. San. Eng. Div.*, 94(1), 85–102.

Temeepattanapongsal, S., Merkley, G. P., Barfuss, S. L., and Smith, B. L. (2013). "Generic Free-Flow Rating for Cutthroat Flumes", *J. Hydraul. Eng.*, 139(7), 727–735.

Tesařík, I. (1967). "Flow in Sludge Blanket Clarifiers", *J. San. Engr. Div.*, 93(6), 105–120.

Thackston, E. L., Hays, J. R., and Krenkel, P. A. (1967). "Least Squares Estimation of Mixing Coefficients", *J. San. Engr. Div.*, 93(3), 47–58.

Thackston, E. L., and Krenkel, P. A. (1967). "Longitudinal Mixing in Natural Streams", *J. San. Engr. Div.*, 93(5), 67–90.

Thackston, E. L., and Schnelle, K. B. (1970). "Predicting Effects of Dead Zones on Stream Mixing", *J. San. Eng. Div.*, 96(2), 319–331.

Theis, C. V. (1935). "The Relation between the Lowering of the Piezometric Surface and the Rate and Duration of Discharge of a Well Using Ground Water Storage", *Trans. AGU*, 16, 519–524.

Theis, C. V. (1952). *The Relation Between the Lowering of the Piezometric Surface and the Rate and Duration of Discharge of a Well Using Ground Water Storage*, U.S. Geological Survey Ground Water Notes No. 5.

Theroux, R. J., and Betz, J. M. (1959). “Sedimentation and Pre-aeration Experiments at Los Angeles”, *Sew. Ind. Wastes*, 31, 1259–1266.

Thielsch, H. (1967). “Manufacture, Fabrication, and Joining of Commercial Piping”, Section 7 in *Piping Handbook*, R. C. King and S. Crocker (eds.), 5th Ed., McGraw-Hill Book Company, New York.

Thirumurthi, D. (1969). “Design Principles of Waste Stabilization Ponds”, *J. San. Eng. Div.*, 95(2), 311–332.

Thirumurthi, D. (1973). “A Break-through in the Tracer Studies of Sedimentation Tanks”, *J. WPCF*, 41(11, Part 2), R405–R418.

Thomas, A. D. (1977). “Particle Size Effects in Turbulent Pipe Flow of Solid-Liquid Suspensions”, *6th Australian Hydraulics and Fluid Mechanics Conference*, Adelaide, Australia, December 5-9, 1977.

Thomas, B. F., and Vogel, R. M. (2012). “Impact of Storm Water Recharge Practices on Boston Groundwater Elevations”, *J. Hydrol. Eng.*, 17(8), 923–932.

Thomas, G. O. (1972). “Determination of Pump Characteristics for a Computerized Pump Transient Analysis”, *International Conference on Pressure Surges*, University of Kent, Canterbury, England, September 6-8, 1972.

Thomas, H. A. (1940). “Discussion of ‘Lateral Spillway Channels’ by T. R. Camp”, *Trans. Am. Soc. Civil Eng.*, 105(1), 627–633.

Thompson, P. C. (2024) “STOMATOPODS: Why Is My Thumb Bleeding and My Mind Blown?”, *How to Influence Friends and Influence Fungi*, C. Balakrishnan and M. Wasowski (eds.), St. Martin’s Press, New York, pp. 17–20.

Thorne, C. R. (1985). “Estimating Mean Velocity in Mountain Rivers”, *J. Hyd. Div.*, 111(4), 612–624. [With discussions April 1987.]

Tippin, D. L. (1978). “New Meters Don’t Cost; They Pay”, *Am. City County*, 93(6), 61–62.

Tobias, A. J. (2012). “Thousand-dollar Water Bills Blamed on Leaks.” *Dayton Daily News*, November 5, 2012. https://www.daytondailynews.com/news/thousand-dollar-water-bills-blamed-leaks/RBx3D

Todd, D. K. (1964). “Groundwater”, Section 13 in *Handbook of Applied Hydrology*, V. T. Chow (ed.), McGraw-Hill Book Company, New York.

Tourbier, J. T., and Westmacott, R. (1981). Water Resources Protection Technology *(#W08)*, Urban Land Institute, Washington, D.C.

Treybal, R. E. (1959). “The Economic Design of Mixer-Settler Extractors”, *Ind. Eng. Chem.*, 53(8), 597–606.

Treybal, R. E. (1961). “Stirred Tanks and Mixers for Liquid Extraction”, *J. Am. Inst. Chem. Eng.*, 5(4), 474–482.

Trieste, D. J. (1992). “Evaluation of Supercritical/Subcritical Flows in High-Gradient Channel”, *J. Hyd. Div.*, 118(8), 1107–1118.

Trieste, D. J. (1994). “Supercritical Flows versus Subcritical Flow in Natural Channels”, *Hydraulic Engineering ’94: Proc. 1994 Conference*, Vol. 2, Buffalo, New York, Am. Soc. Civ. Eng.

Trussell, R. R., and Chao, J.-L., (1977). “Rational Design of Chlorine Contact Facilities”, *J. WPCF*, 49(4), 659–667.

Tuček, F., and Chudoba, J. (1969). “Purification Efficiency in Aeration Tanks with Complete Mixing and Piston Flow”, *Water Res.*, 3(8), 559–570.

Tullis, J. P. (1973). “Cavitation Scale Effects for Valves”, *J. Hyd. Div.*, 99(7), 1109–1128.

Uhlmann, D., Recknagel, F., Sandring, G., Schwaarz, S., and Eckelmann, G. (1983). “A New Design Procedure for Waste Stabilization Ponds”, *J. WPCF*, 55(10), 1252–1255.

Urbonas, B. R., and Roesner, L. A. (1992). “Hydrologic Design for Urban Drainage and Flood Control”, Chapter 28 in *Handbook of Hydrology*, D.R. Maidment (ed.), McGraw-Hill Book Company, New York.

U.S. Congress (1974). *Wetlands: Their Use and Regulation*, Office of Technology Assessment, OTA-O-206, Washington D.C.

U.S. Department of the Army, Corps of Engineers (USACE) (1959). *Backwater Curves in River Channels*, EM 1110-2-1409, Hydrologic Engineering Center, Davis, California.

U.S. Department of the Army, Corps of Engineers (USACE) (1963). *Nonuniform Flow Functions Circular Section*. Miscellaneous Paper No. 2-601, Waterways Experiment Station, Vicksburg, Mississippi.

U.S. Department of the Army, Corps of Engineers (USACE) (1983). *Land Application of Wastewater Research Program*, Special Report 83-9, Cold Regions Research and Engineering Lab, Hanover, New Hampshire.

U.S. Department of the Army, Corps of Engineers (USACE) (1990). *HEC-2 Computer Program: Water Surface Profiles, User’s Manual*, Hydrologic Engineering Center, Davis, California.

U.S. Bureau of Public Roads (USBPR) (1967). *Use of Riprap for Bank Protection*, Hydraulic Engineering Circular No. 11, U.S. Department of Transportation.

U.S. Bureau of Reclamation (USBR) (1963a). *Earth Manual*, 1st Ed. – Revised, U.S. Department of the Interior, Denver, Colorado.

U.S. Bureau of Reclamation (USBR) (1963b). *Linings for Irrigation Canals*, 1st Ed., U.S. Department of the Interior, Denver, Colorado.

U.S. Bureau of Reclamation (USBR) (1965). *Design of Small Dams*, U.S. Department of the Interior, Washington, D.C.

U.S. Department of Agriculture (USDA) (1971). *Soil Conservation Service*, Drainage of Agricultural Land, Section 16, National Engineering Handbook, NRCS, Washington, D.C.

U.S. Department of Agriculture (USDA) (1972). *Culverts*, Section 4, Chapter 14, National Engineering Handbook, NRCS, Washington, D.C.

U.S. Environmental Protection Agency (USEPA) (1972). *Hypochlorite Generator for Treatment of Combined Sewer Overflows. Office of Research and Monitoring*, Program No. 11023 DAA, Water Pollution Control Research Series, Washington, D.C.

U.S. Environmental Protection Agency (USEPA) (1974). *Flow Equalization*, EPA Technology Transfer Seminar Publications, Cincinnati, Ohio.

U.S. Environmental Protection Agency (USEPA) (1994). *Stormwater Management Model (SWMM)*, Version 4.30. Office of Research and Development, Athens, Georgia.

U.S. Environmental Protection Agency (USEPA) (2014). *Water Quality Standards Handbook*, Chapter 5: General Policies, EPA 820-B-14-004, U.S. Environmental Protection Agency, Washington DC.

U.S. Environmental Protection Agency (USEPA) (2020). *Water Quality Analysis Simulation Program (WASP)*. Center for Exposure Assessment Modeling (CEAM), Athens, Georgia, https://www.epa.gov/ceam/water-quality-analysis-simulation-program-wasp, accessed February 3, 2020.

U.S. Green Building Council (2005). "Green Building Rating System for New Construction and Major Renovations, Version 2.2", Washington, D.C.

Vallentine, H. R. (1965). "Rigid Water Column Theory for Uniform Gate Closure", *J. Hyd. Div.*, 91(4), 27–31.

Van der Beken, A. (1969). "Bijdrage tot de hydraulica van drainerbuizen", (Contribution to the Hydraulics of Drain Pipes, in Dutch with English subtitles, Faculty of Agronomy, conclusion, and summary) thesis presented to the University of Ghent, at Belgium, in 1969, in fulfillment of the requirements for the degree of Doctor of Agricultural Engineering. Also published by Research Station of Agricultural Engineering, Merelbeke, Belgium.

Van der Beken, A. (1970). "Discussion of 'Dynamic Equation for Steady Spatially Varied Flow' by B. C. Yen and H. G. Wenzel, Jr.", *J. Hyd. Div.*, 96(12), 2661–2665.

Van der Sterren, M., Rahman, A., and Dennis, G. R. (2013). "Quality and Quantity Monitoring of Five Rainwater Tanks in Western Sydney, Australia", *J. Environ. Eng.*, 139(3), 332–340.

Vanoni, V. A. (Author, ed). (1975). *Sedimentation Engineering*, American Society of Civil Engineers, New York (now Reston, Virginia).

Venkataraman, P., and Subramanya, K. (1973). "A Practical Proportional Weir", *Water Power* 1973, 189–190.

Venkataraman, P., and Rao, P. R. M. (1998). "Darcian, Transitional, and Turbulent Flow through Porous Media", *J. Hydraul. Eng.*, 124(8), 840–846 [with errata 125(4), 437].

Vigander, S., Elder, R. A., and Brooks, N. H. (1970). "Internal Hydraulics of Thermal Discharge Diffusers", *J. Hyd. Div.*, 96(2), 509–527.

Vigander, S., Elder, R. A., and Brooks, N. H. (1971). "Closure to 'Internal Hydraulics of Thermal Discharge Diffusers'", *J. Hyd. Div.*, 97(9), 1518–1520.

Villarreal, E. L. (2007). "Runoff Detention Effect of a Sedum Green-Roof", *Nordic Hydrol.*, 38(1), 99–105.

Villemonte, J. R. (1947). "Submerged Weir Discharge Studies", *Eng. News Rec.*, 193, 866–869.

Villemonte, J. R., and Rohlich, G. A. (1962). "Hydraulic Characteristics of Circular Sedimentation Basins", *Proc. 17th Industrial Waste Conference*, Purdue University, Lafayette, Indiana, pp. 682-702.

Virginia Department of Transportation (VDOT) (2002). *Drainage Manual*, Virginia Department of Transportation, Richmond, Virginia, pp. 9–37.

Vocadlo, J. J., and Sagoo, M. S. (1972). "Slurry Flow in Pipes and Pumps", *Second Symposium on Storage and Flow of Solids*, ASME, Chicago, Illinois, June 1972.

Vogel, S. (1994). *Life in Moving Fluids*, 2nd Ed., Princeton University Press, Princeton, New Jersey.

Vrale, L., and Jorden, R. M. (1971). "Rapid Mixing in Water Treatment", *J. Am. Water Works Assoc.*, 63(1), 52–58.

Wagner, E. G., and Lanoix, J. N. (1959). *Water Supply for Rural Areas and Small Communities*, Monograph Series No. 42, World Health Organization, Geneva, Switzerland.

Wahl, K. L. (1994). "Discussion of 'Evaluation of Supercritical/Subcritical Flows in High-Gradient Channel'", *J. Hyd. Div.*, 120(2), 270–273.

Walker, J. D. (1969). "New Concepts in Aeration-Swirling-Deep Aeration-Channel Stirring", *Presented at the 1969 Annual Meeting North Carolina Section AWWA and WPCA*, Raleigh, North Carolina, November 12, 1969.

Walker, W. R., Skogerboe, G. V., and Bennett, R. S. (1973). "Flow Measuring Flume for Wastewater Treatment Plants", *J. WPCF*, 45(3, Part I), 542–551.

Wallis, G. B. (1961). "Some Hydrodynamic Aspects of Two Phase Flow and Boiling", *International Heat Transfer Conference*, Boulder, Colorado.

Walski, T. M., Barnhart, T. S., Driscoll, J. M., and Yencha, R. M. (1994). "Hydraulics of Corrosive Gas Pockets in Force Mains", *Water Environ. Res.*, 66(6), 772–778.

Wang, C.-T. (1953). *Applied Elasticity*, McGraw-Hill Book Company, Inc., New York.

Ward, J. C. (1964). "Turbulent Flow in Porous Media", *J. Hydraul. Div.*, 90(5), 1–12 [with closure and errata 92(4), 110-121].

Warriner, T. R. (1967). "Field Tests on Chlorination of Poliovirus in Sewage", *J. San. Eng. Div.*, 93(5), 51–65.

Washington State Dept. of Social and Health Services (1975). "Water Distribution Network Analysis by Computer", *Water Sew. Works*, 79–80.

Water Engineering & Management (WE&M) (1990). "California Water Utility Controls Water Hammer", pp. 10, 42.

Water Environment Federation (WEF) (1993). *Design of Wastewater and Stormwater Pumping Stations*, Manual of Practice No. FD-4, Alexandria, Virginia and Reston, Virginia.

Water Environment Federation (WEF) (1998). "Nonmechanical, 'Nonblocking' Stormwater System Traps Debris", *Wastew. Technol. Showcase*, 1(1), 16–17.

Water Environment Federation (WEF) and American Society of Civil Engineers (ASCE) (1998). *Design of Municipal Wastewater Treatment Plants*, Manual of Practice No. 76, Alexandria, Virginia and Reston, Virginia, 10-10 to 10-15 and 11-106 to 11-112.

Water Pollution Control Federation (WPCF) (1952). *Air Diffusion in Sewage Works*, Manual of Practice No. 5, Washington, D.C.

Water Pollution Control Federation (WPCF) (1967). *Sewage Treatment Plant Design*, Manual of Practice No. 8, Washington, D.C.

Water Pollution Control Federation (WPCF) (1971). *Aeration in Wastewater Treatment*, Manual of Practice No. 5, Washington, D.C., Section 6.22.

Water Pollution Control Federation (WPCF) (1978). *Instrumentation in Wastewater Treatment Plants*, Manual of Practice No. 21, Washington, D.C., Chapter 4.

Water Pollution Control Federation (WPCF) and the American Society of Civil Engineers (ASCE) (1959). *Sewage Treatment Plant Design*, Manual of Practice No. 36, Washington, D.C.

Water Pollution Control Federation (WPCF) and the American Society of Civil Engineers (ASCE) (1977). *Wastewater Treatment Plant Design*, Manual of Practice No. 8, Washington, D.C.

Webber, S. (2021). *Excel 2021*, Independently published.

Wehner, J. F., and Wilhelm, R. H. (1956). "Boundary Conditions of Flow Reactor", *Chem. Eng. Sci.*, 6(2), 89–93.

Wells, J., and Schmid, J. H. (1961). "Hold Rain Water on the Roof", *Actual Specif. Eng.*, 6(2).

Wenzel, L. K. (1942). "*Methods for Determining Permeability of Water-Bearing Materials*", U.S. Department of the Interior, Geological Survey Water-Supply Paper 887.

Whipple, W., Jr. (1972). "Corps of Engineers Plans Regional Water Quality Programs", *Civil Engineering*, 42(3), 37–39.

Whipple, W., Jr., Coughlan, F. P., Jr., and Yu, S. L. (1970a). "Aerators for Polluted Rivers", *J. San. Eng. Div.*, 96(5), 1153–1165.

Whipple, W., Jr., and Dittman, F. (1970). "Water Quality and Instream Aeration", *Water Wastes Eng. / Ind.*, September 1970, E-32–E-34.

Whipple, W., Jr., and Hunter, J. V. (1981). "Settleability of Urban Runoff Pollution", *J. WPCF*, 53(12), 1726–1731.

Whipple, W., Jr., Hunter, J. V., Davidson, B., Dittman, F., and Yu, S. L. (1969). *Instream Aeration of Polluted Rivers*, Water Resources Research Institute, Rutgers University, New Brunswick, New Jersey.

Whipple, W., Jr., Hunter, J. V., Dittman, F. W., and Yu, S. L. (1970b). *Oxygen Regeneration of Polluted Rivers: The Delaware River*, Prepared for the Environmental Protection Agency, 16080 DUP, December 1970.

Whipple, W., Jr., Hunter, J. V., Dittman, F. W., and Yu, S. L. (1970c). *Oxygen Regeneration of Polluted Rivers: The Delaware River*, Prepared for the Environmental Protection Agency, 16080 DUP, December 1970.

White, C. M. (1940). "The Equilibrium of Grains on the Bed of a Stream", *Proc. Royal Soc. London*, A174, 322–338.

White, M. H. (1972). "Surge Control for Centrifugal Compressors", *Chem. Eng.* 25, 1972, 54–62.

Whiton, T. B., and Ham, R. (1993). "Troubleshooting Pump Vibration", *Water Environ. Technol.*, 5(1), 45–51.

Wilhelms, S. C., Clark, L., Wallace, J. R., and Smith, D. R. (1981). *Gas Transfer in Hydraulic Jumps*, Technical Report E-81-10, Waterways Experiment Station, Vicksburg, Mississippi.

Wilhelms, S. C., Schneider, M. L., and Howington, S. E. (1987). *Improvement of Hydropower Release Dissolved Oxygen with Turbine Venting*, Technical Report E-87-3, Waterways Experiment Station, Vicksburg, Mississippi.

Williams, G. R. (1950). "Rational Theory of Surface Runoff for Drainage Design", Section IV.1 in *Engineering Hydraulics*, H. Rouse (ed.), John Wiley & Sons, Inc., New York.

Williams, S. (2007). "Langmuir Currents (and Streaks)", *Water Column*, 12(2), 10.

Williamson, J. V., and Rhone, T. J. (1973) "Dividing Flow in Branches and Wyes", *J. Hyd. Div.*, 99(5), 747–769.

Willis, R. F. and Davis, C. (1962). "Flow Patterns in Rectangular Sewage Sedimentation Tanks", *Proc. International Conference on Water Pollution Research*, Paper No. 19, Pergamon Press Ltd., Oxford, England.

Wilson, T. E. (1991). "'A Debate: Rectangular Clarifiers Should be Considered' and 'Rebuttal to The Case for Circular Clarifiers'", *Water Eng. Manage.*, 1991, 26.

Winkler, E., and Guswa, S. (2002). *Technology Assessment Report: Vortechs Stormwater Treatment Systems Vortechnics Inc.* Scarborough, ME, Center for Energy Efficiency and Renewable Energy, University of Massachusetts at Amherst, Amherst, MA.

Winslow, M. (2008). *Lake Champlain: A Natural History*, The Lake Champlain Committee, Burlington, Vermont.

Wislicenus, G. F. (1942). "The Hydraulic Problem of Cavitation in Pumps", *Proc. Second Hydraulics Conference*, arranged by the Iowa Institute of Hydraulic Research, State University of Iowa under the sponsorship of the American Society of Civil Engineers, June 1-4, 1942, Iowa City, Iowa.

Wisner, P. E., Mohsen, F. N., and Kouwen, N. (1975). "Removal of Air from Water Lines by Hydraulic Means", *J. Hydraul. Div.*,101(2), 243–257.

Wolf, D. R., and Phillips, R. (1994). "Retention Basin Reduces CSOs, Increases Treatment", *Water Environ. Technol.*, 6(10), 17.

Wood, D. J. (1966). "An Explicit Friction Factor Relationship", *Civ. Eng.*, 36(12), 60–61.

Wood, D. J. (1970). "Pressure Surge Attenuation Utilizing an Air Chamber", *J. Hyd. Div.*, 96(5), 1143–1156.

Wood, D. J. (1982). *BASIC Computer Program for the Analysis of Pressure and Flow in Pipe Distribution Systems Including Extended Period Simulations*, College of Engineering, University of Kentucky, Lexington, Kentucky.

Wood, D. J. (1991). *Comprehensive Computer Modeling of Pipe Distribution Networks*, Civil Engineering Software Center, College of Engineering, University of Kentucky, Lexington, Kentucky.

Wood, D. J., and Charles, C. O. A. (1972). "Hydraulic Network Analysis Using Linear Theory", *J. Hyd. Div.*, 98(7), 1157–1170.

Wood, D. J., and Funk, J. E. (1988). *Surge User's Manual*, Civil Engineering Software Center, University of Kentucky, Lexington, Kentucky.

Wood, D. J., and Jones, S. E. (1973). "Water-Hammer Charts for Various Types of Valves", *J. Hyd. Div.*, 99(1), 167–178.

Wood, D. J., and Rayes, A. G. (1981). "Reliability of Algorithms for Pipe Network Analysis", *J. Hyd. Div.*, 107(10), 1145–1161.

Woods, W. (1999). "Dowsing-Common Sense or Super Sense?", *Opflow*, 25(9), 4–6.

Wylie, E. B., and Streeter, V. L. (1983). *Fluid Transients*, FEB Press, Ann Arbor, Michigan.

Yang, K. (2001). "Practical Method to Prevent Liquid-Column Separation", *J. Hyd. Eng.*, 127(7), 620–623.

Yao, K. M. (1970). "Theoretical Study of High Rate Sedimentation", *J. WPCF*, 42(1, Part 1), 218–228.

Yao, K. M. (1971). "Nonuniform Flow in Flat Rectangular Channels", *J. Hyd. Div.*, 97(9), 1343–1348.

Yao, K. M. (1972). "Hydraulic Control for Flow Distribution", *J. San. Eng. Div.*, 98(2), 275–285.

Yao, K. M. (1974). "Sewer Line Design Based on Critical Shear Stress", *J. Environ. Eng.*, 100(2), 507–520.

Yen, B. C. (1973). "Open-Channel Flow Equations Revisited", *J. Eng. Mech. Div.*, 99(5), 979–1009.

Yen, B. C. (1975). "Closure to 'Open-Channel Flow Equations Revisited' by B. C. Yen", *J. Eng. Mech. Div.*, 101(4), 485–487.

Yen, B. C. (1992a). "Dimensionally Homogeneous Manning's Formula", *J. Hydraul. Eng.*, 118(9), 1326–1332.

Yen, B. C. (Author, ed.) (1992b). *Channel Flow Resistance: Centennial of Manning's Formula*, Water Resources Publications.

Yen, B. C., and Wenzel, H. G. (1970). "Dynamic Equation for Steady Spatially Varied Flow", *J. Hyd. Div.*, 96(3), 801–814.

Yen, B. C., Wenzel, H. G., Jr., and Yoon, Y. N. (1972). "Resistance Coefficients for Steady Spatially Varied Flow", *J. Hyd. Div.*, 98(8), 1395–1410.

Yen, C. L. (1969). "Effects of Viscous Sublayer on Dispersion", *J. San. Eng. Div.*, 95(6), 1105–1115.

Yih, C. S. (1959). "Two Solutions for Inviscid Rotational Flow with Corner Eddies", *J. Fluid Mech.*, 5(1), 36–40.

Yoon, Y. N., and Wenzel, Jr., H. G. (1971). "Mechanics of Sheet Flow Under Simulated Rainfall." *J. Hyd. Div.*, 97(9), 1367–1386.

Yoshimi, Y., and Stelson, T. E. (1963). "Grit Unbalance in Sewage Flow Division", *J. San. Eng. Div.*, 89(2), 61–83.

Young, J. C. (1985). "Operating Problems with Wastewater Filters", *J. WPCF*, 57(1), 22–29.

Yu, S. L. (1970). "Aerator Performance in Natural Streams", *J. San. Eng. Div.*, 96(5), 1099–1114.

Zhou, H., and Smith, D. W. (1994). "Kinetics of Ozone Disinfection in Completely Mixed Systems", *J. Environ. Eng.*, 120(4), 841–858.

Zic, K., and Stefan, H. G. (1994). *Destratification Induced by Bubble Plumes*, Technical Report W-94-3, Waterways Experiment Station, Vicksburg, Mississippi.

Zsák, E. (1971). "Manifold Flow in Subirrigation Pipes", *J. Hyd. Div.*, 97(10), 1737–1746.

Zuckerman, C. (2011). "Snap, Crackle, Pop", *National Geographic*, Washington, D.C., July 2011.

Zurn Industries Inc. (2007). "Zurn Control-Flo Roof Drainage System", http://www.zurn.com/operations/specdrain/pdfs/miscellaneous/Control-Flo.pdf, accessed July 11, 2007.

Index

Hydraulics and Pneumatics in Environmental Engineering, First Edition. S. David Graber.